THE STATUS OF BIRDS
IN BRITAIN AND IRELAND

David T. Parkin
and
Alan G. Knox

CHRISTOPHER HELM
LONDON

Published 2010 by Christopher Helm, an imprint of A&C Black Publishers Ltd.,
36 Soho Square, London W1D 3QY

www.acblack.com

ISBN 978–1–4081–2500–7

A CIP catalogue record for this book is available from the British Library

This book is produced using paper that is made from wood grown in managed sustainable forests. It is natural, renewable and recyclable. The logging and manufacturing processes conform to the environmental regulations of the country of origin.

Commissioning Editor: Nigel Redman
Project Editor: Jim Martin

Design by Mark Heslington

Printed in Great Britain by Martins the Printers, Berwick Upon Tweed

10 9 8 7 6 5 4 3 2 1

Front cover: Scottish Crossbill *Loxia scotica*. Rothiemurchus, Highland, June (Dean Eades)
Back cover: Common Starlings *Sturnus vulgaris*. Far Ings, Lincolnshire, November (Graham Catley Nyctea Ltd)
Spine: Common Redstart *Phoenicurus phoenicurus*. Clwyd, May (Richard Steel)
Front flap (l-r): Long-tailed Tit *Aegithalos caudatus rosaceus*, Cambridgeshire, February (Rebecca Nason); Willow Ptarmigan/Red Grouse *Lagopus lagopus scotica,* Rosedale Abbey, North Yorkshire, June (Carl Wright); White-throated Dipper *Cinclus cinclus gularis*, Aviemore, Highland, March (Rebecca Nason)

CONTENTS

PREFACE

The birds of Great Britain and Ireland (abbreviated as GB&I) are perhaps the best studied in the world. Countless birdwatchers and ornithologists have observed, monitored and chronicled them for generations. As former Chairmen of the BOU Records Committee (see page 12), we have been aware that a revised account is long overdue. At the time of the last review, published in 1971 as *The Status of Birds in Britain and Ireland*, 466 species had been recorded; today that total has increased by 25% to over 580. Taxonomic realignments account for some of these, but most are due to additional rare migrants. There have been dramatic changes in the distribution and abundance of commoner species, many documented in a series of atlases published by the British Trust for Ornithology (BTO) and the Irish Wild Bird Conservancy (IWC), and in reviews such as the late Chris Mead's *State of the Nation's Birds* (2000) – although this was restricted to Great Britain. After we had repeatedly stressed the importance of documenting these changes, our colleagues across Britain and Ireland encouraged us to undertake this review.

To allow comparison with earlier reviews, we have included the whole of the British and Irish group of islands, though the various national lists are now maintained independently. Despite conspicuous differences, the islands remain a single zoogeographic unit and deserve a comprehensive overview. We have based our researches on an array of literature, although certain sources are of paramount importance. For the commoner species, the excellent (if now dated) atlases of breeding (Sharrock 1976, Gibbons *et al.* 1993), wintering (Lack 1986), historical (Holloway 1996) and migrant (Wernham *et al.* 2002) birds were essential. More recently, the BTO's web-based annual review of populations, published as *Birds in the Wider Countryside* (most recently in 2009), has been invaluable. The national reviews for England, Scotland, Wales and Ireland, published by Poyser, and the most recent *Birds of Scotland* (Forrester *et al.* 2007) have provided much important information. For the rarer species, we have used the reports of rare birds published in *Irish Birds* by the Irish Rare Birds Committee (IRBC) and in *British Birds* by the British Birds Rarities Committee (BBRC), together with supplementary material from Northern Ireland and the Isle of Man. We have been especially fortunate in being allowed access to Keith Naylor's compendium *Rare Birds of Great Britain and Ireland*, which details records from historical times through to the most recent reports of BBRC and IRBC. The Scarce Migrant Reports published in *British Birds*, alongside the more long-standing Rare Breeding Birds Reports have also been helpful sources. For reasons of space, these are rarely cited individually: readers are directed to the appropriate report in the relevant year.

The sequence of species is based on Voous's (1977) *List of Recent Holarctic Bird Species*, varying, sometimes substantially, where later research has indicated that revision is necessary, as noted in the Introduction.

Under each species are listed the subspecies known to have occurred in Great Britain and Ireland, or notes to the effect that the race has not been identified or the species is monotypic. Endemic subspecies are identified as such, and abbreviated codes are given for the status of each race on the list, based on their status in GB (apart from some Scarce Migrants which have occurred only in Ireland or the Isle of Man)

Some of these codes are only generalisations. For example, many forms listed as WM may

RB	Resident Breeder	FB	Former Breeder
NB	Naturalised Breeder	WM	Winter Migrant
MB	Migrant Breeder	PM	Passage Migrant
CB	Casual Breeder	SM	Scarce Migrant
HB	Hybrid Breeder		

sometimes be found in summer. RB and MB together imply that the species is a partial migrant in which a proportion of the population leaves GB&I in winter. CB includes those forms that nest less than annually and those for which there are fewer than 10 breeding attempts each year. SM is generally reserved for rare taxa for which descriptions are required by BBRC and for some of the rarer races not usually identifiable in the field. A small number of PMs (e.g. Ferruginous Duck *Aythya nyroca* and Buff-breasted Sandpiper *Tryngites subruficollis*) are rarer than some SMs.

For each species, we have given three sections, which vary in length depending on the amount of information available. The first of these is a note on the species' taxonomy; here we have concentrated on recent, usually molecular, results, attempting to show how these agree or differ from conventional morphological findings. The second section briefly describes the global distribution and migrations of the species and subspecies; where there are several of the latter, we have concentrated more on those from the Western Palearctic and especially those that have occurred (or might occur) in GB&I. We use 'subspecies' and 'race' interchangeably. We have made extensive use of *Birds of the Western Palearctic* (BWP) and *The Handbook of the Birds of the World* (HBW) for these sections. The third part details the distribution of the species within GB&I. The commoner species are discussed in greater or lesser detail, depending on their status and any changes that might have occurred since the last *Status*. For the rarer species, we have given brief summaries of their occurrence, and for the commoner rarities we have discussed any patterns of appearance.

Because each component part of GB&I (Great Britain, Northern Ireland, the Republic of Ireland and the Isle of Man) now maintains its own national list, it would have been cumbersome to have indicated in the main text the List Categories to which they have been assigned (A, B, C) in each region, and both meaningless and presumptuous to attempt an overall code. We have therefore assigned these to a separate list at the end (Appendix 1). The codes herein are based on the regional authorities and our own judgement; any errors or omissions are our responsibility.

ACKNOWLEDGEMENTS

It is humbling to remember how many people have been generous in helping with the production of this book. Some have read individual species accounts, others have read whole taxonomic groups, while three have read the book more or less in its entirety. We start by acknowledging these three. Patrick Smiddy and Joe Hobbs read the entire book from an Irish perspective; we thank them for giving us the most up-to-date information on Irish birds and pointing out our inaccuracies. George Sangster read the taxonomic sections with his customary eye for detail and pointed us towards several recent references of which we should have been aware. Others who made major contributions include Jeff Baker (Palearctic warblers), Dick Banks (American vagrants), Bill Bourne (tubenoses and other oceanics), Roy Dennis (raptors), Mike Harris (seabirds), Chris Kehoe (rare races), Andy Musgrove (wildfowl and waders), Gavin Siriwardena (common farmland passerines) and Robin Ward (wildfowl and waders). We thank them all most sincerely.

Among the many other friends and colleagues who have helped with information, advice, comments or by reading species accounts, we must especially thank Tommy Aarvak, Per Alström, Phil Atkinson, Allan Baker, Niall Burton, Martin Cade, Mark Collier, Martin Collinson, Greg Conway, Dick Coombes, Peter Cranswick, Humphrey Crick, Pierre-André Crochet, Mark Cubitt, Patrick Cullen, Normand David, Edward Dickinson, Steve Dudley, Tony Fox, Rob Fuller, Bob Furness, Simon Gillings, George Gordon, Jeremy Greenwood, Andrew Grieve, Andrew Harrop, Paul Hebert, Jeff Higgott, Mark Holling, Baz Hughes, Jochen Martens, Mick Marquiss, Bob McGowan, Neville McKee, Tim Melling, Paul Milne, Chris Murphy, Ian Newton, David Noble, Malcolm Ogilvie, Chris Perrins, Håkon Persson, Dick Potts, Norman Ratcliffe, Eileen Rees, Roger Riddington, Jim Rising, Rob Robinson, Deryk Shaw, Ken Smith, Lars Svensson, Don Taylor, Tommy Tyrberg, Keith Vinicombe, Gary Voelker, Steve Votier, Ian Wallace, Roger Wilkinson, Michael Wink, Chris Wormwell, Bob Zink and Frank Zino. Errors or omissions cannot be laid at their door; we confidently assign these to each other.

DTP would like to thank Baz Hughes, Peter Cranswick and Robin Ward at the Slimbridge offices of the Wildfowl & Wetlands Trust for their hospitality and expertise while preparing the wildfowl and wader accounts, and Jeremy Greenwood and the staff at the BTO offices in Thetford for similar kindnesses. DTP also wishes to acknowledge the Leverhulme Trust for the generous award of an Emeritus Fellowship to cover the costs of writing this book.

Finally, Nigel Redman deserves special thanks for his exceptional help in seeing this book through to publication, and his colleagues Jim Martin and Elaine Rose for their patience in converting our work into finished text. DTP also wishes to acknowledge the Leverhulme Trust for the generous award of an Emeritus Fellowship to cover the costs of preparing this book. Finally, words cannot express our thanks to Linda and Ann-Marie for their tolerance and understanding over the past two years.

INTRODUCTION

Great Britain and Ireland ('the British Isles' or 'Britain and Ireland') lie at the western edge of the Palearctic, roughly between 50° and 62°N and between 10°W and 4°E. There are over 6,000 islands in the archipelago, totalling over 315,000 sq km, of which by far the two largest are Britain (almost 217,000 sq km) and Ireland (84,400 sq km). Most of the smaller offshore islands are found along the north and west coasts. The largest groups are the Orkney and Shetland Islands off the north of Scotland and the Inner and Outer Hebrides along the west. Other significantly large islands include the Isle of Man and Anglesey in the Irish Sea and the Isle of Wight off the south of England. There are many smaller islands and groups: the Farne Islands (NE England), Flannan Islands and St Kilda (NW Scotland), Rathlin and Copeland Islands (Northern Ireland), the Saltee group (SE Ireland), and the Blasketts and Skelligs (W Ireland) are amongst those of particular importance to seabirds. Details of many others of ornithological importance can be found in Mitchell *et al.* (2004).

Politically, there are three main administrative units. In order of size, these are the United Kingdom of Great Britain and Northern Ireland; the Republic of Ireland; and the Isle of Man. Great Britain (Britain) includes England, Scotland and Wales. Great Britain, Northern Ireland, the Republic of Ireland and the Isle of Man are each responsible for their own wildlife legislation. Scotland and Wales have devolved government, but the majority of wildlife legislation is common to the whole of Britain. The human population of Britain and Ireland is approximately 65 million, most heavily concentrated in urban areas in the south and east of England.

The whole archipelago lies close to Continental Europe, with the SE corner of England less than 40 km from France. The English Channel to the south and the North Sea on the east form significant barriers to terrestrial animals. There have been several periods of geological history when there were land bridges between SE England and the near Continent, and a relatively free flow of plants and animals was possible at these times. The last land bridge was flooded about 5,000 years ago, since when the purely terrestrial fauna has been isolated. The Channel Islands lie near the shores of France, and their zoological and ornithological affinities are closer to that country; they are not considered here.

GEOGRAPHY AND CLIMATE

The surface geology of Britain comprises ancient rocks in northwest Scotland and northwest Wales through progressively younger formations to the south and east. In Ireland, the older rocks are in the northwest, north and east, with younger deposits in the centre, south and southwest. The upland areas in Britain are in the north and west, compared with the generally lower eastern side. Indeed, a line can be drawn from the River Tees estuary on the Yorkshire/Durham border to the mouth of the River Exe in Devon; north and west of this line is generally upland; south and east is generally lowland. In Ireland, more broken upland areas are found in the north, west and across the south, with an extensive low plain in the centre. The highest peaks in Britain and Ireland are Ben Nevis (Scotland: 1,344 m), Snowdon (Wales: 1,085 m), Corran Tuathail (Republic of Ireland: 1,039 m), Scafell Pike (England: 978m), Slieve Donard (Northern Ireland: 852 m) and Snaefell (Isle of Man: 621 m). The uplands in no way compare with mountains elsewhere in Europe, but nevertheless the climate here can be extreme. There are 284 peaks in Scotland over 914 m (3,000 ft), and many are subject to harsh conditions in winter. The plant and animal life shows closer affinities with boreal habitats of mainland Europe than with adjacent lowlands. Some of the uplands were originally covered with forest, although much of this has long been removed to be replaced by bogs, heather or scrubby woodland. Most of the higher ground does not sustain cultivation and is used for sheep grazing, grouse moor or deer stalking, with human settlements and their associated agriculture largely restricted to the valleys.

Extensive coal measures lie beneath the ground in many parts of the lowlands to the south and east, and these gave the power that drove the industrial revolution. The major cities of lowland Scotland, northern England and South Wales developed around this industry, initiating intensive urbanisation. The concentration of administrative government around the capital cities (especially London and Dublin) boosted the population there, with consequent urban sprawl removing extensive areas from the countryside.

The climate is temperate, with the Gulf Stream bringing comparatively warm water from the tropical North Atlantic to the shores of Britain and Ireland. This is sufficient to prevent the seas around the islands from freezing. Even in winter, sea ice is rarely found south of the Arctic Circle (67°N) in the east Atlantic, although in the west it can extend as far south as Newfoundland (49°N, about the latitude of Paris). This northwards transport of heat by the ocean currents leads to a mild climate for Britain and Ireland, relative to its latitude.

The prevailing westerly winds for much of the year bring weather systems from the Atlantic that are also relatively warm for the latitude and deposit rain, especially over the higher ground and the west. With generally reduced cloud cover and rainfall, the sheltered lowlands are drier and warmer in summer, and it is here that much of the arable agriculture is based. The overall climate can be summarised as four quarters: the north-east is cold in winter and cool in summer; the north-west is mild in winter and cool in summer; the south-east is cold in winter and warm in summer; and Ireland and SW Britain are mild in winter and warm in summer.

The effects of climate change are becoming increasingly apparent. While sceptics may dispute the causes of these changes, there is little doubt among the scientific community that climate change is driven by increased greenhouse gases in the atmosphere and that these are changing the thermal balance of the earth. Recent winters in Britain and Ireland have been milder, with relatively little snow and fewer days with frost. In the highlands of Scotland, the winter sports industry is facing serious hardship. Summers are becoming warmer, with less rainfall. There has been an increase in extremes of weather; storms and short, heavy summer rainfall is commoner than before, with a consequent increase in flooding. Of particular significance to birds, the onset of spring is advancing, with plants flowering and invertebrates emerging two or three weeks earlier than 25 years ago. Many resident bird species are breeding earlier, and some migrants have changed the timing of their migration. The long-term impact on our bird populations is still unclear, but it is likely that we will see species with strong continental influences increase and more northerly specialists become extinct.

FLORA AND VEGETATION

It is unlikely that anywhere in Britain and Ireland has escaped the impact of man. Land use has changed dramatically over several centuries and especially since WWII, with the need for more and cheaper food driving agricultural intensification. From an ornithological standpoint, these changes have been reviewed by Mike Shrubb (O'Connor & Shrubb 1986; Shrubb 2003a, b). The drive for increased efficiency, especially in the lowlands, was initially implemented through larger fields, created by the destruction of thousands of kilometres of hedgerows. Nesting and feeding sites have been lost for a range of farmland birds as varied as Grey Partridge, European Turtle Dove, Lesser Whitethroat and Eurasian Tree Sparrow. Drainage of wet areas has also been widespread within the farmland, with the loss of the marshes, ponds and overgrown ditches essential for Water Rail and Common Grasshopper Warbler. Crops are now grown across the flood plains of our major rivers, replacing lowland wet meadows with cereals and other arable produce. The change to winter wheat has further exacerbated the problem. Soon after harvest, cereal fields are ploughed and seed for next year's crop is sown; this germinates through the autumn, giving it a head start in spring, with an enhanced and early yield. The effects on wildlife are many and varied, from the loss of winter stubble with its associated seeds and invertebrates, to taller spring vegetation reducing the availability of nest sites for grassland birds. Nor have these changes been restricted to the grain fields in the lowlands; in Scotland, the almost complete loss of weed-filled fields of neeps (turnips, or swedes to the English) has had devastating effects on winter seedeaters. Across Britain and Ireland, the increased application of fertiliser and selective herbicides has changed flower-rich meadows to dense grassland for silage. Early and repeated cropping reduces the likelihood of successful breeding by several meadow species, from Yellow Wagtail to Corn Crake, and the loss of weed seeds removes an essential food supply for granivorous birds such as Twite and Common Linnet.

Away from the field species, the loss of mature trees, whether through hedgerow destruction, Dutch elm disease or the selective removal of aged trees for actual or perceived safety reasons or to fuel the wood-burning stoves of the middle classes, has resulted in fewer nest sites for hole-nesting species such as Common Starling, Barn Owl and Stock Dove. Derelict and semi-derelict barns have been modernised, demolished or converted into country cottages (and fitted with wood-burning stoves), again leading to a loss of nest sites for hole-nesters. Even urban dwellers have played their part in impoverishing the landscape, with soffits and fascias sealing openings into roof spaces, and the removal of large trees from streets and gardens. Birds such as Common Swift, Common Starling and House Sparrow are all believed to have suffered from these changes. Rallying to the clarion call of the late Chris Mead ('Britain needs more holes'), some are attempting to counter the trend by providing nest boxes for owls, swifts, sparrows and the like, but long-term problems remain.

Woodlands have also suffered since the 1950s. Increased labour costs have led to reductions in routine management and a reduction in species that depend on secondary growth. Increased numbers of Grey Squirrel are believed to affect the breeding success of open-nesting woodland species, such as Hawfinch, while increased browsing by deer may be reducing the shrub layer required by Common Nightingales and others.

Under pressure from environmentalists, new, and hopefully improved, agri-environment schemes have recently been introduced, changing the methods of farm subsidy to encourage the return (or continuation) of wildlife. Re-creation of hedgerows, the selective planting of trees in field margins and the establishment of crop-free field margins should help our threatened birds and other wildlife to recover. The forest agencies are encouraging the planting of native hardwoods at the expense of faster-growing, but ecologically less justifiable, alien softwoods. Legislation has been introduced to limit, if not eradicate, the tax breaks to timber speculators, thereby reducing the inducement to replace hundreds of hectares of upland and their special plants and animals with non-native conifers.

THE GEOGRAPHIC DIVISIONS AND HABITATS

Some bird species are almost ubiquitous throughout Britain and Ireland, being found just about everywhere there is suitable habitat. During the fieldwork for the Second Breeding Atlas (1988–91), Winter Wrens were found in 3,748 (>97%) of the 3,854 10-km squares that were surveyed, Eurasian Skylarks and Common Blackbirds in 95%, European Robins in 94%, Common Chaffinches in 93%, Willow Warblers in 92% and Common Wood Pigeons and Dunnocks in >90%. Other species were more restricted but still occurred wherever there was suitable habitat. For example, Eurasian Tree Creepers and Goldcrests were only found in 78% and 83% of 10-km squares, but in Britain they were only absent from some northern and western isles of Scotland, treeless areas of the highlands and intensive agricultural land in the English lowlands. Northern Fulmar only bred in 716 (19%) of squares, but there was almost no stretch of cliff around the coasts from which they were absent.

The most dramatic geographic divisions in the ornithological biodiversity across the islands are undoubtedly those between seabird-rich coastal areas and inland, between the uplands and the lowlands, and between Britain and Ireland.

Ireland, whether through historical habitat loss, smaller land-mass or glacial history, has a relatively impoverished fauna and flora compared with Britain, and this extends even to vagile birds. Whatever the cause, the presence of the Irish Sea plays a role in maintaining this impoverishment. Many species that are common and widespread in Britain are absent or rare in Ireland: indeed, only about 60% of the species that breed in Britain also breed 'across the water', with woodland birds especially poorly represented. Amongst other species, there are no Common Nightingales, Marsh or Willow Tits, Eurasian Nuthatches or Tree Pipits. Great Spotted Woodpecker, Common Redstart, Wood Warbler and European Pied Flycatcher breed only occasionally. Short-eared Owl, Yellow Wagtail, Lesser Whitethroat and Carrion Crow are equally rare breeders. On the other hand, Ireland, in common with similarly exposed British west coast locations such as the Isles of Scilly and the Outer Hebrides, receives many trans-Atlantic vagrants and experiences heavy seabird passages.

Boreal or near-boreal species such as Rock Ptarmigan, Snow Bunting and Eurasian Dotterel breed only in the arctic–alpine habitats of the highest summits in the British north and west. The lower slopes in Scotland and the higher parts elsewhere in Britain and Ireland typically comprise heather moor or short grassland. The heather is heavily managed for Willow Ptarmigan (Red Grouse) and supports significant populations of Meadow Pipit, with smaller numbers of Ring Ouzel and Northern Wheatear along the stony valley sides; the upland grass and sheep pasture hold Eurasian Golden Plover, Twite, Eurasian Curlew and Whinchat. Merlin and Peregrine Falcon nest widely in suitable habitat across the uplands, although Golden Eagles are largely restricted to the more remote highlands. Most of the hill regions have a high rainfall, and the wetter areas of blanket bog, especially in Scotland, are breeding sites for waders such as Dunlin, Common Snipe and (less commonly) Common Greenshank and Wood Sandpiper. Many of the lower lakes in the north are oligotrophic and poor for birds, though Ospreys now nest widely among the lochs and river valleys of Scotland, often feeding on Rainbow Trout in managed fishery lochs; a few of the wilder lochs and lochans hold breeding Black- and Red-throated Loons, some of which commute to nearby sea lochs to fish. A few of the eutrophic lakes of central and northern Scotland have breeding Horned and Black-necked Grebes, and a wide range of ducks. The fast-flowing rivers that drain the uplands hold a variety of species: White-throated Dipper and Grey Wagtail frequent the cleanest water, with Common Sandpiper, Eurasian Teal and Sand Martin along the open stretches; Common Goldeneye, Common Merganser, Red-breasted Merganser and Spotted Flycatcher breed in riverine woodlands.

The only native conifers of the uplands are the Scots pine and juniper, and these are now restricted to the mid- and lower altitudes. They were greatly reduced by man and reached their lowest extent in about the mid-1800s, since when there has been a slow recovery. The pine forests of the Scottish highlands hold several species not found elsewhere in Britain and Ireland, including Parrot and Scottish Crossbills, European Crested Tit and Western Capercaillie. Red Crossbill, Northern Goshawk (largely of presumed captive origin), owls and Eurasian Siskin have all taken advantage of the extensive areas of planted conifer in the highlands and elsewhere, especially in lowland Scotland and northern England.

Deciduous woodlands are generally distributed across the lowlands and hillsides, with avian communities that vary in composition with temperature and rainfall. In W Britain there are important areas for breeding Common Redstart, Wood Warbler and European Pied Flycatcher, with Black Grouse hanging on along a few of the woodland margins; further south and east the woods hold Common Nightingale, Hawfinch and Lesser Spotted Woodpecker, all of which are rare or absent further north. A few carefully protected sites in the south hold tiny populations of Common Firecrest and Eurasian Golden Oriole.

Lowland heath was formerly more widespread but has been extensively lost to a combination of agriculture, urbanisation and sheer neglect; 85% has gone from Britain and Ireland since 1800. The areas that are left are chiefly in the south and east and hold some very special birds, including Eurasian Stone-curlew, European Nightjar, Woodlark and Dartford Warbler. Although these can be found away from the heaths, it is here that they are most abundant. Woodlarks and European Nightjars in particular have taken advantage of clear-fell areas in planted conifer forests, with numbers increasing in several counties.

Outside of the urban sprawl, the majority of the lowlands are farmed, more or less intensively depending upon the region. It is in these areas that some of the most profound changes in habitat structure have taken place, especially in the south and east of England, and to a lesser extent in Ireland. These have already been mentioned; the important point being that birds here have suffered some of the most severe declines across the region. Individual losses are discussed in the relevant species accounts, but birds that rely on cereal stubble for winter feeding, hedgerows for nesting and rank vegetation for invertebrate food have all been hard hit. Conversion of lowland pasture, whether to crops or silage, has reduced the habitat for Yellow Wagtail, Meadow Pipit, Northern Lapwing and Corn Crake. Clearance or removal of wet margins, with their rank vegetation, has had a similar effect on Common Grasshopper Warbler, Common Reed Bunting and Water Rail.

Some of the lowland rivers still meander through grassy plains and hold populations of Eurasian Reed Warbler and Common Kingfisher, but others have had their banks cleared of vegetation, with a loss of secure nesting sites. Cetti's Warbler, however, has colonised the denser riverine habitats, especially in the southern third of England but, on the debit side, Marsh Warbler has almost gone. Amongst the general degradation of the landscapes with which we were familiar, there have been some bright spots in the last 50 years. The need for sand and gravel by the construction industry has led to the creation of numerous gravel pits across the lowlands. Some have subsequently been in-filled, while others have gone to water sports. Many have provided a new focus for lowland waterbirds. Great Crested and Little Grebes, Common Shelduck and even Ruddy Duck have increased in numbers and distribution. Eurasian Reed and Sedge Warblers have found breeding sites around the margins; Little Ringed Plover have colonised many of the open gravel sites, and Black-necked Grebe nest in a few secluded places. Canada Geese have benefited enormously; islands in the pits provide safe nesting sites, the open water gives security during the moult, and cereal and silage fields alongside provide grazing through almost all the year. Recently, feral populations of Greylag and Egyptian Geese have also started to take advantage of this habitat.

It may be tempting to dismiss the villages, towns and cities as lost zones for our birds, but this is far from the case. Suburban gardens comprise vast areas in both Britain and Ireland, and provide homes for many tits, thrushes and other species, with European Green and Great Spotted Woodpeckers frequenting lawns and feeders. Rose-ringed Parakeets thrive around the fringes of London. Peregrine Falcons now nest in many of our larger cities, and Eurasian Sparrowhawks patrol the bird tables. Parks hold Common Moorhens and other wildfowl. White (Pied) Wagtails find food and safe places to roost, sometimes in huge numbers. In eastern Scotland, large numbers of Eurasian Oystercatchers nest on rooftops and feed in parks and on roadside verges. Roof-nesting gulls are more widespread and generally far less popular. Together, the urban areas provide for the needs of millions of birds.

The coastline of Britain and Ireland is varied and truly magnificent. Northern Fulmar and Black-legged Kittiwake breed widely on the cliffs, though Fulmars also fly inland to find rocky outcrops or nest in sand dunes in places; in some areas Kittiwake nest on buildings. Nesting seabird diversity increases with distance from the SE corner of England. Great Cormorant, European Shag, Common Murres, Puffin, Razorbill, European Herring and Lesser Black-backed Gulls, and sometimes Northern Gannet nest on cliff colonies from Yorkshire round the coasts of Scotland and Ireland to Dorset and the Isle of Wight. Black Guillemots breed widely in Scotland and Ireland, and Red-billed Choughs use caves or old buildings along the western sea cliffs where there is suitable grassland nearby to forage for invertebrates. Sand and shingle shores are popular with walkers and holidaymakers but less populated by breeding birds. Little Tern and Common Ringed Plover are the most widespread, although sandy foreshores host flocks of scuttling Sanderlings from autumn through to spring. These habitats are frequently backed by coastal grasslands that hold high densities of breeding ducks and waders in summer. Many also have large populations of geese and ducks in winter and provide hunting grounds for harriers and falcons all year round. In a few locations, coastal reed beds provide breeding sites for our small populations of Eurasian Bittern and Bearded Reedling, with larger numbers of Eurasian Reed Warbler in the south.

Estuaries and mudflats are significant winter habitats for waders and wildfowl, but the lack of secure nesting sites limits their importance in summer. These occur all round the coast where major rivers meet the sea; mudflats are also found in sheltered locations where the tidal flow is slow, allowing silt to settle out from the water. Sites such as the North Bull, Strangford Lough, Morecambe Bay, the Wash and Budle Bay all hold thousands of waders in winter, and also Brant Geese and Eurasian Wigeon feeding on the *Zostera* beds.

Our offshore islands, especially if free from terrestrial predators, provide some of the most spectacular colonies of sea birds in Western Europe. As with the sea cliffs, the island colonies with the greatest numbers and diversity generally occur in the north and west, where the birds are closest to rich feeding and disturbance is less. In the northern isles of Orkney and Shetland, the more common terns, gulls and auks are supplemented by Great Skuas and Arctic Jaegers, Black Guillemot and European Storm Petrel, the latter occurring in especially important colonies along the west coast of Ireland. A select few of the islands facing the Atlantic (e.g. Stags of Broadhaven, Rona, St Kilda, Foula) have breeding colonies of Leach's Storm Petrel. Manx Shearwaters breed on islands in the Irish Sea, around the coasts of Ireland and in the Hebrides and St Kilda. Together, these seabird populations are probably the most globally significant avian communities in our archipelago.

STRUCTURE OF ORNITHOLOGY IN GREAT BRITAIN AND IRELAND

SOME KEY ORGANISATIONS

There are currently four main tiers of organised ornithology within Britain and Ireland – with national, regional, county and local organisations that act as foci for ornithological or birding activities. Additionally, there are several universities as centres of research where students gain training for careers in the profession or for lives outside mainstream biology. Not linked to any particular organisation, there are a number of influential birding magazines, including the long-established *British Birds*.

The main national societies in the UK are (in order of age), the British Ornithologists' Union (BOU), the Royal Society for the Protection of Birds (RSPB), the British Trust for Ornithology (BTO) and the Wildfowl & Wetlands Trust (WWT). In Scotland and Ireland, the Scottish Ornithologists' Club (SOC), BirdWatch Ireland (BWI), the Northern Ireland Birdwatchers' Association (NIBA) and the Northern Ireland Ornithologists' Club (NIOC) combine the roles of national and local bird clubs in their regions. In the Isle of Man, the Manx Ornithological Society fulfils the same role. In England and Wales there are also county naturalists' trusts and county and local bird clubs.

The BOU is one of the oldest ornithological societies in the world, founded in 1858. Its membership is international and heavily weighted towards professionals. The BOU publishes *Ibis*, one of the leading journals of ornithology and holds regular scientific conferences on a wide range of topics. It also maintains the British List, recognised by national and regional agencies as the most authoritative compilation of the British avifauna. The list is managed by the BOU's Records Committee (BOURC), which works closely with a subcommittee concentrating on taxonomic matters (the Taxonomic Sub-committee, TSC). The British Ornithologists' Club began life in 1892 as a forum for ornithological discussion – a dining club for London members of the BOU, where they could meet, talk and often exhibit specimens collected around the world, including interesting material from Britain and Ireland. Early in its history it began to publish a bulletin, which initially included a range of material about British birds. This evolved into a leading journal with an international reputation in taxonomy, and, apart from publishing its *Bulletin*, the BOC now plays only a minor role in British ornithology.

The RSPB was, as its name implies, initially a conservation organisation. It was founded in 1889 and has grown over the years until it now numbers over one million members. It still concentrates on conservation, maintaining an estate of over 200 reserves, many with full-time wardens and management teams. Some of the reserves protect particular species (such as Osprey at Loch Garten in Scotland); others are more broadly based (e.g. Minsmere in Suffolk for reedland species). Increasingly, the RSPB is turning to large-scale habitat management and international policy, conserving more than just birds. It also has a dedicated team of investigators who monitor wildlife crime, working closely with police and national conservation agencies to ensure that wildlife legislation is effectively implemented. An increasing amount of the RSPB's efforts are directed towards lobbying local and national government (both here and overseas) on behalf of birds and conservation. As the society has grown in size (and wealth), it has also increased its research investment. This is largely directed towards individual threatened species (e.g. Black-throated Loon, Red-necked Phalarope, Cirl Bunting) but is increasingly being extended to habitat research (e.g. the Scottish flow country). The RSPB is also the UK's formal partner with BirdLife International.

The BTO is a smaller organisation which plays a crucial role in coordinating large-scale fieldwork for bird conservation. It specialises in encouraging volunteer birdwatchers to record birds for individual species' surveys (e.g. European Nightjar, Northern Lapwing), for mapping projects (e.g. three national atlases) or for community recording (e.g. breeding and wetland birds). The BTO also administers the national ringing scheme. Equally valuable surveys involve nest records and garden birds. Some of these will be described in more detail later. The BTO provides professional scientists to assist with the analysis and publication of the results of these surveys, either in its own journals *Bird Study* and *Ringing & Migration* or more widely among the scientific and birding communities.

The WWT is a specialist wildfowl organisation. It is the smallest of the national bird conservation agencies, having been founded in 1946 as the Wildfowl Trust. In addition to undertaking research into wildfowl, it also manages reserves, generally for wintering ducks, geese and swans. Some of these have wildfowl collections where the public can learn about these attractive birds. It produces an annual journal *Wildfowl*, and a more popular magazine, *Waterlife*.

BWI combines the roles of the RSPB and the BTO in the Republic of Ireland, managing a series of reserves, undertaking conservation research, and coordinating surveys, including major exercises such as atlases, often in collaboration with the BTO. It publishes the journal *Irish Birds*.

COORDINATED FIELDWORK

A number of national surveys are used to determine the distribution and abundance of our wild birds. Some of these run in parallel in both Britain and Ireland, coordinated by the BTO and BirdWatch Ireland, while others are country-specific or confined to, for example, Scotland. The more important surveys include the Common Bird Census (CBC), Breeding Bird Survey (BBS), Waterways Bird Survey (WBS), Wetland Birds Survey (WeBS/I-WeBS), Nest Record Scheme (NRS) and the Garden Bird Feeding Survey (GBS). At approximately 20-year intervals, national projects are undertaken to map bird distribution ('Atlases') and, in the Second Breeding Atlas, the abundance of breeding birds. There has also been a single national Winter Atlas. Because data from all of these are important to understanding the status of our wild birds, these surveys are described here. The ringing scheme and the role of bird observatories are also discussed, along with some single-species surveys.

The severe winter of 1962/63 marked a turning point in British and Irish ornithology, for it drew two things to our attention: firstly, that there had been massive mortality among our common birds and, secondly, that it was difficult to assess the extent of this because we did not really know how many birds there had been before. This accelerated the introduction of the Common Bird Census (CBC), the principles of which are relatively straightforward. BTO volunteers choose manageable areas, typically *c.* 70 hectares but varying depending on the complexity of the habitat (woodlands plots are generally *c.* 20 hectares). Visits are made at regular intervals through the breeding season, and every bird that is present is recorded on a visit map; birds deemed unlikely to be breeding (such as gulls overflying a woodland) are excluded from the analysis. At the end of the season, the data for each species are pooled onto a single map, allowing BTO staff to estimate the number of pairs or occupied territories. In the short term, this allows an estimate of the density of birds in that habitat. In the longer term, changes in the number of breeding birds can be monitored. There are two serious drawbacks to this exercise: it is labour and time intensive. To get a full picture of the bird community, it is necessary to make about 10 visits per season to each site, and a visit can last several hours. Because the fieldworkers are largely volunteers, there is little control over the sites chosen. Human nature being what it is, many recorders chose 'interesting' sites: if you are gong to spend 30 or 40 hours a year walking round a piece of countryside, you prefer it to have some birds in it! There was a strong bias towards areas close to towns, because recorders did not want to drive long distances, and sites such as inner cities, plantation conifers and intensive arable farms were all under-recorded (because of an actual or perceived lack of birds).

Nevertheless, this is a valuable exercise, which generates an array of time-series data, but, because of the bias in cover, it does not accurately address the key questions of where our birds are and what is happening to them. Consequently, in 1994, the Breeding Bird Survey (BBS) was launched in Britain and, in 1998, the Countryside Bird Survey in the Republic of Ireland. These similar surveys are based on 1-km squares of the national grids and require two visits each year: one early in the breeding season and one later. Two transects are made across each square, sufficiently far apart as to minimise double-counting of the same individuals. Every bird seen or heard is recorded, again omitting obvious non-breeding species. Because a typical visit lasts about two hours and since only two visits are required each season, this is a much less demanding scheme. Equally important, since the surveys are based upon 1-km squares, it is possible to design them to give random representations of habitats. As with the CBC, observers tend to be concentrated near centres of population. The survey has no density component and the smaller number of visits means that more birds are missed than in CBC fieldwork. In 2005 a total of 2,879 sites were surveyed across the UK by more than 2,300 volunteers, along with about 400 sites in the Republic of Ireland (about 300 volunteers). The increased number of survey locations means that inter-year trends are statistically more robust than the CBC.

The Waterways Bird Survey (WBS) began in 1974, with the intention of providing census data for breeding birds along linear features such as rivers and canals. These habitats, and consequently the species that live there, were poorly covered by CBC censuses and are still under-represented in BBS surveys today. The underlying rationale is essentially a linear CBC; observers make nine visits to the site each year, walking along the water's edge and recording all the waterside birds that are seen and heard. As with CBC, this allows the number of pairs or territorial birds to be mapped, and hence the density and abundance to be tracked on an annual basis. A Waterways Breeding Bird Survey (WBBS) began in 1998 and is currently being assessed as a less intensive project, related to WBS as BBC is to CBC. A similarly named survey with a different methodology was launched in the Republic of Ireland in 2006.

The WeBS project (called I-WeBS in Ireland) began life over 50 years ago as the 'Wildfowl Counts' and involved a series of regular winter visits to inland and coastal wetlands to record the numbers of ducks, geese and swans. Over time, it has become a monthly exercise, and the species recorded have been extended to include waders, herons, egrets, etc. Some sites have now been counted for more than 40 years, and the array of data is impressive. The importance of individual sites for particular species can be assessed; changes in abundance can easily be seen, as can patterns of arrival and departure. Many known important sites are not covered because of a lack of volunteers. As with all regular surveys, sites away from the centres of population tend to be under-represented. Nationally, certain species, such as Common Moorhen, Mallard and Eurasian Teal, are under-recorded because so many occur away from the larger, surveyed areas. Nevertheless, repeated observations from the same sites over many years have produced invaluable data for assessing the distribution and abundance, and especially changes in abundance, of our wetland birds.

The most popular of the BTO surveys is Garden Birdwatch (GBW); this began in 1994, partly in recognition that gardens are an increasingly important habitat and one that is under-recorded by more conventional fieldwork. It also gave the opportunity for a wider range of observers to participate in organised research efforts, and it is certainly popular: currently, about 16,500 recorders send in weekly data on the birds using their gardens. BirdWatch Ireland runs a similar (but not identical) scheme called the Garden Bird Survey, with nearly 1,000 forms returned in 2005–06.

Two further exercises managed by the BTO are important in assessing the status of the birds of Britain and Ireland: nest recording and bird ringing.

The Nest Records Scheme (NRS) is one of the longest-running ornithological surveys in the world, having started life in 1939 as the 'Hatching and Fledging Enquiry'. For nearly 70 years, observers have found nests and noted the fate of their contents, at a current rate of about 30,000 nests per year. These data record the productivity of individual nests, and the pooled data allow overall assessments of annual reproductive effort and success. The number of records is awesome, but inevitably some species are reported more often than others. There is now a total of some 1.4 million records, ranging from tens of thousands of Blue Tits to one each of Eurasian Hoopoe, Common Rosefinch and Short-toed Treecreeper. This does not mean that there has been only one breeding record of these species, rather that only one nest record has been submitted for each. Birds that are scarce and/or declining tend to be under-recorded; this can make the interpretation of population changes hard to assess.

Bird ringing in Britain and Ireland is coordinated by the BTO: training, supervision, the sale of rings and equipment, and annual reporting are all centrally administered. There are over 2,000 ringers, who currently process about 850,000 new birds every year. In Britain, ringing began almost simultaneously in 1909 at the University of Aberdeen (organised by Arthur Landsborough Thomson) and through the fledgling *British Birds* (run by Harry Witherby). It was originally a tool primarily to study migration, but there has been a progressive shift over the years towards the monitoring of populations through a more systematic approach. The early days of 'ring and fling' have been replaced by detailed studies of individual species, groups or habitats, by both amateur and professional ornithologists. While migration studies continue, most notably through the network of coastal bird observatories and migrant ringing sites, there is also a disciplined programme at 'Constant Effort' Sites (CES), providing input into population models for a wide range of species. The CES involves regular (usually weekly) visits to a particular site and the use of the same number of nets in the same locations. A comparison of data over time allows estimates of changes in abundance; furthermore, the identification of different age classes (either through plumage-specific characters or by reading the rings of surviving birds) allows detailed assessments of reproductive success and survival. These data have helped to determine the causes of population changes: declines in breeding success, changes in over-winter survival, etc.

In the early days, the data from both these surveys were all handled as paper copies and stored in files at the BTO headquarters. It has been estimated that the nest records cards alone would make a pile 300 m high, about half as tall again as Canary Wharf! Fortunately, the advent of computers brought a revolution in both input and analysis of the data. Almost all of the ringing information and over half of the nest records data are now submitted electronically.

Reports from all of the surveys are published annually. Data can be extracted rapidly from a diversity of sources for a particular investigation, and integrated approaches to population monitoring, drawing on multiple surveys, are now well established. If a species is in decline, CBC, BBS and WeBS data can indicate when the decrease began and whether it is universal across the regions and habitats. Nest record data can show whether there have been changes in reproductive output. Constant-effort ringing may show whether there are changes in the ratio of adult to juvenile at the end of the breeding season that indicate post-fledging mortality. For migrant species, the analysis of daily counts from coastal observatories may show whether there has been a significant change in the number and date of returning birds, which might indicate problems on the wintering grounds. Unfortunately, some birds (e.g. owls, nightjars) do not lend themselves readily to diurnal survey work. Others (such as Little Ringed Plover, Dartford Warbler, Red-billed Chough) are sufficiently localised that they do not occur in many survey sites, while yet more species are thinly distributed (e.g. Hawfinch, Lesser Spotted Woodpecker) and are only rarely encountered by surveyors. Upland birds are poorly represented in CBC plots and of particular concern in Ireland. For these, special surveys are organised. Although methodologies vary depending on the species, these typically involve a combination of timed visits to known sites to estimate the absolute number of individuals, combined with a more broad-scale (random) survey of the core areas and suitable habitat to determine whether the range is expanding or contracting.

The seabirds of Britain and Ireland are of international importance, not just in their own right but as indicators of the health of our oceans. There are two main strands to the monitoring of their populations. A series of sample sites are censused annually; these are mostly at places such as Fair Isle (Shetland) and the Isle of May (Fife), where there is an ornithological infrastructure that can provide accommodation and support. Occupied nests and breeding success are recorded systematically in designated parts of the colony. These data allow the determination of population trends (albeit at a limited number of sites) and annual productivity of the species involved. Additionally, at approximately 20-year intervals, a more comprehensive survey is undertaken to count every colony around the coasts of the entire archipelago, including the most remote offshore islands. Results from the three most recent surveys were published in book form: Cramp *et al.* (1974) for *Operation Seafarer* in 1969–70, Lloyd *et al.* (1991) for the *Seabird Colony Register* in 1985–88, and Mitchell *et al.* (2004) for *Seabird 2000* in 1998–2002. At a time when the marine environment is subject to a range of challenges, including pollution, overfishing and global warming, these data are of inestimable value for informing science and government of the objective status of our islands' seabirds.

The network of coastal bird observatories has provided information on bird migration since WWII, in addition to offering inexpensive accommodation and training to visiting birders. Their wardens have been in the forefront of developing techniques for identification and the ageing and sexing of birds in the hand, which have subsequently been transferred to other fieldworkers. Daily rounds of recording and trapping have provided data that can be used to monitor migration and vagrancy. The various sites offer different experiences: Portland Bill, Dungeness and Sandwich Bay on the south coast of England record arrivals and departures every year; Cape Clear in SW Ireland is a key site for observing movements of oceanic seabirds and offers a first landfall for many vagrants from North America. Observatories on Copeland, Bardsey and (until its closure) Skokholm provide data on breeding Manx Shearwaters, as well as recording movements of landbirds through the Irish Sea. In the far north, Fair Isle and North Ronaldsay have attracted legions of birders seeking the Siberian vagrants that are found there with such remarkable regularity. With the increased flexibility that mobile phones, pagers, cheap flights and fast cars provide, the need to stay at a bird observatory may be reduced, but to a serious birder nothing can replace the expectation and intensity of an early morning trap round on Fair Isle in late September when the wind has been easterly all night!

The national organisations and surveys are complemented by an astonishing array of specialist, regional and local bird groups that focus on raptors or other particular species, the local county or perhaps individual sites. Most produce reports that are more or less annual and represent a substantial repository of detailed knowledge. Attempts are underway to standardise record-keeping, so that information can be shared and analysed when needed. One specialist group, the North Sea Bird Club, has been coordinating sightings from offshore platforms and ships serving the North Sea oil and gas industry for nearly 30 years. Observations depend on the intermittent presence of birders on any particular structure and thus lack the systematic approach of the bird observatories but, nevertheless, they have yielded much additional information on birds crossing the North Sea, particularly in spring and autumn.

In the preparation of this book, we have used data collected by all of these organisations and surveys. We thank all their participants, from dedicated patch workers counting their sites, through the data handlers who have input the results for computer analysis, to the scientists and researchers who have designed the surveys, encouraged the fieldworkers and seen the analyses through to publication.

ORNITHOLOGICAL PUBLICATIONS

Britain and Ireland are served by a rich ornithological literature. Periodicals include the purely scientific *Ibis* through *Bird Study*, *RSPB Conservation Review*, *Wildfowl*, *Ringing & Migration*, *Irish Birds*, *British Birds*, *Birding World* and many more. These all address different needs and overlapping audiences. *Bird Study* publishes research focused on field ornithology, with a European emphasis. While of international reputation, it is published by the BTO and aims to be readable by the BTO membership. *Conservation Review* and *Ringing & Migration* often carry material of a more technical nature, addressed to their specialist end users – conservationists and ringers. We have already referred to the (usually) annual reports of the county societies, specialist groups and other local organisations. Such groups often undertake their own surveys and projects and have published a number of fine-scale local atlases. There is also an increasingly important 'grey' literature of limited-circulation research and consultancy reports. Whilst locating relevant information has in many ways never been easier, this 'grey' literature is often difficult to find out about and track down unless published on the internet. This literature is often unverified and in general we have been cautious about its use.

British Birds is a monthly not-for-profit journal that has been appearing for over a hundred years. Targeted at the serious birder, it publishes papers on identification, distribution, ecology and behaviour in a format and style that is readily accessible. It also publishes annual reports on rarities and rare breeding birds, accounts of birds new to Britain and changes in status, and articles on taxonomic developments relevant to the British list. The annual *Irish Birds* serves a broadly similar purpose for Ireland, and both are loosely considered 'journals of record'. The monthly *Birding World* specialises in identification and bird-finding articles and rapid reporting of the latest rarities, accompanied by high-quality photographs.

Developments in information technology and digital imaging have led to a revolution in publishing, and ornithology has not lagged behind in its implementation. Books can now be produced quickly and (relatively) cheaply. A glance at the style and presentation of the long-running 'New Naturalist' series published by Collins shows how maps, diagrams and photographs can now be seamlessly incorporated into text, simultaneously adding to the information content and livening up the physical appearance of the finished article. Three styles of contribution will serve as examples.

First, the regional or global books. For years, *The Handbook of British Birds* was the regional *sine qua non* for British and Irish birders, until it was replaced in the latter part of the 20th century by *Birds of the Western Palearctic* (BWP). This was followed by other multi-volume regional texts such as *Birds of Africa (BoA)*, *Handbook of Australian, New Zealand and Antarctic Birds* (HANZAB) and *Birds of North America* (BNA), although the latter was published as a series of single-species booklets. The bar in ornithological publication was raised again by Lynx Edicions with the *Handbook of the Birds of the World* (HBW), a truly magnificent publication with outstanding text and stunning photographs.

Second, in the 1970s, hitherto little-known publishers Trevor and Anna Poyser began a series of specialist accounts of single species or local areas. Often copied but rarely bettered, T & AD Poyser expanded into publishing atlases, national bird books (England, Scotland, Ireland, Wales) and also the results of the seabird surveys that we have already mentioned. They became the ornithological publisher of choice for 'citizen science' by the end of the 20th century. In parallel with Poyser, Christopher Helm created a publishing house that evolved through a series of imprints, focusing more on field guides and identification handbooks. We have called upon the output of both of these publishers extensively throughout this work.

The third publishing development that is particularly significant is the production of high-quality regional avifaunas. Of these, we draw attention especially to Shetland, Norfolk, Suffolk, Scilly and Dorset. Sometimes (e.g. Wiltshire, Isle of Man) these include the results of county atlas surveys; others are more 'traditional'. In all cases they have brought a high quality of publication to regional bird books.

THE BRITISH AND IRISH LISTS

Until the early 20th century, there was no formal structure for maintaining a list of birds that had been recorded in Britain and Ireland. The 'British List' had its genesis in the early work of Yarrell, who produced three editions of his *History of British Birds* between 1837 and 1856 (see Knox 2007). At this time, Ireland was treated as part of 'Great Britain'. While Howard Saunders produced further revisions and editions of Yarrell's work beyond the turn of the century, the first 'official' list was published by the BOU in 1883, based largely on Saunders' editions of Yarrell.

Changes to the avifauna of Britain and Ireland were usually published in *Ibis*, the *Zoologist* and the *Bulletin of the British Ornithologists' Club* and (after its founding in 1907) increasingly in *British Birds*. Updated checklists were published by the BOU in 1915, 1923, 1952, 1971, 1992 and 2006. The 1971 edition included a detailed discussion of distribution and abundance. The two more recent publications have been relatively simple lists.

After its 1915 checklist appeared, the BOU formalised the maintenance of the list by establishing a committee for this purpose. Down the years, many of the leading figures in ornithology from these islands have served on this committee, including Ernst Hartert, Francis Jourdain, Norman Ticehurst and Harry Witherby between the wars; David Bannerman, Cyril Mackworth-Praed, Richard Meinertzhagen, Mick Southern and Bernard Tucker after the Second World War; and James Ferguson-Lees, Peter Grant, Tim Inskipp, Richard Porter, Robert Ruttledge, David Snow, Keith Vinicombe and Ian Wallace since the 1950s. The committee is now named the BOU Records Committee and, to date (August 2009), 59 reports have been published in *Ibis* since 1918. The workings of the committee were described by Marr (1993). In 2000, BOURC recognised the specialised nature of taxonomic advice and the increasing volume of work in this area and re-established a subcommittee with international participation to help deal with this aspect of the list. The Taxonomic Sub-committee's (TSC) reports are also published in *Ibis*.

The BOU's periodic checklists covered Britain and Ireland from 1883 to 1992, and Britain alone in 2006. In 1999, the BOU stopped maintaining a joint list for Britain and Ireland, in response to a request from ornithologists in the Republic. Amongst other reasons, wildlife legislation in the four parts of the islands (Great Britain, Northern Ireland, Isle of Man, Republic of Ireland) are separate and each requires its own faunal list. It was confusing and complicated for the BOU to maintain separate lists for each region as well as a pooled one for the whole of Britain and Ireland, and this latter remnant of 'colonial' ornithology had been inappropriate for some time.

The present volume again covers all of Britain and Ireland and has been prepared with the generous collaboration of colleagues in the Isle of Man, Northern Ireland and the Republic of Ireland, in recognition of the zoogeographic affinities of our avifauna and the value of such a volume to birders and ornithologists.

In the 1950s, although records of birds new to Britain and Ireland were considered by the BOU, assessment of other rare birds (not firsts) was largely by the editors of county bird reports, local avifaunas, *British Birds* and the recently started *Irish Bird Report*. This had worked well up to a point, but it lacked consistency and accountability and was increasingly struggling to cope with the growing number of birdwatchers and their observations. In 1959, a rarities committee was established by *British Birds*, and for the first few years it covered both Britain and Ireland. In time, it became the British Birds Rarities Committee (BBRC), severing its dependence on the journal, though maintaining a close working relationship. The history of this organisation has recently been recounted by Dean (2007a). Ten voting members are elected by the county bird clubs and bird observatory wardens. Retirement is by rotation, the axe falling annually on the longest-serving individual at that time. Stability is maintained by a permanent Chairman and the BBRC Secretary. The list of birds that is considered has changed over the years; some species have been removed as they became commoner or better known. A report is produced every year in *British Birds*, as well as occasional status and identification reviews.

The Irish Rare Birds Committee (IRBC) and the Rarities Committee of the Northern Ireland Birdwatcher's Association (NIBARC) fulfil similar roles in Ireland. Reports are published by both organisations in *Irish Birds* and the *Northern Ireland Bird Report*. For reasons of space, we do not give detailed references to individual reports of rare birds; readers are referred to the relevant year of the appropriate report of the several rarity committees. Similarly, for rare breeding birds, we refer to 'RBBP' for information relating to the publications of the Rare Breeding Birds Panel for Great Britain (also published in *British Birds*).

National lists are important in all four regions within Great Britain and Ireland, as their composition impinges upon conservation policy and legislation. Decision-making is thus required to be independent and objective.

ENGLISH NAMES

The names given to our birds have evolved through the course of time. Local or vernacular use gave us the evocative Peewit (Northern Lapwing), Tystie (Black Guillemot), Cushie Doo (Common Wood Pigeon), Whistler (Ring Ouzel) and even Bum Barrel (Long-tailed Tit). Such names often varied across Britain and Ireland, frequently with confusing consequences. Many countries around the world also had their own English names: the same species might have different names in different countries and different birds might share the same name; three different species were all called 'White Pelican', for example. Although there was a perfectly satisfactory series of unique scientific names, pressure grew for a degree of standardisation, mainly from the growing band of international birders. After much consultation and a long and complicated gestation, the International Ornithological Congress (IOC) recently published recommendations for a world list of English names (Gill & Wright 2006, updated to version 2.1 at www.worldbirdnames.org). In recognition of the international nature of ornithology today, these have been adopted by the BOU and are the names that we have used here. Because some of these new names are unfamiliar within GB&I, in particular cases we have given names in current use as well.

EVOLUTION AND TAXONOMY

EVOLUTION

When we look at birds, we are seeing animals moulded by the forces of evolution. Through countless generations, pairs of birds have reproduced, usually generating more offspring than the environment can sustain. This overproduction leads to a 'struggle for existence', which Charles Darwin and Alfred

Russell Wallace independently recognised would result in the best-adapted individuals surviving at the expense of the rest. Provided that the characters that enabled them to survive are inherited from parent to offspring, there will be gradual evolutionary change through the process of natural selection. The consequence of this continuous process of natural selection is the range of adaptation of birds to their environment that we see in nature. The evolution of these adaptations is mediated by changes in gene frequency. The anatomy and physiology of an animal is largely determined by the array of proteins that are produced through its genes. Many genes have multiple, alternative forms ('alleles') within a population, and an individual inherits one or other of these alleles from each parent. For example, selection for increased migratory performance leads to better survival of those individuals that possess the alleles producing proteins that improve migratory efficiency, and these alleles will increase in frequency from one generation to the next as long as there is genetic variation on which the selection can act. The allele frequencies change and the population evolves.

Not all evolution is driven by selection. Especially in small populations, the random breeding failure of individuals carrying a particular allele (usually a rare one, though not always) may result in the loss of that allele from the population. The population has changed its composition: it has evolved, although not as a result of any form of natural selection. Because these evolutionary changes are random and their effects tend to be individually small, the changes in allele frequency that they generate are called 'random genetic drift'.

In the fullness of time, populations will progressively evolve in these ways until they are recognisably different from their ancestral type. This change may be linear, as a population incrementally changes through a series of intermediate stages, slowly becoming increasingly different from the original in structure and function. The ancestral form gradually transforms into another; this type of progressive linear evolution is called 'anagenesis'.

Alternatively, a population may somehow become separated into two isolates between which there is no gene flow. In European ornithology, this event has often been driven by a glaciation separating birds into refugia in the south-west (Iberia and NW Africa) and south-east (the Balkans and Asia Minor). Over the course of time, these allopatric populations (inhabiting different geographic regions) evolved independently, by a combination of adaptation and selection or genetic drift, and gradually and progressively diverged in genetic structure. Any movement of individuals between these populations would reduce the rate of divergence; alleles that are declining in one population could be reintroduced by this 'gene flow'. In evolutionary biology, individuals that move between populations in this way are termed 'migrants'. They differ from migrants as conventionally recognised in ornithology, which are birds that move seasonally between regions, usually returning to breed in the natal area. Genetic migrants move between populations and become incorporated into a new gene pool, thereby changing its genetic composition. In short, selection and genetic drift promote population differentiation; genetic migration retards this divergence. Ultimately, though, they will become increasingly differentiated, until two recognisably different forms have arisen; such evolution by the splitting of lineages is termed 'cladogenesis'.

Birds evolve and change by these processes of anagenesis and cladogenesis. Because of mankind's need to name things, the science of taxonomy has arisen to identify, name and arrange the multitude of life forms that exist on our planet. The basic unit of taxonomy is the species, and defining what we mean by this has caused great consternation for a long time. Through the second half of the 20th century, the production of species definitions developed into a 'minor industry' (de Queiroz 1998), with at least 25 alternatives being mooted at one time or another. As de Queiroz points out, many of these do not actually tell us what a species is; rather they tell us how we can recognise its existence. He calls them 'special cases' of a more general concept: the General Lineage Species Concept (GLSC), which regards a species as an entity composed of organisms comprising part of an evolutionary lineage. In this, it is closest to the Evolutionary Species Concept (ESC) originally developed by Simpson (1951). It is unfortunate that this is called 'evolutionary', because it might imply that other species concepts are 'not evolutionary'; this is quite untrue: almost all of them are evolutionary in one way or another. The ESC regards a species as 'a single lineage of ancestral descendant populations of organisms which maintains its identity from other such lineages and which has its own evolutionary tendencies and historical fate' (Wiley 1978). Both ESC and GLSC stress the idea of a species being a 'lineage' that is independent from other lineages. The latter is the concept that the BOU Taxonomic Sub-committee has adopted and which is applied to recommendations regarding species-level taxonomy as it affects the British List (Helbig *et al.* 2002).

So, what is a lineage and what does 'maintains its identity' mean? A lineage is a single line of ancestry and descent. When a lineage splits into two or more offshoots, a clade is formed: a clade being a group of lineages with a single common ancestor. If these lineages do not merge back into a single unit, they are said to be maintaining their identity. At some time in the past, before one of the many European glaciations, the ancestor of Icterine and Melodious Warblers was evolving as a lineage. When the ice came down through central Europe, the population was separated into eastern and western isolates. For many generations, there was no exchange of birds between the two populations and they gradually accumulated differences in DNA sequences, plumage and vocalisations, until they are now clearly distinct as separate lineages. Even though Icterine and Melodious Warblers meet along a zone of contact, hybridisation is rare and the offspring have reduced viability, so the species maintain their integrity.

The problem faced by avian taxonomists is that they rarely have much direct data relating to the evolutionary history of their subjects; bird bones are fragile and only rarely found as fossils. The lineages have to be interpreted from a horizontal 'slice' or snapshot of evolution, represented as the contemporary fauna. By comparing the anatomy, physiology, behaviour, vocalisations and now DNA, the existence of independent lineages must be inferred. We identify lineages through their diagnosability; each lineage must be clearly differentiated in characters that are inherited and whose differences imply a prolonged period of independent evolution. In the past, this was largely a matter for anatomists and museum workers; we recognised the lineages by their different appearance, often from skins in a museum tray or bones stored in boxes. In recent years, ornithologists, both amateur and professional, have increasingly worked in the field, observing birds and identifying characters that can be used to differentiate them. Behaviour, ecology and population genetics are now routinely being used to assist in defining the variation within and differentiation between groups of birds, which can then be recognised as separate lineages.

Variation in characters such as plumage pattern and colour, wing and leg length, bill size and even weight have always been used, but now, with the advent of more sophisticated statistics and computers, anatomical variation is being quantified as never before (e.g. waders: Engelmoer & Roselaar 1998). The advent of portable tape recorders has allowed researchers to record vocalisations and either analyse them in the laboratory (e.g. crossbills: Summers *et al.* 2007) or play them back to territorial males and assess their reactions (e.g. leaf warblers: Alström & Olsson 1995). Population biologists have been able to monitor nesting birds in hybrid zones and determine characters such as egg volume, nestling survival or even the fertility of surviving progeny from assortative and dis-assortative pairs (e.g. Pied Flycatchers: Alatalo *et al.* 1990). Data such as these are permitting more detailed analysis of populations than ever before; in particular, comparison of populations allows us to assess the diagnosability of individuals and the degree of genetic isolation: essential parameters in assessing species limits.

MOLECULAR ANALYSIS

The greatest development in recent years has been the application of quick and relatively cheap techniques for DNA sequencing. DNA is the genetic material and is transmitted from parent to offspring in the egg and sperm. It is a linear molecule comprising two strands that are held together by pairs of cross-linking chemicals called bases. There are four of these bases, called guanine, cytosine, thymine and adenine, and they are regularly spaced along the strands. What is unique about them is that guanine only binds to thymine and cytosine only binds to adenine. A gene comprises a stretch of this double-stranded DNA, and the order of the bases along that gene determines the structure of the protein that it codes for. The double-stranded nature of the molecule allows a precise replication of the gene when a cell divides. The molecule simply splits along its length, and each strand is used by the cell as a template to make a new 'complementary' sequence. This process is very precise, but errors inevitably creep in: the 'wrong' base may be substituted; a base may be omitted or an additional one inserted. Such changes in the structure of the gene are mutations and are the ultimate source of genetic variation; their incorporation into or elimination from populations are the consequences of evolutionary change. In general, the longer two populations have been isolated from one another, the more time there has been for the accumulation of mutational changes, and hence the more different they will be in DNA sequence.

The technology now exists to read the sequence of bases along a strand of DNA. If the same gene is analysed from different individuals, it is possible to determine the extent to which they differ; this is sometimes called the 'genetic distance' between the individuals. Individuals that have a recent common ancestor, such as members of the same population, will be genetically more similar than those that have a longer period of independence, such as individuals from populations that are geographically well separated. There is a subculture of evolutionary genetics concerned with the analysis and interpretation of these genetic distances; its researchers construct phylogenetic ('evolutionary') trees to represent the relationships of the various individuals. Sequences that are most similar lie close together in the tree; those that are genetically more different are further apart. These trees are not necessarily the true evolutionary history of the individuals; they are a hypothesis of their relationships based on genetic distances estimated from the sequence data. Of course, the researchers hope that the trees are true representations, and the more independent lines of evidence that support the relationships (sequences of different bits of DNA, vocalisations, plumage, etc.) the more confident they are with its accuracy.

There is now a more sophisticated method of interpreting the sequences. Rather than simply comparing the overall difference between individuals in terms of percentage divergence, it is possible, by looking at the individual bases that differ between the two sequences, to attempt a reconstruction of the route by which one sequence evolved into another. This is much more time-consuming but is closer to the actual evolutionary history of the individuals.

Different genes evolve at different rates. If a gene codes for some vitally important chemical in the bird's body, then any mutational change is likely to be deleterious and even kill its bearer. Genes that are less critical can evolve and differentiate more markedly. This allows the researcher a degree of flexibility in approach. If deep differences are being investigated (such as the separation of passerines and non-passerines), a slowly evolving gene would be chosen. Conversely, if the investigation is into population differentiation among Dunlins, a more rapidly evolving gene would be sought. The most rapidly evolving sequences are generally those in the DNA present in organelles called mitochondria. These structures are transmitted from mother to offspring in the cytoplasm of her eggs. Consequently, every individual receives a single mitochondrial sequence; the mitochondria of males are (generally) not passed on, so a breeding pair has only one transmitted sequence. For a gene in the nucleus, there are typically two copies: one inherited from the two present in the father and one from the two in the mother. There are thus effectively four times as many copies of a nuclear gene in a population as there are mitochondrial. Since evolution occurs more rapidly in small populations, the mitochondrial sequences evolve faster. Additionally, nuclei have mechanisms for repairing mutations; these are not 100% effective, but nevertheless many mutations do get repaired. There are no repair systems in mitochondria, so any mutation in a mitochondrial gene remains to be passed on or lost, depending upon its fate. These two factors result in more rapid divergence of mitochondrial sequences. Added to this, while some mitochondrial genes are under selective control (Zink 2005), other sequences (such as the 'control region') are less constrained and, as a result, can evolve more rapidly.

LINEAGES AND CLADES

Mitochondrial gene sequences are now widely used in avian systematics to assess the degree of molecular divergence or similarity among individuals, populations, subspecies, species or higher taxa. The construction of a molecular phylogeny gives a hypothesis of the routes through which taxa may have evolved to the present day. Groups of closely similar taxa cluster together in the phylogeny; those that can be traced back to a common ancestor are called clades, and frequently molecular clades bear a close resemblance to the relationships postulated by more conventional taxonomy. For example, the two forms of Bonelli's Warbler spring from a common branch: they are 'sister taxa', and Wood Warbler lies within the same clade (Helbig *et al.* 1995). This is in line with the findings of conventional taxonomy: an ancestral lineage split at some point in the past to give rise to two lineages: Wood and Bonelli's Warblers. The latter split again (probably during a glaciation) to give rise to Western and Eastern Bonelli's Warblers; the molecular difference between these two is similar to that between Wood and Bonelli's (*sensu lato*). There are countless examples where molecular data supports morphological taxonomy; the instances where they are not in agreement usually indicate the need for further study.

A species is a segment of an evolutionary lineage, but how do we determine that two lineages are sufficiently divergent to merit being treated as separate 'species'? As noted above, the Evolutionary Species Concept gives an insight into this problem. A species is 'a single lineage of ancestral descendant populations of organisms which maintains its identity from other such lineages and which has its own evolutionary tendencies and historical fate'. It is a group of individuals that are diagnosably distinct from other such groups and that show a level of reproductive isolation that leads us to believe that they are unlikely to merge. The BOU Taxonomic Sub-committee (TSC) has drawn up guidelines to help it determine which lineages could be recognised as separate species, publishing their findings in *Ibis* and at the *International Ornithological Congress* in Beijing (Helbig *et al.* 2002; Parkin *et al.* 2006). These 'guidelines' have been widely used by the TSC in assessing the specific status of avian groups as diverse as shearwaters (Sangster *et al.* 2002b), ducks (Sangster *et al.* 2001) and reed warblers (Parkin *et al.* 2004), and have attracted interest outwith ornithology (e.g. Beasley *et al.* 2005; Li *et al.* 2006).

If two taxa coexist, it is relatively easy to determine whether they are independent. They may never hybridise, or they may hybridise with clear evidence of reduced fitness among the progeny. What happens when populations are allopatric (their ranges do not overlap)? How can we tell whether birds that never meet are sufficiently divergent to merit separate specific status? Molecular data can be especially valuable here; if two taxa whose status we are investigating (e.g. Eastern and Western Bonelli's Warblers) show a similar degree of genetic differentiation to two that we know on other grounds are 'good' species (e.g. Wood and Bonelli's Warblers), then it is hard to avoid the conclusion that the pair under investigation should be given the same status.

Another valuable contribution that molecular data can make is through the identification of clades within the phylogeny. DNA sequences have shown us (Bridge *et al.* 2005) that the white terns ('*Sterna*') comprise several well-differentiated clades. These include groups such as (i) Common, Arctic and Roseate; (ii) Sandwich, Swift and Lesser Crested; (iii) Little, Least and Saunders's; (iv) Bridled, Sooty and Aleutian. The differences among these clades are large, in most cases sufficient to treat them as separate genera: *Sterna*, *Thalasseus*, *Sternula*, *Onychoprion*, respectively, though the evidence for *Thalasseus* is perhaps less strong. The data also suggest that Gull-billed and Caspian Terns are as different from these clades as they are from each other, supporting the view that each of these two belongs in its own genus: *Gelochelidon* and *Hydroprogne*. The merits of separate genera within the terns had been debated for ages; the molecular data helped in resolving the situation. There are many other instances where molecular sequences confirm or resolve generic or higher-level taxonomic placement.

Over the last number of years it has become increasingly apparent that the passerine sequence that has been in use has failed to reflect the evolutionary relationships of birds. The sequence presented here is substantially revised and follows Sangster *et al.* (in press).

SUBSPECIES

Critical to the recognition of species is that they should be diagnosably distinct from other species (Helbig *et al.* 2002). This diagnosability implies that the taxa have diverged in plumage, structure, behaviour, vocalisations and/or genetics to a degree that we can assign an individual to one or other species. However, evolution is a continuous process. Within a species, there may be variations in the environment that are selecting in different directions. For example, northerly populations of European birds are often compelled to migrate to avoid the rigours of winter; more southerly populations are not, and there is a cline of decreasing migratory tendency from north to south. Migrant birds need stronger muscles to drive longer wings, and individuals with these will be favoured in the north. They may also lay more eggs to compensate for the mortality imposed by the migratory journeys. Clines in wing length and clutch size will arise within the species. Alternatively, populations that are geographically separated (e.g. on islands) may diverge as a result of random genetic drift so that they become statistically different, though not necessarily sufficiently so to be diagnosably distinct. Given a few thousand generations, these differences may evolve to a stage where they become diagnosable, even if currently they may not be.

This scenario has been recognised for a long time through the adoption of the concept of 'subspecies'. Some species are fairly uniform in size, appearance, etc. across their full geographic range whereas, within some other species, local populations may show distinct variation from one another. Distinctive populations may formally be described and given a name to identify them and treated as subspecies. The place of subspecies in ornithology has been extensively discussed, e.g. Amadon and Short (1992), who proposed that they are 'spatial subpopulations that are morphologically separable at some reasonable level, e.g. 90%'. Implicit in this definition is that the morphological characters used to define the subspecies are under genetic control. Subspecies may be recognised from metric characters, such as wing or bill length; alternatively, they may be based on plumage details such as colour, barring, spotting, etc. There may be relatively isolated populations (such as the Wrens on island groups around Britain and Ireland) that show quite discontinuous differences; if the differences were much more marked, the various island populations might be treated as separate species. In Continentally distributed species, the variation is often more continuous (clinal), with few major discontinuities. Consequently, identifying the boundary between two subspecies is often not possible; they simply merge into each other.

Many researchers find subspecies useful because they indicate the range of variation within a species. Since evolution is a continuous process, it is possible that some of these subspecies are, in fact, on the way to evolving into a new species. If we came back in 100,000 generations, we might find that some of the subspecies we recognise now have evolved into fully diagnosable species in their own right. Alternatively, changes in the patterns of selection or gene flow may have caused closely related subspecies to merge back into a single type. But defining a subspecies means that, right now, it shows a degree of differentiation, and this differentiation should have an evolutionary cause. They may breed or winter in different regions; they may show adaptations to xeric or mesic environments; they may just be distinctive island forms. But whatever the cause, giving names to distinctive subspecies is valuable, since it allows ornithologists to communicate.

Morphology can evolve very rapidly and differences in anatomy may be quite recent. The patterns revealed by the analysis of these traits may bear no resemblance to the underlying molecular structure of the species. Zink (2004) has shown that the population structure revealed by the molecular analysis of many Nearctic subspecies bears little or no resemblance to the plumage patterns from which the subspecies were originally defined. The implication of this is that populations may be diverging at one level due to historical patterns (of glaciations, for example), but that superimposed upon this is selection for a plumage or metric trait. Eurasian and Nearctic populations of Whimbrels show marked differences in the colour of back and rump and also the underwing; within these two isolates there is clinal variation in size, which shows discontinuities across the Atlantic and Pacific Oceans. One evolutionary force has created the differences in plumage between the Nearctic and Palearctic; a second has resulted in differences in wing and bill length within each of these regions.

The analysis of sequence data also helps with assessing the evolutionary history of the higher groups: genera, families, orders, etc. We have seen that phylogenetic trees constructed from these data do not necessarily represent the true evolutionary history of the species involved; they are merely pictorial representations of the data or hypotheses of the paths that evolution may have taken. The more independent pieces of evidence that taxonomists accumulate which indicate the same relationships, the more confident they become that this is the true evolutionary history of the taxa involved. There is increasing evidence that when a clade branches off early in a phylogeny (is 'basal' to the rest), this indicates that the species included within that clade diverged from the rest of the phylogeny at an early date. While there will be statistical errors associated with these interpretations, they can indicate that these species have a more ancient origin. Such data are now being used to reconstruct not just the relationships of the families of birds but also the order of their evolutionary divergence. DNA sequence data indicate that game birds (galliforms) and wildfowl (anseriforms) (i) are comparatively closely related and (ii) diverged from the rest of the avian lineage at an early date. This supports the views of earlier anatomists that the 'gallo-anserae' grouping is biologically real and also shows that it is older than most other modern birds. The most biologically realistic way to represent the evolution of the various families of modern birds would be through a three-dimensional tree. This cannot easily be presented on paper, and taxonomists are compelled to present their results as an ordered list and, conventionally, the oldest groups are placed first. In the light of this restriction and convention, the gamebird–wildfowl assemblage was recently moved to the beginning of the list of British birds, with divers and grebes (formerly at the start) being relegated to later positions. More substantially, the arrangement of the passerines that has been used for several decades fails to reflect our knowledge of the relationships of these birds and is in need of thorough revision. We have followed the proposals as laid out in Sangster *et al.* (in press). Other changes are likely in the future.

In the species accounts that follow, we give a brief description of any recent molecular information relevant to each species and compare this with results from more traditional data. We do not discuss the latter in any detail: readers are directed towards works such as Vaurie (1959, 1965), BWP or HBW. Where there is evidence for changes in species limits or generic placement, these are discussed. We also give a brief indication of the extent of interpopulation variation and report upon the number of subspecies. Where relevant, the breeding and wintering ranges of those subspecies that have been recorded within Britain and Ireland are given; status codes (see below) for each of the four geo-political regions are given in Appendix 1. Within the species accounts, we use the terms 'subspecies' and 'race' interchangeably.

MIGRATION AND MOVEMENT

Britain and Ireland lie at the western end of the Palearctic, and millions of birds pass through on their way between breeding sites in the Arctic and subarctic and wintering grounds in SW Europe, the Mediterranean or Africa. The routes that these birds use have been recorded both by direct obser-vation and by tracking marked birds. Although visual observation played a major role in establishing migration routes of conspicuous species such as the White Stork, it was with the advent of marking schemes (mainly using rings or individual colour-marks) that the detail began to be filled in. The results of much of the research undertaken in Britain and Ireland have been published by the BTO in their monumental Migration Atlas (Wernham *et al.* 2002).

There are four main components to the migrant populations of our islands: summer visitors, winter visitors, passage migrants, vagrants. The stresses and pressures imposed on these birds are different. While some species may fall into more than one of these groups, they will be considered separately below.

SUMMER VISITORS

The summer visitors are predominantly insectivores that have to leave Britain and Ireland in winter when the invertebrate populations decline. Many migrate to Africa, with an appreciable proportion wintering south of the Sahara. Ringing studies have shown that there are two major migration routes. One tracks through SW Europe, entering NW Africa via the Straits of Gibraltar; the majority of species follow this course. A minority, including Lesser Whitethroat, Marsh Warbler and Red-backed Shrike, head SE across E Europe, through Turkey and the Levant, entering Africa from the NE. Captive-breeding experiments have shown that the initial SW or SE orientation is under genetic control (reviewed in Berthold 2001). More remarkably, Berthold and his colleagues have shown that many small passerines, which predominantly migrate alone and at night, have adapted to change their flight direction during the course of their migration. Eurasian Blackcaps heading for NW Africa change from a SW heading to a more southerly direction at approximately the time that they reach Gibraltar. The survival value of this is obvious; maintaining the initial bearing would have the birds heading for S America – well beyond the endurance of such small animals.

The alternative exit routes from Europe are believed to have evolved during or after one of the glaciations, when the ancestors of today's migrants were restricted to refuges in SW or SE Europe. There is a selective advantage to these routes: they avoid crossing both the Mediterranean Sea and the Sahara Desert. Over the years, problems of aridity and desertification have become increasingly evident. During the late 1960s, there was a drought in the Sahel zone of Africa, where many chats and *Sylvia* warblers winter. This region is immediately south of the Sahara Desert and consists predominantly of grassland and thorn savannah. The soil is generally poor and plant life is critically dependent upon the annual rains. The failure of the rains can put great stress on the migrants that winter there and those that pass through to and from wintering grounds further south. A severe drought in the Sahel during the northern winter of 1968/9 resulted in a heavy mortality of migrant birds. Although most species appear to have recovered from this, there is evidence that increased desertification of the Sahel is reducing the area of habitat suitable for our birds to winter there or feed as they pass through, and the recent decline of some of our summer visitors has been attributed to this.

Research by Peter Berthold has shown that migration distance is also under genetic control and that there is sufficient genetic variation for selection to act on this. This is particularly influencing short-distance migrants, many of which winter in SW Europe and NW Africa. Climate change, however, is bringing warmer winters, and areas suitable for wintering are becoming available further north. The northerly wintering birds are now surviving better; they come into migratory condition earlier; they leave the wintering areas earlier; they arrive back on the breeding grounds earlier; they get the best territories; and they produce more young. Since migratory behaviour is under genetic control, natural selection is working to reduce migration distance.

Long-distance migrants winter well south of the Sahara, where climate change is, so far, having less of an effect, and the survival differences of the more northerly wintering birds are likely less marked. Also, differences in photoperiod will be less in the tropics, so there will be less marked differences in migratory onset. As a result of this, long-distance migrants are not changing their migration patterns to the same extent as more short-distance species; this is having two consequences. Firstly, where there are long- and short-distance migrants with similar ecologies (e.g. Garden Warbler and Eurasian Blackcap), there will be some interspecific competition for territories. Returning first, the short-distance migrants will get the optimal territories, and since these populations are increasing, they will also expand into some of the territories that previously were occupied by the long-distance species. The latter will decline, or at least do less well than the short-distance birds. There is evidence that this is happening. Trans-Saharan migrants, largely unaffected by climate change, are continuing to return at, or closer to, the 'normal' time. The local invertebrates are strongly affected by the warmer, earlier springs, and there is evidence that the breeding of European Pied Flycatchers, for example, is now less synchronised with the primary food sources (Both *et al.* 2006), and breeding success is in decline.

WINTER VISITORS

Having a relatively mild and oceanic climate, Britain and Ireland have long been a wintering area for birds that breed in more arctic or continental regions. Apart from a few thrushes and buntings, these are predominantly wetland species, such as ducks, geese, swans and waders. These birds nest across the northern Palearctic and, in a few cases, Nearctic, taking advantage of vast expanses of wetland and tundra with plentiful invertebrate and plant food, relative security from predation, and long daylight hours to feed and rear their young. The fitness benefits accruing from this behaviour clearly must outweigh the costs incurred by the long migration flights needed to bridge the gap to their wintering grounds. The wintering areas for many of these species (especially the waders and some wildfowl) are our coastal estuaries and mudflats, which provide extensive feeding areas but, in general, few opportunities for breeding. Other species (particularly wildfowl) head for coastal and riverine grasslands and arable fields to forage through the winter, often commuting to secure open water to roost.

Severe winters, especially prolonged periods of freezing, can hit these populations hard, and mortality can rise rapidly. Winter food supplies are also finite. Whether fruits and berries, invertebrates in estuaries and fresh water, or grasses, cereals and arable debris like sugar beet tops, productivity generally decreases at the end of summer and availability to birds declines through natural mortality and predation. Consequently, competition among animals utilising these resources increases, and many birds have evolved flexible responses. As food supply decreases or severe cold arrives, birds wintering in the east relocate further south and west towards milder conditions in Ireland and SW Europe. Observers frequently record inland and coastal 'hard weather' movements of thrushes, finches, ducks, etc. While many of these may be our own breeding birds moving within Britain and Ireland, others are winter visitors moving on as the season progresses.

We have already noted that migratory distance is, at least partly, under genetic control and that there is variation (polymorphism) at the loci that determine this. Some birds are genetically programmed to fly further than others; their survival determines the contribution of the relevant genes to the next generation. As the climate warms up and the winter conditions are ameliorating across Western Europe, many areas are experiencing a progressive rise in temperature and a parallel decline in snowfall and frosts. The northern periphery of the 'traditional' winter range of some species is becoming more suitable for over-winter survival, and birds that do not migrate away are surviving better than before. The distance the birds have to fly in order to survive is reducing. Birds such as Common Ringed Plover and Black-tailed Godwit, which used to be relatively scarce in winter on the east coast of England, are now becoming increasingly common: the rich feeding grounds of the North Sea estuaries are no longer too cold to support wintering populations. Similarly, some ducks and geese (e.g. Mallard, European White-fronted Goose) are becoming scarcer in Britain and Ireland during winter. This decline is not because of enhanced mortality; the birds are now able to survive in wetlands closer to the breeding grounds. This 'short-stopping' is becoming increasingly apparent, especially among species recorded in WeBS surveys.

The short-stopping strategy is not universal, even among wintering wildfowl. Birds that breed in Iceland have less opportunity for short-stopping. Although we may see the emergence of new patterns of wintering distribution, the number of Icelandic birds reaching Britain and Ireland remains buoyant. There is emerging evidence that some species which breed in Greenland are now able to winter in Iceland and are remaining there rather than proceeding further. When a species is in numerical decline within Britain and Ireland, this does not necessarily imply that its populations are in trouble.

PASSAGE MIGRANTS

Birds in this heterogeneous assemblage generally breed to the north of Britain and Ireland and winter to the south, passing through here twice each year. There are relatively few species that are only found here on passage and very few that are common. Waders such as Little Stint and Curlew Sandpiper pass through in spring and autumn, although increasingly they are being found here in winter. Seabirds such as Pomarine Skua and Long-tailed Jaeger skirt the western seaboards en route to and from the Arctic. Surprisingly few passerines fall into this category; perhaps Aquatic Warbler is the only species that neither breeds nor winters in Britain and Ireland, yet makes a regular migration through the islands. There are many northerly species whose range extends southwards to include the northern fringes of Britain and Ireland but do not winter here, and most of these are waders: Whimbrel, Red-necked Phalarope, Temminck's Stint and Wood Sandpiper are examples. Some passerines regularly migrate across Western Europe, between NW Africa and Fennoscandia, and several of these arrive in Britain and Ireland in varying numbers, depending upon the prevailing weather conditions: species such as Bluethroat, Eurasian Wryneck and Icterine Warbler. A final group of birds that are regularly recorded each year but that normally neither breed nor winter here are the classic spring 'overshooting' species. Some of these migrate SW from continental breeding grounds, through Iberia and NW Africa: Alpine Swift, Red-rumped Swallow, Woodchat Shrike, Subalpine and Melodious Warblers, Ortolan Bunting and Yellow Wagtails of the race *flava*. Others leave Europe through the Balkans and Asia Minor, arriving here in spring from the SE, again rarely or never remaining to breed: examples include Red-backed and Lesser Grey Shrikes, Marsh Warbler and Black-headed Bunting.

Much more common are species that breed or winter in Britain and Ireland but have subspecies that are only recorded here on passage. Again, these are mostly long-distance migrants such as waders, examples being *tundrae* Common Ringed Plover, *arctica* Dunlin, *canutus* Red Knot. For the first two of these, the estuaries of Britain and Ireland are essential refuelling sites between the breeding and wintering grounds. Another taxon that falls into this category is the Greenland race *leucorhoa* of the Northern Wheatear.

VAGRANTS

The final, sizeable, component of the non-resident avifauna comprises birds that are recorded here irregularly, including species as diverse as Allen's Gallinule from sub-Saharan Africa, Aleutian Tern and Swinhoe's Storm Petrel from the North Pacific, Hudsonian Godwit, Varied Thrush and Canada Warbler from the Nearctic, Thick-billed Warbler and Yellow-browed Bunting from the E Palearctic, and Moussier's Redstart from NW Africa. The ultimate causes of this vagrancy remain largely a mystery; waders and wildfowl that are powerful fliers and often migrate in flocks probably have different underlying causes than small passerines that typically migrate alone and at night. Nearctic species that cross the Atlantic must usually traverse the ocean in a single stage, and their arrivals can often be associated with particular wind conditions. Palearctic species have the opportunity (although they may not take it) to rest and feed on their way. There are a few common features that may cast light upon the phenomenon of vagrancy.

Many of the autumn arrivals are young birds undertaking their first migration: September and October are key months for rare birds in Britain and Ireland. Those species that migrate alone and by night have the course and distance components determined genetically. Orientation studies using captive birds have shown that both of these are quantitative characters with a mean and variance, with the latter having both genetic and environmental components. When the statistical variation of a character has a normal distribution, there is a clear relationship between the deviation from the mean shown by individuals and the proportion that shows this departure: *c.* 4% of a sample will differ from the mean by more than two standard deviations, 0.25% by three. This may seem a tiny proportion, but with a population numbering tens or hundreds of thousands, increasingly large numbers of birds will show substantial departures from the mean or 'correct' orientation.

Most of these individuals are doomed; there is little evidence that long-distance vagrant juveniles ever make it back to their natal areas. Indeed it is this natural selection against individuals that take the 'wrong' heading that maintains the genetic basis for correct orientation within a population. There is evidence from shorter-distance migrants that changes to the environment may allow the survival of some individuals which deviate from the 'correct' orientation. This could lead to the use of novel wintering grounds and the establishment of new populations, or at least subpopulations. The story of Eurasian Blackcaps wintering in Britain and Ireland has been well documented. This behaviour was unknown 50 years ago but is now widespread. Orientation studies, combined with stable isotope research (Bearhop *et al.* 2005), have shown that these birds come from Eastern Europe, migrating to and returning from Britain and Ireland each winter. This population presumably originated from a few birds whose orientation direction was to the north of the typical south-westerly direction taken by the majority of the population. Instead of heading for Iberia and NW Africa, they ended up in Britain and Ireland. There might be nothing 'abnormal' about these birds, any more than a British (human) male measuring 192 cm (6 ft 3 in). They are simply one tail of the normal distribution, with deviation due to a combination of genetics and environment (though, in the case of the migrant, we do not know what the environmental forces might be). In the past, conditions in Britain and Ireland were too severe in winter and most or all perished. With warmer winters and a greater provision of bird feeders and berry-bearing shrubs in parks and gardens, these birds are now able to survive. Migratory activity is induced by changing photoperiod. Wintering further north, they came into migratory condition earlier and returned to their natal areas earlier and found mates among the other early-returning individuals. These pairs probably obtained better-quality territories and produced more young. It has long been known in theoretical evolutionary genetics that sympatric evolution can occur when there is non-random mating and strong selection (Maynard Smith 1966). Both of these conditions are fulfilled by the Eurasian Blackcaps, and so a subpopulation is evolving that is now genetically programmed to migrate in a more northerly direction.

The key factors here are not that the individuals survive but that, on their return to the breeding grounds, they mate assortatively and have an enhanced fitness. It is possible that some eastern rarities which are increasing here, such as Pallas's Leaf and Yellow-browed Warblers, are evolving along similar lines to the Eurasian Blackcaps. Increasing numbers of these birds are now being found in SW Europe and NW Africa during the winter and surviving until the spring. Their subsequent fate is unknown, but they at least have the opportunity to return and breed.

An alternative explanation for long-distance vagrancy is 'reversed' migration. This phenomenon was first postulated many years ago, but there is still relatively little direct evidence of its existence. The hypothesis is that some birds are 'aberrant' in that they orientate exactly 180° from the norm. Thus, a Lanceolated Warbler that should be orientating SE towards wintering grounds in SE Asia, heads NW and ends up in Europe. While it is intellectually appealing, such aberrant behaviour would have no selective advantage; individuals showing this behaviour would be doomed and whatever genetic forces produced the behaviour would be ruthlessly eliminated every generation. This remains a hypothesis that is almost impossible to disprove.

An alternative hypothesis is 'random scatter', which suggests that some birds disperse randomly. Again, there is little evidence in support. Orientation experiments involving captive birds give no indication of such behaviour; individuals within a population show a directionality that can be related to their behaviour in the wild at that time. This directionality can be quantified and shows a normal distribution, not randomness. As with reverse migration, there would be strong selection against this behaviour and it would be swiftly eliminated from the population. The strength of the 'normal distribution' argument (involving birds in the statistical tails of the species' preferred orientation being the ones that are termed 'vagrants') is that they form a natural part of the population. Their behaviour is controlled by the segregation of genetic variation, the same as for morphological attributes: they are extreme individuals within the population rather than 'aberrants'.

The number of vagrant birds recorded from Britain and Ireland has increased progressively over the decades. Undoubtedly, much of this is due to enhanced skill and experience of birders, combined with excellent field guides, superior optical equipment, greater mobility and better knowledge of the likely sites to find such individuals. Rarity 'hot spots' such as Fair Isle (Shetland), Flamborough Head and Spurn Point (Yorkshire) have been known for generations; others such as Cape Clear (Cork), Portland Bill (Dorset) and the Isles of Scilly (Cornwall) were 'discovered' in the 1950s and 1960s. Birders continue to search out new sites for particular species, leading some to the Bridges of Ross (Clare) for oceanic seabirds, Killybegs (Donegal) for wintering gulls, Islay (Inner Hebrides) for Nearctic geese, and the headlands of Donegal and the Outer Hebrides for seabirds and trans-Atlantic migrants. With better communications and transport systems, birders can head for the 'best' sites at the 'best' times of year – or wait for the pager to bleep with the latest news.

But have there actually been changes in the numbers of rare birds turning up or are we just better at finding them? Logic suggests that there have been no changes; the increasing length of the rare bird reports reflects our skill at finding them. To prove this is more difficult. Certainly, some species are more abundant now than half a century ago. The first Pallas's Leaf Warbler on Fair Isle (Shetland) was found as recently as 1966. It is inconceivable that such a distinctive species was missed by observers as competent as Eagle Clark, and George ('Fieldy') and Jimmy ('Midway') Stout on Fair Isle, who were finding birds like Pechora Pipit, and Lanceolated and Booted Warblers almost routinely. Some birds really are commoner now than previously, and we can use bird observatory data to confirm this.

Observatories such as Fair Isle, Bardsey (Caernarfonshire) and Cape Clear (Cork) have permanent staff (at least outwith the midwinter period) and a corresponding regular and reasonably consistent level of recording over the years. While field skills and binoculars have improved at these locations since the 1950s, the routine of recording has been fairly constant. Consequently, if the number of birds recorded at these sites has remained static over time, or increased less markedly than elsewhere, we can argue that the actual numbers of a particular species have not increased. If the numbers at observatories have increased in parallel with data from non-observatory sites, this might indicate a real increase in occurrence.

BIOGEOGRAPHICAL AFFINITIES OF THE AVIFAUNA
OF GREAT BRITAIN AND IRELAND

ENDEMISM

Isolated populations gradually accumulate genetic differences due to natural selection or random genetic drift. In either case, the populations will eventually become sufficiently divergent to be regarded as separate species. Living in a group of islands separated from Continental Europe, we might expect such divergences among the birds of Britain and Ireland, culminating in the evolution of species or subspecies unique to the region. Such endemics might be found at several levels: on small offshore islands, on the larger land-mass of Ireland, or within the entire archipelago. Since gene flow between populations will retard this divergence, we might also expect endemism to evolve among the less mobile forms, such as Willow Ptarmigan (Red Grouse), Winter Wren or Dunnock. It takes time for significant differences to evolve, and our region has been beset by a succession of glaciations, many of which have rendered the islands uninhabitable to most breeding birds. The most recent of these was at its peak about 20,000 years ago, at which time all of Scotland and the Isle of Man were covered with ice, as were Wales and Ireland, apart from the extreme south, and only the southern parts of England were ice-free (Yalden 1999). Even here the landscape would have been bleak tundra, with sparse vegetation. It is likely that few birds could survive the winter in such inhospitable habitat and, among the passerines, only a very few boreal migrants (Snow Bunting, Lapland Longspur, Arctic Redpoll) would be able to breed successfully. Thus any species or subspecies endemic to all or part of Britain and Ireland may have evolved since the ice retreated about 10,000 years ago and, since most small birds breed at one year of age, this implies within 10,000 generations. Alternatively, distinctive populations may have started to diverge in ice-free areas and moved into Britain and Ireland as the ice retreated.

Although there is no formal link between the extent of genetic divergence and the likelihood of specific distinctness, there are few birds that are regarded as species that differ from their nearest relatives by less than about 2% in their mitochondrial DNA. A widely used estimate for the rate of evolution in this category of DNA is about 2% per million years. Molecular evolution undoubtedly varies among genes and among taxonomic groups (see, for example, Pereira & Baker 2006). Genes that are of major significance to the survival of an individual are likely to be more constrained by natural section than those that are less vital. An example of this apparent lack of evolution occurs in the histones; these proteins are involved with the structure of chromosomes, and some are the same in rodents and humans. Any mutation that changes the structure of a histone runs a serious risk of disrupting the structure of the chromosomes, probably killing the cell. Other proteins are more free to evolve and, although a figure of 2% per million years for avian mitochondrial genes is almost certainly too high (Pereira & Baker 2006), its use will serve to make a point. Divergences of the order of 2% will require a very long period of isolation (about a million years) to evolve through random genetic drift alone: this is certainly far longer than the period that Britain and Ireland have been free of ice. Any forms that show morphological divergence within the archipelago, or between these islands and adjacent parts of Europe, must therefore have evolved through some form of natural selection, or they started diverging elsewhere and moved here as the ice retreated.

There are, in fact, rather few well-marked subspecies or putative species endemic to Britain and Ireland, and several of these are indeed comparatively sedentary, including Willow Ptarmigan (Red Grouse), White-throated Dipper, Winter Wren and Dunnock. Several of the differences in phenotype are stages in clinal variation (e.g. plumage coloration in Wren: BWP), and these reflect the tendencies for mesic homeotherms to be darker coloured with more melanin and less xanthin (Gloger's Rule). This clinal variation is found in many birds, especially passerines, with darker forms frequently occurring in Ireland and the west of Scotland.

In addition to the species already mentioned, Yellow and White Wagtail also differ between Britain and Ireland and elsewhere in Europe; these taxa also show marked inter-regional differentiation across their world range. It is likely that there has been a succession of ecological crises (probably glaciations) for wagtails during their evolutionary history in the Palearctic, and at these times populations became reduced and fragmented. It can be hypothesised that head pattern is of major importance in female choice and/or male–male competition in wagtails (Odeen & Björklund 2003). When a particular population was reduced in numbers, the opportunities for evolutionary change appeared, and the optimum phenotype changed marginally. As the populations expanded and regained contact, female choice had become engrained and alternative phenotypes were disfavoured through sexual selection. There is some preliminary evidence of assortative mating (like with like) in two regions: between *flava* and *flavissima* in Pas de Calais and between *alba* and *yarrellii* in Shetland. Further ecological and behavioural research is needed to confirm these findings.

Two taxa have been strongly advocated as endemic species: Willow Ptarmigan and Scottish Crossbill. Neither of these shows any evidence of molecular divergence from their closest relatives, which is unsurprising given the short time since their isolation from Continental ancestors. The former differs most markedly in its pattern of moult, retaining its red-brown feathering throughout the winter, compared with the European populations, which moult into a white winter plumage. This is presumably an adaptation to deeper, more intense and more prolonged snowfall in the Continental habitat and, while undoubtedly adaptive and selectively based, this distinction is regarded as insufficient evidence for specific separation. Red Grouse is treated here as a well-marked subspecies of Willow Ptarmigan. Scottish Crossbill, on the other hand, coexists with Red and Parrot Crossbills (sometimes both) in conifer forests in the highlands of Scotland. There appear to be habitat and dietary differences, which are associated with differences in bill morphology, and vocal differences as well. There is non-random mating with respect to vocalisations and bill morphology within the crossbill communities across N Scotland. Although these differences must have evolved relatively recently, the bill characters appear to be maintained by strong selection manifest through dietary differences. If hybrid forms are intermediate in bill morphology, and hence (like Darwin's Finches: reviewed by Grant 1986) less efficient in food handling, one can see the selective advantage of vocal differences that reduce the chances of hybridisation. The evidence is compelling that these represent separate and diagnosably different evolutionary lineages, and the Scottish Crossbill is treated here as a distinct species.

CONSERVATION

In 1990, the RSPB and the then Nature Conservancy Council introduced a system for highlighting the conservation plight, or otherwise, of the birds of the United Kingdom, together with the Channel Islands and the Isle of Man. The history of this 'Red List' is reviewed by Gregory *et al.* (2002) and was updated by Eaton *et al.* (2009). Each species is assessed for the likelihood that it will become extinct within the UK, and from this it is placed into one of three categories that identify its conservation importance: red, amber and green. Red List birds have the highest conservation priority; those on the Amber List are less critical but require monitoring in case of any change in their fortunes. Species on the Green List are at no perceived risk.

Birds that are assigned to the Red List are subject to one or more of the following factors:

- They are globally threatened (e.g. Corn Crake);
- There has been a severe decline in the UK population during 1800–1995 (e.g. Red-necked Phalarope);
- There has been a rapid decline of >50% over the last 25 years in the UK breeding population (e.g. Roseate Tern);
- There has been a rapid contraction of >50% in the UK breeding range over last 25 years (e.g. Cirl Bunting).

Amber List species have a wider range of criteria, and these also involve their status in Europe or, for wildfowl and waders, along the relevant flyway. A species is placed on the Amber List if:

- There has been a historical population decline during 1800–1995, but it is now recovering to the extent that the population size has more than doubled over the last 25 years (e.g. Dartford Warbler);

- There has been a moderate (25–49%) decline in the UK breeding population over last 25 years (e.g. Northern Lapwing);
- There has been a moderate (25–49%) contraction of the UK breeding range over last 25 year (e.g. Water Rail);
- There has been a moderate (25–49%) decline in the UK non-breeding population over last 25 years (e.g. Common Ringed Plover);
- It has an unfavourable conservation status in Europe (e.g. Sand Martin);
- There has been a five-year mean of <300 breeding pairs in the UK (e.g. Black-necked Grebe) or of <900 non-breeding individuals (e.g. Smew);
- It is not a rare breeding bird but >50% of the UK breeding population occurs in 10 or fewer sites (e.g. Bearded Reedling);
- It is not a rare bird but >50% of the UK non-breeding population occurs in 10 or fewer sites (e.g. Eurasian Oystercatcher);
- It is not a rare breeding bird but >20% of the European breeding population occurs in the UK (e.g. Great Skua);
- It is not a rare bird but >20% of NW European (wildfowl) or East Atlantic Flyway (waders) or European (others) non-breeding populations occur in the UK (e.g. Common Pochard, Ruddy Turnstone, Great Northern Loon).

Species placed on the Red List are clearly of major conservation concern; urgent conservation action is required, and in many cases national action plans have been developed and implemented. For some species, these have met with success. The decline in Eurasian Stone-curlew has been halted through sympathetic management of its nesting habitat, especially in the Brecklands of East Anglia. Understanding the ecological needs of Eurasian Bittern has led to improvements in the quality of reedbeds and the beginnings of a recovery in the breeding population. Changes in the availability of winter feed through modification of farming practice have slowed the decline in Cirl Bunting, and combining this with the provision of nest boxes seems to be aiding the recovery of Eurasian Tree Sparrow. Indeed, the problems faced by farmland species are generally due to changes in crop production that led to the disappearance of the stubble fields that provided winter food for granivores and insectivores alike; the loss of hedgerows that provided shelter, nest sites and invertebrate food; and the intemperate use of pesticides and herbicides that reduced the populations of invertebrates and their food plants. Attempts by farmers and landowners to reverse these changes are to be welcomed, and the more sympathetic agri-environment support schemes currently in place should further aid the recovery of our farmland birds.

Other species remain at risk. Marsh and Willow Tits have shown serious declines over the past 30 years; studies of their ecology have illuminated some of the problems they face, but reversing the trends will be more difficult and may involve extensive changes in woodland structure and management. Similar declines have been recorded in other woodland birds such as Hawfinch and Common Nightingale. The causes of the declines are not really understood: predation by Eurasian Jay and Grey Squirrel may be threats to the former, and deterioration of the scrub layer by grazing deer for the latter. These are currently untested hypotheses and, in the short term, probably untestable. Similarly, we do not really know why Lesser Spotted Woodpecker is in decline.

Threats to other species may lie overseas. Spotted Flycatcher and European Nightjar may be declining because of a shortage of insect food in Britain and Ireland; alternatively, problems arising from the desertification of the wintering grounds may be equally (or more) significant. There is much to do, and researchers are working hard to identify and alleviate the problems.

1 Fair Isle Bird Observatory building, constructed in 1969 and replaced in 2009. Fair Isle is a major seabird island with a well-justified reputation for vagrants. July (Roger Tidman).

2 Portland Bill (Dorset), June. A Bird Observatory on the south coast of England, ideally located for monitoring movements in and out of southern Britain (Martin Cade).

3 The lush gardens of the Scilly Islands offer refuge to many exhausted migrants from North America as they make their first landfall after an Atlantic crossing. October (Roger Tidman).

4 In addition to monitoring the movement of migrants through the northern Irish Sea, Copeland Bird Observatory in County Down (left island above) hosts an important colony of Manx Shearwater. June (Neville McKee).

5 St Kilda provides a safe nesting place for vast numbers of seabirds and a staging post for migrants between Iceland and the British Isles. July (Chris Gomersall).

6 Bass Rock, June. Many predator-free islands around the coasts of Great Britain and Ireland host major seabird colonies (Roger Tidman).

7 Ouse Washes (Cambridgeshire), February. These are largely protected now, with breeding Black-tailed Godwits and wintering wildfowl including Eurasian Wigeon and Whooper Swan (Chris Gomersall).

8 Minsmere (Suffolk), January. One of the RSPB's flagship reserves, important for reedland species such as Eurasian Bittern, Western Marsh Harrier and Bearded Reedling (Jeff Higgott).

9 Offshore pelagic trips and observations from Atlantic headlands have shown that some seabirds previously considered rare are regular offshore. Bridges of Ross (Clare), September (Tom Shevlin).

10 Ballycotton (Cork). An important resting and feeding site for American ducks and waders after their trans-Atlantic flight. January (Sean Cronin).

11 Fishing ports attract large numbers of gulls; Atlantic trawlers can bring Arctic and Nearctic species inshore. Killybegs Harbour (Donegal), March (Simon Stirrup).

12 Rutland Water, June. A massive reservoir complex with specially managed areas for birds, Rutland Water also hosts the British Birdwatching Fair each August (Anglia Water).

13 Sub-arctic conditions on the highest Scottish summits support Rock Ptarmigan, Eurasian Dotterel and Snow Bunting. Cairngorm (Highland), April (Chris Gomersall).

14 Ancient Scots Pine forest is unique to highland Scotland, holding Western Capercaillie, several crossbills and the endemic race of European Crested Tit. Abernethy (Highland), April (Chris Gomersall).

15 Blanket bogs in the flow country of northern Scotland provide nesting sites for Black-throated Loon and Common Greenshank, though much has been lost to forestry. Sutherland, July (Chris Gomersall).

16 Extensive tracts of the British Isles were afforested through the 20th Century, resulting in the loss of a wide range of habitats; these were largely replaced by a conifer monoculture. Tardree Hill (Antrim), January (Neville McKee).

17 Breeding species in lowland deciduous woods vary with latitude, but include Common Firecrest, Eurasian Hobby and European Honey Buzzard. New Forest (Hampshire), May (Chris Gomersall).

18 Upland deciduous woods, typically containing oak and birch, hold significant numbers of Wood Warbler, Common Redstart and European Pied Flycatcher. Welsh Oak Wood (Powys), October (Chris Gomersall).

19 Later-stage coppice is a scarce resource now, but it provides shelter for species such as Common Nightingale. Kent, May (Chris Gomersall).

20 Many southern heathlands were lost to housing after WWII; those that remain are increasingly important for European Nightjar, Eurasian Stonechat and Dartford Warbler. Purdis Heath (Suffolk), January (Jeff Higgott).

21 Under threat from increased storms and rising sea levels, salt marshes provide feeding and roosting areas for many shore birds. Essex, March (Chris Gomersall).

22 Intertidal mudflats along sheltered shores provide internationally important feeding sites for migrating and wintering waders and wildfowl. Norfolk, November (Chris Gomersall).

23 Many lowland wet meadows have been lost to agriculture, to the detriment of breeding waders such as Northern Lapwing, Common Snipe and Common Redshank. Cambridgeshire, April (Chris Gomersall).

24 Grasslands on largely shell sand substrate still provide lush habitats for breeding waders along the western seaboards of Scotland and Ireland. Argyll, July (Chris Gomersall).

25 Farms with small fields and thick hedgerows have all but disappeared from much of the lowlands, with a consequent loss of many birds. Dorset, June (Chris Gomersall).

26 The post-war drive for cheaper food led to the industrialisation of many farms; the parallel switch to winter cereal devastated populations of many farmland birds. Essex, March (Chris Gomersall).

27 Gravel pits provide breeding and wintering sites for a wide range of wetland species. Dungeness (Kent), June (Chris Gomersall).

28 The tit populations of Wytham Wood in Oxfordshire have been the subject of continuous monitoring for sixty years, giving detailed insights into population biology. May (Chris Perrins).

29 As inner cities are 'regenerated', nesting sites are lost for species like Swifts, although Peregrines are now returning to places such as Derby Cathedral. April (Derby Telegraph).

30 Gardens, cumulatively covering an area the size of Kent, are increasingly recognised as an important habitat; over 15,000 volunteers monitor their gardens every week for the BTO. Thetford (Norfolk), July (Mike Toms).

31 The advent of phone-lines and pagers allows birders to assemble rapidly when a new rarity is found – in this case a Long-billed Murrelet at Dawlish (Devon) in November 2006 (Graham Catley Nyctea Ltd).

32 The backbone of bird-censusing in Great Britain and Ireland is provided by thousands of volunteers working for the BTO, Birdwatch Ireland and the SOC. Derbyshire, September (Linda Parkin).

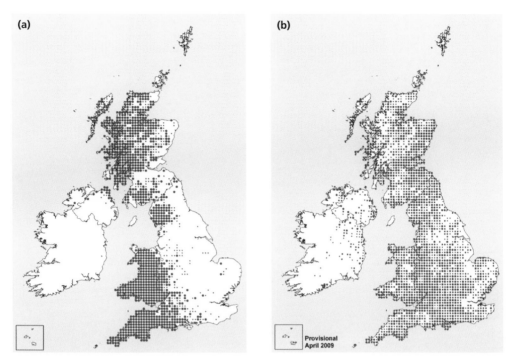

33 Common Buzzards were heavily persecuted until the middle of the 20th Century; a comparison of the BTO Atlas maps of 1968–72 (**a**) and 1988–91 (**b**) shows the range expansion following more sympathetic treatment – an increase that has continued into the 21st Century (BTO).

34 The value of including standardised counts in Atlas surveys is shown by comparing the uniformity of European Robin breeding distribution (**a**) with the non-randomness of its density (**b**); clearly Robins occur almost everywhere, but some regions are much more important than others (BTO).

35 Few things can match the anticipation of an early morning trap round at a bird observatory; the Fair Isle Gully Trap has yielded hundreds of rarities down the years. October (Fair Isle Bird Observatory).

36 Mist nets revolutionised bird ringing, giving ringers flexibility in both space and time, and removing the need for rigid permanent traps. Treswell Wood (Nottinghamshire), January (Chris de Feu).

38 A smaller number of species, such as Lesser Whitethroat, leave Europe via the south-east, passing through the Levant, and entering and leaving Africa through the Rift Valley. This map shows autumn recoveries of birds ringed in Great Britain and Ireland during the summer (BTO).

37 Ringing recoveries have shown that the majority of trans-Saharan migrants leave Europe through Iberia; here the pattern can be seen for autumn recoveries of Eurasian Reed Warblers ringed in Great Britain and Ireland during the summer (BTO).

39 Regular trapping at 'constant effort sites' allows estimation of population changes, breeding success and survivorship of common birds. Treswell Wood (Nottinghamshire), January (Chris de Feu).

40 The use of cannon nets has allowed ornithologists to catch and ring large numbers of ducks, gulls and waders for studies of moult, migration and breeding success. Islay (Argyll), October (Steven Percival).

41 Cannon-netting wildfowl has revealed much about their migrations; here the routes taken by different populations of Brent Goose can be clearly seen (BTO).

42 Transmitting devices are now small enough to be attached to wild birds, allowing their movements to be monitored via satellite links in 'real time'. Wexford Slob (Wexford), March (Tony Fox).

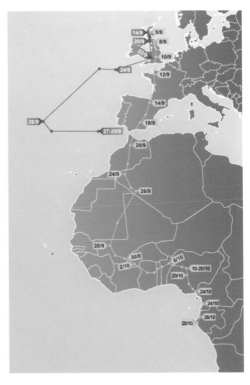

43 Satellite telemetry allows bird migration to be followed in 'real time'; this Osprey flew from its breeding grounds in Scotland to western Africa in only 30 days (Roy Dennis).

44 The fates of an adult (yellow) and juvenile (red) Honey Buzzard; the former successfully reached its wintering grounds: the latter came to grief off SW Spain (Roy Dennis).

45 The use of aircraft for surveys has revealed large accumulations of divers and sea ducks in shallower parts of the Irish and North Seas. Liverpool Bay, August (Peter Cranswick).

46 Improvements in both equipment and analytical tools have revolutionised the study of vocalisations, revealing diagnostic differences among closely related taxa such as crossbills and warblers. Abernethy (Highland), January (Ron Summers).

47 Scottish Crossbill: the only endemic species in Great Britain and Ireland, restricted to mature conifer forests in the highlands of Scotland. Ballater (Aberdeenshire), October (Rab Rae).

48 Red Grouse. This endemic race is regarded as a subspecies of the Willow Ptarmigan, although it has been treated as a separate species in the past. Yorkshire, April (Chris Gomersall).

49 Ecological, behavioural and genetic research has indicated that Hooded (**a**) and Carrion (**b**) Crows are separate evolutionary lineages, and should be treated as separate species (Iain Leach/Oliver Smart).

50 Winter Wren; the endemic Shetland Wren *zetlandicus* is larger and darker than *indigenus*. Shetland, June (Chris Gomersall).

51 Winter Wren; the endemic Fair Isle Wren *fridarensis* differs in plumage and song from the nearby *zetlandicus*. Fair Isle, September (Rebecca Nason).

52 Common Starling; the 'Shetland' race *zetlandicus* has a dark juvenile plumage, and is larger and heavier than the nominate, which breeds over most of Great Britain and Ireland. Lewis (Outer Hebrides), July (Frank Stark).

53 Song Thrush. The endemic Hebridean race *hebridensis* is the darkest form of this species, with a greyer rump and tail than *clarkei*, which occurs in the rest of Great Britain and Ireland and adjacent Europe. Lewis (Outer Hebrides), August (Frank Stark).

54 Coal Tit. The endemic Irish race *hibernicus* has the pale parts suffused with yellow and a brighter cinnamon vent. Antrim, April (Neville McKee).

55 Dartford Warbler. A succession of mild winters has seen a burgeoning of the population of this species, with colonisation of sites in East Anglia and the English Midlands. Suffolk, February (Rebecca Nason).

56 Pied Wagtail, the distinctive race *yarrellii* of the more widespread White Wagtail. This race is a near-endemic to Great Britain and Ireland. Langdon Beck (Durham), March (Iain Leach).

57 Long-tailed Tit. The well-marked race *rosaceus* is endemic to Great Britain and Ireland. Rufford (Nottinghamshire), February (Iain Leach).

58 Great Skua. One of the rarest seabirds in the North Atlantic, 60% of the world population breeds in Scotland. Fair Isle, June (Rebecca Nason).

59 Manx Shearwater; more than 90% of the global population breeds in Great Britain and Ireland. Carnsore Point (Wexford), September (Killian Mullarney).

60 A typical view of passing European Storm Petrels: a quarter of the world population breeds in Ireland, with sizeable numbers in Great Britain. Portland Bill (Dorset), May (Martin Cade).

61 Around 40% of the world's European Shags breed in Great Britain and Ireland. Farne Islands (Northumberland), June (Rebecca Nason).

62 Eurasian Sky Lark. Perhaps more than any other species, losses in Sky Lark populations epitomise the effects of agricultural intensification. Herefordshire, April (Chris Gomersall).

63 European Turtle Dove. The loss of hedgerows and weedy field margins, perhaps allied with problems on the wintering grounds, have led to a dramatic decline in this species. Holme (Norfolk), August (Iain Leach).

64 Eurasian Tree Sparrow. The massive population decline over the last 30 years is widely believed to be due to changes in agricultural practice. Venus Pool (Shropshire), May (Jim Almond).

65 Corn Crake. Frequent mechanical mowing of hay meadows has destroyed many populations; this species is now largely restricted to the Shetland and Orkney Islands, the Western Isles and Ireland. Argyll, July (Chris Gomersall).

66 The breeding distribution of Common Nightingale is retreating southwards through loss of habitat, especially of coppice and woodland with dense understorey. Cambridgeshire, May (Chris Gomersall).

67 Red-backed Shrike is virtually extinct as a breeding bird in Great Britain and Ireland; the reasons for its demise are not clear, but it may relate to a reduction in large invertebrate prey. Norfolk, April (Alan Tate).

68 Marsh Warbler has declined to near extinction as a breeding bird, although it continues to be recorded as a migrant. Hoswick (Shetland), June (Iain Leach).

69 Like many woodland birds, Willow Tit is showing a worrying decrease; the decline may be due to habitat loss, and also predation by woodpeckers. Cannock Chase (Staffordshire), January (Iain Butler).

70 The status of Eurasian Blackcap has changed since WWII, with the adoption of Great Britain and Ireland as a wintering area for birds from central Europe. February (Gary Thoburn).

71 Common Buzzard: this species has seen a dramatic recovery in numbers and distribution, and it is now perhaps Britain's most abundant bird of prey. Gigrin Farm (Powys), June (Ian Butler).

72 Wood Lark is expanding northwards; initially associated with heath and forest clear-fell, this species is now colonising parkland across the English Midlands. Hampshire, April (Alan Tate).

73 Cetti's Warbler arrived in Great Britain relatively recently; it now breeds widely in southern England, and is gradually expanding northwards. Upton Warren (Worcestershire), September (Mark Hancox).

74 Red Kites are now established in England and Scotland following a series of successful re-introductions. The Welsh population is also thriving. Rhayader (Powys), October (Iain Leach).

75 Little Egret; a rare bird only 30 years ago, these herons now breed widely beside wetlands across southern England. Norfolk, May (Roger Tidman).

76 Mediterranean Gull. The breeding population has increased steadily in recent years, and non-breeding birds are increasingly encountered across Great Britain and Ireland. Norfolk, May (Roger Tidman).

77 Roseate Terns. The chicks of these birds are vulnerable to avian predators; provision of shelters has helped reduce losses at several colonies. Coquet Island (Northumberland), June (Paul Morrison).

78 Following severe declines, Bitterns are now recovering thanks to the enhanced protection of existing reed beds and the creation of new ones. Potteric Carr (Yorkshire), December (Iain Leach).

79 Globally threatened, but regularly found on passage along the coasts of southern England, the wintering grounds of Aquatic Warbler have recently been found in West Africa. Djoudj (Senegal), January (Martin Flade).

81 Eastern Bonelli's Warbler; the first 'molecular split', this remains a very rare bird in Great Britain and Ireland. Lundy (Devon), April 2004 (Richard Campey).

80 Lesser Scaup was first recorded as recently as 1987 but is now seen regularly, probably due to increased familiarity among birders. Monkmoor Pool (Shropshire), June (Ian Butler).

82 Long-billed Murrelet: the first for Great Britain and Ireland, a long way from its home in the north Pacific. Dawlish (Devon), November 2006 (Simon Stirrup).

84 Allen's Gallinule; the second for Great Britain and Ireland, this moribund bird died before many birders could see it. Portland Bill (Dorset), February 2002 (Martin Cade).

83 Yellow-nosed Albatross from the southern oceans; the first for Britain was found inland and successfully released. Brean Down (Somerset), June 2007 (Richard Austin).

85 Rufous-tailed Robin: another Fair Isle 'first', this bird from eastern Asia was present for only a few hours. Fair Isle, October 2004 (Rebecca Nason).

86 New birds are still turning up in Britain and Ireland. Days before this book went to press, this Eastern Crowned Warbler *Phylloscopus coronatus* was found in Co. Durham. If accepted, this species will feature in the next edition of this book. South Shields (Durham), October 2009 (Graham Catley Nyctea Ltd).

NON-PASSERINES

FAMILY ANATIDAE

Genus *CYGNUS* Bechstein

Holarctic, Neotropical and Australasian genus of six or seven species, three or four of which are Palearctic; three or four have been recorded in GB&I; one breeds commonly.

Mute Swan

Cygnus olor (J. F. Gmelin)

RB HB monotypic

TAXONOMY Closely related to Black Swan *C. atratus* and Black-necked Swan *C. melancoryphus* (Harvey 2000). There is evidence of genetic differences at single loci between populations (e.g. 'Polish' morph, reviewed in Wieloch *et al.* 2004). Although these indicate the interruption of gene flow, there is no evidence of significant morphological differentiation among populations.

DISTRIBUTION Occurs widely across W and C Europe, both naturally and following human introduction in the 16th and 17th centuries. Spread extensively during the 20th century and marked increase in numbers in second half (Wieloch *et al.* 2004). Now distributed from GB&I and France, east to Caucasus, with isolated populations in Italy, the Balkans and Turkey. Patchily distributed across Asia to NE China. Some populations migratory, leaving areas that freeze in winter for sites further south and/or west. However, evidence of changes in some areas of C and E Europe, possibly because of milder winters and hence more unfrozen water, although increased breeding population probably also plays a part (Wieloch *et al.* 2004). Introduced to Japan, North America, Australia and New Zealand.

STATUS See Whooper Swan in relation to hybrid pairing with that species. Mute Swans are widely distributed across GB&I, except N of Scotland, where more patchily distributed. Breeding population in GB estimated as *c.* 3,500–4,000 pairs in 1955 (Campbell 1960), declining through 1960–80, but increasing to *c.* 5,300 pairs in 1990 (Greenwood *et al.* 1994). A further census in spring 2002 estimated a British total of 6,150 breeding pairs, with a further 19,400 non-breeding individuals, giving a total of *c.* 31,700 birds at the start of the breeding season. Most Mute Swans are highly territorial breeders, but large numbers breed in relatively close proximity in a few locations, e.g. Abbotsbury (Dorset) and Lochs of Harray and Stenness (Orkney) (Ward *et al.* 2007). Note that census methods vary across years, and direct comparisons may be unsafe. Irish populations also show signs of increase from 1,258 pairs in 1978 (Forsyth 1980) to 1,400–1,600 pairs at end of the 1990s (Wieloch *et al.* 2004). WeBS counts indicate progressive increase through the 1990s into the 2000s, with a GB winter population estimate of 37,500 (Kershaw & Cranswick 2003); however, the NI population showed a decline of *c.* 50% from the mid-1990s, largely attributable to falling numbers on Loughs Neagh (NI) and Beg (Londonderry/Antrim), although numbers in Ireland as a whole have been on the increase (Crowe 2005), to a recent estimate of 11,440 (Crowe *et al.* 2008). Adults moult at the breeding site, but non-breeders and failed breeders may undertake seasonal moult migrations to favoured sites (e.g. Abbotsbury, Berwick-upon-Tweed (Northumberland), Montrose (Angus) – see Wernham *et al.* (2002)). Fairly sedentary within GB&I, though juveniles tend to disperse away from the natal site at the end of the summer. Limited exchange between GB and Irish populations, and also between GB and adjacent Continental countries – frequently associated with hard weather.

A steep decline in numbers in the 1960s–80s due to lead poisoning was stemmed when a ban on the use of lead weights for fishing was introduced in 1987. Being conspicuous, many Mute Swans are found after death and causes of mortality are often reported. High on the list are collisions with overhead lines, pollutants (especially oil) and killing by man. Many of the cases of H5N1 avian influenza in Western Europe in 2006 and 2007 were found in dead or dying Mute Swans. There is no evidence that this species is particularly susceptible to the virus, simply that dead swans are more likely to be seen and tested. As swan numbers have increased, so has the potential for conflict with the human population, e.g. winter grazing on grass, cereals or (especially) oil-seed rape, where loss can reach 18% (Parrot & McKay 2001). In some areas, the taking of food from feeders on fish ponds is causing problems, although tapes suspended across the open water can alleviate this. Wieloch *et al.* (2004) suggest that improved management plans are needed to reduce conflicts.

Tundra Swan   *Cygnus columbianus* (Ord)

WM *bewickii* Yarrell
SM *columbianus* (Ord)

TAXONOMY Two subspecies are recognised: broadly speaking, *columbianus* in the Nearctic and *bewickii* in the Palearctic. Eastern Palearctic '*jankowskii*' not now accepted (Vaurie 1965; BWP). There is ongoing debate over the status of *columbianus* and *bewickii*, which may be specifically distinct. They differ diagnosably in the size of the yellow patch at the base of the bill (Evans & Sladen 1980), but all other characters seems to show extensive overlap. Both taxa have expanded their range in recent years, and there is now an area of overlap in NE Siberia (Limpert & Earnst 1994; Syroechkovski 2002) and, since 1970, mixed families have been reported on both breeding and wintering grounds. A preliminary genetic analysis could not resolve the taxa (Harvey 2000). They are thus consistent with phylogenetic species but based on only a single trait. There seems to be little evidence to support their recognition as separate species under other species concepts, and they remain best treated as conspecific.

DISTRIBUTION Nominate breeds across the tundra of Canada and Alaska, the Aleutians into E Siberia (Limpert & Earnst 1994), wintering on the E and W coasts of the USA; *bewickii* breeds from the Kola peninsula, E to Chukotsky. The western populations (roughly west of River Taimyr) winter in W Europe, migrating more-or-less directly along the S coast of the White Sea to Leningrad, and thence along the Baltic coast to Denmark, the Netherlands and GB&I (Wernham *et al.* 2002); those from further east winter along the coasts of China, Korea and Japan (Rees *et al.* 1997).

STATUS *bewickii* ('Bewick's Swan'): Winters chiefly S of a line from the Humber to the Solway, with centres in the Nene/Ouse Washes (Cambs/Norfolk), Lancashire and the Severn Valley; smaller numbers occur in Ireland, especially in SE at Tacumshin, Wexford Slobs, and Cull and Killag (all Wexford), but recent WeBS/I-WeBS data suggest that the RoI wintering population has fallen sharply and contracted to the SE (Crowe 2005). Numbers in Britain increased from the 1960s until *c.* 1990, followed by a steady decline until *c.* 2000, since when numbers have stabilised (Banks *et al.* 2006). Productivity was average or above through the late 1990s, and the decline probably reflects changes in distribution rather than overall abundance. Milder winters across the Continent, resulting from global warming, may have resulted in improved winter feeding and a reduced pressure to fly further west to avoid the harsher weather. An international Bewick's Swan census conducted in mid-Jan 2005 counted 7,216 birds in Britain and Ireland combined. However, two weeks later a roost count was made of 7,491 Bewick's Swans on the Ouse Washes alone.

columbianus ('Whistling Swan'): as with most wildfowl, birds of captive origin mask genuine wild occurrences, although only *c.* 30 were listed as held in captivity in 2001: the most recent year for which census data are available (R. Wilkinson pers. comm.). A few individuals are considered to be genuine vagrants. The first of five Irish birds was present in Co. Kerry, between Dec 1978 and Feb 1979. The first of two in GB was in Hampshire and Somerset, between Dec 1986 and Mar 1990.

Whooper Swan   *Cygnus cygnus* (Linnaeus)

CB HB WM monotypic

TAXONOMY Usually placed as sister taxon to Trumpeter Swan *C. buccinator* and more distant from Tundra Swan. There is some variation in size across the range but insufficient for subspecific differentiation (see Brazil 2003).

DISTRIBUTION Breeds on lakes in the boreal forests, from N Scandinavia to far-eastern Siberia and Kamchatka, with an isolated population in Iceland, where it more usually breeds on open tundra. Winters on lowland lakes and freshwater marshes, from GB&I across Europe and central Asia to Japan, usually beside open water (e.g. Black, Caspian Seas). Part of the Icelandic population is resident.

STATUS A very small number of pairs breed in GB&I, usually in Scotland, and rarely exceeding single figures each year (BS3). One of the better years was 2002, when five pairs laid eggs: two failed and the others were seen with a total of seven young (RBBP). More recently, a pair bred in NI, with two young in June 2006 (RBBP). Mixed pairs with Mute Swans have occasionally been recorded in Scotland and Ireland.

A winter visitor to GB&I, more widely distributed than Tundra Swan and generally more northerly. The majority of birds wintering in GB&I breed in Iceland; only 10 of 370 ringing movements reported by Wernham *et al.* (2002) involved Continental Europe. Large concentrations winter at the Ouse Washes (Cambs/Norfolk) and the Ribble Estuary (Lancs); also N and W Ireland, especially Loughs Erne (Fermanagh), Neagh (NI), Beg (Londonderry/Antrim) and Foyle (Londonderry/Donegal) (J.A. Robinson *et al.* 2004a).

Numbers wintering in Britain remained fairly stable from the mid-1960s to the mid-1980s, but have since risen dramatically to their highest level in 2003–04, whilst numbers in Northern Ireland have also shown a recent strong increase (Banks *et al.* 2006). An international census carried out in Jan 2005 found a total of 24,810 Whooper Swans in Britain and Ireland. The rise of 55% in Britain between censuses in 2000 and 2005 was mostly accounted for by the increase in numbers using the Ouse Washes, where the count of 4,397 in Jan 2005 was the highest ever recorded at a British site (Worden *et al.* in press). It is not clear whether this is due to increase in the Icelandic-breeding birds or swans with Continental breeding origins.

Genus *ANSER* Brisson

Holarctic genus of ten species, eight of which are Palearctic, of which six have been recorded in GB&I; some authorities place Snow, Ross's and Emperor Geese *A. canagicus* in the genus *Chen*.

Bean Goose   *Anser fabalis* (Latham)

WM *fabalis* (Latham)
WM *rossicus* Buturlin

TAXONOMY Five subspecies recognised, comprising two groups, though relations among these still unresolved. Three are largely orange-billed taiga-breeding forms, increasing clinally in size from west to east; *fabalis*: Scandinavia to Ural Mountains; *johanseni*: Urals to L Baikal; *middendorffii*: E of Lake Baikal. The other two are darker-billed, tundra-breeding: *rossicus*: smaller from N and NW Siberia; *serrirostris*: larger from NE Siberia.

A molecular study that examines the genetic structure of the five subspecies is underway, but the results are not yet available. The tundra-breeding forms have relatively shorter, stronger bills than those in the taiga, perhaps reflecting differences in feeding ecology; they are physically smaller and have less orange-yellow on the bill (Roselaar in BWP). These differences, and lack of demonstrated interbreeding, have led some to treat tundra and taiga forms as separate species (Sangster & Oreel 1996; R. C. Banks *et al.* 2007).

DISTRIBUTION Tundra forms nest on islands in lakes or beside pools and rivers in the Arctic, from Scandinavia to Lake Baikal; taiga forms breed further south in the subarctic, in clearings among birch or coniferous forest from N Russia to NE Siberia. Two main wintering areas in temperate lowlands of Europe and E Asia (Wetlands International 2006). In general, *fabalis* and *rossicus* winter in Europe; *johanseni*, *middendorffii* and *serrirostris* in China, although some birds resembling *rossicus* have been seen in the east, and some *johanseni* in the west. However, intergrades between races have caused confusion (Madge & Burn 1988), so the situation is not fully resolved.

STATUS Restricted distribution in GB&I, with main concentrations in the Yare Valley of East Anglia and the Slammanan Plateau near Falkirk (Stirlingshire), with *c.* 200 *fabalis* birds each (Hearn 2004a), though the Yare Valley population has declined since 2001, whilst that at the Slamannan Plateau has increased. Colour rings and neck collars suggest that these birds have different origins (Parslow-Otsu 1991). In the late 1980s, birds were seen in Scotland with rings from a reintroduction scheme in SC Sweden. At about the same time, some birds in the Yare Valley carried neck collars from the Finnish population moulting in N Sweden. Small parties of both *fabalis* and *rossicus* occur elsewhere, predominantly in SE England, being spillover from *c.* 600,000 that winter in Continental Europe. Very few winter in Ireland, and these are mostly *fabalis*, with some records of *rossicus*. Ringing data are very sparse, so little is known of the movements of individual birds.

Pink-footed Goose   *Anser brachyrhynchus* Baillon

WM monotypic

TAXONOMY Now regarded as a separate species from Bean Goose, though analysis of skeletal, anatomical and plumage traits (Livezey 1996) shows them to be closely related. Molecular analyses by Ruokonen *et al.* (2000) confirm this; Bean and Pink-footed Goose are returned as sister taxa, though Lesser White-fronted Goose is in the same clade. Genetic differences are low compared with other birds, supporting a recent evolutionary origin.

DISTRIBUTION Breeds in Iceland, along N part of the E coast of Greenland and W Svalbard. Mt DNA indicates that these populations diverged less than 10,000 years ago, and the present populations have a single origin (Ruokonen *et al.* 2005). Genetic diversity was five times higher in the Svalbard population, suggesting this was the source population from which Iceland/Greenland populations were founded. Ringing recoveries indicate that birds from Greenland and Iceland winter predominantly in Britain, with very small numbers in Ireland. The Svalbard population winters along the Continental coast of the North Sea, chiefly from Denmark to Belgium, with only a few ringing recoveries in GB. The three breeding populations are scattered across difficult terrain, and estimates are subject to error; however, all three seem to be increasing. Svalbard 15,000 to 35,000 between 1965 and 1997; Iceland/Greenland from *c.* 20,000–30,000 in the 1950s to 200,000–250,000 in mid-1990s (Mitchell & Hearn 2004).

STATUS In winter, occurs in E, SE and SW Scotland, Lancashire coast and East Anglia, with seven sites averaging more than 20,000 birds since 1990 (Banks *et al.* 2006). WeBS counts, in conjunction with dedicated roost counts, show an increase from *c.* 10,000 in the 1950s to almost 300,000 birds in 2004–05 (Rowell 2005). This increase was not regular; it was especially steep in the late 1980s, which Pettifor *et al.* (1997) ascribe to relaxation of constraints in the breeding grounds, possibly following the occupation of novel breeding habitat. However, there is little evidence of changes in breeding parameters subsequently, and the population continues to grow. Recent WeBS data (Banks *et al.* 2006) indicate that the increase is fairly uniform across GB, although there is extensive movement between roosting (and presumably feeding) sites. A Population Viability Analysis (Trinder *et al.* 2005) suggested that the population would stabilise at *c.* 220,000, provided that environmental parameters remained constant. The continued increase may suggest that shooting pressure has declined.

Greater White-fronted Goose   *Anser albifrons* (Scopoli)

WM *albifrons* (Scopoli)
WM *flavirostris* Dalgety & Scott

TAXONOMY Morphology, anatomy and plumage (Livezey 1996) indicate sister relationship with Lesser White-fronted Goose, but mt DNA sequences suggest a poorly supported clade also including Bean, Pink-footed and Lesser White-fronted Goose. The lack of clear differentiation indicates a relatively recent origin for these taxa.

Five subspecies are recognised (winter range in parenthesis); *albifrons*: N Russia and NW Siberia (Europe and Middle East); *frontalis*: NE Siberia and N Canada (China, Japan, SW USA); *gambeli*: NW Canada (coastal Texas, Louisiana, Mexico); *elgasi*: SW Alaska (California); *flavirostris*: Greenland (Ireland, W Scotland). A recent review by Ely *et al.* (2005) confirms clinal variation in size variables across the Palearctic, from smallest in west to largest in east, with evidence of a step 'somewhere west of the Lena River'. *Elgasi* and *frontalis*, and *flavirostris* and *albifrons* are (at least partially) sympatric in winter, with the larger (*elgasi, flavirostris*) feeding in marshes or by grubbing for roots and tubers, whereas the smaller forms are adapted to grazing or taking grass seeds. The isolated Greenland population shows no evidence of molecular divergence (Ruokonen *et al.* 2000), but, as Ely *et al.* (2005) suggest, selection on morphological traits can be very strong and we may be seeing the early stages of divergence that, in time, may lead to speciation.

DISTRIBUTION Holarctic. Breeds on the tundra of N Eurasia between 76°N and 64°N, from Kanin Peninsula to Bering Straits, Alaska, NW Canada and W Greenland. Usually by marshes, pools and rivers. Migratory, wintering in open grass or arable land across Europe to Iran, China, Japan, E & SW USA and Mexico.

STATUS Two subspecies winter in GB&I: *albifrons* and *flavirostris*.

Nominate winters principally in S England, especially the Severn and Swale (Kent) Estuaries, with smaller numbers around the coasts of East Anglia to Hampshire (Hearn 2004b). WeBS counts show that numbers increased progressively through the 1970s and early 1980s, peaking at about 10,000 birds. Since then, there has been a fairly steady decline to a minimum of *c.* 1,980 in 2004–05 (Banks *et al.* 2006). This decline is not due to reduced productivity leading to a real decrease in population (Collier *et al.* 2005), but seems to be due to the shifting of wintering areas. Climate change is expected to result in the amelioration of conditions in areas previously too severe for winter survival. Enhanced survival among birds wintering in these sites may result in a progressive reduction in migration distance and reduce or halt the decline in numbers that is now becoming apparent in GB. In support of this, numbers are rising in the Netherlands (van Roomen *et al.* 2004). This race is only a rare visitor to Ireland.

flavirostris ('Greenland White-fronted Goose'): winters mainly in Ireland and W Scotland. Numbers increased to a maximum of over 34,000 in the late 1990s (a little earlier in Ireland), since when there has been a decline at almost every site included in the WeBS/Greenland White-fronted Study survey, to a total of less than 24,000 in spring 2005. Declines have been particularly severe at the key site of Islay (Argyll), where less than 10,000 birds were reported during the 2004–05 winter (Banks *et al.* 2006). Fox *et al.* (2006) suggest that this decrease is real, rather than due to relocation; they postulate falling productivity, with less than 10% young birds in the wintering flocks in every year since 2000. However, mean brood size has not declined, although the proportion of breeders has (Pettifor *et al.* 1999); this must be due to a reduction in the number of pairs breeding, perhaps because of competition with increasing population of Cackling Geese in western Greenland.

Lesser White-fronted Goose   *Anser erythropus* (Linnaeus)

SM monotypic

TAXONOMY Ruokonen *et al.* (2000) found some evidence of affinity with Bean and Pink-footed Geese, though the phylogeny was not strongly supported. This conflicts with conventional taxonomy and the results of Livezey's (1996) analysis of skeletal and morphological traits, which indicate a sister relationship with Greater White-fronted Goose. Further analysis is needed. Ruokonen *et al.* (2004) found evidence of genetic differentiation among some breeding populations across Fennoscandia and N Siberia, indicating that Lesser White-fronted Goose should be treated as three separate conservation units.

DISTRIBUTION N Palearctic, breeding in belt of scrubby tundra, from *c.* 65° to 74°N. Migratory, wintering in grass and arable land in SE Europe, Black and Caspian Seas, and E Asia. Has declined substantially, by >20% between 1990 and 2000, which, coupled with the small global population, led BirdLife International (2004) to define it as 'Endangered' in Europe. Reintroduction programmes in Sweden and Finland have had modest success (Madsen *et al.* 1999); mortality was high among released birds in Finland (Markkola & Tynjälä 1993): in May 2007, there were 10 pairs in Finmark; the releases in the Swedish Lapland scheme resulted in at least 29 broods with 83 immature birds since 2000 (Boere *et al.* 2006).

STATUS Estimating the number of genuinely wild Lesser White-fronted Geese is difficult because of the regular occurrence of birds that are clearly of captive origin; e.g. records from seven WeBS sites in 2003–04 (Collier *et al.* 2005) and two sites in 2004–05 (Banks *et al.* 2006). About 130 birds likely to be wild (120 since 1950), of which the first in GB was an immature shot on Fenham Flats (Northumberland) in Sep 1886, and the only one in Ireland was an adult at North Slob (Wexford) in Mar 1969. There has been a decline in occurrence, with an average of around three per year in the early 1950s to less than one per year through the 1990s. There is no evidence yet that the reintroduction programmes have led to increased records in GB&I, although there have been occasional sightings of neck-collared birds attributable to these schemes, and numbers are increasing in the Netherlands.

Greylag Goose   *Anser anser* (Linnaeus)

RB NB HB WM *anser* (Linnaeus)

TAXONOMY Traditionally, Greylag Goose has been treated as comprising two subspecies, eastern *rubrirostris* and western *anser*. A third subspecies '*sylvestris*' (Iceland, Scotland, coastal Norway, with shorter bill) proposed but has little support (e.g. Vaurie 1965, though see Kampe-Persson 2002). A molecular study (Ruokonen *et al.* 2000) based on 1,180 base pairs of mt DNA showed that differentiation among the lineages of *Anser* was low compared with other birds. Despite the smallness of some sample sizes, the branches revealed in the phylogeny are statistically strongly supported. Subspecies *rubrirostris* and *anser* differ by 1%, and this is similar to other recognisable subspecies within the grey geese.

DISTRIBUTION A globally very widespread species, breeding from Iceland to W China and Mongolia. Kampe-Persson (2002) identifies seven native populations in Europe: Iceland, Faroes and Shetland; Hebrides and adjacent mainland Scotland; coastal Norway; S Sweden to the Netherlands, and east to NW Poland; NW Russia and the Baltic states south to Austria and Croatia; Belarus and Romania, south to Turkey and Greece; and W Siberia. Several of these populations winter in distinct areas, but there is considerable overlap in the winter range of others, supporting the relative lack of genetic divergence.

STATUS See Canada Goose in relation to hybrid pairing with that species. There are four components to the Greylag Goose populations in GB&I, although it is becoming increasingly difficult to differentiate between them in some parts of the range.

(1) A relatively small population of native birds that breeds in N & W Scotland, including the Hebrides. These are generally resident, only moving short distances between breeding and wintering areas. Counts made in Aug at the moulting sites showed a general increase through the 1980s and 1990s, with the latest figures being *c.* 8,500 in Aug 2004 (Banks *et al.* 2006).

(2) Resident population now breeding across much of GB&I. Historically, Greylags bred in S Britain, but in the absence of any biometrics, it is impossible to determine to which source

population they belonged (H Persson pers. comm.). The birds now breeding across most of GB&I originate from a programme of reintroductions for wildfowling, mainly from the Scottish population, that began in the 1930s (Madsen *et al.* 1999). Austin *et al.* (2007) show that re-established Greylag Geese increased by 9.4% per annum over the 1990s, and in the breeding season of 2000 there were thought to be about 25,000 adults in Britain. Re-established Greylags are widespread in Ireland, although in much smaller numbers than Britain (perhaps a few thousand, Crowe 2005). Ringing recoveries (Wernham *et al.* 2002) indicate limited exchange between populations in GB&I and NW Europe, although the latter seem to be increasing sharply (van Roomen *et al.* 2003).

(3) The Icelandic population that winters predominantly in northern Britain. Censuses suggest that late autumn numbers increased from 20,000–30,000 individuals in the 1950s to *c.* 100,000 in the early 1990s, subsequently declining to *c.* 80,000 (Madsen *et al.* 1999), but rising again to *c.* 107,000 (for the UK) in autumn 2004. Orkney is the most important wintering area for the Iceland population, although part winters in Ireland. Numbers here have fluctuated through the years, estimates ranging from 6,000 in 1949, down to 700 (1967), and recovering through 3,800 (1986) and *c.* 4,700 (1993) to >5,000 (Crowe *et al.* 2008).

(4) A few birds from the Dutch subpopulation winter in S England, and occasionally birds from Fennoscandia turn up as vagrants (H Persson pers. comm.).

The main source of mortality in the Icelandic and Scottish populations is hunting; estimates of mortality through this are hard to define because of the uncertainty about both the population sizes and the number of birds killed. However, Madsen *et al.* (1999) suggest that over 30,000 birds were killed in Iceland and *c.* 20,000 in GB each year between 1995 and 1997. These figures comprise a sizeable proportion of the post-breeding population, and an intensive demographic analysis led Trinder *et al.* (2005) to suggest that the Icelandic population will continue to decline unless there is a reduction in hunting pressure. Hunting is less easily monitored in the NW Scottish population because of its more fragmented nature. However, Mitchell *et al.* (1995) suggest that *c.* 20% of the post-breeding population on the Uists (Outer Hebrides) are taken by hunters. Although legally permitted, the re-established populations generally attract less attention from hunters, perhaps due to their relative tameness. Nevertheless, Reynolds & Harradine (1996) estimate that up to 35% of the population of England and Wales may be taken in any particular year.

Snow Goose   *Anser caerulescens* (Linnaeus)

PM *caerulescens* (Linnaeus)
PM *atlanticus* (Kennard)

TAXONOMY Morphological and skeletal characters (Livezey 1996) indicate a clade uniting the sister pair of Snow Goose and Ross's Goose *A. rossii* with Emperor Goose *A. canagicus*, and forming a clade with the 'grey' geese. Ruokonen *et al.* (2000) also found a sister relationship between Snow and Ross's Geese based on mt DNA sequences, but they did not include any *Branta*. In a study of Canada Geese, Paxinos *et al.* (2002) found that Ross's and Emperor Geese were sister to other species in the grey goose clade. Thus, it seems that Snow and Ross's represent an early branch in the lineage leading to modern *Anser*, and these are often placed in a separate genus *Chen*. Molecular studies have documented two hybridisation episodes in Snow and Ross's Geese and support the hypothesis that blue and white Lesser Snow Geese had allopatric breeding distributions in the past (Weckstein *et al.* 2002). Two subspecies of Snow Geese are recognised; *caerulescens* (Lesser) and *atlanticus* (Greater), differing in size and plumage.

DISTRIBUTION Subspecies *caerulescens*, breeding from Wrangel Island in NE Siberia to Baffin Island and Hudson Bay, is polymorphic for plumage; two forms (Blue and Snow) showing clinal variation in frequencies across the range (reviewed by Cooke *et al.* 1995); *atlanticus*, breeding further north in W Greenland and N of Baffin Bay, is almost invariably white. Both subspecies nest on the tundra, usually close to water, and are migratory, *atlantis* wintering in coastal Maryland to N Carolina, while *caerulescens* is more widespread from Washington to California and around coast of Gulf of Mexico.

STATUS Many occurrences of captive origin. There are now small breeding populations in Argyll, Hampshire and East Anglia, of which the first is more or less self-sustaining. This is centred on the islands of Coll and Mull and comprises around 50 individuals. Breeding success is not known accurately, but up to 16 young have been reported in individual years (Dudley 2005). Ringing has recently shown these birds to be highly sedentary. The other populations are more limited and, although breeding is recorded from time to time, productivity appears to be inadequate to maintain the population.

Over 100 individuals thought to be of wild origin have been recorded in Ireland, and these include both blue and white morphs. Most that have been identified to race have been *caerulescens*, but there are five records of *atlanticus* in Scotland and Ireland: one shot from a small flock in Dumfries and Galloway (Feb 1921), one found dead in Lothian (Jan 1954), two in Mayo (Oct 1877; Oct 1886), and one in Wexford (three winters, 1983–86).

There seem to be no records of any in GB that are unquestionably of wild origin; such confirmation can only be through the occurrence of ringed birds. Snow Geese recorded as 'wild' (Category A) in GB&I tend to be supported by circumstantial evidence (e.g. arriving with flocks of wild geese from Greenland/Iceland in winter). Seventeen Snow Geese (including four immatures) were seen flying down the River Thames at Swanscombe (Kent) in Mar 1980 and later the same day at nearby Cliffe. The following month, 17 white-phase birds (incl. four imms) were seen in the Netherlands, with a single blue-phase adult. One of the white adults bore a blue colour ring and a metal ring. The Snow Goose research team at La Pérouse Bay (Manitoba) had used blue rings on male goslings there in 1977.

Genus *BRANTA* Scopoli

Holarctic (and Hawaiian) genus of six species, five of which are Palearctic and have been recorded in GB&I.

Canada Goose/Greater Canada Goose   *Branta canadensis* (Linnaeus)

NB HB *canadensis* (Linnaeus)

TAXONOMY Recently split from Cackling Goose *B. hutchinsii* on basis of morphology, genetics, ecology and behaviour (reviewed by Sangster *et al.* 2005). Canada Goose includes at least seven poorly differentiated subspecies (*canadensis, fulva, interior, maxima, moffitti, occidentalis, parvipes*) and is close to Hawaiian Goose *B. sandvicensis*; Cackling comprises three or four subspecies (*hutchinsii, minima, leucopareia* and perhaps *taverneri*) and is closer to Barnacle Goose *B. leucopsis* (Paxinos *et al.* 2002). Scribner *et al.* (2003a) confirmed the substantial distance between Canada and Cackling Geese.

DISTRIBUTION Breeds in wide range of habitats, from northern tundra to wooded plains of central USA. Some populations strongly migratory, wintering in grasslands, both natural and agricultural. Main distribution (with primary wintering areas in parentheses), from Madge & Burn (1988); *canadensis*: SE Canada, NE USA (Atlantic coast USA); *fulva*: S Alaska, British Columbia (largely resident; some go to N California); *interior*: C & E Canada (SE USA); *maxima*: inland N USA (largely resident); *moffitti*: inland NW USA, SW Canada (W USA); *occidentalis*: coastal S Alaska (Vancouver, Oregon); *parvipes*: Arctic Canada, E Alaska (S USA). Increasing numbers of Canada Geese breeding in W Greenland have been assigned to *interior*, being genetically closer to the Ungava Bay population from Quebec than any other that was analysed (although formal statistical support is low: Scribner *et al.* 2003b). It appears, however, that W Greenland is increasing as a moulting site for Canada Geese, with birds from a variety of races occurring, including individuals that resemble *occidentalis* and *maxima* (AK Fox pers. comm.).

STATUS Canada Goose was introduced to GB&I as an ornamental waterfowl more than 250 years ago and populations maintained a low level until the 1950s. Numbers then grew relentlessly, with only a slight dip around 1990. Austin *et al.* (in press) describe how numbers increased at about 9% per annum during the 1990s, with an estimate of nearly 90,000 adults during the 2000 breeding season. Increase probably facilitated by reservoirs and gravel pits providing suitable breeding habitat, and the transition to autumn-sown cereal providing grazing areas through the winter. The overwhelming majority of this naturalised population was founded from 'large race' birds, now assigned to *B. canadensis*. The Irish population is also increasing, albeit less dramatically, and now exceeds 1,000 birds. Hybridisation with feral Greylag Geese is widespread, and with Barnacle Geese less frequently.

Some birds from from C and N England undertake a moult migration to the Beauly and Cromarty Firths (E Scotland). This formerly involved hundreds of birds annually but smaller numbers now (RH Dennis pers. comm.). More-local movements take place among the wetlands in England. There are a few overseas recoveries, including France, Norway, Sweden and the Faroes (Wernham *et al.* 2002), indicating a small amount of interchange among European countries. There are three long-distance movements: two that involve birds crossing the Atlantic and one from the River Ob in Siberia. One ringed at N Slob (Wexford) in Nov 1993, shot in Maryland (USA) in Jan 1995. At this time, feral Canada Geese were rare in RoI, and the small numbers found each winter in Wexford were often associated with Greenland White-fronted Geese. Another ringed in Maryland (USA) in Feb 1992, seen alive at Muir of Fowlis (Aberdeenshire) in Nov 1992, and shot near Perth in Jan 1993. It belonged to one of the smaller races and other small or medium-sized Canada Geese occur from time to time, but their origin is usually difficult to determine.

Cackling Goose/Lesser Canada Goose   *Branta hutchinsii* (Richardson)

SM *hutchinsii* (Richardson)

TAXONOMY Recently split from Canada Goose *(q.v.)*.

DISTRIBUTION Generally more northerly than Canada Goose, though overlap and sympatry in some areas (summarised by Sangster *et al.* 2005). Three subspecies: *hutchinsii*: Melville Peninsula, islands of Arctic Canada (winters along coastal Gulf of Mexico, Texas to Mexico); *minima*: coastal W Alaska (inland California); *leucopareia*: Aleutian Islands (C California), with a possible fourth *taverneri* N and C Alaska (Washington) – though see Sangster *et al.* (2005).

STATUS Only recently split from *B. canadensis*, and claimed records currently under review. Nevertheless, individuals (especially juveniles) likely to be of this species occur regularly with geese that are known to have arrived from Greenland and NE Canada, and at least some of these are presumed to be wild. Several birds with the wintering goose flocks on Islay (Argyll) have been provisionally identified as *hutchinsii*, as has one wintering with Pink-footed Geese at Holkham (Norfolk) in Feb 1999. Irish records of 'small'-type birds awaiting review (P Smiddy pers. comm.)

Barnacle Goose   *Branta leucopsis* (Bechstein)

HB WM monotypic

TAXONOMY Skeletal and anatomical characters led Livezey (1996) to conclude that Barnacle is sister to Red-breasted Goose, in a clade with Brant Goose. However, molecular analyses (Paxinos *et al.* 2002) indicate a closer relationship with Cackling Goose, placing Red-breasted in a separate subclade with Brant. No subspecific differentiation, despite clear separation into subpopulations.

DISTRIBUTION Three populations, breeding and wintering separately: E Greenland (winters W Scotland, Ireland); Svalbard (Solway Firth); coast and islands of Arctic Russia, now including birds breeding in the Baltic that probably were established from Russian stock in the 1970s (principally in the Netherlands). Frequently nests on cliffs overlooking the sea or on skerries and small islands. In winter, usually on grasslands close to the sea; less often on saltmarsh or estuarine vegetation. All populations increasing sharply, with a consequent expansion in range that has ameliorated its European conservation status (BirdLife International 2004).

STATUS See Canada Goose in relation to hybrid pairing with that species. Naturalised populations of Barnacle Goose are now breeding in various parts of the UK. Collier *et al.* (2005) suggest that many of these originate from free-flying birds from waterfowl collections. Although still comparatively small (WeBS maximum of 1,011 in Dec 2003), numbers are increasing rapidly, especially at ponds and gravel pits in S & E England and at Strangford Lough (Co Down). Naturalised birds in SE England make it difficult to assess the occurrence of wild Barnacle Geese from the Netherlands and Germany, where about 360,000 birds winter.

Two populations winter in GB&I, those from Greenland (Ireland and W Scotland) and Svalbard (Solway Firth). In GB, a sizable proportion of the former occurs on small and relatively remote islands of the Hebrides, and many sites are only censused in special efforts, typically every five years. The most recent international survey was in Mar 2003, when 56,386 birds were counted in GB&I, of which 36,478 were on Islay (Argyll) and 9,034 in Ireland (Worden *et al.* 2004). This population from Greenland has increased from about 10,000 in the late 1950s, although may be showing signs of slowing in recent years; perhaps partly due to a decline in breeding success (Worden *et al.* 2004).

The Svalbard population, which winters almost exclusively in the Solway Firth, has also increased substantially since 1958, when the WWT began its annual census of the winter population, to 28,356 in 2004–05 (Banks *et al.* 2006). In recent years, productivity has been low, declining from *c.* 30% juveniles in 1957–91 to 14.2% in 1993–2003, due to a decrease in both mean brood size and the proportion of adults accompanied by juveniles within the flock. The overall increase in survival from 87 to 95% may compensate for the decline in breeding, allowing the population to continue to rise. However, Trinder *et al.* (2005) show that this is a marginal effect, and even a modest decrease in adult survival could lead to a collapse in this population. Wintering birds from Svalbard usually arrive along the east coast of Scotland and N England, relocating to the Solway after a few days of rest.

Brant Goose/Brent Goose   *Branta bernicla* (Linnaeus)

WM *bernicla* (Linnaeus)
WM *hrota* (O.F. Müller)
WM *nigricans* (Lawrence)

TAXONOMY Morphology suggests affinity with Barnacle and Red-breasted Geese (Livezey 1996), though DNA sequences indicate Barnacle is in a clade with Cackling Goose. Intraspecific situation still unclear, and research underway into species limits. Three (Roselaar in BWP) or four (HBW) races recognised, based on intensity and extent of black on upperparts, colour of belly and flanks, and extent of white collar. As with many geese, strong fidelity to breeding and wintering grounds. Pair formation in late winter combined with strong female philopatry gives potential for rapid mt DNA differentiation. Three races recognised here: dark-bellied *bernicla*: Arctic Russia, Siberia, E to Taimyr (wintering around S North Sea and English Channel); pale-bellied *hrota*: two populations, Greenland and NE Canada (wintering principally in NE N America, Ireland) and Svalbard (wintering Denmark and NE England); 'Black Brant' *nigricans*: breeds from Taimyr through Alaska to Arctic islands of W Canada (wintering China, Japan, W coast of N America to California). Population on Melville Island ('Grey-bellied Brant') intermediate in plumage and may be separate subspecies.

DISTRIBUTION Breeds on coastal and lowland tundra, wintering on estuaries and sandy coasts. The population of *bernicla* declined by *c.* 75% in 1930s, due to disease destroying principal winter food of eel grass *Zostera marina* (Atkinson-Willes & Matthews 1960). Subsequent recovery probably due to a combination of recovering eel grass beds and changed agricultural practices, leading to availability of winter cereals close to the sea.

STATUS Part of the Russian population of dark-bellied *bernicla* winters along the coasts of E and S England, from River Humber (Yorks) to Exe Estuary (Devon), with a few outlying groups, especially at the Burry Inlet (S Wales: Ward 2004), and totalled over 85,000 in Jan 2005. Following a long rise in numbers from the 1960s through the early 1990s, to a peak of *c.* 100,000 birds, there was a clear decline, apparently due largely to low productivity. It is estimated that winter flocks need to hold *c.* 15% juveniles to compensate for annual mortality (Collier *et al.* 2005). This figure has not often been reached since the late 1990s, although changing wintering grounds may contribute to observed decline. Birds leave the Wadden Sea for Britain in severe weather; as winters ameliorate, this behaviour may occur less often.

The E Atlantic (Svalbard and NE Greenland) component of the light-bellied *hrota* race, though increasing in numbers, probably only totals *c.* 6,600 birds and constitutes one of the smallest and most threatened goose populations in the world (Denny *et al.* 2004). Roughly half of the population winters on the mudflats of Lindisfarne in NE England, most of the remainder in Denmark. Numbers at Lindisfarne are relatively stable, with maxima averaging *c.* 3,600 since the turn of the century (Banks *et al.* 2006). However, the size of the flock fluctuates annually depending on conditions in Denmark. The small numbers of light-bellied Brants that appear at other sites along the east coast of Britain are assumed to also be referable to this population.

The E Canadian High Arctic component of *hrota* is also increasing (J.A. Robinson *et al.* 2004b; Banks *et al.* 2006), and now exceeds 33,000 birds, almost all of which winter on estuaries around the coasts of Ireland. In the autumn, most birds make initially for Strangford Lough, where 26,250 in Oct 2004 was the highest on record. As the winter progresses, numbers decline at Strangford and increase elsewhere around Ireland. Recent increases in this population are due to strong breeding performance (Banks *et al.* 2006).

The first 'Black Brant' *B. bernicla nigricans* was recorded at Foulness (Essex) in Feb 1957, and the first Irish birds were at Strangford Lough (Down) in Nov–Dec 1978. Numbers have increased markedly, at least partly due to increased observer awareness. Until the mid-1970s, occurrences were less than annual, but from 1978 they were increasingly found in Irish flocks of Canadian *hrota*. From the early 1980s, birds were regularly found among Brant flocks in Lincolnshire and East Anglia, with up to 18 in some years. Many are probably returning individuals, paired with either *hrota* or other *nigricans*. Hybrid young can appear confusingly similar to Melville Island 'Grey-bellied Brants'.

Red-breasted Goose   *Branta ruficollis* (Pallas)

SM monotypic

TAXONOMY Phylogenetically close to Barnacle Goose (Livezey 1996) or Brant Goose (Paxinos *et al.* 2002).

DISTRIBUTION Rather restricted distribution, principally close to Taimyr Peninsula (Siberia); winters in S Caspian and Black Seas, recently expanding into Romania, Bulgaria, Greece. Serious decline since 1950s probably due to changes in land use (HBW), though population possibly stabilising since 1990 (BirdLife International 2004), albeit at a low level. Nests on tundra, usually close to water, wintering on arable land, and may be particularly vulnerable to avian influenza in coming years.

STATUS As a popular ornamental waterfowl, many individuals of captive origin in the countryside. Birds are seen at all seasons, at a wide range of sites from coastal estuaries to village ponds, and may be timid or approachable. Young birds in autumn along the E and S coast of GB in company with Russian Brants believed more likely to be wild. Many occur with Pink-footed Geese, but whether they arrive with them or transfer from the Brants is not always clear. Numbers vary from none up to six a year, and some of these are likely returning individuals.

The first accepted records for GB both occurred in 1776; one was shot near London early in the year and another at Wycliffe-on-Tees (Cleveland) in winter. Over 70 now recorded in GB in circumstances which suggest a natural origin. There are no records for Ireland (see Appendix 2).

Genus *ALOPOCHEN* Stejneger

A monospecific genus that is widespread across sub-Saharan Africa and has been introduced in GB&I.

Egyptian Goose   *Alopochen aegyptiaca* (Linnaeus)

NB monotypic

TAXONOMY Morphology suggests that affinities lie with Orinoco Goose *Neochen jubata* and perhaps shelducks; molecular data reported by Donne-Goussé *et al.* (2002) supported a close relationship between *Alopochen aegyptiaca* and *Tadorna* but their study did not include *Neochen jubata*.

DISTRIBUTION Widely distributed across African wetlands, only missing from heavily wooded sites. Largely sedentary though may move into areas following rains.

Status Introduced into England probably before the 1700s (Long 1981) and now well established, especially in East Anglia. This species is under-represented in WeBS counts because many occur away from the larger lakes and gravel pits included in this project. A survey in Norfolk in 1991 recorded 846 birds (Sutherland & Allport 1991). The number of sites at which it occurs is increasing, as it spreads away from the traditional heartland of East Anglia. Numbers are increasing at Rutland Water (Leicestershire) and now colonising several midland counties, e.g. Nottinghamshire, Derbyshire, although some of these may be of local captive origin. Not naturalised in Ireland.

Genus *TADORNA* Boie

Palearctic, Ethiopian, Oriental and Australasian genus of seven species, two of which are Palearctic and have been recorded in GB&I.

Ruddy Shelduck   *Tadorna ferruginea* (Pallas)

SM monotypic

TAXONOMY On morphological grounds has been placed in superspecies with S African, Paradise and Australian Shelducks *T. cana*, *T. variegata*, *T. tadornoides* (HBW), but molecular support for this is limited.

DISTRIBUTION Scattered populations from SE Europe across C Asia to Lake Baikal, also NW Africa; Asian populations migratory, wintering in SE Asia and India. Western populations more sedentary, though dispersive in search of waterbodies for breeding. The N African population was estimated as 2,500 individuals and is believed to be stable (Scott & Rose 1996). The European/Russian population is more variable, with (e.g.) decreases in Greece and Turkey but increases in Bulgaria and Russia (Vinicombe & Harrop 1999). There are feral populations in some W European countries; all are small and none is thought to be self-sustaining.

STATUS Clear pattern of arrival in July, peaking in Aug, and declining through the autumn. However, there are regional differences, with higher background level of birds in the SE. Many of these are undoubtedly of captive origin. The species is popular in collections (Vinicombe & Harrop 1999), and young that are neither clipped nor pinioned disperse away from them. These may survive to breed (examples in Vinicombe & Harrop 1999), further adding to the feral population, and (unsurprisingly) such young birds will appear in the wild in June–Sep. Despite breeding in the wild, numbers and productivity are too low for the population to be regarded as self-sustaining.

   In recent years, there have been several attempts to establish whether Ruddy Shelducks occur as natural vagrants to GB&I (e.g. Vinicombe & Harrop 1999; Harrop 2002). These have centred on estimating a baseline for birds of captive origin and then examining periods when the numbers were atypically high. Such an event occurred in 1994, with a substantial influx of birds into NW Europe (especially Scandinavia) in July–Aug and many parties of 5–10 individuals. Numbers of birds were unremarkable in Iberia, suggesting a NW African origin was unlikely and SE Europe the more probable source. Small parties were seen in N England and Scotland more or less simultaneously with the Scandinavian influxes, in accord with a NW movement of wild birds across Europe. Vinicombe &

Harrop (1999, also Vinicombe 2002, Harrop 2002) differ in their conclusions. Vinicombe makes a strong case for a natural origin; Harrop is more conservative, arguing for captive origin or perhaps from the feral population in Askaniya Nova (Ukraine). The species remains in Category B of the British List, last recorded in an apparently wild state in 1946. The situation is similar in Ireland, where the last acceptable record was also in 1946, although a few birds of unknown origin also appear from time to time.

Common Shelduck   *Tadorna tadorna* (Linnaeus)

MB RB WM monotypic

DISTRIBUTION Mostly coastal in W Europe; across inland C Asia and Iran to NE China. Continental populations tend to move south to winter in Mediterranean, NW India, SE China regions.

STATUS Although a common and familiar coastal bird, it has declined in GB in recent years. It breeds predominantly in rabbit burrows and other holes along the coasts; however, increasing numbers breed inland at lakes, especially in gravel pits and sand quarries of the English Midlands. CBC data showed an increase of *c.* 300% from 1966 to 1982, although this was based on a relatively small sample. There followed a period of relative stability until the mid-1990s, since when BBS data indicate a decline of *c.* 38% in the UK. WeBS data support these trends, showing a peak in the 1980s and declining from *c.* 1990, and, despite occasional fluctuations, the trend continues downwards. This is not uniform; sites in N England (e.g. Mersey, Dee and Ribble Estuaries) showed modest increases (although some of these involve moulting birds) whereas some sites in the south (Blackwater, Alde, Severn) appear to be in decline. By contrast, numbers in Northern Ireland are increasing strongly, with the midwinter peak rising steadily since the early 1990s (Banks *et al.* 2006). Recent winter estimates of 78,200 in Britain (Kershaw & Cranswick 2003) and 14,610 in Ireland (Crowe *et al.* 2008).

There is a significant movement of pre- and failed breeders to the Wadden Sea in late summer (Wernham *et al.* 2002), and these birds return in more leisurely fashion during the autumn. Peak numbers generally occur in winter, when the native population is enhanced by Continental immigrants. Whether the decline in S England is due to a higher proportion of the population remaining on the near-Continent is unclear. There is evidence (Wernham *et al.* 2002) that birds are increasingly using British estuaries for moulting, with counts of >1,000 in July–Aug 2000–04 at several sites, including R. Mersey, R. Humber and The Wash.

Genus *AIX* Boie

Holarctic genus of two species: one from the eastern Palearctic, which has been introduced to GB&I (below), the other Nearctic, which is widely held in captivity and frequently escapes to cause confusion over possible natural vagrancy (Wood Duck *Aix aponsa*).

Mandarin Duck   *Aix galericulata* (Linnaeus)

NB monotypic

DISTRIBUTION Eastern Palearctic. NE China, Japan, Ussuriland and Manchuria. Breeds on lakes, ponds or rivers where there are old trees with hollows for nesting; takes readily to nest boxes. The GB population has been isolated from its oriental ancestors for many generations and now forms a small, independent part of the gene pool.

STATUS This species retains its breeding stronghold in SE England, where it was introduced *c.* 250 years ago (Long 1981), although there is an important secondary centre in the Forest of Dean. Numbers at WeBS sites have been increasing in recent years, reaching 550 in 2003–04, although such numbers probably represent only a small fraction of the overall population of this secretive species; Davies (1988) estimated a minimum of 7,000 birds in Britain. Numbers counted by WeBS typically peak in midwinter, when birds are less secretive and apparently accumulate on larger wetlands. Sites in Hampshire, Yorkshire and Derbyshire held record numbers in 2003–04, suggesting that the species is expanding out of its core region. It has been spreading in Scotland in recent years. There is a small, and apparently self-sustaining, population in Co. Down. More or less sedentary, with limited evidence of post-breeding dispersal, though a Dutch-ringed bird was recovered in Highland in Apr 1998.

Genus *ANAS* Linnaeus

Cosmopolitan genus of about 42 species, 11 of which breed in the Palearctic; seven of these have been recorded in GB&I and four from the Nearctic. Because so many wildfowl are kept in captivity, it can be difficult to determine the provenance of many individual birds. Some species which have been recorded in GB&I, such as Chiloe Wigeon *A. sibilatrix* and White-cheeked Pintail *A. bahamensis*, are very unlikely to be of natural origin, although a stronger case can be made for the natural vagrancy of some Falcated Duck *A. falcata*, Baikal Teal *A. formosa* and even Cinnamon Teal *A. cyanoptera*. However, no individuals of these last three species have yet occurred whose credentials are beyond doubt.

Eurasian Wigeon   *Anas penelope* Linnaeus

RB WM monotypic

TAXONOMY A detailed analysis of the dabbling ducks by Johnson & Sorenson (1998, 1999) using cyt-b and ND2 mt DNA sequences showed that the wigeons (American, Eurasian and Chiloe) form a clade with Falcated Duck and Gadwall. This broadly confirmed the findings of Livezey (1996), based on comparative morphology, although there were important differences in detail. Livezey's results suggested that Eurasian and American Wigeon are sister species, with Chiloe Wigeon, Falcated Duck and Gadwall successively less close. Behavioural data (e.g. Johnsgard 1965) support the close affinity of American and Eurasian Wigeons. The wigeons have been analysed by Peters *et al.* (2005) using multiple loci (both nuclear and cytochromal) and relatively large numbers of individuals from across the geographic range of each species. They found that the phylogenies produced from these data were generally in agreement (and with Johnson & Sorenson 1998, 1999), indicating that American and Chiloe Wigeon are sister species, with Eurasian in the same subclade. Falcated Duck and Gadwall are sister species, in a clade rather more separate from the other three. Peters *et al.* (2005) found no evidence of substructure within Eurasian Wigeon to suggest population differentiation. They did, however, find two Eurasian Wigeon whose sequences fell within the American Wigeon clade. These two species hybridise in the wild, especially when one is rare in the range of the other. The two mismatching birds came from the W coast of the USA, where the two species coexist in winter. Pair formation in many wildfowl takes place on the wintering grounds.

DISTRIBUTION A Palearctic species breeding from Iceland, across N Eurasia to Pacific Ocean, S as far as Lake Baikal. Nests beside fresh water, frequently in wooded areas. Most populations migratory, wintering in ice-free temperate regions from GB&I across S Europe and wetlands of N Africa to India, China, Japan and (less commonly) E and W coasts of North America. Winters in areas of grassland adjacent to fresh, brackish or marine habitat.

STATUS Small breeding population (*c.* 300 pairs: RBBP) in GB&I, although many more birds are present in summer than are proved to breed. The Second Breeding Atlas showed a concentration of sites in C and N Scotland, especially in Perth and Kinross, where it breeds widely on alkaline lochs. In recent years, RBBP reports confirmed breeding in Hants, Essex, Northumberland, Fife, Aberdeenshire, Highland and Orkney. Breeding has also been recorded at Lough Neagh (NI, 1933), Rathlin Island (Antrim, 1953) and in Co. Tyrone (1995).

   The winter flocks of Wigeon include birds from across the Palearctic, although the relative contributions of the various breeding areas such as Iceland, Scandinavia and E Siberia are not known. Numbers typically increase through the autumn, peaking in Dec–Jan, declining rapidly thereafter.

WeBS data (Banks *et al.* 2006) give an indication of a steady rise in numbers since the mid-1970s towards a recent winter estimate of about 400,000 (Kershaw & Cranswick 2003). However, the effects of hard weather are superimposed on this trend, with birds moving rapidly both within GB&I and between sites around the North Sea. There are major concentrations on estuarine sites, especially the Ribble (Lancs), with an average of almost 70,000 birds during recent years, and a further nine sites regularly held more than 10,000 birds. Numbers in NI have undergone a decline through the 1990s, which shows no sign of reversing. Numbers peak there in Oct, declining steadily through the winter: perhaps indicative of an Icelandic component to the population, some of which move on to winter elsewhere, possibly in RoI, where a recent estimate totals 82,370 (Crowe *et al.* 2008).

American Wigeon   *Anas americana* Gmelin

SM monotypic

TAXONOMY See Eurasian Wigeon. No evidence from DNA analyses of population substructuring.

DISTRIBUTION A Nearctic species that is widespread across N America, from Alaska to New England, and S to California and Colorado. Breeds in similar habitat to Eurasian and winters in grasslands along Atlantic and Pacific seaboards.

STATUS A Nearctic species that has increased in numbers in recent years (Votier *et al.* 2003), to the extent that it was removed from the list of BBRC species in 2002, by which time *c.* 20 were being found in GB every year. Numbers in Ireland show a similar pattern of increase, with up to nine birds being recorded in some years. Both sexes show this increase, although increase in females began later, when the identification criteria became better known. There is a significant difference in the pattern of finding: more are found in Ireland than in GB during Aug, presumably reflecting the earlier landfall in the more western parts of GB&I. More birds are found in GB during Apr–May than in Ireland, suggesting a different origin of at least some. American Wigeon that cross the Atlantic in autumn to the south of GB&I are likely to associate with Eurasian Wigeon wintering in SW Europe. When the latter begin their northward migration, the American Wigeon apparently move with them through E and C England, where there is a concentration of spring birds. Undoubtedly, there will be new arrivals from N America as well, but there is a preponderance of inland and E coast birds in GB at this time (Votier *et al.* 2003). There are several recoveries or recaptures of Nearctic-ringed American Wigeon in GB&I, all originating from E Canada or NE USA.

Gadwall   *Anas strepera* Linnaeus

RB NB MB WM monotypic

TAXONOMY Molecular data indicate that Gadwall forms a species pair with Falcated Duck (Peters *et al.* 2007); there seems to have been no direct comparison of Nearctic and Palearctic Gadwall to see whether the genetic differentiation evident in teals, wigeons and mallards extends to this species. Peters *et al.* (2005) found one bird from the Netherlands with the mt DNA sequence of an American Wigeon, suggesting past hybridisation.

DISTRIBUTION Holarctic, with populations in N America and Eurasia. Palearctic populations breed from Iceland, GB&I and the Netherlands across Eurasia from 40° to 60°N; those in the Nearctic between 66°N and 30°N. Migratory in both regions, wintering S to north Africa, Arabia, India, S China and to Mexico and Florida, respectively.

STATUS There are two components to the GB&I population: a breeding population enhanced by winter immigrants. Although the RBBP is not especially comprehensive, it is tolerably repeatable in terms of coverage, and the reported figures suggest a steady rise in breeding numbers from *c.* 260 pairs in the 1960s (First Breeding Atlas) to *c.* 1,600 in 2006 (RBBP). The breeding population is greater in the south, the strongholds being in S & E England; however, it breeds as far north as Orkney. It is less common in Ireland, with only small populations, notably in Lough Neagh and Co. Wexford, and the complete demise of a population of over 200 at Ballycotton (Cork) since the early 1990s, due to habitat change.

The winter population has increased unremittingly (at *c.* 4% per annum) since the 1960s, to a recent estimate of 17,100 in winter; WeBS indices show an approximate doubling of winter numbers over the course of the 1990s (Banks *et al.* 2006). This increase parallels elsewhere in Europe (Wetlands International 2002). Maximum numbers occur in Nov–Dec. The most important site is currently Rutland Water (Leicestershire), which held almost 1,100 birds in Aug 2003, and six other sites held average peak numbers of more than 600: the threshold for international importance. The situation in Ireland is slightly different. Crowe *et al.* (2008) estimated 630 in Ireland, with little change apparent in recent years. Ringing recoveries of Gadwall wintering in GB&I are still relatively few (Wernham *et al.* 2002) but suggest that most winter immigrants originate around the S Baltic, although there are a few recoveries from Finland and N Russia.

Eurasian Teal   *Anas crecca* Linnaeus

RB WM PM *crecca* Linnaeus

TAXONOMY Molecular data (e.g. Johnson & Sorenson 1999) suggested, rather surprisingly, that Eurasian and Green-winged Teals are not sister species but that the latter is phylogenetically closer to Speckled Teal *A. flavirostris* of S America, despite the latter having lost its sexual dichromatism. The discovery that molecules may evolve independently of morphology and behaviour in wildfowl has been made several times, e.g. in the wigeons, where Chiloe and American are sister taxa, with Eurasian less close (and again the S American taxon has lost its sexual dimorphism). This represents a challenge for behavioural and evolutionary biologists. Two subspecies are recognised, differentiated by size: *crecca*: Palearctic; *nimia*: Aleutian Islands.

DISTRIBUTION Nominate breeds across N Palearctic from Iceland, GB&I, C Europe N of *c.* 45°N across to Japan and Kamchatka; northern populations migratory, wintering S to tropical Africa, India and Philippines. Nests in swamps and marshes with thick vegetation; winters on wetland, often brackish marshes or coastal mudflats. *Nimia* is a resident race restricted to the Aleutian Islands (Alaska).

STATUS Breeding status difficult to assess due to wide distribution and relatively secretive habits. The Second Breeding Atlas estimated 1,500–2,600 pairs in GB, with 400–675 pairs in Ireland. This suggested a serious decline; indeed, *c.* 25% of 10-km squares were lost in GB&I during the 20 years since the First Breeding Atlas. The breeding population seems to be resident, with some hard-weather movements to France and Iberia in winter; however, relatively few are ringed during the breeding season, so recoveries are correspondingly few.

The number of Teal wintering at GB WeBS sites has increased about fourfold since the 1960s, to a recent winter estimate of 192,000 (Kershaw & Cranswick 2003), with a further 45,000 in Ireland (Crowe *et al.* 2008), although peak numbers fluctuate among years and between sites. The highest numbers are usually in Nov–Jan, and the majority are winter visitors from the near-Continent, Scandinavia and Iceland. There were 14 sites where counts exceeded 4,000: the criterion for international importance. In NI, *c.* 6,000 counted at WeBS sites in 2004–05; although there are significant fluctuations between years, the numbers here seem stable overall.

Green-winged Teal   *Anas carolinensis* Gmelin

SM monotypic

TAXONOMY Recently split from Eurasian Teal (for summary, see Sangster *et al.* 2001).

DISTRIBUTION Breeds across N America between 73°N and 37°N, wintering south to Central America. Habitat similar to Eurasian.

STATUS The number recorded in GB&I has increased steadily since 1950, to an annual total of over 25 in 1990, when Green-winged Teal was removed from the BBRC list and records were no longer collated nationally. Records are still monitored in Ireland, however, and the average number has levelled out at 6–10 per year. All proven records are of male birds; the true figure may be double this.

There are few records in June–Sep, perhaps a combination of low numbers and the difficulty of identifying males in eclipse plumage. However, there are striking differences in patterns of occurrence. In Ireland, birds peak in Nov–Feb, whereas in GB they are later, with the peak in Mar–Apr. The differences are slight but significant. Whether this reflects earlier arrival in the west, with a subsequent dispersal across the Irish Sea, is not easily tested.

Mallard   *Anas platyrhynchos* Linnaeus

RB NB HB WM *platyrhynchos* Linnaeus

TAXONOMY Donne-Goussé *et al.* (2002) examined two mt DNA sequences in a range of wildfowl, including several from the genus *Anas*; they found that Mallard lies within the same clade as Northern Pintail, Eurasian Teal, Northern Shoveler, Chiloe Wigeon and Gadwall. An earlier analysis of a wide range of 'dabbling' ducks (Johnson & Sorenson 1998, 1999) had shown a series of clades, one of which contains 'mallards' from around the world, including Indian and Eastern Spot-bill Ducks *A. poecilorhyncha, A. zonorhyncha,* Laysan and Philippine Ducks *A. laysanensis, A. luzonica,* Pacific and American Black Ducks *A. superciliosa, A. rubripes,* Meller's, 'Mexican', Mottled and Yellow-billed Ducks *A. melleri, A. diazi, A. fulvigula, A. undulata,* with several of which it is known to hybridise (Rhymer *et al.* 1994; Mank *et al.* 2004; Williams *et al.* 2005). Different sequences were found in Mallards (Avise *et al.* 1990), one closely similar to Mexican and American Black Ducks, the other to the Spot-bills and Philippine Duck (Johnson & Sorenson 1999). A more detailed investigation, based on 152 birds from N Russia and Siberia to Alaska (Kulikova *et al.* 2005), confirmed two basic sequences, one of which occurred throughout the studied area, while the other was predominant in Alaska, declining in frequency through the Aleutians into N Asia. They speculate (Kulikova *et al.* 2004) that the existence of an Eastern Spot-billed Duck sequence in some E Asian Mallards is a consequence of the known hybridisation between these taxa; in Europe and W Asia (where Mallard alone occurs) there is only one basic sequence. Hybridisation alone is insufficient to explain their findings from N America, where divergent sequences are found across the continent, not just where Mallards are sympatric with American Black and Mottled Ducks. Interestingly, a small sample of birds from the Aleutians contained a unique DNA sequence; since this population is resident, it demonstrates the importance of pair formation during the winter on the genetic structure of wildfowl populations. Kulikova *et al.* (2005) admit that further research is needed to resolve the evolutionary history of this commonest of ducks.

The intraspecific taxonomy is complex, with a series of taxa that are variously regarded as species or subspecies. We follow the AOU (see Moorman & Gray 1994) in recognising three subspecies of Mallard: nominate *platyrhynchos, conboschas* and *diazi* ('Mexican Duck'); taxa *fulvigula* (including *maculosa*) (S USA/Mexico: Mottled Duck), *wyvilliana* (Hawaii Is: Hawaiian Duck) and *laysanensis* (Laysan Is: Laysan Duck) are treated as full species, albeit very closely related to Mallard.

DISTRIBUTION The most widespread duck in the world, breeding across Eurasia between 40°N and 70°N and N America from 20°N to 70°N. The nominate occurs throughout the species' range, apart from SW Greenland (*conboschas*) and Arizona/Texas to C Mexico (*diazi*). Northern populations are migratory, wintering in S of range, and (in Asia) to S China; southern populations are more resident. Introduced into SE Australia and New Zealand.

STATUS See American Black Duck in relation to hybrid pairing with that species. Mallard is the most widespread breeding waterfowl in GB&I, recorded in *c.* 90% of 10-km squares in Second Breeding Atlas, being absent only from the highest regions where the water is nutrient-poor. Breeding distribution changed little in GB between the two Atlases, although the number of occupied squares declined by *c.* 9% in Ireland. Because of this wide distribution, Mallard is well represented in the Waterways Bird

Survey (WBS), as well as CBC/BBS, so these allow an estimate of population trend. This is generally positive, with an increase of *c.* 185% for both data sets during 1975–2003, although there is evidence of a recent slight decline, especially in Scotland. The breeding population is estimated at *c.* 150,000 pairs, although up to 500,000 hand-reared birds are released each year as hunting quarry, some of which naturalise into the wild population.

Ringing recoveries indicate that the breeding population is largely sedentary, although there is substantial immigration into GB&I from Iceland and N Europe in winter, when Mallards were recorded in *c.* 87% of 10-km squares (Lack 1986). However, WeBS data indicate a progressive decline in British numbers since the early 1980s to a maximum (at counted sites) of 140,000 in Dec 2004 – the lowest ever recorded (Banks *et al.* 2006). This compares with a recent population estimate of about 350,000 (Kershaw & Cranswick 2003), although extrapolation from surveys like WeBS for a very widespread species like Mallard are fraught with difficulty. Recent trends in Northern Ireland also show a decline. Crowe *et al.* (2008) estimate an Irish population of 38,250 but again acknowledge the difficulty of assessing this widespread species. It is interesting that there has been an increase in numbers wintering in the coastal Netherlands since *c.* 1990, perhaps indicating a relocation of some wintering birds. It is tempting to speculate that the decline in GB&I is due to amelioration of winter conditions in the near-Continent. Were this the case, one would predict that the number of European visitors might have declined more than those from Iceland; the latter have less option to undertake a shorter migration and still find suitable wintering grounds. However, it is also possible that the decline stems from a reduction in the number released for shooting and/or a reduction in the breeding populations on small ponds as these have been drained. Despite the relatively high figures in GB&I, there are no sites of international importance, and only one of national importance (the Ouse Washes, Cambridgeshire); the picture is one of widespread distribution rather than close aggregation, as in, for example, Wigeon.

Of some concern are the large urban and rural populations of domestic origin, especially in parks and around settlements. These are a source of genetic variants, such as black or white plumage or bizarre feather structure, that sour the natural phenotype and reduce the fitness of their bearers. Selective removal of these would probably benefit the gene pool.

American Black Duck   *Anas rubripes* Brewster

HB SM monotypic

TAXONOMY Closely similar to Mallard in morphology and behaviour and apparently largely allopatric until European settlement of N America, since when sympatry has led to continuous, and apparently increasing, hybridisation (Mank *et al.* 2004), despite evidence for reduced fitness of hybrids (Mason & Clark 1990). Failure to detect any molecular differences between Mallard and Black Duck (e.g. Avise *et al.* 1990) led to speculation that latter is merely a colour morph (Hepp *et al.* 1988; McCracken *et al.* 2001). However, Mank *et al.* (2004) showed that there are significant differences in genetic structure of older (i.e. pre-1940) museum specimens of Mallard and Black Duck but that these differences had reduced substantially among contemporary wild birds, indicating that ongoing hybridisation has blurred the differences between the two taxa. The future of the Black Duck is uncertain because of this continuing hybridisation.

DISTRIBUTION Breeds in E North America, mainly from Hudson Bay and Labrador to Maryland and Virginia; partially migrant to Louisiana and Florida (Longcore *et al.* 2000).

STATUS There have been *c.* 40 records in GB&I up to 2003, mostly in autumn and winter. The first was an adult female shot at Mullinavat (Kilkenny) in Feb 1954. The first for GB was at Stoke (Kent) in Mar 1967. Annual numbers were fairly low until the late 1990s, since when there have been about three each year. Several of these have stayed for long periods, and some long-staying individuals have paired with local Mallards, rearing hybrid young: e.g. Tresco (Scilly) 1978–84; Aber (Caernarfonshire) 1980–84. As with Green-winged Teal, many female Black Ducks are likely to be overlooked.

Northern Pintail   *Anas acuta* Linnaeus

RB or MB, WM *acuta* Linnaeus

TAXONOMY Analysis of morphology (Livezey 1996) indicates that Northern Pintail forms a subclade with two southern hemisphere pintails: Kerguelen (*eatoni*) and Crozet (*drygalskii*), within a larger clade that includes the remainder of the pintails. DNA sequences confirm the monophyly of the pintails, though the internal structure differs slightly. Northern Pintail has no close relatives in Eurasia. Subspecific taxonomy unclear; some (e.g. HBW) treat *eatoni* and *drygalskii* as races of Northern Pintail, others (e.g. Madge & Burn 1988) consider that they might warrant specific rank; these have not yet been included in any molecular analysis.

DISTRIBUTION Holarctic; breeds across most of USA, Canada and Alaska, and through much of N Eurasia from 50°N to the Arctic Ocean. Migratory, wintering in Central America and Caribbean, S Europe, Africa S to Tanzania, India through to Philippines.

STATUS Breeds only rarely in GB&I (most frequently in Argyll, Orkney and the Outer Hebrides), being recorded in just 94 10-km squares during Second Breeding Atlas. The average number of confirmed pairs reported to the RBBP between 1993 and 2006 was less than 25 pairs each year, with a possible maximum of *c.* 45 (RBBP). Birds sometimes remain well into the breeding season, displaying at apparently suitable sites, but Fox (in Second Breeding Atlas) suggests that unsympathetic water management may be at least partly to blame for their failure to breed.

Regular on passage on their way to S Europe and as a winter visitor from Iceland, Scandinavia and E Europe, especially to coasts of Lancashire, Suffolk/Essex and Hampshire, also inland on washlands of Cambridgeshire. Numbers increase rapidly from Aug, peaking in Nov–Jan, declining through Feb, and most have left by the end of Mar (Banks *et al.* 2006). Numbers at the WeBS sites increased steadily from less than 10,000 in the mid-1960s to peak in the early 1980s at about 40,000. A decline in numbers through the 1980s and 1990s bottomed out at the turn of the century; a slight recovery led to a recent estimate of about 28,000 birds wintering in Britain (Kershaw & Cranswick 2003). NI data are more limited, as numbers are very much smaller (a few hundred birds, mostly on Strangford Lough, Co. Down), but suggest an erratic increase since the early 1990s. Crowe *et al.* (2008) estimated 1,235 in Ireland. During the period 2000–01 to 2004–05, there were 20 sites in GB&NI that regularly held more than 600 birds during the winter, thereby qualifying for international importance.

Garganey   *Anas querquedula* Linnaeus

MB PM monotypic

TAXONOMY Morphology suggests affinity with Eurasian and Green-winged Teals (Livezey 1996) and not the 'blue-winged' ducks (Northern Shoveler, Blue-winged Teal, etc.). However, DNA sequences (Johnson & Sorenson 1998, 1999) indicate that Garganey is closer to the latter group, albeit being basal rather than part of the same clade. More research is needed to confirm the relationships of this group, but Garganey seems to be rather distantly related to most other Eurasian wildfowl.

DISTRIBUTION A Palearctic species, breeding from England and Scandinavia across middle latitudes of Eurasia between 64°N and 35°N to Kamchatka and the Pacific. Strongly migratory, wintering in tropical and subtropical Africa and Asia, sometimes as far SE as Australia.

STATUS Garganey is a summer visitor to GB&I, with a small breeding population. This is difficult to assess because of its secretive habits, but estimated as <115 pairs by Baker *et al.* (2006). On average, only *c.* 20 pairs were confirmed breeding to the RBBP each year, though the possible maximum was *c.* 110 (RBBP). A very rare breeding bird in Scotland and Ireland with only *c.* 10 records in the latter. Favoured breeding habitat is marshy wetland, with patches of rushes along ditch edges. The area of this has been reduced in many river valleys by drainage and agricultural intensification.

Slightly higher numbers occur at WeBS sites during spring and autumn, presumably including migrant birds heading to or from breeding sites elsewhere in GB or adjacent parts of Europe. A handful of birds have also been recorded by WeBS counters during winter months in some years. Amber listed throughout GB&I because of its unfavourable conservation status in Europe and low number of GB&I breeding birds (<300 pairs in GB; <100 in Ireland).

Blue-winged Teal   *Anas discors* Linnaeus

HB SM monotypic

TAXONOMY Mitochondrial genes cyt-b and ND2 (Johnson & Sorenson 1998, 1999) indicate close affinity with Cinnamon Teal *A. cyanoptera;* indeed, there is an indication that species limits might need revision here, since in one study Blue-winged Teal was closer to the S American race of Cinnamon (*cyanoptera*) than these were to the N American race (*septentrionalium*). This might represent a mismatch between the gene tree and the species tree (as also observed in Mallard phylogenies) and merits further investigation.

DISTRIBUTION Breeds from SE Alaska across the prairies of Canada and the central plains of the USA to the Atlantic coasts of N Carolina. Winters S to Peru. Nests in swamp vegetation around shallow pools and lakes, wintering on brackish or seawater lagoons.

STATUS A number of birds have formed mixed pairs with Northern Shovelers, and apparent hybrids have presumably resulted from such pairings – e.g. a female was paired with a male Shoveler at Blagdon Lake (Avon) in summer 1993 and an apparent hybrid male was at Titchwell (Norfolk) in Feb 2002.

About 270 records in GB&I, increasing from *c.* 5 per year pre-1975 to *c.* 10 in the 1990s. Occurs in every month but with peaks in May and Sep–Oct. However, the pattern of occurrence differs between the two islands. In Ireland, there is a clear peak in Sep–Oct, when 50% of the birds are found; these are likely freshly arrived from N America. In GB, the pattern is more uniform; the maximum counts come from Sep–Oct, with a secondary peak in Mar–June. These may be birds that crossed the Atlantic further south in the autumn and wintered in SW Europe in association with Palearctic ducks, then being found in GB on their way north.

Northern Shoveler   *Anas clypeata* (Linnaeus)

MB HB WM PM monotypic

TAXONOMY A member of the 'blue-winged' clade, being sister species to Australasian Shoveler *A. rhynchotis* and in same subclade as Cape Shoveler *A. smithii.*

DISTRIBUTION Holarctic, distributed from Alaska through W and C North America; also Iceland and across Eurasia between about 73°N and 35°N. Strongly migratory, wintering S to Central America, tropical Africa, India and SE Asia. Breeds beside shallow, usually mud-fringed, pools and lakes; in winter, also found in saline habitats such as mudflats and estuaries.

STATUS See Blue-winged Teal in relation to hybrid pairing with that species. The breeding population of Shoveler is estimated at 1,000–1,500 pairs (Baker *et al.* 2006), most of which breed on marshy pools in S and E England, though there are appreciable numbers in lowland Scotland, Orkney, Outer Hebrides and N Ireland. GB&I birds migrate to SW Europe or W Africa in winter (Wernham *et al.* 2002), to be replaced by birds from Scandinavia and NE Russia; nine sites hold internationally important numbers (Banks *et al.* 2006). Numbers at British WeBS sites increased steadily from the 1960s until the turn of the century, after which there was evidence of a slight decline. Kershaw & Cranswick (2003) estimated a non-breeding population of about 15,000 birds, and Crowe *et al.* (2008) estimate about 2,500 birds wintering in Ireland. As with some other dabbling ducks (e.g. Mallard), this may reflect amelioration of winters closer to the breeding grounds reducing the selective pressure to migrate to GB&I. Certainly, numbers wintering in the Wadden Sea are increasing, although not in the delta areas further south in the Netherlands (van Roomen *et al.* 2003).

Genus *NETTA* Kaup

Palearctic, Ethiopian and Neotropical genus of three species, one of which is Palearctic and has been recorded in GB&I.

Red-crested Pochard   *Netta rufina* (Pallas)

NB SM monotypic

TAXONOMY Conventionally treated as monotypic; however, Gay *et al.* (2004) found differences in both morphology and mitochondrial gene sequences, indicating that W European and C Asian populations are genetically distinct and should be treated as separate conservation units. Nuclear genes showed less differentiation, supporting the view that the divergence is due to low female dispersal. They also show that the European populations were probably established by a recent colonisation.

DISTRIBUTION Breeds in scattered populations across Europe from Iberia, France to Turkey, and then more abundantly through C Asia to NW China. N populations are migratory, wintering as far south as Israel, N India, NW Indo-China.

STATUS A naturalised population is becoming established in GB, centred especially on Cotswold Water Park (Gloucestershire) (Dudley 2005). Numbers here were stabilised at 50–60 birds through the early 1990s but have now begun to increase, reaching at 129 in Oct 2004. Smaller numbers occur at sites elsewhere, particularly further down the Thames Valley, for example on the gravel pits at Lower Windrush (Oxfordshire), Wraysbury (Berkshire), etc., but also notably at Baston and Langtoft Gravel Pits (Lincolnshire). There are >50 records from Ireland, but some probably refer to birds of captive origin, especially those in summer–early autumn. Natural vagrancy from the Continent no doubt occurs but is generally difficult to substantiate.

Genus *AYTHYA* Boie

Holarctic, Australasian, Ethiopian and Oriental genus of 12 species, five of which are Palearctic and four of these, plus four Nearctic species, have been recorded in GB&I. Several species will hybridise, both in captivity and in the wild, to produce progeny than can resemble other species in the genus.

Canvasback   *Aythya valisineria* (Wilson)

SM monotypic

TAXONOMY Sorenson & Fleischer (1996) show this to be sister to allopatric Common Pochard.

DISTRIBUTION As with Redhead, a species predominantly of W North America, breeding from Alaska to Winnipeg, and then S into central USA (Mowbray 2002). Some populations resident but mostly migratory, wintering along coasts of USA and through southern states to Mexico. W populations use the Pacific flyway; those more central use the Mississippi. However, sizable proportions migrate to the Atlantic coast, although approaching it from inland rather than the north. In spring, migration takes them from coastal Atlantic wintering grounds NW towards Great Lakes and thence to prairies.

STATUS There have been six records (all of males) in GB since the first was found at Cliffe (Kent) in Dec 1996, before moving to the Ouse Washes (Cambridgeshire). Most records in midwinter but also in spring and through to July. No autumn records. There are no records from Ireland.

Common Pochard   *Aythya ferina* (Linnaeus)

MB or RB, WM PM monotypic

TAXONOMY Morphological analysis suggests a clade with Redhead and Canvasback; mt DNA supports the relationship with Canvasback but suggests that Redhead is closer to Ring-necked.

DISTRIBUTION A Palearctic species that breeds in Iceland, across Europe from S Finland to the Mediterranean, and east to L Baikal. W populations generally resident (except Iceland) or partial migrant; eastern populations strongly migratory. Winters around Western Europe, the Mediterranean, Nile Valley, Black, Caspian and Aral Seas, and from Pakistan across N India to S China.

STATUS Small numbers (*c.* 500 pairs) of Pochard breed in GB&I, especially on pools and lakes with extensive emergent vegetation (RBBP). In a review of breeding distribution and abundance, Fox (1991) showed that the increase in the 20 years since 1970 owed much to the colonisation of gravel pits in the S and E of England. Relatively small numbers breed in Scotland, Wales (apart from Anglesey) and Ireland, where over 20% fewer 10-km squares occupied between the two Breeding Atlases.

GB&I is much more important as a winter habitat, hosting up to 25% of the NW European population, with about 60,000 in Britain (Kershaw & Cranswick 2003) and about 38,000 in Ireland (Crowe *et al.* 2008). Although Loughs Neagh and Beg (Londonderry/Antrim; by far the most important single site) still hold internationally important concentrations, it is a matter of concern that the number wintering at this site has declined sharply since surveys began in the 1980s (Allen *et al.* 2004). The reasons for this are unknown, but the parallel declines in Tufted Duck and Goldeneye, compared with relatively stability in Mallard, Teal and Gadwall, suggest a common cause with these diving ducks, perhaps related to molluscan or (possibly) chironomid food in the lakes. A slight decline is also apparent in GB, though this may be within surveying error, but this masks a severe decline in Scotland in the late 1970s. As a species that feeds on open fresh water, rising winter temperatures might be expected to lead to a reduced pressure to move S and W from sites in the Netherlands and S Scandinavia.

Redhead   *Aythya americana* (Eyton)

SM monotypic

TAXONOMY Conventional taxonomy places this close to Common Pochard and Canvasback, a position supported by Livezey (1996). A study of transpositions of mt DNA sequences to the nucleus by Sorenson & Fleischer (1996) used *Aythya* as its target group. Albeit based on a limited array of species, they found strong evidence that Redhead is sister to Ring-necked, and Canvasback to Pochard.

DISTRIBUTION Breeds in W North America, in Alaska, across Canada to the prairies and in the marshlands of the western USA (Woodin & Michot 2002). Partial migrant, wintering from S of breeding range to Mexico, especially along the gulf coast. It seems that only a small proportion of the population migrates along the Atlantic seaboard; most birds use Great Plains or Mississippi flyways.

STATUS There are three records from GB, and this rarity compared with Ring-necked is unsurprising in view of the differences in distribution and migration routes. The first was a male at Bleasby (Nottinghamshire) in Mar 1996, and probably the same bird at Rutland Water (Leicestershire) in Feb of the following year. A similarly mobile male was at various sites in S Wales in 2001–04, and a first-winter female was present on Barra (Hebrides) during winter 2003–04. There is one record for Ireland, a male at Cape Clear Island (Cork) in July 2003.

Ring-necked Duck   *Aythya collaris* (Donovan)

HB SM monotypic

TAXONOMY Morphology suggests affinity with Tufted Duck and scaups (Livezey 1996), but Sorenson & Fleischer (1996) found evidence that it is sister to Redhead, in a clade that includes Pochard and Canvasback, but not the black-and-white *Aythya*. A more extensive review is needed, but the results seem statistically robust.

DISTRIBUTION Breeds in shallow freshwater wetlands in a relatively narrow band across N America, from Yukon and Mackenzie Rivers to James Bay, and N Great lakes to Newfoundland and Canadian Maritimes. Extends W into C Alaska, and S into N central USA. Migratory, wintering across much of S North America, from British Columbia across southern Great Plains to New England, and S to Mexico, including Caribbean, Bermuda.

STATUS A male was paired with a Tufted Duck in the Outer Hebrides in 2004. Two young were hatched but are not known to have fledged.

The first Ring-necked Duck in GB&I was in Mar 1955 at Slimbridge (Gloucestershire), yet by 1994 this species, by then with >250 records, was removed from the list of BBRC birds. The trend towards increasing numbers is paralleled in Ireland, where there were over 130 records by 2003 since the first at Lurgan Park Lake (Armagh) in Mar 1960. Multiple arrivals include a party of five at Loch Leven (Perth and Kinross) in Sep 2003. It seems very unlikely that this change in occurrence reflects earlier lack of familiarity with the species; it is not hard to identify and is unlikely to have been overlooked.

As with Blue-winged Teal, there are more records from Ireland in Jan–Feb, whereas in GB the peak occurs in Mar–Apr. It is tempting to speculate that some of the latter are birds that crossed the Atlantic further south during the previous autumn or winter and are recorded in GB during their passage north.

Ferruginous Duck   *Aythya nyroca* (Güldenstädt)

PM monotypic

TAXONOMY Morphological data indicate that this is part of a clade with Baer's Pochard *A. baeri*, Hardhead *A. australis* and Madagascar Pochard *A. innotata* (Livezey 1996), and this is supported by a molecular analysis (Sorenson & Fleischer 1996), although Baer's was not included. This clade is sister to one including the rest of the Holarctic *Aythya*, and some believe that it should be treated as a superspecies.

DISTRIBUTION Has a scattered and fragmented distribution from W Europe into C Asia, with isolated populations south into N Africa and Iran and Pakistan. Migratory, wintering range equally fragmentary, with sites in sub-Saharan Africa, Rift Valley, coastal Mediterranean, S Caspian, NW Indian subcontinent. The nearest breeding sites to GB&I are the Netherlands, where it breeds irregularly.

STATUS Ferruginous Duck has remained a scarce bird (rare in Ireland) since the 1950s, with an average of about five per year, apart from two periods, 1976–82 (10 per year) and 1998–2003 (13). It is predominantly a winter visitor, most records in Oct–Jan, though with a secondary peak in Apr, presumably reflecting vagrancy associated with migration. The majority of records are from S & E England, with some instances where the same bird probably returned for several successive winters. There is a substantial excess of males among those where the sex is reported; this will likely reflect the difficulty of identifying females (or separating these from eclipse or immature males). However, there are early records of flocks of males: for example, 10 at Hickling Broad (Norfolk) in 1903 (Taylor *et al.* 1999), so the bias may indicate different dispersal patterns.

Tufted Duck   *Aythya fuligula* (Linnaeus)

RB HB WM PM monotypic

TAXONOMY Through morphological analyses, Livezey (1996) found this to be part of a clade that included Ring-necked Duck and Greater, Lesser and New Zealand *A. novaeseelandiae* Scaups, a finding that Sorenson & Fleischer (1996) supported (apart from Ring-necked, q.v.) through molecular means. Since some authorities regard Tufted and Ring-necked as part of the same superspecies, this situation needs resolution.

DISTRIBUTION Breeds across the Palearctic in a band between 45°N and 70°N, from Iceland, GB&I to Kamchatka and Japan. Northern populations migratory, but those breeding in W Europe generally resident. Winters in wetlands of N Africa, Mediterranean, Black and Caspian Seas, and across S Asia from Pakistan to S Japan.

STATUS See under Ring-necked Duck for comment on hybrid breeding with that species. In the breeding season Tufted Ducks are widely distributed across GB&I, and during Second Breeding Atlas were recorded in *c.* 45% of 10-km squares. This represented a slight increase (*c.* 7%) over the First Breeding Atlas but masks apparent differences across the region. The number of 10-km squares in Ireland declined by *c.* 24%, perhaps partly due to reduced coverage, whereas occupancy in GB increased by *c.* 15%. Occupied squares tended to be in the lowlands, and in Ireland it was scarce S of a line from Dundalk (Louth) to Cape Clear (Cork, Second Breeding Atlas). As with Mallard, Tufted Ducks are sufficiently widespread to be represented in WBS and BBS surveys. The former indicate a rise of almost 50% since 1975, mostly since 1993; BBS results suggest a rise of *c.* 35% since the scheme began in 1994. Data from elsewhere are too few for meaningful analysis.

The WeBS monthly indices show peak numbers in late summer, followed by a gradual decline through the winter; the peak may reflect the post-breeding maximum or possibly birds arriving to moult in GB waters. WeBS data indicate a modest increase in GB since *c.* 1980, with a recent estimate of about 90,000 birds (Kershaw & Cranswick 2003). In NI, Crowe *et al.* (2008) estimated a winter total of about 37,000, although there has been a serious decline at Loughs Neagh and Beg (Londonderry/Antrim), where numbers in 2004–05 were about half those before 2000. This must be a matter for concern, since this is the only site in GB&NI supporting mean peak numbers in excess of the threshold for international importance for Tufted Duck. Since Pochard and Goldeneye have also declined at this site, but dabbling species have not, this suggests a common cause.

Ringing data indicate that Icelandic birds winter in Scotland and Ireland, whereas those wintering in England and Wales are predominantly from Scandinavia, E Europe and NW Russia. There may be further hard-weather movements to Ireland and France when conditions become especially severe.

Greater Scaup   *Aythya marila* (Linnaeus)

CB WM PM *marila*

TAXONOMY A limited phylogeny based on the analysis of mt DNA transpositions shows that Greater and Lesser Scaup are sister taxa, in a separate clade to Pochard, Redhead, Ring-necked Duck and Canvasback. However, morphological analyses conflict with this; Livezey (1996) places them in a clade with Tufted and Ring-necked Duck; clearly there is need for a more extensive analysis. Two subspecies are recognised: *marila* and *nearctica*, differing in the vermiculations on the back.

DISTRIBUTION One of the few circumpolar ducks, it breeds in a narrow band of lakes and marshes across the tundra. Subspecies: *marila* breeds across the Palearctic from Iceland to about River Lena, wintering in coastal NW Europe and beside Black and Caspian Seas; *nearctica* breeds from the Lena, east across Alaska and Canada to Labrador, wintering south to China and California in the Pacific, and coastal Canada and USA in the Atlantic. Some birds winter inland: e.g. Great Lakes (USA/Canada), E Europe.

STATUS Very rare breeding bird, averaging less than one record per year, usually in Scotland (most often Orkney and Outer Hebrides) or N Ireland. Many of these are only 'possible breeding', as for example in 2002, when a pair was seen displaying in Argyll and a female remained until July, but with no firm evidence of breeding (RBBP). There were only 29 occupied 10-km squares in Second Breeding Atlas, of which just 7 showed evidence of breeding.

The number of Scaup wintering in GB&I fell dramatically in the 1970s, when about 75% of the population disappeared. This coincided with pressure to clean up estuaries and coastal waters and seems to be linked to the reduction in untreated sewage outflows at sites such as Seafield (Midlothian), which alone held *c.* 30,000 birds in 1970–80 but <50 in 2004. Since then, numbers in GB have stabilised, with a recent estimate of about 7,600 birds (Kershaw & Cranswick 2003), including sizable flocks in SW Scotland, Islay (Argyll) and in the Moray Firth. Crowe *et al.* (2008) estimated an Irish total of 4,430, the majority of which occur at Loughs Neagh & Beg (Londonderry/Antrim). WeBS counts in NI did not start until the late 1980s, after the population crash in GB. Numbers there rose to a peak at around the turn of the century, appeared to crash dramatically but then bounced back to a record high in 2004–05.

Lesser Scaup   *Aythya affinis* (Eyton)

SM monotypic

TAXONOMY Closely related to Greater Scaup.

DISTRIBUTION A widespread and abundant species in N America, ranging from Alaska and the southern shores of Hudson Bay, south to the Great Lakes in the east and Wyoming in the west (Austin *et al.* 1998). Strongly migratory, it winters across the southern USA and south to Guatemala, including the Caribbean. Eastern populations migrate through Great Lakes and across to Atlantic coast and thence to Florida and Bermuda; this probably source of GB&I vagrants.

STATUS There have been *c.* 100 Lesser Scaup recorded in GB and a further 12 in Ireland since the first at Chasewater (W Midlands) in Mar 1987. The first Irish bird soon followed, being found on Lough Neagh in Dec 1989, with what was probably the same bird being seen at a range of sites through into 2001. This increase is no doubt partly due to greater awareness of identification criteria for what was previously regarded as a difficult, if not intractable, problem (e.g. Perrins 1961). Records come from all months except Aug, though the majority are in winter. As for several Nearctic ducks there is the possibility that some cross the Atlantic in autumn or during harsh winter weather and winter in W Africa or SW Europe. Birds from NE USA that head out to sea to avoid adverse weather will tend to arrive in Iberia or Morocco, and there are now increasing numbers of records of Nearctic ducks wintering in the Azores. As the days begin to lengthen, these birds might join parties of Eurasian *Aythya* ducks and enter GB from the SW, moving up the flyways of the Severn and Trent. Of the 15 records from non-coastal English counties, 12 were in the period Feb–June, compared with only 14 of 33 coastal records, a significant difference.

Genus *SOMATERIA* Leach

Holarctic genus of three species, all occur in the Palearctic and two have been recorded in GB&I.

Common Eider   *Somateria mollissima* (Linnaeus)

RB WM *mollissima* (Linnaeus)

TAXONOMY Conventionally included with King and Spectacled Eiders in *Somateria*, Common Eider is a polytypic species found across the Holarctic; six subspecies are recognised: five in the Atlantic (*mollissima, faeroeensis, borealis, dresseri, sedentaria*) and one in the Pacific (*v-nigrum*). These differ in bill shape and colour, and to a lesser extent in plumage of the head and neck; European and N Pacific Eiders also differ in display repertoire (McKinney 1961). Taken together, these findings suggest that species limits would repay further investigation. Tiedemann *et al.* (2004) examined genetic variation in Eiders and found substantial differentiation in mt DNA sequences among colonies but less at nuclear genes. This finding supports strong female philopatry, with more extensive male-mediated gene flow among colonies. Tiedemann *et al.* (2004) also found genetic evidence in support of post-glacial colonisation of NW Europe by a series of step-wise events. Pennington *et al.* (2004) report that 133 males and females caught in Shetland were significantly smaller than birds wintering on the River Ythan (Aberdeenshire), suggesting that this population may be closer to *faeroeensis*. There are few other data relating to metric variation within GB&I. See also below.

DISTRIBUTION Breeds along the coasts of both Atlantic and Pacific Oceans, from *c.* 45°N, but absent from extreme north. Subspecies: *mollisssima*: NW Europe to Novaya Zemlya; *faeroeensis*: Faroe Is; *borealis*: N Atlantic, from Baffin Is, through Greenland, Iceland to Franz Josef Land; *dresseri*: NE North America from Labrador to Maine; *v-nigrum*: NE Siberia to NW North America; *sedentaria*: Hudson Bay. Partially migrant, northern populations move south to avoid extreme winter conditions.

STATUS The Second Breeding Atlas revealed a continuous distribution around the northern coast of Britain from Coquet Island (Northumberland) to Mull of Galloway, and from Carlingford Loch (Louth/Down) to Donegal Bay in Ireland, with a few isolated sites outside this, notably around Walney Island (Cumbria). The largest colony is at the Sands of Forvie (Aberdeenshire), with *c.* 1,350 nests in the 1990s. There was a slight decline in occupied 10-km squares with confirmed breeding in GB since the first atlas, but these were mostly in the more remote areas and likely reflect poor coverage rather than genuine loss. The population was estimated in Second Breeding Atlas as *c.* 31,000 pairs in GB and *c.* 600 in Ireland; this is our second commonest breeding duck (after Mallard).

Counts of Eiders are most reliable during the autumn, when birds congregate in flocks at moulting, feeding or roosting sites. Numbers at WeBS sites in GB showed a steady increase from the 1960s until the late 1990s, when the non-breeding British population was estimated at 73,000 birds (Kershaw & Cranswick 2003); since then, there seems to have been a levelling out or even a slight decline (Heubeck 1993, Banks *et al.* 2006). Irish populations (most in NI) have also been increasing since the mid-1990s; Crowe *et al.* (2008) estimated a total of 2,890 for Ireland. A detailed annual survey in the Firth of Clyde carried out each Sep shows that this region holds substantial numbers of birds, although a decline has been apparent over recent years.

Although large numbers have been ringed in GB, our knowledge of movement within GB&I is biased since most ringing occurred at only a few sites on the E coast. Recoveries suggest that part of the populations from Northumberland and Aberdeenshire winter on the estuaries of the Forth and Tay, and there are other data indicating similar movement around the North Sea to areas with better food supplies. However, some of the emigrants from Forvie (Aberdeenshire) are replaced by birds from elsewhere; an early study of genetic variation by Milne & Robertson (1965) revealed differences between the sedentary and migratory components at Forvie. There is exchange between breeding populations at Forvie and the S Baltic; the majority of these are males, and Baillie (in Wernham *et al.* 2002) postulates that these are birds that paired in winter with philopatric females from elsewhere and returned with them to their natal colonies.

There are a number of records of birds resembling *borealis* from GB, including one at Musselburgh (Lothian) in Feb 1978, and others in NE Scotland and the N & W Isles. However, identification of this race remains problematic, and birds showing at least some of the characters of *borealis* have been noted in British breeding colonies. Such records require to be reviewed before this taxon can be added to the British List. Birds resembling *borealis* were seen in Wicklow in Mar 1998 and in Donegal in Mar and June 2004; small numbers may be resident in Donegal (P Smiddy pers. comm.).

King Eider   *Somateria spectabilis* (Linnaeus)

SM monotypic

TAXONOMY Closely related to Common Eider, with which it regularly hybridises.

DISTRIBUTION Breeds along the coast all round the Arctic, apart from Iceland and Fennoscandia. Migratory, wintering Iceland, Scandinavia, NW Atlantic, NE Pacific from Alaska to British Columbia, Kamchatka.

STATUS King Eider has been recorded in every month of the year, although most commonly during the winter, with over 80% in Scotland and generally less common in the S & W. Assessing trends is difficult because of the number of long-staying (and returning) individuals that also move between adjacent sites. However, it appears to be increasing in frequency, though this may reflect increased interest in wintering sea-ducks and an awareness of the relative ease with which females can be identified (Suddaby *et al.* 1994). Thus, in 1950–1980, less than 5% of birds seen were reported as female, compared with *c.* 25% subsequently.

Genus *POLYSTICTA* Eyton

Holarctic monospecific genus that has been recorded in GB&I.

Steller's Eider   *Polysticta stelleri* (Pallas)

SM monotypic

TAXONOMY Not included in any molecular analysis of phylogeny; Pearce *et al.* (2005) found evidence of slight differentiation among breeding and wintering groups. Traditionally placed in a monotypic genus, though closely related to the other eiders in *Somateria*.

DISTRIBUTION Population apparently separated into two regions along the coast of the Arctic Ocean: western and eastern (Solovieva *et al.* 1998). The former breeds from central Taimyr in the west, to the Yamal Peninsula in the east; there are scattered breeding records from further west to Norway (Haftorn 1971). These birds are presumed to be the source of the wintering flocks in the Barents and Baltic Seas, which totalled between 30,000 and 50,000 in the early 1990s (Nygård *et al.* 1995). The eastern population breeds between the Khatanga and Kolyma Rivers of Siberia, on the New Siberian Islands, and east to the N and W coasts of Alaska. These birds are thought to winter in the region centred on the Commander and Kurile Islands (30,000 birds: Solovieva *et al.* 1998) and the Aleutian Islands and coastal Alaska (*c.* 150,000 birds: Solovieva *et al.* 1998). Trends are difficult to assess because of the weakness of earlier data, but it is generally thought that the eastern population shows evidence of a serious decline, whereas the western population may be modestly increasing (Solovieva *et al.* 1998).

STATUS The first record for GB&I was a sub-adult male shot at Caister (Norfolk) in Feb 1830; perhaps not surprisingly, 8 of the 15 records are from the N & W Isles of Scotland. There have been two especially long-staying individuals: a male on S Uist (Outer Hebrides) from May 1972 to Aug 1984, and another male on Westray/Papa Westray (Orkney) in Oct–Nov 1974 and then erratically from July 1978 until July 1982. There are no records from Ireland.

Genus *HISTRIONICUS* Lesson

Monospecific Holarctic (circumpolar) genus that has been recorded in GB&I.

Harlequin Duck   *Histrionicus histrionicus* (Linnaeus)

SM monotypic

TAXONOMY Minor morphological differences between Atlantic and Pacific populations but generally considered too slight for subspecific recognition. A genetic investigation of some Pacific populations following the Exxon Valdez oil spill (Lanctot *et al.* 1999) found little evidence of population differentiation, although a radio-telemetry exercise in NE Canada suggested that populations in Greenland and E Canada are demographically separate (Brodeur *et al.* 2002). This study also suggests there are slight genetic differences between populations in E and W North America.

DISTRIBUTION Breeds in clean and fast-flowing rivers in Iceland, E & W Greenland, NE Canada from Labrador to Hudson Bay, and around the N Pacific from N Rockies and Alaska to Seas of Okhotsk and Japan. Pacific populations winter south to NW USA and Japan. Icelandic population generally resident; some birds from NE Canada migrate to moult and winter in SW Greenland (Brodeur *et al.* 2002).

STATUS The first bird was collected at Filey (N Yorks) in autumn 1862. Of the 16 records since 1950, 11 have been in Scotland; there are no records from Ireland. Most birds were found in Dec–Feb, though there are two modern records each for Apr and Oct, and one for June. Several of the records are of more than one individual: e.g. three together (1886), two (1965), two (1996).

Genus *CLANGULA* Leach

Monospecific Holarctic (circumpolar) genus that has been recorded in GB&I.

Long-tailed Duck   *Clangula hyemalis* (Linnaeus)

WM monotypic

TAXONOMY Morphology suggests the Long-tailed Duck is closer to the goldeneyes than scoters (Livezey 1995); little studied at a molecular level.

DISTRIBUTION Breeds along the Arctic coasts across the Holarctic, from James Bay in Nearctic, Greenland, Iceland and Fennoscandia in Palearctic. Migratory, wintering along the temperate coasts of Atlantic and Pacific and inland on Great Lakes, and Black and Caspian Seas and Lake Baikal.

STATUS Although birds are regularly seen displaying offshore in spring, there are no fully satisfactory breeding records for GB&I (see review in BS3).

A winter visitor that occurs along the shallow coasts and shoals of Scotland, the north of Ireland and NE England (Lack 1986), with major concentrations over richer feeding grounds in the Moray Firth, Forth Estuary, Orkney and Shetland. WeBS counts for this species are not entirely reliable due to the problems associated with counting seaducks: heavy seas can render accurate counts difficult, while calm conditions can lead to birds being distributed beyond the range of binoculars and telescopes. Dedicated surveys may reveal substantially higher counts then the regular WeBS counter. Numbers appear to be fairly stable, the British estimate being about 16,000 birds (Kershaw & Cranswick 2003). Crowe (2005) suggests that Irish numbers are unlikely to exceed 2,000 birds, perhaps far fewer. These numbers pale into insignificance in comparison with the populations elsewhere in Europe (half a million birds in N Norway and the Baltic), and no site in GB&I reaches international importance.

Genus *MELANITTA* Boie

Holarctic genus of five species, of which four have been recorded in GB&I.

Black Scoter/Common Scoter   *Melanitta nigra* (Linnaeus)

RB or MB, WM PM monotypic

TAXONOMY A recent review (Collinson *et al.* 2006) recommended that *M. nigra* should be split into *M. nigra* and *M. americana*, on the basis of differences in bill pattern and shape and in vocalisations, which indicate they are separate evolutionary lineages and have been reproductively isolated for a long time. There is anecdotal evidence that vagrant *M. americana* 'ignore' *M. nigra* in winter flocks in GB.

DISTRIBUTION Breeds on fresh water on tundra and taiga bogs across the north coasts of the Palearctic, from Iceland, GB, through Fennoscandia to River Olenek (Siberia). The extent of contact between this and *M. americana* in Siberia is unclear, as is the existence of hybridisation in any contact zone.

STATUS Evidence of breeding was found in 24 10-km squares during the fieldwork for Second Breeding Atlas; these were scattered from Shetland to Islay (Argyll) in Scotland, and Fermanagh to Galway in Ireland. Full breeding data are usually only available after targeted surveys; thus in 1995 195 pairs were estimated for GB&I (Underhill *et al.* 1998), but since then there have only been casual records (RBBP).

Gathers in large moulting flocks in summer and early autumn. In winter, distributed along shallow, sandy coasts and shoals around much of GB&I, and Lack (1986) estimated 25,000–30,000 birds, but this was from land-based counts and is now known to be a serious underestimate. Aerial surveys in the late 1990s revealed *c.* 20,000 birds wintering in Carmarthen Bay (Wales), suggesting this to be a key site for the species. In Aug 2002, 6,000 moulting birds were found in Cardigan and Liverpool Bays. Further surveys during winter 2002–03 gave a rough estimate of >25,000 at the latter site alone, increasing to an estimated 79,000 in Feb 2003 – more than the previously known total for the whole of GB&I (Cranswick *et al.* 2004). These aggregations make Common Scoter especially vulnerable to pollution, as was shown in the *Sea Empress* oil spill in Pembrokeshire Feb 1996, and additional offshore surveys have revealed similarly unsuspected concentrations in the Moray, Forth and Solway Firths and off N Norfolk. Crowe *et al.* (2008) estimate an Irish wintering total of 23,190.

Although the discrepancies between shore-based and aerial counts may cast doubt over the value of the former for this species, the continuity of records under WeBS is valuable for assessing trends. Although these massively underestimte the total population, they suggest that Common Scoter is increasing modestly in GB&I. Liverpool and Carmarthen Bays host internationally important numbers. There are regular reports from inland waters, especially in the Midlands of England, often in late summer, suggesting that a proportion of the population migrates from S Wales along the Severn/Trent flyway.

American Scoter/Black Scoter   *Melanitta americana* (Swainson)

SM monotypic

TAXONOMY Now treated as a separate species from *M. nigra* (summarised by Collinson *et al.* 2006).

DISTRIBUTION Two apparently disjunct populations breed along the Arctic coastal regions of E and W Nearctic; E Siberia, Kamchatka to W Alaska, and E Canada, Newfoundland. W birds winter along Pacific coast to California; E populations along Atlantic coast of Canada, USA.

STATUS Following a recent review, there have been eight records from GB since the first at Gosford Bay (Lothian) from Dec 1987 to Jan 1988, returning in Mar and Nov 1989. The pattern seems to be one of occasional, long-staying individuals: e.g. the Moray Firth 1989–93; Llanfairfechan (Caernarvonshire) 1999–2007. There are no records from Ireland.

Surf Scoter   *Melanitta perspicillata* (Linnaeus)

WM monotypic

TAXONOMY Morphological analyses (Livezey 1995) indicate that Surf and 'Velvet' scoters are sister taxa, more distant from Black and American Scoter.

DISTRIBUTION Nearctic. Breeds discontinuously from Labrador to Alaska; missing from NC of region and Arctic islands. Winters along both coasts of N America, from Aleutians to California and Labrador to Florida.

STATUS The first record from GB was in the Firth of Forth some time before 1837; by 1990, there were 230 records, and the species was removed from the BBRC list. The first for Ireland was shot in Belfast Bay in Sep 1846, and by 2003, there were 135 individuals. Many of the records from GB&I involve multiple sightings, and some birds appear to wander widely over prolonged periods of time, making numbers difficult to assess.

   The majority of records are from Oct–Mar, with no evidence of passage movement, although there are slight differences between GB and Ireland. In GB, most often recorded off Aberdeenshire in mid-summer. There is a peak of records elsewhere in Oct, the build-up to which starts in Aug; in Ireland, however, the peak is later (Nov) and does not begin until Oct. This pattern is quite different to all other Nearctic ducks and raises the possibility that these birds are arriving not from the west but from the NE, perhaps moving north with Common or Velvet Scoters in spring, returning in late summer. Since multiple sightings of adult birds are not unusual, and some appear with juveniles in mid-summer, this raises the possibility that Surf Scoters occasionally breed in the Palearctic. The largest numbers of records occur between the Moray Firth and the Firth of Forth, very similar to the core area for Velvet Scoter in the UK, and relatively few in Liverpool Bay and Carmarthen Bay, which host the largest flocks of Common Scoters. To some extent, however, this may be due to difficulties in viewing birds at these latter sites.

Velvet Scoter   *Melanitta fusca* (Linnaeus)

WM PM monotypic

TAXONOMY Three subspecies were recognised, but Collinson *et al.* (2006) recommend that these be split into *M. fusca* and *M. deglandi* (including *stejnegeri*) on the basis of diagnostic differences in bill structure and pattern, male plumage, tracheal structure and vocalisations. They comment that a case could be made for further splitting *deglandi* and *stejnegeri* but recommend further study.

DISTRIBUTION Breeds on lakes and large ponds in NW Eurasia, from Scandinavia to about River Yenesei (Siberia). Largely coastal in winter, from N Norway, around North Sea, but also inland in Black and Caspian Seas. The two subspecies of *M. deglandi* have not been recorded from GB&I; *deglandi* breeds from Alaska to W Hudson Bay, wintering along both coasts of N America; *stejnegeri* breeds in Siberia, E of River Yenesei, and winters along the Pacific coast from Kamchatka to China. It is not certain whether *stejnegeri* and *fusca* are ever sympatric; further study in the River Yenesei is desirable.

STATUS Regular winter visitor but much more thinly scattered around the coasts than Common Scoter and recorded in only 109 10-km squares during the Winter Atlas, compared with 457 for Common (Lack 1986). Rare in Ireland and recorded in only 13 squares, chiefly along the E coast. Despite the limited distribution, there are significant concentrations in the Moray Firth, St Andrews Bay and the Firth of Forth, where >1,000 regularly winter, and parties of a few hundred winter elsewhere in SE Scotland (Banks *et al.* 2006). More typically, small numbers are found in Common Scoter flocks along the E and S coasts of England, although within these they frequently associate more closely with each other than with their congenerics. The number counted by WeBS surveys varies depending on the sea conditions, but winter numbers were recently estimated at *c.* 3,000 (Kershaw &

Cranswick 2003). Velvet Scoters occasionally occur inland but less frequently than Common Scoter, supporting the theory that the records of Common Scoter in the English Midlands are on route to South Wales (where few Velvets occur).

Genus *BUCEPHALA* Baird

Holarctic genus of three species, all of which have been recorded in GB&I.

Bufflehead   *Bucephala albeola* (Linnaeus)

SM monotypic

TAXONOMY Little studied recently, but Livezey (1986) suggested that it is close to *Mergus* and Smew. Perhaps unusually for a duck, almost no evidence of natural hybridisation, perhaps due to its size and distinctive display (Gauthier 1993).

DISTRIBUTION Breeds by shallow lakes from Alaska to Hudson Bay and Great Lakes; migratory, along the coasts from Aleutians to Mexico, Newfoundland to Gulf of Mexico and inland across S USA to Mexico. With east coast wintering grounds along the Atlantic coasts of USA and breeding areas west of this, there is little opportunity for trans-Atlantic vagrancy as far north as GB&I.

STATUS This is a popular species in captivity and some records are believed to be of captive origin. The first presumed wild bird in GB was shot on Tresco (Scilly) in Jan 1920. There were *c.* 11 more to 2007, widely scattered across GB and, apart from birds that arrived (or were found) in Devon and Shetland in Nov, all have occurred in late winter or spring. Five inland records are all from Feb–June. The possibility of these having crossed the Atlantic further south and being found on their northwards migration is appealing. There is a single Irish record at the Gearagh (Cork) in Jan–Mar 1998. The population is increasing rapidly in N America, so more records might be expected.

Barrow's Goldeneye   *Bucephala islandica* (Gmelin)

SM monotypic

TAXONOMY Based on morphology and behaviour, believed to be sister taxon to Common Goldeneye (Livezey 1995).

DISTRIBUTION Four centres of population: Iceland, SW Greenland, Labrador, Alaska to California. Iceland population resident; remainder variably migratory, wintering on Pacific coast close to breeding grounds and along Atlantic coast of N America.

STATUS A popular ornamental duck, there have been many records that undoubtedly relate to escapes. The first regarded as wild was an adult male at Irvine (Ayrshire) in Nov–Dec 1979. Further apparently wild birds were found in Aberdeenshire (May–June 2005), probably with the same one returning to C Scotland the following winter, and in Co. Down (Nov 2005–Apr 2006), again returning in Nov 2006.

Common Goldeneye   *Bucephala clangula* (Linnaeus)

RB WM PM *clangula* (Linnaeus)

TAXONOMY Close to Barrow's Goldeneye in morphology (Livezey 1995) but detailed molecular phylogeny lacking. Two subspecies recognised (*clangula*, *americana*), based on size and bill thickness, but not all authorities recognise these.

DISTRIBUTION Breeds in tree holes beside freshwater lakes and rivers across the temperate Holarctic, wintering south of breeding range, on both fresh and salt water. Subspecies *clangula*: GB&I,

Scandinavia, C Europe to Kamchatka; *americana*: Alaska to Newfoundland. N limits determined by presence of suitable trees for nesting.

STATUS Breeding has continued to increase, especially in nest boxes, since the first pairs in Highland during the 1970s. Accurate censuses are not made every year, so the data in the RBBP reports is usually incomplete. The latest survey of the core area of Strathspey (Highland) in 2002 found at least 91 clutches, although 33 were not incubated, leading to an estimate of *c.* 150 egg-laying females in Scotland (RBBP) though *c.* 200 laying females were reported to RBBP in 2006. Outside Strathspey, only two nests were found in 2002, though numbers have been increasing since, especially in Aberdeenshire. Goldeneyes summer at many sites across GB&I and have been recorded breeding S to the Scottish Borders. A pair bred at Lough Neagh (NI) in 2000, the first for Ireland.

   Wintering birds begin to arrive in Oct, later than for many of our other wintering wildfowl, rapidly increasing through the autumn to reach a peak in Jan, before declining again through Feb–Apr. Ringing data indicate that the majority of winter immigrants come from Scandinavia. Recorded in 50% of 10-km squares during the Winter Atlas. Present at hundreds of lakes, pools, rivers and estuaries across GB&I, with concentrations of over 500 at several sites both inland e.g. Rutland Water (Leicestershire), Abberton Res (Essex), Loughs Neagh (NI) & Beg (Londonderry/Antrim) and coastal (e.g. Firths of Forth, Moray and Clyde). Numbers have declined at some sewage outfalls following the introduction of improved water treatment (Campbell 1984). The total for GB has increased fairly steadily from the early 1970s (Banks *et al.* 2006), to an estimate of about 25,000 birds (Kershaw & Cranswick 2003), though with a slight decline from the turn of the century. The WeBS data for NI reveal a less successful scenario: numbers have more than halved since the early 1990s; Crowe *et al.* (2008) estimate an Irish total of 9,665 for the winter between 1999 and 2004, a very large proportion of which are at Loughs Neagh and Beg. Although still the only site in GB&I holding numbers close to international importance, the population at Loughs Neagh and Beg has declined from >10,000 in 1993–94 to less than 4,500 in 2003–04. As with Pochard and Tufted Ducks, the reasons for this are unknown; changes in food supply seem the most likely, but whether this is due to declines in molluscs or chironomids (or both) is still unresolved.

Genus *LOPHODYTES* Reichenbach

Nearctic monotypic genus which has been recorded in GB&I.

Hooded Merganser   *Lophodytes cucullatus* (Linnaeus)

SM monotypic

TAXONOMY Previously treated as a member of the genus *Mergus*.

DISTRIBUTION Two populations in N America: one from SE Alaska through Rocky Mts to Oregon; the other from SE Canada, through Great Lakes and Canadian Maritimes south to Mississippi. Former population partially migrant, inland birds wintering S to California; latter more completely migratory, wintering along Atlantic and Gulf coasts, from Maine to Mexico.

STATUS Because of its popularity as an ornamental waterbird, this species had a long and chequered history (reviewed in BBRC Report for 2007), before finally being admitted to the British List in 2008 on the basis of three records, from Outer Hebrides (2000), Northumberland (2002) and Shetland (2006). Four individuals in Ireland; two obtained in Cork Harbour in Dec 1878; one obtained in Kerry in Jan 1881; one seen in Armagh in Dec 1957. No recent reords from Ireland.

Genus *MERGELLUS* Selby

Holarctic monotypic genus which has been recorded in GB&I.

Smew   *Mergellus albellus* (Linnaeus)

WM monotypic

TAXONOMY Previously treated as a member of the genus *Mergus*. Systematics of the merganser group (*Lophodytes, Mergellus, Mergus*) complex and not fully resolved. However, most analyses (e.g. Livezey 1986, 1995) show that Smew is not especially closely related to the others, apart perhaps from Hooded Merganser.

DISTRIBUTION Breeds in hollow trees (and nest boxes) close to lakes and rivers across Palearctic, from Finland, Sweden to E Siberia, Kamchatka. Strongly migratory, wintering from W & C Europe, across C Asia to China, Japan. W populations may move further S & W in response to severe winter weather.

STATUS Smew visit GB&I in winter from breeding grounds in N Russia and Fennoscandia, but their distribution is relatively localised at mainly inland sites, and most birds occur over a relatively short period in midwinter. Fieldwork for the Winter Atlas only identified 243 occupied 10-km squares (mostly in GB), and only 45 held more than two birds. WeBS counts indicate that the number of Smew wintering in GB&I rose in the mid-1990s, when it reached *c.* 450 in 1996–97. However, numbers are dependent on winter conditions in the near-Continent, where much larger numbers winter (e.g. recent average peak of over 600 on the IJsselmeer in the Netherlands (van Roomen *et al.* 2005)). In severe weather, such as Jan 1979, numbers can exceed 400, compared with a more typical figure at that time of <100 (Chandler 1981). The recent milder winters have seen a decline towards 250, the majority of which are on sand and gravel pits in SE England, with Wraysbury Gravel Pits (Berkshire) the most important site each winter since 1996–97 (to 2004–05 at least). Smew is a scarcer bird in Scotland, Wales and Ireland.

Genus *MERGUS* Linnaeus

Palearctic, Nearctic and Neotropical genus of five species, three of which have been recorded in GB&I.

Red-breasted Merganser   *Mergus serrator* Linnaeus

RB WM monotypic

TAXONOMY Close to Common Merganser but detailed systematics unresolved. Greenland population has been assigned to separate subspecies but not generally recognised.

DISTRIBUTION Holarctic; more northerly than Common Merganser, breeding from Arctic Ocean to Great Lakes in N America and to GB&I, Baltic and Lake Baikal in Palearctic. Partial migrant, most northerly populations wintering along ice-free coasts of N Atlantic and N Pacific and on the lakes of temperate Asia.

STATUS The Second Breeding Atlas recorded evidence of breeding in *c.* 450 10-km squares in GB&I, a decline from the First Breeding Atlas, especially in Scotland and Ireland. Distribution is predominantly in N & W Scotland, the English Lakes, N & W Wales and coastal Ireland (where inland breeding seems to be in decline). GB&I birds are relatively sedentary, although there is a tendency for males and immatures to moult in coastal waters. The breeding population was estimated at *c.* 2,850 pairs during Second Breeding Atlas.

    The winter population is supplemented by immigrants from elsewhere in N Europe but the precise totals are hard to determine because of problems censusing the numbers at sea. Abundance not well understood due to widespread winter distribution in NW Scotland and the N & W Isles, which are not well covered by WeBS. Ringing recoveries suggest that many immigrants come from Iceland, and Wernham *et al.* (2002) suggest that a sizeable concentration of birds from Greenland and Iceland winters off the north of Scotland. It also seems probable that some Continental birds winter in SE

England, though ringing data are sparse. Regular WeBS surveys indicate that the population increased fairly steadily from the 1960s, to a peak in the late 1990s and then fluctuated around a similar level since. Kershaw & Cranswick (2003) estimated the British population at 9,800 birds, whilst Crowe *et al.* (2008) estimate 3,390 in Ireland. The NI population has remained fairly static since the early 1990s. No sites within the islands hold internationally important numbers of birds.

Common Merganser/Goosander   *Mergus merganser* Linnaeus

RB WM *merganser* Linnaeus

TAXONOMY Appears to be close to Red-breasted Merganser (Livezey 1986, 1995) but molecular data limited. Three subspecies recognised on basis of size, bill pattern and colour of greater coverts: *merganser, orientalis, americanus.*

DISTRIBUTION Breeds across Holarctic, beside lakes and rivers, usually S of tree line. Migratory; in winter, much less common on coast than Red-breasted Merganser. Subspecies *merganser*: N Palearctic from Iceland to Kamchatka; *orientalis*: S of nominate, from Afghanistan to China; *americanus*: across N America, from Alaska to Newfoundland, and S to N USA. Northern populations migratory, moving S of breeding range to avoid frozen lakes and rivers.

STATUS Breeds in wooded river valleys, especially in N and W GB. There was a marked expansion between the two Atlases, especially in N England and Wales; in Scotland, the picture was one of gains and losses, with little overall change. Overall, the number of occupied squares increased by *c.* 65%; there is only a handful of breeding records from Ireland. WBS data indicates a steady increase of *c.* 97% during 1981–2003. From June–Oct, males leave the breeding grounds and migrate to fjords of N Norway to moult alongside most other drakes from W Europe (Little & Furness 1985; Rehfisch *et al.* 1999). Exceptionally rare breeding bird in Ireland (occasional pair).

The species is more widely dispersed in winter in Britain, where most birds are thought to derive from the breeding population but an unknown proportion (probably mostly in the south-east) suspected to come from Continental populations. There was a steady increase from the 1960s to the 1990s, which probably reflects an increase in the GB breeding population, followed by a slight downturn in the late 1990s. Numbers sharply up during the hard winters of 1995–96 and 1996–97, especially, but not exclusively, in S and E England, and possibly due to a increase in the number of Continental immigrants into this part of the region. No sites hold numbers above the international threshold, and only three (all in S Scotland) reach national significance (Banks *et al.* 2006). Kershaw & Cranswick (2003) estimated the British winter population at 16,100 birds. Although several hundred birds are shot under legal licence (and an unknown number illegally) in Scotland, this appears only to have local effects on the population (e.g. on the Tweed). Very scarce in Ireland.

Genus *OXYURA* Bonaparte

Cosmopolitan genus of six species, of which an American species has been introduced to GB&I.

Ruddy Duck   *Oxyura jamaicensis* (J. F. Gmelin)

NB *jamaicensis* (J. F. Gmelin)

TAXONOMY McCracken and his colleagues used morphology and molecules to reveal evidence of two clades within *Oxyura*, broadly Old World and New World (see McCracken & Sorenson 2005 for refs). Thus, White-headed Duck *O. leucocephala* closer to African Maccoa and Australian Blue-billed Ducks *O. maccoa, O. australis*; Ruddy closer to Argentine Blue-billed Duck *O. vittata*. Brua (2001) follows AOU in treating Andean Duck *O. ferruginea* as separate species, leaving two weakly differentiated subspecies of Ruddy: *jamaicensis* and *rubida*, which differ in size and body colouration.

DISTRIBUTION Breeds from British Columbia to Manitoba, and south to California and Texas (*rubida*); also in W Indies (*jamaicensis*). Separate populations along Andean chain from Colombia to Tierra del Fuego now treated as separate species. Inland populations from N America move south to winter mainly along the Pacific coast of USA and Mexico, with lesser numbers along Atlantic and Gulf of Mexico.

STATUS Introduced to Britain following accidental escape of birds that avoided pinioning in the 1950s (Hudson 1976). First bred in Avon (1960) and Stafford (1961). There followed a rapid expansion in range and numbers (Banks *et al.* 2006) and now distributed across much of lowland GB and relatively isolated populations in NI and Anglesey. The annual rate of increase seemed to be declining slightly from 8–9% in the 1980s to 6–7% more recently (Pollitt *et al.* 2003), and the species seems prone to heavy mortality during hard winters (Vinicombe & Chandler 1982). Ruddy Ducks have bred in NI since 1973, but, although they have occurred in many counties in RoI, they have not become established. As the GB population increased, birds began to be found elsewhere in Europe, having been recorded in 19 countries by 1996 (Hughes 1998). Many of these are presumed to originate from the GB population, although there are no recoveries of British-ringed birds overseas (Wernham *et al.* 2002). As the numbers increased, so the species began to show evidence of migratory behaviour that mirrors some of the N American populations. A migratory population became established in Iceland, and Ruddy Ducks began to be found in Scandinavia.

Of more concern, however, was the regular arrival of Ruddy Ducks in Iberia, believed to put at additional risk the globally threatened populations of White-headed Ducks *Oxyura leucocephala*, with which Ruddy Duck appears to hybridise freely (Green & Hughes 1996). Green & Hughes (2001) believe that this is the most important threat to the survival of the White-headed Duck. Research in the early 1990s indicated that it should be possible to reduce the number of Ruddy Ducks in GB by selective shooting and that this should in turn reduce the level of immigration into the Iberian peninsula. A programme of culling was initiated in 1999 and was incorporated into the International Species Action Plan for White-headed Duck (Hughes *et al.* 2005), since when several thousand birds have been removed from the GB population. There may be slight evidence that this is bearing fruit, since the WeBS index showed a 15% drop from its peak value in 2002–03 (Collier *et al.* 2005).

Natural vagrancy has been postulated but, in the absence of any ringing recoveries, this cannot be confirmed. A recent molecular study (Muñoz-Fuentes *et al.* 2006), using both mitochondrial and nuclear genes, has revealed that European (and Icelandic) birds are genetically very homogeneous; N American Ruddy Ducks are more variable, and the lack of variation in Europe indicates a single founding event. Were natural vagrancy to occur regularly (and these birds to become incorporated into the gene pool), more variation might be expected in the European population.

FAMILY TETRAONIDAE

Genus *LAGOPUS* Brisson

Holarctic genus of three species; two of which breed in GB&I.

Willow Ptarmigan/Red Grouse   *Lagopus lagopus* (Linnaeus)

Endemic: RB *scotica* (Latham)

TAXONOMY Up to 19 subspecies recognised (HBW; Madge & McGowan 2002), based on geography and summer plumage, and Irish birds ('*hibernica*') sometimes separated as a 20th. The picture is confused; for example, Sœther (1989) summarised evidence that subspecific boundaries between *lagopus* on the Norwegian mainland and *variegata* on adjacent islands were not supported by molecular analyses and that the plumage similarities on which these were based were more likely to be due to parallel selection than phylogenetic history. Using the mt DNA control region, Freeland *et al.* (2006)

found no genetic differentiation between British and Irish populations, concluding that there was little molecular support for *hibernica*. The 'Red Grouse' populations of GB&I named *scotica* have sometimes been treated as specifically distinct. This is based on plumage traits that vary clinally across NW Europe, and microsatellite loci showed similarly little evidence of differentiation. Molecular analysis of cyt-b (Lucchini *et al.* 2001) shows 3.1% difference between GB and Scandinavia, similar to specific differences elsewhere in the galliforms; there is a pressing need for a thorough review of Willow Ptarmigan population genetics, akin to that undertaken for Rock Ptarmigan by Holder *et al.* (1999, 2000).

DISTRIBUTION Holarctic, breeding from *c.* 47°N to the coasts of the Arctic Ocean. The most northerly populations are partially migratory, moving *c.* 200 km from the tundra to woodland habitats in winter (HBW).

STATUS The majority of the population of the endemic race of Willow Ptarmigan ('Red Grouse') occurs in GB, with only limited numbers in Ireland (749 10-km squares in the Second Breeding Atlas compared with only 64 in Ireland). This distribution follows declines of *c.* 13% in GB and *c.* 66% in Ireland since the First Breeding Atlas, although data from the BBS indicates that the situation in GB may have stabilised since 1994. Numbers on grouse moors tend to be better monitored, and it is known that *c.* 2.5 million birds were shot annually in GB&I early in the 20th century (Lack 1986). Bags declined progressively from the 1920s (Hudson 1992), especially in Ireland, and by the Second Breeding Atlas, the population of GB&I was estimated at only 250,000 pairs. The decline has been assigned to loss or degradation of heather moor, increases in predation and infection by the parasitic nematode *Trichostrongylus tenuis*. A study in N England did not implicate predation by raptors in the declining populations, although it might have had limited effects on post-breeding abundance (Redpath & Thirgood 1997); however, predation by crows and foxes is a major cause of mortality, especially on less well-keepered moors.

Within individual populations, there is evidence of cycling of abundance, with a periodicity ranging from 4–5 to 10 years. The causes of this are unclear (Moss *et al.* 1996). They may relate to the densities of nematodes (Hudson *et al.* 1999); however, a near 50-year study by Watson *et al.* (2000) showed that peak numbers tended to occur one or two years after high June temperatures and that these in turn showed the cyclicity. Another population on lower ground did not show such a correlation (Moss *et al.* 1996), so although a link between warm springs and enhanced breeding success is intellectually attractive, it seems that there is more to Red Grouse cycles than parasites and/or microclimate.

Rock Ptarmigan   *Lagopus muta* (Montin)

Endemic: RB *millaisi* Hartert

TAXONOMY Up to 30 subspecies recognised across the arctic (four from W Europe), based on variations in size and plumage that are 'usually slight' (Vaurie 1965) and depend much on geography. Holder *et al.* (1999, 2000) examined morphological and molecular variation among 10 subspecies around the Bering Sea and found evidence that recent glaciations had led to genetic divergence among geographical isolates; contemporary patterns of variation reflect post-glacial redistribution combined with restricted gene flow among the island populations. A related study from W Europe (not including Scottish birds) also showed striking genetic differentiation among populations from the Pyrenees, the Alps and Norway, in accordance with the limited powers of dispersal (Caizergues *et al.* 2003). On a more local scale, there was evidence that males are more philopatric.

DISTRIBUTION Holarctic; generally a High Arctic species but extending S from Fennoscandia (*muta*) into Scotland (*millaisi*), the Alps (*helvetias*) and the Pyrenees (*pyrenaica*). Further subspecies occur from the Urals to the Bering Sea, across the Aleutians to Newfoundland, Greenland and Iceland. Generally sedentary, moving short distances in response to adverse weather, though longer, perhaps irruptive, movements have been recorded in the High Arctic.

STATUS Occurred widely across N Scotland in the 19th century, including Hoy (Orkney) and Outer Hebrides, but it had been lost from the former of these by 1831 and the latter by 1938, though unclear if occasional more recent records to Outer Hebrides are immigrants or a relict population at very low density. Numbers were low in Scotland in the 1940s, as with several other populations of European tetraonids, but showed a recovery through the second half of the 20th century. During the fieldwork for the First Breeding Atlas, Ptarmigan were found breeding in 162 10-km squares, mostly in the Grampian Mountains and the W Highlands, and this changed little by the Second Breeding Atlas. However, abundance within populations showed considerable cyclic variation, with a periodicity usually close to 10 years, but occasionally less (Watson *et al.* 1998). Peak numbers tended to occur one to two years after high June temperatures, but the fine detail varied depending on the substrate, and synchrony between two populations may have been due to migration between study sites (Watson *et al.* 2000).

Population numbers and breeding success were adversely affected by the introduction of skiing in the Cairngorm Mountains. This was partly direct mortality due to birds being killed by striking wires associated with the ski lifts. There was also mortality due to the increased number of corvids attracted to the area by the development; these birds predated nests and young. In general, cable deaths were more significant in the area of highest ski activity, whereas corvid predation extended away from the ski centre.

Genus *TETRAO* Linnaeus

Palearctic genus of four species; two of which breed in GB&I, one following reintroduction.

Black Grouse   *Tetrao tetrix* Linnaeus

Endemic: RB *britannicus* (Witherby & Lönnberg)

TAXONOMY DNA sequences from both nuclear and mitochondrial genes (e.g. Drovetski 2002) show that Black Grouse and Capercaillie comprise a well-supported clade, leading these to be combined in the genus *Tetrao*. Within this, Black Grouse is sister to Caucasian Grouse *T. mlokosiewiczi*. Variation within the former is clinal; six or seven subspecies are recognised (Madge & McGowan 2002; HBW), of which three occur in the W Palearctic.

DISTRIBUTION Breeds in forests across the Palearctic, from GB to China, between 68°N and 48°N. Subspecies *britannicus* occurs in GB; *tetrix*: Europe from Scandinavia and the Alps to NE Siberia; *viridanus*: SE Russia, east to the River Irtysh. Most populations are resident, though there are occasional irruptions, and Siberian birds move to avoid harsh weather conditions.

STATUS At the end of the 19th century had already been lost from parts of Wales and S England, but still relatively widespread in mainland Britain north of Cumbria and Durham, and C Wales, SW & C England, with pockets remaining from Norfolk to Yorkshire (Holloway 1996). Declines set in from early in the 20th century, with populations becoming increasingly fragmented, especially in England and Wales. Here, afforestation initially provided quality habitat, but as the trees matured, this became increasingly unsuitable, especially since *c.* 1980.

The population of GB fell from 25,000 displaying males (1990) through 6,500 (1995–96) to less than 5,100 in 2005 (Warren & Baines 2004, 2008a, b). The pattern was not, however, one of uniform decline. Numbers fell by over 50% in S Scotland between the latter two surveys, although those in N England rose slightly from *c.* 770 to *c.* 1,030. In this area, the population became divided into two subunits, with limited opportunities for gene flow between them. Despite the collapse in numbers, it is widely, if thinly, distributed in Scotland and relatively common in parts of the Borders, Perth and Kinross, Aberdeenshire and restricted areas of Highland. Absent from Ireland.

Ecological studies (reviewed in Lindström *et al.* 1998) have shown that habitat requirements are a mosaic of woodland edges, with grass meadows for display, cover for nesting and a patchwork of low vegetation for adults and young to feed. Efforts by the RSPB and the national conservation agencies are underway to create or renew this habitat mix in several parts of GB. Restriction of grazing in selected areas and predator control appear to be successful in raising the population, although cold, wet weather during the chick-rearing phase can still adversely affect recruitment. Reintroduction is being tested in England to determine its feasibility more widely, but predation can be high and more needs to be learned. Woodland management in Wales led to an 85% increase of lekking males during 1997–2002, and research is underway in Scotland to determine methods of designing woodland to maximise habitat for this species.

Western Capercaillie   *Tetrao urogallus* Linnaeus

NB FB *urogallus* Linnaeus

TAXONOMY Molecular data (e.g. Drovetski 2002) confirm that this is sister to Black-billed Capercaillie *T. parvirostris*. Shows considerable interpopulation variation in colour saturation and female spotting, though much of this is clinal (Madge & McGowan 2002). From five (Vaurie 1965) to 12 (HBW) subspecies have been recognised, but Madge & McGowan (2002) limit this to eight, seven of which occur in the W Palearctic. Segelbacher *et al.* (2003) investigated genetic differentiation among Capercaillies from 14 sites from W Europe, finding that the populations in the Pyrenees (long-isolated), C Europe (recently isolated) and Scotland (introduced) were genetically depauperate, compared with the more extensive and interconnected populations in the Alps and the boreal forests. Recent studies show that Cantabrian/Pyrenean birds are distinct from populations elsewhere in Eurasia, with limited hybridisation (see Segelbacher & Piertney 2008 for refs).

DISTRIBUTION Breeds in conifer woodland, between 69°N and 43°N, from Iberia and Scandinavia, east to the Altai in NW Mongolia. Subspecies *cantabricus*: Cantabrian Mts, Spain; *aquitanicus*: Pyrenees; *major*: Alps to Carpathian Mts, S to Balkans; *urogallus*: Scandinavia to W Siberia; *kureikensis*: N Russia; *volgensis*: C and SE Russia; *uralensis*: S Urals to Novosibirsk; *taczanowskii*: C Siberia to the Altai (Madge & McGowan 2002). Birds in the S Carpathians are sometimes separated as *rudolfi*, but this is not generally recognised.

STATUS Probably occurred in England in historical times and there are sub-fossil remains from even earlier, but long since extinct; extinct in Ireland by 17th century; reintroductions have been unsuccessful (Hutchinson 1989; Brown & Grice 2005). Became extinct in Scotland in the 17th century, probably due to the widespread felling of the native pine forests. Replanting during the 19th century recreated habitat to the extent that birds were successfully reintroduced during the 1830s and probably reached a peak in numbers early in the 20th century (Storch 2001). Further felling during WWI and WWII resulted in the loss of this habitat and numbers declined again, until by the end of the century there were probably only *c.* 1,000 birds left, mainly in Perth and Kinross, Aberdeenshire and Highland. A Species Action Plan was instituted and additional factors leading to the latest decline were identified (Moss *et al.* 2001). A particular problem was mortality of adults flying into deer fences in the forests; however, low productivity due to poor breeding success was also significant. More chicks were reared when early Apr was warm, perhaps because of accelerated plant growth, providing superior food when eggs were being formed; broods were also more successful if late May to early June was warm and dry, probably improving foraging conditions for young chicks. During the period 1975–99, mid-Apr was cooler than normal, leading to a delay in the warming of the environment, reducing the availability of insect food for the young. Research also showed that increases in both predation by corvids and foxes, and disturbance by people may have added to the problems of breeding birds. Predator control, removing fences (or making them more visible) and more sympathetic forest management seem to be having an effect, and numbers began to rise through the early 21st century. Improved survey methods in 2003–04 suggested a population of 1,980 individuals (95% limits: 549–2041), but the change in methodology means that direct comparison with earlier surveys is unsafe (Eaton *et al.* 2007b). Alternative data suggest there may have been a recent increase in the Strathspey population, but habitat is still limited and Capercaillie remains one of the most threatened members of Scotland's avifauna.

FAMILY PHASIANIDAE

Genus *ALECTORIS* Kaup

Palearctic and marginally Oriental and Ethiopian genus of about seven species; one of which has been introduced into GB&I.

Red-legged Partridge
Alectoris rufa (Linnaeus)

NB *rufa* (Linnaeus)

TAXONOMY The *Alectoris* partridges form a tightly knit group of seven species, sharing many anatomical and morphological characters, and molecular analyses (Randi 1996; Randi & Lucchini 1998; Kimball *et al.* 1999) supported earlier enzyme results (Randi *et al.* 1992) indicating three separate clades: Arabian and Barbary *A. melanocephala, A. barbara*; Philby's, Chukar and Przewalski's *A. phylbyi, A. chukar, A. magna*; Red-legged and Rock *A. graeca*. Although hybridisation occurs between the last two of these, Randi & Bernard-Laurent (1999) found circumstantial evidence of natural selection against hybrids, supporting their status as parapatric species. Three subspecies are recognised by Madge & McGowan (2002), though variation is slight and clinal.

This species has interbred widely with released Chukar *A. chukar* (E*), and Red-legged Partridge x Chukar hybrids have been recorded in GB.

DISTRIBUTION W Europe, particularly Iberia, France, GB; also adjacent island groups. Subspecies *rufa*: France, NW Italy, Corsica, GB (introduced); *hispanica*: Portugal, NW Spain, Madeira (introduced); *intercedens*: E & S Spain, Balearics, Canaries, Azores (introduced).

STATUS Red-legged Partridge is not a native species; according to Lever (2005), it was probably introduced into GB&I during the reign of Charles II (*c.* 1670). Large numbers were imported from France during the 18th and 19th centuries, with varied success; releases by gun clubs in Ireland have not resulted in permanent establishment. At the time of the First Breeding Atlas, it was distributed widely in S & E England and E Wales, with fewer records from lowland Scotland. It was not recorded in Ireland in the First Breeding Atlas. By the time of the Second Breeding Atlas, it had spread north and west, with an increase of over 30% in the number of 10-km squares in GB. It had also been introduced to Ireland, though was only recorded at a very few sites (see Appendix 2). Numbers recorded in CBC fieldwork declined through the 1980s but recovered into the late 1990s; according to Baillie *et al.* (2006), this is largely driven by a substantial increase in the number released onto shooting estates. Populations fluctuate considerably through the year, as birds are released to supplement the breeding population and are then subsequently shot.

Genus *PERDIX* Brisson

Palearctic genus of three species; one of which breeds in GB&I.

Grey Partridge
Perdix perdix (Linnaeus)

RB NB *perdix* (Linnaeus)

TAXONOMY Many subspecies have been described, but only six (below) are currently recognised (Potts 1986; Madge & McGowan 2002). Variation is mostly clinal, but overall situation confused by repeated releases in some areas. Luikkonen-Anttilla *et al.* (2002) analysed a short sequence (390 bp) of a mitochondrial gene in wild Grey Partridges from a series of sites across Europe. The detected sequences fell into western and eastern groups or clades, differing by over 3%. Only the western sequence was found from Poland through to France (including England), and only the eastern sequence in S Finland and Greece; both sequences were found in birds from Bulgaria, N Finland and

Ireland. The implication is that the two sequences evolved during past glaciations, in Iberian (or Italian) and Balkan (or Caucasian) refugia; the mixed populations likely result from release of captive-bred birds, e.g. birds in game farms in Finland consist entirely of the western type. Native birds are subspecies *perdix*, and there is little evidence of any other subspecies being imported (Potts 1986).

DISTRIBUTION Occurs from N Iberia, GB&I, across Europe to N Finland, then east to W China. Subspecies *hispaniensis*: Iberia; *armoricana*: N France; *sphagnetorum*: Netherlands, NE Germany; *perdix*: GB&I, Scandinavia, S to Alps and Balkans; *lucida*: E of nominate, from Finland to Black Sea, and E to Urals and Caucasus; *canescens*: Turkey to NW Iran; *robusta*: Asia, from Urals to China. Most populations resident, though those in the Far East may move S in winter.

STATUS During the fieldwork for the the First Breeding Atlas, Grey Partridges were found across much of England and Wales, apart from the uplands and Pembroke peninsula. In Scotland, they were absent from most of the highlands and islands, and in Ireland their distribution was limited to the central belt and the N coastal fringes. However, a decline was already evident (First Breeding Atlas), and this was confirmed by a serious range contraction by the time of the Second Breeding Atlas. Birds had been lost from much of Ireland, the Cornish peninsula, and parts of Wales and SW Scotland. In general, losses in GB were from areas where they had been less abundant, but the decrease was serious nevertheless. In Ireland, the decline has been 'catastrophic' (P Smiddy pers. comm.), despite continued releases by gun clubs, including birds from eastern Europe.

CBC and BBS data confirmed the population declined suddenly and dramatically in the late 1970s; over 85% of birds were lost, and the evidence (reviewed by Potts 1986) suggested that agricultural intensification, and specifically the use of herbicides, had devastated the plants on which the insects (which formed the mainstay of early chick diet) depended. Attempts to modify agricultural practice seem to be having limited success (Baillie *et al.* 2006), and local populations are being sustained at their present low levels by ongoing releases of captive-bred stock for shooting purposes.

Genus *COTURNIX* Bonnaterre

Palearctic, Oriental and Ethiopian genus of at least eight species; one of which has been recorded in GB&I.

Common Quail   *Coturnix coturnix* (Linnaeus)

MB *coturnix* (Linnaeus)

TAXONOMY Kimball *et al.* (1999) found *Coturnix* to be sister to *Alectoris*, and Dyke *et al.* (2003) found evidence of similar phylogenetic affinity from an extensive morphological study. Up to six subspecies recognised on the basis of size, plumage and distribution (Vaurie 1965; HBW). However, Puigcerver *et al.* (2001) were unable to discriminate among these using two each of biometric and plumage traits, and cast doubt on the diagnosability of most. They recommend treatment as monotypic, although their samples were small, and a more extensive analysis is needed.

DISTRIBUTION A widespread summer visitor to much of the W Palearctic and adjacent areas of C Asia, south of *c.* 64°N to the Mediterranean in the west, and the Himalayas in the east; also resident populations in E Africa, from Ethiopia to the Cape. Described subspecies from the W Palearctic include *coturnix*: N Africa, Europe, C Asia; *confisa*: Canaries, Maderia; *conturbans*: Azores; *inopinata*: Cape Verdes. Most birds breeding in W Palearctic are migratory, probably wintering in sub-Saharan Africa, though small numbers winter in W Europe.

STATUS A bird of cereal and hay fields, especially on chalk. Secretive nature and difficult habitat combine to make confirmation of breeding difficult, and the vast majority of records are of calling males. The distribution mapped in the Atlases largely reflects this, but gives a false impression of abundance since there can be marked differences in numbers and distribution between consecutive years. The Second Breeding Atlas gives separate maps for 1988–1990. The middle year was a 'Quail year' and birds were heard calling as far north as Inverness, with a few records to Shetland; the flanking years showed a much thinner scatter of records, with only a handful north of Yorkshire. Although a much scarcer bird in Ireland, a similar pattern was evident, with many more records in the middle year. There is good evidence that they form aggregations, with later-arriving birds being attracted by calling males; this means that distribution can be very non-random, with areas of apparently suitable habitat being unoccupied even in good years.

The pattern of abundance may be cyclical, with peak numbers occurring every four to six years (see e.g. Pennington *et al.* 2004), although the underlying causes are unclear. Brown & Grice (2005) suggest that arrivals are correlated with successful breeding in N Africa. Since Quail can breed at three months old, birds reaching GB&I may be the offspring of Quail that bred earlier in the year. There is some evidence of a severe decline across much of Europe in 1970–90 (BirdLife International 2004); this continues in SE Europe, although N & W populations show signs of limited recovery. Quail are probably affected by overgrazing and desertification of the wintering grounds, but a near absence of ringing recoveries from GB&I means that the location of these are not known with certainty: a female shot in Spain in Sep was probably still on migration south.

Genus *PHASIANUS* Linnaeus

Palearctic and Oriental genus of two species; one of which has been introduced into GB&I.

Common Pheasant   *Phasianus colchicus* Linnaeus

NB *colchicus* Linnaeus
NB *torquatus* J. F. Gmelin
NB *principalis* P. L. Sclater
NB *mongolicus* J. F. Brandt
NB *pallasi* Rothschild
NB *satschuensis* Pleske

– the population consists largely of intraspecific hybrids

TAXONOMY Closely related to Green Pheasant *P. versicolor*, with which it has sometimes been regarded as conspecific; molecular comparisons lacking. Mt DNA sequences suggest that Common Pheasant is sister to Reeves' Pheasant *Syrmaticus reevesii* (Kimball *et al.* 1999; Bush & Strobeck 2003) although the morphological analyses of Dyke *et al.* (2003) suggested a more distant relationship. Thirty subspecies are listed by Madge & McGowan (2002).

DISTRIBUTION The native range extends from Korea, China and Vietnam, west to Black Sea (Hill & Robertson 1988), though Madge & McGowan (2002) suggest that birds as far west as NE Greece and SE Bulgaria may be of natural origin. Introductions are recorded for *c.* 50 countries (Long 1981), and viable populations occur on all continents except Antarctica. Hill & Robertson (1988) suggest that at least six subspecies have been released in GB&I (distributions from Madge & McGowan 2002): *colchicus* (E Georgia to NE Azerbaijan, possibly including the Balkan populations), *torquatus* (Henan to Guangdong), *mongolicus* (E Kazakhstan to NW Xingjiang), *principalis* (S Turkmenistan to NW Afghanistan), *pallasi* (far east of Siberia to N Korea, NE China) and *satscheuensis* (W Sichuan). In severe weather, may undertake extensive movements to lower or more sheltered ground.

STATUS Johnsgard (1988) suggests that pheasants were introduced to GB by the Romans between 55 BC and 400 AD, though Hill & Robertson (1988) are less convinced. According to Lever (2005), it has occurred in GB since at least 1058, when it was being eaten at a monastery at Walton Abbey (Essex); in Scotland, certainly occurred in 1578 and in Ireland since the 1580s. The introduced races have hybridised to produced a Pheasant that combines the characteristics of several races. Baillie *et al.* (2006) suggest that *c.* 12 million birds are released each year for sporting purposes and that *c.* 2 million of these probably survive to breed naturally. Brown & Grice (2005) report that productivity of the feral stock is low, perhaps because herbicides remove many of the plants needed by the insects which the chicks eat.

The Second Breeding Atlas showed that Pheasants occur throughout GB&I, apart from the highlands and islands of Scotland, where they are more limited in range, and parts of the extreme W of Ireland. The range had expanded slightly northwards since the First Breeding Atlas, and CBC/BBS data confirm a steady increase in density, especially in England. There is very little ringing data (Pheasants are not normally ringed); however, the few recoveries are extremely short-range, suggesting that the species is very sedentary.

Genus *CHRYSOLOPHUS* Gray

Palearctic genus of two species; both of which have been introduced into GB&I.

Golden Pheasant   *Chrysolophus pictus* (Linnaeus)

NB monotypic

TAXONOMY Golden and Lady Amherst's Pheasants are the only members of the genus *Chrysolophus*, which Bush & Strobeck (2003) found to be part of a large unresolved clade that included most other 'pheasants', and similarly weak resolution had been recovered by Kimball *et al.* (1999). Evidently the pheasants have evolved through a rapid radiation and molecular differentiation is slight. Morphological and plumage characters, which are likely to be driven by a combination of sexual and natural selection, show greater differentiation, although a cladistic analysis based largely on skeletal characters was equally unsuccessful in producing a phylogeny that was both well differentiated and well supported (Dyke *et al.* 2003).

DISTRIBUTION Bamboo forests and adjacent woodlands of Central China; introduced to France, New Zealand and GB&I but only established in last of these.

STATUS Released at a range of sites in GB from 1880 to 1975 (listed in Lever 2005) but only seem to have become well established in SW Scotland and the Norfolk/Suffolk Brecklands. Now believed extinct in SW Scotland, where none recorded since 1996 (BS3). Because of its secretive behaviour, estimating numbers (or even establishing continued existence) can be problematic. Numbers may be declining in Breckland as a consequence of reduced keepering and the cessation of releases; inbreeding may also be an increasing problem (Taylor *et al.* 1999, Piotrowski 2003).

Lady Amherst's Pheasant   *Chrysolophus amherstiae* (Leadbeater)

NB monotypic

TAXONOMY Closely related to Golden Pheasant, with which it will hybridise.

DISTRIBUTION Similar areas of Central China to Golden, though at higher altitudes; however, information on current distribution in China is scanty (Madge & McGowan 2002). Introduced into lowland England.

STATUS Brought to GB by Lady Sarah Amherst in 1828, although these birds died soon after arrival (Nightingale 2005). Subsequent attempts were more successful at sustaining a population, although several were blighted by the simultaneous release of Golden Pheasants, with which Lady Amherst's hybridise freely, and the stock rapidly lost the desired phenotype. Pure Lady Amherst's were released at Woburn (Bedfordshire) in the 1890s and at several other sites in S England and N Wales. Most of these staggered along for a few years before finally expiring. The population in Bedfordshire increased to peak, probably in 1970–80, at around 100–200 pairs. A decline set in at key sites shortly after this as the habitat changed; the trees matured, understorey was removed and disturbance increased. Nightingale (2005) believed that, by 2004, the key counties of Bedfordshire and Buckinghamshire held no more than 20 males with an unknown number of females. He suggested that, without intervention, these populations are unlikely to be sustainable. Since releases of non-native birds are no longer permissible, it seems probable that this species will die out in GB.

FAMILY GAVIIDAE

Genus *GAVIA* Forster

Holarctic genus of five species, two of which are only marginally Palearctic. All have been recorded in GB&I, two as breeding species, one as a regular visitor and two as vagrants.

Red-throated Loon/Red-throated Diver   *Gavia stellata* (Pontoppidan)

RB MB WM PM monotypic

TAXONOMY Divers have been relatively little studied using DNA sequences; however, Wink *et al.* (2002a) examined a small sample while establishing the identity of an unusual Great Northern Loon. They sequenced *c.* 1,000 base pairs from the mt cyt-b gene. Unsurprisingly, in view of its differences in morphology, ecology and behaviour, they found that Red-throated is sister to the other divers, differing by *c.* 4%. Despite its wide geographic distribution, there is little evidence of interpopulation variation, although some birds from Spitzbergen and Franz Josef Land have greyish edges to mantle feathers in breeding plumage. However, this character is variable (de Korte 1972) and the species is regarded as monotypic.

DISTRIBUTION Found across the Holarctic, N of *c.* 58°N in the Palearctic, slightly further S in the Nearctic. In Europe, distributed around the coasts of Greenland, through Iceland, NW Ireland, Scotland and from S Scandinavia across Eurasia to the Pacific. Winters primarily at sea, from the S parts of the breeding area, S to California and Florida in Nearctic, Iberia, Black Sea and coastal China in Palearctic.

STATUS Breeds on lochans and small pools, often far from the sea, across much of N and W Scotland, especially Orkney, Shetland, Outer Hebrides and Caithness to Wester Ross (Second Breeding Atlas). Assessing exact numbers is difficult because of remoteness of sites and frequent presence on water where breeding has not actually occurred (Gomersall *et al.* 1984). Long-term monitoring of the Shetland population revealed a decline through the 1980s, which has been ascribed (at least partly: Pennington *et al.* 2004) to fluctuations in food supply – principally sandeels. The total population there was probably at an all-time high of *c.* 700 pairs in 1983, although this had fallen to *c.* 424 pairs in 1994 (Gibbons *et al.* 1997). In Orkney, the 1994 survey revealed a minimum of 94 breeding pairs and an estimated total of 301 adults. Since then, numbers in the Northern Isles have remained stable but elsewhere there has been an increase. Overall, the British breeding population grew significantly (by 34%) between the 1994 and 2006 surveys. The latest survey indicated a total British population of about 1,250 breeding pairs and a total adult population of about 4,150 birds (Dillon *et at.* 2009). Shetland held 33% of the breeding population and the Outer Hebrides 26%. Fewer than 10 pairs breed in Ireland (all in Donegal) and none in England and Wales.

In winter, this is the commonest and most widespread diver in GB&I, being present around most of the coasts, occasionally in large numbers. Until the late 1990s, it was thought that perhaps 5,000 wintered in GB&I. However, aerial counts in the S North Sea have revealed that perhaps up to 11,000 occur in the shallow waters of the outer Thames Estuary alone and that up to 1,500 winter between Anglesey and Morecambe Bay (Cranswick *et al.* 2005). Over 1,000 were estimated in Cardigan Bay in Mar 2004 (Hall *et al.* 2005). Other shallow coastal areas, such as the Firth of Forth, Orkney and Lincolnshire, also harbour significant accumulations, although probably not of a size comparable with these. Pooling data from 2001 to 2006, O'Brien *et al.* (2008) estimated a total wintering population of at least 17,000 individuals. Since many of these birds are up to 20 km from land, they are rarely observed by shore-based observers. However, coastal watchers have sometimes recorded hundreds, or even thousands, of birds, particularly in the SE. Ringing recoveries indicate that many of these wintering birds are from Fennoscandia, although individuals ringed in Greenland have also been found. Birds ringed as nestlings in Orkney and Shetland have been recovered as far south as Biscay, but more usually around the coasts of Scotland and Ireland, and in the S North Sea.

Black-throated Loon/Black-throated Diver   *Gavia arctica* (Linnaeus)

RB or MB, WM *arctica* (Linnaeus)

TAXONOMY Now usually treated as a separate species from the Nearctic Pacific Loon *G. pacifica*, on the basis of differences in size, colouration and bill structure, and the fact that they coexist without hybridising (HBW). Wink *et al.* (2002a) found no sequence variation between Black-throated and Pacific Loons and that these two formed a clade separate from the two larger species. Two subspecies are recognised, *arctica* and *viridigularis*, on the basis of a greener throat patch in the latter.

DISTRIBUTION Somewhat similar distribution to Red-throated, though extends further south, and absent from Greenland, Iceland. Nominate breeds across W Palearctic from Scotland and Scandinavia, E to about the River Lena, wintering around the coasts of W Europe, and also in Mediterranean, Black and Caspian Seas; *viridigularis* breeds from the River Lena across Siberia to the Pacific, and winters in the NW Pacific as far south as China.

STATUS Within GB&I, a much rarer breeding bird than Red-throated, being absent from the N Isles and scarcer elsewhere in Scotland. Main concentrations are north of the Caledonian Canal, especially in Sutherland, Wester Ross and Lewis (Outer Hebrides). The installation of floating rafts (e.g. Hancock 2000) has improved breeding success, as fewer nests are lost to flooding and terrestrial predators: e.g. in 2002, 18 chicks were fledged by 35 raft-nesting pairs, compared with only 6 by 32 on raft-free sites (RBBP). An RSPB survey in 2006 estimated the population of GB&I to be *c.* 217 occupied territories, an increase of *c.* 16% over 1994, suggesting there is space for further increase. There are no breeding records from England and Wales and only sporadic attempts in Ireland.

Little is known about this bird outside the breeding season. The major wintering grounds are along the east shores of the North Sea, and birds are thinly distributed around the coasts of GB&I, with modest concentrations in the north and west and only small numbers inland. It is not well represented by WeBS counts, though these have shown that Moray and Cornwall hold larger numbers, and these may be enhanced by cold weather on the near-Continent. It does not seem to form any significant proportion of the wintering flocks of divers in the S North Sea. A recent aerial survey to assess possible impacts of wind farm development have not uncovered any major new concentrations of this species comparable with those found for Red-throated. Only small numbers winter in Ireland.

Pacific Loon/Pacific Diver   *Gavia pacifica* (Lawrence)

SM monotypic

TAXONOMY Recently split from Black-throated Loon *(q. v.)*.

DISTRIBUTION A largely Nearctic species that breeds from Alaska across N Canada to Hudson Bay; it also occurs in NE Siberia, where it is sympatric with the *viridigularis* race of Black-throated Loon. Migratory, leaving the frozen waters of N America to winter along the Pacific coasts as far S as China and California. Small numbers are also found along the Atlantic coast of the USA as far S as New York.

STATUS The first records were in 2007 when three were found in GB: the first was at Farnham Gravel Pits (N Yorkshire) in Jan–Feb 2007; the second at Llys-y-Fran Reservoir (Pembrokeshire) and the third at Mount's Bay (Cornwall), both in Feb–Mar 2007, returning the following year. A bird found off Eriksay (Outer Hebrides) in Oct 2005 is under consideration.

Great Northern Loon/Great Northern Diver   *Gavia immer* (Brünnich)

HB WM monotypic

TAXONOMY Molecular data (Wink *et al.* 2002a) indicate that this forms a species pair with Yellow-billed Loon, differing by *c.* 1% from the latter, although only one Yellow-billed was sampled. However, this supports morphological and behavioural data indicating a close affinity between the two largest divers.

DISTRIBUTION Predominantly Nearctic, only breeding in Iceland and Bear Island in W Palearctic. Winters along both coasts of N America S to California and Florida; in Palearctic, winters at sea from N Norway to Iberia.

STATUS A species that has held attractions as a potential breeding bird for many years (Ransome 1947). A recent review of claimed breeding records in Scotland found none completely satisfactory, though there is one well-documented case of a hybrid Great Northern Diver x Black-throated Diver paired with a Black-throated Diver producing one young in Loch Maree (Highland) in 1971, and possibly the same birds in the same place with two young in 1970 (BS3).

The N Isles and the western coasts of GB&I are important wintering areas for Great Northern Loons (Lack 1986); however, because of the isolated nature of much of this coastline, surveys and censuses are rarely possible on an annual basis. The highest site total in recent years was of 781 in Scapa Flow (Orkney) in Mar 1999 from a dedicated survey. Aerial surveys through the winter of 2003–04 (Dean *et al.* 2004) recorded over 750 birds at a series of sites between Aberdeen and Mull, including Scapa Flow, although very few were seen S of Inverness. As with Red-throated Loon, numbers recorded from land-based surveys are usually much smaller, so that, while having the benefit of continuity, WeBS data are likely to be less accurate. Although these show annual fluctuations, there is no evidence of profound changes in abundance, although standardised boat-based surveys in Shetland in recent years have revealed a decline (AJ Musgrove pers. comm.).

Yellow-billed Loon/White-billed Diver   *Gavia adamsii* (G. R. Gray)

SM monotypic

TAXONOMY Only one was sampled by Wink *et al.* (2002a), but this was sister to Great Northern Loon, forming a clade separate from Black-throated and Pacific Loons.

DISTRIBUTION A Holarctic species, breeding in Arctic Canada and Siberia from the Kola Peninsula to the Pacific. Majority winter in N Pacific but small numbers off NW Norway. Still relatively little known, and breeding status rather vague, although apparently fairly common in NE Siberia (HBW).

STATUS The first record for GB was a bird shot at Embleton (Northumberland) in Dec 1829. By 1960, there were still barely 20 records, but with improvement in skills and equipment, the number of records increased rapidly, so that by 2007 over 300 birds had been recorded. Birds have been seen in every month, although there are fewer in summer; there is now evidence of a peak of records in May, from the N & W Isles of Scotland, suggesting a previously unsuspected migration from the Atlantic towards the breeding grounds in Siberia.

It remains a rare bird in Ireland, with the first record as recently as Feb 1974, and a total of only eight birds to 2004. Three of these are from May, suggesting that observations from the NW coast at this time might strengthen the theory of a spring passage.

FAMILY DIOMEDEIDAE

Genus *THALASSARCHE* Reichenbach

Southern ocean genus of five species, two of which have been recorded as a vagrant to GB&I.

Black-browed Albatross   *Thalassarche melanophris* Temminck

SM *melanophris* Temminck

TAXONOMY Morphological and molecular analyses (Sangster *et al.* 2002a; Penhallurick & Wink 2004) have confirmed that there are four distinct clades of albatrosses, with Black-browed (*sensu lato*) being sister taxon to Grey-headed Albatross *T. chrysostoma*. Campbell Albatross *T. m. impavida* Mathews 1912 may be a distinct species (Robertson & Nunn 1998), but Penhallurick & Wink (2004) argue that the molecular difference between this and the nominate (Burg & Croxall 2001) is lower than between most other pairs of mollymawks so the evidence for specific status for *impavida* is weak. The sequence of mt DNA from *impavida* on Campbell Island differs markedly from those of other populations of Black-browed Albatross from across the southern oceans (Alderman *et al.* 2005). Hybridisation occurs between *impavida* and *melanophris* on Campbell Island (Burg & Croxall 2001), and the *impavida* sequence has been identified among the latter at this site. These data support Rheindt & Austin (2005), who suggest that *impavida* may be better treated as a separate species and that *T. melanophris* should therefore be monotypic.

DISTRIBUTION *T. melanophris* breeds on sub-Antarctic islands: Falklands and S Georgia, Kerguelen, Heard, Antipodes, Macquarie Islands. A few also breed on Campbell I (Burg & Croxall 2001) alongside *T. (m.) impavida*. Outside the breeding season, migrates northwards into cold-water areas of Benguela and Humboldt Currents.

STATUS The first record for GB was a sub-adult found exhausted at Linton (Cambridgeshire) in July 1897, and there have been about 20 further records. Most of these are in the period June–Sep and often involve repeated sightings of the same individual, which complicates determining the number of birds. There is a remarkable series of records from Hermaness (Shetland), where a bird returned almost every year between 1972 and 1995 and might even have been the same individual as on Bass Rock in 1967–1969. The first Irish record was a bird off Cape Clear (Cork) in Sep 1963; there have been about 11 individuals in total, including a report of two adults off the Old Head of Kinsale (Cork) in Sep 1976.

Atlantic Yellow-nosed Albatross   *Thalassarche chlororhynchos* (J. F. Gmelin)

SM monotypic

TAXONOMY Molecular analysis (Nunn *et al.* 1996; Penhallurick & Wink 2004) show this to be part of a clade that includes the other 'mollymawks': Black-browed, Buller's *D. bulleri*, Shy *D. cauta*, etc., supporting earlier morphological studies. Traditionally considered as comprising two subspecies: the nominate in the Atlantic and *carteri* (sometimes known as '*bassi*') in the S Indian Ocean. Adults differ consistently in plumage (sometimes obscured by wear and moult) and bill morphology (Robertson 2002); juveniles and immatures are more difficult to distinguish but may also differ in bill morphology. Molecular analysis reported by Robertson & Nunn (1998) and summarised in Penhallurick & Wink (2004) showed the genetic distance between these to be similar (*c.* 3%) to that between taxa generally regarded as full species and led the former to recommend these to be treated as separate species as well. We follow these recommendations here.

DISTRIBUTION *T. chlororhynchos* breeds on Tristan da Cunha and Gough Island in the S Atlantic. *T. carteri* breeds on islands in the S Indian Ocean, including Crozet, Amsterdam and Kerguelen. Both species disperse away form the breeding grounds after breeding, though they appear to remain generally in their respective oceans.

STATUS One, an immature, Brean (Somerset) in June 2007; taken into care and released the following day, subsequently relocated in July at Carsington Water (Derbyshire) and near Messingham (Lincolnshire) (Gantlett & Pym 2007). Photographs suggest that a Yellow-nosed Albatross seen in Sweden on 8 July 2007 was the same individual, but birds recorded in Norway in June and July 2007 were different. A bird resembling this species was also seen off the Faroes in mid-July 2007.

FAMILY PROCELLARIIDAE

Genus *FULMARUS* Stephens

Holarctic and Antarctic genus of two species, one of which breeds widely in GB&I.

Northern Fulmar	*Fulmarus glacialis* (Linnaeus)

RB MB PM *glacialis* (Linnaeus)

TAXONOMY Two subspecies are recognised: *glacialis* in North Atlantic and *rogersii* in North Pacific, with a more slender bill. Molecular analysis (Penhallurick & Wink 2004) confirm that these are sister taxa, forming a clade with the Southern Fulmar *F. glacialoides*. Only nominate *glacialis* recorded in GB&I. Plumage and bill sizes of *glacialis* vary; dark ('blue') forms occur at high frequencies in some northern populations (e.g. Spitzbergen, Bear Island) but not all.

A recent genetic analysis (Burg *et al.* 2003) compared mt DNA sequences from seven colonies across the North Atlantic in an attempt to identify the source population for the colonisation of GB&I. The data do not conform with a pattern of sequential colonisation of sites by a limited number of individuals as there is too much intracolony variation. The results suggest that the source of most of the birds in GB&I was Iceland, rather than St Kilda, although the limited variability of the latter colony made reaching a firm conclusion difficult. The low level of variation at St Kilda perhaps indicates that this population has itself been through a relatively recent bottleneck, and Burg *et al.* (2003) suggest that this could be due to heavy and sustained killing of animals for food.

DISTRIBUTION Oceanic. Outside breeding season, adults disperse from breeding grounds to return early in the New Year. Nominate race breeds from Newfoundland in the west, through Svalbard and Novaya Zemlya in the north and east, to GB&I and northern France in the south; *rogersii* breeds in the North Pacific from Eastern Russia to Alaska.

STATUS In GB&I, breeds on all coasts where there are suitable cliffs with wide ledges, apart from Lincolnshire and most of East Anglia. Scarce along south coast of England, apart from Cornwall. Especially abundant on Scottish islands, where it may nest on flat tops of those free of mammalian predators.

The Fulmar attracted constant interest through the 20th century, and its numbers were monitored more or less regularly. It has increased dramatically since it began its population expansion around the coasts of GB&I in the middle of the 18th century (Fisher 1952). The first breeding record for Ireland was rather later, in County Mayo in 1911, again spreading round the coasts to colonise Co. Down in 1954 (Hutchinson 1989). Fisher (1952) believed that the colonisation of GB&I began from St Kilda (see above). The rate of increase has been estimated as *c.* 15% prior to WWII, since when it has slowed (Tasker 2004a). The three major seabird surveys estimated numbers in GB&I as *c.* 310,000 apparently occupied nests in 1969–70, *c.* 535,000 in 1985–88, and 538,000 in 1998–2002, suggesting that it is now more stable. Tasker (2004a) discusses the reasons for the population changes, concluding that there is no clear single cause. The increase in food, through the expansion of the commercial whaling industry and subsequently from offshore trawlers, has been postulated as a factor (Fisher 1952), as has a genetic change in the northern populations (Wynne-Edwards 1962), for which there is no firm evidence.

Fledglings depart the natal ledges and spend the next four or five years at sea, returning then to colonies during the breeding season until the birds are sexually mature at about nine years (Dunnet *et al.* 1979). Ringing recoveries during the pre-breeding period come from Newfoundland to the White Sea (Wernham *et al.* 2002). Birds ringed as nestlings and subsequently found in colonies when of breeding age range widely across GB&I, with no obvious pattern of movement between adjacent colonies (Wernham *et al.* 2002). This supports the finding of Burg *et al.* (2003) that the spread around the coasts of GB&I was not necessarily by 'stepping stone' colonisation.

As with many pelagic tube-noses, the increase in long-line fisheries in the North Atlantic is of serious concern. Tasker (2004a) reports that between 50,000 and 100,000 Fulmars may be killed annually but stresses the need for more detailed study. Another potential threat comes from the collapse of some marine stocks, though this may not be entirely anthropogenic. The decline in the planktonic copepod *Calanus finmarchicus* since the 1950s may be linked to rising sea temperature, and Tasker discusses this at length. Almost all major fish species have shown recent declines, leading to reduced levels of commercial fishing and hence fishery discards. However, research has shown that the food of Fulmars varies across their range, both seasonally and geographically, so a single linking factor may not be apparent. At present, Fulmars seem to be secure.

Blue forms occur regularly in small numbers off coasts of GB&I in autumn and winter; often following severe weather.

Genus *PTERODROMA* Bonaparte

Holarctic genus of over 30 species, two of which have been recorded as vagrants to GB&I.

Fea's Petrel   *Pterodroma feae* (Salvadori)

SM race undetermined

TAXONOMY Molecular analysis of the mitochondrial cyt-b gene (Penhallurick & Wink 2004) shows that the *Pterodroma* petrels comprise a single clade. Morphologically, Fea's Petrel *P. feae* and Zino's Petrel *P. madeira* are closely similar to Soft-plumaged *P. mollis* and have traditionally been grouped with them (e.g. Sibley & Monroe 1990; Warham 1990, 1996). Recent molecular data from Zino *et al.* (2008) suggest that these are not especially close, supporting the views of Bourne (1983) that they are separate species: Fea's Petrel is sister-taxon to Cahow *P. cahow*, with 2.9% divergence, whereas it differs from Soft-plumaged by 4.6% (F Zino pers. comm.; Zino's Petrel was not included by Penhallurick & Wink). There is no evidence of intraspecific variation in *P. madeira*, but *P. feae* has been split into two subspecies, based largely on size, more recently supported by molecular differences (Jesus *et al.* in press).

DISTRIBUTION Soft-plumaged Petrel breeds in the S Atlantic at Gough and Tristan da Cunha. Fea's is restricted to Cape Verde (*feae*) and the Desertas (Madeira, *deserta*); Zino's only breeds in the highlands of Madeira. On abundance alone, Fea's is much more likely to occur in GB&I than Zino's Petrel.

STATUS 'Soft-plumaged' Petrels were first recorded at Cape Clear (Cork) in Sep 1974 and Dungeness (Kent) in Oct 1983, but identification was regarded as inconclusive by IRBC and BOURC respectively. With increasing skills and awareness, there are now over 30 records and they are almost annual, the majority during Aug–Sep, coming either from pelagics (in GB waters) or from well-watched Atlantic headlands in W Ireland such as the Bridges of Ross (Clare), Brandon Point (Kerry) and Galley Head (Cork), though with further records N to the N Isles. While the global abundance of the two forms means that Fea's is much more likely than Zino's, the difficulties of separating the two species meant that it took until Aug 2001 for the first definite Fea's record for GB of a bird seen, and more importantly photographed, *c.* 60 miles SW of Scilly (Cornwall). In Ireland, *c.* 30 individuals of either *feae* or *madeira* have occurred, with none identified to species level.

Black-capped Petrel/Capped Petrel   *Pterodroma hasitata* **(Kuhl)**

SM *hasitata* (Kuhl)

TAXONOMY Molecular analyses of Penhallurick & Wink (2004) and Zino *et al.* (2008) show this species to be sister taxon to Cahow and Fea's Petrels.

DISTRIBUTION Black-capped Petrel, more evocatively called 'Diablotin' from its devilish nocturnal call, is much reduced in distribution. Formerly, relatively widespread in Jamaica (race *caribbaea*, possibly extinct), now only breeds in the mountains of Hispaniola, possibly Cuba, Dominica and Martinique. Outside the breeding season, it ranges along the east coasts of N and S America from Maryland to Brazil.

STATUS The first record is of a bird found at Swaffham (Norfolk) in Mar or Apr 1850. A tideline corpse found at Barmston (Yorkshire) in Dec 1984 was originally placed in Category D but, following the redefinition of this category, it was elevated to Category A in 2006. There are no records from Ireland.

Genus *BULWERIA* Bonaparte

An oceanic genus of two species, one of which has been recorded as a vagrant to GB&I.

Bulwer's Petrel   *Bulweria bulwerii* **(Jardine & Selby)**

SM monotypic

TAXONOMY Forms a species pair with Jouanin's Petrel *B. fallax* of the Indian ocean. Penhallurick & Wink (2004) included Bulwer's in their molecular analysis of Procellariiiformes and found it to be sister to a clade that included *Procellaria*; however, the genetic distances were large and *Bulweria* is clearly a rather distinct genus.

DISTRIBUTION A tropical and subtropical seabird that breeds on islands in the Atlantic and Pacific Oceans; there has been no genetic comparison of birds from these two regions. In the Atlantic, Bulwer's nests from the Azores to Cape Verde, dispersing away form the breeding islands into the S & SW Atlantic outside the breeding season.

STATUS One Irish record of a bird off Cape Clear Island (Cork) in Aug 1975. Previously accepted records from Britain were recently reviewed and found no longer to be acceptable.

Genus *CALONECTRIS* Mathews and Iredale

Holarctic genus of two, three or four species, one of which occurs in the seas around GB&I.

Cory's Shearwater   *Calonectris diomedea* **(Scopoli)**

SM *borealis* (Cory)
SM *diomedea* (Scopoli)

TAXONOMY Molecular analysis (Penhallurick & Wink 2004) suggests that *Calonectris diomedea* (*sensu lato*) is sister to Streaked Shearwater *C. leucomelas* of the W Pacific and part of a clade that also includes most of the *Puffinus* shearwaters (though see Rheindt & Austin 2005). Three taxa are recognised, separated on size and colouration: *diomedea* (Mediterranean); *borealis* (E Atlantic); *edwardsii* (Cape Verde Is). The Cape Verde populations are now widely regarded as a separate species *C. edwardsii* Cape Verde Shearwater (see Hazevoet 1995; Hillcoat *et al.* 1997). Molecular analysis reveals that *diomedea* and *borealis* are distinct (Gómez-Díaz *et al.* 2006). There is also a sharp discontinuity in metrics (see Gómez-Díaz *et al*. 2006 for refs), and playback experiments provide evidence of the potential for assortive mating (Bretagnolle & Lequette 1990). It is likely that this group is best treated as three separate species.

DISTRIBUTION Nominate *diomedea* ('Scopoli's Shearwater') breeds on islands in N and W Mediterranean; *borealis* breeds in islands of Azores, Canaries and Madeira group; *edwardsii* breeds Cape Verde Is. Migratory, most wintering at sea off W coast of southern Africa, but small numbers reach the coasts of Uruguay and Brazil (BWP). Some Mediterranean birds apparently winter in the Indian Ocean

STATUS Regularly seen in modest numbers off headlands and pelagic boat trips in S and W of GB&I in July–Oct, with occasional large, concentrated movements. Smaller numbers recorded elsewhere, including the North Sea, usually after strong NW gales. The average number each year increased around 1980 (Fraser & Rogers 2005), probably due to the increase in skills and sea-watching. Whether these birds are foraging individuals from breeding colonies, non-breeders or sub-adults is unclear.

One record of the nominate race at sea, 6 miles S of St Mary's (Scilly) in Aug 2004. An earlier record, seen from Bridges of Ross (Clare), in Aug 2003 is under consideration by IRBC (P Milne pers. comm.).

Genus *PUFFINUS* Brisson

Holarctic genus of perhaps 20 species, five of which have been recorded in GB&I.

Great Shearwater   *Puffinus gravis* (O'Reilly)

PM monotypic

TAXONOMY Molecular analysis (Penhallurick & Wink 2004) suggests a close affininty with Sooty Shearwater *P. griseus*, also Flesh-footed *P. carneipes*, Short-tailed *P. tenuirostris* and other Pacific shearwaters (though see Rheindt & Austin 2005).

DISTRIBUTION Breeds more or less exclusively on Nightingale and Inaccessible Islands (Tristan da Cunha) and Gough Island. Visits North Atlantic during austral winter, moving northwards along coasts of Americas to wintering grounds between Newfoundland and Greenland, then east and south, before returning to South Atlantic with a more easterly bias.

STATUS Recorded regularly in western approaches to GB&I from headlands and pelagics from July to Oct, though occasional in winter. Numbers vary seasonally and annually, with occasional strong movements, especially off SW Ireland. Scarce in North Sea, usually after NW gales.

Sooty Shearwater   *Puffinus griseus* (Gmelin)

PM monotypic

TAXONOMY Closely allied with Great Shearwater *(q.v.)*.

DISTRIBUTION Breeds on sub-Antarctic islands in S Pacific (especially off New Zealand); also islands off S tip of South America. Majority spend austral winter in N Pacific, but smaller (though still sizable) numbers move north along E coast of South America to winter in North Atlantic. Crosses Atlantic in July and present in European waters from Aug to Nov, feeding especially in Rockall and Faroese fishing grounds (BWP).

STATUS Occurs in most months from Apr to Nov but principally Aug and Sep. Largest concentrations recorded in SW, when heavy movements can be recorded off headlands and on pelagics. Smaller numbers around northern coasts of GB&I; commoner than Great Shearwater in North Sea, where several hundred sometimes recorded in a day moving past promontories.

Recent studies in the E Pacific suggest that there have been dramatic changes in distribution that can be linked to climate-induced changes in the marine ecosystem (Robinson *et al.* 2005). Whether similar patterns have occurred in the N Atlantic seems not to have been examined.

Manx Shearwater   *Puffinus puffinus* (Brünnich)

MB monotypic

TAXONOMY The 'Manx Shearwater' complex has attracted much study recently. Morphological, behavioural and ecological analyses have indicated that *P. puffinus*, *P. yelkouan* and *P. mauretanicus* should all be regarded as separate species (see Sangster *et al.* 2002b). Austin (1996) and Austin *et al.* (2004) examined the molecular genetics of the complex and confirmed the three taxa in a single clade, somewhat separate from the rest of the small black-and-white *Puffinus* shearwaters.

DISTRIBUTION Oceanic. Outside breeding season, adults disperse away from the breeding grounds. Breeds predominantly in the north-east Atlantic, with only a tiny and apparently precarious population off Newfoundland. The main centres of population are in Iceland, the Faroe Islands and, especially, GB&I, with much smaller populations in France, the Azores, the Canaries and Madeira.

STATUS *Seabird 2000* estimated the population of GB as 280,000–310,000 apparently occupied sites, with a further 27,000–61,000 in Ireland; this comprises >75% of the global population. The major colonies within the archipelago are Rum (Hebrides): *c.* 120,000; Skomer (Pembrokeshire): 101,800; Skokholm (Pembrokeshire): 46,200; Bardsey (Caernarvonshire): 16,183; Inishtooskert (Kerry): 9,696; Puffin Island (Kerry): 6,329; and Inishabro (Kerry): 5,611. Because of problems with accurate censusing and the differences in survey technique down the years, it is difficult to determine what changes may have taken place since since *Operation Seafarer* (1969–70). There was evidence of a slight decrease in the colony on Rum between 1985 and 1995, but this was not statistically significant (Furness 1997). In view of the international responsibility for this species and the potential for loss due to climatic or anthropogenic effects, close and continued monitoring of Manx Shearwaters should be a priority.

A sustained programme of ringing was initiated by RM Lockley on Skokholm in the late 1920s, and continued there by Oxford University until work was transferred to the nearby island of Skomer in 1976. This has been supplemented by studies at other colonies in GB&I (e.g. Copeland (Down), Rum, Bardsey, Kerry) and laid the foundations of our knowledge of Manx Shearwater distribution both inside and outside the breeding season. Breeding birds were formerly believed to range widely in search of food for their young, but Brooke's view (1990) that this was unlikely is supported by recent research using global positioning technology mediated through a device attached to the bird's back (Guilford *et al.* 2008). Females may forage hundreds of km from the nest during the pre-laying period, but during the egg and chick stages most ring recoveries and GPS records are close to the colony.

As with many ground-nesting seabirds, terrestrial mammalian predators pose a significant threat to breeding birds. It is likely that the population on Canna was lost due to Brown Rats *Rattus rattus* and feral cats *Felis silvestris*, and Newton *et al.* (2004) discuss other possible examples of predation by species as diverse as Ferrets *Mustela furo* and Red Deer *Cervus elaphus* and, of course, man. There also exists the possibility of climate change adversely affecting breeding success, whether through prolonged periods of heavy rain flooding the burrow or the reduction in food species adversely affecting growth and survival.

At the end of the breeding season, both adults and juveniles cross the Atlantic to winter off the coasts of Brazil, Uruguay and Argentina (Hamer 2003). Return to breeding area in Feb–Apr, probably further to the east in the S Atlantic, then further to the west in the N Atlantic, overall completing a figure-of-eight migration (Brooke 1990). Sub-adults may remain in the S Atlantic or summer in the Caribbean or on the rich feeding grounds off E Canada.

Balearic Shearwater   *Puffinus mauretanicus* (Lowe)

PM monotypic

TAXONOMY Austin (1996) and Austin *et al.* (2004) analysed the mitochondrial cyt-b sequence of this and related taxa and found that *mauretanicus* formed a sister group to *P. yelkouan*, differing at about 2.2% base pairs, compared with *c.* 0.5% variation within the taxa. Plumage, morphological and behavioural data (Yésou & Paterson 1999; summarised in Sangster *et al.* 2002b) indicated that they were also distinct in other characters and formed separate evolutionary lineages in the W and E Mediterranean. On the strength of this, the BOU TSC recommended that they be treated as separate species.

DISTRIBUTION Restricted to the Balearic Islands, where the population was estimated at less than 2,000 pairs and declining (BirdLife International 2004). This makes it one of Europe's most threatened seabirds. The majority leave the Mediterranean in mid-summer to feed in the Atlantic, especially in Biscay, but recorded from Norway to South Africa.

STATUS Regular visitor to waters around GB&I in summer and autumn, especially in the S and W, but also enters North Sea and is frequently seen in small numbers from vantage points on the E coast of England and Scotland.

Macaronesian Shearwater   *Puffinus baroli* (Bonaparte)

SM *baroli* (Bonaparte)

TAXONOMY Austin *et al.* (2004) report a molecular analysis of the phylogeny of the 'small' shearwaters, based on 917 bp of cyt-b. They found no evidence to support the conventional taxonomy, instead identifying three clades that reflect geographical distribution. A North Atlantic clade comprises *lherminieri*, *baroli* and *boydi*, with the latter two being sister taxa; *loyemilleri* from Panama was identical to *lherminieri*. The N Atlantic forms are clearly distinct from *assimilis*, so three options are available: *lherminieri*, *baroli* and *boydi* are treated as conspecific; *baroli* and *boydi* are regarded as one species and *lherminieri* as a second; the three taxa are treated as separate species. The molecular differentation between *baroli* and *boydi* is modest, but that between these and *lherminieri* is broadly similar to differences between other taxa of *Puffinus* that are treated as full species. Until further data becomes available, the BOU Taxonomic Sub-committee recommended that they be treated as two species: Audubon's Shearwater *P. lherminieri* and Macaronesian Shearwater *P. baroli*.

DISTRIBUTION Macaronesian Shearwater is confined to the N Atlantic, where *baroli* breeds from the Azores south to the Canaries and *boydi* breeds in Cape Verde. Both forms seem to show rather limited dispersal away from the breeding areas and appear not to cross the equator (Sinclair *et al.* 1982).

STATUS The first record for Britain and Ireland was off the Bull Rock (Cork) in May 1853, when a 'Little Shearwater' settled on board the sloop *Olive* (Ussher & Warren 1900). The first for GB was found dead near Earsham (Norfolk) in Apr 1858. It is now evident that a small, but significant, number of Macaronesian Shearwaters wander northwards, with individuals being recorded from GB&I in most years, usually in June–Sep. Records are increasing slightly, probably as observers gain familiarity with the key features; however, there is a distinct possibility that some of the multi-date records within years relate to the same individuals associating with feeding flocks of Manx and Sooty Shearwaters. Remarkably, in 1982 and 1983, the same male (on the basis of vocalisations) was captured beside a boulder pile on Skomer (Pembrokeshire) – on one occasion landing on the chest of a former President of the BOU, who was 'resting' after a strenuous day in the field. It is possible that a female was also present in 1983 (James 1986). All those identified to race have been *baroli*.

FAMILY HYDROBATIDAE

Genus *OCEANITES* Keyserling & Blasius

Antarctic and sub-Antarctic genus of two species, one of which has been recorded as a vagrant to GB&I.

Wilson's Storm Petrel
Oceanites oceanicus (Kuhl)

PM *exasperatus* Mathews

TAXONOMY Penhallurick & Wink (2004) attempted to resolve the taxonomy of the Procellariiformes using cyt-b data from a wide range of species. In particular, they report divergences that led them to recommend a series of changes to the generic names within the storm petrels. Rheindt & Austin (2005) have challenged these results on theoretical and logical grounds. Until this difference is resolved, we leave the taxonomic relations within the storm petrels unchanged. Two subspecies are generally recognised, differing in size: the smaller more northerly breeding *oceanicus* and the larger more southerly *exasperatus*.

DISTRIBUTION Breeds widely on sub-Antarctic islands from Cape Horn and Kerguelen, south to the coast of Antarctic continent. Many colonies number hundreds of thousands of pairs. Migrates north after breeding to spend austral winter in northern parts of Indian, Atlantic and Pacific Oceans.

STATUS The first for GB was found in a field near Polperro (Cornwall) in Aug 1838 and there was only one more in the next 130 years. Irish records follow a similar pattern: the first two were in 1891 and no more for *c.* 70 years. It remained a rare bird until quite recently, when they started to be found on pelagic trips (first off RoI and then off SW England). There have been over 400 records off England between 1950 and 2005, all but four of which date since 1986. Increasingly being identified from land, especially from sites in W Ireland such as the Bridges of Ross (Clare) and Cape Clear (Cork). It is now clear that large numbers winter in Atlantic waters close to GB&I.

The two Irish birds from 1891 were long-winged and within the size range of *exasperatus* (details in Ruttledge 1975); only these two have been racially identified.

Genus *PELAGODROMA* Reichenbach

Widespread monospecific oceanic genus that has been recorded as a vagrant to GB&I.

White-faced Storm Petrel
Pelagodroma marina (Latham)

SM *hypoleuca* (Webb, Berthelot & Moquin-Tandon)

TAXONOMY The data reported by Penhallurick & Wink (2004) suggest that this species is sister to Grey-backed Storm Petrel *Garrodia nereis*. However, the genetic distances are large and the bootstrap support low. Several races have been described, of which *hypoleuca* (Salvage and Canary Islands), *eadesi* (Cape Verde) and *marina* (Tristan/Gough) are potential visitors to GB&I.

DISTRIBUTION In addition to the Atlantic populations, it also breeds on islands round the coast of Australia and New Zealand. The European populations seem to be stable (BirdLife International 2004) but very limited in area.

STATUS There is one record for GB, of an immature female caught alive on Colonsay (Argyll) in Jan 1897. There are no records from Ireland.

Genus *HYDROBATES* Boie

Largely Palearctic monospecific genus that breeds in GB&I.

European Storm Petrel   *Hydrobates pelagicus* (Linnaeus)

MB monotypic

TAXONOMY Regarded as monotypic by BWP and here, though support for the separation of Mediterranean populations as subspecies '*melitensis*' has been proposed on basis of larger size, especially bill (Lalanne *et al.* 2001). DNA sequences (and vocalisations) support divergence of Mediterranean and Atlantic populations (Cagnon *et al.* 2004).

DISTRIBUTION Oceanic. Globally, the species is truly European, with only a few (if any) colonies in NW Africa. There are a handful of colonies in the Mediterranean, principally in Italy, Malta and the Balearics. The rest are on islands along the Atlantic seaboard. Outside breeding season, adults disperse away from the breeding grounds.

The majority of the colonies outside GB&I have not been assessed by playback (currently regarded as the most accurate way to identify occupied sites) (Mitchell 2004a), and estimates are probably less accurate. The largest populations are likely to be in the Westermann Islands off Iceland (50,000–100,000 pairs) and the Faroe Islands (150,000–400,000 pairs), with somewhere between 2,000 and 13,000 pairs elsewhere. The Mediterranean population '*melitensis*' totals only 7,000–18,000 pairs.

STATUS This species is present on a number of the smaller offshore islands along the Atlantic coast of GB&I and during *Seabird 2000* only one mainland colony was found from a total of 95 active colonies surveyed (Mitchell 2004a). The 95 active colonies surveyed in this way ranged in size from a handful of individuals to over 27,000 (at Inishtooskert, Kerry). The total estimate of the GB&I population was 99,000 apparently occupied sites, of which the majority were in Ireland where there were *c.* 57,000 on 18 islands.

Tape-luring of birds at and away from breeding colonies has revealed a large population of sub-adults wandering around the coasts of GB&I, with individuals often moving >200 km in two or three days (Wernham *et al.* 2002). There is some evidence of substructuring, with birds from the N Isles being less commonly trapped in the Irish Sea and vice versa. There is also some evidence that these two subpopulations may winter in different areas: Irish and Irish Sea birds tend to be recovered in Biscay and off Mauritania, whereas many Scottish recoveries are from coastal S Africa. Recaptures also suggest that sub-adult birds do not return to the natal colonies for two to three years (Wernham *et al.* 2002), but data are sparse, and there is little evidence of where these birds might be in the meantime.

As with Leach's Storm Petrel and Manx Shearwater, terrestrial predators pose a serious threat to burrow-nesting seabirds. Thus, European Storm Petrels are absent from all islands with rats in the Scilly archipelago (Heaney *et al.* 2002) and were almost totally restricted to rat-free islands in Orkney and Shetland (de León *et al.* 2006), and the population of Foula was reduced to a few pairs through predation by feral cats (BWP). Avian predators are now recognised as an equal threat; Phillips *et al.* (1999) showed that Great Skuas from St Kilda took *c.* 7,500 of this species every year. As Mitchell (2004a) points out, this total is greater than the estimated population of St Kilda, and such high levels of predation clearly cannot be sustained.

Genus *OCEANODROMA* Reichenbach

Widespread oceanic genus of 11–13 species, one breeding in GB&I and two others occurring as vagrants.

Leach's Storm Petrel   *Oceanodroma leucorhoa* (Vieillot)

MB PM *leucorhoa* (Vieillot)

TAXONOMY Several subspecies described from Pacific but little variation among Atlantic populations. In view of the presumably long period of isolation, the lack of morphological differentiation between N Atlantic and N Pacific populations is a little surprising. Molecular analyses may reveal greater differences.

DISTRIBUTION Oceanic. Outside the breeding season, adults disperse from the breeding grounds. Mitchell (2004b) shows that this species is much more restricted in breeding range than the European Storm Petrel. He argues that Leach's feeds on macro-zooplankton, which tend to be found beyond the edge of the continental shelf, and so it breeds as close to these areas as possible. Indeed, he found a significant correlation between colony size and proximity to the 1,000 m isobath, and all colonies in GB&I were within 70 km of the edge of the shelf.

STATUS Leach's Storm Petrel has a limited distribution in GB&I, where it is known to breed only at few remote sites off the N & W coasts of Scotland and off the W of Ireland. The survey teams of *Seabird 2000* found the vast majority on the islands of St Kilda (45,433 apparently occupied sites), with smaller numbers on the Flannan Islands (1,425), North Rona (1,132) and the Stags of Broadhaven, Co. Mayo (310). The remaining sites where birds were found mustered a total of less than 60 between them (Mitchell 2004b). As with other burrow-nesting oceanic seabirds, it is difficult to determine whether these numbers are stable. Census methods are still being developed, and *Seabird 2000* estimates were based on playback. Previous surveys generated results now regarded as insecure. For example, the population on part of North Rona was estimated as 328 apparently occupied sites during *Seabird 2000*, remarkably close to the figure of 327 burrows found in 1936 (Ainslie & Atkinson 1937). Intervening surveys (Robson 1968; Love 1978) supported these figures, but all differ markedly from the 2,000 to 3,000 pairs reported in 1958 (Bagenal & Baird 1958 – based on mark-recapture). A 1972 survey suggested the population of the whole island to be 500 pairs (Lloyd *et al.* 1991), less than half the figure in 2000.

   The population of GB&I, although important, is minor compared with numbers from Atlantic Canada (>4.75 million pairs) and Iceland (80,000–150,000). Pacific populations are also substantial, probably in excess of 10 million pairs.

   The recapture/recovery rate of Leach's Storm Petrel is very low (Wernham *et al.* 2002), and ringing data are of little use in determining movements. Present off the coast of W Africa during late autumn and winter (Bourne 1992), and large numbers have been estimated in Biscay at the same time. These are not all from GB&I, however, for the numbers are too large even for the E Atlantic population. They presumably include birds from the W Atlantic, and there is limited ringing data to support this (Wernham *et al.* 2002). Circumstantial evidence thus suggests that birds from the W Atlantic probably occur in GB&I waters during the late summer, particularly after storms. Little is known of the movements of sub-adult birds.

   Mitchell (2004b) suggests that predatory mammals are a serious threat, and in GB&I Leach's Storm Petrel currently nests entirely on predator-free islands. Mink and Brown Rats are perhaps the biggest threat, and it is imperative that the greatest care is taken to ensure that these do not gain access to key areas such as St Kilda. Serious threats are also posed by avian predators, including gulls and crows, and especially Great Skuas, which are increasing rapidly in many areas. Phillips *et al.* (1999) estimate that about 14,000 Leach's Storm Petrels may be taken by skuas each year on St Kilda and that this may be a serious threat to the local populations. Indeed, one subcolony (cited by Mitchell) seems to have declined by 48% between 1999 and 2003 and this decline matches that predicted by Phillips *et al.* (1999) based on their observations of skua predation. These figures relate solely to breeding birds, and non-breeders visiting the colonies at night are probably also at risk.

Swinhoe's Storm Petrel   *Oceanodroma monorhis* (Swinhoe)

SM monotypic

TAXONOMY Penhallurick & Wink (2004) show that this species is part of a clade that includes Leach's *O. leucorhoa*, Tristram's *O. tristrami*, Band-rumped *O. castro* and European *H. pelagicus* Storm Petrels (Matsudaira's *O. matsudairae* was not included in their analysis). However, relationships within this clade were not clear, so further discrimination is not possible.

DISTRIBUTION Breeds on islands of the Yellow Sea from Japan, China and Korea. Migrates SW to winter in the N Indian Ocean and Arabian Gulf, but movements otherwise little known. There has been a series of records from the Salvage Islands (Madeira); these are reviewed by Zino in Hagemeijer & Blair (1997). In brief, a male was found in a Band-rumped Storm Petrel burrow in 1983; another male was captured in a wall in 1988; between 1993 and 1996, a female with a vascularised brood patch was regularly caught on a nest in a collapsed wall. The possibility of breeding in the Salvage Islands is unlikely but cannot be discounted.

STATUS A remarkable series of events began in July 1989, when a dark-rumped petrel was trapped and ringed at Tynemouth, Northumberland (Cubitt 1995). Three nights later, also in July, a second bird was trapped. The following year, a third bird was trapped and, even more remarkably, this one was re-trapped in 1991 (once), 1992 (once), 1993 (three dates) and 1994 (two dates). There are further records from Cove (Aberdeenshire) in Aug 2000 and Great Skellig Rock (Kerry) in July 2000. Apart from a bird seen off the Bridges of Ross (Clare) in Aug 1985, all confirmed records involve individuals attracted to tape lures at night.

Band-rumped Storm Petrel/   *Oceanodroma castro* (Harcourt)
Madeiran Storm Petrel

SM monotypic

TAXONOMY Part of an unresolved clade recovered by Penhallurick & Wink (2004) – see Swinhoe's Storm Petrel. For several years, Band-rumped Storm Petrel has been known to show out-of-season breeding at Ascension Island (Allan 1962) and to have two breeding seasons within a single year in the Galapagos (Snow & Snow 1966; Harris 1969). This has now been confirmed at Baixo and Praia in the Azores (Monteiro & Furness 1998) and also in the Desertas (Madeira; Nunes 2000). In both these archipelagos, birds breeding in the 'hot season' (Apr–Aug) are physically smaller than those breeding in the 'cool season' (Sep–Jan). On Vila Islet (Azores), breeding only occurs in the cool season, and here the birds are large: similar in size to the cool-season birds from Baixo and Praia (Monteiro & Furness 1998). Ringing studies revealed a low level of mixing of the two seasonal populations, indicating seasonal fidelity and consequently little gene flow between them. Prospecting hot-season birds in the Azores showed no more response to the vocalisations of cool-season birds than they did to Cory's Shearwater calls. Egg dimensions and nestling mass differ similarly between the seasonal populations. It seems likely that the hot-season birds in the Azores should be treated as a separate species and that the seasonal populations in Madeira should be regarded as separate management units (see Friesen *et al.* 2007 for recent refs). Bolton *et al.* (2008) recently named the hot-season birds *O. monteiroi*.

DISTRIBUTION Disjunct distribution ranging from Berlengas (Portugal), Canaries, Azores, Madeiran archipelago, Cape Verde, Ascension and St Helena in Atlantic, Galapagos, Hawaii, and islands off Japan. The European populations appear to be small but stable.

STATUS One: a female found dead at Blackrock Lighthouse (Mayo) in Oct 1931. Subject to acceptance: single birds, 6 miles S of Scilly, July 2007 and off Pendeen (Cornwall), Sep 2007.

FAMILY PHAETHONTIDAE

Genus *PHAETHON* Linnaeus

Widespread tropical and subtropical genus of three or four species, one of which has occurred as a vagrant to GB&I.

Red-billed Tropicbird   *Phaethon aethereus* Linnaeus

SM race undetermined but likely to have been *mesonauta* Peters

TAXONOMY Kennedy & Spencer (2003) examined the phylogenetic relationships of frigatebirds and tropicbirds using sequence data from four mitochondrial genes totalling 1,756 bp. They found that clades for these two groups were very strongly supported but that these were not closely related either to each other or to other members of the Pelecaniformes such as gannets, cormorants or pelicans.

Three races are recognised on a combination of facial pattern, intensity of black on primaries and secondaries, and barring on upperparts: nominate *aethereus* (South Atlantic), *mesonauta* (E Pacific, Caribbean and tropical Atlantic), *indicus* (Arabian Gulf and Red Sea).

DISTRIBUTION Nominate breeds on islands in tropical Atlantic, including Ascension, St Helena and Fernando Noronha. Race *mesonauta* breeds in the Caribbean, Cape Verde and islands off Senegal; also on tropical islands off Pacific coast of N, C and S America. Race *indicus* is restricted to islands in Red Sea, Arabian Gulf and Arabian Sea. Disperses away from colonies during non-breeding seasons.

STATUS The first record is of a bird photographed 20 miles (32 km) SSE of Scilly (Cornwall) in June 2001. Race undetermined but not *indicus*. Two more records (though possibly of the same individual) followed in the same general area during Mar–Apr 2002. There are no records from Ireland.

FAMILY SULIDAE

Genus *MORUS* Vieillot

Temperate oceanic genus of three species, one of which breeds in GB&I.

Northern Gannet   *Morus bassanus* (Linnaeus)

RB MB PM monotypic

TAXONOMY Analysis of cyt-b sequences (Friesen & Anderson 1997) confirmed that the gannets and boobies represent separate evolutionary lineages. Conventional wisdom (e.g. Nelson 1978) regarded Northern Gannet as sister to Cape Gannet *M. capensis*, but the same molecular study showed that Cape Gannet is closer to Australasian *M. serrator* with Northern having separated earlier.

DISTRIBUTION Widely distributed across the North Atlantic with the principal colonies being in GB&I, Canada, Iceland and France, with smaller numbers in Norway, the Faroes, Germany and Russia. Since 1900, numbers have increased at about 2% per year through six global surveys and show no signs of levelling off (Wanless & Harris 2004a). However, this apparent uniformity masks individual variation across countries and colonies.

STATUS Breeding populations of Gannets in GB&I are dispersed across a small number of colonies scattered around our coasts, mostly on offshore islands off Scotland, but including one each in England and Wales and several in Ireland. These have been surveyed at regular intervals and most are increasing. Of 14 colonies surveyed during *Seabird 2000*, 13 showed an increase from the previous surveys. Most sites not included in this survey also showed signs of stasis or an increase (Wanless & Harris 2004a).

GB&I birds disperse away from the colonies after breeding, generally moving southwards. Although immature birds range further south than older individuals (Wernham *et al.* 2002), birds of all ages can be seen in all parts of the range at any time. There is evidence from ringing recoveries that birds from east Atlantic colonies mix in the winter and that some even cross the Atlantic (Wernham *et al.* 2002). Ringing data also show that there is considerable colony fidelity but that birds may move between sites, either to newly established colonies or to those of longer standing. Whether these individuals are first-time or established breeders is not reported.

Northern Gannet is included in the 'Amber' category because >50% of the UK breeding population is concentrated at 10 or fewer sites and because >20% of the European population breeds in the UK. However, the general and continuing increase in numbers of breeding Gannets suggests that the species is in a healthy state in GB&I. Annual survival rises from 30–40% (depending on the colony) in the first year of life to 92% at adulthood (Wanless *et al.* 2006). The main threats seem to be from human activity. Harvesting for food only occurs at one site (Sula Sgeir, Highland), and population growth here is less than at similar colonies elsewhere. Environmental pollutants are known to occur in Gannet tissues but apparently at levels below those likely to adversely affect reproductive success (Newton *et al.* 1990). Oil spills can and do kill Gannets but do not appear to have significant effects on population size. Indeed, Nelson (2002) has suggested that the increase in numbers reflected recovery from persecution rather than expansion in numbers or extension of range.

The detailed knowledge of population size, dispersal, feeding rates, reproductive success and other demographic parameters in this species has led to the theoretical modelling of east Atlantic gannetries. This supported the observation that smaller colonies increased more rapidly than larger ones, expansion perhaps limited through competition for food at the latter (Moss *et al.* 2002).

FAMILY PHALACROCORACIDAE

Genus *PHALACROCORAX* Brisson

Cosmopolitan genus of 35–40 species; two breed in GB&I and one has occurred as a vagrant.

Great Cormorant   *Phalacrocorax carbo* (Linnaeus)

RB MB *carbo* (Linnaeus)
RB WM PM *sinensis* (Blumenbach)

TAXONOMY The taxonomic relationships of the shags and cormorants were confused until the 1980s, when a combination of behaviour (van Tets 1976) and morphology (Siegel-Causey 1988) were drawn together (see Johnsgard 1993), although there remain some contentious taxa. A molecular study (Kennedy *et al.* 2000) based on three mt DNA genes found some sections of the phylogeny were robust, others lacked statistical support. We follow the recommendations of Kennedy *et al.* (2000) in placing all shags and cormorants in the single genus *Phalacrocorax* in the meantime.

The Great Cormorant shows morphological differentiation around the world, which has resulted in six or seven races being generally accepted: nominate *carbo* (E Canada across N Atlantic to Norway and GB&I), *sinensis* (the rest of Europe, E to India and China), *hanedae* (Japan), *maroccanus* (NW Africa), *lucidus* (W, S and E Africa), *novaehollandiae* (Australasia). Molecular studies (Goostrey *et al.* 1998; Winney *et al.* 2001) established that *sinensis* and *carbo* differ in their genetic architecture and that this could be used to resolve the identity or origin where subspecies was in doubt. Winney *et al.* (2001) also found evidence that birds from Norway and Scotland were genetically different from those in England, Wales and NW France. Newson *et al.* (2004) developed the proposal of Alström (1985) that gular angle could be used to differentiate between *sinensis* and *carbo*.

DISTRIBUTION Widespread from E Canada to Australasia, and from the Arctic to the tropics. Predominantly, but not exclusively, coastal, occupying areas of open water, either fresh or saline. Usually close to shore, nesting on cliffs, flat islands or in trees, depending on availability.

STATUS Great Cormorants in GB were historically coastal at all seasons (Holloway 1996), but since the 1950s an increasing proportion has taken to breeding inland, especially in England: Great Cormorants have bred inland in Ireland for over 100 years (Holloway 1996). Both coastal and inland populations have increased in numbers (see Sellers 2004), although this is not uniform across GB&I. For example, coastal populations in Ireland, England, Wales and IoM have probably increased, but those in N and W Scotland decreased. Inland breeding has increased steadily since *c.* 1950, principally in SE England. DNA studies and gular angle analyses confirm that a variable proportion of this is due to the colonisation of lowland England by *sinensis*. However, ringing and DNA data indicate that coastal *carbo* are also present at inland colonies during the breeding season, so inland colonists have two origins. Further research is needed to test whether the birds of different origins show any evidence of assortative mating. There are still only *c.* 20 records of *sinensis* from Ireland, with the first as recently as 1985 (NI) and 2000 (RoI).

WeBS data from GB&NI show a steady increase since monitoring Great Cormorants began in the late 1980s, and the most recent winter estimates are *c.* 23,000 in Britain (Kershaw & Cranswick 2003) and *c.* 13,700 in Ireland (Crowe *et al.* 2008).

Great Cormorant is included in the UK 'Amber' category because >50% of the breeding population is concentrated at 10 or fewer sites and because >20% of the European population winters here. Oil spills kill small numbers of Great Cormorants but do not appear to affect the populations; chemical contaminants have been found in some populations and linked with reproductive failure in the Netherlands (Boudewijn & Dirksen 1995) but have not been found to have had an adverse effect in GB&I. Declines in colonies in NW Scotland appear to be linked with increased adult mortality rather than food shortages during the breeding season (Budworth *et al.* 2000) and may be partly due to shooting along salmon rivers during the winter.

Great Cormorants have long been persecuted by man, though the extent of this declined through the second half of the 20th century (Sellers 2004) and no doubt is partially responsible for many of the increases in numbers. The colonisation of lowland England is likely due to the increase in lakes, reservoirs and gravel pits, with their attendant fish stocks, although reduction of marine fish stocks may also have played a part. Pressure is now increasingly exerted by the fishing industry for control measures to be imposed: the 'black plague' is an unwelcome addition to many freshwater fisheries.

Double-crested Cormorant   *Phalacrocorax auritus* (Lesson)

SM race undetermined

TAXONOMY Traditionally placed fairly close to Great Cormorant in systematic lists based on morphology and behaviour (reviewed by Johnsgard 1993), although the molecular analysis of Kennedy *et al.* (2000) indicates that these two species are members of different, well-supported clades. These might be best treated as separate genera, but presently it seems safer to leave them in *Phalacrocorax*. Generally partitioned into five subspecies: *cincinatus* (Alaska), *albociliatus* (Pacific seaboard from S British Columbia to Baja California), *floridanus* (Atlantic and Gulf coasts of North America), *auritus* (Atlantic coasts of NE North America, inland from central Canada to N Texas and Kansas), *heuretus* (Bahamas, Cuba).

DISTRIBUTION Breeds across much of North America, from Alaska and Newfoundland to Baha California and Florida. Nominate race is strongly migratory, wintering south to Gulf of Mexico and Florida. Other races show less movement: *cincinatus* S to British Columbia, interior populations of *albociliatus* move towards the sea, *floridanus* and *heuretus* are almost sedentary.

STATUS The first record for GB&I was a long-staying bird at Billingham (Cleveland) from Jan to Apr 1989. A similarly long-staying individual was present at Nimmo's Pier (Galway) from Nov 1995 to Jan

1996. The only other record is of a bird found in the hold of a cargo ship in Glasgow Docks (Clyde) in Dec 1963; the ship was newly arrived from Newfoundland. However, since the bird had been in confinement during the crossing, it cannot be considered for the British List.

European Shag   *Phalacrocorax aristotelis* (Linnaeus)

RB *aristotelis* (Linnaeus)

TAXONOMY Kennedy *et al.* (2000) show that *P. aristotelis* is part of the same clade as *P. carbo*, though its position relative to the other species is not well resolved. Three subspecies are recognised on size and the extent of yellow at the base of bill: *aristotelis* (Iceland and coasts of W Europe) is slightly larger and with less yellow than *desmarestii* (Mediterranean and Black Seas), *riggenbachi* (coastal Morocco).

DISTRIBUTION Essentially a marine species, rarely inland. Breeds on rocky coasts and offshore islands; much more restricted to this habitat than Great Cormorant. Nominate is distributed along coasts from Iceland, the Faroes in the NW, and Russia and Norway in the NE, south through GB&I and France to Spain and Portugal. Replaced in Mediterranean by *desmarestii*, with largest populations in Croatia, Greece, Sardinia and the Balearics (Wanless & Harris 2004b); breeds as far east as Turkey and Ukraine. In N Africa, the race *riggenbachi* is now extremely rare, with less than 50 pairs in Morocco (Thévenot *et al.* 2003). Largely sedentary, though juveniles may disperse locally. Some colonies may be deserted in winter due to lack of food (HBW).

STATUS About 50% of the nominate race breeds in GB&I and this represents *c.* 45% of the global population of the species. The main colonies are in the N and W, from the Farne Islands (Northumberland) to S Cornwall, with small numbers in Yorkshire, S Devon and Dorset. *Seabird 2000* estimated the GB&I population at *c.* 32,300 apparently occupied nests, distributed in Scotland (66%), Ireland (12%), England (19%) and Wales (3%). This may well be a significant underestimate (BS3). Most of the principal colonies in Scotland and Ireland declined in numbers between the Seabird Colony Register Census of 1985–88 and that of *Seabird 2000* (1998–2002), whereas those in England and Wales increased over the same period. Wanless & Harris (2004b), who report these figures, suggest that, for methodological reasons, many of the declines recorded at individual colonies could be more severe than the data indicate.

Many adults seem to remain within 100 km of the breeding colony, only moving away in response to bad weather or food shortage (Wernham *et al.* 2002). Younger birds disperse more widely outside the breeding season and are more likely to be found inland (Potts 1969), usually following wrecks. The direction of dispersal varies among regions, and open sea appears to be a significant barrier. Birds from the north-east move north and south along the coast; the Irish Sea seems to be a barrier, for few Irish birds have been recovered in Wales, or vice versa (Wernham *et al.* 2002). Interchange among colonies is limited: adult birds show almost no movement between breeding colonies and most juveniles recruit within 12 km of their birthplace.

Populations of Shags around the coasts of GB&I increased in numbers through the 20th century following the reduction of human exploitation for food (Potts 1969) and the removal of a bounty directed at Great Cormorants in 1981. Wanless & Harris (2004b) argue persuasively that some of the decreases in numbers observed at individual colonies may be due to adults deferring breeding through periods of local unfavourability. Severe adult mortality can also occur due to adverse weather (Harris & Wanless 1996), oil spills (Heubeck 1997) and paralytic shellfish poisoning (Armstrong *et al.* 1978). It seems that recovery from these catastrophic, though usually local, events can occur and that reasons for the widespread progressive declines (especially across Scotland and Ireland) probably lie elsewhere. Wanless & Harris (2004b) conclude that this remains a mystery but that, because GB&I host such a high proportion of the global population of this species, efforts to resolve the issue should be given high priority.

FAMILY FREGATIDAE

Genus *FREGATA* Lacépède

Pantropical genus of five species, two of which have been recorded as vagrants in GB&I.

Ascension Frigatebird · *Fregata aquila* Linnaeus

SM monotypic

TAXONOMY A molecular analysis of frigatebirds and tropicbirds (Kennedy & Spencer 2003) revealed that these groups are monophyletic but not especially closely related. Magnificent and Ascension Frigatebirds are each other's closest relatives, with strong statistical support.

DISTRIBUTION A rare and very localised species, now restricted to Boatswainbird Islet off Ascension in the S Atlantic. Population recently estimated as 1,000–1,500 pairs (HBW). Sedentary, as far as is known, with few records far from Ascension.

STATUS One record: an immature female was caught exhausted on Tiree (Argyll) in July 1953. For 50 years this bird was accepted as a Magnificent Frigatebird, but a review of 1950–1958 GB records revealed its true identity (Walbridge *et al.* 2003).

Magnificent Frigatebird · *Fregata magnificens* Mathews

SM monotypic

TAXONOMY Recent study has indicated that this forms a clade with Ascension *(q.v.)*. Some (e.g. Vaurie 1965) have suggested that the Cape Verde and Galapagos populations of Magnificent Frigatebird comprise separate subspecies (*lowei, magnificens*, respectively) to elsewhere (*rothschildi*), but these divisions are not generally accepted.

DISTRIBUTION Occurs on both Atlantic and Pacific coasts of the Americas, from Florida to Brazil, and California to Ecuador, respectively, and across to Cape Verde in Atlantic and Galapagos in Pacific. Little evidence of movement, apart from dispersal of immatures away from the natal colony.

STATUS Two records: an adult female was found in an exhausted state at Scarlett Point (Isle of Man) in Dec 1998. Taken into care, it died 10 months later in Oct 1999. A second individual (an adult male) was found exhausted near Whitchurch (Shropshire) in Nov 2005; taken to Chester Zoo, it died a few days later. Both birds occurred following the passage of Category 5 hurricanes from the Caribbean NE across the Atlantic. There have been around eight further records of frigatebirds in GB&I that were not identified to species; remarkably, in view of the IoM record, three of these have been in County Dublin (1988, 1989 & 1995).

FAMILY ARDEIDAE

Genus *BOTAURUS* Stephens

Cosmopolitan genus of four species; two of which have been recorded in GB&I, one of which breeds.

Eurasian Bittern · *Botaurus stellaris* (Linnaeus)

RB WM *stellaris* (Linnaeus)

TAXONOMY Data from cyt-b and DNA/DNA hybridisation (Sheldon *et al.* 2000) both indicated that *Botaurus* is sister taxon to *Ixobrychus*, supporting an earlier analysis based on osteology (McCracken

& Sheldon 1998). More detailed investigations of species limits within these genera have not been published, but conventional wisdom (e.g. Vaurie 1965; HBW) recognises two subspecies: *stellaris* and *capensis*, differing in darkness of upperparts.

DISTRIBUTION The nominate breeds across the Palearctic from the Atlantic to Pacific, between 61°N and 34°N. Populations in W Europe, from Denmark, GB to Morocco, and SE to Black and Caspian Seas, are generally resident; northern and eastern populations, from Fennoscandia to China, Japan are more migratory, wintering in S Eurasian wetlands and through Nile Valley to C Africa. Replaced in SE Africa by resident race *capensis*.

STATUS Formerly widespread in reedbeds, mires, fens and similar flooded habitats with tall emergent vegetation. Drainage for agriculture in the 17th–19th centuries progressively reduced the number and extent of breeding sites; extinct as a breeding bird in GB&I since the late 19th century, though recolonised England from 1911, mainly in areas of extensive reedbeds. The population grew steadily to the 1950s, when the population reached *c*. 80 booming males and a decline set in. A reduction in water quality through eutrophication, combined with ageing of the reedbeds and loss or overgrowth of feeding ditches, led to a rapid fall in the breeding population through to the 1990s, when the number of males fell to <20. An intensive research programme into the ecological needs of breeding birds resulted in improved management of existing reedbeds and the creation of new ones. This appears to have been successful, and there are signs of recovery, especially in Suffolk (refs in Brown & Grice 2005). In recent years, booming males have been heard away from the key areas of East Anglia and NW England, in locations abandoned since the 1950s. The number of booming males in 2006 was estimated at 44–65 (RBBP). The need for continued management of freshwater reedbeds is now recognised as imperative; Bitterns do not seem to like saline conditions, and if the reeds become choked and overgrown, their suitability as a nesting habitat is soon lost.

Outside the breeding season, Bitterns are scarce winter visitors to suitable habitat across GB&I. Though much less common in Scotland and Ireland, birds are regularly found from late autumn in small reedbeds across the Midlands and southern England and into Wales. Ringing recoveries suggest most come from the Netherlands, Belgium, Germany and S Sweden (Wernham *et al.* 2002), often arriving following hard weather in these areas. Ringing data from within GB (mostly East Anglia) indicates that birds also disperse away from the breeding areas in the autumn.

American Bittern   *Botaurus lentiginosus* (Rackett)

SM monotypic

TAXONOMY Closely related to Eurasian Bittern and to species in S America and Australasia; sometimes included with S American *B. pinnatus* in *B. stellaris*, but detailed phylogeny lacking. However, DNA/DNA hybridisation (Sheldon 1987) suggests Eurasian and American Bitterns are genetically distinct.

DISTRIBUTION Breeds across Canada, south of *c*. 55°N (Hancock & Kushlan 1984; Gibbs *et al.* 1992), through USA, though discontinuously in southern states. Northern populations are migratory, to avoid frozen wetlands, wintering as far south as Panama.

STATUS There are 37 records from GB and 22 from Ireland, but now occurs much less often than formerly. The first were birds shot at Puddletown (Dorset) in autumn 1804 and near Armagh (Armagh) in Nov 1845; the former of these is the type specimen for this Nearctic species. Birds have been recorded in most months from Nov to May, with rather more in Oct–Nov, but no obvious peak and no differences across the archipelago.

Genus *IXOBRYCHUS* Billberg

Cosmopolitan genus of eight species, one of which has been recorded in GB&I.

Little Bittern   *Ixobrychus minutus* (Linnaeus)

CB SM *minutus* (Linnaeus)

TAXONOMY Closely related to Least *I. exilis* and Yellow *I. sinensis* Bitterns but genetic data inconclusive; morphological differences currently regarded as adequate for specific differentiation. Four widely allopatric races recognised: *minutus* in Eurasia and others in sub-Saharan Africa, Madagascar and Australia.

DISTRIBUTION Nominate breeds across much of Europe, apart from Fennoscandia and GB&I, to W Siberia, and scattered wetlands across the Middle East to N India. Migratory, wintering in southern half of Africa; Indian populations probably resident (HBW).

STATUS Occasional presence of more than one bird at some sites has led to the suspicion of breeding, and this was confirmed in 1984 when a pair raised three young at Potteric Carr (S Yorks).

Over 475 records of Little Bittern in GB since the first was shot at Christchurch (Dorset) in 1773 and 55 records from Ireland to 2004. Although annual totals vary somewhat, numbers rose fairly steadily through to the mid-1970s, though now reduced to an average of *c.* 5 per year. There are records from every month, with a distinct peak in May that corresponds with their arrival in C Europe, and some males may remain, calling for prolonged periods. Significantly fewer birds reported from Ireland in autumn compared with the rest of GB; presumably, spring arrivals are birds that 'overshoot' from France and Iberia, whereas autumn birds are dispersants/migrants from breeding grounds in NW Europe.

Genus *NYCTICORAX* Forster

Nearly cosmopolitan genus of two species, one of which has been recorded in GB&I.

Black-crowned Night Heron   *Nycticorax nycticorax* (Linnaeus)

PM *nycticorax* (Linnaeus)

TAXONOMY Relationships of *Nycticorax* to other herons not clear; molecular analyses inconclusive and cyt-b does not resolve situation. A near-cosmopolitan species, breeding on all continents, apart from Australasia and Antarctica, and separated into four subspecies, of which three (*hoactli, obscurus, falklandicus*) are from the Americas (Vaurie 1965; HBW).

DISTRIBUTION Nominate breeds across Europe, from Iberia to the Netherlands, S and E to Kazakhstan, and from Indian subcontinent through SE Asia to Japan, Philippines; also much of wetland Africa and Madagascar. Western birds migratory, wintering to tropical Africa; northern component of eastern populations also migratory, to SW Pacific. N American race *hoactli* breeds across Americas, from S Canada to N Chile, N Argentina; northern birds migratory, wintering in southern N America and Caribbean.

STATUS The number recorded each year varies from zero (1959) to 65 (1990), with a fairly steady increase since 1960. The distributional picture has long been muddied by free-flying birds of captive origin and even transient breeding populations established from this source. A free-flying colony of American *hoactli* bred around Edinburgh Zoo from 1951 (Thom 1986), though eliminated by 2004 (BS3). Some of the past records in C Scotland were probably from this source. Similarly, birds in East Anglia are sometimes 'tainted' as escapes from a collection at Great Witchingham (Norfolk, Brown & Grice 2005). There are records from every month of the year, and clear peaks in Apr–May and late autumn (Oct–Nov) are in accord with natural vagrancy, so the majority of the records from GB are likely to be genuine. There are 62 Irish records to 2004, and birds tend to occur here in years when numbers are also high in England.

Genus *BUTORIDES* Blyth

Cosmopolitan genus of two or three species; one of which has been recorded as a vagrant to GB&I.

Green Heron   *Butorides virescens* (Linnaeus)

SM monotypic

TAXONOMY Has a varied taxonomic history, sometimes (e.g. Vaurie 1965; HBW) being combined with Striated *B. striata* and Lava *B. sundevalli* Herons but treated as separate species by others (e.g. Monroe & Browning 1992; Davis & Kushlan 1994). Several poorly defined subspecies proposed.

DISTRIBUTION SE Canada and USA (apart from more arid parts of NW), south to Panama and Caribbean. Northern populations migratory; wintering in southern parts of range.

STATUS Five records in GB; the first was shot near St Austell (Cornwall) in Oct 1889. The subsequent birds were also recorded in autumn; perhaps surprisingly, three were from eastern counties. The only record for Ireland was a bird at Schull (Cork) in Oct 2005.

Genus *ARDEOLA* Boie

Palearctic, Oriental and Ethiopian genus of four to six species, only one of which has been recorded in GB&I.

Squacco Heron   *Ardeola ralloides* (Scopoli)

SM monotypic

TAXONOMY *Ardeola* was not included in osteological (McCracken & Sheldon 1998) or molecular (Sheldon *et al.* 2000) phylogenetic reconstructions, so its taxonomic position has not been recently reassessed. Generally regarded as close to Cattle Egret, and these placed in the same genus in the past.

DISTRIBUTION Breeds across S Europe and N Morocco, east to Aral Sea and south to Israel, Iraq marshes and SE Iran. Migratory, wintering through sub-Saharan Africa, apart from the rainforests and SW Cape. There are also small, generally resident populations in Africa, from the wetlands of the west to Kenya and south to the Cape.

STATUS Recorded in every month except Dec, though with a clear peak of occurrence in May–June, and <13% from the autumn. The first for GB was shot near Madron (Cornwall) in spring 1834; the first for Ireland was obtained near Youghal (Cork) in May 1849. All the rare herons have shown increases in frequency of records since 1950, although Squacco rather less than Cattle, Little and Great Egrets. Numbers of Squacco Herons reported by BBRC per decade rose from 4 (1950s) to 28 (1990s), compared with 0 to 61 (Cattle Egret) and 2 to 80 (Great Egret). Still a very rare bird in Ireland, with only 17 individuals to 2004.

Genus *BUBULCUS* Bonaparte

Nearly cosmopolitan monospecific genus that has been recorded in GB&I.

Cattle Egret   *Bubulcus ibis* (Linnaeus)

CB SM *ibis* (Linnaeus)

TAXONOMY Taxonomic position of Cattle Egret has been much disputed, with authorities variously placing it in *Ardeola* and *Egretta*; molecular studies (Sheldon *et al.* 2000) are inconclusive and is now generally placed in its own genus. Two (or three) subspecies recognised, of which *coromandus* is sometimes treated as a species.

DISTRIBUTION Nominate race is widespread throughout Africa, apart from desert regions, also along N Mediterranean from Iberia to Turkey, across suitable habitat in Middle East, and on many islands of Indian Ocean. Replaced by *coromandus* in SE Asia, from Indian subcontinent to Australasia. Most populations resident, though those in the north may move south: e.g. from C France to Iberia and NW Africa. Nominate has colonised the Americas in historical times (see Telfair 1994).

STATUS Nested for the first time in 2008 (see below).

Well over 100 have been recorded in GB, at all times of the year, though with a clear peak in Apr–May, presumably reflecting over-shooting of migrants returning from N Africa. It is surprisingly rare in Scotland and Ireland; the first was recorded in the latter as recently as Mar 1976 and still less than 10 records. Numbers in GB have increased in recent years and multiple arrivals are not unusual: a party of eight was recorded in Hertfordshire in May 1992. This paled into insignificance in 2007–08, when a major influx occurred, especially in SW England. A party of 18 and several smaller groups were found In Cornwall, Devon and Dorset. Perhaps unsurprisingly (in view of the Little Egret scenario), some remained into the spring, and breeding was confirmed at two sites in Somerset and possibly elsewhere.

Genus *EGRETTA* Forster

Cosmopolitan genus of *c.* 12 species, one of which breeds in GB&I and two have occurred as rare vagrants.

Little Blue Heron   *Egretta caerulea* (Linnaeus)

SM monotypic

TAXONOMY Molecular analysis of cyt-b from some American herons (Sheldon *et al.* 2000) showed Little Blue to be sister to Tricolored *E. tricolour*, in a clade that also included Snowy Egret *E. thula* and the S American Whistling Heron *Syrigma sibilatrix*. These and other ongoing studies suggest that generic limits may need revision. Some geographical variation in plumage but insufficient for subspecific differentiation (Rodgers & Smith 1995).

DISTRIBUTION Breeds along the coasts of southern N America, C America and northern S America. Occurs from Massachusetts in the east and California in the west, through coastal Mexico and the Caribbean into much of northern S America, apart from the Andes. Northern populations migratory and juveniles especially show extensive dispersal away from the breeding sites, with birds sometimes occurring north to Canada and south to Argentina.

STATUS A juvenile was found at Letterfrack (Co. Galway) in Sep–Oct 2008 (subject to acceptance).

Snowy Egret   *Egretta thula* (Molina)

SM race undetermined

TAXONOMY Osteological and molecular analyses (McCracken & Sheldon 1998; Sheldon *et al.* 2000) only included three N American species of *Egretta* but indicate that these form a clade, sister to the S American Whistling Heron *Syrigma sibilatrix*. Two races of Snowy are recognised, but these are poorly differentiated (Parsons & Master 2000).

DISTRIBUTION The nominate race breeds from the NE USA to E Mexico, and south through the Caribbean to Chile, Argentina. Northern populations migratory, wintering from the Gulf coast to Central America. Localised race *brewsteri* W of the Rocky Mountains in the USA.

STATUS The only record for GB&I is an individual that was found at Balvicar (Argyll) in Nov 2001, subsequently wandering around SW Scotland until last seen in Sep 2002 in Dumfries and Galloway.

Little Egret   *Egretta garzetta* (Linnaeus)

RB PM *garzetta* (Linnaeus)

TAXONOMY Relationships among various small egrets still debated, reflecting problems of assessing closely related allopatric/parapatric taxa (Hafner *et al.* 2002). Pacific Reef Egret *E. sacra* is clearly distinct, being structurally different from and sympatric with Little Egret in the Far East. More complex issues relate to the various taxa of Little (*garzetta, nigripes, immaculata*), 'Dimorphic' (*dimorpha*) and 'Western Reef' (*gularis, schistacea*). There is morphological and plumage variation, and the taxa fall into three groups that are generally given specific rank; a molecular analysis would be valuable to confirm the phylogeny of this group.

DISTRIBUTION Widely distributed on wetlands across southern Eurasia through to Australasia; also Africa, apart from the rainforests and deserts. Nominate: Palearctic, including NW Africa, through to India (largely inland) and SE Asia; *nigripes*: islands of SE Asia and SW Pacific; *immaculata*: Australasia; replaced by (and sometimes sympatric with, *E. g. gularis*: coasts of W Africa, S of Mauritania and *E. g. schistacea*: coasts of E Africa, Arabia to SE India; and by *E. dimorpha*: Madagascar and adjacent coasts of continental Africa.

STATUS Just one recorded in Ireland before 1950, none in Scotland and less than a dozen in England and Wales. From the 1950s onwards, Little Egrets extended their range northwards in both France and Iberia, and birds were increasingly recorded in S England. Initially, small numbers occurred in spring, but from 1989, parties began to arrive at the end of the breeding season; autumn roosts at S coast estuaries grew every year, with increasing numbers staying through the winter. Breeding was confirmed in Dorset in 1996, and in subsequent years more colonies were established until, by 2002, breeding occurred in at least 11 counties in England and Wales (Brown & Grice 2005). The increase has continued and, by 2006, a minimum of *c.* 60 colonies totalling *c.* 450 pairs (RBBP). The first breeding occurred in Ireland in 1996, when 12 pairs bred at a site that held only a single summering individual the year before (Smiddy & Duffy 1997); in 2006, it bred in at least five counties.

At the end of the 20th century, the winter total reached 1,650 (Sep 1999), with over 800 wintering birds still present in Jan 2000 (Musgrove 2002); peak in Sep 2003 could have been as high as 4,400 birds (Collier *et al.* 2005). Although Little Egrets can be hit hard by severe winters, it seems likely that this pattern of post-breeding dispersal, overwintering and colonisation will continue, at least in the milder parts of GB&I.

Genus *ARDEA* Linnaeus

Cosmopolitan genus of about 10–14 species, three of which have been recorded in GB&I, one as a common breeding bird.

Great Egret   *Ardea alba* Linnaeus

PM *alba* (Linnaeus)

TAXONOMY Variously placed in *Egretta* and *Casmerodius*, molecular data (Sheldon *et al.* 2000) support osteological analysis (McCracken & Sheldon 1998) in revealing a closer affinity with *Ardea*, although position relative to *Bubulcus* remains unresolved. Four subspecies recognised, largely on size and colour of bill, legs; *modesta* sometimes treated as a separate species.

DISTRIBUTION Across all continents, apart from Antarctica, south of *c.* 50°N. Nominate: Europe to C Asia and across Middle East to Iran; *modesta*: Japan, Korea to India and through SE Asia to

Australasia; *melanorhynchus*: wetland areas of sub-Saharan Africa; *egretta*: N & S America, from Great Lakes to C Argentina. Northern populations are migratory to avoid winter ice.

STATUS At one time, a seriously rare bird, with less than a dozen in the 150 years following the first at Hornsea mere (Yorkshire) in the winter of 1825. Only two recorded during 1950–73; numbers have risen to >25 per year since 2000. Recorded in every month; there is no obvious peak, though records are fewer in Jan–Mar. Not recorded in Ireland until a bird was present at Moneygold (Sligo) in May–June 1984. Since then, there have been a further 13 records, scattered across the island. The Netherlands population is increasing rapidly, and Dutch colour-ringed birds have been seen here on several occasions. Colonisation just seems a matter of time.

Grey Heron   *Ardea cinerea* Linnaeus

RB WM *cinerea* Linnaeus

TAXONOMY Not included in recent molecular and osteological analyses but closely related to Great Blue Heron *A. herodias* and Cocoi Heron *A. cocoi*. Four subspecies recognised on basis of leg length and plumage, but much variation is clinal (Hancock & Kushlan 1984).

DISTRIBUTION Breeds across Eurasia, south of *c.* 64°N, into SE Asia and Africa (apart from desert areas). Nominate: occurs through most of range, being replaced by *jouyi* in Far East, from Japan to Java, *firasa* in Madagascar and *monicae* on islands off Banc d'Arguin, Mauritania. Most northerly populations are migratory, also many birds across C Asia.

STATUS Being a large, colonial bird at the top of the freshwater aquatic food chain, Grey Herons have attracted continuous interest down the years. This was one of the first birds to be surveyed nationally, and censuses have continued at regular intervals since 1928. The pattern that emerges is one of increase from *c.* 8,000 pairs in 1928 to an estimate of >14,000 in 2003. The increase is not, however, continuous; harsh winters can take a heavy toll of birds: the population dipped markedly in the years following 1947–48 and 1962–63 and, to a lesser extent, 1985–86. The increase is due both to increases at existing colonies and the establishment of new ones. The Second Breeding Atlas shows that Grey Heron occurs almost everywhere in GB&I, apart from the most mountainous regions, but that the breeding density is highest along major rivers and at coastal sites, especially in Scotland and Ireland. Marquiss (in Wernham *et al.* 2002) speculates that a run of milder winters has allowed birds to survive the winter at sites which were previously sub-optimal, thus boosting the breeding population. Certainly, the mean laying date has advanced markedly; Grey Herons now breed a month earlier than in the late 1960s. Birds raised in GB&I show two peaks of mortality: one shortly after fledging, when young, inexperienced birds are especially susceptible, and during Dec–Feb, when competition for feeding sites becomes intense.

The population increase has taken place despite persecution at some fish farms, where lack of simple protective measures allows the birds to take quantities of young trout and salmon (Carss & Marquiss 1992). At one or two sites, a condition akin to 'brittle bones' affects nestlings, leading the long bones to fracture and causing crippling deformity and death. Circumstantial evidence points to environmental pollutants as a possible cause, but research into this is continuing.

Ringing data show that many Grey Herons come to GB&I each winter, especially from S Scandinavia and the Netherlands; Norwegian birds winter mainly in Scotland and Ireland; Netherlands birds in England. GB&I Grey Herons move between colonies, with many recoveries of birds ringed as nestlings at one site found elsewhere in a subsequent breeding season. Ringing recoveries outside GB&I indicate the movement of some birds south, to Iberia and even as far as W Africa.

Great Blue Heron   *Ardea herodias* Linnaeus

SM race undetermined.

TAXONOMY Molecular analysis of a limited number of American herons (Sheldon *et al.* 2000) showed *Ardea* (Great Blue Heron) to be sister to Great Egret and in a clade that also included Cattle Egret. This parallels results from DNA–DNA hybridisation. Polytypic, with five races recognised by HBW.

DISTRIBUTION Breeds through much of C and S North America, from Canadian prairies to Caribbean and just into northernmost S America; also along Pacific coast of Canada to S Alaska. Northern populations partially migrant, those further south resident or dispersive in hard weather.

STATUS One in the Scillies (Cornwall) in Dec 2007 was the first for GB&I.

Purple Heron   *Ardea purpurea* Linnaeus

PM *purpurea* Linnaeus

TAXONOMY Three subspecies recognised, largely on basis of neck and gular streaking (Hancock & Kushlan 1984): *purpurea, madagascariensis, manilensis.* The critically endangered *bournei* (Cape Verde Islands) is sometimes treated as a separate species (e.g. Hazevoet 1995).

DISTRIBUTION Three centres of distribution in W Eurasia & Africa, Madagascar and SE Asia. W Eurasian *purpurea* breeds across Europe and into adjacent parts of NW Africa including Cape Verde, south of c. 53°N, and into Asia to Kazakhstan and Iran; these populations largely migratory, wintering W Africa, S of Sahara to Red Sea, also Nile Valley, Arabian Gulf. Africa populations, also *purpurea*, largely resident in East Africa. Replaced in SE Asia by *manilensis*; again, northern birds are migratory, wintering south to Malaysia. Madagascar population (*madagascariensis*) believed resident (Hancock & Kushlan 1984).

STATUS The commonest of the rare herons, with over 450 records to 1982, when BBRC stopped collecting data. Since then, there have been *c.* 20 birds each year (Fraser & Rogers 2006), though it remains rare in Ireland, with only 18 individuals. Recorded in every month, with a peak in Apr–June and a lesser one in Aug–Sep, corresponding with migration times. Many late-summer records involve juvenile birds undergoing post-fledging dispersal. Adults have been reported summering in suitable habitat (e.g. Minsmere, Suffolk) on several occasions, but breeding has not yet been confirmed.

FAMILY CICONIIDAE

Genus *CICONIA* Brisson

Near-cosmopolitan (not Nearctic, Australasia) genus of seven species, two of which have been recorded in GB&I.

Black Stork   *Ciconia nigra* (Linnaeus)

SM monotypic

DISTRIBUTION Breeds from Iberia (partially resident), across Eurasia from France to the Yellow Sea, between 61°N and 30°N, with isolated populations in Iran and southern Africa. Most northern hemisphere populations are migratory, wintering in tropical Africa, N of Indian subcontinent and S China. Apparently avoids sea crossings less than White Stork.

STATUS Recorded over 150 times in GB, since the first on West Sedgemoor (Somerset) in May 1814. A rare bird until the 1970s, becoming more common, with *c.* 10 each year in the early 1990s, though now fewer. Reported from Mar to Nov, with most records Apr–Sep, and generally in the S & E of England. Many individuals remain for only one day, but some stay longer or wander round the country: e.g. a first-summer bird in Aberdeenshire, Northumberland and Suffolk between July and Sep 1998. There is only a single Irish record: a bird at Foxrock (Dublin) in Aug 1987.

White Stork   *Ciconia ciconia* (Linnaeus)

FB PM *ciconia* (Linnaeus)

TAXONOMY Closely related to Oriental Stork *C. boyciana*, with which it was long considered conspecific. Now treated separately (Vaurie 1965; HBW) on basis of differences in size, plumage, bill structure and colour, and behaviour. Two subspecies recognised on basis of size (especially of bill): *ciconia* and *asiatica*.

DISTRIBUTION Breeds in W Eurasia, from NW Africa N to Estonia and E to Turkey; then discontinuously E to Lake Balkash. Migratory, wintering in sub-Saharan Africa, except rainforests and deserts of SW; also Indian subcontinent.

STATUS A pair nested on St Giles Cathedral, Edinburgh (Lothian), in 1416 (though see Bourne 2008). Hundreds of records down the years, but interpreting these is bedevilled by a combination of multiple sightings of mobile birds and free-flying individuals of captive origin. However, wild birds occur regularly, including ringed individuals from natural and reintroduced populations on the near-Continent. Recorded in every month of the year but with maximum numbers in Mar–Oct; much less common in Ireland (31 individuals), as would be expected from a species so averse to crossing open sea.

Some individuals remain for prolonged periods, leading to the hope that further breeding will occur. In 2004, a pair was found in W Yorkshire, displaying and carrying nest material onto an electricity pylon. Remarkably, both were ringed: the female was a wild bird that had been taken into care in S France during winter 2002–03 and the male had escaped from a wildlife park in Belgium. The nest site was regarded as inappropriate by the local electricity board, which erected an alternative platform and transferred the nest material. Unfortunately, this appeared not to the birds' liking and they moved on after a few days.

FAMILY THRESKIORNITHIDAE

Genus *PLEGADIS* Kaup

Old World and Australasian genus of three or four species, one of which is a rare visitor to GB&I.

Glossy Ibis   *Plegadis falcinellus* (Linnaeus)

SM *falcinellus* (Linnaeus)

TAXONOMY Closely related to Puna and White-faced Ibises *P. ridgwayi* and *P. chihi.*

DISTRIBUTION Nominate race widely distributed, breeding along the Atlantic coast of N America S to Venezuela, NW Africa and across Eurasia from Iberia to Kazakhstan, S of *c.* 50°N; also populations through sub-Saharan Africa, south C Asia through Indo-China to Australia. Latter ascribed to race *peregrinus.* European populations disperse widely after breeding, before migrating to sub-Saharan Africa. Eastern populations more dispersive than migratory, probably related to rainfall (HBW).

STATUS The first in GB were two near Reading (Berkshire) in Sep 1793; and for Ireland, 'several' in Wexford in summer 1818. There are now over 400 reports, but determining totals is difficult since many involve more than one individual, e.g. eight at Budleigh Salterton (Devon) Sep 2002. Occurred much more frequently in the past (<1950). Seventeen together around Slimbrige (Gloucestershire) in Apr–May 2007 were part of an influx of at least 36 birds into SW GB&I. Some individuals remain for extended periods, such as one to two birds that wandered around Kent, Norfolk, Suffolk and Essex from 1975 to 1992. There are records for every month of the year, with a minor peak in May and a major one in Aug–Nov. Autumn birds are earlier in GB, with >55% of reports in Aug–Sep, compared with only <25% of Irish; perhaps suggesting an arrival from the east. However, a bird ringed as a nestling in Spain in 2006 was seen in Lincolnshire in Feb 2008, having arrived via Wexford in Jan.

Genus *PLATALEA* Linnaeus

Palearctic, Oriental, Ethiopian and Australasian genus of five or six species, one of which has been recorded in GB&I.

Eurasian Spoonbill   *Platalea leucorodia* Linnaeus

CB FB PM *leucorodia* Linnaeus

TAXONOMY Little studied recently; believed to be closely related to Black-faced *P. minor*, African *P. alba* and Royal *P. regia*. Three subspecies are recognised (Vaurie 1965; HBW), but two are very restricted in range.

DISTRIBUTION Nominate breeds at scattered sites in Europe, from Iberia to the Netherlands, and then E to Manchuria, Korea and S to Persian Gulf and Indian subcontinent; northern populations are migratory, wintering in wetlands of N Africa, Nile Valley and coastal wetlands of Middle East. Indian populations largely resident. Subspecies *balsaci* breeds in islands off Mauritania and *archeri* along the coasts of Somalia and Red Sea.

STATUS Formerly bred in East Anglia, probably until 17th century (Brown & Grice 2005), but seemingly extinct as a breeding bird until the 1990s. In 1999, two young were raised in NW England, and nest platforms have been built at several sites and eggs occasionally laid. Colonisation seems imminent, but establishment has not yet occurred. Has never bred in Ireland, where it is a rare visitor. Occurs widely as a non-breeding bird, with spring and summer birds most frequent in the east and small numbers of wintering birds primarily in the south-west. A party of 10 juveniles in County Mayo in Oct 2005 included two colour-ringed birds from the Netherlands.

FAMILY PODICIPEDIDAE

Genus *PODILYMBUS* Lesson

New World genus of two species, one of which has been recorded as a vagrant in GB&I.

Pied-billed Grebe   *Podilymbus podiceps* (Linnaeus)

HB SM race undetermined but unlikly to have been other than *podiceps* (Linnaeus)

TAXONOMY A Nearctic species, apparently closely related to the Atitlán Grebe *P. gigas* of Guatemala, which is now probably extinct. Three subspecies recognised: *antarcticus* of S America, *antillarum* of the Gtr and Lsr Antilles and nominate from N America.

DISTRIBUTION Widespread across the Americas, from C Canada to Tierra del Fuego, though absent from inland S America, east of the Andes. Nominate breeds from Manitoba to Panama and is mostly sedentary. However, birds from Canada and N USA migrate S to winter as far south as California and Central America.

STATUS An adult, present at Stithians Res (Cornwall) between Apr and Sep 1993, was paired with a Little Grebe and hatched three young.

The first record was at Blagdon Lake (Somerset) in Dec 1963; this remained in the area for several years, certainly until June 1968. Up to 2003, there had been about 37 records in GB, ranging from Caithness and the Hebrides to Scilly. The actual total is a little imprecise; because of the longevity of some individuals, there is a possibility of duplication of some records. First Irish bird was at Lady's Island Lake (Wexford) in May 1987; a further five records to 2000.

Birds have been newly found in all months, apart from July and Sep, with peaks in Jan, Apr and Nov. Two of these coincide with times of migration; the midwinter peak may be related to hard-weather movements along the Atlantic seaboard of N America.

Genus *TACHYBAPTUS* Reichenbach

Near-cosmopolitan genus of five species, one of which breeds commonly in GB&I.

Little Grebe   *Tachybaptus ruficollis* (Pallas)

RB MB HB WM *ruficollis* (Pallas)

TAXONOMY This and Least Grebe *T. dominicus* are the most variable members of a cosmopolitan genus of small grebes, distributed across most major land-masses. Isolated forms from Madagascar (two) andAustralasia have been given specific rank. There are up to nine subspecies of *T. ruficollis* (Vaurie 1965; HBW); a comprehensive phylogenetic analysis is overdue. Races in the W Palearctic include *ruficollis*, *iraqensis* and *capensis*, differing in size and wing colouration.

DISTRIBUTION S and C Europe, from GB&I to Aralo-Caspian, all of Africa (except arid Saharan zones), India, Indo-China, Malay Peninsula, coastal China to Korea. Nominate race breeds NW Africa and across Europe to the Urals (Vaurie 1965); *iraqensis* breeds Iraq, Iran; *capensis* breeds C and S Africa, Caucasus to Burma. Populations generally sedentary, except more northerly birds that migrate to avoid freezing conditions in winter.

STATUS See Pied-billed Grebe in relation to hybrid pairing with that species. The breeding atlases showed Little Grebe to be a widespread species across lowland GB&I, being absent only from sites that lack standing water with emergent vegetation. There was evidence of a decline between the two surveys, especially in Ireland (*c.* 35%); however, this could be partly due to differences in methodology. This species is covered by three different regular surveys, although these give conflicting results. The Waterways Breeding Bird Survey, which concentrates on linear habitats such as streams and canals, shows signs of an irregular decline of *c.* 75% since the late 1970s. BBS data does not confirm this, indicating an increase of *c.* 25% since the mid-1990s. Both are based on relatively small data sets. The WeBS counts are more robust and show that numbers have increased steadily since 1990. Through the year, numbers peak in Sep–Oct in GB (Dec in NI). While some of this may be due to local dispersal of breeding birds from small uncounted sites to larger sites that are included in the WeBS survey, there may also be immigration into GB&I from the Continent. This is supported by records of birds from Fair Isle, where almost all records are of birds in Mar–Apr and Sep–Nov (Dymond 1991). Little Grebes have declined along rivers and canals, perhaps due to a combination of Mink and the loss of waterside vegetation.

Kershaw & Cranswick (2003) estimated a British winter population of 7,770 birds, whilst Crowe *et al.* (2008) estimated an Irish winter population of 2,345.

Genus *PODICEPS* Latham

Near-cosmopolitan genus of nine species, of which four have been recorded in GB&I, where all breed, one commonly, three less so.

Great Crested Grebe   *Podiceps cristatus* (Linnaeus)

RB WM *cristatus* (Linnaeus)

TAXONOMY Great Crested, Red-necked and Slavonian Grebes are believed to be closely related, on the basis of morphology and behaviour, and are sympatric across large areas of the Palearctic (see Vlug 2002). Great Crested Grebe is treated as three subspecies (*cristatus, infuscatus, australis*); the two southern hemisphere forms differing in a tendency to retain breeding plumage throughout the year (HBW).

DISTRIBUTION Occurs across the Palearctic, in S and E Africa and Australasia. Nominate breeds from Ireland to Lake Baikal and N China, south to Mediterranean, Iran to N India; remains in breeding areas where these are ice-free, elsewhere migrates S to winter in coastal waters; *infuscatus* breeds in E and S Africa; *australis* in Australasia.

STATUS A bird of larger waterbodies that, especially in England and Wales, benefited from the creation of reservoirs and gravel pits in the 20th century. Less widespread in Scotland, where it is largely restricted to relatively undisturbed eutrophic lowland lakes; especially common in the wet lowlands of the centre of Ireland. Heavily persecuted for its plumage during the 19th century (Holloway 1996), its recovery has been well documented through a series of surveys since 1932 (see Hughes *et al.* 1979). The 1975 census of GB revealed *c.* 7,000 breeding adults, with a further 1,400 in Ireland. Analysis of count data for the Second Breeding Atlas suggested higher numbers, with 8,000 in GB and 4,100 in Ireland. BBS data since 1994 suggests a further modest increase in GB&NI.

As with Little Grebe, there is evidence of immigration into GB&I during the winter. Birds are recorded in spring and autumn at E coast sites that lack a breeding population. There are also a few ringing recoveries that show exchange with the near-Continent and S Scandinavia; however, data are limited since so few birds have been ringed. WeBS counts outside the breeding season show a steady increase in GB&NI since 1985, with large concentrations at estuaries, major reservoirs and gravel pits of S England, such as Lade Sands (Kent), Rutland Water (Leicestershire), Chew Valley (Somerset). More recent data suggest that numbers in NI may be stabilising; several sites hold in excess of 1,000 birds in the autumn; since the peaks occur in different months, there may be movement among these waterbodies.

Kershaw & Cranswick (2003) estimated a British winter population of 15,900 birds, whilst Crowe *et al.* (2008) estimated an Irish winter population of 5,385.

Red-necked Grebe   *Podiceps grisegena* (Boddaert)

CB WM *grisegena* (Boddaert)

SM *holboellii* Reinhardt

TAXONOMY Two subspecies recognised: nominate and *holboellii*, differing in plumage and size: the latter being larger, with almost no overlap in wing and bill lengths. On the strength of these differences, some authorities (e.g. Bocheński 1994) believe they are better regarded as separate species; there is clearly a need for detailed molecular analysis.

DISTRIBUTION A Holarctic species. The nominate breeds across E Europe, from Finland and SE Scandinavia to Black Sea, and east to Kirgiz Steppes; migrates south to ice-free waters from N Sea to Caspian and Aral Seas (HBW). A second group of populations (*holboellii*) breeds from E Siberia and Manchuria, across Alaska and W Canada to N USA; these winter S along both sides of Pacific and also on Atlantic coast of USA as far south as Florida.

STATUS Individuals have been regularly recorded summering across GB&I, with sporadic (and usually unsuccessful) attempts to breed. A pair nested successful in 2001, hatching two young and rearing one, on a loch in the Scottish Borders.

Winters regularly in small numbers, especially along the E & S coasts of England, and S Scotland. WeBS counts show that the Firth of Forth is particularly important in regional terms, with up to 50 birds recorded in some winters. While these data do not reflect the total numbers, they have the advantage of being repeatable across years and indicate relative stability in the winter population. In some winters, there may be significantly higher numbers as birds cross the North Sea to escape severe weather in NW Europe. Chandler (1981) documents such an event in 1978–79, when over 480 birds were found in GB&I during a prolonged period of strong easterly winds and associated low temperature during Feb. Rarely reported in WeBS surveys from NI and only a few records each year from Ireland as a whole.

There is a single record of the Nearctic race *holboellii* shot at Gruinard Bay (Highland) in Sep 1925 (McGowan 2006).

Horned Grebe/Slavonian Grebe   *Podiceps auritus* (Linnaeus)

RB WM *auritus* (Linnaeus)

TAXONOMY Two or three subspecies recognised: *auritus*, *arcticus* and *cornutus*. Variation in plumage and metrics is slight and statistical. Treated as monotypic by BWP, but populations from Iceland and North America differ in haplotype frequencies (Boulet *et al*. 2005), so their retention seems justifiable.

DISTRIBUTION An Holarctic species. Nominate breeds in a narrow band of the Palearctic from Iceland, Scotland and Scandinavia across Eurasia between 63°N and 50°N to Kamchatka and the Pacific, though Atlantic populations sometimes called *arcticus*. In the Nearctic, *cornutus* has a broader latitudinal range, from Alaska and NW Canada to the Great Lakes. As with other grebes, migrates south to avoid frozen lakes and estuaries, wintering chiefly at sea south of the breeding range, along the Atlantic and Pacific coasts of Eurasia and N America, but also on inland waters such as the Caspian Sea.

STATUS The commonest of the rarer breeding grebes, though still very restricted in breeding distribution and largely confined to Scotland. The main centre remains at Loch Ruthven (Highland) with *c.* 50% of the GB population, but ongoing decline in the number of sites where breeding occurs, falling from 33 in 1993 to 16 in 2002 and 19 in 2006 (RBBP). Productivity at Ruthven remains lower than elsewhere in Europe, possibly due to loss of chicks after they leave the nest, although accidental disturbance and egg-collecting have been implicated at some sites (Second Breeding Atlas). Elsewhere in GB, productivity seems to be even lower, perhaps driving down the number of sites.

Winters around the coasts of GB&I, from Shetland to Cornwall, with larger numbers in Scotland and Ireland than elsewhere. Evidence suggests that birds from Fennoscandia winter in the south, whereas those further north are either local birds or from Iceland, Greenland. WeBS data indicate that numbers are currently buoyant, with increases at several favoured sites, including Lough Foyle (Londonderry/Antrim), Blackwater Estuary (Kent) and Pagham Harbour (W Sussex). As with Red-necked Grebe, numbers may be inflated in severe winters as birds cross the North Sea to avoid harsh conditions on the near-Continent.

Black-necked Grebe   *Podiceps nigricollis* C. L. Brehm

RB or MB, WM PM *nigricollis* C. L. Brehm

TAXONOMY As with Red-necked and Slavonian Grebes, two subspecies that are essentially Nearctic and Palearctic, but also a third from S Africa: *nigricollis*, *californicus*, *gurneyi* respectively. The differences in plumage are, however, only slight (Koop 2003).

DISTRIBUTION Four main centres of distribution: *nigricollis* breeds in two regions, W & C Europe into C Asia, and in E Asia, wintering in SW Europe and SE Asia, respectively; *californicus* breeds in western N America, from SW Canada to California and NW Mexico, wintering in Central America to Guatemala; *gurneyi* is largely resident in southern Africa.

STATUS Breeding distribution shows a pattern of scattered colonisation, increase, decline and extinction; relatively few sites have a prolonged period of occupancy. Breeds at a scatter of locations across GB, with typically less than 10 pairs at a single site. Only occasionally do numbers at any site rise above this; the totals reported are in the range 40–80, depending on coverage (RBBP). Poor productivity has been ascribed to predation (Mink *Mustela vison*, Brown Rats *Rattus norvegicus*, terrapins), but deliberate or accidental disturbance and egg-collecting remain threats.

Because of the overlap between breeding and non-breeding seasons in WeBS surveys, published data from this source sometimes lack site details. Nevertheless, it is clear that the number of wintering birds is lower than for Slavonian and was estimated at *c.* 120 by Lack (1986), who also showed a strong bias in distribution towards the S of GB&I. In 2004–05, over 75% of wintering birds recorded by WeBS were found on just four key sites. Recently there seems to have been a decline in SW England, though whether this will be sustained or is due to random effects of small numbers remains to be confirmed. Very scarce in Ireland.

FAMILY ACCIPITRIDAE

Genus *PERNIS* Cuvier

Palearctic and Oriental genus of at least three species, one of which is a scarce breeder in GB&I.

European Honey Buzzard   *Pernis apivorus* (Linnaeus)

MB PM monotypic

TAXONOMY Two extensive analyses of raptor phylogeny (Lerner & Mindell 2005; Griffiths *et al.* 2007) have shown that the honey buzzards are part of a clade that also includes Egyptian Vulture and Bearded Vulture *Gypaetus barbatus*, although these are not closely related. More limited study of the honey buzzards themselves (Gamauf & Haring 2004) revealed an oriental group more closely similar to each other than to the European, which showed little molecular variation among the individuals screened. There may be a case for realigning species limits within the eastern taxa, but *P. apivorus* is monotypic.

DISTRIBUTION Breeds across Europe, from Iberia and GB through Scandinavia to about the River Ob in W Siberia, and S to Asia Minor and the Caspian Sea. Strongly migratory; W populations pass chiefly through Gibraltar, though a small number use Italy and Sicily; E populations cross Bosphorus or fly round E Black Sea, thence through Turkey to Suez. Adult birds tend to migrate on a narrow front, whereas juveniles are more widespread. Winters in equatorial Africa, possibly as far S as Tanzania.

STATUS A scarce, or even rare, breeding bird whose secretive habits and those of their observers have combined to generate an incomplete picture of its distribution and abundance. Tubbs (in Second Breeding Atlas) suggests that observer caution may be justified, since deliberate and accidental disturbance affected the density of European Honey Buzzards at a site in the Netherlands. The Second Breeding Atlas recorded evidence of breeding in 27 10-km squares, an increase of 16 over the First Breeding Atlas. Tubbs also suggests that, since both surveys were incomplete, these data are of little value except to show a widespread, though fragmentary, distribution RBBP data on breeding European Honey Buzzards is supplemented by occasional special surveys. Some birds may be present without breeding, so observed pairs may not be a good estimator of the actual population. Although often incomplete, the RBBP data have the advantage of being objective and suggest a modest increase (in the areas that make returns) from <10 breeding pairs in the early 1990s to a peak at 33 pairs in

2000; however, this was a year of a species survey so these records may be inflated. Clements (2005) suggested that there are, in fact, many more pairs that go unrecorded, especially in the more remote areas. He reports that >60 sites have held breeding European Honey Buzzards over the last 30 years, with *c.* 30 more sites where breeding has not been proved, and proposes that the lack of earlier records from regions such as Wales and Cumbria (which now hold significant numbers) probably indicates relatively recent colonisation. Nests as far N as highlands of Scotland, though main areas in S of England. The creation of areas of clear fell adjacent to mature woodland in upland forest provides both feeding and breeding areas, from which European Honey Buzzards have benefited in recent years. Gibbons *et al.* (2007) suggest that less than 5% of potential range is presently occupied.

One adult and six young have been tracked by satellite from their breeding site near Inverness (Highland). The adult wintered in Gabon; four juveniles wintered in W Africa from Nigeria to Liberia, and two were lost in the Atlantic Ocean.

Rare in Ireland, with only 32 records and no evidence that it has ever bred.

Genus *MILVUS* Lacépède

Palearctic, Oriental, Ethiopian and Australasian genus of two to five species, two of which have been recorded in GB&I.

Black Kite   *Milvus migrans* (Boddaert)

HB PM *migrans* (Boddaert)

TAXONOMY Traditionally treated as seven subspecies: *migrans, lineatus, formosanus, govinda, affinis, aegyptius, parasitus.* Molecular data (Lerner & Mindell 2005) support morphological analysis in placing *Milvus* as sister to Brahminy and Whistling Kites *Haliastur indus, H. sphenurus.* Johnson *et al.* (2005) investigated the relationships of *Milvus* from across its range (apart from *formosanus*), from Cape Verde to Australia, using three sequences of mt DNA. They found that kites fell into four groups: Eurasian Black (*migrans, lineatus, govinda, affinis*), C African Black (*aegyptius, parasitus*), S African Black (*parasitus*) and Red (*M. milvus*). The African Yellow-billed Kites (*aegyptius, parasitus*) were clearly differentiated from the other Black Kite taxa but comprised two well-supported clades from South Africa/Madagascar and Senegal–Uganda that did not agree with traditional subspecies limits (birds identified as '*parasitus*' occurred in both clades). Furthermore, these clades were not sister groups: *parasitus* from S Africa was sister to Red Kite, albeit with only weak support. Eurasian/Australasian Black Kites also comprised two clades: *migrans* and *affinis/lineatus/govinda*, and the latter in turn comprised two subclades: *affinis* and *lineatus/govinda.* This all implies that Eurasian Black Kites are a distinct evolutionary lineage, quite separate from the African Yellow-billed. However, in view of the paraphyly in the latter, further analysis is required to differentiate the subspecies *aegyptius* and *parasitus.* A case could be made for splitting the Eurasian taxa into western (*affinis*) and eastern (*lineatus, govinda*), but as the number of individuals analysed by Johnson *et al.* was rather limited, further analysis is necessary.

These data also revealed that individuals collected in Cape Verde in the 1920s and identified as *fasciicauda* were genetically indistinguishable from Red Kites. Birds sampled more recently (2002) were genetically indistinguishable from N African Black Kites. There is thus no molecular evidence supporting the specific status of the (historical) Cape Verde Kite. Contemporary birds in Cape Verde appear to have suffered introgressive hybridisation from Black Kites, and they appear to have been replaced by Black Kites in the eastern islands. The continued existence of Cape Verde Kite needs further investigation.

DISTRIBUTION Widespread across the Old World, from *c.* 64°N to S tip of Africa and Australia; absent only from desert regions of Africa and C Asia. European and N Asian populations migratory, wintering in sub-Saharan Africa and SE Asia. Subspecies: *migrans* (NW Africa, Europe, W/C Asia), *lineatus* (E Asia, from Siberia to N India, N China), *formosanus* (Taiwan, Hainan), *govinda* (C, S India,

Malay peninsula, Indo-China), *affinis* (Sulawesi to N, E Australia), *aegyptius* (NE, E Africa), *parasitus* (sub-Saharan Africa, Madagascar).

STATUS One report of mixed breeding pair with Red Kite in N Scotland in 2006, producing two hybrid young (RBBP).

Surprisingly for a large raptor, there seems to have been only a single record from the 19th century: a bird killed at Alnwick (Northumberland) in May 1866. The next was a bird shot at Aberdeen (Aberdeenshire) in Apr 1901. There was then a further gap of 37 years until the third, on Tresco (Scilly) in Sep 1938 – also shot. Records increased through the second half of the 20th century, with annual averages of <1 (1960s), 3 (1970s), 10 (1980s), 15 (1990s) and 11 (since 2000), probably being correlated with increased numbers and awareness of birders. There are only eight records from Ireland (to 2004), since the first near Garryvoe and Ballycotton (Cork) in Apr–May 1980.

There is a strong peak in numbers in Apr–May, with almost 70% of records in this period. Numbers tail off through the summer, with no evidence of a second peak in the autumn.

A bird resembling subspecies *lineatus* ('Black-eared Kite') wintered in Lincolnshire and Norfolk during 2006–07; the record is under consideration.

Red Kite   *Milvus milvus* (Linnaeus)

RB NB WM PM *milvus* (Linnaeus)

TAXONOMY Molecular analysis (see Black Kite) reveals a close relationship with (historical) 'Cape Verde Kite' *fasciicauda*, which has sometimes been treated as separate species.

DISTRIBUTION Nominate breeds from Iberia and parts of adjacent NW Africa, E to Italy and N to GB, S Scandinavia. In C Europe, breeds to Caucasus and Ukraine. Populations in N & C Europe are migratory, wintering in S France and Iberia. However, increasingly birds are remaining on or close to breeding grounds, especially in S Sweden. Race *fasciicauda* restricted to Vape Verde Is.

STATUS Historically distributed throughout GB and probably also in Ireland. Its decline and ultimate demise in England and Scotland began during the 18th and 19th centuries (Holloway 1996). Persecution, poisoning, cold winters and (in Ireland) felling the woodlands played a significant part, and by the 1870s it had been lost from England and the last birds bred in Scotland in 1898. Only in Wales did it continue to breed, albeit in ever-decreasing numbers, probably reaching its nadir in the 1930s, when Davis (1993) could find evidence of only 14 broods reared during the decade. This population faced threats from egg-collectors, predation and persecution by shepherds and gamekeepers but survived through a combination of secrecy and the protection of a small group of dedicated conservationists. At this time, it was restricted to unproductive land at the heads of the valleys that dissect the Cambrian mountains, feeding on carrion from the sheep hills of mid-Wales.

The protection afforded through the second half of the 20th century saw a slow recovery of the population, hampered by low fertility and low survival of young that could be ascribed to a combination of poor climate, unsuitable habitat and inbreeding depression. Mid-Wales is not ideal habitat for Red Kites; it just happened to be the place where they were not exterminated. Recovery accelerated in the mid-1980s as the population expanded into the more productive lowlands, and in 2002 217 pairs were known to have bred, rearing at least 179 young from 128 nests (RBBP). The population has since expanded west into Pembroke, and the first pair crossed the border into England in 2006.

For most of the 20th century, Red Kite was a scarce passage migrant in England, occurring predominantly in the S and E, and mostly in Mar–Apr. In Scotland, it was very rare, with only a handful of records.

During the 1980s, a plan was developed for the reintroduction of Red Kites into England and Scotland, with a success that surpassed all expectations. The initial releases were in the Chiltern Hills (Buckinghamshire/Oxfordshire) and on the Black Isle (Highland). A total of 93 young Red Kites were released in each area during 1989–94, and breeding took place in both regions in 1992 (Evans *et al.* 1997). The source populations were predominantly sedentary birds from Spain (England) and migratory birds from Sweden (Scotland); a few Welsh birds were released into the Chilterns. The Chiltern release succeeded spectacularly, with >300 breeding pairs being located in 2006, although that in Scotland has been less successful, reaching 40 pairs in the same year. Breeding productivity is similar in the two areas, but in Scotland post-fledging survival is much reduced by illegal poisoning on adjacent grouse moors (RBBP). Poisoning is also the major cause of death of free-flying Red Kites in England (Carter & Grice 2000), but levels are lower than in N Scotland, probably because most releases have been away from grouse moors and gamekeepers understand better that Red Kites are not a threat to healthy pheasants and partridges.

Further reintroductions were undertaken, using birds from Germany and Spain, as well as England and Scotland. These were in Northamptonshire (releases in 1995–98; 74 pairs by 2006); two sites in N England: Yorkshire (1999–03; 40 pairs by 2006) and Durham (2004–06; 5 pairs by 2006); three additional Scottish sites: Stirlingshire (1996–01; 28 pairs by 2006), Dumfries/Galloway (2001–05; 17 pairs by 2006) and Aberdeenshire (2007). Releases are also planned for Ireland, where records have increased in recent years and tagged birds are regularly seen. A pair nested unsuccessfully in County Antrim in 2002. Large areas of GB&I provide suitable habitat and Gibbons *et al.* (2007) suggest that only *c.* 5% of potential range is occupied. Colonisation of these vacant areas is, at least partly, dependent on sympathetic treatment by landowners, and continued reports of poisoned birds is a cause for concern to the conservation movement.

Genus *HALIAEETUS* Savigny

Holarctic, Oriental, Ethiopian and Australasian genus of eight species, two of which have been recorded in GB&I.

White-tailed Eagle   *Haliaeetus albicilla* (Linnaeus)

NB FB PM monotypic

TAXONOMY Palearctic sister taxon to Bald Eagle *(q.v.)*. Larger birds of resident population in Greenland sometimes treated as separate subspecies *groenlandicus*. A molecular study (Seibold & Helbig 1996) found two distinct haplogroups with a predominantly eastern and western distribution and extensive overlap in Europe.

DISTRIBUTION Breeds from S Greenland, W Iceland, across Palearctic from Scandinavia and Baltic States to Bering Straits (once in Alaska), and S to about latitude of Lake Balkash at 35°N. Most populations migratory (not Greenland, Iceland, Norway), wintering C Europe, through coastal Middle East to Pakistan; also coastal China, Japan.

STATUS In the Middle Ages, White-tailed Eagles were widespread through GB&I but suffered extensive persecution and habitat loss. By the 19th century, they were still distributed as a breeding bird from Shetland and Orkney through the Hebrides to SW Scotland and around the coasts of Ireland, especially in the NW (Love 1983, 2003). Habitat loss and persecution led to extinction in Ireland by 1898 and Scotland by 1916. Changes in attitude to birds of prey through the 20th century led to hopes of reintroduction. The first attempts in Glen Etive (Argyll) in 1959 and on Fair Isle (Shetland) in 1968 were both unsuccessful, in that they did not result in establishment, although the experience gained was valuable (Love 1983).

The island of Rum (Inner Hebrides) was identified as a more appropriate venue, and a reintroduction programme began there in 1975, when three females were released. Over the next 10 years, nearly a hundred Norwegian birds were released on Rum and about 60 in Wester Ross during 1993–98. In 1985 the first eggs hatched successfully. Since then, the number of breeding pairs (sometimes trios) has increased steadily, so that by 2006 36 pairs bred, of which 17 raised 29 young (RBBP). Although there are occasional instances of egg robbery, most failures are natural and the population is now self-sustaining. A comparison of the demography of released and wild-bred birds (Evans *et al.*

2009) found that survival and breeding success were higher for the latter, suggesting that, in future releases, efforts should be directed towards early breeding attempts to maximise the proportion of wild-bred birds in the population. The Scottish birds are mainly confined to the Inner and Outer Hebrides, though young birds roam widely, usually in the N and W of Scotland. One young bird moved to Norway and subsequently bred there. Further releases have taken place in E Scotland (Fife, from 2007) and Ireland (Killarney, also from 2007). There are extensive areas of potential habitat within the UK, of which less than 5% is currently occupied (Gibbons *et al.* 2007).

In England, many individuals fell to the gun throughout the 19th and early 20th centuries. The first authenticated specimen seems to be in Grantham Museum, shot near Nocton (Lincolnshire) in Jan 1732. There are over 400 records through to 1950, presumably from the Baltic population, since when numbers have declined so that there seem to be only about 35 from that date. Assessing totals is problematic since many individuals remained for days, or even weeks, roaming around and being reported from widely different localities: for example, in 1985, what was presumably the same bird was recorded in Cleveland, N Yorkshire, E Yorkshire and Lincolnshire on 26 and 27 Oct. A bird that had been present in Suffolk since 14 Apr 1984 was found shot at Warham (Norfolk) on 11 May that year; it was taken into care but died the following day. It had been ringed in a nest in Schleswig-Holstein (Germany) in June 1983. There were <10 records from Ireland in the 20th century, with the first (perhaps inevitably) being shot at Clare Island (Mayo) on 27 Nov 1935. White-tailed Eagle has been recorded in every month but, in England and Wales, it is predominantly a winter visitor, with most records being between Oct and Mar.

Bald Eagle   *Haliaeetus leucocephalus* (Linnaeus)

SM race undetermined.

TAXONOMY Lerner & Mindell (2005) found *Haliaeetus* (including the SE Asian fishing-eagles *Ichthyophaga*) to be monophyletic and sister to a large speciose clade that included both *Buteo* and the New World *Buteogallus*. Within *Haliaeetus*, Bald Eagle was sister to White-tailed, in a subclade that also included Steller's Sea Eagle *H. pelagicus* and Pallas's Fish Eagle *H. leucoryphus* (see also Seibold & Helbig 1996). Two subspecies are recognised: *leucocephalus* and *washingtoniensis*, differing in darkness of plumage.

DISTRIBUTION Subspecies *washingtoniensis* breeds from the Aleutian Is, across Alaska and much of subarctic Canada to the N USA; nominate *leucocephala* is more southerly, breeding S to Florida, Baja California. Inland populations of *washingtoniensis* are migratory, wintering in S of range; coastal populations of Alaska, Aleutians less so, wintering along the Pacific seaboard.

STATUS There are two records of juvenile birds from Ireland. The first was shot in Jan 1973 near Garrison (Fermanagh); the second was found exhausted at Ballymacelligot (Kerry) in Nov 1987 and was subsequently returned to the USA. There are no records from GB. A single record from Llyn Coron (Anglesey) in Oct 1978 was placed in Category D (see Appendix 2).

Genus *NEOPHRON* Savigny

Monospecific genus breeding in the Palearctic, Oriental and Ethiopian regions that is a vagrant to GB&I.

Egyptian Vulture   *Neophron percnopterus* (Linnaeus)

SM *percnopterus* (Linnaeus)

TAXONOMY In their study of the molecular phylogeny of Old World vultures, Lerner & Mindell (2005) found that Egyptian Vulture was sister to Bearded Vulture, although genetically divergent, indicating a long period of separate evolution and justifying their placement in separate monotypic genera. In the Western Palearctic, two subspecies are recognised, *percnopterus* and *majorensis*, based on size differences. A third subspecies (based largely on bill colour), *ginginianus*, occurs in Asia. Recognition of subspecies is supported by genetic analyses of mt DNA control region (Donázar *et al.* 2002) and microsatellites (Kretzmann *et al.* 2003).

DISTRIBUTION Nominate race breeds from Iberia and the Balkans, south to the rainforests of W Africa and to Tanzania in E Africa; also across Asia to Lake Balkash and the Himalayas. Subspecies *ginginianus* breeds from Nepal and the Himalayas, south through peninsula India. Populations N of Sahara and Himalayas are migratory, wintering in the southern parts of its range.

STATUS There are two old records of Egyptian Vulture, both shot. The first was at Bridgwater Bay (Somerset) in Oct 1825; the second at Peldon (Essex) in Sep 1868. Birds recorded in Hampshire in 1968 and 1969 were both thought to have been of captive origin. There are no records from Ireland.

Genus *GYPS* Savigny

Old World genus of at least eight species, one of which has been recorded in GB&I.

Griffon Vulture   *Gyps fulvus* (Brisson)

SM race undetermined.

TAXONOMY In view of the conservation concern at the collapse of many populations of vultures, especially in the Indian subcontinent (refs in Johnson *et al.* 2006), a study of the phylogenetics of *Gyps* was undertaken as a matter of urgency. Johnson *et al.* (2006) provided strong evidence in support of splitting Long-billed Vulture into *G. tenuirostris* and *G. indicus*. Species limits within Griffon Vulture, which is traditionally treated as two subspecies, also appeared to need revision. The European and C Asian race *fulvus* was closer to Rüppell's *G. rueppellii* and the race from the Indian subcontinent *fulvescens* closer to Himalayan *G. himalayensis*. They stress, however, that these findings are preliminary and warrant further study.

DISTRIBUTION Taxon *fulvus* breeds in scattered populations from Iberia and NW Africa across S Europe and the Balkans, through Turkey and the Middle East to the Altai; *fulvescens* breeds from Afghanistan through the N of the Indian subcontinent to Assam. Most adults are fairly sedentary, with juveniles dispersing from Europe into the arid regions of sub-Saharan Africa.

STATUS In view of their generally sedentary nature and aversion to sea crossings (see e.g. Bildstein *et al.* 2009), records of vultures in GB&I have historically been viewed unfavourably. The only accepted record is of one obtained in Cork harbour in spring 1843.

Genus *CIRCAETUS* Vieillot

Palearctic, African and Oriental genus of six species, one of which is a vagrant to GB&I.

Short-toed Snake Eagle/Short-toed Eagle   *Circaetus gallicus* (J. F. Gmelin)

SM monotypic

TAXONOMY A member of a clade of snake eagles and serpent eagles, whose substructure more or less matches conventional taxonomy, except that Congo Serpent Eagle *Dryotriorchis spectabilis* is nested within *Circaetus* (Lerner & Mindell 2005).

DISTRIBUTION SW Europe, through Alps, Italy to Balkans, Asia Minor east to Lakes Balkash and Baikal; also NW Africa. Isolated populations in Indian subcontinent and Lombok to Timor (HBW). W populations migratory, wintering across Africa in Sahel.

STATUS The only record for GB&I is of a juvenile present on several islands of the Scilly group in Oct 1999.

Genus *CIRCUS* Lacépède

Cosmopolitan genus of 9–13 species, four of which have been recorded in GB&I, one as a vagrant.

Western Marsh Harrier   *Circus aeruginosus* (Linnaeus)

RB MB PM *aeruginosus* (Linnaeus)

TAXONOMY The only harriers included in the study by Lerner & Mindell (2005) were Western Marsh Harrier, and African Marsh Harrier *C. ranivorus*; perhaps unsurprisingly, these proved to be extremely close. Western Marsh is now generally separated from Eastern Marsh *C. spilonotus*, though Fefelov (2001) found only slight differences. Two subspecies recognised, *aeruginosus* and *harterti*, based on colour of underparts. A preliminary analysis in Simmons & Simmons (2000) found the various forms of 'Marsh Harrier' to be monophyletic, though more detailed analysis is needed to supplement the morphological and behavioural data.

DISTRIBUTION Nominate breeds across Palearctic from Atlantic to W of Lake Baikal and from S Finland to the Mediterranean. Westernmost populations are resident, those from elsewhere migratory, wintering across India and the savannahs of sub-Saharan Africa. Race *harterti* in NW Africa. Replaced by taxa in E Asia, E Africa, Australasia and islands of Indian Ocean that are usually now given specific status (HBW).

STATUS Probably once widespread (Holloway 1996), but land drainage for agriculture in the 19th century must have destroyed much habitat and persecution finished the job, so that by the end of that century, it was likely extinct as a breeding bird in GB. A similar fate befell the Irish population, which appears to have gone by the end of WWI. The extinction era in GB lasted only a few years, and breeding was again confirmed in 1911, when a pair bred at Horsey (Norfolk). Numbers built up slowly until the 1950s, when *c.* 15 pairs nested in England and Wales. Only four 10-km squares held breeding birds during the First Breeding Atlas, with a further two 'probables', all in East Anglia. This represented a serious decline, which was ascribed to a combination of (unintentional) disturbance to nest sites by the boating public and poisoning by toxic pesticides.

The period since the First Breeding Atlas has seen a remarkable growth and expansion of the population; evidence of breeding was obtained in 32 10-km squares in the Second Breeding Atlas, an increase of >300%, and Marsh Harrier had recolonised Scotland, Wales and Ireland. Underhill-Day (1998) reviewed the history of the UK population through 1983–95 and found a steady increase, such that, by 1995, 156 females reared *c.* 350 chicks. By 2006, the RBBP recorded 265 females rearing 453 young, though both of these are probably underestimates. This growth has involved a change in nesting habitat. Formerly largely restricted to extensive reedbeds, they have taken to crops, ditches and arable dykes, although 80% were still within 5 km of salt water. Birds are now nesting regularly in NW England, Scotland and on Humberside and, since there is little conflict with game-rearing (Underhill-Day 1998), this expansion should continue. Gibbons *et al.* (2007) report that only *c.* 10% of potential habitat is presently occupied. Ireland has not yet been colonised; birds are now recorded regularly in suitable habitat though there may be a shortage of males (P Smiddy pers. comm.).

Many young Marsh Harriers are ringed. Limited recovery data indicates migration through W France and Iberia into NW Africa; a young bird from Scotland was tracked by satellite to Senegal in 2004. There are a few recoveries of birds ringed overseas, indicating possible exchange between breeding populations. In recent years, more Marsh Harriers (predominantly females) have been wintering in GB&I; with the continued amelioration of winter climate, this trend is likely to continue. Most British birds are still migratory.

Northern Harrier/Hen Harrier   *Circus cyaneus* (Linnaeus)

RB MB HB WM PM *cyaneus* (Linnaeus)
SM *hudsonius* (Linnaeus)

TAXONOMY Nearctic *hudsonius*, sometimes regarded as separate species. Closely related to Cinereous Harrier *C. cinereus*, with which it has been considered conspecific. The analysis in Simmons (2000) suggested that Black Harrier *C. maurus* is sister to Cinereous and in a clade with Pallid, thus separate from Hen Harrier. As with Marsh Harrier complex further investigation is called for.

DISTRIBUTION Holarctic (nominate), breeding from N Iberia, GB&I and Scandinavia, across N & C Europe, Siberia to Kamchatka, and South to Manchuria and Lake Baikal. Nearctic *hudsonicus*, from Alaska, California to Newfoundland and Texas. Northern populations are strongly migratory, less so further south. Palearctic populations winter to Mediterranean, through Iran and N Himalayas to S China and Japan; Nearctic birds to C America.

STATUS See Pallid Harrier in relation to hybrid pairing with that species. The Hen Harrier is one of Britain's most persecuted birds. According to Holloway (1996), it was much more widespread and probably not especially threatened prior to 1830. The development of game-shooting estates led to direct conflict with grouse and pheasant interests, and declines seem to have set in around this date. The combination of ground-nesting and fidelity to the territory made them an easy target, and serious efforts were made to eradicate them, especially from grouse moors across much of the uplands. By 1900, Hen Harriers are believed to have been largely restricted to Orkney, Outer Hebrides and Ireland (First Breeding Atlas). By the time of the First Breeding Atlas, they had returned to large parts of Scotland, especially in the SW, probably following a decline in keepering during the two World Wars; the reduction in moorland management allowed heather to grow tall enough for successful nesting. Increased afforestation of the uplands led to young plantations that were also suitable nesting sites.

   The Second Breeding Atlas revealed the colonisation of I of Man, and a continued spread in Scotland, centred on Orkney, the southern Hebrides and parts of the Highlands. Etheridge (in Second Breeding Atlas) discusses the reasons for this, suggesting that they avoid sheep walk, deer forest, wetlands and coverless high ground. However, at the same time, Hen Harriers had disappeared from other parts of GB&I, especially S Ireland, N & C England and parts of E Scotland. The reasons for these losses differ. In some areas (e.g. Ireland) agricultural practices were changing (O'Flynn 1983) and areas of afforestation were reaching maturity; both of these led to changes in habitat, with a loss of the deep ground cover required for nesting. In Scotland and England, there was systematic persecution on grouse moors; productivity in this habitat was significantly below that on unmanaged upland. Concern for the Northern Harrier's welfare led to special surveys in 1988–89, 1998 and 2004 (Bibby & Etheridge 1993, Sim *et al.* 2001, 2007a). The first two of these revealed little overall change in numbers, a local decline in Orkney and an increase in N Ireland. There was, however, evidence of differences among regions and habitats between these two surveys. In Scotland, numbers increased on grouse moor and decreased in young plantations, probably because of a degradation of the latter habitat as the trees matured. In Ireland, there were no records from grouse moors, but harriers were found breeding in open spaces within mature plantations; in some sites, birds had taken to nesting in the crowns of deformed conifers. The decline in Orkney represented a continuing trend, with numbers decreasing from *c.* 110 breeding females in the late 1970s to 34 in 1998. This was puzzling since there is little persecution in Orkney, and other factors (such as changes in land use or persecution in the winter range) must be at work. The small Welsh population appeared to be stable, but in England almost all of the birds were on grouse moor, where there was evidence of continued persecution. The most recent survey (2004) revealed a marked increase in most regions, probably associated with a move away from moorland. However, in regions where this habitat continued to be used (S & E Scotland, N England) numbers continued to decline.

   There is little doubt that Hen Harrier numbers in many areas are constrained by illegal killing of breeding birds and equally little doubt that, where harriers (and Peregrines) are common, they can adversely affect grouse numbers and hence shooting bags. A major study in N England during 1992–96 (Redpath & Thirgood 1997) showed that eliminating persecution of Hen Harriers resulted in a significant increase in their numbers. On one moor, the number of breeding females rose from 2 to 20 over five years, although this coincided with a peak in a vole cycle and voles are key prey of harriers. Grouse numbers had been in decline on this moor for almost 50 years, but there was still a decrease of 50% in the bag, which could be ascribed to raptor predation. Since most grouse populations fluctuate cyclically, raptor predation might have significant effects during trough years, and estate managers are concerned that stocks might then be driven to uneconomically low levels. It has been recommended that less of grouse moor should be devoted to grassland, as this is the preferred habitat of voles and Meadow Pipits; reducing their populations might limit the appeal of grouse moors to Hen Harriers.

Many Hen Harriers leave the breeding grounds in autumn to spend the winter in the lowlands and coasts of southern GB&I. Ringing recoveries show that some move as far as W France and Iberia, but the majority remain within the region. There are also a few recoveries of birds ringed overseas that indicate immigration to GB&I from Scandinavia and the Low Countries.

One record of *hudsonius,* a juvenile on the Isles of Scilly from Oct 1982 to June 1983.

Pallid Harrier   *Circus macrourus* (Gmelin)

HB SM monotypic

TAXONOMY Relationship with Hen and Montagu's Harriers unclear; has hybridised with both. Structurally and ecologically more akin to latter, but investigation required to disentangle evolutionary history. Molecular data summarised in Simmons (2000) indicates Pallid is sister to Cinereous and Black.

DISTRIBUTION Extensively sympatric with Montagu's, breeding from Ukraine to NW China and south to S Caspian. A bird of the steppes that periodically extends W; has bred Germany, Sweden, Denmark. Migratory, wintering in savannahs of sub-Saharan Africa almost to Cape, Indian subcontinent, S & E China. Migration recorded through Italy, Cyprus, more so in spring.

STATUS The first was present on Fair Isle (Shetland) in Apr–May 1931, when it was shot. There had been 25 records to 2007, including a wintering juvenile in Norfolk that frequented the area around Stiffkey and Warham Greens between Dec 2002 and Mar 2003. Birds have been seen in most months, including a male near Aberfeldy (Perthshire) in May 1993, which was seen displaying to a female Hen Harrier, and (even more remarkably) a second summer male in the Durkindale area (Orkney) in Apr–June 1995 (also Sep and possibly Nov) that bred unsuccessfully with a female Hen Harrier. There are no records from Ireland.

Montagu's Harrier   *Circus pygargus* (Linnaeus)

MB PM monotypic

TAXONOMY Despite superficial morphological similarities, not especially closely related to Pallid Harrier. Molecular analysis reported in Simmons (2000) reveals that *pygargus* is sister to the entire 'Marsh Harrier' clade.

DISTRIBUTION Sympatric with Pallid Harrier in W Asia but extends further west to breed Iberia, NW Africa. Totally migratory, wintering in sympatry with Pallid in savannahs of Africa, India, Sri Lanka.

STATUS In the late 19th century, bred more or less regularly in, or closely adjacent to, Cornwall, Hampshire, Kent and East Anglia, but numbers were low, probably rarely more than *c.* 20 pairs (Holloway 1996). There seemed to be an increase through the mid-20th century, with an estimated 40–50 pairs spread across 22 counties of England and Wales, with a few further pairs in Scotland and Ireland (Second Breeding Atlas). Fieldwork for the First Breeding Atlas showed a scatter of breeding pairs, with small concentrations in Cornwall and around The Wash, but pairs were erratic in their return, so that annual numbers were lower than the maps suggested. A pair bred in Ireland in 1971 but not subsequently, although the species continues to be recorded on a regular basis.

The Second Breeding Atlas painted a gloomier picture; breeding was confirmed in only five 10-km squares, although birds were present in suitable habitat at a further 27. Gibbons *et al.* (2007) report that only about 5% of potential range is occupied, so expansion is possible. The Second Breeding Atlas comments that nesting habitat has changed over the years, with more birds using winter cereals or oilseed rape. Within these, protection has often been possible, increasing nest success to *c.* 85%. This protection has allowed a modest increase, and the RBBP reported an average of *c.* 10 breeding females per year in the decade to 2006 (RBBP). While these data are likely incomplete, they have the advantage of being collected in a fairly consistent manner and indicate a very low, but apparently stable, population. Colour-ringing of nestlings has shown that these return close to the natal site, often being paired with unringed individuals, suggesting that new birds are entering the breeding pool from elsewhere in Europe, limiting the possible effects of inbreeding in an otherwise small population. Most overseas recoveries are from France, although there is a recovery from Senegal at the western end of the species' wintering grounds.

Genus *ACCIPITER* Brisson

Cosmopolitan genus of *c.* 45 species, two of which breed in GB&I.

Northern Goshawk   *Accipiter gentilis* (Linnaeus)

NB FB *gentilis* (Linnaeus)
SM *atricapillus* (Wilson)

TAXONOMY Relationships within *Accipiter* have been little studied using DNA sequences, so phylogenetics are based on morphology; may be closely allied to Henst's (Madagascar) and Meyer's (New Guinea) Goshawks *A. henstii, meyerianus*, but little direct evidence. Interpopulation variation largely clinal and 8–10 subspecies recognised, some of which are poorly differentiated. Those from W Palearctic include *gentilis* (Europe), *arrigonii* (Corsica, Sardinia), *buteoides* (northern Palearctic, from N Sweden to River Lena); at least three more in E Asia, plus *atricapillus* (most of Nearctic range), *laingi* (British Columbia seaboard).

DISTRIBUTION Holarctic. In Nearctic, breeds from C USA north to tree line; in Palearctic, from tree line south to Mediterranean, Asia Minor, Caspian, Lake Baikal and C China. Most northerly populations are migratory to an extent that depends on harshness of winter and cyclical status of prey (e.g. Snowshoe Hare: Squires & Reynolds 1997). Palearctic birds seem to be less dispersive (HBW).

STATUS The Northern Goshawk was an extremely rare breeding bird in GB&I at the end of the 19th century and may even have been extinct. Holloway (1996) could only find five counties where breeding was confirmed during 1875–1900, and it is generally believed that the causes of the decline, on both sides of the Irish Sea, were deforestation and persecution. During the First Breeding Atlas, evidence of breeding was found in 31 10-km squares, but this increased to 91 during the Second Breeding Atlas, with presence being recorded in a further 156. In a review of distribution and abundance published between the Atlases, Marquiss & Newton (1982) comment on the difficulty of confirming breeding and the overenthusiasm of observers in assessing presence. Despite these reservations, there is no doubt that a period of expansion was under way at this time. They identified 22 areas that had been colonised during the 15 years following 1965; most, if not all, of these originated from falconers' birds that either escaped or were deliberately released. In support of this, they note that adjacent European stocks were at a low ebb at this time due to pesticide poisoning; most colonisations were in the west, rather than the east, as would be expected were they naturally established from Europe; the birds were physically larger than adjacent populations; several birds were seen to be carrying jesses, anklets and bells. Where identifiable, they were mostly birds from Fennoscandia, with a few smaller birds from C Europe; consequently, most are likely to have been subspecies *gentilis* (Marquiss 1981). Populations are now well established in Wales, the Scottish Borders, Aberdeenshire, parts of the Midlands and East Anglia and in smaller numbers elsewhere. Population reported as 272 confirmed pairs in 2006, but this is definitely a big underestimate as almost entirely from monitored study areas (RBBP). Gibbons *et al.* (2007) report that only about 15% of potential range is currently occupied in the UK. Further expansion is possible, provided that colonisation is tolerated. In Ireland, believed to be a former breeder but disappeared due to deforestation (and persecution?), probably during the 17th and 18th centuries (P Smiddy pers. comm.). In the 20th century only a rare visitor but signs of increase in recent times, with evidence of breeding in NI since mid-1990s.

Breeding is usually in large areas of mature woodland; a wide range of trees may be adopted, probably related to availability. Most birds are ringed as nestlings, and Wernham *et al.* (2002) state that reports of recoveries are lower than for raptors of similar size, suggesting illegal killing and failure to report the ring. Most recoveries were in the area where the bird was ringed, with relatively few moving between these, although a bird from S Wales was found in Ireland. There is only one overseas bird that has been recorded in GB: one from Norway that was retrapped in Lincolnshire; there are no recoveries from the Netherlands, despite many birds being ringed there, supporting the view that colonisation from the Continent is unlikely.

There are several records of the Nearctic race *atricapillus* from Ireland (Hutchinson 1989), of which the first was shot in County Tipperary in Feb 1870. The only record from GB was a bird on Tresco (Scilly) on 28 Dec 1935.

Eurasian Sparrowhawk   *Accipiter nisus* (Linnaeus)

RB WM PM *nisus* (Linnaeus)

TAXONOMY The relationships within *Accipiter* have not been investigated using DNA, so resolving the lineages within some species' groups is difficult. Eurasian is close to the African Rufous-breasted *A. rufiventris*, and possibly Sharp-shinned Hawk *A. striatus*, but phylogenetic resolution is incomplete. Shows interpopulation variation in size, colour and barring of males. Six subspecies recognised, four from the W Palearctic: *nisus* (Europe to W Siberia), *wolterstorffi* (Corsica, Sardinia), *granti* (Maderia, Canaries), *punicus* (NW Africa).

DISTRIBUTION Nominate breeds in wooded areas across the Palearctic, from the treeline south to Mediterranean, Caspian. Northern populations migratory, wintering in W Europe, Red Sea and Nile Valley. Eastern populations winter south of Himalayas to Sri Lanka, Indo-China.

STATUS Consistently persecuted down the years and one of the species hardest hit by organochlorines in the 1950s and 1960s. These were taken at low levels by insectivorous birds and accumulated through the food chain to accumulate in Eurasian Sparrowhawks, either killing them directly or severely affecting their reproductive performance (Prestt 1965, Ratcliffe 1970, Newton 1974). Many parts of GB&I, especially in lowland arable areas, lost almost all of their breeding birds and Eurasian Sparrowhawks became rare. Bans and restrictions on the use of these toxic chemicals were introduced in the 1960s and, by the time of the First Breeding Atlas, recovery was under way in Ireland, and even in C & E England numbers were beginning to increase. The fieldwork for the Second Breeding Atlas confirmed that the recovery was almost complete; Eurasian Sparrowhawks were found in almost every 10-km square in England and Wales and most of the wooded areas of Scotland and Ireland. Newton (in Second Breeding Atlas) suggested that it might have been commoner in GB&I than at any time in the 20th century, because of the opportunities for nesting provided by extensive areas of afforestation in the uplands.

Before the mid-1970s, Eurasian Sparrowhawks were too rare to register in the CBC, but as they recovered, birds were increasingly recorded in both this and the subsequent BBS surveys. There was a steady increase in numbers to the early 1990s, at which time the population reached a plateau. Much of this increase can be ascribed to improved breeding success; failures at the egg stage declined and brood sizes increased, presumably as the declining pesticide residues reduced the adverse effects on egg viability. Since 1990, the population has shown signs of a modest decline, as has brood size. Whether this is a temporary abberation or indication of something more serious, such as a decline in woodland passerines, will need to be monitored closely.

Ringing recoveries indicate that movement between natal and subsequent breeding sites is modest, with females dispersing more (median = 13 km) than males (6 km). There are few overseas recoveries, indicating that the population of GB&I is resident. There is, however, a substantial immigration of birds into GB&I, especially from Scandinavia. Newton (in Wernham *et al.* 2002) points out that the pesticides that so affected birds in GB&I would likely have had similar effects on Scandinavian birds wintering in these islands. They do not breed in Shetland, and Pennington *et al.* (2004) comment that autumn numbers have increased in recent years, which may reflect recovery of the Scandinavian populations.

Genus *BUTEO* Lacépède

Holarctic, Ethiopian and Neotropical genus of *c.* 27 species, one of which breeds in GB&I and another occurs as a winter visitor.

Common Buzzard   *Buteo buteo* (Linnaeus)

RB WM PM *buteo* (Linnaeus)

TAXONOMY A variable taxon with up to 11 subspecies recognised (more by some authorities): *buteo* (most of Europe), *arrigonii* (Corsica, Sardinia), *rothschildi* (Azores), *insularum* (Canaries), *bannermani* (Cape Verdes), *vulpinus* (N Scandinavia to Caucasus, east to C Asia), *menetriesi* (Crimea, Caucasus to Iran), *japonicus* (E Palearctic, from Lake Baikal to Japan), *refectus* (W China), *toyoshimai* (Izu, Bonin Is), *oshiroi* (Ryukyu Is). Differences based on size, colouration and plumage pattern, but much variation is clinal and there is extensive variability within populations (HBW). Kruckenhauser *et al.* (2004) attempted to resolve this using multivariate morphological analysis and variable DNA sequences from the mt genome, with mixed results. The former indicated three clusters, related to habitat (islands, dry open and mixed habitats, forests), presumably reflecting parallel adaptation to life in these biomes. The molecular data gave a series of clades, all poorly supported, indicating a rapid divergence with incomplete lineage sorting. However, they suggest that *refectus* should be raised to specific rank, as well as *japonicus* (including *toyoshimai*). They also found that *bannermani* is phylogenetically closer to Long-legged Buzzard *B. rufinus*, with which it may be conspecific. The others should remain as subspecies of *B. buteo*. Accepting these recommendations would result in Common Buzzard having about seven races.

DISTRIBUTION Breeds across the wooded temperate areas of the Palearctic, from the Atlantic to China. W & C European and Japanese populations are resident, but those from across the N Palearctic are strongly migratory, wintering in SE Asia, Indian subcontinent or S & E Africa. These migrations follow hills and mountain chains, with large concentrations (e.g. >400,000 in spring at Eilat) at geographic bottlenecks such as Gibraltar, Bosphorus, Bab el Mandab and Suez.

STATUS Has shown a remarkable recovery since middle of 20th century. Two hundred years ago, it bred throughout GB&I (Newton 1979), yet by 1860 had been eliminated from all but the western areas. Two World Wars resulted in a decrease in keepering activities and Common Buzzards began to spread, although this was limited in the 1950s and 1960s due to effects of myxomatosis on the rabbit population and organochlorines affecting reproductive success. By the end of the 1950s, there was still a strong negative association between Common Buzzards and gamekeepers (Fig. 36; Newton 1979). Atlas fieldwork in the late 1980s documented a significant expansion in N Ireland, and expansion was also apparent eastwards out of the heartlands of the SW, Wales and the Lake District and into more areas in E & S Scotland. The next decade saw further major expansion in both range and abundance (Clements 2002), so that Buzzards bred in most areas of suitable habitat across GB, apart from the counties bordering the North Sea, and even here there were significant populations in Lincolnshire and Norfolk. A similar expansion has occurred in Ireland since *c.* 1990, with breeding being confirmed in many counties, including Cork and Waterford on the south coast. Gibbons *et al.* (2007) suggest that, even now, only 70% of potential range is occupied.

Ringing data indicate that Common Buzzards are relatively sedentary, with only three recoveries involving exchange between GB and Continental Europe. However, there is movement across the North Sea, for Common Buzzards are regularly recorded in Shetland, where they do not breed (Pennington *et al.* 2004). Radio-tracking shows that young birds disperse away from the natal site, usually, but not always, returning to breed nearby.

CBC/BBS data confirmed the increase in population; during the period 1967–2003, counts increased by >400% (Baillie *et al.* 2006). These were most marked in England, but even in the previously comparatively well-populated Scotland and Wales, there has been a doubling in numbers. The reasons for the most recent resurgence are likely to be complex. Undoubtedly, diminution of toxic residues has led to an increase in productivity: nest failures have declined since the 1960s. An increase in the rabbit population is also a likely factor; in many areas, these remain an important source of food. A overall reduction in keepering, the threat of prosecution, and perhaps a more sympathetic attitude towards Common Buzzards by many land managers may all be playing their part. Clements (2002) estimated the population of GB at 44,000–61,000 territorial pairs and suggested, quite plausibly, that it may be the most abundant diurnal raptor.

Rough-legged Buzzard   *Buteo lagopus* (Pontoppidan)

WM PM *lagopus* (Pontoppidan)
SM *sanctijohannis* (J. F. Gmelin)

TAXONOMY Has not been included in any molecular phylogenies, so affinities based on morphology and behaviour. Four subspecies recognised: *sanctijohannis* (Nearctic), *lagopus* (Scandinavia to NW Asia), *menzbieri* (NE Asia), *kamtschatkensis* (Kamchatka), based on size and intensity of colour.

DISTRIBUTION Holarctic. Breeds in narrow belt of essentially treeless tundra, across Alaska, N Canada, N Scandinavia to Bering Straits; missing from Greenland. Migratory, wintering south into USA, C Europe, C Asia to N China. Timing and distance of migration depend on availability of food supplies, as does breeding success.

STATUS A winter visitor, although there are occasional records of birds summering. Sharrock (1974) analysed data for 1958–67, but there has not been an assessment of abundance in recent years. Sharrock showed that most birds arrive along the east coast of England and Scotland in Oct–Nov, with a gradual spread inland, occasionally as far as S Ireland. Numbers rose along the east coast again in May, presumably reflecting emigration back to N Eurasia. Much less common in Ireland, with only 43 records to 2004.

The breeding populations fluctuate with the density of arctic rodents, such as Lemmings, and the occasional higher numbers in GB are undoubtedly related to these. It is tempting to speculate that large numbers occur in GB&I in years following peak Lemming populations; as the rodents crash, the raptors are compelled to emigrate in search of food. Interestingly, Taylor *et al.* (1999) report peak numbers in Norfolk during the winters of 1974–75, 1978–79, 1985–86 and 1994–95, and Pennington *et al.* (2004) found peaks in Shetland in the autumns of 1982 and 1988. These dates do not coincide, perhaps suggesting that the source populations were different. However, changes in the pattern of Lemming cycles has led to changes in breeding dispersal across the tundra. There are a few recoveries of birds ringed in Scandinavia and found in GB.

One record of North American race *sanctijohannis*: found exhausted at Tulla (Clare) in Oct 2005, taken into care and subsequently released in Co Wexford.

Genus *AQUILA* Brisson

Holarctic, Ethiopian and Neotropical genus of 9–11 species, two of which have been recorded in GB&I, one breeding and the other as a vagrant.

Greater Spotted Eagle   *Aquila clanga* Pallas

SM monotypic

TAXONOMY Helbig *et al.* (2005a) investigated the phylogeny of the aquiline eagles using both nuclear and mitochondrial gene sequences. They found extensive paraphyly within the eagles: conventional genera were not monophyletic using the molecular data. Indeed, their data suggest that all genera with more than one species are non-monophyletic. Greater Spotted Eagle and Lesser Spotted Eagle *A. pomarina* were sister taxa and formed a clade with Long-crested Eagle *Lophaetus occipitalis* of sub-Saharan Africa, traditionally placed in its own monospecific genus. Although Helbig *et al.* (2005a) proposed rearranging the *Aquila* and *Hieraaetus* eagles into several genera, relationships among these groups are as yet poorly resolved, and they are, for now, best retained in an enlarged *Aquila*. A second study (Helbig *et al.* 2005b) compared the genetics of Greater and Lesser Spotted Eagles in more detail. They found evidence of hybridisation; several birds that were phenotypically Lesser contained mt DNA from Greater. Nuclear markers were intermediate, suggesting this to be a recent hybridisation rather than incomplete lineage sorting, and hybridisation was asymmetric since no Greater Spotted Eagles were found with Lesser Spotted Eagle mt DNA. Furthermore, estimates of gene flow between the two species indicated this to be much stronger for nuclear genes than mt sequences, suggesting that female hybrids are less likely to survive: a potential example of Haldane's Rule (reviewed by Coyne 1985).

DISTRIBUTION Breeds in a relatively narrow belt of extensive wet forest between 64°N and 45°N, from the Baltic to China. Migratory, wintering from Nile Valley and Arabia across N India to Indo-China and coastal China.

STATUS The only Irish record is of two birds present near Youghal (Cork) for several weeks through the winter of 1844–45, until they were shot in Jan 1845. There have been 13 records from GB (all in England), since the first at Cheesering (Cornwall) in Dec 1850. Most were shot, and all except one were between Oct and Jan. The most recent was in 1915.

Golden Eagle   *Aquila chrysaetos* (Linnaeus)

RB *chrysaetos* (Linnaeus)

TAXONOMY See Greater Spotted Eagle. Helbig *et al.* (2005a) showed the spotted eagles to be distinct from the rest of *Aquila* and also showed that Golden Eagle is a member of a clade that also includes the African Verreaux's *A. verreauxii* and Australian Wedge-tailed *A. audax* Eagles; perhaps more surprisingly, African Hawk-eagle *A. spilogaster* and Bonelli's *A. fasciata* are also part of this clade. Four members of this clade shared a four-base pair deletion in a nuclear sequence (strongly indicating shared ancestry), but Golden Eagle lacked this. Overall, the nuclear and mt DNA data are in good agreement, but they are certainly at variance here. Helbig *et al.* (2005a) propose that the generally close agreement between nuclear and mitochondrial data imply that Golden Eagle shares its ancestry with the other four species and has regained the deletion since it evolved into a separate lineage.

Six subspecies are recognised, based on size and plumage colour, including the extent and balance of yellow/rufous to crown and nape. Two of these (*chrysaetos, homeyeri*) occur in the W Palearctic; the remaining four are Nearctic (*canadensis*) or E Palearctic (*daphanea, japonica, kamtschatica*). A limited study of Japanse and Korean birds (Masuda *et al.* 1998) found little evidence of interpopulation divergence in mt DNA sequences. Wink *et al.* (2004a) extended this study and found two major clades within Golden Eagles; the first comprised European birds (*chrysaetos, homeyeri*) and the second Nearctic/E Palearctic (*canadensis, japonica, kamtschatica*); however, they reported on few birds from C & W Siberia. Within the two clades, there was little geographic structure, except that a few African birds (Mali, Ethiopia) clustered together.

DISTRIBUTION Holarctic, breeding in mountains and steppe, from limits of tundra to semi-desert. Five subspecies range across Palearctic; *chrysaetos*: N & C Europe across N Palearctic to Altai Mts;

homeyeri: more southerly, from Iberia, N Africa to Caucasus; *daphanea*: C Asia to Himalayas and S China; *kamtschatica*: from Altai Mta to Kamchatka; *japonica*: Japan. Korea; *canadensis*: Nearctic. N populations migratory where habitat is inhospitable in winter; elsewhere resident, with some local movements in severe weather.

STATUS The earliest records are open to doubt since there was often confusion with White-tailed Eagle. Extinct from most parts of Ireland by the end of the 19th century and from England and Wales before then. The last breeding outside Scotland seems to have been in Ireland (Hutchinson 1989) in Donegal (1910) and Mayo (1912). In Scotland, Golden Eagles also declined: persecution by shepherds and gamekeepers was followed by collectors of eggs and skins, leading to extinction in the SW and serious reductions elsewhere. Baxter & Rintoul (1953) estimate a population of <100 pairs by the 1870s. After this date, conversion of sheep walk to deer forest resulted in reduced persecution, and the population began a recovery that was enhanced by WWI & WWII, when keepering activity was reduced.

At the time of the First Breeding Atlas, Golden Eagles were established in most of the N, W & C Highlands, and most of the Hebrides. There were a few occupied eyries in SW Scotland, and in 1969 a pair bred in England. There was little change in the overall population by the Second Breeding Atlas; minor changes in distribution could be ascribed to changes in the level of persecution and the indirect effects of land management, e.g. deer culling in the Grampians led to a reduction of winter carrion (Watson 1997). Disturbance, whether deliberate or accidental, can also affect eagles; Watson & Dennis (1992) showed that breeding failure was higher at sites where this was greater.

In the 1980s, Golden Eagles had occupied further sites in SW Scotland and young birds were regularly seen in Cumbria, away from the initial nest site. Watson (1997) estimated the population of Scotland during the late 1990s to be *c.* 420 pairs, though persecution continues, with nests destroyed and birds being shot and poisoned on estates across the Highlands (Watson 1997). Indeed, some sites appear to be 'black holes': juvenile birds move into apparently vacant territories and disappear, to be replaced by yet more juveniles. Persecution is higher on grouse moors in the E of Scotland than in deer forests, where Golden Eagles are tolerated better, and Whitfield *et al.* (2007) conclude that this remains a major influence upon territory occupancy. The latest estimate (2003) of the British population was 442 pairs, a slight increase since the 1990s (Eaton *et al.* 2007a).

In Cumbria, nesting was regular at Haweswater (details listed by Brown & Grice 2005), but only one young was reared since 1992. From 1976 to 1982, a second pair nested in Cumbria, and occasionally birds nest elsewhere in N England; however, successful colonisation seems as far off as ever. A pair bred in Antrim from 1953 to 1960, and a subsequent reintroduction programme in Donegal is under way: nests have been built in the wild and eggs laid, but apparently none have hatched to 2006.

Ringing data indicate that the population of GB&I is self-contained; there are no records of Golden Eagles being recovered after crossing the North Sea, and this lack of dispersal is supported by the almost total absence of records from the N and E coast observatories.

FAMILY PANDIONIDAE

Genus *PANDION* Savigny

Cosmopolitan monospecific genus that breeds in GB&I.

Osprey   *Pandion haliaetus* (Linnaeus)

MB PM *haliaetus* (Linnaeus)

TAXONOMY A unique genus whose systematic position is still not fully resolved; most studies (see Seibold & Helbig 1995) indicate affinities with hawks and eagles but well differentiated from these. Four subspecies (*haliaetus, carolinensis, cristatus, ridgwayi*) generally recognised; the first three of these were analysed by Wink *et al.* (2004b) using mt cyt-b sequences. They found little differentiation within subspecies, but genetic differences between them ranged from 1.9 to 3.8%, which are large enough to indicate a prolonged period of independent evolution and are similar to the differences between closely related eagles such as Greater and Lesser Spotted (1.7%) and Bonelli's and African Hawk-eagle (2.1%). A strong case can be made for splitting Osprey into two or three distinct species, as suggested by taxonomists over 150 years ago.

DISTRIBUTION One of the most widely distributed raptors, breeding across middle latitudes of the Palearctic, and through SE Asia to Australia. Northern populations are strongly migratory; those from S North America, Mediterranean, S China are resident. Distribution of subspecies: *haliaetus*: whole of Palearctic range, northern populations wintering to sub-Saharan Africa as far as the Cape, India, islands of SE Asia; *carolinensis*: continetal N America, from Alaska to Labrador, and S through Rocky Mts, migrates to S America, populations in California, Florida are resident; *ridgwayi*: Caribbean, resident; *cristatus*: Indonesian islands, E of Java/Sulawesi to Australia, partially migrant.

STATUS Historically, Ospreys were widespread across GB&I, but persecution, initially in the Middle Ages because of their depredations at fish farms and later in the 19th century for eggs and skins, led to a near-total decline until the last breeding attempt at Loch Loyne (Highland) in 1916. It remained a scarce bird of passage. Dennis (2007) considers that it was never entirely lost but that occasional breeding continued through 1920–1950, before the first pair 'recolonised' and bred in 1954. Success stuttered for a year or two (Brown & Waterston 1962), through disturbance and egg-collecting, but gradually the number of pairs built up to 26 (1980), 52 (1987), passing 100 in the late 1990s to 182 pairs in 2004 (BS3). The majority of these are in Highland, Perth and Kinross, and NE Scotland. The species was recorded in 39 10-km squares during the fieldwork for the Second Breeding Atlas. Persecution continues to be a problem, with nests robbed of their eggs most years; Dennis (1991) recorded that *c.* 10% of nests were robbed during 1954–90. However, the population has now reached the stage where losses to this and other factors such as disturbance, bad weather and food shortages can be compensated by the reproductive success of the population as a whole.

In Scotland, Ospreys were encouraged into new areas through the provision of artificial nest platforms; the success of these was instrumental in a reintroduction programme centred on Rutland Water (Leicestershire) that started in 1996. Because males were thought to be the philopatric sex, from 1996 to 2001, more were released than females. However, it became clear that females also returned close to the natal site, so in 2005, nine females and two males were released to redress the gender imbalance. The first young hatched in 2001 and 14 were reared to 2006. Birds from this project were instrumental in successful breeding in Wales in 2004; two pairs hatched young, though one nest was lost in a storm. The males had both been released at Rutland Water as part of the reintroduction project. In 2000, a pair of unmanipulated birds bred successfully in Cumbria (RBBP); two or three pairs nested subsequently but only a single pair in 2006. Ospreys total over 150 pairs in GB, but probably only occupy *c.* 20% of their potential range in the UK (Gibbons *et al.* 2007). Ospreys have not yet bred in Ireland, though they are increasing as a passage migrant.

Most young Scottish Ospreys head south through W France and Iberia, although a few are recovered further east in the Mediterranean (Wernham *et al.* 2002). Many winter in coastal W Africa, although there are also recoveries further north in Africa, suggesting that not all birds go so far. Re-sighting of colour rings has shown that birds may winter in exactly the same estuaries in successive years. Satellite tracking has allowed individual birds to be followed in real time. A female moved 4,878 km to S Senegal in only 20 days and another flew from N France to Guinea-Bissau in 18 days, though most move at a more leisurely pace. However, it has also illuminated diffculties for a few Scottish birds on their first migration; some that orientate to the west of south transit through Ireland and are then faced with a hazardous journey across the Atlantic to France or Iberia. Ringing has also pointed to the origins of the original Scottish colonists; most recoveries have been of birds ringed in Scandinavia and several of these have been found breeding in Scotland.

FAMILY FALCONIDAE

Genus *FALCO* Linnaeus

Cosmopolitan genus of *c.* 39 species, nine of which have been recorded in GB&I, four as breeding birds and the rest with varying degrees of rarity.

Lesser Kestrel	*Falco naumanni* Fleischer

SM monotypic

TAXONOMY Closely related to Common Kestrel but, unlike the latter, shows little interpopulation variation. Wink *et al.* (2004c) examined genetic variation among birds from Spain, Italy, Israel and Kazakhstan, and these formed a distinct clade, separate from Common Kestrel. There was evidence of differentiation between Kazakh birds and the rest, but genetic differences were slight and not diagnostic; one individual from each of Italy and Israel was outside the 'European' clade. None of a series of birds sampled at roosts in South Africa possessed the 'European' sequences, indicating that birds from this region do not winter there.

DISTRIBUTION Breeds in a narrow belt of warm temperate steppe or agricultural land from Iberia, NW Africa to Lakes Balkash and Baikal. Isolated population in N China. Most populations migratory, wintering across sub-Saharan Africa, outside the rainforests, and in coastal Arabia. Some birds remain in NW Africa, Azerbaijan over winter.

STATUS There are 16 accepted records up to 2002, of which the first was caught alive near Dover (Kent) in May 1877. There is a broad geographic spread to the records, from Shetland to Scilly, but a clear bias towards the south. The only Irish record (to 2004) is an adult male near Shankhill (Dublin) early Nov 1890–Feb 1891, when it was shot. Birds have been recorded in most months from Feb to Nov, with rather more in Apr–May.

Common Kestrel	*Falco tinnunculus* Linnaeus

RB MB WM PM *tinnunculus* Linnaeus

TAXONOMY Extensive interpopulation variation, leading to at least 11 named subspecies (Vaurie 1965; HBW); differences based on size, colouration and extent of markings. Five subspecies in W Palearctic, four on islands. Hille *et al.* (2003) investigated genetic differentiation across the Cape Verde Is using microsatellite loci and found, as expected for island populations, less genetic variability than among continental (Austrian) birds. There was strong genetic support for two subspecies in Cape Verde, within which adjacent islands were genetically more similar. There is a need for a more extensive analysis along these lines to compare the various subspecies of Common Kestrel and also their relationships with the species on Madagascar, Mauritius and Seychelles in the Indian Ocean, Spotted Kestrel *F. moluccensis* from SE Asia, and even Nankeen Kestrel *F. cenchroides* and American Kestrel, which have long been thought to be closely related allopatric taxa.

DISTRIBUTION Widespread across Europe and Asia, breeding from the tundra of N Norway to South Africa, and from Siberia to the Himalayas, missing only from rainforests and deserts. Most northerly populations are migratory, generally wintering within the southern range of the species but also across India and Indo-China. Palearctic subspecies are *tinnunculus*: entire N Palearctic, from Iberia, NW Africa to Sea of Okhotsk; *dacotiae*: E Canary Is; *canariensis*: W Canary Is, Madeira; *neglectus*: NW Cape Verde Is; *alexandri*: SE Cape Verde Is; four more races in Africa and two in Asia.

STATUS At the time of the Second Breeding Atlas, this was the most widespread diurnal raptor in GB&I, occurring in >85% of 10-km squares and with especially high densities in eastern England. However, this picture hides an overall decrease of 7% from the First Breeding Atlas, more significantly in Ireland but also in the W Highlands of Scotland and the Pembroke peninsula. The reasons for this decline were not apparent at the time but may be due to Buzzard predation.

Population estimates from CBC/BBS data indicate that numbers increased rapidly during 1966–76, as the levels of organochlorine pesticides declined in the environment; certainly egg and chick failures decreased through this period (Baillie *et al.* 2006). However, populations in GB declined in the decade following their peak in the mid-1970s, perhaps due to the effects of agricultural intensification on the populations of small mammals in the countryside. However, this seems unlikely to be the sole cause of the loss of breeding birds reported in the Second Breeding Atlas since densities increased in some intensively farmed regions. There may be a slight reversal of this since 2000, but it remains a matter for concern.

Strong immigration into GB&I in winter; ringing data show that much of this is from Scandinavia and the Baltic states. Many are juveniles that arrive in Sep–Oct, with passage recorded at most N & E coast observatories at these times. There is some evidence of emigration of GB&I breeding birds, either locally to lower ground or more distantly, occasionally even to N France. This is more pronounced among birds from N England and Scotland, presumably to avoid harsh winter conditions in these areas, for Village (in Wernham *et al.* 2002) reports increased mortality in cold winters, especially of sub-adults.

American Kestrel   *Falco sparverius* Linnaeus

SM *sparverius* Linnaeus

TAXONOMY Believed to be closely related to Common Kestrel complex (Smallwood & Bird 2002), but molecular investigation of the affinities of this group needs investigation. No less than 17 subspecies have been recognised, based on size, colour pattern and saturation, barring on upperparts.

DISTRIBUTION Nominate *sparverius* is most widespread, breeding from the treeline, south through Canada and much of USA. Northern populations migratory, wintering generally within range in N & C America. Other races breed across Caribbean, C & S America to Tierra del Fuego.

STATUS The American Kestrel has only been recorded twice in GB and never in Ireland. The first was a male on Fair Isle (Shetland) in May 1976; this was followed a few days later by a female on Bodmin Moor (Cornwall) in June1976. This species is regularly kept in captivity and other records are assumed (or known) to be of captive origin.

Red-footed Falcon   *Falco vespertinus* Linnaeus

PM monotypic

TAXONOMY Recently separated from Amur Falcon *F. amurensis* on basis of diagnostic differences in male and female plumage.

DISTRIBUTION Breeds in E Europe from Baltic States and Hungary to Lake Baikal and extreme NW China (HBW). Strongly migratory, wintering in Africa, south of Namibia and Zambia.

STATUS There are hundreds of records of this attractive bird from across GB but only 22 from Ireland. The first for GB was a first-summer male shot at Great Yarmouth (Norfolk) in May 1830, and the first for Ireland soon followed, when an adult male was shot in Co. Wicklow in the summer of 1832. The number of records has increased since the mid-20th century, from an annual average of <4 (1950–69) to 14 (1970–89) and, more recently, *c.* 23 (1990–2003), although this last includes an exceptional figure of *c.* 130 in 1992. Birds have been recorded in every month from Mar to Nov but half have been in May.

Amur Falcon   *Falco amurensis* Radde

SM monotypic.

TAXONOMY Until recently, this distinctive taxon was regarded as a race of Red-footed Falcon. Differences in behaviour and diagnostic characteristics of plumage, especially in the female, indicate these are separate evolutionary lineages and merit specific rank.

DISTRIBUTION Breeds from Lake Baikal, E to Sea of Japan and S to C China. Strongly migratory, wintering in SE Africa, from Malawi to Transvaal (HBW). Several records from W Europe in recent years, perhaps as observers gain familiarity with the identification criteria.

STATUS One second-winter male present at Tophill Low Nature Reserve (E Yorkshire) in Sep–Oct 2008 (subject to acceptance).

Merlin   *Falco columbarius* Linnaeus

RB MB WM PM *aesalon* Tunstall
WM PM *subaesalon* Brehm

TAXONOMY Variable taxon, with nine subspecies recognised by Vaurie (1965), HBW; differences largely clinal and based on intensity of pigmentation, being darkest in humid biomes and paler in xeric. Two races in W Palearctic (*aesalon*: Britain, Faroes east to C Siberia; *subaesalon*: Iceland), four in C & E Palearctic, two in N America. Gradation between *aesalon* and *subaesalon*, with birds from Faroes and N Britain intermediate and dividing line suggested above somewhat arbitrary.

DISTRIBUTION Widespread across Holarctic, breeding in a range of habitats from open tundra and moorland to boreal forests. Nearctic range extends from taiga forest of Alaska, Hudson Bay, Labrador to NW USA, wintering to N South America. In Palearctic, from Iceland to Pacific, generally between 72°N and 43°N; migratory across much of range, wintering from Mediterranean basin, across C Asia to China.

STATUS Merlins were widespread across N & W GB&I during the 19th century, being absent only from lowland England (Holloway 1996). Although probably never common, they suffered similar persecution to other raptors, especially by collectors of eggs and skins. Merlins did not recover to the same extent as the others during the first half of the 20th century and by the First Breeding Atlas were restricted to the remoter uplands of GB and the coastal fringes of Ireland. Prestt (1965) suggested that the failure to recover might have been due to the loss of nesting habitat, with afforestation and disturbance both playing a part in this. Newton (1973) showed that breeding failure was prevalent and could be ascribed to egg breakage due to the disruption of calcium metabolism by toxic chemicals. Since they breed on high ground, it was presumed that this chemical load was acquired in winter, when many Merlins move to lower agricultural land to feed on small passerines.

In a later study, Newton & Haas (1988) found that organochlorine levels had declined during 1964–86 and egg shell thickness recovered. However, they then found high levels of other contaminants, including mercury, and showed that breeding success was lowest in birds with high egg mercury. Whether this was driving the decline was (and still is) unclear. The highest levels of mercury were found in populations from Orkney and Shetland, which winter along the shores, feeding on waders, which have higher levels of mercury than passerines. However, Merlins from the N Isles did not show the correlation with breeding success, which was limited to the mainland of Britain.

Because of their small size and unobtrusive habits, Merlins are difficult to census, but a dedicated survey in 1982–83 (Bibby & Nattrass 1986) revealed that numbers had continued to decline in those areas where the data were reliable and repeatable and estimated the population as 550–650 pairs. They reported that breeding success was especially low in areas where the declines were most severe. They did, however, find that Merlins tended to utilise tree nests in areas where the density was low; although this was because optimum habitat tends to be in the relatively treeless mix of grass and heather in the uplands, where avian prey is more accessible. There was little difference in breeding success between tree and ground nests. A survey in 1993–94 indicated a further increase in the population to *c.* 1,300 pairs (Rebecca & Bainbridge 1998). Subsequent reports from RBBP are serious underestimates, coming largely from a few sample sites.

Subspecies *aesalon* breeds across GB&I and moves to lower ground in winter, when found along coasts and farmland. Some birds leave GB&I to winter in W France and Iberia, but the majority appear to remain within the islands.

Ringing shows that Icelandic *subaesalon* is a winter visitor to GB&I, particularly in the W and Ireland. Birds in the N Isles and N mainland of Scotland are similar in wing length to Icelandic birds.

The Nearctic race *columbarius* was formerly on the British List on the basis of a Meinertzhagen specimen from the Oouter Hebrides (Nov 1920); this record was subsequently reviewed and found no longer to be acceptable (Knox 1993).

Eurasian Hobby   *Falco subbuteo* Linnaeus

MB PM *subbuteo* Linnaeus

TAXONOMY Probably forms a species complex with African, Oriental and Australian Hobbies *F. cuvierii*, *F. severus*, *F. longipennis*. However, this has not been investigated using DNA sequences. Two subspecies are generally recognised: *subbuteo* and *streichi*, differing particularly in wing length (*streichi* is shorter).

DISTRIBUTION Breeds across the Palearctic, from Iberia, GB&I to Sea of Okhotsk, Kamchatka and from *c.* 68°N to Mediterranean, Iran, N China and Japan. Populations in SE China, Indo-China (*streichi*) generally resident, hence shorter wings. Other populations (*subbuteo*) migratory, wintering well south of breeding areas into C & S Africa, C Asia.

STATUS Traditionally regarded as one of the most difficult raptors to census because of its low density and unobtrusive behaviour; recent research has suggested that it is much more abundant than was previously believed. Holloway (1996) recorded breeding in 36 counties in the 19th century, predominantly S of a line between the Rivers Humber and Severn. Little more definitive data emerged before the Atlases; the first of which suggested a marked contraction in range since 1900, with few records in East Anglia or N of a line from the Severn to the Wash. By the Second Breeding Atlas, there had been a substantial expansion, and the total of 628 10-km squares in which Eurasian Hobbies were recorded represented an increase of >140% on the First Breeding Atlas. This was especially noticeable in the eastern parts of Wales, East Anglia and the N Midlands. Many of these were sites where breeding was not confirmed but nevertheless indicated a substantial range increase. Hobbies have bred successfully as far north as Strathspey (Highland) since 2001, though breeding in Scotland has long been suspected (BS3). The species is scarce in Ireland; 126 individuals have been recorded and there is no evidence of breeding.

Estimating actual numbers was more problematic: densities are too low for birds to be accurately recorded in CBC or BBS and most data come from specialist work by raptor study groups. Clements (2001) reviews some of these: across C & S England there were densities of between 1.3 and 4.9 pairs per 10-km square. Based on these, Clements estimated the population to exceed 2,500 pairs. While there is no evidence that clutch size changed during 1970–2000, BTO nest records indicated an increase in brood size from 1.96 to 2.44 fledged young per pair. This success will have boosted the number of birds entering the breeding pool, and an increase in corvids at the same time will have increased the number of available nests.

There are few ringing recoveries of Eurasian Hobbies ringed in GB&I; most are relatively local, although there are several from W France, also from Belgium, Germany and the Netherlands. They appear to migrate on a broad front, crossing the Mediterranean and the Sahara to winter in the savannahs of the Zambesi basin (Wernham *et al.* 2002). However, this aspect of the biology of Eurasian Hobbies from GB&I is little known.

Eleonora's Falcon   *Falco eleonorae* Géné

SM monotypic

TAXONOMY Probably closely related to Sooty Falcon *F. concolor*, with similar breeding biology, replacing Eleonora's in oases of the Libyan, Egyptian deserts and through the Red Sea and Persian Gulf to coastal Pakistan. Polymorphic for plumage pattern, with all-dark and pale 'typical falcon' forms. The frequency of each morph varies little among colonies, although Walter (1979) reported evidence of non-random mating, with mixed pairs more common than expected. Has been suggested that there is a selective advantage to this, since prey have to learn two images of predators.

DISTRIBUTION Has a remarkable distribution, usually nesting on cliff-bound islands through the Mediterranean and off the coast of NW Africa. Migratory, wintering chiefly in Madagascar and the S Rift Valley. A late migrant, timing its breeding so that young are in the nest during the autumn migration of the northern passerines on which it feeds.

STATUS There are only five records of this species from GB&I. The first was seen at Formby Point (Lancashire) in Aug 1977. The second was found freshly dead in E Yorkshire in 1981, with three more subsequently, none of which has remained for more than a day. There is no clear pattern of arrival, with records from June, July, Aug and Oct (2). There are no records from Ireland.

Gyrfalcon   *Falco rusticolus* Linnaeus

SM monotypic

TAXONOMY A preliminary study by Wink *et al.* (2004d) confirms that Gyrfalcon is a member of the *Hierofalco* complex, being closely similar to Saker and Lanner Falcons *F. cherrug, F. biarmicus*. Only a limited series of Gyrfalcon was included in this analysis, with, significantly, none from Siberia or America. Wink *et al.* report genetic distances among this group are very small, and resolution of species limits are not feasible using their sequences. A more detailed analysis (Nittinger *et al.* 2005, 2007) confirms the poor differentiation among, and also indicates possible non-monophyly of, Gyrfalcon, Saker and Lanner Falcons. Intraspecific taxonomy unclear; now treated as monotypic but several subspecies have been recognised in the past. Shows variation in plumage and colour, more or less resolvable to three forms: white, grey and dark. Potapov & Sale (2004) discuss the inheritance of this; there seems to be a major locus with two alleles that determine dark (black/brown) or white, with grey being heterozygous. However, there may be modifying genes affecting pattern and barring. Although there is considerable interpopulation variation, in general white forms are most abundant in the extreme north: Kamchatka/Chukotka and Canadian Arctic/Greenland (Potapov & Sale 2004).

DISTRIBUTION Holarctic, breeding along the coasts and mountainous regions of the Arctic. Found across tundra and taiga in most of N Canada, across Greenland, Iceland, N Scandinavia and N Siberia to the Bering Straits. A summer visitor to the extreme north, wintering S of range in N America, Asia but relatively scarce in W Europe.

STATUS About 150 records of Gyrfalcon from GB since 1950 and at least 200 prior to that date; about two-thirds from Scotland. There are also *c.* 120 records from Ireland to 2004. The number recorded has increased slightly since the 1950s, but the average remains below five each year. Gyrfalcon has been recorded in every month of the year, but the majority are in winter and almost none June–Aug. There seems to be a peak in Mar–Apr in England and Wales, Scotland, and Ireland, with most in the N & W Isles. Grey birds are commoner in Oct–Jan, whereas white predominate in Feb–Apr. In the 25 years from 1950 to 1975, less than 40% of birds were white phase, since then the figure has risen to over 65%. Presumably, the white birds come from Greenland (certainly more white birds are recorded from Ireland than elsewhere in GB&I), but why there should be an increase in these is unclear. Perhaps the situation is the reverse and the impression of an increase in white birds is due to fewer non-white ones coming to GB&I as warming of the Arctic reduces the need to leave the breeding grounds in N Russia and Scandinavia.

Peregrine Falcon   *Falco peregrinus* Tunstall

RB WM PM *peregrinus* Tunstall

TAXONOMY A variable taxon that has been split into *c.* 20 subspecies, many restricted to small island groups. Four occur in the W Palearctic: *peregrinus*, *brookei*, *pelegrinoides* (sometimes treated as a separate species), *madens*; it is likely that the Nearctic race *tundrius* from Arctic N America and Greenland occurs in winter. White & Boyce (1988) reviewed morphological variation among Peregrine Falcon subspecies and presented a phylogeny. A molecular study by Nesje *et al.* (2000) used microsatellite loci to show that Scandinavian birds were genetically distinct from Scottish birds and that there was evidence of genetic differences across Scandinavia. A small sample from the Nearctic (including *peali*, *tundrius*, *anatum*) was also clearly distinct from the Scandinavian birds, indicating a long period of separation; however, Brown *et al.* (2007) recommended that the subspecies *tundrius* be merged with *anatum*, based on lack of molecular differentiation.

DISTRIBUTION Cosmopolitan, breeding in every continent (except Antarctica) wherever there is suitable nesting habitat and sufficient food. Most northerly populations are migratory, wintering across temperate latitudes. The nominate race breeds across much of the N Palearctic, N of the Pyrenees; *brookei* breeds south of this, from Iberia, N Africa to the Caucasus Mts; *pelegrinodes*, from Canary Is through inland N Africa to Iran/Iraq; *madens* resident on Cape Verde Is (ranges from HBW).

STATUS Probably no large bird has attracted as much attention down the years as the Peregrine Falcon, culminating in two comprehensive monographs (Ratcliffe 1980, 1993). Widespread across Scotland, N England, Wales and coastal areas of Ireland and SW England in the 19th century, it suffered persecution from game-rearing interests and also from egg-collectors and falconers to feed their respective activities (Holloway 1996). Nevertheless, it was abundant in the wilder part of GB&I through until WWII, when killing of birds and destruction of nests was permitted to reduce the mortality of carrier pigeons, especially along the coast of the English Channel; perhaps 50% of the population of England was destroyed (First Breeding Atlas). Numbers began to recover after the legislation expired in 1946, but a far more serious problem arose in the 1950s with the toxic effects of organochlorine pesticides, which accumulated in Peregrine Falcons from their avian prey. Beside direct poisoning of birds by these insecticides, DDT was also widely used through to the 1950s. A side effect of this was to affect calcium metabolism; this led to thinning of eggshells and the collapse of many eggs during incubation. The combination of direct poisoning and reduced breeding success meant that populations fell markedly, and Peregrine Falcons were virtually extirpated from wide areas of S Britain, remaining only in areas such as Cumbria, W Wales and the hills of SW Scotland, where their prey was less contaminated. The Irish populations also fell, although less dramatically (First Breeding Atlas). The restrictions on the use of these toxic materials came into force through the 1960s and recovery began; the resilience of the populations is remarkable. By the time of the Second Breeding Atlas, the number of 10-km squares occupied had more than doubled, although agricultural England was still poorly populated.

Although Peregrine Falcons are poorly represented in CBC/BBS surveys, there are regular national censuses, and these show a continued climb in numbers in GB&NI from 385 pairs in 1961 to *c.* 1,500 in 2002, with particular growth at coastal sites (A.N. Banks *et al.* 2003). However, there is evidence of a decline again, this time in N & W Scotland, N Wales and N Ireland. Although sample sizes are small, there are indications that clutch sizes are declining (Baillie *et al.* 2006). No territories have been occupied in Shetland since 1999, and this has been linked to high levels of PCBs and mercury and competition for nest sites with Fulmars (BS3). Declines on grouse moors in E Scotland have been attributed to persecution.

Movements are not well understood because of the non-randomness of ringing effort; only two birds ringed in GB&I have been recovered overseas (W France, S Spain), although there is considerable exchange among regions within GB&I. Ringing data do, however, demonstrate that there is extensive immigration from Scandinavia in autumn. A crash in the Scandinavian populations was ascribed to the acquisition of toxic chemicals on the wintering grounds in lowland England; the observation that recoveries are predominantly from this part of GB&I may support the hypothesis. Lifjeld *et al.* (2002) found that Norwegian birds were genetically more similar to one another than were Hobbies and Merlins, further indicating the possibility of a recent population bottleneck.

Scottish Peregrine Falcons have been used for reintroductions in Sweden, Germany and the USA.

FAMILY RALLIDAE

Genus *RALLUS* Linnaeus

Cosmopolitan genus of 9–14 species (and several species on Pacific islands believed to be extinct), one of which breeds in GB&I.

Water Rail	*Rallus aquaticus* Linnaeus

RB WM PM *aquaticus* Linnaeus

TAXONOMY Closely related to African and Madagascar Rails *R. caerulescens*, *R. madagascariensis*, although molecular phylogenetic data are lacking. Four subspecies listed by Taylor & van Perlo (1998), based on colour and size, of which *indicus* is sometimes treated as a separate species.

DISTRIBUTION Icelandic populations sometimes separated as *hibernans* but not generally accepted; nominate *aquaticus* is widely distributed across Europe and N Africa, as far east as N Kazakhstan; *indicus* replaces nominate in E Palearctic, from Rivers Yenesei and Lena to Japan; *korejewi* is the southern form, breeding discontinuously from Iran and Aral Sea to Szechwan (Taylor & van Perlo 1998). Northern populations are migratory, to avoid frozen wetlands, wintering in south of breeding range or beyond, e.g. Libya, Nile Valley, Burma, coastal China.

STATUS As a breeding bird, widely but thinly distributed across GB&I, being scarcer in Scotland than elsewhere. Densest populations are in Ireland, especially around Lough Neagh and in the Shannon river system. In GB, breeding was widespread but avoided upland areas and concentrated in permanent wetlands, particularly close to the coasts of S & E England. Overall, between the two Breeding Atlases, distribution declined by *c.* 35% in both GB and Ireland, being variously ascribed to loss of habitat such as vegetated river margins and coastal marshland, 'improvement' of field ditches and drains, and canalisation of river systems.

Results from ringing are biased because of the non-random nature of trapping: relatively few sites catch a large proportion of the birds handled. Most native birds seem to be fairly sedentary, with recoveries typically within 20 km of the ringing site, though occasional long-distance movements (e.g. to Germany) suggest that the GB&I population is not entirely resident. It is evident, both from ringing data and direct observations, that the population is boosted by substantial numbers of immigrants during the winter. Birds are regularly seen and trapped in autumn at coastal observatories where they rarely occur in summer. Recoveries indicate that these originate from middle latitudes of Europe: the Low Countries, N Germany and S Scandinavia. WeBS data confirm that the winter population is higher than that in summer, although numbers counted are often higher when waterbodies freeze and the birds are forced out into the open to feed. At most sites, numbers peak from Nov to Feb (Cranswick *et al.* 2005). Because the counts tend to be low, it is difficult to assess whether there are any changes in abundance; however, several counts in the S & W of GB, such as Loe Pool (Cornwall), Rye Harbour and Pett Level (Sussex), and Chew Valley (Avon), seem to have recorded fewer birds in recent years.

There are no ringing recoveries to indicate exchange between GB&I and Iceland.

Genus *PORZANA* Vieillot

Cosmopolitan genus of up to 17 species (some extinct); four of these have been recorded in GB&I, one as a scarce breeder, the rest as vagrants.

Spotted Crake   *Porzana porzana* (Linnaeus)

MB PM monotypic

DISTRIBUTION Widespread across Eurasia, between 64°N and 40°N, from GB, Iberia to River Yenesei and south to Lake Baikal. Strongly migratory though wintering areas are still not well known; the western populations winter in Africa from the Nile and Rift Valleys to N South Africa, with smaller numbers in isolated wetlands across the continent. Eastern birds mostly winter in the Indian subcontinent (not extreme south).

STATUS A rare breeding bird in GB and only one 10-km square held birds in Ireland during the Second Breeding Atlas. Restricted to areas of wetlands. An instability in the population is reflected in the almost total lack of consistency in occupation of 10-km squares in the two Breeding Atlases; of approximately 30 occupied squares in each survey, only two were common to both. Evidence of breeding was found at 11 sites in GB, 9 of which were in Scotland: mostly eutrophic wetlands in the north. However, a dedicated survey in May–June 1999 (Gilbert 2002) found 73 singing males at 29 sites. Standing water and tall vegetation were key requirements, and within sites crakes selected a mosaic habitat including wet grass and standing water. Previous and subsequent counts reported by RBBP are consistently below this figure and suggest that Spotted Crakes are generally underrecorded.

There is clear evidence of passage to or through GB&I during the autumn, with birds being recorded at many wetland sites, especially along the E coast. These presumably originate from N and E European countries such as Germany and Poland, although there are no recoveries to support this contention. Away from breeding areas, the annual total in GB is typically *c.* 60 birds.

Sora   *Porzana carolina* (Linnaeus)

SM monotypic

DISTRIBUTION Widely but locally distributed across N America from SE Alaska and Northwest Territories to S Hudson Bay and Canadian Maritimes, south to Pennsylvania in the east and Arizona in the west. Migratory, wintering from southern edges of breeding range through Central America to Colombia and Venezuela. Not averse to crossing sea, with passage records from Bermuda and wintering birds recorded on most Caribbean islands.

STATUS The first Sora for GB&I was shot near Newbury (Berkshire) in Oct 1864; there were three more prior to 1913, then a gap of 60 years before the series of modern records that began with a bird found in Scilly in 1973. The first Irish bird was caught on *HMS Dragon* in sea area Shannon/Rockall during the winter of 1919; the location was 160 km W of Ireland and technically lies outside the recording area. The next, which was recorded at Galway in Apr 1920, is strictly the first for Ireland. As with GB, there was a long gap to the next record at Tacumshin (Wexford) in Aug 1998, with only one more since. Most of the *c.* 20 birds in GB&I have been found in autumn, but there are records from Jan and Apr. The majority have been along the western fringes, but individuals have been found in Sussex, Nottinghamshire and Lincolnshire.

Little Crake   *Porzana parva* (Scopoli)

SM monotypic

DISTRIBUTION Breeds at scattered sites in SW Europe, then more continuously from C Europe, E Baltic through Romania, Turkey to Caspian, and across Russia to NW China. Absent from most arid regions of central Asia. Migratory but wintering sites are poorly known (Taylor & van Perlo 1995). In Africa, winters in wetlands along the Mediterranean, through the Rift Valley and across W Africa; eastern populations winter at scattered sites from Arabia and S Iran, across Pakistan to NW India.

STATUS The first recorded Little Crake seems to have been one 'obtained' at Catsfield (E Sussex) in Mar 1791. There have now been over 100 in GB&I, recorded in all months except June but with a definite peak in Mar–May. There is a clear bias towards the south and east, but individuals have been recorded inland and along the west coast. There are only two Irish records, both over one hundred years old; the first was shot at Balbriggan (Dublin) in Mar 1854, the second in Offaly in Nov 1903.

Baillon's Crake   *Porzana pusilla* (Pallas)

FB SM *intermedia* (Hermann)

TAXONOMY Six subspecies recognised by Taylor & van Perlo (1995): *pusilla, intermedia, mira, mayri, palustris, affinis*; only two of which occur in Palearctic.

DISTRIBUTION Very widespread across Europe, southern Africa, Asia through to Australasia. Most subspecies are relatively local, but *intermedia* breeds at scattered sites across Europe and in wetlands through E Africa to the Cape, including Madagascar; *pusilla*: breeds from C Russia and Ukraine across Siberia to Japan and at wetland sites to Iran, N India, Indo-china. European birds winter in N Africa to Senegal and Sudan; wintering areas of Asiatic birds less well known but has been recorded from Malaysia to Philippines (Taylor & van Perlo 1995).

STATUS Holloway (1996) summarises reported instances of breeding in the mid-19th century in Cambridgeshire and Norfolk, but there are no more recent breeding records. Baillon's Crake is currently slightly rarer than Little Crake, with about 75 records from GB since the first near Beccles (Suffolk) in 1819. Frequent records in 19th and early 20th century but much rarer in recent decades. The first Irish bird was obtained near Youghal (Cork) on 30 Oct 1845 and the only other record was near Tramore (Waterford) in Apr 1858.

There is a clear difference in the seasonality of this and Little Crake. Only 15% of the latter have been recorded in summer (June–Aug), compared with almost 40% of Baillon's. Conversely, almost 70% of Little Crakes have been recorded in winter (Nov–Apr) compared with less than 40% of the Baillon's

Genus *CREX* Bechstein

Palearctic and African genus of two species, one of which breeds locally in GB&I.

Corn Crake   *Crex crex* (Linnaeus)

MB PM monotypic

TAXONOMY African Crake *C. egregia* sometimes placed in *Porzana* or monospecific genus *Cecropsis*. There is clinal variation in plumage colour, with eastern populations being less saturated; however, this is insufficient for racial separation.

DISTRIBUTION Breeds across Europe between 64°N and 33°N, from Faroes, Scandinavia, GB&I, N Spain through W Asia to Lake Baikal; absent from Italy, Greece. Winters sparsely in NW Africa and

Egypt (Taylor & van Perlo 1995) and more regularly down the east side from Sudan to South Africa. However, abundance and distribution are little known. Many records exist of individual birds across the continent; these may be migrants on passage or wintering individuals.

STATUS According to Holloway (1996), in the late 19th century, Corn Crakes bred in every county in GB&I and most offshore islands where there was suitable habitat. Fieldwork for the First Breeding Atlas revealed that it had been lost from much of lowland England and S Wales, and by the Second Breeding Atlas it was virtually gone from mainland GB and from three-quarters of the 10-km squares in Ireland occupied in the First Breeding Atlas. National surveys put numbers to this decline, showing that the maximum in GB and I of Man declined from 746 in 1978–79 to 596 in 1988 and that 525 of the latter were in the Hebrides (Hudson *et al.* 1990). Largely restricted now to Orkney, Outer Hebrides, more remote parts of Inner Hebrides and limited numbers on Scottish mainland. A comprehensive survey of Britain and I of Man in 2003 found 831 callng males, a rise of 40% since 1998. Most of these were in areas where active habitat management was under way (O'Brien *et al.* 2006). In Ireland, there had also been declines (Casey 1998); the number of singing males fell by *c.* 75% during 1988–1993; while there was a further loss of 7% by 1998, there were signs of recovery in N Donegal.

The dramatic fall in population across GB&I led to an intensive programme of research (see Green 1996). Intensification of grassland management seemed to lie at the root of the problem, combined with early and unsympathetic cutting of the hay crops. Experiments on some Hebridean crofts that introduced simple changes in husbandry, such as delaying the first cut until after the eggs had hatched (or finding and avoiding the nests and eggs) and cutting the crop from the inside out (allowing adults and chicks to escape) reduced chick mortality substantially. Following the implementation of these practices, the number of singing males increased in the Hebrides and Orkney from 446 in 1993 to >1,000 in 2006; much smaller numbers breed on the mainland of Scotland.

In 2001, a reintroduction programme was initiated at Nene Washes (Cambridgeshire). Young, captive-bred birds were imported from Germany and reared to fledging before being released into the wild at about five weeks of age (Newbery *et al.* 2004); others were retained as captive breeding stock. By 2006, over 200 had been released, and in 2004 young birds were seen in the wild for the first time. A further pair bred in 2005, and in 2006 three released males and one unringed individual were recorded calling in the washes, increasing to 12 calling males in 2008.

Genus *GALLINULA* Brisson

Cosmopolitan genus of eight or nine species, one of which breeds commonly in GB&I.

Common Moorhen   *Gallinula chloropus* (Linnaeus)

RB WM *chloropus* (Linnaeus)

TAXONOMY A variable taxon, with 12 subspecies recognised by Taylor & van Perlo (1995); the underlying variation is clinal, involving size, shape and size of the frontal shield, and the colour of the upperparts. The only race in the W Palearctic is the nominate.

DISTRIBUTION Globally very widespread, occurring in all major geographical regions, apart from Australasia. Subspecies *chloropus* breeds commonly throughout Europe, apart from northern parts of Fennoscandia; also found in N Morocco and Algeria, the Nile Valley and across into Arabia. Northern populations are migratory to escape frozen waters, wintering as far as southern Europe and N Africa.

STATUS Widely distributed across GB&I, only really missing from the bleaker upland areas of Britain, especially the N & W Highlands of Scotland; occurrence is patchy in the N & W Isles, apart from Orkney and the Uists. Especially common in lowlands beside ponds, streams, lakes and rivers where there is suitable emergent vegetation for nesting. The distribution contracted between the Atlases, particularly in NE Scotland, C Wales and the Cornish peninsula; land drainage may have played a part in this, but there is little firm evidence. Similarly, in Ireland, there was a contraction from the far NE and NW, which again might be due to changes in agricultural practices.

The numbers of Common Moorhens are monitored through CBC/BBS and especially by the Waterbirds Survey. The former began in the mid-1960s, and there was immediate evidence of an increase in numbers, which doubtless reflected a recovery following losses in the severe winters early in the decade. Numbers peaked around 1970 and then showed a modest decline through to the 1990s, followed by a slight increase. Baillie *et al.* (2006) postulate that the decline may be due to a combination of loss of farm ponds for nesting and increases in the abundance of Mink. However, breeding performance has deteriorated since the 1960s. Clutch size has declined and nest failure has increased, although average brood size has increased, perhaps because the failing nests are from young birds and/or in poor-quality habitat. The population of GB&I is largely sedentary (Wernham *et al.* 2002), but ringing recoveries show that there is significant immigration from the Netherlands and Denmark. WeBS data support this immigration, with peak numbers being recorded during the winter months, although local birds may accumulate on larger waterbodies outside the breeding season. There is evidence that the number of birds is declining at some sites, especially in the west of Britain; this may reflect a reduction in migratory behaviour as climatic conditions ameliorate further east.

Genus *PORPHYRIO* Brisson

Cosmopolitan genus of six species, two of which are vagrants to GB&I.

Allen's Gallinule   *Porphyrio alleni* Thomson

SM monotypic

TAXONOMY Previously placed with Purple Gallinule *P. martinica* and S American Azure Gallinule *P. flavirostris*, with which monophyletic, in genus *Porphyrula*, but this group better merged with similar *Porphyrio*.

DISTRIBUTION Breeds across sub-Saharan Africa, apart from the extreme south-west; also Madagascar. Breeding season varies across continent, depending on latitude; e.g. S of equator, usually during boreal winter; northern populations (Nigeria, Cameroon) during boreal summer. Movements are little known (Taylor & van Perlo 1995). It appears in arid zones following rains, but whether this is genuine migration or merely dispersal out of permanent wetlands such as swamps and river valleys may vary between regions. Taylor & van Perlo (1995) report that 16 of 17 W Palearctic records were in the period Oct to Feb (chiefly Dec), presumably reflecting post-breeding dispersion from populations in the northern tropics.

STATUS There are only two records of this African species from GB&I. The first was caught on a fishing boat off Hopton (Suffolk) in Jan 1902; the second was found moribund on Portland (Dorset) almost exactly 100 years later in Feb 2002.

Purple Gallinule   *Porphyrio martinica* (Linnaeus)

SM monotypic

TAXONOMY See Allen's Gallinule. Some variation in plumage colour but not sufficient for racial differentiation.

DISTRIBUTION Breeds in the eastern USA, from Pennsylvania south through Central America to Uruguay. Most northerly and southerly populations migrate towards the tropics to avoid their respective winters. Has a remarkable tendency to vagrancy, with birds being recorded from most islands in the Atlantic and also SW Africa, where they tend to arrive when birds from the most southerly populations are migrating north to avoid the austral winter (Silbernagl 1982).

STATUS The only record for GB&I is of an immature found exhausted on St Mary's (Scilly) in Nov 1958.

Genus *FULICA* Linnaeus

Cosmopolitan genus of 10–11 species, two of which have been recorded in GB&I, one as a common breeder, the other as a vagrant.

Eurasian Coot   *Fulica atra* Linnaeus

RB WM *atra* Linnaeus

TAXONOMY Morphological analyses indicate this to be closely related to American and Red-knobbed Coots *F. americana*, *F. cristata* and several other taxa across the world, but molecular phylogenetic analysis not available. Four subspecies recognised by Taylor & van Perlo (1995), based largely on size and the extent of white in the secondaries. Only one race (*atra*) occurs in the Palearctic.

DISTRIBUTION Very widely distributed across Eurasia south of 65°N, from the Azores and Canaries to Japan, though absent from arid regions of Asia; in Africa, found in N Morocco, Algeria and Nile Delta; across India, Sri Lanka; SE Asia from S China through Malaysia to Australasia. Northern populations are migratory, especially from Finland and NE Europe, wintering around the North Sea and into Iberia.

STATUS Widespread though generally more restricted than Common Moorhen, being absent from smaller waterbodies such as ponds and streams. The Second Breeding Atlas showed it occurred across much of lowland GB&I, away from the uplands and their associated oligotrophic waters. As with Common Moorhen, scarce in the N & W Isles, though present on eutrophic lochs in Orkney and the Uists. Again, as with Common Moorhen, a contraction of range in Ireland between the Atlases may be due to drainage of wetlands associated with changes in agricultural practice.

CBC and BBS data indicate a steady increase in numbers since the 1960s, possibly representing a recovery from mortality associated with the hard winters early in that decade. The Waterbirds Survey (WBS), which started in 1974, also shows an increase in population through to about 1990 and a slight decrease since. WBS recorders tend to concentrate on linear bodies such as rivers and canals, and the decline that they report may be due to the increase in Mink in these habitats. Ringing data indicate that a proportion of the GB population is migratory, moving through W France to winter in wetlands beside Biscay and the Mediterranean (Wernham *et al.* 2002).

The population of Eurasian Coots is enhanced in the winter by immigrants. WeBS data show that numbers increase through the autumn, generally peaking in Oct–Dec, and this can be further enhanced when hard weather hits the Continent. The winter maximum at counted waters remained fairly static in both GB and NI through the 1990s but has shown signs of a decline since the turn of the century. There is no obvious geographic pattern to this, though the decline in NI is largely due to a fall in the winter population on Lough Neagh (NI) (Collier *et al.* 2005). Ringing recoveries indicate that many of the immigrant birds come from the S Baltic, from Denmark to W Russia, predominantly to winter in SE England.

American Coot   *Fulica americana* J. F. Gmelin

SM race undetermined

TAXONOMY American Coots likely evolved in the New World (Brisbin *et al.* 2002), and this species is closely related to several other taxa, including Eurasian, Hawaiian *F. alai*, Red-knobbed and Caribbean *F. caribaea* Coots. Two subspecies: *americana* N and C America; *columbiana* Colombia.

DISTRIBUTION Nominate breeds across N America from SW Alaska, central prairie provinces and Nova Scotia, south to Panama. Most populations are migratory, wintering along the Pacific coast, Gulf states and Central America.

STATUS American Coot has been recorded six times in GB and once in Ireland. The first was a bird at Ballycotton (Cork) in Feb–Apr 1981. The GB records began with one at Stodmarsh (Kent) in Apr 1996; the second was at South Walney (Cumbria) in Apr 1999; subsequently, there have been records from Dumfries and Galloway (2004), Outer Hebrides (2004, 2005), Shetland (2004, returning in 2005).

FAMILY GRUIDAE

Genus *GRUS* Brisson

Near-cosmopolitan genus of 10 or 11 species, two of which have been recorded in GB&I, one breeding locally, the other as a vagrant.

Common Crane   *Grus grus* **(Linnaeus)**

CB PM *grus* (Linnaeus)

TAXONOMY Part of a clade that also includes Red-crowned *G. japonensis*, Whooping *G. americana*, Hooded *G. monacha* and Black-necked *G. nigricollis* Cranes; the phylogeny based on mt cyt-b supports conventional taxonomic arrangements (Krajewski & King 1996). Two subspecies have been described for some time, based on geography and plumage colour (BWP, but see HBW). Third subspecies described recently (Ilyashenko 2008), lacking red patch on back of head.

DISTRIBUTION Breeds in a wide range of wetlands, including forest pools and swamps, across Palearctic from Scandinavia to E Siberia and S to Caspian Sea and Lake Balkash. Nominate is replaced E of Turkestan by *lilfordi* (BWP); isolated population *archibaldi* in Trans-Caucasus. Traditionally strongly migratory, wintering well S of breeding range in S Europe, E Africa, Tibet, E China, although recently increasing numbers have remained in or around southern breeding grounds. This may be, at least partly, due to climatic amelioration.

STATUS Became extinct as a breeding bird several hundred years ago, although probably abundant to the Middle Ages. Archaelogical evidence suggests early man ate Common Cranes, and there are records of them featuring at royal banquets in the 13th century (Brown & Grice 2005). Until 1981, it was a scarce or even rare visitor, with <50 birds being recorded in the average year across GB&I. There were occasional influxes into SE England, usually in autumn when birds were displaced from their migration route by adverse weather in the southern North Sea.

Historical evidence indicates that Common Cranes formerly bred in the fenland of E England and probably the broads of Norfolk. In 1981, they bred again in the latter area and have continued to do so since. Success rate varies; in some years no young are reared, with foxes and Marsh Harriers being blamed for losses. Breeding has been successful in most years since 1990 and the population has grown slowly. Most recently (2007), eight pairs bred in broadland, with other pairs in Suffolk, Yorkshire and possibly Lincolnshire (A Grieve pers. comm.). Although some young birds disperse away form the natal area after fledging, a wintering flock exists near Hickling Broad (Norfolk), with up to 35 individuals in 2007.

In Ireland, it remains a rare bird, with only 80 individuals since mid-19th century and there is no evidence of recent breeding.

Sandhill Crane   *Grus canadensis* **(Linnaeus)**

SM *canadensis* (Linnaeus)

TAXONOMY Not especially close to any other *Grus*, Sandhill Crane seems to have undergone a prolonged period of independent evolution (Krajewski & King 1996). Six subspecies are recognised, though a study utilising microsatellite and mitochondrial markers (Jones *et al.* 2005) indicated that there is significant gene flow among these. Since nuclear genes were more clinal than mt DNA sequences, this implies a higher degree of philopatry among females compared with males.

DISTRIBUTION Now has a remarkably fragmented distribution across Canada and USA and into extreme E of Siberia. Most southerly populations are resident but majority show strong migratory behaviour, wintering in S USA and N Mexico. Nominate race breeds across Arctic and subarctic from NE Siberia to Baffin Is and S to N Ontario; subspecies *rowani* replaces this in subarctic Canada from British Columbia to Ontario. Other races in continental USA to Cuba.

STATUS The first for GB&I was a bird of the race *canadensis* shot near Galley Head (Cork) in Sep 1905. There are two modern records (race undetermined), both from Shetland: Fair Isle in Apr 1981 and Exnaboe/Sumburgh in Sep 1991; the latter bird subsequently moved to the Netherlands.

FAMILY OTIDIDAE

Genus *TETRAX* Forster

Palearctic monospecific genus that is a vagrant to GB&I.

Little Bustard   *Tetrax tetrax* (Linnaeus)

SM monotypic

TAXONOMY Molecular analysis of nuclear and mt DNA sequences (Pitra *et al.* 2002) support the placing of Little Bustard in a monospecific genus; it has no close relatives within the phylogeny and it seems not to be especially close to Great Bustard, as suggested in the past. There is no evidence of intraspecific variation, so it is treated as monotypic.

DISTRIBUTION Breeds in scattered grasslands from Morocoo and Iberia, through France and N Italy; then from Ukraine to Kazakhstan and extreme NW China (HBW). W populations are sedentary, moving only in response to adverse weather and food shortage. Further E, most birds are migratory, moving from the breeding areas on the steppes to winter Iran, Azerbaijan.

STATUS This species has shown a substantial decline in records over the past 200 years; the number in the four 50-year periods from 1800 are (approximately) 51, 72, 39 and 22. Although more limited in numbers, the 10 Irish records show a similar pattern (2, 6, 2, 0). The majority of records (>70%) are from Oct to Jan, though there is a slight indication that more may have come outside this period since WWII (35%; 21% before WWII). However, the numbers are small and the difference is only marginally significant. It is much rarer in Ireland, with only *c.* 10 records and none since the 1930s.

Genus *CHLAMYDOTIS* Lesson

Palearctic and marginally Oriental genus of two species, one of which is a vagrant to GB&I.

Macqueen's Bustard   *Chlamydotis macqueenii* (J.E. Gray)

SM monotypic

TAXONOMY Molecular, morphological and behavioural data differentiate E and W populations of 'Houbara' Bustard, which is now treated as two species (summarised by Sangster *et al.* 2004a).

DISTRIBUTION Macqueen's Bustard breeds from the Middle East across Arabia to Iran. W populations are sedentary, moving only when climate becomes too hostile; those further E are truly migratory, crossing the Himalayas to winter from Persian Gulf to NW India.

STATUS There are five records from GB, though only one in the modern era. The first was shot in Oct 1847 at Kirton-in-Lindsey (Lincolnshire); three others followed in the 1890s and one was present at Hinton (Suffolk) in Nov–Dec 1962. There are no records from Ireland.

Genus *OTIS* Linnaeus

Palearctic monospecific genus that is a vagrant to GB&I.

Great Bustard   *Otis tarda* Linnaeus

FB SM *tarda* Linnaeus

TAXONOMY Pitra *et al.* (2002) showed that Great Bustard is sister to, but well differentiated from, *Chlamydotis*. They also showed that the two subspecies are genetically distinct but less so than other bustards that are treated as separate species. Two races are thus recognised: *tarda* and *dybowskii*.

DISTRIBUTION Nominate race occurs in a relatively narrow band of steppe from SW Russia through Kazakhstan; more fragmented populations occur in Europe, in Morocco, Iberia, Germany, Hungary to Turkey, with remnants elsewhere. These largely resident, with only dispersive or irruptive migrations in response to impacts of winter weather on food availability.Pitra *et al.* (2000) examined nuclear and mt DNA sequences among European birds and found evidence that Iberian birds were genetically differentiated from those in C Europe. They conclude that birds in these two regions should be treated as separate evolutionary significant units and any management should take account of their differences. The eastern race *dybowskii* is migratory, breeding from SE Russia to NW China and wintering in N China.

STATUS Common in the past and formerly bred as far N as Scotland. Last wild breeding in GB was in Suffolk in 1832. A reintroduction programme was started in 2004; eggs were collected from the Saratov region of Russia and between 6 and 32 young birds reared in captivity each year prior to release on Salisbury Plain (Wiltshire). Of 32 birds released in 2005; at least 12 survived to 2006. This survival rate compares favourably with 78% mortality reported in a natural population. Some of the birds have wandered away from the release site, e.g. one seen Portland (Dorset) Oct 2005. Two pairs of released birds hatched chicks for the first time in 2009.

Just *c.* 25 individuals in GB since Dec 1949, including a party of up to 9 in Suffolk in Jan 1987. Only three in Ireland; two near Thurles (Tipperary) in Dec 1902 and one near Castletownbere (Cork) in Dec 1925. As with Little Bustard, there has been a marked decline in records through the 20th century. Over 75% of records (rather than birds) in midwinter (Dec–Feb). There is a slight trend for fewer midwinter records since WWII: 78% before 1940; 64% since.

FAMILY HAEMATOPODIDAE

Genus *HAEMATOPUS* Linnaeus

Holarctic genus of about 11 species; one Palearctic, which breeds commonly in GB&I.

Eurasian Oystercatcher   *Haematopus ostralegus* Linnaeus

RB MB WM PM *ostralegus* Linnaeus

TAXONOMY Baker *et al.* (2007) show that oystercatchers form a clade with stilts, avocets, Ibisbill *Ibidorhyncha* and (perhaps surprisingly) *Pluvialis* plovers – although they recommend prudence until confirmatory data become available. Four subspecies recognised by Hockey (1996) and HBW, although the race from New Zealand is sometimes treated as a separate species *H. finschi*. The others are *ostralegus*: Europe and W Russia; *longipes*: C Russia to W Siberia and Kazakhstan; *osculans*: E Asia. Subspecies *buturlini* sometimes recognised (e.g. Dickinson 2003).

DISTRIBUTION Widespread across NW Europe; predominantly coastal in Iceland, France, Scandinavia and Baltic States; also inland in GB&I, Russia E to Caspian and Aral Seas. Isolated populations in Orient: Kamchatka, Korea, coasts and major rivers of China. Most populations partially migrant; winters along coasts of Europe, Africa, N of equator, Red Sea, Arabia, India.

STATUS The majority of those breeding in GB&I now do so in fields, often either adjacent to rivers and other wetlands, where they feed largely on earthworms and Tipulid larvae, or along shorelines, where the adults can collect shellfish for their young. Inland breeding largely confined to NE Scotland until late 19th century but has spread W and S since then, especially through 20th century. In the First Breeding Atlas, inland breeding throughout most of Scotland and in NW England S to Lancashire. Also breeding along or near coasts from Humber to Kent and around W & S coasts to W Sussex. Breeding at that time also round W, N & E coasts of Ireland. By the Second Breeding Atlas, continued expansion of inland breeding in N England and East Anglia and more scattered across Midlands to mid-Wales, with losses in Devon and Cornwall. Little change in Ireland. Notable population of urban, roof-nesting birds in Aberdeen. In common with other parts of W Europe, the number of birds breeding in GB&I continued to increase through the 1990s, although declines were reported elsewhere (e.g. Netherlands), perhaps reflecting decreased food supplies due to overfishing on adjacent estuaries.

Ringing data confirm that there is a general southwards movement in winter, especially towards the rich feeding grounds of W Britain. Sub-adult birds may remain here for several years, before returning to the breeding grounds as mature birds.

Wetlands International (2002) recognise five relatively distinct subpopulations of *ostralegus*, of which four occur in GB&I. These comprise birds breeding in Iceland, the Faroes and Scotland (wintering in Ireland and W Britain); Norway (wintering North Sea); Baltic and NW Russia (Wadden Sea); S Britain and Ireland (wintering Atlantic coast S to Morocco). In the 1990s, the global population of *ostralegus* was estimated as 1.02M (Stroud *et al.* 2004), of which about a third wintered in GB&I; the proportion of this total in each subpopulation is not known with any accuracy. The most important sites are Morecambe Bay (*c.* 40,000), Solway Firth (*c.* 30,000) and the Dee Estuary of NW England (*c.* 18,000), but flocks in excess of 1,000 occur on many intertidal mudflats around the coast of GB and to a lesser extent dispersed along rocky shores, where they take crabs and mussels rather than cockles and worms. Numbers in Ireland are somewhat less, partly reflecting the lower number of sandy estuaries, but Strangford Lough (Co. Down) has held winter populations in excess of 8,000 (Banks *et al.* 2006).

WeBS data probably represent good coverage for this species and are valuable in identifying changes in abundance both at individual locations and nationally. These show a progressive increase from the 1970s through to 1990, then a decline in the early 1990s, since when numbers have remained fairly stable; in 1994–95 to 1998–99, the winter population of GB, Ireland and IoM were estimated at *c.* 315,200, 67,620 and 3,990 respectively (Rehfisch *et al.* 2003a). The first two represent 31 and 7% of the flyway population. In general, these flocks include birds from Iceland, the Faroes, Scandinavia and the Netherlands, though in different proportions around the coasts of GB&I.

FAMILY RECURVIROSTRIDAE

Genus *HIMANTOPUS* Brisson

Near-cosmopolitan genus of two to five species; one from the Palearctic, which is a rare visitor to GB&I.

Black-winged Stilt   *Himantopus himantopus* (Linnaeus)

CB SM *himantopus* (Linnaeus)

TAXONOMY Molecular data from RAG-1 nuclear gene (Paton *et al.* 2003) and various mt genes (Baker *et al.* 2007) show that avocets, stilts and oystercatchers form a clade; indications that stilts may be sister group to thick-knees Burhinidae, Magellanic Plover *Pluvianellus* and sheathbills *Chionis* spp., but this is not strongly supported. At least three species of stilts, of which two are localised to Australia (*Cladorhynchus leucocephalus*) and New Zealand (*H. novaezelandiae*). Several races of Black-winged Stilt recognised (HBW), although only one (nominate) in Palearctic, and others are sometimes treated as full species: e.g. *H. mexicanus* (USA to N South America) and *H. melanurus* (S South America). Taxonomy in need of review: morphological and behavioural differences may or may not justify this realignment of species limits.

DISTRIBUTION Widely distributed across suitable habitat in all continents. Nominate breeds in shallow pools and marshes across temperate zone of Palearctic from Iberia to N China, wintering in Africa; African populations generally resident, from sub-Saharan zone to the Cape, except tropical forests; populations from Indian and Malay peninsula also generally non-migratory.

STATUS Breeds sporadically, the first fully documented occasion being in 1945, when three pairs bred at Nottingham Sewage Works, raising four young (Staton 1945). Subsequent attempts have met with varying success, although these appear to be increasing in frequency, a pleasing side-effect of climate change.

Being a conspicuous bird, there are many records from earlier centuries, with the first specimen possibly being a bird collected in Dumfries and Galloway before 1684. The number recorded annually has increased slightly since 1950 but fluctuates among years, with peak numbers for the period in 1965 (20), 1987 (40) and 1990 (26). Much less common in Ireland, with a total of only 47 individuals; the first was at Youghal (Cork) in winter 1823–24. There is, however, a record of six birds together at Ballycotton (Cork) in Mar 1990.

Recorded in every month of the year, though over half of all records are in spring. Interestingly, the peak months differ among the regions: Ireland (Mar), Wales (Apr), England (May), suggesting an arrival from the SW (there are too few records from Scotland for analysis). There is also a much smaller secondary peak in the autumn: Sep (England) and Sep–Oct (Ireland).

Occasional individuals remain for a long time; a bird arrived in N Norfolk in Aug 1993, staying in the Titchwell area for over a decade before finally disappearing in May 2005.

Genus *RECURVIROSTRA* Linnaeus

Near-cosmopolitan genus of four species; one Palearctic, which breeds in GB&I.

Pied Avocet   *Recurvirostra avosetta* Linnaeus

RB MB WM PM monotypic

TAXONOMY Avocets and stilts are sister groups on both morphological (Björklund 1994; Chu 1995) and molecular (Paton *et al.* 2003; Baker *et al.* 2007) grounds. Four species of avocets, two in Americas and one in Australia. Pied is most widespread but shows little variation across range, apart from clinal increase in size from Europe into Asia. As with stilts, more detailed molecular analysis desirable.

DISTRIBUTION Breeds on temperate wetlands from GB&I to N China. Also, in Africa from Rift Valley lakes to the Cape. Palearctic populations migratory, wintering around coasts of Europe, Africa, N India, China. African populations generally resident.

STATUS The story of the re-colonisation of England following WWII is well documented (e.g. Cadbury *et al.* 1989). Birds moved into areas of East Anglia that had been flooded as part of GB's wartime coastal defences. Subsequent protection and management of the sites led to a steady expansion and increase, so that a total of 1,570 pairs bred at 66 sites in 2006 (RBBP). The spread northwards has reached Lancashire and County Durham, perhaps due to a combination of protection and climate change. The total number of pairs (>1,200) is still only a tiny fraction of the 65,000 birds that breed in W Europe. Bred at Tacumshin (Wexford) in 1938 but otherwise a rare visitor to Ireland, with 133 individuals in total.

The number of birds recorded at WeBS sites in winter continues to rise (Banks *et al.* 2006). Concentrations of *c.* 1,000 birds were recorded in midwinter 2004–05 at Poole Harbour (Dorset) and on the Thames Estuary, with similar totals on the Swale (Kent) and Alde (Suffolk) in Mar. Significant numbers are now found regularly further north in late summer (e.g. River Humber), which probably reflect post-breeding aggregations. There is a clear movement towards SW Europe in autumn. Although many of the birds that winter on the estuaries of Devon and Cornwall are British, there are some Continental individuals among them; British-ringed birds have been recovered as far south as Morocco (Wernham *et al.* 2002).

FAMILY BURHINIDAE

Genus *BURHINUS* Illiger

Near-cosmopolitan (not Nearctic) genus of seven species; one Palearctic, which breeds in GB&I.

Eurasian Stone-curlew

Burhinus oedicnemus (Linnaeus)

MB *oedicnemus* (Linnaeus)

TAXONOMY Strong molecular support for a sister group relationship between thick-knees and sheath-bills (Paton *et al.* 2003), though the morphological data are less convincing (Björklund 1994; Chu 1995). Mostly tropical or subtropical, there are nine species, and representatives occur on all continents (except Antarctica). Eurasian Stone-curlew is polytypic, with at least six races across the Palearctic, two (*distinctus, insularum*) as island forms from the Canaries. Nominate race breeds from Britain and Iberia, across S Europe to Ukraine and the Caucasus. Replaced by *indicus* and *harterti* in Indian subcontinent and northern parts of SW Asia, respectively; *saharae* occurs from N Africa through Turkey to Iran.

DISTRIBUTION Breeds on steppe and dry grassland, also in arable land with open patches of ground. Northern populations migratory; those from N Europe winter round Mediterranean and Africa S to Sahel zone. Extent of movements little known for many populations (HBW).

STATUS Stone-curlews are in decline across Europe and most populations are now fragmented and discontinuous. The declines in GB&I began long ago with the reduction in active rabbit-warrening and the conversion of permanent grassland to arable. The range contracted, and by the Second Breeding Atlas, there were only two populations remaining, in the brecklands of East Anglia and the downland around Salisbury Plain. At that time, the breeding population was estimated to be *c.* 160 pairs. Conservation efforts centred on reversing the loss of habitat by the creation of new, or sympathetic management of existing, permanent and semi-permanent grassland and the preparation of areas of bare earth within arable fields, where the birds can nest. These developments are improving the situation for Stone-curlews and the numbers in both isolates are increasing steadily; in 2006 there were almost 350 pairs, of which 222 were in Norfolk and Suffolk (RBBP).

Away from S England, it remains a rare bird, with only a handful of records from elsewhere in GB and 23 individuals in Ireland, the first being at Clontarf (Dublin) in Jan 1829. Recently, colour-ringed birds from the two GB populations have been found outside these centres. If the agricultural environment is suitably encouraging, the possibility exists of Stone-curlews reoccupying their historical range: they bred as far north as Yorkshire and Nottinghamshire until the end of the 19th century (Holloway 1996). Ringing recoveries show that many of our birds migrate through W France to winter in Iberia and NW Africa, though there is a recovery from Sierra Leone that suggests some may move much further.

FAMILY GLAREOLIDAE

Genus *CURSORIUS* Latham

Ethiopian, Oriental and Palearctic genus of four or five species; the Palearctic species is a vagrant to GB&I.

Cream-coloured Courser	*Cursorius cursor* (Latham)

SM *cursor* (Latham)

TAXONOMY Morphologically, coursers and pratincoles have always been treated as close allies; recent analyses of morphology (e.g. Chu 1995) and a molecular analysis using part of the nuclear RAG-1 gene (Paton *et al.* 2003) support this and suggest the courser/pratincole clade is sister to that of the gulls, terns, auks and skuas. Although not part of the GB&I fauna, we note the support given for regarding buttonquails Turnicidae as part of the Charadriiformes through their sister relationship to this complex.

Cream-coloured Courser is normally treated (e.g. HBW) as five subspecies; *exsul*: Cape Verde Is.; *cursor*: Canaries, N Africa to Arabia; *bogolubovi*: SE Turkey and Iran to Pakistan and NW India; *somalensis*: Ethiopia, Horn of Africa; *littoralis*: N Kenya, S Somalia. Pearson & Ash (1996) suggest that last two of these are better regarded as races of a separate species *C. somalensis*.

DISTRIBUTION Desert and semi-desert across N Africa from Mauritania to N Egypt; isolated populations in SE Turkey, N Iran, S Pakistan and NW India; E African populations extend from E Sudan through the Horn of Africa to N Kenya. Island populations on Cape Verde, Canaries. Populations in N Africa migratory, wintering to S of Sahara; those from Turkey, Iran probably winter NW India/Pakistan but detailed movements little known.

STATUS About 35 records but only seven since 1950; this reduction parallels rest of Europe and may be due to population decline since 19th century (HBW). Records widely scattered, N to Central Scotland. Most records in Oct, though some Sep to Dec. The first for Britain was an immature, near Wingham (Kent) 1785. There is only a single Irish bird, recorded at Raven Point (Wexford) sometime during Dec 1952 or Jan 1953.

Genus *GLAREOLA* Brisson

Ethiopian, Oriental and Palearctic genus of seven species; three Palearctic, all of which have been recorded as vagrants in GB&I.

Collared Pratincole	*Glareola pratincola* (Linnaeus)

SM *pratincola* (Linnaeus)

TAXONOMY No recent analyses but traditionally treated as polytypic with up to five subspecies, all of which occur in Africa and Western Eurasia. Nominate *pratincola*: dispersed populations from Iberia to Aral Sea, Iraq, Pakistan; *fuelleborni*: scattered across Africa from Senegal to Kenya, S along Rift Valley to Cape, with isolated populations Zaire, Namibia; *erlangeri*: restricted to coastal Somalia and Kenya.

DISTRIBUTION Nests on flat areas such as steppe or salt pans, usually close to water. Widely distributed across S Europe to Turkey and Black, Caspian and Aral Seas to Lake Balkash; isolated populations in Iraq, Pakistan. Most (all?) populations migratory, wintering S of Sahara and along Rift Valley.

STATUS About 95 records, since the first at Bowness (Cumbria) Oct 1807; *c.* 60 since 1950. Despite relative frequency in GB, only a single record from Ireland, on Bann Estuary (Londonderry/Antrim) in Oct 1970. Recorded in every month except Jan, but one peak in May–July and a lesser one in Sep–Oct. Decrease from around three per year in 1970s to around one a year since 1980 may be due to severe decline in Spanish population.

Oriental Pratincole   *Glareola maldivarum* J.R. Forster

SM monotypic

TAXONOMY Exact affinities within pratincoles uncertain but usually (e.g. HBW) regarded as close to Collared and Black-winged. Despite relatively wide distribution, no evidence of subspecific differentiation.

DISTRIBUTION Breeds N India and Indo-China, N and E to Mongolia and Japan. Migrates to India, Indonesia to N and W Australia.

STATUS First record: first-summer, Dunwich (Suffolk) June–July 1981. Four more since, all in SE England from Norfolk to Sussex between May and Sep. Not recorded in Ireland.

Black-winged Pratincole   *Glareola nordmanni* Fischer

SM monotypic

TAXONOMY Closely related to Collared Pratincole.

DISTRIBUTION More northerly than Collared, breeding on steppe, grassland and arable from N of Black Sea to Kazakhstan. Strongly migratory, wintering in S and W Africa. Populations declining, perhaps due to intensification of agriculture on breeding grounds and/or use of pesticides in Africa (HBW).

STATUS First record was a bird shot near Northallerton (N Yorks) in Aug 1909. Since then, *c.* 36 records, mostly in Aug but recorded May–Nov. Mostly unaged but several juveniles in Aug–Sep, suggesting post-natal dispersal. The first of only two Irish birds was an immature female shot at Belmullet (Mayo) in Aug 1935; the other was at Larne (Antrim) in Aug 1974.

FAMILY CHARADRIIDAE

Genus *CHARADRIUS* Linnaeus

Cosmopolitan genus of *c.* 32 species; three of the nine Palearctic species breed in GB&I and four more occur with varying degrees of regularity. There are five Nearctic species, two of which are vagrants to GB&I. *Charadrius* is part of a clade that includes lapwings and Diademed Plover *Phegornis mitchelli* and is sister to thick-knees, avocets and stilts (Paton *et al.* 2003).

Little Ringed Plover   *Charadrius dubius* Scopoli

MB PM *curonicus* J. F. Gmelin

TAXONOMY No recent detailed investigation, though Chu (1995) indicates close to Kentish Plover; however, few *Charadrius* included in his analysis. Three subspecies are recognised: *curonicus* across Eurasia from GB&I to China; *jerdoni* from India to Indo-China; *dubius* from SW Pacific islands: Philippines to New Guinea.

DISTRIBUTION Widespread across Eurasian and Oriental regions, from Iberia and NW Africa to New Guinea. Northern population migratory, wintering in tropical Africa, India and Malay archipelago. Other races generally resident or local migrants.

STATUS Colonised GB in 1938 when the first pair nested at Tring Reservoirs (Hertfordshire); being a bird of freshwater shingle banks in rivers and lakes, they were pre-adapted to nesting in the gravel pits and sand quarries that burgeoned across lowland England following WWII, reaching Scotland in 1968 and Wales in 1970. There are no confirmed breeding records from Ireland. Its numbers in GB have been monitored regularly, and by the time of the Second Breeding Atlas, the total was approaching 1,000 pairs, probably surpassing this during the 1990s (Brown & Grice 2005). Latterly, birds have nested on a wider range of substrates: disturbed ground around building developments, reclaimed mine spoil heaps, even supermarket car parks have all held breeding birds. Remains a rare bird in Ireland and scarce in Scotland away from the main breeding areas in the south. Many of these nesting habitats are essentially transient or unstable, and it is easy to envisage a decline setting in unless suitable sites are retained.

In Ireland, only 54 individuals since the first record of four together in Dublin in Sep 1953; there has been a recent surge in records including winter ones. Considered a probable for future breeding; may even have done so (Co. Cork 2006).

The bulk of Little Ringed Plovers arrive back in GB&I during Mar–Apr and most have gone by the end of Aug; there are, however, a few recent wintering records. They are rarely recorded in large numbers; post-breeding parties of up to 12 birds presumably involve two or three families associating at or close to the breeding site. There is a single ringing recovery from Togo; whether this is a true indication of the wintering area of British birds is unclear though it is within the known wintering grounds for the species.

Common Ringed Plover   *Charadrius hiaticula* Linnaeus

RB MB WM PM *hiaticula* Linnaeus
PM *tundrae* (Lowe)

TAXONOMY Two (BWP, HBW) or three (Engelmoer & Roselaar 1998) subspecies recognised, though some hang on to the belief that Semipalmated Plover is also conspecific, despite clear evidence to the contrary. Nominate *hiaticula* breeds N Canada, Greenland, Iceland, Jan Mayen, S Scandinavia, GB&I and France, wintering in Europe and N Africa (many are effectively resident); *tundrae* across Arctic from N Scandinavia to E Siberia, wintering in SW Asia and E & S Africa.

DISTRIBUTION Breeds on sand and shingle close to the sea and along rivers, from islands off NE Canada, across Arctic Russia and Siberia to Pacific Ocean. Demonstrates leap-frog migration: northern populations strongly migratory, wintering in Africa as far as the Cape; those breeding further south are less migratory, and some of the most southerly breeding birds may be resident.

STATUS Both subspecies occur in GB&I; one chiefly on passage, while the other breeds and winters here. The breeding population was monitored regularly during 1970–90 (refs in Prater 1989), but more recent national trends are not known (Thorup 2006). About 1,100 pairs are estimated to breed in RoI, with a further 8,540 pairs in GB&NI; the largest proportion is in Scotland. Many birds nest on sand and shingle banks close to the sea, although there has been an increasing tendency to utilise habitat beside lakes and rivers or on sand and gravel works, especially in England and S Scotland. Ringing data indicate that these birds are comparatively sedentary, with most recoveries from within GB&I, though a few reach as far south as the Mediterranean. There is also clear evidence that *hiaticula* from S Scandinavia, Denmark and N Germany come to S GB&I for the winter (Wernham *et al.* 2002).

Birds passing through GB&I on migration to wintering grounds in Africa will include *tundrae*; the number involved cannot easily be estimated because of the difficulty of separating the races in the field. Perhaps only small numbers (<100 each year) of *tundrae* pass through Scotland, though birds thought to be of this race have occurred at high altitude in the Cairngorms in summer (BS3). The latest population estimates of passage birds (1997–98) for GB, Ireland and IoM are 32,450, 14,580 and 949 (Rehfisch *et al.* 2003a, Crowe *et al.* 2008), representing 44%, 20% and >1% of the flyway population. WeBS counts indicate a progressive decline in wintering numbers since the early 1990s; whether this is a real decrease or the movement of birds away from surveyed sites is not clear. Only about 27% of the GB birds were on estuaries, so WeBS trends may be unreliable. Repeat surveys of non-estuarine shorelines in 1984–85 and then 1997–98 showed a 15% decline (Rehfisch *et al.* 2003b). There is evidence of a shifting of the winter population, with increasing numbers being found along the North Sea coast of England; this presumably reflects the trend towards warmer winters (Rehfisch & Crick 2003), allowing birds to remain on the relatively richer feeding grounds of the east coast estuaries through the less severe winter weather.

Substantial numbers winter in W Africa, where *c.* 98,000 were counted on Banc D'Arguin in 1980; a repeat survey in 1997 recorded only 40,000 (Zwarts *et al.* 1998). Coupled with the declines in GB&I, these data suggest that there is currently extreme pliability in the numbers of Ringed Plovers, extensive and comparatively unrecorded movement among sites or (most worryingly) a serious decline in this species across its range.

Semipalmated Plover   *Charadrius semipalmatus* Bonaparte

SM monotypic

TAXONOMY Replaces Common Ringed Plover *C. hiaticula* in Canada, with some sympatry. Included by Joseph *et al.* (1999) in phylogenetic analysis of *Charadrius*; Palearctic taxa generally excluded from this so little help in resolving relationships. Nol & Blanken (1999) report that Common Ringed and Semipalmated Plovers are monophyletic but with 8% divergence in mt cyt-b, confirming both their close relatedness and their specific identity.

DISTRIBUTION Breeds across Alaska and N Canada from Aleutians to Nova Scotia; winters along coasts from S USA to Patagonia. Evidence from birds ringed in James Bay (Canada) suggests easterly bias during autumn migration, with a prolonged migratory period from July to Nov.

STATUS The first record was a juvenile on St Agnes (Scilly) Oct–Nov 1978. One further record of a bird at Dawlish Warren (Devon), through summer 1997 and probably the same bird in Mar–May 1998. One Irish record, a juvenile at Arranmore Island (Donegal) in Oct 2003.

Killdeer   *Charadrius vociferus* Linnaeus

SM *vociferus* Linnaeus

TAXONOMY A member of the monophyletic Ringed Plover complex that includes Common Ringed *C. hiaticula*, Little Ringed *C. dubius* and Kentish *C. alexandrinus*. Three subspecies; *vociferus* from continental N America; *peruvianus* in coastal Peru and N Chile; *ternominatus* from Antilles. Latter two races believed resident; nominate is migratory (Jackson & Jackson 2000).

DISTRIBUTION Breeds in a wide variety of short vegetation. Antilles and S American races probably resident; nominate found across N America in suitable habitat S of boreal forests; winters from southern N America into coastal Ecuador, Colombia and Venezuela. An early spring migrant, usually deserting winter grounds by Feb–Mar (Jackson & Jackson 2000); earliest birds may arrive in northern breeding areas by end Mar. Prone to long-distance hard-weather movements, possibly accounting for many of the records from GB&I.

STATUS A female apparently paired with a male Common Ringed Plover in Shetland in 2007; although distraction display was seen on several occasions, there appears to have been no firm evidence of breeding.

First record: near Christchurch (Dorset, then Hampshire) Apr 1859. First Irish record was a bird shot near Balbriggan (Dublin) in Jan 1928. A total of *c.* 60 records, including 17 from Ireland, and only 7 before 1950. Mostly Nov–Mar and widespread, though unsurprisingly more in the west. Annual pattern of records is unusual: of 56 winters from 1950 to 2006, no Killdeers were recorded in 29, with multiple arrivals in 14. Only 13 years with a single record is less than would be expected and, since relatively few were parties, this supports the idea of a common cause for their arrival such as adverse weather conditions in N America.

Kentish Plover   *Charadrius alexandrinus* Linnaeus

FB PM *alexandrinus* Linnaeus

TAXONOMY Five subspecies usually recognised; *alexandrinus* from S Eurasia E to Korea, N Africa; *dealbatus*: E China, Japan; *seebohmi*: S India, Sri Lanka; *nivosus*: USA and Caribbean; *occidentalis*: coasts of Peru, Chile. Last two often called Snowy Plover and may be a separate species.

DISTRIBUTION Usually nests on sand, by sea or saline lakes and lagoons. Breeds across temperate Eurasia to Pacific, also N Africa, Nile and Rift Valleys, coastal Arabia. N populations migratory, wintering across N Africa, India, Indo-China to Philippines. Separate populations in N and S America, where inland birds migratory.

STATUS Once bred along the coast of SE England at sites in Kent and Sussex such as Dungeness, Pegwell and Sandwich Bays and Rye Harbour, with up to 40 pairs during 1907–1911 (Brown & Grice 2005). Numbers declined through the 20th century and now only sporadic breeding occurs, with the last successful pair being in Lincolnshire in 1979. The loss seems to have been due to a combination of persecution for eggs and skins, disturbance by shingle extraction and habitat loss to coastal housing.

Currently, it is a scarce passage bird and numbers seem to be falling; Fraser & Rogers (2006) report an average of *c.* 30 per year since 1986 but comment that this includes a fairly steady decline since 1999. This mirrors the scenario along the Atlantic and North Sea coasts of Europe, where decreases have been recorded in all countries apart from France, Portugal and the very small population in Denmark (BirdLife International 2004). In line with the former breeding distribution, most records come from SE England. Elsewhere in GB&I, it is extremely rare, with less than 20 records each from Scotland and Ireland and not many more from Wales.

Lesser Sand Plover   *Charadrius mongolus* Pallas

SM *atrifrons* group (*atrifrons* Wagler, *pamirensis* (Richmond), *schaeferi* de Schauensee)
SM *mongolus* group (*mongolus* Pallas, *stegmanni* Portenko)

TAXONOMY Five races generally recognised based on breeding plumages and relative sizes (see Roselaar in BWP); *mongolus*: inland in E Siberia; *stegmanni*: Kamchatka, Chukotsky; *pamirensis*: Kazakhstan, W China; *atrifrons*: Himalayas, Tibet; *schaeferi*: E Tibet. These usually divided into two groups: NE *mongolus/stegmanni* and SW *atrifrons/pamirensis/schaeferi*.

DISTRIBUTION Disjunct across C and E Asia, with populations from Kazakhstan to Himalayas, E Siberia, Kamchatka, Chukotsky. All populations migratory, wintering in coastal E Africa, from Red Sea to Cape, also coasts of Arabia, India to Australia.

STATUS Four or five records. The first was an adult at Pagham Harbour (Sussex) in Aug 1997; this was identified as member of the *atrifrons* group from photographs. The second at Rimac (Lincs) in May 2002 was a female of the same race. The third at Keyhaven Marshes (Hampshire) in July 2003 belonged to the *mongolus* group, as did a bird at Aberlady (Lothian) in July 2004. The identification of bird at the Don Estuary (Aberdeenshire) in Aug 1991 is being reassessed. Not recorded in Ireland.

Greater Sand Plover   *Charadrius leschenaultii* Lesson

SM race undetermined

TAXONOMY Three subspecies; *columbinus*: Turkey and Syria to S Caspian; *crassirostris*: Caspian to Kazakhstan; *leschenaultii*: Altai Mts, across Asia S of Lake Baikal to W China.

DISTRIBUTION Scattered populations breed in desert or semi-desert from Turkey to Lake Baikal. W populations of *columbinus* winter NE Africa, Red Sea; *crassirostris* winters E Africa, S to Cape; *leschenaultii* winters S and E to Australia. Generally coastal in winter.

STATUS The first was a first-winter bird, recorded at Pagham Harbour (W Sussex) in Dec 1978–Jan 1979. There have been about 12 further birds, including one that wandered round SE England for a week in Aug 1992. Recorded fairly widely across GB from Apr to Dec. Not recorded in Ireland.

Caspian Plover   *Charadrius asiaticus* Pallas

SM monotypic

TAXONOMY Believed to be closely related to Oriental Plover *C. veredus* (Hayman *et al.* 1986). No morphological variation.

DISTRIBUTION Caspian Sea to Kazakhstan; usually breeds close to water. Migratory, wintering in E half of Africa from Nile Delta to Cape.

STATUS The first record was, remarkably, a pair shot near Great Yarmouth (Norfolk) in May 1890. Then a gap of almost 100 years, until two widely separate individuals in 1988 (Scilly in May, Midlothian in July), followed by a further bird in Shetland in 1996. There are no records from Ireland.

Eurasian Dotterel   *Charadrius morinellus* Linnaeus

MB PM monotypic

TAXONOMY Now treated as member of *Charadius*; previously sometimes in the monospecific genus *Eudromias* (e.g. Hayman *et al.* 1986).

DISTRIBUTION Arctic and subarctic from GB&I across N Eurasia to Bering Straits; isolated populations in Pyrenees, Alps, Carpathians, Altai. All populations migrate to Africa and Middle East, wintering in narrow zone from Morocco to Iran.

STATUS As a breeding bird, Eurasian Dotterels are restricted to the summits of some our highest and most remote hills; most of these are in Scotland, with a few pairs nesting in N England, the Lake District and S Scotland and one breeding record from Ireland. Numbers declined, especially in N England, through the latter part of 19th century, probably due to habitat degradation (refs in Thompson & Brown 1992). The Scottish population has been more stable through the 19th and 20th centuries and seems to have increased between the two Breeding Atlases, with densities increasing and new hills colonised. Whether this due to genuine increases or improved reporting is unknown (e.g. Watson & Rae 1987). Subsequent decline in the Scottish population since the late 1980s (Whitfield 2002). The reasons for this are unclear; Eurasian Dotterels appear to move relatively freely between regions, with records of birds failing in one massif and relocating to another for a second attempt (Wemham *et al.* 2002). Climate change may be playing a part but further study is needed. In 1999, the population of GB&I was estimated as *c.* 630 incubating males, probably indicating a decline since a peak in the late 1980s.

Elsewhere in GB&I, Eurasian Dotterels are more familiar as passage migrants, showing remarkable fidelity to their staging areas – sometimes even to the same fields. Relatively few are detected by WeBS counters, as the preferred habitat seems to be grass or arable fields. The numbers seen in spring are markedly fewer than the breeding population, suggesting that many fly direct to the hills. Ringing recoveries of Scottish birds indicate that they winter in NW Africa and some birds pass through Scottish hills in spring on way to Scandinavia.

Genus *PLUVIALIS* Brisson

Holarctic genus of four species, one Nearctic, all of which have been recorded in GB&I.

American Golden Plover   *Pluvialis dominica* (P.L.S. Müller)

PM monotypic

TAXONOMY Coexists with Pacific Golden Plover in parts of Alaska without interbreeding. Difference in size and plumage between this and *P. fulva* is marked (Connors 1983) but slightly less in area of sympatry (BWP); now regarded as specifically distinct (Knox 1987, Connors *et al.* 1993).

DISTRIBUTION Nests on tundra, preferring drier areas, across Alaska and N Canada to Hudson Bay and Baffin Island. Winters in South America, on grasslands of Chile and Argentina. Route for many involves crossing W Atlantic to S America, possibly in single journey from James Bay, Canada to S America; this probably source of birds in GB&I. Failed breeders may leave breeding grounds as early as late June but juveniles rarely depart before late Aug. Spring passage seems to be more continental, with relatively few along Atlantic coast (Johnson & Connors 1996).

STATUS There are three records from the 1800s, including the first for GB at Perth (Perth and Kinross) in Aug 1883 and the first for Ireland at Belmullet (Mayo) in Sep 1894. There were then no records for over 50 years, until the first modern record of a bird shot near Kells (Meath) in Nov 1952, and the first modern record for GB was an adult on Fair Isle (Shetland) in Sep 1956. Since then, no doubt with improved observer awareness, the species has become increasingly recorded, with over 300 individuals in GB&I. Peaks occur in May and Aug–Nov, especially Sep–Oct. In autumn, adults are significantly earlier than juveniles: 75% of the latter arrive in Oct–Nov whereas 60% of adults are found in Aug–Sep. Presumably, some of the earlier adults are failed breeders. There are several records of first-summer birds in Apr–June; perhaps birds from the previous autumn that have wintered in Africa or Europe, especially since spring passage in N America is less oceanic than in the autumn.

Pacific Golden Plover   *Pluvialis fulva* (Gmelin)

SM monotypic

TAXONOMY Closely related to *P. dominica* and *P. apricaria*, coexisting with former in Alaska and latter around Yamal Peninsula (Siberia), with little or no interbreeding (Connors *et al.* 1993). Birds breeding in Alaska have longer wings than those from Siberia (Johnson & Connors 1996).

DISTRIBUTION Yamal Peninsula to Alaska, breeding on dry tundra. Strongly migratory, with populations wintering in Horn of Africa, coastal SE Asia to Australia and many islands of S Pacific. Some birds make the journey from Alaska to Pacific islands in single stage but route north in spring is less well known. Studies of fat levels indicate that this may comprise >35% body weight, giving flight range of as much as 10,000 km (Johnson *et al.* 1989). Williams & Williams (1990) report that migrant *P.*

dominica may fly as high as 6 km over the ocean; parallel studies not available for *P. fulva* but probably similar, perhaps accounting for regular occurrence in GB&I.

STATUS The first for GB was a bird shot at Epsom (Surrey) in Nov 1870, but first Irish record at Tacumshin (Wexford) not until Aug 1986. A much rarer bird than *P. dominca*, with *c.* 80 records (only 10 in Ireland to 2004) and showing quite a different pattern of occurrence. Only 16% of aged individuals were juveniles, compared with 62% *P. dominica*. Adults peak in July–Aug, a month earlier than *dominica*; juveniles in Oct–Nov. Wide scatter of locations but a distinct bias towards the north and east of GB and S of Ireland. 80% of Pacific Golden Plovers have been recorded since 1990, compared with only *c.* 57% of American Golden Plovers. This striking significant difference ($P<0.01$) may reflect increasing observer awareness (broadly coinciding with treatment as separate species) or genuine changes in patterns of vagrancy.

Relating to the latter, there are anecdotal reports (summarised in Jukema *et al.* 2001) suggesting Pacific Golden Plovers probably occurred regularly in the Netherlands prior to the flooding of the Ijsselmeer in the 1930s. The local trappers ('wilsternetters') reported that flocks of small dark plovers came to roost inland, usually appearing after several days of frost which drove the Eurasian Golden Plovers away. These birds often associated with Dunlin, something Eurasian Golden Plover rarely did. There are nine skins of these birds, obtained by the wilsternetters, that are clearly Pacifics. This indicates the possibility, or even likelihood, that these birds migrated SW across Asia to W Europe; the subcutaneous fat and dense plumage reported by the wilsternetters would then be adaptations to severe winters; birds wintering in the Pacific show neither of these features. Whether these birds were a separate subspecies as suggested by Jukema *et al.* (2001), the population is now extinct; there seem to be no records from GB&I at the relevant times that indicate our islands were involved in this remarkable story.

European Golden Plover   *Pluvialis apricaria* (Linnaeus)

RB MB WM PM monotypic

TAXONOMY One of a group of closely similar forms (*P. apricaria, P. dominica, P. fulva*). Variable in plumage across range, but BWP and Byrkjedal & Thompson (1998) suggest not consistently enough for racial separation, although HBW recognise two subspecies. This is a species that would merit a phylogenetic investigation.

DISTRIBUTION Breeds on moorland and coastal tundra, from Iceland, GB&I, Scandinavia and Baltic States to Taimyr. Winters GB&I, France and around Mediterranean Sea to S Caspian. Southern populations less migratory.

STATUS Showed a decline of *c.* 10% in occupancy of 10-km squares between the Atlases, the losses being especially severe across Scotland, from Galloway and the border counties to the S & E Highlands. Much of this was ascribed to afforestation (Second Breeding Atlas), with 19% of the population of Caithness and Sutherland being displaced (Thompson *et al.* 1988). Some losses also occurred in Ireland and Wales, though the species was scarce there to begin with. Concerns about human disturbance, especially in the Pennines, currently seem unfounded (e.g. Yalden & Yalden 1988). The main threats to the breeding population remain habitat loss and increased predation, especially in areas where keepering is reduced.

The winter population of both GB and Ireland is poorly known (Gillings 2005a). WeBS and IWBS data give an indication of numbers at key sites but many occur on unsurveyed farmland. In 2005, the GB winter population was estimated at 250,000, excluding farmland birds, but a more systematic survey in 2006–07 across all habitats suggested the figure to be *c.* 400,000. In Ireland, peak numbers occur in Dec–Jan but this might include immigrants from GB. The latest population estimate for Ireland is *c.* 166,700 (Crowe *et al.* 2008) and represents *c.* 18% of the flyway population; the majority of these are from Iceland (Wernham *et al.* 2002). While some of those wintering in GB are also from Iceland, others are either local birds or immigrants from the Continent. There are ringing records (and sightings) of birds that have been trapped on passage through the Netherlands.

For such a conspicuous member of our avifauna, it is surprising how little we know about its abundance and distribution; there is a clear need for more systematic surveying (Gillings 2005a). Certainly, numbers at individual sites seem to be very labile, with counts fluctuating markedly within and among years, and many observers report localised declines from the interior of Britain. Regional trends in coastal wetlands (in GB) show a marked increase on the east coast and reduction elsewhere (Gillings 2005b; Gillings *et al.* 2006). Sometimes, this can be correlated with changes in local weather; for example, an increase across east coast sites in Dec 2004 and Jan 2005 coincided with severe weather in the Netherlands and a fall in numbers there (Banks *et al.* 2006), but there may also be a general trend towards the (now) milder eastern regions.

Grey Plover   *Pluvialis squatarola* (Linnaeus)

WM PM monotypic

TAXONOMY Phylogeny of *Pluvialis* unresolved, so relationships with rest of genus unclear. Slight clinal variation in size (BWP) but, despite wide distribution, no evidence of subspecific differentiation (though see Engelmoer & Roselaar 1998).

DISTRIBUTION Breeds on lowland tundra in High Arctic, from White Sea E to Bering Straits, coastal Alaska and islands off N Canada. Strongly migratory, wintering along coasts of all major continents, south of areas where shoreline freezes.

STATUS Winters on muddy and sandy shores and estuaries around the coasts of GB&I, especially in the S & E, with small numbers of sub-adults remaining through the summer months. The pattern of occurrence differs between GB and Ireland. Peak numbers are recorded in GB during the autumn, decreasing to a plateau through the winter, presumably indicating the southwards movement of passage migrants, followed by a more stable winter population. In Ireland, this autumnal peak is not evident; numbers increase through the autumn to a maximum in midwinter, suggesting that there is less passage though this region. Crowe (2005) shows that the number wintering in the Irish Republic has declined quite markedly since the mid-1990s. These data are more limited than those for GB and NI, where wetland bird censuses began 25 and *c.* 5 years earlier, respectively. Both sets of data show an increase to *c.* 1995, since when there has been a decline that parallels the picture in the Republic, and these changes are also detectable at a local level (Ward *et al.* 2003).

The increases through the early 1990s are paralleled across W Europe, from Spain to Germany. Further south, in Mauritania and Tunisia, winter populations have declined, suggesting a northwards shift to the wintering areas (Stroud *et al.* 2004). However, the increases in W Europe may have been partly driven by rising numbers of birds in the arctic regions of European Russia (Stroud *et al.* 2004); whether this expansion also resulted in increased productivity seems not to be known. It is possible that the more recent decline in the winter populations in GB&I is due to warmer winters on the European side of the North Sea reducing the need for birds to leave the rich feeding grounds in N Germany and Denmark; certainly there is direct evidence that the number wintering in the Netherlands has increased steadily in recent years (van Roomen *et al.* 2005).

The latest population estimates are 52,750 (GB), 6,315 (Ireland), 52 (IoM); the former represent 21% and 3% of the flyway, and 96% of these were counted on estuaries (Rehfish *et al.* 2003a; Crowe *et al.* 2008).

Genus *VANELLUS* Brisson

Near-cosmopolitan (not Nearctic) genus of 24 species. Six occur in the Palearctic, of which three have been recorded in GB&I, one as a common breeder.

Sociable Lapwing   *Vanellus gregarius* (Pallas)

SM monotypic

TAXONOMY Reanalysing an osteological data set derived by Strauch (1978), Chu (1995) showed that the lapwings form a clade with oystercatchers, stilts and avocets, although statistical support is not strong. Using sequences from the nuclear gene RAG-1, Paton *et al.* (2003) found a similar relationship, although again with only limited statistical support. However, these two independent data sets both suggest this relationship may be valid.

DISTRIBUTION Nests in dry steppe and stubble across Kazakhstan and adjacent parts of Russia but declining, perhaps due to intensification of agriculture on breeding grounds. Strongly migratory, wintering NW India and Pakistan, Iraq and Sudan, though now uncommon in the latter areas. Global population estimated at less than 10,000 mature individuals (HBW). However, recent reports of parties in excess of 1,000 individuals in SE Turkey and NW Syria, giving hope of larger population. A review of the ecology in NW Kazakhstan (Belik 2005) suggests the decline there is due to dense ground vegetation because of fewer livestock and new shelter belts providing nest sites for corvids (primarily Rooks), which intensively predate nests and eggs.

STATUS The first record was near St Michael's-on-Wye (Lancashire) in autumn *c.* 1860. The first of only four Irish individuals was near Navan (Meath) in Aug 1899, with the other three records in Co. Waterford (1909), Kerry (1985) and Offaly (1998); the last three were all in Dec. Very rare in first half of 20th century, with only five records before 1950, but increased in frequency and almost annual since late 1960s. Has occurred in most months but a distinct peak Sep–Oct. The regularity of occurrence is surprising in view of its global rarity.

White-tailed Lapwing   *Vanellus leucurus* (Lichtenstein)

SM monotypic

TAXONOMY No recent phylogenetic study, so relationships to other lapwings unclear.

DISTRIBUTION Usually nests close to water, along rivers or beside lakes from S Turkey across Middle East to Caspian and Aral Seas. Mostly non-migratory, though C Asian populations move S to winter in Pakistan and NW India. Birds wintering in Sudan perhaps from Middle East populations.

STATUS The first record was of an adult male at Packington (Warwickshire) in July 1975. Four more records subsequently: in 1979 (Dorset), in 1984 (Durham and Shropshire, both May) although the latter two could relate to the same individual, and 2007 in Dumfries and Galloway, and Lancashire (the same individual). In both 1975 and 1984, vagrant birds were recorded elsewhere in Europe, suggesting a possible influx from the Middle East breeding grounds. Not recorded in Ireland.

Northern Lapwing   *Vanellus vanellus* (Linnaeus)

RB MB WM PM monotypic

DISTRIBUTION Widespread across temperate Eurasia, from Atlantic to Pacific, nesting in short vegetation in open grassland and arable crops. N populations migratory, wintering (often in large flocks) S of the 3°C isotherm; distribution determined partly by severity of winter weather but as far south as Mediterranean Sea in W, across N India to S China.

STATUS Breeding range contracted by 9% (GB) and 29% (Ireland) between the two Breeding Atlases, and there was a further decline in GB of 49% through 1990–2001 (Chamberlain & Crick 2003); similar losses have occurred in Ireland. Losses were especially noticeable in arable areas, where the conversion to autumn sowing and the increased cultivation of silage both resulted in a sward too long for successful nesting. Intensive grazing, especially of marginal uplands, also reduced productivity; clutch sizes were lower and nest failures higher in these habitats (Chamberlain & Crick 2003). A recent study has shown that spring/summer fallow enhanced nest survival and (perhaps unsurprisingly) that nest predation was more likely when they were close to field borders or predator perches (Sheldon *et al.* 2007). Stroud *et al.* (2004) report a contraction in range in over half the countries of Europe during 1970–90 and ascribe this to agricultural intensification. Although this was less evident in eastern Europe, the same effect is likely to happen as farming practices change there too.

WeBS data from survey sites in GB indicate an increase in winter numbers from 1980 to 2004, whereas in Ireland there has been a decrease since 1990 (Crowe *et al.* 2008); a recent survey of wintering flocks (Gillings and Fuller 2009) suggested a decline in GB, but it is recognised that earlier surveys may have been hampered by the volatility of Northern Lapwing flocks and the likelihood of missing some and counting others more than once. Ringing data indicate a substantial immigration to Ireland, especially from NW GB. There is also emigration to SW Europe from all parts of GB, and this is increased during periods of winter cold; Peach *et al.* (1994) showed that ringing recoveries from Iberia rose during severe winters.

Numbers at individual wintering sites in RoI tend to be higher than those in GB, with 34 censused sites holding more than 2,000 birds, compared with only 8 in GB and none in NI. The decline in NI may be due to fewer birds wintering there, and the resulting shift eastwards might, at least partly, account for changes in distribution within GB. There is evidence of a movement towards more easterly localities, and some observers have reported localised disappearance from inland sites, although this is less noticeable than for European Golden Plovers (Gillings 2005b). An alternative scenario is that, in GB, a worsening of the suitability of agricultural land for wintering Northern Lapwings has forced them onto wetland sites, where they are more amenable to census. A further scenario is that recent milder winters have permitted Northern Lapwings to winter further east and closer to the breeding grounds (Gillings *et al.* 2006). The wintering population in GB is *c.* 620,000, while that for Ireland is *c.* 207,700 (Crowe *et al.* 2008, Gillings and Fuller 2009).

The status of Northern Lapwings is in need of careful assessment in both winter and summer, and the analysis of habitat usage in winter should be urgently addressed.

FAMILY SCOLOPACIDAE

Genus *CALIDRIS* Merrem

Holarctic genus of 19 species, 14 of which breed in the Palearctic and 12 in the Nearctic; all except Rock Sandpiper *C. ptilocnemis* have been recorded in GB&I, though only one breeds here commonly and another three rarely.

Great Knot   *Calidris tenuirostris* (Horsfield)

SM monotypic

TAXONOMY Chu's (1995) reanalysis of Strauch's (1978) osteological data is rather uninformative, placing Great Knot in a large and unresolved clade that includes almost 30 other taxa. In their supertree analysis, Thomas *et al.* (2004b) found that it was sister to Red Knot, largely based on morphological data. A phylogeny based on the mt DNA sequences for cyt-b and ATPase showed sister species with Red Knot/Surfbird *Aphriza* (Borowik & McLennan 1999).

DISTRIBUTION Breeds on montane tundra, with or without scattered shrubs, in NE Siberia but distribution still poorly known (HBW). Long-distance migrant, wintering principally in SE Asia and Australia, with small numbers in coastal N India, Pakistan and Arabian Gulf.

STATUS The first record for GB was at Scatness and Pool of Virkie (Shetland) in Sep 1989. This was followed by a second at Seal Sands (Cleveland) in Oct–Nov 1996 and a third on the Wyre Estuary

(Lancashire) in July–Aug 2004. There is a single record from Ireland: on the Swords Estuary (Dublin) in July 2004. All were adults.

Red Knot   *Calidris canutus* (Linnaeus)

WM PM *canutus* (Linnaeus)
WM PM *islandica* (Linnaeus)

TAXONOMY From their analysis of cytochrome oxidase 1, Hebert *et al.* (2004) found Red Knot to be closely similar to Surfbird, a conclusion supported by Thomas *et al.* (2004b).

Piersma & Davidson (1992) recognise six subspecies, based on body size, plumage and migratory flyways. From W to E, these are (wintering areas in parentheses): *canutus*: Taimyr Peninsula (W & SW Africa); *piersmai*: New Siberian Islands (NW Australia); *rogersi*: Chukotsky Peninsula (SE Australia, New Zealand); *roselaari*: NW Alaska, Wrangel Island (SE USA); *rufa*: central Canadian Arctic (S Patagonia, Tierra del Fuego); *islandica*: NE Canada, N Greenland (NW Europe).

Buehler & Baker (2005) examined the rapidly evolving mt control region and found evidence of recent divergence, perhaps during or after the last glacial maximum (*c.* 20,000 yrs BP). They suggest a sequential eastwards expansion, perhaps from a refuge in C or E Eurasia (*canutus*). Genetic divergence indicates that *piersmai* and *roselaari* became established as the ice receded, the latter having crossed Behringia as the ice sheets melted. Subsequent expansion westwards led to the establishment of *rogersi* in the Palearctic and *rufa* and *islandica* in the Nearctic. This hypothesis is supported by a progressive decrease in genetic diversity: *canutus* has the highest level, consistent with its basal position (and hence likely the oldest taxon); *piersmai* and *roselaari* are intermediate; *rogersi*, *rufa* and *islandica* are least variable, as expected if they diverged most recently, with insufficient time to recover variation lost through a succession of founding effects.

DISTRIBUTION Holarctic and circumpolar in High Arctic latitudes, nesting on tundra close to water and usually near sea between 83°N and 64°N. Winters along coasts on tidal mud or sand flats of bays and lagoons.

STATUS The race wintering in GB&I is *islandica*. Numbers fell across the range in the early 1970s, probably due to three summers when weather conditions in the Arctic were very poor, leading to extensive adult mortality as well as widespread breeding failure (refs in Stroud *et al.* 2004). Following a decline of *c.* 40% in the wintering population, there was a steady recovery through to the 1990s, after which numbers stabilised at *c.* 450,000, of which *c.* 300,000 winter in GB&I (Crowe 2005, Banks *et al.* 2006). The most important sites, holding recent mean peaks in excess of 20,000 birds each, are the Wash, Morecambe Bay and the estuaries of the Humber and Thames (in E England), and the Ribble, Alt and Dee (in NW England), with another seven sites holding internationally significant numbers (>1% of the racial population). Numbers are lower in Ireland, though 8% of the total population winters there, with internationally important concentrations at Dundalk Bay (Louth) and Strangford Lough (Co. Down). The latest total winter population estimates are *c.* 283,600 (GB), 18,970 (Ireland), 172 (IoM); the first of these represents 63% of the flyway population; >95% were on estuaries.

In the 1990s, the global population of the nominate race was estimated at *c.* 340,000 individuals (Stroud *et al.* 2004), breeding on the Siberian tundra and wintering around the coasts of Africa from Mauritania to the Cape. This reflected a decrease of *c.* 34% during the first half of that decade. The cause of this decrease seems unclear. Migration seems to comprise an initial flight from the breeding grounds to the Wadden Sea, where perhaps the entire population rests and feeds for two to three weeks during Aug (adults) and Sep (juveniles) prior to the next stage south, again perhaps in a single stage to NW Africa. Relatively few of these birds cross the North Sea to GB&I, though juveniles are found in some years. Returning birds again predominantly use the Wadden Sea in May, remaining there to feed for two to three weeks; adverse weather may force these birds further west, and parties are more often found in GB&I in spring than autumn.

Sanderling   *Calidris alba* (Pallas)

WM PM monotypic

TAXONOMY Osteology (Chu 1995) and supertree analysis (Thomas *et al.* 2004b) both indicate part of a large, relatively unresolved clade of *Calidris*. Borowik & McLennan (1999) identified membership of a clade with Dunlin, and Purple and Buff-breasted Sandpipers, but Hebert *et al.* (2004) suggested greater similarity with White-rumped Sandpiper. Only slight geographical variation in plumage and size, though Englemoer & Roselaar (1998) suggest two subspecies. However, generally treated as monotypic: extensive morphological overlap between populations.

DISTRIBUTION Holarctic and circumpolar. Breeds in very high Arctic from 83°N and 64°N on stony tundra. Migratory, wintering along most sandy or muddy coasts that are ice-free in the boreal winter.

STATUS Regularly seen along sandy shores around the coasts of GB&I, its habit of persistently chasing the ripples being both endearing and unique. The E Atlantic population has been estimated at *c.* 123,000 (Stroud *et al.* 2004), and WeBS data indicate that the Ribble estuary (Lancashire) holds internationally significant numbers (>2,300) during the winter, and that site plus the Wash and the estuaries of the Thames and the Alt (Lancashire) all reach international significance on passage. Generally, two populations of Sanderlings use the beaches of GB&I. One of these comprises migrants staging in GB&I for two to three weeks in Aug and May, on their way between the breeding and wintering grounds. Ringing data suggest that many (most?) of these birds originate in Greenland and some continue as far as South Africa, although others remain here throughout the winter. The second group are primarily winter visitors, originating in W Siberia; although many of these remain in W Europe, some continue south into W Africa (refs in Stroud *et al.* 2004).

The number recorded in GB has fluctuated over the years; a slight decline through 1975–95 was followed by an increase to 2000, followed by another decline. Numbers in Ireland, though smaller, have been more stable since the mid-1990s; as with GB, showing peaks in Aug and May. The latest winter estimates are *c.* 20,540 (GB) and 6,680 (Ireland), representing 17 and 5% of the flyway population (Rehfisch *et al.* 2003a; Crowe *et al.* 2008). About 65% of the GB birds were on non-estuarine habitats, which are poorly covered by WeBS fieldwork. The decline reported by this is paralleled by a decline of >20% on non-estuary habitats between 1984–85 and 1997–98 (Rehfisch *et al.* 2003a): perhaps another instance of changes in distribution in reponse to warmer winters.

Semipalmated Sandpiper   *Calidris pusilla* (Linnaeus)

SM monotypic

TAXONOMY Borowik & McLennan (1999) report that this forms a clade with Baird's and Western Sandpipers; although supporting this for Western, Hebert *et al.* (2004) found Baird's to be more different. The evolutionary history of the small stints still needs resolving. Semipalmated Sandpiper shows clinal variation in size, especially in bill length, which is longer in the east. Harrington & Morrison (1979) reported a step in this cline, and Haig *et al.* (1997) found evidence of genetic differences across this boundary but insufficient for racial separation.

DISTRIBUTION Northern Nearctic, breeding on wet tundra from W Alaska to N Labrador between 71°N and 55°N; also in dryer areas near lakes and deltas. Long-distance migrant, taking 4,000-km flights to winter in Caribbean and along coast of N South America.

STATUS The first record for GB was an adult at Cley (Norfolk) in July 1953, and the first for Ireland was at Ballycotton (Cork) in Oct 1966. Subsequently, there have been over 140 records in GB&I, all between May and Nov. 65% of the aged birds were juveniles and 60% of these occurred in Sep. Adults occurred from July, presumably the earliest involve dispersing failed breeders. The handful of spring birds may include individuals that crossed the Atlantic the previous autumn and wintered in Africa or the W Palearctic.

Western Sandpiper   *Calidris mauri* (Cabanis)

SM monotypic

TAXONOMY Cytochrome oxidase 1 and cyt-b both suggest this is close to Semipalmated Sandpiper (Borowik & McLennan 1999, Hebert *et al.* 2004), in accord with morphology-based taxonomies (e.g. HBW).

DISTRIBUTION Restricted to coastal Alaska and Chukotsky Peninsula, wintering along E and W coasts of USA south to N South America.

STATUS First recorded at Tresco (Scilly) in Aug 1969, with five subsequent records from GB. There have been three in Ireland, with the first at North Slob (Wexford) in Sep 1992. Of the birds that were aged, three adults were in Aug and the four juveniles were in Sep.

Red-necked Stint   *Calidris ruficollis* (Pallas)

SM monotypic

TAXONOMY Omitted by Borowik & McLennan (1999) and Hebert *et al.* (2004), though Thomas *et al.* (2004b) placed it in a clade with Long-toed Stint and Pectoral Sandpiper. Conventional taxonomy places this close to Little Stint, as did Sibley & Monroe (1990). Another species whose phylogenetics warrant further investigation.

DISTRIBUTION C Taimyr discontinuously through to Chukotsky Peninsula, also occasionally in N and W Alaska. Winters in SE Asia to Australasia. Appears to be declining in Australia and Tasmania, though not in New Zealand (HBW)

STATUS Eight records to 2003, six in GB and two in Ireland. The first record was of an adult at Blacktoft Sands (Humberside) in July 1986. Four of the five subsequent birds have also been adults (in July–Sep), as were the two Irish birds. Both of the latter were at Ballycotton, Cork, the first in July 1998. It is remarkable that this species was not recorded before 1986, especially since almost all have been adults and were relatively easy to identify. The pattern of low frequencies of juveniles being recorded in GB&I is typical of other Siberian vagrant waders, such as Pacific Golden Plover, Great Knot and the sand plovers.

Little Stint   *Calidris minuta* (Leisler)

PM monotypic

TAXONOMY Placed in a superspecies with Red-necked Stint by Sibley & Monroe (1990), but Thomas *et al.* (2004b) found close affinity with Semipalmated Sandpiper. Omitted by Hebert *et al.* (2004), though Borowik & McLennan (1999) found it comprised a clade with Least and White-rumped Sandpipers. Warrants further analysis.

DISTRIBUTION N Palearctic, from NE Norway across tundra to New Siberian islands. Winters from Mediterranean and Africa, to Middle East and Indian subcontinent.

STATUS The population wintering in Europe and W Africa has been estimated as *c.* 200,000; of which only a handful winter in GB&I. More often recorded on passage, especially on the E & S coasts, and typically along the margins of fresh or brackish water. Scarce in the west of GB and in Ireland. The numbers can fluctuate markedly among years, although there does not appear to be any regularity in the cycles of abundance. Most are in autumn, when juveniles far outnumber adults. Good years are often (although not always) associated with higher than average numbers of Curlew Sandpipers, and local weather conditions can sometimes play a part in the influx as birds may arrive with easterly winds.

Temminck's Stint   *Calidris temminckii* (Leisler)

CB PM monotypic

TAXONOMY Has not been included in any extensive molecular phylogenies but usually placed close to Long-toed Stint (e.g. Vaurie 1965; Sibley & Monroe 1990).

DISTRIBUTION Across N Palearctic, between 74°N and 62°N, from C Norway to Bering Straits; generally S of Little Stint, though some distributional overlap. Winters in C Africa, Middle East, coastal India and Indo-China. According to BirdLife International (2004), Temminck's Stints are declining in Finland and Sweden, though stable in Norway.

STATUS Temminck's Stint is a very rare breeding species in GB. Birds have been present at a small number of sites in Scotland since the 1970s, but successful breeding is not often confirmed; the average number of possible/probable pairs through the 1990s varied between two and four, but last confirmed breeding was in 1993 (RBBP). There is a single (unsuccessful) breeding record from England: in Yorkshire in 1951 (see Mather 1986). It seems unlikely that the hoped-for colonisation of Scotland is imminent.

A scarce migrant through GB&I, usually singly on fresh water. Birds are generally commoner in spring than autumn, and peak numbers occur in May and Sep–Oct. Typically they appear singly or in very small numbers, only rarely away from fresh or brackish water margins. Fraser & Rogers (2006) find little evidence of change in abundance from 1968 to 2003 (averaging 105, 95, 112 per year in the decades since 1980). However, the authors note an approximately five-year periodicity in numbers, which makes identifying trends more problematic. It is much less common in Ireland, with only 33 records.

Long-toed Stint   *Calidris subminuta* (Middendorff)

SM monotypic

TAXONOMY Not included by Hebert *et al.* (2004) or Thomas *et al.* (2004b), but Borowik & McLennan (1999) found this to be sister to Red-necked Stint; however, they omitted Temminck's. Disjunct populations but no morphological evidence for subspecies.

DISTRIBUTION Scattered locations across taiga and S tundra, from SW Siberia to S Chukotski Peninsula. Winters SE Asia to Australia.

STATUS Two records from GB and one from Ireland. The first was an adult at Marazion (Cornwall) in June 1970; this remained unconfirmed for many years because so few observers were familiar with the species in adult plumage (Round 1996). The second in GB, a juvenile at Saltholme Pools (Cleveland) in Aug–Sep was, by contrast, relatively straightforward. The only Irish record is of an adult at Ballycotton (Cork) in June 1996.

Least Sandpiper   *Calidris minutilla* (Vieillot)

SM monotypic

TAXONOMY Borowik & McLennan (1999) identified a clade with Little Stint and White-rumped Sandpiper, but Sibley & Monroe (1990) placed in superspecies with Long-toed Stint.

DISTRIBUTION Widely distributed across Alaska and boreal Canada. Winters from S USA to Brazil, migrating overland in broad front.

STATUS This species is much rarer in W Europe than Semipalmated Sandpiper, despite extensive overlap in their breeding ranges. The first for GB was at Marazion (Cornwall) in Oct 1853; there were three more records in the 19th century and then none for over 50 years. The modern series of records began in 1955, and there have been *c.* 30 birds since then. The first for Ireland was in Aug 1955 at

Akeragh Lough (Kerry). Since then, there have been a further eight to 2004. One record in Feb; birds in July–Aug were all adult, whereas those in Sep–Oct were juveniles. 70% of the aged Least Sandpipers have been adult, compared with only *c.* 35% of Semipalmated Sandpipers.

White-rumped Sandpiper   *Calidris fuscicollis* (Vieillot)

PM monotypic

TAXONOMY Hebert *et al.* (2004) found the cytochrome oxidase I sequence to be closely similar to Sanderling, with Dunlin rather more different. This differs from Borowik & McLennan (1999), who found White-rumped in the same clade as Little and Least Sandpipers, with Sanderling and Dunlin well separated, both from White-rumped and from each other. Since both of these studies involved mt genes, further analysis is clearly required. Geographic variation has been reported in bill, wing and tarsus length (Wennerberg *et al.* 2002), although no subspecies currently recognised.

DISTRIBUTION A High Arctic species, restricted to Canada north of 63°N. Nests in marshy tundra, with good vegetation. Migrates in long stages, up to 4,000 km. Route apparently direct from NE North America to N South America, and thence in shorter stages to winter in Argentina and S Chile (HBW).

STATUS This is one of the commonest Nearctic waders in GB&I with up to 30 records each year. The first for GB was at Stoke Heath (Shropshire) sometime before 1839; by 2006, there had been over 400. The first Irish record is much later, having been seen at Gt Saltee (Wexford) in Oct 1956, although a skin in the Ulster Museum is believed to have been obtained in Ireland (probably Belfast Lough) before Apr 1836. There has now been a total of 184 individuals in Ireland. It has been reported in every month, although over 97% were in July–Nov. The first long, trans-oceanic, stage of the autumn migration is likely the source of many of the birds that turn up in W Europe.

Baird's Sandpiper   *Calidris bairdii* (Coues)

SM monotypic

TAXONOMY Another species whose phylogenetics need resolution. Cytochrome oxidase 1 (Hebert *et al.* 2004) showed this to be close to Buff-breasted Sandpiper, whereas cyt-b (Borowik & McLennan 1999) indicated a closer affinity with Western and Semipalmated Sandpipers. Based largely on morphology, Sibley & Monroe (1990) regarded it as close to White-rumped and Pectoral Sandpipers.

DISTRIBUTION Northern belt of Canada and Alaska, beyond 61°N, to Ellesmere Island; also extreme NW Greenland and Chukotsky Peninsula. Nests on dry tundra and stony ridges. Usually migrates inland, across N America to winter beside high Andes grasslands and lakes in Chile and Patagonia.

STATUS The first in GB was an adult female on St Kilda (Outer Hebrides) in Sep 1911; there have now been over 200 records. The first for Ireland was at Akeragh Lough (Kerry), in Oct 1962; by 2004, there had been 79 birds in Ireland. There are a few records for May–June but highest numbers in Sep, with *c.* 60% of all records in this month and, as with most vagrant waders, juveniles account for most of the Sep records.

Pectoral Sandpiper   *Calidris melanotos* (Vieillot)

CB PM monotypic

TAXONOMY Usually regarded as close to Sharp-tailed Sandpiper (e.g. Vaurie 1965; Sibley & Monroe 1990), but Borowik & McLennan (1999) found them to be widely separate within *Calidris*, as did Thomas *et al.* (2004b), though the latter included similar data.

DISTRIBUTION Breeds in wet, well-vegetated habitats on N coast of Siberia from Taimyr to Chukotsky, and from W Coast of Alaska to Hudson Bay. Winters in Chile and Argentina, also small numbers in SE Australia, New Zealand.

STATUS Occasional reports of birds displaying and holding territories in Scotland; 'very likely' breeding in NE Scotland in 2004 and possible attempt in W Isles same year.

The first confirmed record for GB was a female at Breydon Water (Norfolk) in Oct 1830, and the first for Ireland was in a Dublin market, from Portumna (Galway) in Oct 1888. Records increased so fast in the 1950s that the species was removed from BBRC scrutiny in 1962. There are a very few spring records; most birds are found in Aug–Oct. Multiple sightings are not unusual, with up to four being found together. Although there is a distinct bias towards the S & W, this species has been recorded in almost every county in GB&I. This is the commonest Nearctic wader in both GB and Ireland.

Sharp-tailed Sandpiper   *Calidris acuminata* (Horsfield)

SM monotypic

TAXONOMY Borowik & McLennan (1999) and Thomas *et al.* (2004b) both found a clade linking this with Broad-billed Sandpiper and Ruff; however, the former's data may weigh heavily in the derivation of the latter's supertree.

DISTRIBUTION More restricted than Pectoral Sandpiper, only breeding from Lena Delta to River Kolyma; more restricted in habitat than Pectoral as well (HBW). Most adults migrate overland or along the coast, to New Guinea and islands S and E to Australasia, but recent research has shown that many (if not most) juveniles initially migrate east from the Siberian breeding grounds to Alaska, where they fatten up before the long flight across the Pacific.

STATUS The first for GB was an adult shot in late Sep 1848 in Great Yarmouth (Norfolk); 25 records to 2003. There are only three Irish records, the first being present at Tacumshin (Wexford) in Aug 1994. Unusually for a vagrant wader, 22 of the 27 that have been aged were adult, and the species occurs slightly earlier than many, with a peak in Aug–Sep. The adults that are found in GB&I may be failed breeders or post-breeding dispersal, whereas the deficieny of juveniles may stem from apparently different migration route.

Curlew Sandpiper   *Calidris ferruginea* (Pontoppidan)

PM monotypic

TAXONOMY Vaurie (1965) and Sibley & Monroe (1990) regard this as phylogenetically close to Dunlin and Purple Sandpiper, but cyt-b suggests that it is distant from most other *Calidris* (Borowik & McLennan 1999).

DISTRIBUTION Breeds along N coast of Siberia from Yamal Peninsula to N Chukotsky, also on offshore islands. Usually nests on damp parts of open tundra. Winters widely from W Africa to Australasia, chiefly along coast but also inland in W Africa and Rift Valley.

STATUS Birds passing through GB&I are a tiny part of the population that breeds in W Siberia and migrates through NW Europe to the wintering grounds in SW Europe and W Africa; this has been estimated at *c.* 740,000 individuals. This population increased markedly through the last quarter of the 20th century. Numbers recorded in GB &I peak in Sep but fluctuate annually, with occasional invasion years, sometimes in association with Little Stints. Unlike the latter, there are almost no records of wintering birds. Larger numbers occur on E and S coast saline pools and estuaries, but birds can occur almost anywhere. In Ireland, shows a similar pattern of occurrence to Little Stint but commoner.

Underhill (1987) showed how the age structure of Curlew Sandpiper populations in South Africa correlated with Siberian Lemming cycles; in years when Lemmings were scarce, the proportion of juvenile Curlew Sandpipers was low, suggesting that predation was higher on the breeding grounds (as proposed by Roselaar 1979). Kirby *et al.* (1989) examined data from influxes during 1969–87, comparing the breeding success with Lemming abundance; they found that influxes occurred in years when breeding success had been high but that in other years with high success, numbers in GB&I were low. They concluded that simple correlations were not evident and that local weather conditions were also significant in determining the number of birds reaching our shores.

Stilt Sandpiper   *Calidris himantopus* (Bonaparte)

SM monotypic

TAXONOMY Molecular data suggest an independent lineage within *Calidris*, though until recently (e.g. HBW) placed in monospecific genus *Micropalama* (see Sangster *et al.* 2004b).

DISTRIBUTION Northern coast of Alaska and Canada to N Hudson Bay; isolated populations in N Manitoba. Migrates through inland Americas to winter in C South America.

STATUS The first record for GB was an adult at Kilnsea (E Yorkshire) in Aug–Sep 1954, and the first for Ireland was found at Akeragh Lough in Oct 1968. Despite its relatively wide distribution in N America, this remains a rare bird in GB&I, with only 30 records to 2003. There is no obvious geographical pattern to their occurrence and birds have been found in most months from Apr to Nov but chiefly in Aug and Sep. It is, however, interesting that several of the spring birds were on the E coast of England: possibly these are birds that either crossed the Atlantic the previous autumn and survived the winter further south or crossed further south earlier in the spring and are following their instincts to head north.

Purple Sandpiper   *Calidris maritima* (Brünnich)

CB WM PM monotypic

TAXONOMY Mt DNA suggests this to be sister species to Rock Sandpiper and in a clade with Dunlin and Buff-breasted Sandpiper. Included in few other molecular analyses, so confirmation lacking, but this is in accord with Sibley and Monroe (1990), who treat Rock and Purple Sandpipers as members of a superspecies. Suggestion that considerable variation in size perhaps 'warranting recognition of at least two races' (Roselaar in BWP) has subsequently been reinforced (Engelmoer & Roselaar 1998), with the recommendation that Icelandic birds are sufficiently larger to merit separate subspecific status (*littoralis*).

DISTRIBUTION Holarctic, breeding from Ellesmere and Baffin Islands, around coast of Greenland through N Scandinavia to Taimyr. Some populations (e.g. Iceland, W Greenland) only partially migrant but most winter S along rocky coasts of NW Europe and E North America.

STATUS A few pairs have bred in the highlands of Scotland since 1978, nesting at >1,000 m.

Purple Sandpipers winter on rocky shores around the coasts of GB&I, especially in the N and W, with very few on the east coast of England S of Yorkshire. They generally arrive in late Sep and depart by the beginning of May. There are at least three components to the wintering population of GB&I (Stroud *et al.* 2004): a small proportion of the population that breeds in NE Canada, most of which winter in N America; Scandinavian breeders wintering in east GB; birds from Svalbard and NW Russia that winter in SE England as well as the Netherlands. Icelandic birds appear to be almost (if not entirely) resident. The use of biometrics has permitted the dissection of wintering populations (refs in Summers in Wernham *et al.* 2002); Norwegian birds (with the shortest bills and wings) comprise *c.* 25% of the winter population, mostly along the E coast of England and Scotland; sightings of colour-ringed birds support this. Most of the remainder of the winter population have long bills and wings; since Icelandic birds differ in wing length and are resident, these must be from the Canadian population and again there are ringing recoveries that link Canadian birds with a flyway through Iceland to GB (and presumably Ireland, although there are no recoveries to substantiate this).

Census data from WeBS counts are difficult to interpret because many birds winter in small numbers scattered along rocky shores all round GB&I. Data from regularly monitored sites indicates a steady decline since the mid-1980s (Banks *et al.* 2006), which supports a 21% decline in numbers at non-estuarine sites between 1984–85 and 1997–98 (Rehfisch *et al.* 2003a); believed to be due to a decrease in rocky shore invertebrates following the improvement in coastal water quality. The winter population has been estimated at 17,530 (GB), 3,330 (Ireland) and 10 (IoM); the former of these represent 9 and 4% of the flyway population (Rehfisch *et al.* 2003a; Crowe *et al.* 2008).

Dunlin   *Calidris alpina* (Linnaeus)

MB WM PM *schinzii* (C. L. Brehm)
WM PM *alpina* (Linnaeus)
PM *arctica* (Schiøler)

TAXONOMY Mt DNA indicates close affinity with Rock and Purple Sandpipers (Borowik & McLennan 1999) and with Sanderling and White-rumped Sandpiper (Hebert *et al.* 2004), suggesting the need for more research. A variable number of subspecies have been described, from five (Glutz von Blotzheim *et al.* 1975), to six (BWP), or even nine (HBW) or more. Greenwood (1986) attempted to clarify the situation through multivariate analysis of six metric traits and detected six groups that he proposed as subspecies (*pacifica, alpina, arctica, schinzii, sakhalina, arcticola*). Although birds breeding in the central Canadian Arctic could not be separated on morphology, they were geographically isolated and Greenwood proposed that these should also be accorded subspecific status (*hudsonia*). C Siberian '*centralis*' recognised by some authorities (e.g. Tomkovich 1986).

Dunlins have now been analysed using rapidly evolving mt control region sequences, and five separate lineages have been identified closely related to the morphological subspecies (Wenink *et al.* 1993, 1996). These lineages are European: *arctica, schinzii, alpina*; C Siberian: '*centralis*'; Beringian: *sakhalina*; Alaskan: *arcticola, pacifica*; Canadian: *hudsonia*. Buehler & Baker (2005) extended this analysis, confirming that Canadian Dunlins are the most divergent (supporting Greenwood's view that they should be recognised as a distinct subspecies); European birds are differentiated from Siberian, Beringian and Alaskan, and these last from each other (supporting the recognition of '*centralis*'). Buehler & Baker (2005) relate these differences to evolutionary time and suggest that Dunlins went through a bottleneck some 200,000 yrs BP but that since then the populations have been larger and the major lineages have diverged through restricted gene flow due to their fidelity to both breeding and wintering areas. Wenink & Baker (1996) and Wennerberg (2001) have shown the potential of these differences for determining the composition of flocks at stopover or wintering sites, although there is overlap between European and Siberian phylogroups – perhaps due to continuing hybridisation.

DISTRIBUTION Widely distributed across N Holarctic between 77°N and 50°N. Generally nests on wet or boggy sites, close to pools or open water; also wet grassland, both coastal and upland. Not such a long-distance migrant as many *Calidris*; in winter, found extensively on estuaries, mudflats, fresh and brackish pools, along ice-free coasts, almost exclusively in the N Hemisphere.

STATUS Three subspecies occur regularly; there is a ringing recovery from the Yamal peninsula, which is well within the range of '*centralis*'.

The breeding population (*schinzii*) of GB&I has recently been estimated at *c.* 9,500 pairs (Baker *et al.* 2006), although Thorup (2006) revised this to 18,300–33,500 with a further 150 in Ireland; most of this race (984,000 birds) breeds in Iceland and SE Greenland and a further 2,000 pairs in the Baltic. Numbers breeding in GB&I have declined, probably as a result of afforestation, especially in the flow country of N Scotland, although a pair recently bred by the Menai Straits (N Wales). GB&I breeders migrate into SW Europe and NW Africa (Wernham *et al.* 2002), to winter alongside the rest of the subspecies, many of which pass through GB&I on migration. Winter counts suggest that *schinzii* has increased in numbers, although the extent to which this is due to better coverage in Mauritania is not clear (Zwarts *et al.* 1998).

The relatively small population of *arctica*, breeding in NE Greenland, has been estimated at no more than 7,000–15,000 pairs and passes through GB&I on its way to the wintering grounds in W Africa. However, it is difficult to identify with confidence and the limited information on its status here is derived from ringing (Wernham *et al.* 2002).

In winter, the bulk of the Dunlins in GB&I are *alpina*. There are few data from the breeding grounds relating to abundance, but winter censuses suggest a total of *c.* 1.37M birds, of which *c.* 555,800 occur in GB (95% on estuaries), *c.* 88,480 in Ireland and 704 in IoM (Rehfisch *et al.* 2003a; Crowe *et al.* 2008). Numbers increase through the autumn to peak in midwinter, declining subsequently; however, hard weather can result in movements around the Atlantic seaboard of Europe, with rapid, short-term changes in numbers. Census data began in the 1970s in GB, and the totals declined until the late 1980s; a subsequent increase to the mid-1990s was paralleled in NI, where censuses began in 1988, and both regions show a further decline since then (more markedly in NI). Wetland censuses began in 1994 in RoI, and after reaching a peak in 1996, Dunlin numbers have declined here also. Although numbers fluctuate among years, there is evidence of a more marked decline at west coast sites, which might be interpreted as 'short stopping', the phenomenon whereby birds fly a shorter distance to the wintering grounds in response to climatic amelioration.

Genus *LIMICOLA* Koch

Monospecific genus from the Palearctic that is a scarce migrant through GB&I.

Broad-billed Sandpiper   *Limicola falcinellus* (Pontoppidan)

SM *falcinellus* (Pontoppidan)

TAXONOMY Mt DNA suggests this forms a clade with Sharp-tailed Sandpiper and Ruff, which, if confirmed, will require a reassessment of generic limits. However, most taxonomies (e.g. Vaurie 1965; Sibley & Monroe 1990) place Broad-billed Sandpiper in a monospecific genus. Two subspecies recognised, based on colour of upperparts; *falcinellus*: Scandinavia, NW Russia; *sibiricus*: Taimyr, River Lena to River Kolyma.

DISTRIBUTION W subspecies breeds in bogs and peatlands of subarctic Scandinavia and NW Russia, wintering on intertidal mudflats at scattered sites from SW Africa to Sri Lanka. E subspecies breeds on wet tundra, winters in similar habitat to *falcinellus* from NE India to Australia.

STATUS The first record for GB was shot at Breydon Broad (Norfolk) in May 1836, and the first for Ireland suffered the same fate at Belfast Lough (Antrim) in Oct 1844. Since then, it has been recorded in all months from Mar to Oct, although there is a strong peak of records in May, with almost half of some 200 birds being found in this month. In only five years were more than 10 individuals recorded, all in the 1980s. The annual average for 1980–2000 was significantly higher than before or since.

Genus *TRYNGITES* Cabanis

Monospecific Nearctic genus that has been recorded as a vagrant in GB&I.

Buff-breasted Sandpiper   *Tryngites subruficollis* (Vieillot)

PM monotypic

TAXONOMY Traditionally placed in a monospecific genus (e.g. Sibley & Monroe 1990), but recent molecular data suggest sister species to Baird's Sandpiper and in same subclade of *Calidris* as Pectoral Sandpiper. Borowik & McLennan (1999) also found it nested within *Calidris* but in a separate clade. Further investigation required to confirm position and validity of *Tryngites*.

DISTRIBUTION Breeds in High Arctic of Canada and Alaska, also N Chukotsky, nesting on dry tundra with thin vegetation. Migrates through interior N America to winter in C South America.

STATUS The first record for GB was near Melbourne (Cambridgeshire) in Sep 1826, and the first for Ireland was near Dublin at some time previous to, or in, 1845. Since those early birds, there have been hundreds of records, involving parties of up to seven individuals. The number recorded annually increased through the 1960s and 1970s to a maximum of *c.* 70 individuals in 1977. After 1982, BBRC stopped assessing records, though the IRBC continued. The annual totals fluctuated with rather fewer through the early 1990s, though this remains one of the commoner American waders in GB and the second commonest in Ireland, with 251 records to 2004. Recorded in every month from Apr to Nov but >70% in Sep; widely distributed across the islands but more often in the S & W.

Genus *PHILOMACHUS* Merrem

Palearctic monospecific genus, recorded regularly in GB&I; breeds in small numbers.

Ruff	*Philomachus pugnax* (Linnaeus)

CB WM PM monotypic

TAXONOMY Traditionally placed in monospecific genus (e.g. Sibley & Monroe 1990), but molecular data suggest sister species to *Limicola falcinellus* and *Calidris acuminata* (Borowik & McLennan 1999).

DISTRIBUTION Occurs in scattered isolated populations across C Europe, then continuously across N Eurasia from Scandinavia to Bering Straits. Breeds adjacent to water on meadows and marshes where suitable lekking areas exist. Strongly migratory, wintering widely in sub-Saharan Africa; also round periphery of Arabia and Indian subcontinent. Zöckler (2002) estimated the Eurasian population at 2.2–2.8 million birds, of which 98% now breed on the Arctic tundra. He argues that populations have moved north and east, resulting in a decrease in numbers (especially in wet grassland) west of a line from Finland and the Baltic states to Hungary but an increase along parts of the northern coast of Siberia. Whether this shift in distribution has included a global decline or merely involves a relocation is unclear.

STATUS The breeding population of Europe has been estimated at 244,000–526,000 females (Thorup 2006), of which less than 10 usually occur in GB&I, as far N as Scotland. Leks continue to be found across GB, though they are rarely utilised for more than a few years before the birds abandon the site; recently, two leks in Cambridgeshire hosted *c.* 30 males, and 7 males and 3 females, in 2002 (RBBP). A female was seen with two small young in Lancashire in 2006, but successful breeding is rarely recorded. The preferred habitat of fresh and fresh/brackish water meadows are still available, and reserve management is re-creating more of these; furthermore, in parts of northern Europe, birds breed successfully on peatland and such habitat also exists in GB&I. However, there has been a substantial decline in most countries of NW Europe, which has been attributed to land drainage removing the desired wetlands and increased use of pesticides reducing the availability of soil invertebrates. Stroud *et al.* (2004) report that Ruffs have completely disappeared from areas of intensive agriculture, and since the total population from Latvia, Estonia and Poland to GB and France is now <2,000 breeding females (Thorup 2006), the likelihood of successful re-colonisation seems remote. Only recorded as passage migrant and winter visitor to Ireland, with no evidence of any breeding attempts.

Present in all months and fairly common passage migrant in Apr and Aug–Sep; also occurs in winter, and WeBS data show that wintering numbers have increased steadily since the mid-1980s. Often found on freshwater pools in the S & E of GB&I; numbers peak in Aug–Sep, when the juveniles move through en route to the wintering grounds in W Africa.

Genus *LYMNOCRYPTES* Boie

Palearctic monospecific genus, occurring as a winter visitor to GB&I.

Jack Snipe   *Lymnocryptes minimus* (Brünnich)

WM PM monotypic

TAXONOMY Included in phylogenetic analysis by Thomas *et al.* (2004a), who found it to lie in a strongly supported clade that included stints, snipes, woodcocks, shanks and curlews. The internal structure was less well resolved, and Jack Snipe's position in a separate subclade to Common Snipe needs confirming. Analysis of morphology (Chu 1995) indicates Jack is sister to snipes and woodcocks, though Baker *et al.* (2007) placed it sister to *Limnodromus*.

DISTRIBUTION Breeds from N Scandinavia to E Siberia, with isolated populations in Poland, Baltic States, S Sweden. Winters in S and W Europe, N and sub-Saharan Africa, scattered wetlands across Middle east to India.

STATUS A passage migrant and winter visitor to GB&I, Jack Snipe are widely distributed in treeless wetlands but in small numbers; only a handful of WeBS sites record >10 individuals, with only Chat Moss (Gt Manchester) and Doxey Marshes (Staffordshire) holding >30. However, higher numbers are usually found on sites where special efforts are made to record the species, and these are not made during standard WeBS counts. Numbers reported are too slight in either GB or Ireland to seek trends but appear to be generally stable (Crowe 2005, Banks *et al.* 2006). Movements detected through ringing recoveries indicate that some birds move onwards from GB into France and N Iberia, but Ireland appears to be a wintering area since no birds ringed there have been recovered further south. The origin of birds found in GB&I is unclear, though recoveries in GB&I of individuals ringed in Scandinavia indicate Fennoscandia as a likely source.

Genus *GALLINAGO* Brisson

Near-cosmopolitan genus (not Australasia) of 15–16 species, seven in Palearctic, of which three have been recorded in GB&I: one breeds and two are vagrants.

Common Snipe   *Gallinago gallinago* (Linnaeus)

RB MB WM PM *gallinago* (Linnaeus)
RB MB WM PM *faeroeensis* (C. L. Brehm)

TAXONOMY Molecular analysis of cyt-b (Thomas *et al.* 2004a) found *Gallinago* to be part of a weakly supported subclade with curlews, within a larger clade that included stints, shanks and turnstones. Nuclear genes RAG-1 (Paton *et al.* 2003) and myoglobin (Ericson *et al.* 2003) confirmed the larger clade but suggested a closer relationship between woodcocks and snipes, a finding supported by osteological analysis (Chu 1995).

Two subspecies generally recognised in northern hemisphere (see also under Wilson's Snipe); *gallinago*: Palearctic (except Iceland–Orkney); *faeroeensis*: Iceland, Faroes, Orkney and Shetland. Several races in southern hemisphere recognised by BWP now regarded as separate species (HBW). Status of whole group needs further study.

DISTRIBUTION Widely distributed across arctic and temperate Eurasia. Migratory, with most northern habitats vacated in winter, as birds relocate to ice-free regions S to equator.

STATUS The European population of *gallinago* has been estimated as 800,000–1,300,000 pairs (Thorup 2006). Of these, *c.* 48,000 (<5%) were in GB and a further 19,000 pairs in Ireland. In common with many other parts of W Europe, Common Snipe are declining as a breeding bird in GB&I,

probably due to habitat loss through drainage and reclamation of wetlands, and it has been suggested that relatively few now breed away from nature reserves.

Stroud *et al.* (2004) estimate the population of *faeroeensis* at *c.* 190,000 pairs, of which 180,000 breed in Iceland; Pennington *et al.* (2004) estimated the breeding population of *faeroeensis* in Shetland at *c.* 5,000 pairs, although recognising that accurate censusing was very difficult.

Ringing data suggest that birds from northern GB winter in Ireland, whereas those from further south migrate S to W France, Iberia and even NW Africa. However, some appear to be resident and these are joined in winter by perhaps a million birds from countries round the Baltic and NW Russia.

Winter numbers are monitored through WeBS counts, but the consistency of these is questionable; the number recorded may reflect observer effort as much as actual numbers present. Consequently, there can be extensive fluctuations between years (Banks *et al.* 2006), even at the same site; for example, maximum winter numbers on the Somerset Levels were 972, 308 and 1,513 during 2002–05. With data as variable as this, overall trends are hard to elucidate.

Ringing recoveries indicate that most *faeroeensis* winter in Ireland, although there are a few from the Hebrides (Wernham *et al.* 2002); Shetland birds have been recovered from Cornwall and Spain.

Wilson's Snipe   *Gallinago delicata* (Ord)

SM monotypic

TAXONOMY Morphological and vocal/acoustic evidence indicates that *delicata* should be treated as a separate species from Common Snipe, though Zink *et al.* (1995) found levels of molecular differentiation intermediate between populations and species.

DISTRIBUTION Widely distributed across arctic and temperate N America. Northern populations are migratory; birds relocate to ice-free regions S as far as N South America.

STATUS The first was one near Coleraine (Londonderry/Antrim) in Oct 1991. The first from GB was a juvenile, St Mary's (Scilly), Oct 1998–Apr 1999. Another (or more than one), St Mary's (Scilly), from Oct 2007 into 2008.

Great Snipe   *Gallinago media* (Latham)

SM monotypic

TAXONOMY Has not been included in any molecular analysis of phylogeny. Conventionally treated as close to Common Snipe. European birds distributed in two main areas: W Scandinavia, where found in earthworm-rich mountain fens around the treeline, and in E Europe, where patchily distributed in annually flooded lowland meadows. Some morphological and genetic variation between these populations but treated as monotypic.

DISTRIBUTION Tends to breed in more wooded habitat than Common, from Poland and Baltic States to Yenesei; also N Scandinavia. Winters in tropical and southern Africa. Many populations have declined and become fragmented, primarily due to loss of wetland habitat, but over-hunting also possible cause.

STATUS The first confirmed record is from Kent, prior to 1787, and there have been over 630 subsequently. These are widely scattered through GB, with a bias towards the N Isles and the east coast. Surprisingly few in Ireland, with only *c.* 21 since the first two were shot in Kildare in Oct 1827.

Recorded every month of the year, mostly Aug–Oct with *c.* 45% in Sep. Annual records have declined since the 1950s, despite an increase in observer cover and awareness; this may reflect the declining European population or the reduction in hunting records from GB&I in the latter part of 20th century. It is now a very rare bird.

Genus *LIMNODROMUS* Wied-Neuwied

Holarctic genus of three species, two of which have been recorded as vagrants to GB&I.

Short-billed Dowitcher   *Limnodromus griseus* (Gmelin)

SM race undetermined

TAXONOMY Ecological, behavioural and morphological differences justified the separation of this from Long-billed Dowitcher; enzyme and molecular data have substantiated the independence of their lineages (Avise & Zink 1988; Hebert *et al.* 2004). Chu's (1995) osteological analysis indicates Short-billed is sister to godwits, though Long-billed was not included. However, nuclear DNA sequences suggest dowitchers are closer to snipes and woodcocks (Paton *et al.* 2003). This conflict with morphology may be due to convergent evolution and habitat specialisation (Paton *et al.* 2003).

Three disjunct subspecies based on size, breeding plumage and vocalisations (Miller *et al.* 1983); *griseus*: Labrador and Quebec; *hendersoni*: Canadian interior, from British Columbia to Manitoba; *caurinus*: S Alaska.

DISTRIBUTION Breeds in three separate populations (Labrador/Quebec, Canadian interior, Alaska), nesting in bog and swamp vegetation. Migratory, wintering generally along E and W coasts of Americas, S to Brazil and Peru. Autumn migration of adults starts at end of June, with passage through Atlantic Canada from mid-July–Aug, thence south to wintering grounds; juveniles migrate later. Timing and routes of *hendersoni* and *griseus* differ markedly (Jehl *et al.* 2001).

STATUS One record from GB: a juvenile near Fraserburgh (Aberdeenshire) in Sep 1999 and the same bird at Seal Sands (Teesside) in late Sep–Oct 1999. Three records from Ireland: the first in Sep–Oct 1985 at Tacumshin, Wexford and a second on various dates around Dublin and Meath in Mar–Sep 2000 and again in Co. Dublin in May 2001; a third in Wexford and Dublin from June–Aug 2004 and Oct 2004–Mar 2005.

Long-billed Dowitcher   *Limnodromus scolopaceus* (Say)

SM monotypic

TAXONOMY See Short-billed Dowitcher.

DISTRIBUTION More northerly than Short-billed; similar habitat, though breeds closer to sea. Populations occur in NE Siberia, coastal Alaska and (more isolated) N Inuvik. Migratory; some move S along Pacific coast to Mexico, others cross Canada and head S along Atlantic coast of N America to winter in coastal USA, except NE, and Mexico. Latter route probable source of many European birds. Failed breeders depart mid-July; juveniles much later, usually not leaving Alaska before Sep.

STATUS The first record for GB was a first-winter male from an unknown locality in coastal Devon in Oct 1801; the first Irish bird was at Port Laoise (Laois) in Sep 1893. Birds have been found in GB&I most years since 1950, and 1–16 individuals every year since 1962. Before *c.* 1985, over 100 dowitchers were recorded that cannot now be safely ascribed to species (but were presumably mostly Long-billed), and the apparent increase in numbers latterly is partly due to this. Long-billed Dowitcher has been recorded in every month of the year but peaks in Sep–Oct, with a secondary peak in Apr–May. Since the identification criteria for separating dowitchers was established in the 1980s (reviewed by Chandler 1998), more Short-billed might be expected; however, the latter remains a very rare bird. Long-billed initially move SE in autumn on a trajectory that could lead them over the sea; Short-billed has a more north–south route and is less likely to stray over the Atlantic and hence to W Europe.

Genus *SCOLOPAX* Linnaeus

Holarctic and Australasian genus of around eight species, two from the Palearctic, one of which breeds in GB&I.

Eurasian Woodcock   *Scolopax rusticola* Linnaeus

RB MB WM PM monotypic

TAXONOMY Morphology suggests woodcocks are allied to snipes (Chu 1995), and analyses of nuclear RAG-1 gene (Paton *et al.* 2003) and combined RAG-1 and mitochondrial DNA (Baker *et al.* 2007) support this.

DISTRIBUTION Breeds in thick, damp forests across Eurasia, from GB&I, and Scandinavia to China. Northern populations migratory, wintering around Mediterranean, N India and Indo-China.

STATUS Breeds widely in woodlands across GB, though less abundant in Ireland. The Second Breeding Atlas reported a substantial decrease in distribution compared with the First Breeding Atlas, with birds lost from *c.* 30% of 10-km squares in GB and from *c.* 65% in Ireland, although some of this might have been due to limited night-time coverage. Eurasian Woodcock prefer an open understorey and tend to abandon woodlands as they approach the thicket stage (Second Breeding Atlas); these habitats were rarely included in the CBC study, so the generality of a decline of *c.* 75% at these sites is unclear (Baillie *et al.* 2006). BBS censuses are also poor at recording this species so a special survey was undertaken in 2003, the results of which suggested in excess of 78,000 roding males in Britain, far exceeding all previous estimes (Hoodless *et al.* 2009). They also found evidence that mixed woodland was more favoured and that numbers varied markedly among the regions of GB. Fuller *et al.* (2005) suggest that local decreases may be due to any or all of recreational disturbance, overgrazing of the ground cover by deer, drying out of the substrate through climate change or adjacent land management.

Ringing data indicate that many Eurasian Woodcock from GB&I are resident and fairly site faithful. There are a few longer-distance recoveries; birds from N Britain are recovered either in Ireland or along the Atlantic coast of France and Iberia. Birds from further south are less likely to winter in Ireland.

There is a substantial immigration into GB&I during the autumn, which may account for >90% of the winter population (Wernham *et al.* 2002). These birds arrive from Oct to Dec, generally remaining until Feb–Mar. Those wintering in the north of GB mostly come from Scandinavia, whereas those further south seem to have a more easterly origin, from the Baltic republics and NW Russia. Irish wintering Woodcock come from a wider range, encompassing both of the above regional origins.

Genus *LIMOSA* Brisson

Holarctic genus of four species, two in the Palearctic; three have been recorded in GB&I; one breeds. Paton *et al.* (2003) used the nuclear RAG-1 gene to show that *Limosa* is part of a major wader clade that also includes stints, shanks, snipes and woodcock; mt DNA supports this conclusion (e.g. Hebert *et al.* 2004).

Black-tailed Godwit   *Limosa limosa* (Linnaeus)

MB *limosa* (Linnaeus)
MB WM PM *islandica* C. L. Brehm

TAXONOMY Of the four godwits, three were included in the barcode study of Hebert *et al.* (2004); Black-tailed was the one omitted. Höglund *et al.* (2009) examined the mt DNA of Black-tailed and Hudsonian, which differed by 5%, supporting treatment as separate species. Closely allied to

Hudsonian; Sibley & Monroe (1990) place it in the same superspecies. Three subspecies recognised; *islandica*: Iceland, Faroes, Shetland and Lofoten Islands (HBW); *limosa*: W Europe across Eurasia to N of Lake Balkash; *melanuroides*: a series of isolated populations across E Asia, including NE China, Kamchatka, E Mongolia. All subspecies have unique haplotypes but small genetic distances (Höglund *et al.* 2009) and no substantial gene flow between them.

DISTRIBUTION Nests in wet grassland and marshes, with soft substrate for feeding. Winters in GB&I, France, round the Mediterranean, sub-Saharan Africa, Middle East, Pakistan and N India, Malay Peninsula through to Australia. More often around coasts in winter, favouring lagoons and sheltered estuaries, etc. but less marine than Bar-tailed Godwit.

STATUS The small population of the nominate race breeding in E England was estimated at *c.* 50 pairs in 2002 (RBBP), mostly in the Cambridgeshire fens. These birds are restricted to a few sites and are vulnerable to both predation and flooding. Ringing has only generated a single recovery: from sub-Saharan Africa. No colour-ringed individuals from the fens have been seen anywhere during the winter, implying that they leave Europe, and hence are *limosa* rather than *islandica*. The birds that breed in N & W Scotland are *islandica* and totalled up to 10 pairs in 2002–06 (RBBP). This race also breeds sporadically in Ireland, usually in the midlands.

Three subpopulations of race *limosa* are recognised by Stroud *et al.* (2004), only one of which occurs in GB&I; this breeds in W Europe, from the Netherlands (*c.* 47%) through to Ukraine (including GB&I). The population has decreased markedly over the past 30–40 years; as with Ruffs, this seems to be due to drainage of wetlands and intensification of agriculture, including earlier mowing of fertilised grasslands. Also seriously prone to flooding and predation thus, in 2006, one site in Cambridgeshire produced only four young from 48 pairs. Nominate does not occur in Scotland.

WeBS counts from GB show a progressive increase in the wintering population since the census began in 1975; this show no signs of levelling out. Ringing data indicate that virtually all of these birds are *islandica*. Numbers peak rapidly in Aug–Sep, declining slightly through the winter, perhaps as birds die or relocate further south into France and Iberia or west into Ireland. Counts from NI indicate an increase in numbers since the early 1990s, but the pattern within a winter is somewhat different. Numbers peak in Sep–Oct, then fall rapidly to a low level that is maintained until Apr, when numbers increase slightly. These data presumably reflect a passage of Icelandic birds through NI in the autumn to wintering areas further south in the Republic, where perhaps half of the population winters (Crowe 2005); the smaller peak in spring perhaps indicating that much of the population bypasses (or overflies) NI on its way back to Iceland. The winter census data from the Republic show a different pattern again; only a modest increase since the mid-1990s and fairly constant numbers through the winter. Rehfisch *et al.* (2003a) estimated the winter population in GB as 15,360 (44% of the flyway total); however, this was in 1998–99, and numbers have continued to rise since then, so the figure is unlikely to be accurate. Comparable figures for Ireland were collected more recently (2003–04); estimated as 13,880, or 40% of the flyway population (Crowe *et al.* 2008). Clearly, GB&I is a major wintering area for this species.

The current population size of *islandica* is *c.* 47,000 and has increased over the last 100 years (Gunnarsson *et al.* 2005), without any obvious signs of levelling out. It has been shown (Gill *et al.* 2001) that as the population rose, birds wintering in W Europe expanded into estuaries where food was more difficult to acquire and Black-tailed Godwit survival was consequently reduced. The breeding population showed a parallel expansion in Iceland to smaller marshes with more dwarf birch, where food supply and breeding success were both lower than in the extensive bogs that were colonised in earlier years.

Hudsonian Godwit   *Limosa haemastica* (Linnaeus)

SM monotypic

TAXONOMY Phylogenetic analysis based on cytochrome oxidase 1 gene (Hebert *et al.* 2004) suggested sister species to Marbled *L. fedoa*, but Black-tailed omitted from this study. Höglund *et al.* (2009)

found 5% mt DNA difference from Black-tailed. Chu (1995) could not resolve Black-tailed, Bar-tailed and Hudsonian from Strauch's (1978) morphological data. Disjunct distribution across Canada and Alaska but no subspecific differentiation.

DISTRIBUTION Breeds on marshes along edge of taiga forest at widely scattered sites from Alaska to Hudson Bay. Haig *et al.* (1997) found evidence that birds from Saskatchewan and Mackenzie Delta were genetically distinct, though their sample sizes were small and the sampling range was limited. Long-distance migrant, with many crossing W Atlantic directly from James Bay to S America (Morrison 1984). Winters in S Argentina but still relatively little is known about this species; e.g. staging posts in S America not clearly defined (Elphick & Klima 2002).

STATUS The first record for GB&I was an adult at Blacktoft Sands (E Yorkshire) in Sep–Oct 1981; this individual was subsequently reported from Countess Wear (Devon) in Nov 1981–Jan 1982 and (presumably the same bird) back at Blacktoft in Apr–May 1983. A second bird was at Collieston (Aberdeenshire) in Sep 1988. In view of the autumn migration route being down the W Atlantic to S America, it is a little surprising that so few have been found in W Europe. Not recorded in Ireland.

Bar-tailed Godwit   *Limosa lapponica* (Linnaeus)

WM PM *lapponica* (Linnaeus)

TAXONOMY Cytochrome oxidase 1 sequences indicate this is the most divergent of godwits (Hebert *et al.* 2004). Three subspecies recognised, though variation in size and plumage is clinal (BWP); *baueri* largest: NE Siberia and Alaska; *menzbieri* intermediate: between Khatanga and Kolyma Rivers of N Siberia, though Vaurie (1965) doubts this taxon, believing it should be merged with *baueri*; *lapponica* smallest: N Europe and N Siberia W of Khatanga.

DISTRIBUTION Breeds on tundra along coasts of Arctic Oceans from Lapland to Alaska. Long-distance migrant; populations of *lapponica* from W of Yamal Peninsula winter along coasts of Europe; those from further east winter along Atlantic coasts of Africa and Arabia. Eastern subspecies winter in Malay Peninsula and islands to Australasia. Majority winter on intertidal areas; only rarely inland (Hayman *et al.* 1986). A powerful migrant, satellite telemetry has shown that Bar-tailed Godwits are capable of flying across the Pacific Ocean in a single stage: one bird flew from North Island (New Zealand) to Yalu Jiang (China) without stopping; the shortest distance between these sites is 9,575 km, but the actual route that it followed was 11,026 km and took approximately nine days.

STATUS A passage migrant and winter visitor to GB&I, these two components comprising different subpopulations. Biometric data suggest that birds breeding west of the River Yamal in Siberia winter in W Europe, whereas those from here to the Taimyr migrate through the North Sea on their way to wintering grounds in W Africa (refs in Stroud *et al.* 2004). Although most of these stage in the Wadden Sea, other sites in the S North Sea are used on both migrations; there is a widespread influx of Bar-tailed Godwits into GB&I in May of most years and many of these have bleached primaries, indicating a winter spent in the strong sunlight of W Africa. These birds seem to be poorly recorded by WeBS survey, probably because their passage is coastal, with birds being seen from sea-watches as they pass directly to the Wadden Sea

In winter 2004–05, WeBS counts peaked at *c.* 37,500 in Jan, with a further 1,500 in NI (Banks *et al.* 2006); such a pattern is fairly typical, although it represented a decrease on previous years. Despite numbers being relatively stable over the past 20 years, there may be the beginnings of a decline. Certainly, the winter population in the Netherlands has increased in recent years (van Roomen *et al.* 2005); this is perhaps another example of a species shifting its wintering range in response to climate changing towards warmer winters. Latest population estimate for GB (Rehfisch *et al.* 2003a) is of 61,590 (51% of flyway total), with a further 16,280 (14%) in Ireland.

Genus *NUMENIUS* Brisson

Holarctic genus of eight species, five in the Palearctic; five have been recorded in GB&I, two as breeding species.

Little Curlew   *Numenius minutus* Gould

SM monotypic

TAXONOMY Usually regarded as close to Eskimo Curlew *N. borealis*, but Roselaar (in BWP) points out resemblances to Upland Sandpiper *Bartramia longicauda*. Not included in any molecular analyses; relationship to Eskimo Curlew would bear investigation.

DISTRIBUTION Formerly believed to be rare, but breeding range wider than earlier reports (HBW) and abundance in N Australia in winter also suggests otherwise. Breeds chiefly inland on upland taiga of Yakutia, in clearings and forest edges. Migrates to winter principally in dry grasslands of N Australia, lesser numbers in E and in New Guinea.

STATUS First record: adult, Kenfig (Mid-Glamorgan) in Aug–Sep 1982. A second bird was in the Cley area (Norfolk) in Aug–Sep 1985. There are no records from Ireland.

Eskimo Curlew   *Numenius borealis* (J.R. Forster)

SM monotypic

TAXONOMY The ecological and taxonomic replacement of Little Curlew in the Nearctic; formerly regarded as conspecific.

DISTRIBUTION Only known breeding sites from Beaufort Sea along Amundsen Inlet to N of Great Slave Lake, although probably also N Alaska (Gollop *et al.* 1986). Migrated SE in Aug and Sep towards coastal Labrador and New England, thence trans-oceanic across W Atlantic to S America, wintering in Uruguay and Argentina. Return route further west, with spring records from Texas and SE USA (Gollop *et al.* 1986). Declined rapidly between 1870s and 1890s due to over-hunting in USA. Now probably extinct, with no confirmed records since birds in Texas in 1962 and Barbados in 1963.

STATUS First record was of two shot near Woodbridge (Suffolk) Nov 1852. Of the four further records between 1855 and 1887, three in Sep in Aberdeenshire. There is a single record from Ireland, obtained in a Dublin market in Oct 1870, probably shot in Sligo. Most (or all) occurrences followed storms or hurricanes off NE USA.

Eurasian Whimbrel   *Numenius phaeopus* (Linnaeus)

MB PM *phaeopus* (Linnaeus)
SM *hudsonicus* Latham

TAXONOMY Thomas *et al.* (2004a) report that Whimbrel and Curlew are part of a poorly supported clade with Common Snipe, based on cyt-b. Four subspecies generally recognised: *phaeopus*: Iceland and N Europe to Taimyr; *alboaxillaris*: N Caspian; *variegatus*: Taimyr to Kolyma River. Roselaar (in BWP) suggests that variation is clinal and that intergrades occur between these. Zink *et al.* (1995) compared *variegatus* with *hudsonicus* and found differences in mt DNA fragment polymorphism consistent with specific distinctness. Although European birds were not included in their analysis, these results support the morphological differences and the case for specific separation of the taxa. Engelmoer & Roselaar (1998) have recommended that the birds breeding in Iceland, the Faroes and Scotland should be treated as an additional race *icelandicus* and that the birds breeding in Alaska and NW Territories should be treated as subspecies *rufiventris*. Tomkovich (2008) recently named a further race, *rogachevae*, from E Evenkia, in C Siberia.

DISTRIBUTION Widely, though patchily, distributed across boreal and subarctic regions; Iceland, Faroes, N Scandinavia, N Russia and Siberia to Far East. Migratory, wintering along coasts of Africa, and Arabia to Australia. In N America, patchily and discontinuously distributed from Alaska and NW Canada (*rufiventris*) to Hudson Bay (*hudsonicus*). Migratory, wintering along coasts of Middle and S America.

STATUS Small numbers breeding in Scotland. The majority of these are in Orkney, Shetland and Outer Hebrides. Breeding in Shetland is restricted to heathland; the adults rarely take their chicks away from this habitat, unlike Curlews, many of which utilise more fertile grassland or plough. The Shetland population is estimated to be *c.* 500 pairs (Pennington *et al.* 2004), but there is some evidence of a recent decline. There was a possible breeding attempt in NW Ireland in 2005.

Although Whimbrel may be found in most months, there is a distinct pulse of birds moving through the region in Apr–May and a more protracted one in July–Sep. The numbers involved are too high to represent just the local breeding population, and many must be en route from the wintering grounds in W & S Africa to breeding sites further north and comprise a mix of birds from breeding populations in Iceland and Fennoscandia. There appears to be an west/east split in migration, with western localities (e.g. Somerset Levels) having larger numbers in spring and eastern ones (e.g. East Anglia) in autumn. The Icelandic population has been estimated as *c.* 200,000 pairs; although Stroud *et al.* (2004) suggest that this estimate is 'imprecise', it is clear that the population is large. The component of *phaeopus* breeding from Fennoscandia and the Baltic states to European Russia probably winters alongside Icelandic and Scottish birds in W Africa and has been estimated as between 50,000 and 100,000 pairs. In view of the large numbers breeding in the W Palearctic, it is surprising that more are not seen on passage through GB&I. Ringing recoveries indicate that Shetland birds winter along S coast of W Africa.

Subspecies *hudsonicus*: recorded five to six times in GB and twice in Ireland. The first in GB was on Fair Isle (Shetland) in May 1955, and the Irish birds were at Tralee (Kerry) in Oct 1957 and Tacumshin (Wexford) in Sep 1980.

Slender-billed Curlew   *Numenius tenuirostris* Vieillot

SM monotypic

TAXONOMY Little studied; affinities unclear but probably phylogenetically close to Eurasian Curlew.

DISTRIBUTION Breeding grounds unknown; probably in taiga, E of Urals. Formerly relatively common in winter in NW Africa but now very rare; the last record from Morocco was in winter 1994–95. Decline may be due to some or all of: loss of steppe habitat to agriculture, drainage of staging posts, excessive hunting. Parties of *c.* 50 in Iran (1994) and *c.* 20 in S Italy (winter 1995), 4 on the Evros Estuary (Greece) on 7 Apr 1999 and a succession of records from Bulgaria to 2002 (Nankinov *et al.* 2003) give hope for continued existence. A first-year bird in Britain (below) seems to indicate that Slender-billed Curlew must have bred somewhere in 1997. The search for the breeding area remains an ornithological Holy Grail (Gretton *et al.* 2002) – if, indeed, it still exists.

STATUS One record: a first-summer bird at Druridge Bay (Northumberland) in May 1998. It had just started primary moult, as expected for Eurasian Curlew at this age. This plumage had not been described in the literature and identification proved far from straightforward (Steele & Vangeluwe 2002). No records from Ireland.

Eurasian Curlew   *Numenius arquata* (Linnaeus)

RB MB WM PM *arquata* (Linnaeus)

TAXONOMY Cyt-b sequences (Thomas *et al.* 2004a) indicate that Eurasian is sister to Eastern Curlew *N. madagascariensis*, though few curlew species were included in their analysis. Chu (1995) found that

Eurasian is sister to Long-billed Curlew *N. americanus*, with Whimbrel rather more distant, but these only curlews included. Variation clinal, with two subspecies: *arquata*: GB&I to Ural Mts; *orientalis*: Urals through C Russia to Manchuria.

DISTRIBUTION Breeds in damp or boggy areas with suitable feeding sites, from Iceland, across temperate and subarctic Eurasia to Lake Baikal. Both races migratory, though W populations less so; the nominate winters along the coasts of S Europe, Africa and the Middle East; *orientalis* further east through S Asia to the Philippines.

STATUS Irish Eurasian Curlews showed a serious breeding decline between the Atlases, being lost from *c.* 20% of 10-km squares; although the British and Manx populations were more stable, *c.* 3% of squares were still abandoned in GB. The declines are probably chiefly due to loss of nesting habitat, although the causes of this differ among regions. In parts of Ireland, afforestation and re-seeding of moorland have both been implicated in the loss of grassland. In GB, birds disappeared from a belt of lowland between the Cornish peninsula and the Humber, along the southern fringes of the range occupied during 1968–72. Much of the habitat was lowland grassland; drainage and conversion to cereal may have resulted in a worsening of habitat quality for Eurasian Curlews.

As with the rest of N Europe, the populations in GB&I are declining, although Eurasian Curlews remain a fairly common breeding bird of the uplands; Stroud *et al.* (2004) give an estimate of *c.* 200,000 pairs for the European population (*arquata*), with a further 100,000+ in NW Russia. The breeding population of GB is *c.* 103,000 pairs, with a further 7,000 in Ireland, representing *c.* 30% of the nominate race. There is a substantial movement of British birds to Ireland in autumn, especially from Scotland and N England; Eurasian Curlews from S England and Wales also winter in Ireland, but a proportion migrate to the coasts of Biscay and Iberia.

Whilst many birds winter at well-monitored wetland sites, a sizeable proportion winters in a more dispersed fashion along non-estuarine coasts, as well as inland in some areas. Similarly for Ireland, the maximum census figures of *c.* 34,000–40,000 are likely to be unrepresentative of the true total. The number counted each winter has increased progressively since the mid-1970s in GB, though numbers in Ireland have been more fluid. Only two sites in GB&I hold internationally important concentrations (>4,200): Morecambe Bay (*c.* 11,000) and the Solway Estuary (*c.* 4,500). Some of these birds are immigrants, with recoveries in GB&I of birds ringed from the Netherlands, Scandinavia, Finland and W Russia. In general, birds wintering along the eastern side of GB&I are immigrant; those further west are from the native population. The latest winter population estimate (Rehfisch *et al.* 2003a) is of 147,100 (35% of flyway total). Only 52% of these were on estuaries; a substantial number winter on non-estuarine coasts, especially in Scotland and particularly in Orkney, where *c.* 29,000 were recorded in winter 1997–98 (Rehfisch *et al.* 2003b); comparable figures for Ireland were 54,650 (13% of flyway) and 8,925 (>2%) for IoM.

Genus *BARTRAMIA* Lesson

Nearctic monospecific genus that has been recorded as a vagrant in GB&I.

Upland Sandpiper   *Bartramia longicauda* (Bechstein)

SM monotypic

TAXONOMY In a study of N American waders, cytochrome oxidase 1 sequences showed that Upland Sandpiper was closely similar to Long-billed Curlew and Whimbrel (Hebert *et al.* 2004), though statistical support is not specified. This supports Chu's (1995) reanalysis of Strauch's (1978) osteological data, which showed Upland Sandpiper is sister to a curlew clade.

DISTRIBUTION Prairie grassland and wet meadows from S Alaska, across middle Canada to E seaboard of USA. Migrates to grasslands of Argentina and Uruguay. Range and numbers contracted with cultivation of prairies, though now seems stable (HBW).

STATUS A total of *c.* 45 recorded in GB since the first shot near Warwick (Warwickshire) in Oct 1851; 11 more in Ireland since one near Ballinasloe (Galway) in autumn 1855. All but two have been in Sep–Dec and predominantly in the S and W or the N Isles.

Genus *XENUS* Kaup

A monospecific Palearctic genus that is vagrant to GB&I.

Terek Sandpiper   *Xenus cinereus* (Güldenstädt)

SM monotypic

TAXONOMY Position of *Xenus* was not resolved by Chu (1995), but nuclear and mt DNA sequences (Pereira & Baker 2005) indicate sister to clade with all *Tringa* species.

DISTRIBUTION Breeds in forest tundra from Finland and W Russia to E Siberia, between 70°N and 52°N. Migratory, wintering along coasts of SW Africa and around Indian Ocean to Australia.

STATUS The first bird reported in GB was at The Midrips (East Sussex) in May 1951 and a total of 63 up to 2003. Only two Irish records (to 2004); the first at Rosslare Backstrand (Wexford) in Aug–Sep 1996 and a second at Blennerville (Kerry) in Sep–Oct 2004. There is a clear peak in reports for May–June, with a strong bias towards the S and E; presumably many are overshooting spring migrants. However, one bird wintered in SE Northumberland from Nov 1989–May 1990, returning (or relocated) Aug 1990–Jan 1991.

It is curious that there were no records prior to the 1950s; the bird is not hard to identify and they are unlikely to have been overlooked.

Genus *ACTITIS* Illiger

Holarctic genus of two species; the Palearctic representative breeds in GB&I, the Nearctic form occurs as a vagrant.

Common Sandpiper   *Actitis hypoleucos* (Linnaeus)

MB WM PM monotypic

TAXONOMY Morphology, osteology and DNA all confirm sister relationship between Common and Spotted Sandpipers (Chu 1995; Pereira & Baker 2005). However, node linking these to *Tringa* is weak for both data sets, so relationships rather ill-defined. Not included by Paton *et al.* (2003).

DISTRIBUTION Widespread across Palearctic from Atlantic to Kamchatka and Japan, between 71°N and 33°N. Typically nests along riverbanks and lakesides, sometimes seashores. Winters beside streams, rivers, lakes, estuaries, from Mediterranean across Africa to Cape, and through Middle East and India to Australia, avoiding only desert areas.

STATUS Two populations in W Eurasia are recognised by Stroud *et al.* (2004); one breeding in W & C Europe and wintering in W Africa; the other in E Europe and W Asia and wintering in E Africa. Common Sandpipers recorded in GB&I are part of the former, which is estimated as 464,000–683,000 pairs (Thorup 2006). Breeds locally in Ireland but unquantified. The population of GB is rather uncertain, having been estimated as *c.* 18,000 pairs during the 1990s (Stroud *et al.* 2004), although this was revised to *c.* 24,000 pairs by Dougall *et al.* (2004). Data from the Waterbirds Survey indicates a progressive decrease in numbers since the mid-1980s, supported latterly by BBS data. A study in the Peak District (Derbyshire) has shown birds to be highly site faithful, allowing detailed analysis of their population dynamics (refs in Wernham *et al.* 2002), and this has shown that survival and productivity can vary depending on climatic conditions. There also appears to have been a reduction in average clutch size since 1968. The reasons for these declines are not apparent (Baillie *et al.* 2006).

Migrants move through S GB&I during Mar–May, returning from July to Oct. Most of the migrant population passes through southern Britain on its way between breeding sites in Scandinavia and wintering grounds in Iberia and NW Africa (Wernham *et al.* 2002); British breeding birds have been recovered as far south as W Africa. There is evidence that birds may show fidelity to their wintering sites; birds have been trapped in successive winters in Senegal wetlands, and there is a similar record from Hampshire (Wernham *et al.* 2002).

Small numbers overwinter in GB&I; e.g. WeBS reports of 154 individuals at 105 sites between Nov 2004 and Mar 2005 (Banks 2006).

Spotted Sandpiper   *Actitis macularius* (Linnaeus)

CB SM monotypic

TAXONOMY Closely related to Common Sandpiper.

DISTRIBUTION Breeds across N America from *c.* 30°N, excluding only Keewatin and Arctic islands. Winters S of breeding area to Brazil.

STATUS A pair was found with a nest and four eggs at Uig Bay, Skye (Highland) in 1975 (Wilson 1976). Unfortunately, the breeding attempt failed. A returning bird was seen paired with a Common Sandpiper in Yorkshire in 1991, in the company of three large young.

Following a recent review by BOURC, some earlier records are not now accepted, and the first record for GB was a bird seen at Helston (Cornwall) in June 1924; the first for Ireland was in Feb 1899, near the River Finnea (Westmeath). This species shows a striking pattern of occurrence. Only four were recorded in GB&I before 1950. The numbers increased rapidly thereafter, to a total of over 130 by 2003, of which 14 were in Ireland. Up to eight have been recorded in a year, and the species is now annual. It has been recorded in most months but shows clear peaks in May–June and Sep–Oct. There are also several records of birds wintering. Increase in recent records followed a paper on their identification by Wallace (1970). Whilst adults are striking, juveniles are much more subtly different.

Genus *TRINGA* Linnaeus

Holarctic genus of 13 species; three breed in GB&I and a further seven have been recorded with varying degrees of rarity. The shanks have been the subject of a molecular study by Pereira & Baker (2005), based on mt and nuclear DNA sequences and compared with a morphological analysis. Mt DNA evolves faster than nuclear and has been used for more closely related species; nuclear sequences are more suitable for examining deeper levels of a phylogenetic tree. Since a combination of data optimises tree resolution, their findings are based on this approach.

Green Sandpiper   *Tringa ochropus* Linnaeus

CB WM PM monotypic

TAXONOMY Osteology (Chu 1995) cannot resolve relationships with other *Tringa*, but Pereira & Baker (2005) show sister species to Solitary, with which once thought conspecific.

DISTRIBUTION Breeds in wet woodlands, usually close to open water, from Scandinavia to Yellow Sea, between 68°N and 49°N. Migratory, wintering from GB&I through Mediterranean to tropical Africa, Indian, SE Asia.

STATUS Breeding was first confirmed in the Highlands of Scotland in 1959, and in 1999–2006 one or two pairs bred. Summering individuals and possible breeding on many other occasions (BS3).

GB&I lie on the extreme western fringes of the range. Population numbers are very vague, with the European and W Russian component estimated as 335,000–630,000 pairs (Thorup 2006): breeding in boreal forests and utilising the old nests of songbirds and squirrel dreys does not make for easy censuses.

Most records from GB&I are of birds on autumn passage, arriving in early July and peaking through Sep, usually along the margins of pools and marshes. Some remain through the winter, moving to sites with flowing fresh water, usually singly, and often taking up territories that they defend quite vigorously (refs in Smith *et al.* 1999). Given its liking for small ditches and creeks, it is difficult to assess numbers. WeBS counts peak in autumn at *c.* 500, declining to c. 100 in midwinter, but these will be a fraction of the true figure: it was noted in 593 10-km squares in the winter atlas, often with more than one bird per square. It is scarce in Ireland and rare in Scotland in winter.

The limited ringing data suggest that these birds come from Scandinavia and those that continue their migrations move through SW Europe into NW Africa.

Solitary Sandpiper   *Tringa solitaria* Wilson

SM race undetermined but see below.

TAXONOMY Osteological analysis was unable to resolve position within *Tringa* (Chu 1995) but, by pooling mt and nuclear sequences, Pereira & Baker (2005) showed it to be sister to Green Sandpiper. Two subspecies recognised based on size, colour of spots on upperparts, face pattern: *cinnamomea*: N and W of range, from Alaska, NW Territories, to NE Manitoba; *solitaria*: S and E, from E British Columbia across S Canada to Labrador. DNA sequences of cytochrome oxidase 1 gene show substantial differences (Hebert *et al.* 2004), which indicate a prolonged period of reproductive isolation and suggest that the two races might be better treated as separate species.

DISTRIBUTION Boreal forests, breeding in marshes and swamps from S Alaska to NW Territories and British Columbia to Labrador and Newfoundland. Migrates overland, to winter from Mexico to Argentina.

STATUS A total of *c.* 30 recorded in GB since the first by the River Clyde (Lanarkshire) prior to 1870; 11 of these have been in Scilly. There had only been three in Ireland up to 2001, and the first was as recent as Sep 1968 at Akeragh Lough (Kerry); the other two were in County Cork, in Sep 1971 and 1974. All have been in period July–Dec. A bird seen and photographed on Fair Isle in Sep 1992 showed clear features of the nominate race (McGowan & Weir 2002).

Grey-tailed Tattler   *Tringa brevipes* (Vieillot)

SM monotypic

TAXONOMY Osteological analysis Chu (1995) and mt DNA sequences of Pereira & Baker (2005) show sister relationship with Wandering Tattler *T. incanus*; nuclear sequences give different result, though with little statistical support. Former show this clade sister to Wood Sandpiper but latter shows basal to clade including all *Tringa* except Solitary and Green Sandpipers. Exact affinities need resolving, though nested position within *Tringa* clade indicates that the tattlers belong within that genus.

DISTRIBUTION Principally NE Siberia alongside streams and lakes in montane tundra and taiga. Isolated population further west in Putorana Mountains. Migratory, wintering in islands of SW Pacific to Australasia.

STATUS Two records for GB&I, both long-staying individuals. The first was on the Dyfi Estuary (Ceredigion) in Oct–Nov 1981; the second in Nov–Dec 1994 at Burghead Bay (Moray). Not recorded in Ireland.

Spotted Redshank
Tringa erythropus (Pallas)

WM PM monotypic

TAXONOMY Molecular sequences indicate that Spotted Redshank is sister to a clade comprising Greater Yellowlegs and Greenshank; leg colour appears not to be a good taxonomic character and to have evolved independently in different clades. Chu's (1995) analysis of osteology also returns Greenshank as sister to Spotted Redshank, within a larger unresolved clade that also included Redshank, Lesser Yellowlegs and Marsh Sandpiper, further indicating the limited value of leg colour in the taxonomy of *Tringa*.

DISTRIBUTION Breeds across High Arctic from Lapland to Chukotskaya; winters in tropics, with staging posts in temperate zones across Eurasia.

STATUS Spotted Redshanks occur regularly on passage and in winter, usually in estuaries and coastal wetlands, and preferring muddy sites to sand. Parties rarely larger than 10–20. Spring passage peaks in Apr–June, with the return birds occurring during July–Oct. The latest population estimate (Rehfisch *et al.* 2003a) is of 136: just 1% of the flyway total. In Ireland, before the 1950s, was 'as rare as many Nearctic waders' (P Smiddy pers. comm.), then became much commoner with parties of 100+ recorded; now seems to be declining again.

Greater Yellowlegs
Tringa melanoleuca (Gmelin)

SM monotypic

TAXONOMY Sister species to Greenshank.

DISTRIBUTION Nearctic, breeding in swamps and marshes in belt from Alaska, across Canada to Labrador, Newfoundland. Winters from S USA to Patagonia, many using inland route, with relatively few overflying Atlantic (HBW). This perhaps accounts for scarcity in GB&I compared with Lesser Yellowlegs.

STATUS The first record was a bird shot on Tresco (Scilly) in Sep 1906, with 28 records in GB to 2006. The first Irish bird was also shot, near Skibbereen (Cork) in Jan 1940, with a total of 12 records to 2004. Reported every month except Feb and June, with peaks in May, Aug–Sep.

Common Greenshank
Tringa nebularia (Gunnerus)

RB MB WM PM monotypic

TAXONOMY Forms strongly supported clade with Greater Yellowlegs (Pereira & Baker 2005), indicating weakness of leg colour as taxonomic character.

DISTRIBUTION Widespread in marshes across taiga from N Scotland, Scandinavia to Kamchatka and Amurland, mainly between 65°N and 55°N. Winters GB&I, Mediterranean to S Africa, India, Indo-China, Australia.

STATUS The number of birds breeding in Scotland was estimated at *c.* 1,200 pairs in the 1990s, with a further one or two pairs in Ireland (Stroud *et al.* 2004); they comprise a small fraction of the population that breeds across N Europe, from S Scandinavia through Finland to NW Russia and estimated as 78,000–132,000 pairs (Thorup 2006). The stability of the European population is not known. Since European birds winter widely beside fresh and brackish water in W Africa, they do not lend themselves to easy census. Numbers in Scotland are subject to fluctuations and some losses have been attributed to afforestation of the nesting habitat (refs in Thompson *et al.* 1988). Christian & Hancock (2009) examined the structure of a population in N Scotland over a 25-year period; they found little effects of adjacent forestry cover but stress that the plantings were new and the situation might change as the trees matured. Single pairs bred in Co. Mayo in 1972 and 1974, possibly also in 1971.

Three birds ringed on the breeding grounds in Scotland have been recovered, one in Iceland and two in SW France, suggesting them to be relatively short-distance migrants (Wernham *et al.* 2002). Recoveries of birds trapped on passage in S Britain come from as far as S Nigeria. Others remain closer to home, and small numbers regularly winter in GB&I, especially on estuaries in S England, Wales and Ireland. Rehfisch *et al.* (2003a) estimated a winter population of 597 in GB and a further 1,265 in Ireland (Crowe *et al.* 2008). The trend in recent years has been for an increase (Banks *et al.* 2006), and several sites hold >25 birds during Nov–Feb. Of course, these numbers pale into insignificance alongside the global population but indicate that climatic amelioration may see further increases.

Lesser Yellowlegs   *Tringa flavipes* (J. F. Gmelin)

SM monotypic

TAXONOMY Osteology could not resolve relationships with other shanks (Chu 1995), but Pereira & Baker (2005) show strongly supported clade with Willet *T. semipalmata*; since this is nested within many other shanks, they recommend that *Catoptrophorus* (in which Willet formerly placed) should be merged with *Tringa*, as earlier proposed by Vaurie (1965).

DISTRIBUTION Breeds in swamps and bogs, usually in or close to taiga forest, from Alaska, N and W Canada to Manitoba and Quebec. Migratory, wintering from S USA to Patagonia; some move E and follow the Atlantic coast, or even overfly the ocean to Lesser Antilles and S America.

STATUS There have been over 250 in GB since the first at Misson (Nottinghamshire) in winter 1854–55. The first for Ireland was at Lady's Island Lake (Wexford) in Oct 1955 and 87 by 2001. A far more abundant bird in GB&I than Greater Yellowlegs, with about seven times as many records, probably due to the more coastal autumn migration route in North America. Similar distribution through the year, and there are several records of bird wintering in GB&I.

Marsh Sandpiper   *Tringa stagnatilis* (Bechstein)

SM monotypic

TAXONOMY Pereira & Baker (2005) showed sister relation to Redshank/Wood Sandpiper clade, though morphology and osteology (Chu 1995) could not resolve this.

DISTRIBUTION Breeds in marshes of steppe and boreal regions of Eurasia; predominantly between 57°N and 45°N, from W Russia to Lake Baikal. Migratory, widely distributed in winter across S Africa, India, Australia; also smaller numbers in E Mediterranean, Arabian Gulf. According to BirdLife International (2004), the European population increased substantially between 1970 and 1990.

STATUS The first record was of two at The Midrips (East Sussex) in Sep 1937; there have now been over 120 records, mostly from the eastern side of GB. Five birds have been reported from Ireland since the first at Tacumshin (Wexford) in Aug 1982. Recorded in all months from Apr to Oct, with peaks in May and July–Aug. Never more than two in a year during 1950–76, but this happened 18 times from 1977 to 2003 – in agreement with the increased European population but also with the increased number of observers.

Wood Sandpiper   *Tringa glareola* Linnaeus

MB PM monotypic

TAXONOMY Morphology and osteology suggest affinities with tattlers (Chu 1995) but only weakly supported, and molecular evidence indicates sister to Redshank (Pereira & Baker 1995).

DISTRIBUTION Breeds in swamps and bogs along woodland edges and in scrub from Scandinavia to Kamchatka and Amurland, between 71°N and 51°N. Migratory, wintering in sub-Saharan Africa down to the Cape, Nile Valley, India and SE Asia to Australia.

STATUS There is a small breeding population in Scotland, with birds regularly found in suitable habitat, and with evidence of an increase since the 1980s. Numbers are rarely large, 4–12 pairs at seven localities in 2002 being about average for recent years, although the number of birds and occupied sites varies. Most in Sutherland and Caithness, with fewer as far S as Perth and Kinross (Chisholm 2007). Confirmed breeding is less often reported due to the risk of disturbance, the remoteness of many sites and the scarcity of observers in N Scotland. It is possible, even likely, that more birds remain to be discovered.

Wood Sandpiper has bred in England on at least one occasion; John Hancock obtained a nest and eggs and one adult at Prestwick Carr (Northumberland) in 1874. Other claims of breeding are less well authenticated (Brown & Grice 2005).

Much less common than Green Sandpiper on passage throughout GB&I, the maximum in GB of 72 during Aug 2004 (which was unusually high) compares with 478 Green Sandpipers in the same month. Chisholm (2007) shows that the number of breeding attempts correlates with the number of birds recorded in Scotland on spring passage. Rather few have been ringed in GB&I, and recoveries are correspondingly few; however, a bird ringed in Lincolnshire was found dead three years later only *c.* 47 km away, suggesting the possibility of fidelity to stop-over sites (Wernham *et al.* 2002). Very rare indeed in winter.

Common Redshank   *Tringa totanus* (Linnaeus)

RB MB WM PM *totanus* (Linnaeus)
WM PM *robusta* (Schiøler)

TAXONOMY Chu (1995) found this to be allied to Lesser Yellowlegs and other shanks (see Spotted Redshank), but could not resolve the internal structure of the clade. Pereira & Baker (2005) had more success, finding that Redshank is sister to Wood Sandpiper, in a clade that also included Marsh Sandpiper. Redshank is conventionally divided into six subspecies: *robusta*: Iceland, Faroes; *totanus*: N Scandinavia to W Siberia, S to Iberia and Mediterranean, including GB&I; *ussuriensis*: S Siberia, Mongolia to Lake Baikal; three other races in Oriental region. '*Britannica*' previously recognised from S Scandinavia, GB&I and coastal Netherlands (e.g. Vaurie 1965); validity doubted by BWP but supported by Engelmoer & Roselaar (1998), who recommend its reinstatement; Ottvall *et al.* (2005) found differences in nuclear and mt DNA between *robusta* and Scandinavian birds, but less difference between N Scandinavian *totanus* and S Scandinavian '*britannica*'.

DISTRIBUTION Breeds widely across S and W Palearctic, from Iceland to coastal China; absent from tundra and taiga of Siberia. Most populations migratory, though those from S and W less so. Winters along coasts of Europe and Africa, predominantly N of equator, also Arabia, India, Malay Peninsula, Indo-China.

STATUS The numbers of breeding Redshanks have been surveyed several times (refs in Thorup 2006), most recently suggesting *c.* 37,000 pairs in GB and 4,000–5,000 in Ireland. However, BBS data indicate that the breeding population in GB declined by 12% during 1994–2005, and there is also evidence of a northwards shift in distribution between the Atlases, which mirrors the picture across Europe. The decline, which occurred in both uplands and lowlands, has been attributed to changes in grassland management in the former and drainage of wetlands in the latter.

Size of global population uncertain, but estimated at *c.* 2.4 million (Robinson *et al.* 2005); within this, Stroud *et al.* (2004) identify several components, three of which occur in GB&I: the breeding population (treated as '*britannica*' by Stroud *et al.* 2004) is generally resident, though some migrate to Iberia in winter; *robusta* breeds in Iceland and comes to GB&I for the winter, though Engelmoer & Roselaar (1998) suggest that birds breeding in N Scotland, Orkney and Shetland may also be of this form; W populations of *totanus*, breeding from Fennoscandia and the Baltic states to Poland and the western parts of C Europe and wintering in NW Africa, passing through GB&I on both migrations.

The size of the Icelandic/Faroese populations of *robusta* is less well known, and the estimate of *c.* 64,000 pairs seems to be based on either 'informed assessment' or guesses of the relative proportions of these and the resident population in winter flocks (Stroud *et al.* 2004). Since 1980, the winter data from WeBS counts indicate increases, modest in GB and sharper in Ireland, both N and S (Crowe 2005, Banks *et al.* 2006). It is interesting that WeBS counts from NI show distinct peaks in Mar and Oct that are not seen in GB; these are presumably the Icelandic birds on their way to and from the breeding grounds.

The importance of our islands for wintering Redshank is highlighted by there being 40 sites that held internationally important concentrations, with totals of 116,100 in GB, 31,090 in Ireland and 418 in IoM (Rehfisch *et al.* 2003a; Crowe *et al.* 2008), probably representing *c.* 60% of the combined populations of migratory *robusta* and the resident breeding population ('*britannica*'). As with many species of waders, Redshanks historically have tended to winter along the western coasts, where temperatures are higher and cold stress is reduced. However, climate change is leading to a progressive rise in the temperature of east coast sites, enhancing the survival of birds wintering here and apparently resulting in an eastward shift of wintering birds, at least on non-estuarine sites.

The third component comprises birds from the W populations of *totanus* that pass through GB&I en route to wintering grounds in W Africa. The number of birds involved is still not known (see Stroud *et al.* 2004) but is at least 250,000 individuals. Small Redshanks are trapped on the E coast of GB that are representatives of this race.

Genus *ARENARIA* Brisson

Holarctic genus of two species, one from the Palearctic that is a common visitor to GB&I.

Ruddy Turnstone   *Arenaria interpres* **(Linnaeus)**

WM PM *interpres* (Linnaeus)

TAXONOMY Ericson *et al.* (2003) review results from three nuclear DNA sequences and report that *Arenaria* is sister to *Calidris*, a finding supported by Thomas *et al.* (2004a) and Hebert *et al.* (2004), based on different mt DNA genes, and in agreement with the osteological analyses of Chu (1995). Two subspecies recognised: *interpres*: Canadian Arctic, from Ellesmere and Axel Heiberg islands through Greenland, N Eurasia to NW Alaska; *morinella*: most of Arctic Canada, W of Axel Heiberg Island, to NE Alaska. Wenink *et al.* (1994) examined rapidly evolving mt DNA sequences from widely separate populations and found a lack of geographical structure. They argue that extensive gene flow is unlikely, since Turnstones are confined to different migratory flyways and exhibit strong site fidelity. They also suggest that the species has more likely undergone rapid expansion from a recently bottle-necked refugial population, with insufficient time for molecular differentiation, and that the racial differences arose through strong phenotypic selection.

DISTRIBUTION Predominantly a circumpolar species, breeding on boreal tundra, only extending S into Scandinavia. Winters along ice-free coasts almost everywhere, from Iceland to Africa, through Indian Ocean to Australia; coastal USA to Chile and Argentina.

STATUS Summering birds occur regularly, and breeding strongly suspected in N and W Isles and N Scottish mainland, though none considered conclusive (BS3).

Stroud *et al.* (2004) identify three components to the global population of Turnstones, two of which occur regularly in GB&I, as either passage migrants or winter visitors; these are birds from NE Canada and E Greenland, which winter chiefly in W Europe (including GB&I), and from N Europe, which winter mainly in W Africa, passing through GB&I en route.

The Nearctic breeding population comprises 40,000–80,000 pairs, which implies a post-breeding population of up to 240,000 birds (refs in Stroud *et al.* 2004). This is higher than the numbers estimated to winter along the shores of Europe (94,000 in the 1990s), but this figure itself may be inaccurate because of the difficulty of surveying all of the expanses of rocky shore that host small flocks in winter. WeBS counts include only a small proportion of the available habitat. They show, however, that Turnstones arrive in Aug, and numbers remain at a fairly even level through until Apr, when most depart for the breeding grounds. Birds ringed in GB&I during the winter are found in or on their way to Greenland and NE Canada. There are recoveries of birds ringed on Ellesmere Island, but none from further west where *morinella* breeds.

The second component is more difficult to quantify; the population breeding in N Europe and European Russia has been estimated as 15,000–40,000 pairs (refs in Thorup 2006). Some of the birds trapped in GB&I during the autumn belong to this population, with summer recoveries from the Gulf of Bothnia and winter ones from the coasts of W Africa, between Morocco and Congo. Although Thorup's estimate is broad, together with an add-on for juveniles, even a lower estimate of 45,000 individuals is well above the census data (32,000) from W Africa; further indication of the difficulty of surveying a species that typically winters in a multitude of relatively small, widely dispersed, flocks. Rehfisch *et al.* (2003a) estimated 49,550 in GB (53% of flyway total), 11,810 in Ireland and 429 in IoM. Only 22% of these were on estuaries, and surveys of non-estuarine coasts between 1984–85 and 1997–98 showed a 16% decline over this period (Rehfisch *et al.* 2003b). WeBS data also showed a modest decline since the late 1980s in GB&I; this is similar to Purple Sandpiper and, in view of the similar rocky shore habitat of many birds, may have the same cause.

Ringing data also confirm the striking site fidelity of winter Turnstones, occupying the same stretch of coastline, in company with the same individuals, as long as they survive.

FAMILY PHALAROPIDAE

Genus *PHALAROPUS* Brisson

Holarctic genus of three species, two breeding in Palearctic and all recorded in GB&I.

Wilson's Phalarope   *Phalaropus tricolor* (Vieillot)

SM monotypic

TAXONOMY From his reanalysis of Strauch's (1978) osteological data, Chu (1995) found five 'sandpiper' lineages: snipes, calidrines, tringines, phalaropes and curlews, but the relations among these were poorly resolved. Paton *et al.* (2003), Ericson *et al.* (2003) and Baker *et al.* (2007) all found evidence of sister relationship between *Tringa* and *Phalaropus*, though statistical support for latter was also low. Osteology showed Wilson's to be sister to the other two species, as did Dittmann & Zink (1991) and Thomas *et al.* (2004b).

DISTRIBUTION Breeds on prairie wetlands and marshes from Alberta and California to Great Lakes. Migratory, wintering down W side of S America from Peru to Patagonia.

STATUS First recorded in GB&I in Sep 1954, when one at Rosyth (Fife), within 50 miles of the birthplace of Alexander Wilson, after whom the species was named. The first Irish record was at Lady's Island Lake (Wexford) in Aug 1961. The species is now annual with a small peak of numbers in May and a larger one in Aug–Oct. Numbers increased steadily through the 1970s to a remarkable spell from 1977 to 1999, when there was an average of *c.* 11 a year. This coincided with increased autumn migration through eastern N America, which Colwell & Jehl (1994) suggest may have been due to drought conditions in Quebec and Ontario. Latterly, numbers have declined to around three a year, perhaps as the drought eased and reduced the need for this eastern movement.

Red-necked Phalarope   *Phalaropus lobatus* (Linnaeus)

MB PM monotypic

TAXONOMY Morphological and molecular data support sister status with Red Phalarope.

DISTRIBUTION Widespread across northern tundra, further inland in Nearctic, closer to Arctic Ocean in Palearctic. Winters at sea, especially on rich fishing grounds of Humboldt Current off Ecuador, Peru; also off Arabia and among islands of SW Pacific.

STATUS A very rare breeding species, largely restricted to Shetland (mainly Fetlar) and Outer Hebrides, with occasional pairs elsewhere in GB&I. Number of pairs varies, but usually within range of 15–35 breeding males present. Bred at several sites in Ireland in the past, but not recently. BirdWatch Ireland continue with habitat management work in Co. Mayo, hoping birds will return to breed in that area.

Red-necked Phalarope is otherwise a scarce migrant through GB&I, with most records in autumn, coming from wetlands along the North Sea coasts, possibly being Scottish birds on their way to wintering grounds in the Arabian Sea. Rarely more than 50 recorded away from the breeding grounds in GB during a year.

Red Phalarope/Grey Phalarope   *Phalaropus fulicarius* (Linnaeus)

PM monotypic

TAXONOMY Sister species to Red-necked Phalarope.

DISTRIBUTION Circumpolar; more northerly than Red-necked Phalarope in Nearctic. Breeds in Siberia E of Taimyr, coastal Alaska, most N Canadian islands to Greenland and Iceland. Winters further south in Humboldt Current, also Benguela Current off SW Africa and upwellings off W Africa.

STATUS More abundant than Red-necked Phalarope on passage, with an average of *c.* 200 being recorded each year in GB, almost always in autumn. The number has increased with the rising popularity of sea-watching and pelagics. Severe westerly gales in Sep–Oct can bring substantial movements, especially along the west coast of Ireland, when hundreds of birds are occasionally recorded, sometimes followed by records far inland across GB. More usually, they occur as individuals passing coastal headlands or stopping to feed in sheltered coves or coastal wetlands.

FAMILY STERCORARIIDAE

Genus *STERCORARIUS* Brisson

Bipolar genus of about five or six species, four of which have been recorded in GB&I, two as breeding birds.

Pomarine Skua   *Stercorarius pomarinus* (Temminck)

PM monotypic

TAXONOMY Several lines of evidence (mt and nuclear DNA, enzyme variations, feather lice, behaviour and calls) suggest that Pomarine Skua is sister to the Great Skua complex rather than to

Arctic and/or Long-tailed Jaegers (Andersson 1973, 1999a; Cohen *et al.* 1997), leading to two possible conclusions: Pomarine Skua is a *Catharacta* or the Great Skuas are *Stercorarius*. Braun & Brumfield (1998) reanalysed the mt DNA data and supported treating all skuas as congeneric; *Stercorarius* has priority. Pomarine shows similar plumage polymorphism to Arctic, and genetics presumably the same. Pale morph commonest everywhere, over 90%.

DISTRIBUTION Breeds across the Arctic tundra, except Iceland, and breeding not proven in Scandinavia (BWP). Winters off coasts of Middle and Central America, Caribbean, Argentina, Atlantic coast of Africa, Arabia, and the Pacific and Indian Oceans from Malaysia to Australia.

STATUS Passage migrant through GB&I, some wintering. In spring, regularly recorded off W coasts Apr–June, especially W Ireland and Hebrides. Some pass through English Channel, less often seen on North Sea coasts in spring. In the autumn, passage starts end July, and birds are usually commoner on the E side of GB&I. Earliest birds generally adult, presumably failed or non-breeders. Fox & Aspinall (1987) showed that most autumn birds are juveniles, with evidence of a three-year cycle, with larger numbers coinciding with peak Lemming years in the Russian High Arctic.

Parasitic Jaeger/Arctic Skua   *Stercorarius parasiticus* (Linnaeus)

MB PM monotypic

TAXONOMY Morphology, plumage, behaviour, ectoparasites, enzymes and molecular data all support sister group relationship of Arctic and Long-tailed Jaegers (see Pomarine Skua for refs). Well-known plumage polymorphism controlled by single locus, with the dark allele showing incomplete dominance. Non-random mating occurs in some colonies, with evidence of female preference for dark morph at Fair Isle (O'Donald 1983). Fitness differences also reported there, but not supported by data from other colonies (Phillips & Furness 1998).

DISTRIBUTION Widespread across Arctic tundra: predominantly coastal in Palearctic, but also breeds inland in Nearctic (Olsen & Larsson 1997). Less widespread than Pomarine in winter, principally around southern half of S America, SW Africa and SE Australia. On passage, may occur anywhere between breeding and wintering sites. Plumage polymorphism clinal in Atlantic. Colonies at Spitzbergen and Bear Island are almost 100% pale, declining to *c.* 25% in N Britain. In Finland, only *c.* 4% are pale. Frequencies may vary among colonies, e.g. Fair Isle vs Shetland; N vs S Iceland (O'Donald 1983). Dark morphs reach higher frequencies in coastal areas than tundra at comparable latitudes (see Furness 1987). Also, evidence of a consistent increase in frequency of dark morph between 1935 and 1970 in several Atlantic populations (Furness 1987).

STATUS Breeds in N and W Scotland, especially Orkney and Shetland; distribution hardly changed since 1850, though relative size of colonies has (Furness & Ratcliffe 2004). Overall, numbers increased from 1,039 apparently occupied territories (aot) in Seafarer (1969–70) to 3,388 aot during the SCR Census (1985–88), and then declined to 2,136 aot during *Seabird 2000* (1998–2002). This decline was especially severe in Shetland, where every known colony decreased. There was an overall decrease in Orkney, though some colonies increased, declines that were slightly offset by increases in the populations on N Lewis (Outer Hebrides) and new colonies in Sutherland and St Kilda. The decline is partly due to habitat loss through agricultural change but also competition with, and predation by, Great Skuas. Food availability may also be involved; Furness & Ratcliffe (2004) report that since *Seabird 2000* nest occupancy has declined further, and the continued loss of sandeel stocks and fishing discards makes the immediate future look bleak.

The commonest migrant skua along the coasts of GB&I, occurring regularly in spring, though much commoner in autumn. Spring passage begins in Mar and is fairly rapid, usually in smaller flocks than Pomarine Skua or Long-tailed Jaeger (Olsen & Larsson 1997). In the autumn, passage starts in Aug, extending to Nov; adult birds tend to be earlier than juveniles, and (unsurprisingly, in view of the relative distribution) dark morphs decline through the autumn.

A 10-year study of summer seabird distribution in the North Sea (Camphuysen 2005) showed concentrations in coastal zones, predominantly near Orkney but also south to the Farnes (Northumberland), and mostly associated with terns and Kittiwakes. Less than 1% sub-adult, supporting observations that these do not come north but remain on the wintering grounds.

Long-tailed Jaeger/Long-tailed Skua   *Stercorarius longicaudus* Vieillot

CB PM *longicaudus* Vieillot

TAXONOMY Sister taxon to Arctic Jaeger. Two subspecies recognised, though doubt has been cast on validity of *pallescens* (N America and Greenland) by Vaurie (1965) and de Korte (1972). Shares plumage polymorphism with Arctic Jaeger and Pomarine Skua, but dark morph is exceptionally rare.

DISTRIBUTION Breeds from C Scandinavia, E across Arctic tundra to Pacific, and S to Sea of Okhotsk; in Nearctic, from W Alaska, across N coast of Canada and Arctic islands to Greenland. Winters at sea in sub-Antarctic, N to S Africa and S tip of South America. Migrates offshore through mid-Atlantic and across the Pacific.

STATUS Noted inland in apparently suitable upland habitat in Scotland from time to time, and nest with eggs at a coastal site in Angus and Dundee in 1980, though eggs predated.

A scarce, but regular, passage migrant through seas off GB&I. Numbers have increased since 1960s, with >1,000 in most years since the late 1980s (Fraser & Rogers 2006). In spring, regularly seen from western headlands of Ireland (Hutchinson 1989) and off Hebrides, especially with onshore winds, but less common in N Sea. In autumn, more regular in North Sea, sometimes in numbers. Also small numbers recorded inland, suggesting that overland migration occurs.

Great Skua   *Stercorarius skua* (Brünnich)

MB PM *skua* (Brünnich)
SM race undetermined but one of the subspecies *maccormicki* Saunders/*antarcticus* (Lesson)/*hamiltoni* (Hagen)/*lonnbergi* (Mathews)

TAXONOMY Extensive evidence now supports the sister status of Great and Pomarine Skuas (see Pomarine Skua). These show affinities in behaviour, vocalisations, ectoparasites, enzymes and both mt and nuclear DNA (Cohen *et al.* 1997, Ritz *et al.* 2008). Great and Pomarine Skuas closest to S hemisphere 'brown' skuas, but species limits in latter ambiguous due to current and historical gene flow (Ritz *et al.* 2008). The extent of divergence in populations of *lonnbergi* also needs further study (Ritz *et al.* 2008). Morphological and behavioural data suggest that the southern skuas are closely related, a view supported by molecular analysis. Mt DNA sequences indicate that northern hemisphere *skua* differs more from the five Antarctic taxa (*antarctica, chilensis, hamiltoni, lonnbergi, maccormicki*) than these do from each other (Cohen *et al.* 1997, Ritz *et al.* 2008). The molecular divergence is probably adequate to support separate specific status for *skua*, but possibly not for any of the Antarctic forms (Votier *et al.* 2007). Hybridisation has been recorded among some of these (e.g. Ritz *et al.* 2006), and further morphological, behavioural and molecular data are required to resolve their taxonomy.

Convergence of Great and Pomarine Skuas may be due to hybridisation. Furness (1987) summarises the evidence that large skuas evolved in the Antarctic and that Great Skua arose through colonisation of the North Atlantic. Andersson (1999a) discusses this in detail, concluding that individually rare male colonists might have hybridised with female Pomarines, thereby transferring the (maternal) mt DNA into the former's gene pool. Which of the Antarctic skuas is the source population is presently not resolvable.

Andersson (1999b) further questioned why the morphology of the Great Skuas differs from the three small species. He suggests that they have retained a juvenile-like plumage into maturity and that this 'neoteny' was selectively favoured both because it made individuals more cryptic, facilitating a kleptoparasitic lifestyle, and because juvenile plumage reduces inter-individual aggression, facilitating colonial nesting. He notes that southern populations of Arctic Jaegers are more colonial and kleptoparasitic than those further north and include higher proportions of dark-morph birds.

DISTRIBUTION Breeds from Iceland, Faroes, Scotland and Norway, north to Bear Island, Jan Mayen and Svalbard. Migrates S along Atlantic seaboard to winter from Western Approaches through Biscay and Iberia to NE Africa. Smaller numbers cross central Atlantic. Young birds tend to remain in wintering grounds, with some wandering across the Atlantic between Greenland and Labrador.

STATUS During *Seabird 2000*, a total of 9,635 apparently occupied territories was estimated for Scotland and Ireland, of which *c.* 23% were in Orkney and *c.* 71% in Shetland (Furness & Ratcliffe 2004), and comprised *c.* 60% of the global population. This represents an increase of 26% from the SCR Survey of 1992 and >200% from 1982. The increases occurred across most colonies, although some (e.g. N Mainland, Shetland) declined sharply. Some increases were too sharp for natural recruitment and must reflect immigration. Some populations seem to be limited by shooting on lambing areas (Furness & Ratcliffe 2004). The only known breeding in Ireland has been of a single pair (successfully) in 2001–2004 in the NW, although others have been seen prospecting. As with other seabirds, populations face stress following the reductions in available sandeel stocks and declining availability of white-fish discards (Hamer 2001). During summer months, Camphuysen (2005) found Great Skuas to be abundant close to breeding colonies in Orkney and Shetland, but further south was more widespread across North Sea than Arctic. Ringing recoveries indicate that adults winter south to Iberia, with a few entering the Mediterranean; immatures winter further south, with some recovered in Brazil (Wernham *et al.* 2002).

Two records of 'southern skuas' belonging to one (or more) of the races *maccormicki* ('South Polar Skua'), *antarcticus, hamiltoni* or *lonnbergi* ('Brown Skua'): St Agnes (Scilly) Oct 2001; Aberavon (Glamorgan) Feb 2002.

FAMILY LARIDAE

Traditional taxonomy based on anatomical characters has most recently been examined in 58 taxa by Chu (1998), who included 117 skeletal traits and 64 from the integument (plumage and soft parts). He found that skeletal traits were of value in determining higher levels of phylogeny, but that the integument was important for lower-level relationships. Moynihan (1959) had previously developed a phylogeny based on behavioural and phenetic traits. Similar studies using mt DNA sequences have been undertaken on 32 gull species (Crochet *et al.* 2000), 86 Charadrii, including 30 gulls (Thomas *et al.* 2004a), and all 53 larids (Pons *et al.* 2005). These approaches allow a comparison of morphology, behaviour and molecules in assessing gull systematics, but raise as many questions as they resolve. Gulls are monophyletic; studies that suggest otherwise are usually (e.g. Schnell 1970a, b) based entirely on phenetic characters that are subject to strong natural selection, which can confound history and adaptation, or which may be primitive and therefore unreliable for establishing phylogenetic relationships.

Behaviour and morphology divide the gulls into two broad groups: large white-headed gulls and small, often dark-headed ones, but there are a number of exceptions that have traditionally been assigned to small or monotypic genera (e.g. *Pagophila, Xema, Rhodostethia, Rissa*). Molecular studies and Chu's morphology-based analysis indicate that the genus *Larus*, as we usually think of it, is not monophyletic, and that the 'hooded' gulls are not homogeneous. The molecular and morphological data are not entirely in agreement and some groups identified by the former are not supported by the latter. They do agree that, of the species that have been recorded in GB&I, *Rissa, Xema* and *Pagophila* are distinct from the rest of the gulls, and these should be retained in separate genera. Both sets of data indicate that Black-headed, Bonaparte's and Slender-billed Gulls are in a clade that is distinct from the rest of the gulls. Along with some other species that have not been recorded in GB&I, they are placed in their own genus *Chroicocephalus*. Finally, Little and Ross's Gulls are sister taxa, and again separate from the remaining gulls; they too are placed in their own genera. Although there are indications that the Nearctic 'hooded' gulls, such as Franklin's and Laughing, may be in a separate clade, this is not strongly supported, and it seems wiser, for now, to retain the remaining species in a single (still rather large) genus *Larus*.

These findings are discussed in depth by Pons *et al.* (2005). They stress the importance of inter-preting morphology with caution: plumage, bill and leg colour and bill structure are all potentially subject to strong natural or sexual selection, and these can confound true phylogenetic relationships. In several of the species accounts that follow, we cite Pons *et al.* (2005) for brevity; in many cases, other studies (e.g. Crochet *et al.* 2000; Thomas *et al.* 2004a) reach similar conclusions.

Genus *PAGOPHILA* Kaup

Holarctic genus containing one species, which is a vagrant in GB&I.

Ivory Gull   *Pagophila eburnea* (Phipps)

SM monotypic

TAXONOMY Molecular results (e.g. Pons *et al.* 2005) show this as sister taxon to Sabine's Gull, in accord with morphology (Chu 1998). Latter's postulated affinity to Black-legged Kittiwake not strongly supported.

DISTRIBUTION Probably the most northerly bird, breeding on sea cliffs and rocky outcrops in extreme north of Russian and Canadian Arctic, from Novaya Zemlya, Svalbard, Franz Josef Land; islands off N Canada to NE and NW Greenland. Winters across High Arctic, S into Bering Sea in Pacific, and Labrador coast in Atlantic.

STATUS The first record for GB was a bird shot at Unst (Shetland) in Dec 1822; first Irish: two at Blennerville (Kerry) in Feb 1847. Now, *c.* 150 records, from every month but chiefly Nov–Feb. No evidence of any change over time; most records from northern coasts, especially Orkney and Shetland, and more scarce in Ireland, with only 16 records.

Genus *XEMA* Leach

Holarctic genus containing one species, which is a regular migrant in GB&I.

Sabine's Gull   *Xema sabini* (Sabine)

PM monotypic

TAXONOMY Molecular data indicate that this forms a species pair with Ivory Gull (Pons *et al.* 2005), a finding unexpected from traditional analysis, which has usually (tentatively) placed Sabine's with Swallow-tailed *Creagrus furcatus* (e.g. HBW). Ivory/Sabine's pair forms a clade with the kittiwakes, but this is less well supported.

DISTRIBUTION A High Arctic species, breeding along the northern shores of N America and Asia; also E and W Greenland, Svalbard. Migrates through Atlantic and E Pacific to winter on the rich feeding grounds off W coast of southern Africa and S America.

STATUS The first for GB&I was shot in Belfast Lough (Antrim) in Sep 1822, and the first for GB was obtained at Milford Haven (Pembrokeshire) in autumn 1839. Although it is now annual with *c.* 100 records per year, it is rare in spring; e.g. in 2002, only five recorded in late May from the annual total of *c.* 108 in GB (Fraser & Rogers 2005). Autumn passage usually Aug–Nov, especially along the S & W coasts of Ireland. Small numbers pass through North Sea and are regularly seen off headlands such as Flamborough and Filey (Yorkshire); more use the W coastline of GB, sometimes in quite large numbers. Not infrequent inland, usually after storms.

Genus *RISSA* Stephens

Holarctic genus of two species, one of which breeds in GB&I.

Black-legged Kittiwake

Rissa tridactyla (Linnaeus)

RB MB WM PM *tridactyla* (Linnaeus)

TAXONOMY Closely allied to Red-legged Kittiwake *R. brevirostris* of Aleutian and Commander Islands, but differs in plumage and mt DNA (Olsen & Larsson 2003; Pons *et al.* 2005). Comprises part of a poorly defined clade with Ivory and Sabine's Gulls. Two well-defined races recognised: *tridactyla* N Atlantic, N Siberia; *pollicaris*: N Pacific (McCoy *et al.* 2005). Lack of morphological differentiation within the two oceans may be due in part to an apparently low level of natal philopatry: less than a quarter of new entrants to a colony had been hatched there (Couson & Coulson 2008). Initial evidence suggests vocal differences between Atlantic and Pacific populations as well (Mulard *et al.* 2009).

DISTRIBUTION Nominate breeds along rocky coasts of N Atlantic, from NE Canada, Greenland, Iceland, GB&I to Iberia; also east along northern coast and islands of Siberia; *pollicaris* breeds around N Pacific from Sea of Okhotsk to N British Columbia. Both races winter at sea, across temperate N Pacific and N Atlantic.

STATUS Breeds on sea cliffs around GB&I, and less often buildings and harbours, though rare or absent from low-lying coasts of S and E of England. Populations increased throughout first half of 20th century, perhaps due to reduced persecution; this increase continued from *Operation Seafarer* (1969–70) to SCR census (1985–88), but then declined to *Seabird 2000* (1998–2002), when estimated at *c.* 416,000 apparently occupied nests (Heubeck 2004). These declines evident at most major colonies, especially in Shetland (down 69%), but less severe from Orkney to Moray (7%), though rising to *c.* 30% over much of rest of GB&I. This decline probably reflected poor breeding success (Heubeck 2004). At one colony, reduced numbers were shown to be due to low adult survival and poor breeding success; numbers were unlikely to recover if a local sandeel fishery, which had been closed due to its presumed adverse effects on top predators, was allowed to be re-activated or if sea temperatures increased further (Frederiksen *et al.* 2004). In long-lived species, adult survival is critical for maintaining numbers, and this declined at some sites (Harris *et al.* 2000).

Both these findings were ascribed partly to poor sandeel stocks, but other factors may also be involved. Predation by Great Skuas has increased in Shetland (Heubeck *et al.* 1997), perhaps also as a response to the declining sandeels. A further factor that has been implicated (at least locally in NE England) is toxin produced by algal blooms (Coulson & Strowger 1999). Heubeck (2004) suggests the relative contributions of fisheries, climate change or pollution on this species are unclear, but that declines in key food fish have played a major role. Unlike diving species, such as Guillemots, Kittiwakes feed from the surface water and cannot easily change to demersal species; indeed, evidence is now accumulating that some alternative food stocks such as pipefish are unsuitable as food for nestlings, being too long and bony for ingestion and of low energy content (Harris *et al.* 2008).

Outside the breeding season, Kittiwakes spread widely across the N Atlantic, with recoveries of juvenile birds from Greenland within six weeks of fledging (Wernham *et al.* 2002); some adults are found here as well. Later in the autumn, the pattern changes, with fewer recoveries from W Geenland and more from NE Canada, suggesting that birds move south to winter in the richer fishing grounds off Newfoundland. Not all young birds cross the Atlantic, and a similar pattern of southwards movements are seen in recoveries from the coasts of W Europe. The distribution of young birds during the second year is less well known, but it seems that most do not return to the breeding colonies until their second or third year.

Genus *CHROICOCEPHALUS* Eyton

Holarctic genus of 11 species, one of which breeds and two of which are vagrants to GB&I.

Slender-billed Gull   *Chroicocephalus genei* (Brême)

SM monotypic

TAXONOMY Pons *et al.* (2005) found Slender-billed Gull to be part of a clade of 'masked' gulls that included Black-headed, Bonaparte's and an array of other species from around the world, although its position within this clade was unclear. Given *et al.* (2005) found a poorly supported sister relationship with Bonaparte's, and its sister status to Black-headed Gull (based on anatomy and plumage e.g. Chu 1998) was not supported. The loss of the dark head/mask has been proposed as an adaptation to taking live fish (Isenmann 1976).

DISTRIBUTION A localised breeder around the Mediterranean and E into Aralo-Caspian. Among the major countries, populations are declining in Russia (BirdLife International 2004), apparently stable in Italy and Ukraine, and increasing in Spain, France, Senegal, Mauritania and Turkey (Olsen & Larsson 2003). European birds migrate to Mediterranean coasts and NW Africa (from Spain). Eastern populations winter in Red Sea, Arabian Gulf and N coasts of Indian Ocean.

STATUS Nine individuals. First at Langney Point (E Sussex) on four dates in June–July 1960. Subsequently, most in May (including two, twice), but also Apr, Aug and Sep. There are no records from Ireland.

Bonaparte's Gull   *Chroicocephalus philadelphia* (Ord)

SM monotypic

TAXONOMY Chu's (1998) analysis of larid osteology placed Bonaparte's in a clade with Little and Ross's, whereas integumentary traits suggested a larger and more loosely defined assemblage. Molecular studies (Given *et al.* 2005, Pons *et al.* 2005) were rather inconclusive, identifying Slender-billed as the sister taxon, but with little support. More investigation desirable. Some authorities (e.g. HBW) recognise four subspecies, though we follow Burger & Gochfeld (2002) and regard it as monotypic.

DISTRIBUTION Breeds widely across Canada in the taiga forest from Hudson Bay to the Pacific, and N into Alaska, and is the only gull that almost invariably nests in trees (Burger & Gochfeld 2002). Migrates S to winter along the Atlantic, Pacific and Gulf of Mexico coasts of N America, and along the River Mississippi. May occur in large flocks (>30,000 birds) on northern waters prior to freezing.

STATUS First records were adult birds in Belfast in Feb 1848 (the next Irish record was not for over 130 years) and Loch Lomond (Clyde) Apr 1850; the Belfast record was also the first for the W Palearctic. Numbers have increased markedly in recent years, perhaps as observers have become more familiar with identification. Over 150 recorded in GB to 2003, with a further 33 in Ireland; more in spring than autumn, with peak numbers in Mar–May.

Black-headed Gull   *Chroicocephalus ridibundus* (Linnaeus)

RB MB HB WM PM monotypic

TAXONOMY Morphological data suggest is sister species to Slender-billed Gull (Chu 1998) or Brown-hooded Gull *C. maculipennis*, but molecular data indicate closer affinity with Brown-headed Gull *C. brunnicephalus* of SC Asia (Given *et al.* 2005, Pons *et al.* 2005). Eastern birds ('*sibiricus*') reputedly have longer wings and bill, but Vaurie (1965), BWP and HBW do not recognise this race.

DISTRIBUTION Breeds widely through temperate zone, north to boreal forests, from Iceland across Europe and Asia to NE China and Kamchatka; small population in NE N America. Northern populations migratory to avoid frozen water, wintering along rivers and coasts of S and W Europe, N, W and E Africa, India and Far East. Global population estimated as 2.1–2.8 million pairs (Dunn 2004).

STATUS See Mediterranean Gull in relation to hybrid pairing with that species. Black-headed Gulls in GB&I nest colonially on and around fresh- and salt-water lagoons, sand and shingle banks, etc., both inland and close to the sea. Numbers estimated at 141,890 apparently occupied nests during the *Seabird 2000* survey, representing a decrease of *c.* 16% since the SCR Census (1985–88). The changes were not uniform, either in distribution or abundance. Overall, Scottish populations increased, although not everywhere; there were declines at inland colonies in England and Wales, and especially Ireland (>50% since 1980s). Mink are a proven threat in W Scotland (Craik 1998) and possibly elsewhere. Changes in agricultural practices and the decline of invertebrates through increased pesticide usage may also be involved.

Populations in GB&I are largely resident, though there is exchange between Britain and Ireland, and some birds from the south move to W France and Iberia in winter (Wernham *et al.* 2002). The winter population is enhanced by extensive immigration, especially from Scandinavia and colonies adjacent to the Baltic, often with a striking winter-site fidelity between years. The Winter Gull Roost Survey (A.N. Banks *et al.* 2007) estimated a wintering population of 2.1 million birds. British-ringed nestlings may relocate to different colonies to breed, though the extent of this is hard to determine. Some have been found at colonies in the Netherlands, Germany and Scandinavia during the breeding season: again the extent is not known, but this gene flow would help explain the relatively low morphological differentiation across the range.

Genus *HYDROCOLOEUS* Kaup

Holarctic genus containing one species, which has been recorded in GB&I.

Little Gull   *Hydrocoloeus minutus* (Pallas)

CB WM PM monotypic

TAXONOMY Morphological analyses have usually indicated that Little Gull shows affinities with Ross's, and the molecular data support this (Pons *et al.* 2005).

DISTRIBUTION Breeds around the E Baltic and discontinuously across Russia to E Siberia. Migrates S and W to winter around Caspian, Black and Mediterranean Seas. E populations winter along Pacific coast of China. Small numbers breed in N America, especially along the shores of Hudson Bay.

STATUS First confirmed breeding at the Ouse Washes (Cambridgeshire) in 1975, though nest predated and one adult killed (Carson *et al.* 1977); subsequent attempts elsewhere in GB in 1978 and 1987 were also unsuccessful (Messenger 2001). Records of juveniles with down on the head and primaries incompletely grown from NE England (1980) and Scotland (1988, 1991) may represent successful breeding (Messenger 2001).

A scarce bird in GB&I until the 1950s, predominantly in the SE, since when it has increased markedly (Hutchinson & Neath 1978). Initially, this only involved the autumn, but then spring records increased too, culminating in birds remaining through the summer (see above). In recent years, records indicate the increasing importance of the E coast of England as moulting area for Baltic breeding birds (Hartley 2004). The largest numbers in recent years have been recorded at Hornsea Mere, Yorkshire, where 7,000 counted in Sep 2004, but Seaforth at the mouth of the Mersey remains a key site in spring.

Numbers have also increased in Ireland. Until 1952, it was regarded as a vagrant, but since then has been increasingly regular (Ruttledge 1975), and hundreds are now recorded annually.

Genus *RHODOSTETHIA* MacGillivray

Holarctic genus containing one species, which is a vagrant in GB&I.

Ross's Gull   *Rhodostethia rosea* (MacGillivray)

SM monotypic

TAXONOMY Molecular and morphological data support a sister relationship between Ross's and Little Gulls (Chu 1998; Pons *et al.* 2005), but exact placement within the gull phylogeny is unclear.

DISTRIBUTION A disjunct breeding distribution, with centres in High Arctic of Siberia, E of Taimyr. Smaller numbers in Arctic Canada and NE Greenland. Main population migrates E to winter in along the pack ice of Bering Sea, S to Kamchatka. Small numbers, perhaps from Greenland and W Siberian colonies seem to winter in N Atlantic, with records from Iceland to the Netherlands, and off NE coast of N America.

STATUS First record in GB was an immature, between Whalsay and Skerries (Shetland) in Apr 1936, which was caught but died later same day. The first for Ireland was in Jan 1981 at Portavogie (Co. Down). Now, *c.* 100 records, including 17 from Ireland, predominantly from N and E coasts, with a steady increase in numbers since first records, and averaging four to five per year since early 1990s. Recorded in every month except July, but peaks in Jan–Feb.

Genus *LARUS* Brisson

Cosmopolitan genus of about 35 species; 16 have been recorded in GB&I, of which six have bred.

Laughing Gull   *Larus atricilla* Linnaeus

SM monotypic

TAXONOMY Morphological data (Chu 1998) suggest that Laughing Gull is part of a large clade whose internal structure is poorly resolved. The DNA results place it in a clade with Franklin's Gull and the S American Grey, Lava and Dolphin Gulls *L. modestus, L. fuliginosus, L. scoresbii* (Crochet *et al.* 2000; Pons *et al.* 2005), though internal relationships are unresolved.

DISTRIBUTION Breeds on Atlantic coast of N America, from Nova Scotia to Gulf of Mexico; also W coast from S California to Mexico; *atricilla* breeds on islands in Caribbean. Migrates to winter on coasts of S America.

STATUS First record at The Crumbles (E Sussex) in July 1923, and the first for Ireland at Tivoli (Cork) in Aug 1968. There are now over 150 records from GB&I, including >50 in autumn 2005, predominantly in the SW, following trans-Atlantic passage of the remnants of Hurricane Wilma. Over half of these were aged as first-winter birds, similar to previous years, but markedly different in 2006, when 60% of reports referred to adults. Some of these may have been 'Wilma' birds that survived in European waters and were re-discovered the following year. Apart from the events of 2005, records are fairly uniform through the year, though slightly fewer in Feb–Apr.

Franklin's Gull   *Larus pipixcan* Wagler

SM monotypic

TAXONOMY In the phylogeny of Pons *et al.* (2005), closely related to Laughing Gull.

DISTRIBUTION Breeds across inland N America from W Canada to Montana and Minnesota. Winters along Pacific coast of C and S America.

STATUS First record for GB&I was an adult on Farlington Marshes (Hampshire) in Feb–May 1970. Subsequently, there have been *c.* 50 records in GB, though there is some uncertainty since individuals move around. Peaks in May–Aug and Nov–Jan. The first for Ireland was a first-summer bird at Ballyheigue (Kerry) in May 1993; this is surprisingly recent in view of the number and variety of

Nearctic seabirds recorded in Ireland. The total in Ireland now stands at seven (to 2004), with birds found in many coastal counties; once again records are difficult to assess because of potentially long-lived, long-staying and mobile individuals.

Mediterranean Gull   *Larus melanocephalus* Temminck

RB MB HB WM PM monotypic

TAXONOMY Behavioural, morphological and molecular data all indicate this to be closely related to Pallas's Gull, though Pons *et al.* (2005) found an unresolved polytomy that included a variety of other species (e.g. *audouini, relictus*) as well.

DISTRIBUTION Breeds at scattered sites across S Europe, also Netherlands, Belguim and France. Outside Europe, common on coasts of Ukraine and E to Azerbaijan. Migrates to Mediterranean, especially Spain. Previously globally rare and highly restricted in distribution in SE Europe, to the extent that Voous (1960) commented 'an unmistakable relict ... probably in the course of becoming completely extinct'.

STATUS With a significant range expansion through the 1960s, new colonies were established in the Netherlands, Belgium and France. Birds were present at Needs Oar Point (Hampshire), in the summers of 1966, 1967, and breeding was proved in 1968 (Taverner 1970). Initially (both here and elsewhere), some birds paired with Black-headed Gulls and, once in Scotland, with a Common Gull. As numbers increased, mixed pairs became rarer. Numbers rose to almost 500 pairs at 34 sites by 2006, although *c.* 90% of these were at just three sites on the south coast of England (RBBP). The first (unsuccessful) breeding attempt in Ireland was in Antrim in 1995. Bred successfully in 1996 at Lady's Island Lake (Wexford), and most years since then, with a maximum of seven pairs at all sites in 2002.

   A very rare bird in GB&I prior to the 1950s, and still less than 50 records per year in the late 1960s (Sharrock 1974). Now, widespread, especially along the coast of SE England, but regularly seen in smaller numbers at inland gull roosts and elsewhere throughout GB&I. Three-figure flocks are regularly reported from Folkestone (Kent) and from the Isle of Wight. The first record for Ireland was Sep 1956, when one was present at Belfast Lough (Antrim).

Audouin's Gull   *Larus audouinii* Payraudeau

SM monotypic

TAXONOMY Morphology and plumage (Chu 1998) suggest this to be in a clade with Ring-billed and Mew Gulls, but molecular data (Thomas *et al.* 2004a, Pons *et al.* 2005) include it with Mediterranean and Pallas's. As with Slender-billed, the loss of the dark hood may be an adaptation to taking live fish.

DISTRIBUTION A Mediterranean endemic, breeding on islets and deltas from Spain to Turkey and Cyprus, also Morocco, Algeria, Tunisia (Olsen & Larsson 2003). Marked increase in population since 1960s, especially in S Spain, e.g. Ebro Delta – none in 1980 to 10,500 in 2000. Winters around coasts of Mediterranean and along Atlantic coast of N Africa as far as Gambia.

STATUS The first was a second-summer bird at Dungeness (Kent) in May 2003, closely followed by a second at Beacon Ponds (Yorkshire) in June 2005. Three more quickly followed in 2007, 2008. The second bird had been ringed, but the number could not be read. Since many thousands have been marked in the Ebro Delta (Spain), it is most likely to have come from there. There are no records from Ireland.

Pallas's Gull/Great Black-headed Gull   *Larus ichthyaetus* Pallas

SM monotypic

TAXONOMY Behavioural (Moynihan 1959), morphological (Chu 1998) and molecular data (Crochet *et al.* 2000) indicate this to be in a clade with Mediterranean and several other species including Relict *L. relictus* and Audouin's. Recent molecular analyses (e.g. Pons *et al.* 2005) extend this and confirm inclusion of other hooded species such as Relict, Sooty and White-eyed Gulls, *L. hemprichii, L. leucophthalmus*, and the white-headed Audouin's Gull. This clade is well separate from the other dark-headed gulls.

DISTRIBUTION An Asiatic species, breeding in a series of scattered sites from Crimea to Mongolia. Winters on E coast of Mediterranean and around the coasts of Arabia and India to Myanmar.

STATUS One record: adult, Exmouth, Devon, in late May or early June 1859. No records from Ireland.

Mew Gull/Common Gull   *Larus canus* Linnaeus

RB MB HB WM PM *canus* Linnaeus
SM *heinei* Homeyer.

TAXONOMY On the basis of plumage and osteology, Chu (1998) found that this formed a clade with Ring-billed Gull. N American and NE Siberian birds were each other's nearest neighbours, with European *canus* lying further apart; however, statistical support was low. Pons *et al.* (2005) confirmed earlier molecular findings that Ring-billed and Mew Gulls are early offshoots from the lineage leading to the *L. argentatus* complex, but they did not investigate the relationships of the different populations.

Four subspecies generally recognised (e.g. BWP): *canus*: from Iceland and GB&I, across N Europe to Moscow and Kola Peninsula; *heinei*: N Russia and Siberia possibly to Sea of Okhotsk; *kamtschatschensis*: far east of Siberia and Kamchatka; *brachyrhynchus*: Alaska and NW Canada. *Canus* and *heinei* intergrade in W Russia (Olsen & Larsson 2003); *heinei* and *kamtschatschensis* may overlap, with evidence of clinal variation in NE Siberia. Zink *et al.* (1995) found a difference of *c.* 2% in mt DNA between *kamtschatschensis* and *brachyrhynchus*, based on restriction fragment analysis rather than sequencing, and suggested that this supported specific differentiation: a conclusion followed by Olsen & Larsson (2003).

DISTRIBUTION Breeds across Eurasia from Iceland to the Pacific, south of the tundra zone. Northern birds strongly migratory. W Eurasian populations winter in NE Atlantic, and E Mediterranean, Black and S Caspian Seas; eastern populations winter along Pacific coast from Japan to S China Sea.

STATUS See Ring-billed Gull in relation to hybrid pairing with that species. Breeds in coastal and inland sites across Scotland, N England and N & W Ireland; less common further south. Has recently taken to rooftop nesting in Scotland (Tasker 2004b). Populations have increased across GB&I, by *c.* 2% per year, to *c.* 48,000 apparently occupied nests, although some sites are adversely affected by terrestrial predators. GB&I breeders tend to move S and SW in winter, but the majority remain within GB&I. Wernham *et al.* (2002) suggest that GB&I is a major wintering focus for European birds, especially from Scandinavia and the Baltic states. The Winter Gull Roost Survey (A.N. Banks *et al.* 2007) estimated a wintering population of almost 700,000 birds.

There are ringing records that document exchanges of birds between GB&I and the breeding range of *heinei*; also birds have been trapped whose metrics are closer to this race than to the nominate (BOU 1994).

Ring-billed Gull   *Larus delawarensis* Ord

HB SM monotypic

TAXONOMY Morphology and plumage have traditionally led to placement with Mew and Audouin's Gulls (e.g. Chu 1998). This former is supported by molecular analysis of mt DNA (Pons *et al.* 2005), although the clade also included a range of other white-headed gulls.

DISTRIBUTION A common, predominantly inland-breeding gull of temperate N America. After a decline in the late 19th century, has increased substantially, especially in Great Lakes, where some colonies have increased by >400% (Olsen & Larsson 2003). Winters along Atlantic and Pacific coasts S to Mexico and Caribbean.

STATUS One mixed pair with a Common Gull on Copeland Island (Co. Down) in 2004, discovered when apparent hybrid (ringed as a pullus Common Gull in the Copeland colony in 2004) was located at Millisle (Co. Down) in Feb 2008 (N McKee pers. comm).

One of the more dramatic changes in status of birds in GB&I. The first record in GB involved an adult at Blackpill (W Glamorgan) in Mar 1973, and in Ireland two were present at Belmullet (Co. Mayo) in Feb 1979. Since then, the number of records has increased until it is now widely recorded, especially in the S and W. It is difficult to assess the exact numbers of newly arrived birds because of their mobility and propensity to remain for months, or even years, at particular sites. Fraser & Rogers (2005) estimate *c.* 70 in 2002 (GB only), pointing out that this included only 14 first-year birds and proposing that many of the older birds might be passage migrants, returning winterers or birds that had initially arrived elsewhere in Europe or Africa. Although probably overlooked in the past, certainly in sub-adult plumages, this rise likely reflects the increased N American population, and there may even be a regular trans-Atlantic passage of birds wintering along the W fringes of Europe and NW Africa.

Lesser Black-backed Gull   *Larus fuscus* Linnaeus

RB MB PM *graellsii* A.E. Brehm
PM *intermedius* Schiøler
SM *fuscus* Linnaeus

TAXONOMY Complex; recent molecular and morphological studies indicate clinal variation from *graellsii*, through *intermedius* to *fuscus* (Liebers *et al.* 2001, Liebers & Helbig 2002). However, mt DNA shows cline continues (albeit with slight step) through *heuglini*, *barabensis*, *taimyrensis* to *vegae* (Liebers *et al.* 2004). Further research needed in the zone of contact between *fuscus* and *heuglini*, in particular into the extent of hybridisation and introgression; this may indicate that the European and Siberian taxa merit separate specific rank.

DISTRIBUTION Main breeding range of *graellsii*, Iceland, Faroes, GB&I, Continental Europe from W Iberia to Denmark. Merges into *intermedius*, which occurs Denmark and Atlantic coast of Norway. Populations in Baltic assigned to *fuscus*, also possibly N coast of Norway (Olsen & Larsson 2003). Other races mostly nest along coast and islands of Arctic Siberia: *heuglini* breeds Kola to Taimyr Peninsulas, chiefly wintering Sri Lanka to Red Sea and Persian Gulf; *barabensis* breeds Urals to Kazakhstan, winters principally around Persian Gulf; *taimyrensis* breeds E of *heuglini* from Taimyr Peninsula to Kara Sea, winters chiefly Japan, Korea, Taiwan; *vegae* breeds from Kara Sea to Pacific, wintering S along seaboard of Japan, Korea, China. However, the validity of *taimyrensis* has been questioned (Yésou 2002).

STATUS See Yellow-legged Gull in relation to hybrid pairing with that species. Population of Lesser Black-backed Gulls in GB&I estimated to be 116,684 during *Seabird 2000* (Calladine 2004), over half being in England and about 22% inland. Coastal populations have increased markedly since *Operation Seafarer* (1969–70): England (+85%), Scotland (+83%), Wales (+79%) and Ireland (+77%). Numbers have also increased at many inland sites, especially on rooftops, where 30 times more birds nest than in the 1970s. The global population of *graellsii* has been estimated as *c.* 179,000 pairs (Stroud *et al.* 2001), of which almost 70% nest in GB&I.

GB&I populations mainly migrate south to winter on the coasts of Iberia and NW Africa (Wernham *et al.* 2002), though there is evidence of a change since the 1960s. Over the last 20 years, many birds of all ages have taken to wintering in GB&I, especially in S Ireland. The Winter Gull Roost Survey (A.N. Banks *et al.* 2007) estimated a wintering population of 125,000 birds. Birds come to GB&I in winter from Iceland, Faroes and Scandinavia. Some of these come from the breeding grounds of *intermedius*, and there are two records of *fuscus* ringed in Finland.

European Herring Gull   *Larus argentatus* Pontoppidan

RB HB *argenteus* C. L. Brehm
WM PM *argentatus* Pontoppidan

TAXONOMY Nuclear and mt DNA data indicate that one of the two ancestral stocks of the 'Herring Gull' complex originated in the NE Atlantic. This ('pre-*argentatus*') was clearly distinct from *cachinnans* and gave rise to the ancestors of both Glaucous and Yellow-legged Gulls, as well as contemporary European Herring Gulls. In coastal France, pink-legged *argenteus* breeds alongside yellow-legged *michahellis* with strongly assortive pair formation. These two forms are now distinct and largely reproductively isolated. Birds from Scandinavia differ sufficiently in plumage and metrics to be treated as a separate subspecies: *argentatus*. In fact, European Herring Gulls can be found containing DNA characteristic of both Atlantic and Aralo-Caspian isolates. This could be due to the retention of molecular lineages by this taxon from before their separation into two isolates. Alternatively (and perhaps more likely) hybridisation has occurred between European Herring Gull and either Lesser Black-backed or Caspian Gulls; certainly, the evidence indicates that the Aralo-Caspian sequences found in some European Herring Gulls are more recently evolved than those characteristic of the Atlantic isolate. Hybridises with Glaucous Gull in Iceland (Pálsson *et al.* 2009).

DISTRIBUTION The nominate race breeds in Germany and Fennoscandia, and is dispersive rather than migratory, though it winters regularly as far south as GB&I and France. Grades into subspecies *argenteus*, which breeds from Iceland, GB&I, Netherlands through France; as with *argentatus*, more dispersive than migratory, wintering south to Biscay.

STATUS See Yellow-legged Gull in relation to hybrid pairing with that species. European Herring Gulls of the race *argenteus* breed widely around the coasts of GB&I, although the number of birds involved has changed markedly over the past 75 years. Chabrzyk & Coulson (1976) calculated that there was an annual increase of *c.* 13% from the 1930s to the 1970s, although numbers have fallen since then. Coastal populations declined from *c.* 344,000 apparently occupied nests (*Operation Seafarer*) through *c.* 177,000 (SCR census) to 147,000 (*Seabird 2000*), an overall drop of 43% (Madden & Newton 2004). The decline has been partly offset by an increase in rooftop nesting, from *c.* 3,000 to *c.* 20,000 over the same period, although this by no means compensates for the coastal losses. Urban-nesting European Herring Gulls, particularly common in NE Scotland, appear to have a higher survival and higher fledging success than in nearby coastal colonies.

Chabrzyk & Coulson (1976) ascribed the increase through the middle of the 20th century to a combination of heightened protection and the exploitation of new food resources, such as fishing discards, landfill sites and sewage outfalls. The subsequent decline has been blamed on outbreaks of botulism in the dense breeding colonies (see Madden & Newton 2004 for a discussion of this). However, botulism is unlikely be the only cause; a reduction in unprocessed sewage, a closer control of landfill sites and the decline in the fishing industry have also resulted in a diminution of food resources.

Many thousands of birds utilise inland lakes and reservoirs as nocturnal roosts, and the Winter Gull Roost Survey (A.N. Banks *et al.* 2007) estimated a wintering population of 730,000 birds.

Yellow-legged Gull   *Larus michahellis* J. F. Naumann

CB HB WM PM *michahellis* J. F. Naumann

TAXONOMY Nuclear and mt DNA show that this taxon originated as a southern offshoot of an Atlantic isolate. This population colonised the Mediterranean basin, where it perhaps evolved into *L. armenicus*; these taxa are now reproductively isolated, genetically differentiated and specifically distinct. A second colonisation of the Mediterranean by *atlantis* gave rise to *michahellis*; however, this taxon is not diagnosable or reproductively isolated from *atlantis* and these are retained as conspecific. There are forms along the coasts of Iberia and NW Africa that may or may not be distinct from *atlantis* and are retained within that race.

DISTRIBUTION Subspecies *atlantis* breeds on the Atlantic archipelagos of the Azores, Madeira and the Canaries; largely sedentary but wanders south along coast of NW Africa, perhaps as far as Nigeria (Olsson & Larsson 2003). Nominate race breeds along the coasts of SW Europe, from Biscay and Iberia, through the Mediterranean to Black Sea. Small, isolated populations from N Italy to Poland. Disperses widely in winter, being recorded as far N as GB, Denmark and S Baltic. The populations breeding along the N & W coasts of Iberia have been described as '*lusitanius*', but this taxon is not generally recognised.

STATUS A few breeding records in SW England (especially Dorset) since 1995, involving both pure *michahellis* and mixed pairs with both *L. argenteus* and *L. fuscus* (Green 2004). As with other newly colonising gulls (e.g. Mediterranean Gull), mixed pairs have also been recorded elsewhere.

The first record for GB was a male collected at Breydon Water (Norfolk) in Nov 1886. Numbers have increased markedly over the last 20 years, with flocks of up to 100 being recorded across GB&I. While this is likely partly due to increased observer awareness, the numbers involved will include birds from the rapidly increasing populations in France. Much less common in winter gull roosts than Caspian: the BTO survey (2003/4–2005/6) only recorded 5, compared with 100 Caspian Gulls (N Burton pers. comm.).

Caspian Gull   *Larus cachinnans* Pallas

WM monotypic

TAXONOMY The evolutionary history of the white-headed gulls of the Herring/Lesser Black-backed complex has attracted a great deal of interest for over 80 years, since Dwight (1925). Once regarded as the classic ring species (Mayr 1942), recent molecular studies, utilising both nuclear and mt DNA, have revealed that this not the case. The traditional 'Herring Gull' is paraphyletic, with two sister clades; one of these is broadly 'Atlantic' and includes Yellow-legged, Armenian and most individual European Herring Gulls (plus Great Black-backed Gull and Palearctic individuals of Glaucous Gull); the other centres on the 'Aralo-Caspian' and includes *cachinnans*, *barabensis*, *heuglini*, *taimyrensis*, Lesser Black-backed Gull, Kelp Gull *L. dominicanus* and some individuals of European Herring Gull, an 'Arctic/Pacific' grouping of *vegae*, *smithsonianus*, *mongolicus* (plus Iceland, Glaucous-winged and Nearctic individuals of Glaucous Gull and some Slaty-backed Gulls *L. schistisagus*). Western Gull *L. occidentalis* is not included in these clades and forms an outgroup. Genetic differences among the taxa are generally well defined, except within the Arctic/Pacific grouping of the Aralo-Caspian clade; these are poorly separated with some haplotype sharing. As a result of these researches, which have been reviewed and summarised by Yésou (2002) and Collinson *et al.* (2008), species limits within the complex have been revised, and the present accounts are based on these changes. To save space, individual citations are not given; reference shoud be made to the reviews by Yésou (2002) and Collinson *et al.* (2008).

Molecular data suggest that *cachinnans* is one of the oldest members of the complex, originating in the Caspian region before expanding northwards towards the Arctic Ocean, and giving rise to the dark-mantled group that includes, amongst other taxa, *fuscus*, *heuglini*, *barabensis*. Range expansion has led to sympatry with *argentatus*, with a degree of hybridisation that may account for the morphological variability of *cachinnans*. However, the hybrid zone appears to be narrow compared with the dispersal ability of both forms. Its distinctive molecular structure led to this being recognised as a separate species.

DISTRIBUTION Breeds from north of the Black Sea across Kazakhstan to Lake Balkash, with populations around the Caspian. Now spreading W through European Russia into Hungary, Czech Republic and Germany (Olsson & Larsson 2003). Migratory, wintering to S Caspian, Arabian Gulf, Israel and E Mediterranean. There seems to be a post-breeding dispersal westwards, with Black Sea birds being recovered as far as S Scandinavia, Netherlands and France.

STATUS The first accepted record for GB was a bird at Muckin (Essex) in Sep 1995; the first for Ireland was a second-winter near Belfast Lough (Antrim) in Feb 1998. As observers have become more familiar with the species, the number of records has increased, and in the BTO survey of winter gull roosts (2003/04–2005/06) a total of 110 were identified. This is clearly an underestimate, but puts a minimum figure on the species' abundance (N Burton pers. comm.). The majority of birds are found in the S & E, generally in winter, and often in feeding areas such as landfill sites or at inland gull roosts.

American Herring Gull   *Larus smithsonianus* Coues

SM *smithsonianus* Coues

TAXONOMY Taxa *smithsonianus*, *mongolicus* and *vegae* fall in the Aralo-Caspian clade that also includes Iceland, Glaucous-winged, Slaty-backed and Nearctic Glaucous Gulls. The last four of these are clearly distinct from each other, on the basis of molecular data, morphology and reproductive isolation, and their specific rank can be in little doubt. The first three differ from the others in the same array of characteristics, but their internal relations are less clear cut. There are clear molecular differences between *smithsonianus* and European Herring Gulls, and they apparently vary in their responses to each other's vocalisations, although they are not fully diagnosable on plumage characters. Nevertheless, the DNA and vocal evidence indicate that these are separate evolutionary lineages, and American and European Herring Gulls are treated as specifically distinct. Although *vegae* differs sharply from *mongolicus* and *smithsonianus*, full diagnosability has not yet been proven. The lack of conclusive differences between *smithsonianus*, *vegae* and *mongolicus* led the BOU TSC to retain these as conspecific (Collinson *et al.* 2008) in the meantime. As with the divide between 'Siberian' and 'European' Lesser Black-baked Gulls, these three may well prove to warrant specific separation in the future.

DISTRIBUTION Nominate is widespread across northern N America, from Alaska to Newfoundland, and S as far as New England, where population is increasing rapidly (Olsson & Larsson 2003). Southern populations generally resident, but northern birds move south to winter as far as Panama. Subspecies *mongolicus* breeds in C Asia from the Altai to NE Mongolia, wintering from C Japan along the coast of China to Vietnam. Subspecies *vegae* is more northerly, breeding from W Taimyr to Bering Straits; western populations, which more often have yellow legs, are sometimes separated as '*birulai*'. Migratory, wintering from Sea of Okhotsk through Japan to C China.

STATUS The first record for GB&I was a first-year at Cobh Harbour (Cork) in Nov–Dec 1986. This was the second record for the Western Palearctic; the first was in Nov 1937, when a second-winter bird was caught on a ship 480 km off the coast of Spain. It had been ringed in Aug of the previous year as a chick on Kent Island (New Brunswick). The first for GB was also a first-year, Neumann's Flash (Cheshire) Feb–Mar 1994. Now that the identification features are being established, more of these birds are being recorded, with a total of 16 to 2007. They are especially found around harbours, such as Killibegs (Donegal), where deep-sea Atlantic trawlers come to unload their catches; 62 individuals had been found in Ireland to 2004.

Iceland Gull   *Larus glaucoides* Meyer

WM *glaucoides* Meyer
WM *kumlieni* Brewster
SM *thayeri* Brooks

TAXONOMY The complex of Iceland, 'Kumlien's'and 'Thayer'sGulls' is not fully understood. On the basis of plumage differences, carefully assigned to breeding locality, Weir *et al.* (2000) concluded that *kumlieni* (E Baffin Is, N Hudson Bay) is a product of hybridisation between *glaucoides* (S Greenland) and *thayeri* (Canadian Arctic, W of Hudson Bay). Chu (1998) found that these three formed a clade with Glaucous. A re-assessment of their earlier results, led Gay *et al.* (2005) to conclude that they comprised a clade with Glaucous and Slaty-backed Gulls, although bootstrap support was minimal. There seems to be little evidence to support specific status for any of these three forms.

DISTRIBUTION This is in a state of flux, with apparent changes in range of all three taxa since the 1800s (Weir *et al.* 2000). Iceland Gull now breeds around the coasts of S Greenland; 'Thayer's' on the coasts and islands of the Canadian Arctic, from Southampton Island N and W to Banks Island; 'Kumlien's' Gull along the N fringes of Hudson Bay and the E end of Baffin Island.

In winter, 'Thayer's' occurs principally along the Pacific coasts of Canada and USA, but with scattered records from all across S Canada and the USA. 'Kumlien's' winters along the Atlantic coasts of Canada, again with scattered records from the eastern USA. Nominate Iceland Gull winters E and S to Norway, Finland, GB&I, sporadically to E Canada.

STATUS Iceland Gull is a regular winter visitor, usually in small numbers, especially in N and W; also regular at gull roosts and refuse tips inland across GB&I.

Over 100 records of 'Kumlien's Gull' since first for GB at Blackness (Orkney) Nov 1869, and for Ireland at Ballycotton (Cork) in Oct 1971. Most of the 63 Irish records are from Jan to Mar.

The first accepted record of 'Thayer's Gull' was around Cork city in Feb–Mar 1990. There were five more records to 2005, all in Feb–Mar. There are no accepted records from GB.

Glaucous-winged Gull   *Larus glaucescens* Naumann

SM monotypic

TAXONOMY Pons *et al.* (2005) included this in their molecular phylogeny of the Laridae, finding it part of a poorly resolved clade that also included California *L. californicus*, Iceland and Slaty-backed Gulls; it was, however, closest to Glaucous.

DISTRIBUTION Breeds along the coasts of the N Pacific, from the Commander Is through the Pribilof and Aleutian Is to S Alaska and Oregon. N birds move south in winter to avoid the ice, sometimes as far as S California.

STATUS Two records from GB&I: a bird found and trapped at Gloucester landfill site (Glos) in Dec 2006. Then wearing a colour ring, it was relocated on the Tywi estuary (Carmarthenshire), Hempsted (Glos) and at Beddington Sewage Farm (Surrey), in Mar–Apr 2007; another seen Saltholme Pools (Cleveland) Dec 2008–Jan 2009.

Glaucous Gull   *Larus hyperboreus* Gunnerus

HB WM *hyperboreus* Gunnerus

TAXONOMY From osteology and plumage, forms part of the Iceland Gull complex (Chu 1998), a finding supported by the molecular analyses of Pons *et al.* (2005). However, Liebers *et al.* (2004) report that Palearctic and Nearctic Glaucous Gulls have different mt DNA. The former share sequences with European *argentatus*; the latter possess a variety of sequences also found in other Nearctic gulls. Hybridisation with European and American Herring Gull has been observed (refs in Olsen & Larsson

2003; Pennington *et al.* 2004). The possibility exists that mt DNA transfer has occurred between Glaucous and other arctic gulls as well.

Three subspecies recognised by BWP: *hyperboreus*: Arctic Canada to Taimyr; *pallidissimus*: E Siberia from Taimyr to Pribilofs; *barrovianus* Alaska and adjacent N Canada. In view of the molecular differences summarised by Liebers *et al.* (2004), the limits of these need review.

DISTRIBUTION Breeds on sea cliffs and islands around the Arctic. Partial migrant, with individuals wintering as far south as California, New England, English Channel and Japan.

STATUS One female, paired with a European Herring Gull in Shetland 1975–79, raising chicks to fledging each year (Pennington *et al.* 2004).

Winter visitor to GB&I, decreasing in abundance further south. Some winters, numbers may be high (especially after northerly gales), with parties of up to 50 birds recorded in Shetland (Pennington *et al.* 2004). Regular in fishing ports such as Scalloway, Lerwick (Shetland), Peterhead, Fraserburgh (Aberdeenshire), Killybegs (Donegal). Frequently in and around harbours and estuaries, but also inland on refuse tips and reservoir roosts. Ringing recoveries indicate that birds come from Iceland, Bear Island and Norway (Wernham *et al.* 2002).

Great Black-backed Gull   *Larus marinus* Linnaeus

RB WM monotypic

TAXONOMY Chu (1998) suggests that Great Black-backed Gull is part of a large clade of 'white-headed' gulls; molecular studies support this, though statistical confidence in some parts of the phylogeny is weak (Pons *et al.* 2005). Liebers *et al.* (2004) confirm that Great Black-backed lies within the 'Herring Gull' complex. They did not include Nearctic samples in their analyses, and these may show a different pattern.

DISTRIBUTION Breeds on rocky shores and islands in NE Canada and W Greenland, and from Iceland to White Sea, including Svalbard, GB&I to NW France. Dispersive, and partial migrant S to Biscay and Florida.

STATUS Breeds around coasts of GB&I, but scarce or absent between Forth and Portland. Main concentrations are in the N Isles and along the Atlantic seaboard, but also some in Irish Sea (Reid 2004). Total population for GB&I estimated at 19,713 during *Seabird 2000*, down by *c.* 12% since *Operation Seafarer* (1985–88). However, trends variable across GB, with declines of *c.* 80% in Caithness, but increase of almost 200% in Argyll and Bute. In Ireland, declines in almost all regions, but increase of >110% in Co. Dublin. Increases attributed to enhanced protection (Reid 2004); decreases, at least partly, to Mink predation: Craik (2002) reports that only seven young fledged at Scottish sites where Mink were active predators.

Birds breeding in N and E of GB generally disperse S and E along North Sea coasts in winter; many from the NW to winter in the Irish Sea; those breeding in the S and W migrate towards Biscay. Numbers in winter are enhanced by strong immigration, especially from Scandinavia. The Winter Gull Roost Survey (A.N. Banks *et al.* 2007) estimated a wintering population of 76,000 birds.

FAMILY STERNIDAE

Genus *ONYCHOPRION* Wagler

Cosmopolitan genus of four species, three of which are vagrants to GB&I.

Aleutian Tern

Onychoprion aleuticus (S. F. Baird)

SM monotypic

TAXONOMY Molecular analysis (Bridge *et al.* 2005) confirms the view of Roselaar (in BWP) that Aleutian Tern is phylogenetically close to Spectacled *O. lunatus* and Bridled *O. anaethetus*. The DNA study indicates that Sooty Tern *O. fuscatus* is also part of this complex, and the wide separation of these from the rest of *Sterna* and *Chlidonias* led Bridge *et al.* (2005) to propose resurrecting *Onychoprion* as a genus for these four species.

DISTRIBUTION Breeds in coastal SW Alaska, Aleutian Islands, Kamchatka Peninsula, occasionally Japan (Brazil 1991). The wintering grounds are still largely unknown, though it has been seen on passage in Hong Kong and in winter in the Philippines.

STATUS The only record in GB&I is of a bird seen and photographed on the Farne Islands (Northumberland) in May 1979 (Dixey *et al.* 1981).

Sooty Tern

Onychoprion fuscatus (Linnaeus)

SM *fuscatus* (Linnaeus)

TAXONOMY Bridge *et al.* (2005) show, from mt DNA sequences, that Sooty, Bridled and Aleutian form a clade, and recommend placing these in *Onychoprion* (see Bridled Tern). Several races recognised, though some are of doubtful validity (see Roselaar in BWP). Of relevance to GB&I are *fuscatus* (Atlantic) and perhaps *nubilosus* (Red Sea, Indian Ocean, W Pacific), differing slightly in fresh plumage but similar in size. Molecular analysis indicates that Atlantic populations are poorly differentiated from Indo-Pacific colonies (Avise *et al.* 2000).

DISTRIBUTION Strongly pelagic; more so than Bridled Tern. Breeds throughout the tropics, usually on islands, only avoiding areas of cold-water currents (e.g. west coast of S America, Angola, Namibia). Colonies in Atlantic include St Helena, Ascension and islands in Gulf of Guinea. Has bred Madeira (HBW). Winters at sea. Juveniles from Caribbean regularly cross the Atlantic to feed on the rich fishing grounds at the northern end of the Benguela Current. Presumably at least some of the records from GB&I involve birds displaced northwards from these areas.

STATUS GB: 25 or 26 individuals recorded. The first was killed with a stone near Tutbury (Staffordshire) in Oct 1852; two records from Ireland: an adult at Bridges of Ross (Clare) in July 2002 and a mobile bird in Dublin and Down in July–Aug 2005. The pattern of occurrence is generally from Apr to Oct, with evidence of a recent distributional bias at or close to active tern colonies.

Bridled Tern

Onychoprion anaethetus (Scopoli)

SM *antarcticus* (Lesson)

TAXONOMY Molecular analysis (Bridge *et al.* 2005) shows that this species forms a clade with Sooty and Aleutian (also *O. lunatus* – W Pacific), well separated from all the remaining terns, apart from White Tern *Gygis alba* and the noddies. The paraphyly of the terns means that the conventional placement of this clade in *Sterna* cannot be sustained, and Bridge *et al.* (2005) recommend resurrecting *Onychoprion*. Four races are recognised by Roselaar (in BWP): *anaethetus* from Indonesia, Australia and W Pacific; *melanopterus* from the Caribbean and W Africa; *antarcticus* from Red Sea, Persian Gulf, and W Indian Ocean; *nelsoni* from the Pacific coast of Central America. The two forms most likely to occur in GB&I (*melanopterus* and *antarcticus*) differ in plumage details (BWP; Olsen & Larsson 1995).

DISTRIBUTION Breeds on tropical and subtropical islands around the world, extending further north than Sooty. Less pelagic than Sooty Tern, typically feeding within *c.* 15 km of land. Winters at sea.

Migrations little known; however, some juveniles may cross the Atlantic from the Caribbean to winter off W coast of Southern Africa.

STATUS There have been *c.* 20 individuals recorded in GB since the first was found dead at Dungeness (Kent) in Nov 1931; this individual has been assigned to the race *antarcticus*. The only Irish record is one of the race *melanopterus* (Swainson) found dead on the shore of the North Bull Island (Dublin) in Nov 1953; being a tideline corpse, this record is assigned to the Irish Category D3 (see Appendix 2). As with Sooty Tern, most records are between May and Oct, with a similar distributional trend towards some recent birds being seen at or close to active tern colonies.

Genus *STERNULA* Boie

Cosmopolitan genus of about seven species, one of which breeds in GB&I.

Little Tern   *Sternula albifrons* (Pallas)

MB PM *albifrons* (Pallas)
SM *antillarum* Lesson/*athalassos* (Burleigh & Lowery)/*browni* (Mearns)

TAXONOMY Roselaar (in BWP) discusses seven races: *albifrons* (Europe and C Asia, to N India); *guineae* (W and C Africa); *sinensis* (SE and E Asia); *antillarum* (coastal USA and Caribbean); *athalassos* (Mississippi basin); *mexicana* (Sonora, Mexico); *staebleri* (Chiapas, Mexico), but regards the validity of some of the N American taxa as uncertain, as do Thompson *et al.* (1997) and Whittier *et al.* (2006). Bridge *et al.* (2005) show strong support for sister relationship between *albifrons* and *S. nereis* (Fairy Tern) and between *antillarum* and *S. superciliaris* (Yellow-billed Tern). They also found that these four taxa form a clade well separate from the rest of the terns, on the strength of which they recommend including these 'small terns' in *Sternula*. The '*albifrons*' that they used were from Australia, which might explain the affinity with Fairy Tern. The relationships between *antillarum*, *guineae* and European *albifrons* are not resolved by this study. Olsen & Larsson (1995) suggest that *guineae* is smaller than European *albifrons*, the bill is frequently all yellow, and the rump and tail are often grey, but with overlap in all characters so that some *guineae* are indistinguishable from *albifrons*. Taxon *antillarum* is also slightly smaller than *albifrons*, with marginally less black at the bill tip, and a grey rump and central tail. Diagnostic contact calls have been described, but most other vocalisations of *antillarum* are indistinguishable from *albifrons*. This group is in need of analysis, with birds being examined genetically from right across the range.

DISTRIBUTION Widely distributed outside Arctic regions of Eurasia, and around the coasts of Africa and Australasia. Breeds along coasts and also inland (Fasola *et al.* 2002), especially in Europe and Asia. European populations migrate south to winter around the coasts of Africa and Arabia.

STATUS Breeds exclusively on coastal sand or shingle at a limited number of sites round GB&I, though absent from Shetland and N Ireland and only one colony in Wales. The number of occupied colonies declined from 154 (1969–70) and 168 (1983–88) to 130 (2000), as did the number of apparently occupied nests, from *c.* 2,800 in the mid-1970s, to 2,153 during *Seabird 2000* (Pickerell 2004). Little Terns show limited site fidelity, moving both within and between years, and so, perhaps not surprisingly, colonies did not show a uniform decrease in size. While more than a dozen regions lost their Little Terns, others increased dramatically. Because of its vulnerability, many colonies are monitored carefully each year, and the latest figures (2006) indicate evidence of slight recovery in some regions.

Annual fledging success varied from 0.19 to 0.70, and this is not enough to maintain the population (Ratcliffe *et al.* 2000). The failure to fledge chicks successfully has several causes, of which disturbance and predation are highest. Disturbance may be natural: high tides may flood the nests and increasing populations of other birds may force Little Terns to use sites closer to the water's edge. Terrestrial predators include foxes, hedgehogs and domestic dogs; avian predators include Kestrels, Carrion

Crows, gulls and Magpies. Electric fencing and effective wardening have reduced the effects of the former, but avian predation remains a problem. Carrion Crows and Magpies take eggs, and Kestrels in particular are a major predator of chicks at some sites. Attempts to provide shelters, under which the chicks may hide, are not always a success: sometimes the chicks may not use them; alternatively, Kestrels have been seen to take chicks from inside the shelter.

Restoration of breeding sites by creating or renewing shingle banks has proved a success at some places, especially where islands have been created that isolate the birds from some terrestrial predators, but, as Pickerell (2004) points out, until deterrents to aerial predators can be established, the problem of low productivity will remain.

Those breeding in GB&I are *albifrons*; a bird seen and recorded at Rye Harbour (1983–1992) belonged to the *antillarum/athalassos/browni* group.

Genus *GELOCHELIDON* Brehm

Near-cosmopolitan monospecific genus that is a scarce visitor to GB&I.

Gull-billed Tern   *Gelochelidon nilotica* (J. F. Gmelin)

CB SM *nilotica* (J. F. Gmelin)

TAXONOMY A molecular study (Bridge *et al.* 2005) shows that Gull-billed Tern is sister species to Caspian *H. caspia*, and well separated from the rest of the 'white terns'. Since the divergence between these two species is large, they recommend placing them in separate monospecific genera *Gelochelidon* and *Hydroprogne*, respectively. Roselaar (in BWP) recognises six races: *nilotica* (Europe and W Asia), *affinis* (E Asia), *aranea* (E North America), plus three further races from Australia and W & S America These are differentiated on size, and colour of the upperparts. European and N American taxa probably only separable in the hand on bill depth at base.

DISTRIBUTION Breeds in southern latitudes of Europe and across central Asia to China; in Americas, on both coasts of USA south to tropical S America. Also in parts of the Indian subcontinent and Australia. Northern populations migrate south to winter inland in Africa and India; also through India and the Malay peninsula to Australia. Less marine than most terns, apart from *Chlidonias*, in many areas preferring terrestrial prey such as grasshoppers and dragonflies.

STATUS The only breeding record is of a pair which nested at Abberton (Essex) in 1950, although the single chick died before fledging.

Over 400 individuals in GB to 2003 (Rogers *et al.* 2004). The European populations of the Gull-billed Tern declined substantially between 1970 and 1990 (Tucker & Heath 1994), and this decrease continued through into 2000 (BirdLife International 2004). In parallel with this, records in GB have shown a marked decline. An annual average of 6.1 birds was seen in the 27 years between 1950 and 1976, compared with only 3.8 in the 28 years to 2003 ($P<0.05$), and this is despite the increase in observer effort in the second period.

Only 12 Irish records to 2004, with the first being in July 1969, when an adult was found at Ballyconneely (Galway). In view of recent records of Nearctic race Black Tern in GB&I, there is the intriguing possibility that some of these (especially those in Sep–Oct) might have been *aranea*. The peak month for records is May, but birds are seen through the summer to Sep, with only small numbers outside this period.

Genus *HYDROPROGNE* Kaup

Near-cosmopolitan (not S American) monospecific genus that is a scarce visitor to GB&I.

Caspian Tern   *Hydroprogne caspia* (Pallas)

PM monotypic

TAXONOMY Bridge *et al.* (2005) show that it forms a clade with Gull-billed Tern, and this is sister to the clade including the medium to large white terns and *Chlidonias*. Consequently, neither Caspian nor Gull-billed should be retained in *Sterna*, and, since they differ from each other by *c.* 7% in *c.* 2,800 bp of mt DNA sequence, they recommend placing them in different genera, retaining Caspian in the monospecific genus *Hydroprogne*. Roselaar (in BWP) suggests that subspecies previously recognised on size variation are not substantiated by larger samples and that Caspian Tern is best treated as monotypic.

DISTRIBUTION Breeds on coasts and larger lakes across Europe, Africa, Asia, Australasia and N America. Baltic population migrates south across central and eastern Europe to winter in west and central Africa (including the Nile Valley) and Arabia. N American populations winter in C America, south to Panama.

STATUS Over 250 individuals to 2006. Despite the increased number of observers in GB&I, number of records has remained around five each year for the last 40 years. The European population declined during 1970–1990, but has shown signs of recent recovery (BirdLife International 2004); the apparent stasis in the number of records could stem from this decline. The overwhelming majority of records are from May to Aug, but small numbers occur earlier and later than this; most are from the S and E of England, but individuals may turn up anywhere. It remains, however, a scarce bird in Ireland with only around eight records to 2005, and none from NI.

Genus *CHLIDONIAS* Rafinesque

Cosmopolitan genus of three (possibly four) species, three of which have been recorded in GB&I.

Whiskered Tern   *Chlidonias hybrida* (Pallas)

SM *hybrida* (Pallas)

TAXONOMY Molecular data (Bridge *et al.* 2005) indicates that Whiskered forms a clade with the other marsh terns. Roselaar (in BWP) recognises three races: *hybrida* (Eurasia), *delalandii* (S and E Africa) and *javanicus* (Australia). These differ slightly in size and more consistently in plumage colouration, with the African form being darkest and the Australian lightest.

DISTRIBUTION Freshwater lakes, marshes and rivers across S and C Europe, through C Asia to China; also S and E Africa, N India and southern Australia. Winters across C Africa and Nile Valley; African breeders are relatively sedentary. Eastern populations winter in Malay Peninsula south to Australia.

STATUS Over 150 individuals to 2006. There was an increase in numbers from the 1960s, following Williamson's (1960) key paper on the identification of juvenile and winter marsh terns, but has remained a scarce species with rarely more than six individuals each year. Much more regular in S and E England; remains very rare in Scotland. Usually singletons, but occasional flock, most spectacularly 11 at Willington (Derbyshire) in April 2009. There are less than 20 records from Ireland, all in the Republic, with a gap of >120 years between the first (1839) and second (1961). The first for NI occurred the same weekend as the Willington flock. Recorded in all months from Apr to Nov, but with a distinct peak in May (43%) and June (25%).

Black Tern   *Chlidonias niger* (Linnaeus)

CB FB PM *niger* (Linnaeus)
SM *surinamensis* J. F. Gmelin

TAXONOMY Holarctic. Bridge *et al.* (2005) only included the Eurasian form in their analysis and found that it was sister to White-winged Tern. In view of the separation of these taxa from *Sterna* and *Thalasseus*, the molecular evidence supports the retention of the genus *Chlidonias* for the marsh terns. Two races recognised: *niger* from Eurasia and *surinamensis* from N America. Differences in juvenile plumage apparently diagnostic (Olsen & Larsson 1995).

DISTRIBUTION Breeds across C North America (*surinamensis*) and C and E Europe through to Russian Altai (*niger*). Nearctic birds move south to winter on both coasts of C America. Most Old World birds seem to winter in coastal W & S Africa and the Nile Valley.

STATUS An abundant breeding bird in East Anglia and elsewhere in S and E England until the middle of the 19th century. Now only attempts to breed sporadically (and usually unsuccessfully) – e.g. Norfolk 1969, 1975; Fermanagh 1967, 1975; Nottinghamshire 1975. On passage, occurs regularly in the spring, typically under anticyclonic conditions in May, and often in considerable numbers. Most abundant in the S and E, but recorded in most counties in most years. In autumn tends to be more coastal, again chiefly in the S and E.

Subspecies *surinamensis* has been recorded three times in Ireland to 2006, with the first (a juvenile) at Sandymount Strand (Dublin) in Sep 1999. There is a single record from GB: also a juvenile, at Weston-super-Mare (Somerset) in Oct 1999.

White-winged Tern   *Chlidonias leucopterus* (Temminck)

PM monotypic

TAXONOMY Molecular study of Bridge *et al.* (2005) showed that it is the sister species to Black Tern and forms a clade with Whiskered and the New Zealand Black-fronted Tern *Onychoprion albostriatus*. The latter might therefore be more appropriately regarded as a fourth species in *Chlidonias*; further research is needed.

DISTRIBUTION Predominantly an inland breeder, on freshwater lakes and marshes, from SE and E Europe across Russia and C Asia to China. Moves south in autumn, to winter in southern half of Africa, Persian Gulf, coastal China, Malay Peninsula to Australasia.

STATUS Over 800 individuals to 2003 (Rogers *et al.* 2004), of which 73 in Ireland. The average number each year was <10 in the 1950s, increasing to *c.* 20 p.a.; as with Whiskered, this was likely a result of Williamson's identification paper (1960). Birds usually occur singly, between Mar and Nov, but occasionally there may be two or three together. Most records are in May–June and Aug–Sep, and they have been recorded in most parts of GB&I, but the majority are in the S and E.

Genus *STERNA* Linnaeus

Cosmopolitan genus of 17 species, four of which breed in GB&I and four occur as vagrants.

Elegant Tern   *Sterna elegans* Gambel

SM monotypic

TAXONOMY Forms a clade with Sandwich Tern that is sister to the clade including Swift *S. bergii*, Lesser Crested and Royal Terns.

DISTRIBUTION Breeding apparently restricted to Pacific coasts of California and Mexico. Winters as far south as Peru, but has occurred on eastern seaboard of N America in Texas. Vagrancy to NE Atlantic may follow either moving too far south in winter and then returning up the 'wrong' ocean, or crossing Central America to Caribbean. In either case, an easterly component to the northward migration could lead vagrants across the Atlantic to W Europe.

STATUS There are five records of this Pacific tern in Ireland, including the first at Carlingford Lough (Down) in June–July 1982, and subsequently at Ballymacoda (Cork) in Aug 1982. Subsequent birds have been found in 1999, 2001, 2002 and 2005. However, in view of the multiple sightings of other vagrant terns (e.g. Lesser Crasted and Forster's), there is a possibility that some of these refer to the same individual. A record from Devon in July 2002 is currently under review.

Sandwich Tern   *Sterna sandvicensis* Latham

MB HB PM *sandvicensis* Latham
SM *acuflavida* Cabot

TAXONOMY Slight variation in size, with N American form *acuflavida* averaging a little smaller, with differences in juvenile plumage in particular. Cayenne Tern *eurygnatha* occurs from the Caribbean along the coasts of E South America; it is similar in size to *acuflavida* but with all-yellow bill. Much overlap in measurements. Bridge *et al.* (2005) only examined North American birds, from range of *acuflavida* and *eurygnatha*, and found 0.29% divergence in mt DNA sequences, compared with 1% for Sandwich/Elegant comparison. Thus, no molecular support for specific status of Cayenne Tern, though these may have evolved very recently. The American races of Sandwich Tern are closer to Elegant than to European *S. s. sandvicensis* (Efe *at al.* 2009), and *acuflavida/eurygnatha* should perhaps be treated as a separate species (Cabot's Tern *S. acuflavida*). The 'crested' terns may be better placed in genus *Thalasseus* (Bridge *et al.* 2005).

DISTRIBUTION European race breeds predominantly in GB&I, with smaller numbers in coastal areas from S Baltic, Denmark, Germany and France. Small numbers in Mediterranean, then east through Turkey to Black and Caspian Seas. Migrates south to winter round coasts of Africa (less so on E side) to Arabia. American race breeds on Atlantic and Gulf coast to Mexico, also Bahamas. As with many terns, sizable components of population may move between colonies from year to year.

STATUS See Lesser-crested Tern in relation to hybrid pairing with that species. The Sandwich Tern is almost exclusively coastal in its breeding (Ratcliffe 2004a). Unlike Common and Arctic, whose colonies tend to be widely dispersed, Sandwich Terns breed at a more limited number of sites. The largest colonies are at Scolt Head (Norfolk) (4,200 pairs), the Farne Islands (1,900) and Coquet Island (1,700) (both in Northumberland), Sands of Forvie (800–1,000, Aberdeenshire), Lady's Island Lake (Wexford) (1,000) and Strangford Lough (Co. Down) (900).

Overall, the population of GB&I (14,252) rose by *c.* 18% between *Operation Seafarer* (1969–70) and *Seabird 2000*, although it was higher (16,047) during the intervening survey in 1985–88 (Lloyd *et al.* 1991). The picture is slightly confused, however, because of the propensity of Sandwich Terns to move almost entire colonies between sites from one year to the next. Thirty-three sites were abandoned between 1985 and 2000, and 9 new sites were occupied – a net loss of 24. Some sites have been counted annually since the late 1960s (Ratcliffe *et al.* 2000), and these show a slow increase through the late 1970s and 1980s, followed by a sharp decline until 1995, and a subsequent partial recovery. As Ratcliffe (2004a) comments, this pattern of fluctuating population seems to be 'not wholly atypical' of Sandwich Tern.

Terrestrial predators are a severe threat to many Sandwich Tern colonies. Foxes are especially serious, as there is a tendency for these animals to return repeatedly, year after year, until the colony is destroyed or moves elsewhere. Adult terns may defer breeding for several seasons while searching for new and more secure sites, and this could account for the interannual fluctuations in population size. Competition with gulls for nesting space can lead to wholesale colony desertion, as, for example, at

Inchmickery (Firth of Forth), with birds failing to relocate that season. Persistence of the Northumberland colonies is largely due to gull control. Equally serious might be the killing of immature birds in West Africa for food or recreation. The loss of sub-adult birds from a population during a period when reproduction is also suppressed could have severe consequences for its long-term stability.

A bird found in Herefordshire in Nov 1984 had been ringed in North Carolina, USA (Mead & Hudson 1986), confirming the occurrence of *acuflavida* in GB&I.

Royal Tern   *Sterna maxima* Boddaert

SM *maxima* Boddaert

TAXONOMY Two races currently recognised, though Roselaar (in BWP) casts doubt on their validity. Nominate *maxima* occurs from southern United States of America through to northern South America; *albididorsalis* breeds in West Africa. Molecular analysis (Bridge *et al.* 2005) confirms its close affinities with Swift and Lesser Crested Terns. These three form a clade (sometimes placed in a separate genus *Thalasseus*) that is sister to Sandwich and Elegant, and the whole group is well separated from the Arctic, Roseate, Common Tern clade.

DISTRIBUTION Tropical and subtropical. Western populations breed on both coasts of USA, Gulf of Mexico and Caribbean; small numbers on Atlantic coast of southern S America. In winter, largely coastal, tropical regions of the Americas. Eastern populations breed Mauritania, Senegal and Gambia, dispersing both north to Morocco and south as far as Namibia in winter.

STATUS The first in GB was seen at Kenfig (Mid-Glamorgan) in Nov 1979 (Moon 1983); it had been ringed in North Carolina as a nestling in 1978–79, establishing the occurrence of ssp. *maxima* in GB&I. Birds from America or Africa seem equally likely to occur in GB&I. In view of the slight differences in morphology, it seems likely that confirmation of African birds here will depend on recovery of a ringed individual. One found dead on North Bull Island (Dublin) in Mar 1954 is the only record from Ireland and is placed in the Irish Category D3 (see Appendix 2; Sharrock & Grant 1982).

Lesser Crested Tern   *Sterna bengalensis* Lesson

HB SM *torresii* (Gould)

TAXONOMY Two or three races currently recognised, largely on size, but distribution of these is patchy and slightly confused. Larger birds *torresii* occur in Mediterranean and also Australasia and Indonesia; smaller birds *bengalensis* breed in E and S India, Sri Lanka and E Africa. Thus, *bengalensis* occurs in centre of range, splitting *torresii* into two isolates. Further confusion since Red Sea breeders are small and hence *bengalensis*, while Persian Gulf birds are large and hence *torresii*. Mediterranean Sea birds have been named '*emigrata*', but probably not separable from *torresii* (Roselaar in BWP). Further analysis of interpopulation differentiation is called for. Molecular study by Bridge *et al.* (2005) shows that Lesser-crested Tern is sister taxon to Royal and forms a clade with Swift; this trio is closely allied to clade of Sandwich/Elegant Terns.

DISTRIBUTION Disjunct breeding range includes Mediterranean and Red Seas, Arabian Gulf, NW and NE coasts of Australia, Indonesian islands. Individuals occasionally found outside range, sometimes paired with Sandwich Terns (e.g. Banc d'Arguin, Ebro Delta, Farne Islands). More widely distributed outside breeding season, around shores of Indian Ocean and throughout Malaysian/Indonesian/Philippine archipelagos.

STATUS First GB record from Anglesey in July 1982. A total of nine records but perhaps not nine individuals because of the longevity of the species. What was almost certainly the same bird was seen at a wide range of sites between Lothian and Sussex, on a succession of dates from 1984 to 1997. This individual paired with one or more Sandwich Terns on the Farne Islands, producing hybrid young in several years between 1992 and 1997. Some of these survived, and one that fledged in July 1997 was subsequently seen in the Vendee, France, in Sep that year. There is only one record from Ireland, of a bird seen at Ballycotton (Cork) in Aug 1996; this was possibly the same individual that had been seen in Scilly earlier that month.

Forster's Tern   *Sterna forsteri* Nuttall

SM monotypic

TAXONOMY The position of this taxon was not fully resolved by the molecular analyses of Bridge *et al.* (2005). It was sister species to Snowy-crowned Tern *S. trudeaui* of South America, with a divergence of 4.9%, but the affinities of this pair could not be determined. They were closest to the Common, Roseate, Arctic group, but without strong support, and a case could equally be made for their being more closely allied with subgenus *Thalasseus*. Morphology and behaviour suggest that their affinities lie closer to Common Tern, but more analysis is needed of this pair.

DISTRIBUTION Mostly at freshwater lakes and marshes, coastal saltmarshes and pools from SW Canada across the USA to Atlantic and Gulf coasts. Nests on floating vegetation, occasionally on sand or mud. Migrates to southern half of N America through to Panama. Some northwards dispersal after breeding, and sometimes winters north to New England.

STATUS The first was recorded in Feb–Mar 1980, at Falmouth (Cornwall). The first for Ireland was at Dublin Bay (Dublin) in Nov 1982–Feb 1983. It is difficult to be precise about the number of birds subsequently, since there have been several occasions when individuals have been seen at the same site in successive years. In Flintshire, two birds were seen in 1987 close to places where a singleton had been recorded previously. Whether these represent two or three individuals is not known. About 20 have been recorded in GB and about 21 in Ireland to 2004, though the possibility of double recording of wandering individuals cannot be discounted. The majority of records are from the Irish Sea basin, from Oct to Mar, although long-staying individuals have summered.

Common Tern   *Sterna hirundo* Linnaeus

MB PM *hirundo* Linnaeus

TAXONOMY Polytypic, varying in colour of bill and feet; also length of wing and bill, and colour of body. Nominate *hirundo* breeds across E North America to Caribbean, Europe south to Mediterranean, across Middle East to W Siberia; also isolated populations in N and W Africa. Replaced in E Siberia and NE China by *longipennis*, and by *tibetana* from Kashmir through Tibet to E China. Racial differences described by Roselaar (in BWP) and Olsen & Larsson (1995). Recent molecular analysis of mt DNA sequences (Bridge *et al.* 2005) suggests that Common Tern is rather differentiated from Arctic and Roseate groups, although Zink *et al.* (1995) found no evidence of divergence between subspecies *hirundo* (USA) and *longipennis* from the Russian Far East.

DISTRIBUTION Much less marine than Arctic or Roseate Terns, breeding along coasts but also well inland on islands and edges of lakes and rivers, and on grassy and tundra sites away from open water. Winters along coasts of C and S America, southern Africa, India, and the Malay Peninsula through to Australia; also on the great rivers of S America.

STATUS The Common Tern is widely distributed around the coasts of GB&I, only being absent from the mainland of Britain between Anglesey and Portland, although there are very few in coastal Yorkshire or Lincolnshire. It differs from other terns of GB&I in breeding inland on islands and sandbars of rivers and lakes. Ratcliffe (2004b) reports that these comprise only about 8% (Britain) and 19% (Ireland) of the total, and the number of inland colonies seems to be in decline. The total population was estimated as *c.* 14,500 apparently occupied nests during *Seabird 2000*, reasonably similar to *Operation Seafarer* (14,890: BWP) and the 1985–88 census (14,861: Lloyd *et al.* 1991).

However, this masks some regional differences; Wales and Ireland showed increases, England stability and Scotland a decline over recent years.

Distribution and status have remained broadly stable since the 1985–88 survey, despite these local differences. Mammalian predators, especially American Mink, are a serious threat at some locations. These have escaped from fur farms from the 1950s (Dunstone 1993). They spread rapidly through coastal and riverine habitats and caused serious depredations of many of the native fauna of GB&I. They reduced the number of tern colonies and forced many birds to relocate to offshore islands, where the environment was more secure (Ratcliffe 2004b). Foxes and gulls can also wreak havoc in colonies of Common Terns, especially now that gamekeepers are fewer in number. Efforts to control these predators at or around colonies are underway in many regions.

Other threats to breeding colonies, especially inland, stem from disturbance by human activity. Water recreation is increasingly popular, and disturbance (often unintentional) to breeding birds is frequent. Nesting habitat can also become damaged or destroyed by direct or indirect means. Examples of the former include draining and canalisation of waterways, erosion of banks and islands, creation of sailing and windsurfing facilities; of the latter, neglect of island vegetation resulting in scrub encroachment onto breeding areas. Artificial islands have proved valuable as alternative nesting sites in several areas; combined with electric fences, they can almost eliminate terrestrial predation.

Roseate Tern   *Sterna dougallii* Montagu

MB PM *dougallii* Montagu

TAXONOMY Polytypic, varying principally in colour of bill during breeding, also length of bill and wing. Roselaar (in BWP) suggests that the subspecies are in need of review and revision, as there is overlap between many populations. Atlantic form *dougallii* varies in bill colour from almost entirely black to mainly red. Other races currently recognised are *korustes* (India, Sri Lanka, Burma), *gracilis* (Australia, New Caledonia), *bangsi* (Arabia, W Indian Ocean to Japan) and *arideensis* (E coast of Africa and Seychelles), though some authorities (e.g. Roselaar in BWP) believe the last two should be merged. Lashko (2004, in Ratcliffe *et al.* 2004) used mt DNA to identify distinct Atlantic and Indo-Pacific lineages, but little evidence of differentiation within these two regions, and suggests that there is no molecular justification for the continued recognition of *korustes*, *gracilis*, *bangsi* and *arideensis*, and that these should be merged into a single taxon. A geographically more limited study of nuclear microsatellites (Szczys *et al.* 2005) also found strong differences between N Atlantic and W Australian birds. A molecular analysis of mt DNA sequences (Bridge *et al.* 2005) suggests that Roseate is sister species to White-fronted *S. striata* and that these form a clade with Black-naped *S. sumatrana*.

DISTRIBUTION Breeds in all continents except Antarctica. Major centre of distribution is W Pacific, especially Japan, Taiwan and Australia; sizable colonies in Kenya, Somalia and Seychelles; smaller numbers in NE USA and Caribbean; European populations comprise only about 2% of global figure and are concentrated in Azores and GB&I, with only a handful in France and elsewhere. The extent of interchange between regions is not known. There is considerable local movement between the well-studied colonies in GB&I. For example, the colonies in Anglesey and N Ireland decreased due to relocation to Rockabill Island (Dublin) (Newton 2004). Similarly, in the Azores, variation in both size and location of colonies suggests extensive movement among sites, and Lebreton *et al.* (2003) believed that up to 42% of birds breeding for the first time may do so at a site different from the natal colony.

STATUS Roseate Tern is the rarest tern that breeds regularly in GB&I, and the number of pairs has declined substantially since the 1960s. At the time of *Operation Seafarer* (1969–70), the total population was about 2,384 apparently occupied nests (BWP), of which 107 were in the Firth of Forth (Scotland), 202 on Anglesey, 332 in Northumberland (principally on Coquet Island), 251 were in Co. Down (Carlingford and Strangford Loughs) and 1,352 were in Co Wexford (especially Lady's Island Lake). By the end of the century, almost all these colonies had shown a decline, and several of the smaller ones were extinct (Newton 2004). Numbers in Northumberland were down to 34 and in Co. Wexford to 116; colonies in the Firth of Forth and Co. Down had gone entirely, and even Anglesey was down to just 2 pairs. Part of this is undoubtedly due to the relocation of many birds to Rockabill Island, where the numbers had risen to 618 but, even so, the total number had declined by 67% to only *c.* 790 nests. Since the fieldwork for *Seabird 2000*, numbers have increased at Coquet Island, perhaps due to conservation effort but, at *c.* 90 nests, are still below Seafarer values.

Newton (2004) argues that the reasons for this decline probably lie on the wintering grounds. It has been shown that the capture of immature Roseate Terns, especially in Ghana, was likely to be a significant cause of the decline of adult birds at European colonies during the 1980s. Education programmes alleviated this, with an increase in recruitment, but recently it seems that the practice is on the increase again. Winter food is important off the coast of W Africa, and there is a risk that more efficient fishery technology may reduce the stocks available, with a consequent adverse effect on recruitment.

Recent efforts to improve the quality of nesting habitat show some signs of success. Roseate Terns are as susceptible as other terns to nest loss through flooding and egg/chick loss from predation. The reduction of gull numbers (Rockabill) and rat control (Lady's Island Lake) are thought to have led to an increase in breeding success, and the provision of nest boxes increased the area suitable for nesting (Casey *et al.* 1995). Similar efforts on Coquet Island may be one reason for the increase in Northumberland in recent years. Roseate Terns remain susceptible, and their status in GB&I continues to give cause for concern.

Arctic Tern   *Sterna paradisaea* Pontoppidan

MB PM monotypic

TAXONOMY A molecular analysis of mt DNA sequences (Bridge *et al.* 2005) indicated that Arctic Tern forms a clade with the sister species Antarctic *S. vittata* and South American *S. hirundinacea*. This clade is closer to the Roseate group than to Common Tern.

DISTRIBUTION Holarctic, from GB&I northwards along coasts of N Europe, across Siberia, Alaska, Canada, E coast of USA to Massachusetts, Greenland, Iceland and Faroes. Oceanic migrant, wintering at sea from southern tips of Africa and S America down to pack ice.

STATUS Arctic Tern is widely distributed around the coasts of Scotland and Ireland, with only small numbers in England and Wales, except for the large colonies in Northumberland and Anglesey. *Seabird 2000* (Ratcliffe 2004c) found that 85% of the population of GB&I occurred in Scotland and that 66% were in Orkney and Shetland. While the majority of colonies were of less than 100 apparently occupied nests, colonies in excess of 1,000 were recorded at Dalsetter, Papa Stour and Fair Isle (Shetland), Papa Westray (Orkney) and the Farne Islands (Northumberland). The biggest Irish colonies were at Rock Island (388) and Illaunamid (329), both in Co. Galway, and Cockle Island (308) and Strangford Lough (281), both in Co. Down.

There seems to have been a decline across GB&I since *Operation Seafarer* in 1969–70, although Ratcliffe (2004c) is cautious about extrapolating across all the sites because of variation in both methodology and the year of census. Where there are repeat counts from the same colony, the evidence is of interannual fluctuation superimposed on decrease. Thus, overall, the population of Orkney and Shetland seems to have declined from about 40,000 pairs in 1980 to about 38,000 in 2000, although there are variations in relative abundance in the regions within this period. Changes are equally variable elsewhere in GB&I. Increases in the Outer Hebrides and on the east coast of Scotland are at variance with widespread declines elsewhere in Scotland and NE England. The Welsh and Irish populations both showed a steep increase, though again this varied among colonies.

It seems to be generally agreed that these changes are, at least partially, linked to food supply. Northern populations are strongly dependent on sandeels during the breeding season. Decreases in sandeel stocks in the 1980s were mirrored by almost total breeding failure in the tern colonies, accompanied by indications of severe food shortage (Monaghan *et al.* 1992). There seems to have been a slight improvement through the early 1990s, with an increase in sandeel availability leading to at least a stabilisation in numbers (Brindley *et al.* 1999). However, this situation appears to have reversed through the later 1990s, with a combination of reduced sandeel availability and bad weather during hatching. It is clear that populations of the Arctic Tern in northern GB&I remain at risk from further fluctuations in sandeel abundance.

As with many seabirds, terrestrial predators are a serious threat. American Mink, Fox, Brown Rat and gulls have all been implicated in colony failure, but control measures have been successfully implemented at some sites. Continued vigilance is imperative: recovery of populations is likely to involve both predator control and increased stocks of sandeels within flight range of the breeding colonies.

FAMILY ALCIDAE

Genus *URIA* Brisson

Holarctic genus of two species, one of which breeds in GB&I; the other is a vagrant.

Common Murre/Common Guillemot   *Uria aalge* (Pontoppidan)

RB MB WM *aalge* (Pontoppidan)
RB MB WM *albionis* Witherby

TAXONOMY Friesen *et al.* (1996a) used mt cyt-b to show that Common Murre is sister to Thick-billed, in a less well-supported clade that includes Razorbill and Little Auk; this relationship was supported by enzyme genetics. Five subspecies recognised by Sluys (in BWP), but much of the variation in size and plumage colour is clinal. Nominate *aalge* is chocolate brown, blacker on the head and neck. In SW Europe, birds have been separated as *albionis*, which is slightly smaller and browner-mantled. The dividing line between these is approximately N England and SW Scotland. In the NE Atlantic, *hyperborea* occurs; this is larger than *aalge*, with a longer and deeper bill, but the variation is clinal, with birds increasing in size with latitude (see Cadiou *et al.* 2004). Populations in the North Atlantic are weakly differentiated at microsatellite loci and the majority of birds could not be assigned to their true provenance (Riffaut *et al.* 2005): a case can be made to merge all N Atlantic populations in a single subspecies. Birds from the Pacific have been separated as *californica* in the USA and *inornata* elsewhere in the N Pacific.

DISTRIBUTION Nominate *aalge* breeds from Canada across Greenland, Iceland, the Faroes, Scotland, the Baltic and Norway north to about the Lofoten Islands. It is replaced by *albionis* in England, Ireland and the Atlantic coast of Europe south of Heligoland. The clinal nature of the colouration adds to the difficulty in defining the boundary. From N Norway through Spitzbergen to Novaya Zemlya, *hyperborea* occurs. Oceanic in winter, dispersing away from the breeding colonies out to sea. Populations of *inornata* breed from Hokkaido, Japan, around the Pacific rim through the Bering Sea, and across Canada. South from about Washington State, *californica* is found.

There is also a genetic variant ('bridled') with a thin white eye-ring, extending backwards as fine streak to back of ear coverts. This is controlled by a single gene, with the bridled allele recessive (Jefferies & Parslow 1976). Bridling is restricted to Atlantic populations and increases in frequency with latitude from 0% in Iberia to >50% in Bear Island (BWP). There are reports of quite dramatic changes in frequency of bridling over a relatively short period (Southern 1962), but since there is also significant variation within individual colonies (Southern 1966), these data need interpreting with care. Indeed, it is evident that there was no change reported when the same observers made repeated visits to the colonies; change in frequency was only recorded when different observers were involved (Parkin 1980). A study of the demography of a polymorphic population on the Isle of May (Fife) showed a significantly higher breeding success for pairs that included at least one bridled bird (83.8% vs 79.5%: Harris *et al.* 2003). The frequency of bridling in the colony increased from 1946 to 2000, but selection was not necessary to explain this.

STATUS Common Guillemot is the most numerous seabird in Britain (Harris & Wanless 2004), with an estimated total population of about a million pairs. The majority of these are in Scotland, especially Caithness, Sutherland, Orkney, Shetland and the W Isles. During *Seabird 2000*, the largest colonies were Handa (Sutherland) (112,676 birds), Rathlin Island (Antrim) (95,117), Berriedale (Caithness) (79,071), Lambay Island (Dublin) (60,754). Populations in England and Wales were smaller, with the largest being Bempton (E Yorkshire) (46,685) and Skomer Island (Pembrokeshire) (13,852).

The total population of GB&I rose from *c.* 652,000 during *Operation Seafarer* (BWP) through *c.* 1,183,000 in 1985–88 (Lloyd *et al.* 1991) to *c.* 1,559,000 in *Seabird 2000*. There was thus a striking increase of 139% between the first two censuses and a lesser one of 32% between the second and third. Of the 28 major sites in GB&I, 20 showed an increase; the smallest colonies showed the largest increases, suggesting a degree of density-dependence to population control (Harris & Wanless 2004). The small colonies were also in the south, making it difficult to separate out latitudinal effects. Continued monitoring indicates that numbers in the N Isles are now declining, perhaps as a consequnce of overfishing or climate change (see below). Common Guillemots are known to visit colonies during the autumn and winter, apparently to compete for the best nest sites. There has been little recent change in return dates to colonies in S Scotland, but Shetland colonies have shown dramatic variation (Harris *et al.* 2006). During the 1960s, return dates advanced to early Oct at a time when populations were expanding. As the population fluctuated from 1970 to 2005, so the return date moved, almost in mirror image, supporting the importance of competition for sites. Recently, Harris *et al.* (2007) have shown that winter survival of young birds varies markedly among years, with two periods of high mortality coinciding with sharp changes in the plankton community ('regime shift'), and also that post-fledging survival was much higher for the progeny of early-breeding, high-quality adults.

During the breeding season, Common Guillemots have been shown to be susceptible to drowning following entrapment in fishing nets. This seems to be less of a problem in GB&I than elsewhere (Carter *et al.* 2001), although Whilde (1979) reported possible effects to the populations of N and W Ireland. Oiling incidents can cause dramatic mortality and, if severe, can destroy individual colonies. Harris & Wanless (2004) discuss this in detail, and conclude that the biology of the species, to some extent, buffers many populations against the most severe consequences. Nevertheless, oiling remains a very visible threat, though perhaps less significant in the long term than overfishing of sandeels and climate change. The former may remove an important component of chick diet, and the latter may increase summer storms and simultaneously increase both the risk of eggs and chicks being washed off the ledges and the problems associated with food gathering by the parents.

The race *hyperborea* has been suspected of occurring on several occasions; large birds have been found dead in Shetland (Pennington *et al.* 2004) and elsewhere in N GB; a dead bird was found in NE England that had been ringed at a colony in Murmansk (Wernham *et al.* 2002). These records require to be reviewed before this taxon can be added to the British List.

Thick-billed Murre/Brünnich's Guillemot   *Uria lomvia* (Linnaeus)

SM race undetermined; all likely to have been *lomvia* (Linnaeus)

TAXONOMY Polytypic; four races defined from morphology. Variation in plumage, with *eleonorae* (E Taimyr to New Siberian Islands) and *heckeri* (Wrangel Island to Chukotsky Peninsula) averaging paler, and *arra* (N Pacific) darker than nominate *lomvia*. Morphological variation mostly clinal, increasing in size from west to east, but Sluys (in BWP) suggests the evidence for *heckeri* and *eleonorae* is weak and they should be merged into *arra*.

DISTRIBUTION Widely distributed in Arctic waters of Pacific and Atlantic Oceans, breeding from 46°N to 82°N (Nettleship & Evans in Nettleship & Birkhead 1985). Nests on islands and sea cliffs from Newfoundland along the Labrador coast to Baffin Island, down the west coast of Greenland (less abundant on east), round Iceland, Arctic coasts of Norway, Spitzbergen, Jan Mayan, across N coast of Siberia to Pacific, and south to Aleutian Islands and N Japan. Disperses away from breeding sites in winter, leaving those areas where the sea freezes. Ringing recoveries suggest that Spitzbergen birds winter towards Greenland (BWP). Main wintering area for N Siberian birds seems to be in the Barents Sea, and west to N Norway. This is probably the source of the British records.

STATUS The first record is of a female found dead at Craigielaw Point (Lothian) in Dec 1908; there is now a total of *c.* 40 individuals. One or two have been recorded in most recent years; many found dead, sometimes associated with oil spills. The majority of records are from N and E Scotland; and one was present in a Shetland Common Guillemot colony in June–July 1989. The only record from Ireland was a bird at Ballyteigue Bay (Wexford) in Dec 1986. Birds usually occur associated with severe Atlantic weather systems (BBRC 2009).

Genus *ALCA* Linnaeus

Holarctic monospecific genus that breeds in GB&I.

Razorbill
Alca torda Linnaeus

RB MB WM *islandica* Brehm
WM *torda* Linnaeus

TAXONOMY Polytypic; two races. Nominate *torda* from NE North America, Greenland and NE Atlantic. Replaced in Iceland, Faroes, GB&I and Brittany by *islandica*, which is smaller; BWP suggests that there is no overlap in wing length or weight (though sex differences may mask this).

DISTRIBUTION Nominate breeds along rocky coasts and islands from Maine to N Hudson Bay, west coast of Greenland, around the Baltic Sea, and north from S Sweden, coastal Norway and Finland to the White Sea. Subspecies *islandica* breeds from Iceland through the Faroes, GB&I (except SE) to Brittany and the Channel Islands (Nettleship & Birkhead 1985).

Winters on marine inshore waters, adults usually fairly close to breeding sites but immatures wander further, south to New England in west Atlantic, and Morocco in east.

STATUS Razorbills breed round the coast of GB&I, wherever there are islands or cliffs and rocky headlands, except from Flamborough Head to the Scilly and Channel Islands, with a few small isolates in S Cornwall and Dorset. The fieldwork for *Seabird 2000* found a total of *c.* 216,000 Razorbills, an increase of 30% from *Operation Seafarer* (Merne & Mitchell 2004). However, the increase was not uniform across the region. Orkney populations remained fairly static, and those in Shetland, which had increased between 1969–70 and 1985–88 (Lloyd *et al.* 1991), fell back to Seafarer levels again. This reflected a decline through the period 1986–88 at several Shetland sites, though numbers are recovering (Mavor *et al.* 2002). Numbers held steady on the NW coasts of Scotland and in the W Isles, though this masks significant differences among colonies: Caithness colonies declined; some in Sutherland increased; Mingulay increased; the Shiants and Flannans decreased. Away from Scotland, numbers rose along the North Sea coasts, in Wales and SW Britain. Irish colonies generally increased from *c.* 41,000 in Seafarer to *c.* 51,500 in *Seabird 2000*, but again there were major differences among colonies. Some of the changes are doubtless genuine, but others may reflect the difficulties of census work at some of these sites. Merne & Mitchell (2004) discuss this in more detail.

Razorbills and Guillemots generally show a consistent pattern of change around the coasts of GB&I. Increases or decreases in one species are mirrored in the other (Merne & Mitchell 2004), suggesting that similar controlling factors may be at work. Certainly, there is evidence of birds being killed in fishing nets in some regions, and both bad weather and oil spills have had adverse effects on localised populations. Availability of food for nestlings has been implicated at some sites, and problems with stocks of sandeels continue to pose threats. Since Razorbill is a diving predator, it has a wider range of prey species, including demersal fish, and is not as reliant on one prey as (e.g.) Arctic Tern or Kittiwake. Razorbill is also a long-lived species, spending up to five years before breeding, so mortality of individual cohorts is less of a challenge than for species with a shorter life expectancy.

Subspecies *torda* occurs in winter; the first authenticated instance for GB seems to be from Dungeness (Kent) in 1937. However, the first for Ireland was much earlier: a first-summer individual was shot on The Tearaght (Kerry) in June 1885.

Genus *PINGUINUS* Bonaterre

Holarctic monospecific genus that has been recorded in GB&I. Extinct.

Great Auk   *Pinguinus impennis* (Linnaeus)

FB (extinct) monotypic

TAXONOMY Analysis of mt DNA sequences (Moum *et al.* 2002) confirm that Razorbill was the closest N Atlantic alcid to Great Auk.

DISTRIBUTION N Atlantic. Probably on many low-lying islands from Newfoundland to Norway.

STATUS Extinct. Once common in N & W Scotland (see BS3 for detailed review). Reports from Lundy 1835 and St Kilda 1840 suggest extinction in GB&I occurred about 1840. Archaeological remains from England and Isle of Man. Once abundant in Ireland, with remains from around the coasts; the only known Irish specimen is an immature female taken in Co. Waterford in May 1834.

Genus *CEPPHUS* Pallas

Holarctic genus of three species, one of which breeds in GB&I.

Black Guillemot   *Cepphus grylle* (Linnaeus)

RB *arcticus* (C. L. Brehm)

TAXONOMY Molecular studies show that the 'black' guillemots *Cepphus* form a strongly supported clade (Friesen *et al.* 1996a, Kidd & Friesen 1998a). Five subspecies (*grylle, mandtii, arcticus, faeroeensis, islandicus*) listed by Sluys (in BWP) but only three (*grylle, mandtii, islandicus*) recognised by Vaurie (1965), who regards *arcticus* and *faeroeensis* as stages in a cline between High Arctic *mandtii* and the lower-latitude *grylle*. Differences are based largely on size and the relative amounts of black and white in upperparts (especially wing patch) of adults in both summer and winter plumage; however, Kidd & Friesen (1998b) showed that the extensive molecular variation among populations does not reflect morphologically based subspecies boundaries. Adult winter *mandtii* have more white on upperparts than other races, though this is clinal. *Islandicus* is most distinct, with a dark brown streak across the white speculum and without white markings on the back in winter plumage (Vaurie 1965). *Faeroeensis* has smaller streak (but still more than *arcticus*) and primary 10 is all white (Sluys in BWP).

DISTRIBUTION Circumpolar; breeding between 43°N and 82°N. Subspecies *mandtii* breeds Hudson Bay and Southampton Island, south to Labrador, where it grades into *grylle*; along both coasts of Greenland south to about 72°N, through Jan Mayen, Bear Island, Spitzbergen and east to Novaya Zemlya, the New Siberian Islands, Wrangel and Herald Islands, and along the N coast of Alaska. Further south, except Iceland, *mandtii* is replaced by *arcticus*, from Labrador to New England, southern Greenland, GB from the Llyn Peninsula of N Wales, N around Scotland and S to Aberdeenshire, Ireland, around the Baltic coasts of Finland, Sweden and Denmark, and north along the Atlantic coasts of Norway and Finland to the Kola Peninsula and the White Sea. The three subspecies *islandicus, faeroeensis* and *grylle* are restricted to Iceland, the Faroe Islands and the Baltic, respectively (Sluys in BWP).

STATUS Mitchell (2004c) regarded this to be the hardest seabird to survey in GB&I because it breeds in small numbers along rocky coasts and low-lying islands. Many are either hidden in the nest or away feeding when survey teams arrive. Nevertheless, techniques were developed (Ewins 1985), and these were used in both the *Seabird Census Register* and *Seabird 2000*. Not all counts were directly comparable (especially from Ireland) because of differences in date that influence the birds' behaviour, but some comparisons were possible. The total population was estimated to be 42,683, 87% of which in Scotland. The British figures are broadly comparable with data from the 1980s, while the population in N Ireland seems to have more than doubled. The similarity of the two data sets for Britain hides considerable heterogeneity. Populations in some parts of Orkney and Shetland show evidence of increases, while others have decreased. Those in the Outer Hebrides increased overall, but the majority of colonies on mainland Scotland and through the Inner Hebrides had declined. Further south, into both sides of the Irish Sea, the pattern was reversed, and the majority of sites showed an increase, although the number of birds at these was relatively small. Differences in census methods made comparison in Ireland impractical.

Food supply is not a major determining factor in these changes in abundance (Mitchell 2004c). There can be heavy, though localised, mortality following inshore oil spills (e.g. Sullom Voe 1978: Heubeck & Richardson 1980). This and other events led to substantial losses of adult birds, but recovery was encouragingly rapid; usually within five years the population was back to pre-spill levels. Mitchell suggests that a limiting factor for Black Guillemots is the availability of nest sites, and they have expanded in numbers at sites where artificial nest cavities have been made available (Greenwood 1988). Mammalian predation is serious at nests that are easily accessed; birds adapt by moving to sites higher off the ground, away from the more typical boulder beach sites. However, American Mink, cats and rats remain a serious threat; at some sites, eradication has been followed by population increase or even colonisation. Mitchell (2004c) also stresses the long-term problem that a reduction of numbers by, for example, Mink could lead to a fragmented population, and the limited dispersal of Black Guillemots might adversely affect the possibility of recolonisation.

Genus *BRACHYRAMPHUS* Brandt

Holarctic (N Pacific) genus of three species, one of which is a vagrant to Britain.

Long-billed Murrelet   *Brachyramphus perdix* (Pallas)

SM monotypic

TAXONOMY Friesen *et al.* (1996b) used a combination of sequence data from mt cyt-b and allele frequencies from 37 enzyme loci to analyse the relationships of all extant alcids. They found that Marbled *B. marmoratus* and Kittlitz's *B. brevirostris* were sister taxa, with Long-billed the third member of the *Brachyramphus* clade. This confirmed the morphological data that suggested Marbled and Long-billed should be treated as separate species.

DISTRIBUTION Breeds in Russia on the Kamchatka Peninsula, Kuril Islands, Sakhalin Island, western coast of the Sea of Okhotsk, north-eastern Hokkaido and possibly in the Commander Islands. Most birds winter in the seas surrounding Hokkaido and Honshu Japan, with a few reaching South Korea and southern Japan. Vagrant to Switzerland, Dec 1997. The British bird (below) was closely followed by another in Romania in Dec 2006. It arrived at the same time as an influx of Little Auks. Much commoner in inland N America than Marbled Murrelet, Long-billed has been recorded as a vagrant along the Atlantic coast from Newfoundland to Florida, usually in late autumn or early winter.

STATUS One record of a juvenile at Dawlish (Devon) in Nov 2006. Not recorded in Ireland.

Genus *SYNTHLIBORAMPHUS* Brandt

Holarctic (N Pacific) genus of four species, one of which is a vagrant to Britain.

Ancient Murrelet   *Synthliboramphus antiquus* **(Gmelin)**

SM monotypic

TAXONOMY Close relationship to Japanese Murrelet *S.* wumizusume confirmed by molecular and enzymatic study of Friesen *et al.* (1996a), and part of a clade also including Xantus's *S. hypoleucus* and Craveri's *S. craveri.*

DISTRIBUTION Pacific coasts from Korea and Hokkaido north through Kamchatka, the islands of the Bering Sea, south through Alaska into British Columbia. Migratory, wintering south to China in the west and California in the east.

STATUS Adult present on Lundy (Devon) in May–June1990, returning in spring 1991 and 1992. Not recorded in Ireland.

Genus *ALLE* Link

Monospecific Holarctic genus that is a winter visitor to GB&I.

Little Auk   *Alle alle* **(Linnaeus)**

WM PM *alle* (Linnaeus)

TAXONOMY Friesen *et al.* (1996a) provide strong evidence from enzymes and mt DNA that this is part of a clade with Razorbill, Common and Thick-billed Murres. Two subspecies recognised by Vaurie (1965) and Sluys (in BWP): no variation in plumage; nominate *alle* smaller than *polaris*. However, age-related differences are considerable with sub-adult (brown-winged) birds being significantly smaller than breeding adults (Stempniewicz 2001).

DISTRIBUTION Circumpolar; breeds on Arctic islands close to feeding grounds, between 60°N and 82°N. Nominate *alle* breeds from Arctic Canada (E Baffin and Ellesmere Islands) through Greenland, Jan Mayen, Spitzbergen to Bear Island. Replaced in Franz Josef Land by *polaris*. Little Auk also breeds in Novaya Zemlya and Severnaya Zemlya, but the racial identity of these birds seems unclear. Recent evidence of breeding by birds believed to be *alle* in Bering Straits (Day *et al.* 1988). Formerly bred Iceland (Stempniewicz 2001).

In E Atlantic, winters in the N North Sea, Skagerrak and Kattegat, with numbers increasing from *c.* 180,000 in autumn to *c.* 850,000 in midwinter (Stempniewicz 2001).

STATUS Occurs regularly off the coasts of the N Isles and E Scotland, principally Sep–Mar, and in smaller numbers further south. Occasionally found inland, usually associated with winter wrecks. Strong movements may occur off E coasts, usually following periods of strong northerly winds. Numbers can be substantial, with hundreds or thousands of birds passing coastal headlands such as Filey and Flamborough (E Yorkshire). Less common off Ireland, but movements involving low hundreds of birds have been recorded from the west coast.

Long-winged specimens from Shetland, Orkney (Dec–Jan) and Aberdeenshire (Apr) have been attributed to ssp. *polaris* (BS3), but require to be reviewed before this taxon can be added to the British List.

Genus *FRATERCULA* Brisson

Holarctic genus of three species, one of which breeds in GB&I.

Atlantic Puffin   *Fratercula arctica* (Linnaeus)

RB MB WM PM monotypic

TAXONOMY Molecular evidence (Friesen *et al.* 1996a) confirms that Atlantic Puffin and Horned Puffin *F. corniculata* are sister taxa, with close affinities to Tufted Puffin *F. cirrhata*. In the past, has been regarded as polytypic, but most variation is in size and is largely clinal (reviewed by Moen 1991). Northern populations tend to be larger and southern smaller, but these size differences are paralleled at more local levels: birds from north Iceland have longer wings than those from the south (Petersen 1976, BWP); birds from NW Scotland are bigger than those from the SE (Harris 1979), wing length increases clinally up the Norwegian coast (Pethon 1967). The functional link of these correlations is not clear, but they are consistent.

DISTRIBUTION Breeds from E North America across SW Greenland to Iceland, Spitzbergen, Bear Island and east to Novaya Zemlya. Extends south through Scandinavia, GB&I to Brittany and the Channel Islands.

STATUS Fieldwork for *Seabird 2000* led to an estimate of 600,751 apparently occupied burrows in GB&I, of which 82% were in Scotland and 13% in England (Harris 2004). This represented an increase from 452,069 during *Operation Seafarer* (1969–70) and 506,626 in the *Seabird Colony Register* (1985–88). The main centres of population were in the N & W Isles, especially St Kilda (141,000), the Shiant Islands (65,000), Sule Skerry (59,000), the Farne Islands (55,000), the Isle of May (42,000) and Fair Isle (40,000), with lesser (but still substantial) numbers at Foula (Shetland), Coquet Island (Northumberland) and Skomer (Pembrokeshire).

Census methods changed between Seafarer and the SCR, so total numbers are not directly comparable, but there was consistency in technique at many colonies between SCR and *Seabird 2000*, so direct comparisons are more meaningful. Harris (2004) lists 22 comparisons and shows wide fluctuations. For example, the Isle of May colony, which has been monitored regularly for many years has increased from 2,000 burrows (1970) through 12,000 (1984) and 42,000 (1988) to 69,000 (2003) (Wanless *et al.* 2003). Coquet Island and the Farne Islands (Northumberland) showed similarly impressive increases, but the Foula population declined from 70,000 burrows (1969–70), through 48,000 (1987) to 22,500 (2000).

As with many oceanic species that breed in burrows, clear threats are posed by predators, both mammalian and avian; rats, gulls and skuas have all been implicated. Great Black-backed and Herring Gulls and Great Skuas kill Atlantic Puffins at some colonies, and Arctic Skuas and Herring and Lesser Black-backed Gulls steal fish from the returning adults as they come to the colony to feed their nestlings. Although predation is generally thought to be unimportant, it can have significant effects at individual sites. At Ailsa Craig, the number of breeding Atlantic Puffins declined rapidly following the arrival of Brown Rats in the latter part of the 19th century, and they were almost extinct as a breeding bird by the 1930s. The rats were eradicated in 1990–91 (Zonfrillo & Monaghan 1995), and the birds rapidly returned (Zonfrillo 2002). A similar recovery was observed following rat eradication from Handa (Sutherland), although at some sites, there seems to be coexistence (Harris 2004).

Oil spills have been implicated in severe losses to some smaller colonies; Harris (2004) reported that several colonies in Brittany may have been reduced due to oil-based mortality following wrecks, but believed that food supply is equally (if not more) important as a limiting factor in Puffin population biology. Several studies have shown that breeding success is reduced when food supply is compromised (e.g. Durant *et al.* 2003). As fish stocks are sensitive to both overexploitation and changes in sea temperature, the greatest long-term threats to the Puffin may come from man and climate change (Harris 2004).

FAMILY PTEROCLIDIDAE

Genus *SYRRHAPTES* Illiger

Palearctic genus of two species, one of which is a vagrant to GB&I.

Pallas's Sandgrouse   *Syrrhaptes paradoxus* (Pallas)

CB SM monotypic

TAXONOMY The affinities of sandgrouse have long been debated. At one time or another, they have been suggested as allies of partridges and grouse, pigeons and waders (HBW). Molecular data are beginning to resolve the debate. Using nuclear sequences, Paton *et al.* (2003) found no support for a position within the shorebirds, though they might be sister to these. Closely related to Tibetan Sandgrouse *S. tibetanus*, but apparently no recent analysis.

DISTRIBUTION An Asiatic species, breeding in semi-desert and dry steppe from Aral Sea across Kazakhstan to Mongolia, NW China. Generally resident, though most northerly birds move S to avoid snow. Otherwise, occasionally irruptive, though reasons unclear: perhaps related to snow cover or failure of food supply (HBW).

STATUS There have been three major invasions into GB&I: in 1863, 1888 and 1908. The first of these totalled over 900 individuals, with individual parties of up to 50 birds, many of which were shot. The influx of 1888 began in May and was much larger, with a total of over 4,300 birds being recorded. Several parties of over one hundred were reported, predominantly along the eastern side of GB, but birds were recorded as far as W Ireland. Breeding occurred at several sites. For example, two eggs were collected at each of two sites near Beverly (E Yorks) in June and July 1888, and five young were reported from Kinshaldy (Fife) in Aug of the same year. Five young were also recorded on Newmarket Heath (Suffolk) in Aug, and broods were reported from Findhorn (Moray and Nairn) on more than one occasion in Aug. This influx extended for a couple of years, but the largest numbers were in 1888, declining through to about May 1890. The third invasion occurred in 1908, with a few birds being recorded through to 1909. This was smaller than the previous two influxes, totalling *c.* 100 individuals, and did not reach Ireland. There were no reports of breeding on this occasion. There have been six records in GB since 1950, along the east coast from Shetland to Kent.

The totals for Ireland are: 18+ in 1863, two in 1876 and 110+ in 1888. The only modern Irish record is of two at Wexford Harbour in May 1954, although the possibility of Black-bellied Sandgrouse *P. orientalis* could not ruled out in the 1954 Wexford record.

FAMILY COLUMBIDAE

Genus *COLUMBA* Linnaeus

Cosmopolitan genus of 35–50 (or more) species, three of which breed commonly in GB&I.

Common Pigeon/Rock Dove   *Columba livia* J. F. Gmelin

RB *livia* J. F. Gmelin

TAXONOMY Molecular phylogenies derived from mitochondrial and nuclear genes showed that *Columba* is paraphyletic; Old and New World *Columba* species are both monophyletic, and in different parts of the gene tree (Johnson *et al.* 2001, Pereira *et al.* 2007). The AOU now treat the New World clade as a separate genus *Patagioenas* (R.C. Banks *et al.* 2003). Within the Old World clade, Common Pigeon is sister to Hill Pigeon *C. rupestris* of C & SE Asia; Snow Pigeon *C. leuconota* of Himalayas is also generally believed to be close to these but was not included in the analysis. Over 12 subspecies are recognised based on variations in size and colouration (HBW, Gibbs *et al.* 2001); three of these occur in W Europe, two as island races: *atlantis, canariensis*. There are large urban and rural populations of feral pigeons in many regions; these originate from escaped and released domesticated birds and lost racing pigeons, and native gene pools in many adjacent areas have been heavily contaminated.

DISTRIBUTION Almost cosmopolitan, occurring naturally across Europe and much of C and SE Asia, but also after introduction across the S half of N America. Typically nests in cliffs and crevices in rock faces; this naturally adapts it to populated urban sites and is now found in many (most?) major cities of the world Nominate race breeds in S & W Europe and the Mediterranean coast of Africa to the Caucasus Mts; subspecies *atlantis* breeds on Madeira, Azores and Cape Verde; *canariensis* on Canaries and islands off Morocco. Five other races occur in Africa and four in Asia. Most populations appear to be sedentary, despite pigeon fanciers being able to breed racing pigeons selectively for medium-distance navigation ability.

STATUS Until comparatively recently, urban populations ('Feral Pigeon') were omitted from many surveys and censuses. Consequently, analyses are biased in the earlier years. Holloway (1996) was unable to assess the status of feral birds in his review of 19th-century records, limiting this to 'pure' Common Pigeons, which he concluded had occurred around the coasts of Scotland, including most of the islands, and round Ireland, except for the E coast of the Republic. The only places in England and Wales where he was confident wild birds occurred were the Lleyn Peninsula (including Anglesey), coastal Pembroke and the sea cliffs of Yorkshire. By the time of the Atlases, data were pooled for both feral and wild birds, although Feare (in Second Breeding Atlas) attempted to differentiate between the two populations. It appeared that wild Common Pigeons continued to decline as Ferals increased; the former appeared to have been lost from coastal areas of N & W Ireland, and many populations were becoming 'contaminated', with birds showing characteristics of feral stock. Near-pure Common Pigeons now confined to Outer Hebrides and more remote areas of Inner Hebrides (BS3). In lowland England, Common Pigeons continued to increase in numbers in urban areas. Populations were expanding in rural areas, and overall there was an increase of *c.* 40% in the number of 10-km squares occupied between the two Breeding Atlases.

The origins of the feral populations are a combination of escaped or released domestic birds and lost racing pigeons. Ringed individuals from the latter source are often seen in urban flocks. Both these sources include birds with a variety of plumage types, ranging from almost black through to orange, with varying amounts of white. These traits are largely under genetic control, and Murton *et al.* (1973) found differences in reproductive characters among these. The solid dark ('dark blue checker', 'velvet') phenotypes are controlled by separate alleles, and males of these have larger testes and higher numbers of sperm-generating cells. The testes increase in size through the breeding cycle but, at all seasons, Murton's team found that testis volume was larger among melanics. They also found evidence of sexual selection, with females choosing melanic males preferentially, presumably because of their higher sexual activity. The high frequency of dark birds in urban sites is thus accounted for, and their gradual replacement of 'pure' Common Pigeons can also be understood. A recent study in Vienna showed an increase in the proportion of darker phenotypes with age, further supporting a selective advantage for these types in urban environments (Haag-Wackernagel *et al.* 2006). Any selective disadvantage of the melanic alleles, such as differential visibility to predators in flight (Pielowski 1959) seems inadequate to stem their inexorable advance through urban populations, though in rural areas, especially the more remote sea cliffs, a reversion to wild-type plumages appears to take place.

Stock Dove   *Columba oenas* Linnaeus

RB WM PM *oenas* Linnaeus

TAXONOMY Not included by Johnson *et al.* (2001) in their study of pigeon phylogenetics, but generally accepted as evolutionarily close to Common and Wood Pigeons. Two subspecies recognised, nominate and the disjunct *yarkandensis*.

DISTRIBUTION Nominate breeds in woods and agricultural land where there are old trees with nest holes, but if habitat is otherwise suitable will nest in holes in cliffs. Occurs from Iberia and GB&I across most of W Eurasia to Caspian Sea and the uplands of C Turkey and N Iran. Isolated population in Kazakhstan to Tien Shan is paler and larger and has been described as *yarkandensis*. N & E populations are migratory; western birds winter in milder parts of Europe, eastern ones move S into Iran.

STATUS Widespread across GB, except in N & W Scotland and the N & W Isles. The Atlases revealed a slight contraction in range, with loss of birds from parts of Cornwall, Wales and Scotland. Now found in woods, parks and farmland with trees, though may breed in buildings and cliffs where tree holes are lacking. The situation in Ireland was less encouraging; >30% of 10-km squares occupied in the First Breeding Atlas were abandoned by the Second Breeding Atlas. Stock Doves spread and increased throughout the early 20th century, benefiting from the less intensive arable agriculture of the time. Indeed, the first breeding records for Ireland were not until the 1870s, and birds were not widespread for 100 years. Their populations were hit hard by chemical seed dressings, initially cyclodienes in 1950, and then organochlorines. When these were banned in the early 1960s, the populations began to recover. However, O'Connor & Shrubb (1986) suggest that the recovery was not complete since agricultural practices had changed in the interim, and the introduction of herbicides, absence of fallow and loss of winter stubbles all reduced the amount of available seed, especially in winter.

CBC and BBS data are only available subsequent to the chemical ban, so are restricted to the recovery period. It is clear that numbers rose to the 1980s and, after a period of stability, are increasing again to a level approximately three times that of the 1960s. Clutch and brood sizes have changed little since then, but nest failures are declining. Interestingly, Stock Dove is one of the few species that is nesting later than 30 years ago, with nest initiation seemingly two weeks later than in the 1970s (though sample sizes are small; Baillie *et al.* 2006). This might be artefactual: as conditions improve for the species, it will nest longer through the season and the average initiation date will become later.

Ringing data indicate that Stock Dove is fairly sedentary, with only limited dispersal from the natal site. There are a few recoveries of birds from W France and Iberia, suggesting a limited migration, and there are regular movements in varying numbers through south coast watch-points such as Portland Bill, though whether these are British or Continental birds is unclear. Immigration from Fennoscandia is reported to have declined in recent years (Wernham *et al.* 2002), but there is little quantitative data to support this.

Common Wood Pigeon   *Columba palumbus* Linnaeus

RB WM *palumbus* Linnaeus

TAXONOMY A member of the Old World clade of *Columba* (Johnson *et al.* 2001), and generally regarded as closely allied with Trocaz *C. trocaz*, Bolle's *C. bollii* and Laurel *C. junoniae* Pigeons of Madeira and Canaries. Six subspecies recognised, two restricted to Atlantic islands (one now extinct), differing especially in colour and size of neck patches.

DISTRIBUTION Restricted to W Palearctic, breeding in woods and farmland from NW Africa, Iberia, GB&I across all of Europe except extreme north to Caspian Sea and Persian Gulf. Nominate in majority of range; separate subspecies in Madeira (*maderensis*), Azores (*azorica*), N Africa (*excelsa*), S Caspian (*iranica*). Disjunct populations in Kazakhstan through to Himalayas have been described as *casiotis*. N Africa and island races are sedentary; populations in SW Europe are largely resident, undertaking hard-weather movements in extreme cold; rest of Continental birds are migratory, moving SE or SW to winter in milder regions.

STATUS Common Wood Pigeons were recorded in over 90% of 10-km squares during the two Atlases, being absent only from the highest hills and parts of Shetland and Outer Hebrides. CBC/BBS fieldwork has revealed a steady increase in population of *c.* 150% since 1965, more steeply in England and Wales than elsewhere, with a slight decline in Scotland during 1994–2004, and is now perhaps the most economically serious avian pest species in GB&I. The Repeat Woodland Survey confirmed these increases, indeed some data suggest that it may have been more pronounced than the BBS data indicate. There was evidence of a slight decline in numbers through the early 1970s (Baillie *et al.* 2006), and Inglis *et al.* (1990) showed that this was likely due to the decrease in acreage under grass and clover, on which Common Wood Pigeons depended in winter. The introduction and spread of oil seed rape reversed this trend; winter food became more plentiful, and Common Wood Pigeons responded rapidly. Data on survival and reproduction are too meagre for analysis, but over-winter survival seems key to the present increase.

Large numbers of Common Wood Pigeons are ringed as nestlings, and there is evidence from recoveries that some of these emigrate to W France in winter. Heavy and prolonged visible migration of south-bound Common Wood Pigeons commonly seen along S coast in late autumn, with thousands of birds passing sites such as Portland Bill in a single day. Whether Common Wood Pigeons come to GB&I from the Continent is still a matter for debate (see Inglis in Wernham *et al.* 2002), although birds recorded on Fair Isle (Shetland) in autumn. Very few ringing recoveries of Continental birds in GB.

Genus *STREPTOPELIA* Bonaparte

Palearctic, Oriental and Ethiopian genus of about 15–17 species; three which have been recorded in GB&I, two as breeding birds and one as a vagrant.

Eurasian Collared Dove	*Streptopelia decaocto* (Frivaldszky)

RB *decaocto* (Frivaldszky)

TAXONOMY Johnson *et al.* (2001) showed *Streptopelia* to comprise three subclades, which were all part of a poorly supported major clade within the pigeons and doves. Two smaller clades comprised Pink Pigeon *Nesoenas mayeri* (Mauritius) and Malagasy Turtle Dove *S. picturata,* and Spotted Dove *S. chinensis* (SE Asia) and Laughing Dove *S. senegalensis* (Africa). Within the larger subclade, Eurasian and African Collared Doves *S. decaocto* and *S. roseogrisea* were sister taxa; these have been regarded as conspecific in the past, but the molecular data indicate that they are sufficiently divergent to be separate evolutionary lineages. Two subspecies of Eurasian are recognised based on darkness of plumage and colour of orbital skin: *decaocto* and *xanthocycla.*

DISTRIBUTION Very widespread across Eurasia, including most of Europe, Middle East, Iran through Indian subcontinent to E China. Nominate occurs across most of range, replaced by *xanthocycla* in Burma and S China. Most populations resident, though high-altitude populations move to lower ground in winter; however, has shown remarkable expansion of range during 20th century, so dispersal distances are potentially very large.

STATUS As is now well known, Collared Doves arrived in GB&I in the 1950s; the colonisation is well documented by Hudson (1965). Following a 1952 record from Lincolnshire that was deemed a possible escape, a pair bred in Norfolk in 1955 and reared two young. The next few years saw a rapid increase, with confirmed breeding in Moray, Lincolnshire and Kent (1957), Northumberland and Hertfordshire (1958), and spread more widely in subsequent years, with the first breeding in Ireland in 1959 and in Wales in 1960. The patchwork pattern of settlement suggests that initially there were several independent colonisations, but latterly new sites were occupied by expansion. The spread across W Europe was initiated from the Balkans in about 1930. It has been suggested (e.g. Mayr 1951) that it stemmed from some kind of behavioural mutation, causing a more rapid dispersal; there is, however, no evidence to support this, and the reasons remain a mystery. Within little more than a decade, its numbers had increased to the extent that it was added to the Schedule of 'pest species'.

By the time of the First Breeding Atlas, Eurasian Collared Dove bred throughout GB&I, from Shetland to Scilly, and from westernmost Ireland to Kent. There were gaps in the upland areas of all four countries, some of which were infilled by the Second Breeding Atlas. CBC and BBS data show that numbers continue to climb, although, as with Stock Dove, there was a plateau in the early 1980s, perhaps associated with the run of cold winters early in the decade, which affected several other species. Nest record data indicate that brood sizes continue to increase, and nest failure at the chick stage is declining, so the population looks likely to remain buoyant.

There are 44 ringing recoveries of foreign-ringed birds prior to 1979, during the initial period of colonisation; most of these involve birds from Belgium and the Netherlands, supporting the spread from the near-Continent. Dispersal distance within GB&I was higher in the early days of settlement than later, and this tended to be in a W or NW direction (Wernham *et al.* 2002).

European Turtle Dove   *Streptopelia turtur* (Linnaeus)

MB PM *turtur* (Linnaeus)

TAXONOMY Molecular data (Johnson *et al.* 2001) indicate this to be sister to Adamawa Turtle Dove *S. hypopyrrha*, a little-known species from W Africa; however, Dusky Turtle Dove *S. lugens* of E Africa, which is often regarded as conspecific with this, was not analysed. European Turtle Dove is treated as four subspecies on basis of size and plumage tones (Gibbs *et al.* 2001): *turtur, arenicola, hoggara, rufescens.*

DISTRIBUTION Breeds across the W Palearctic, from the Sahara to *c.* 58°N, generally excluding Scandinavia, and through Siberia to about Lake Baikal. Nominate occurs across N of range, N of line from Mediterranean to Kazakhstan; replaced by *arenicola* in N Africa, Middle East to Afghanistan and N China; other two races restricted to smaller areas of N Africa. Most populations migratory, only those in extreme S of range resident. Others winter in Sahel, from Senegal to Red Sea.

STATUS In GB&I, this is a bird of the lowlands, typically nesting in hedgerows on agricultural land and feeding on open ground. The two Atlases revealed a marked contraction in range, with no breeding records in 1988–91 from *c.* 25% of the 10-km squares that held birds in 1968–72, and evidence of breeding being found in only *c.* 65% of squares that had previously been utilised. The declines were most marked in Wales, N, SW & C England and Ireland, although it could never have been regarded as a regular breeder in the last of these, and is now rarely more than a passage migrant in NW England. In parallel with this, the index of abundance based on CBC data showed a progressive decline from the late 1970s, which continued until the mid-1990s, and was also found elsewhere in W Europe.

These declines prompted a study by the Game Conservancy (reviewed by Browne & Aebischer 2005). Their findings were that changes in land management were a major factor, in particular the loss of tall, overgrown hedges for nesting, and short, weed-rich areas for feeding. There had been a marked change in nest productivity, both the number of clutches and the fledging success per pair had fallen since the 1960s. They also recorded changes in migratory behaviour; Turtle Doves leave earlier in the autumn than in the 1960s, and they speculate that this may be due to climate change: warmer, earlier springs bring the plants to fruition earlier in the year, but the Turtle Dove's breeding season has not evolved fast enough to remain in synchrony. Foraging behaviour in GB has changed; there has been a marked increase in reliance on cultivated seed, such as wheat and oilseed rape. Browne & Aebischer (2005) suggest that the provision of tall hedges for nesting and weedy margins to crop fields will be essential to the recovery.

That conditions on the wintering grounds may also be significant has been indicated by an admittedly small study of Turtle Doves breeding on Oléron Island, France (Eraud *et al.* 2009); over-winter survival, measured from the recapture of ringed birds, correlated significantly with a measure of cereal production in Mali–Senegal. This might indicate the sensitivity of this bird (and presumably other migrant and non-migrant granivores) to agricultural changes in sub-Saharan Africa.

Oriental Turtle Dove   *Streptopelia orientalis* (Latham)

SM *orientalis* (Latham)
SM *meena* (Sykes)

TAXONOMY In the phylogeny of Johnson *et al.* (2001), Oriental was sister to the Eurasian/Adamawa Turtle Dove pairing, supporting conventional taxonomic placements (HBW). Six subspecies described, differing in size, richness of colour, and breadth and tone of terminal tail band; much variation is clinal and some races probably not recognisable.

DISTRIBUTION Occurs in SE Asia, roughly S of 64°N and E of 65°E. Subspecies *meena* and *orientalis* are most northerly, and northern populations of both are migratory; former breeds from SW Siberia and Pakistan, to Altai Mts and Nepal, wintering in India, latter breeds from C Siberia to Kamchatka and Japan, S through China to IndoChina, wintering in SE Asia. Other races breed in India, Ryu Kyu Is and Taiwan.

STATUS Two subspecies have been recorded in GB totalling eight individuals; of those positively identified, there have been three of each. The first record was of *orientalis* at Scarborough (N Yorks) in Oct 1889; the first *meena* was at Spurn (E Yorks)in Nov 1985. All records have been Oct–Jan, apart from a bird on St Agnes (Scilly) in May 1960. There are no records from Ireland.

Genus *ZENAIDA* Bonaparte

New World genus of seven species, one of which has been recorded as a vagrant in GB&I.

Mourning Dove   *Zenaida macroura* (Linnaeus)

SM race undetermined

TAXONOMY Five subspecies recognised by AOU, plus a sixth (Socorro Dove), now regarded as a full species *Z. graysoni*, (Johnson & Clayton 2000), but now extinct in the wild.

DISTRIBUTION Extremely abundant and widespread in N America, breeding across the continent S of Great Lakes to C Mexico. According to Mirarchi & Baskett (1994), it is the most hunted bird in N America, with close to 70 million birds shot every year. Most populations are migratory, but some individuals resident year round. Majority of range is occupied by two races (*marginella, carolinensis*), in the W and E of N America, respectively. Other races have very restricted range.

STATUS Three records from GB and one from Ireland. The first was trapped on Calf of Man in Oct 1989 and found dead the following day. The second and third were both found on N Uist (Outer Hebrides) in Nov 1999 and 2007. All three were first-year birds. The first record for Ireland was at Inishbofin (Galway), Nov 2007.

FAMILY PSITTACIDAE

Genus *PSITTACULA* Cuvier

Cosmopolitan genus of about 14 species, one of which has been introduced into GB&I.

Rose-ringed Parakeet   *Psittacula krameri* (Scopoli)

NB race(s) undetermined; closest to *borealis* (Neumann) and *manillensis* (Bechstein)

TAXONOMY Apart from the endemic and endangered Mauritius (or Echo) Parakeet *P. echo*, Rose-ringed is the only *Psittacula* to occur outside SE Asia. Four subspecies are generally recognised, two each from Asia and Africa, differing chiefly in size and bill colour.

DISTRIBUTION Occurs in two regions: Africa and Asia. African populations extend from Senegal to the Red Sea coasts of Djibouti and N Somalia, inhabiting woodland north of the rainforests; two races: *krameri* and *parvirostris*, found W and E of Sudan/Uganda. Asian populations breed across most of the Indian subcontinent, Sri Lanka and Burma; subspecies *borealis* more northerly, distributed from Pakistan through Nepal to Burma; *manillensis* occurs across peninsular India S of 20°N (Juniper & Parr 1998). Birds have been introduced into Europe, N America and the Orient, and sizeable populations have built up in several places.

STATUS May have been present in GB for >100 years (see Morgan 1993, Chandler 2003), but birds began to be seen regularly in 1969, and breeding was confirmed in 1971 (Lever 2005). By the mid-1980s, the population had spread to most counties in SE England, along with Yorkshire, Manchester, Liverpool, Norfolk, Renfrew and Clwyd. The Second Breeding Atlas confirmed the status in SE England, although the more northerly populations were not recorded. The number of individuals is difficult to assess, but there are at least several thousand: a roost at Esher (Surrey) regularly holds >2,000 individuals. The rapid increase and spread has been ascribed to a combination of breeding at a young age, longevity and lack of competition. There are no records from Ireland.

  In an attempt to identify the source of the GB populations, Pithon & Dytham (2001) analysed the biometrics of a series of wild and captive birds, comparing these with museum specimens. They concluded that British colonists are closer to the (larger) Indian subspecies *borealis* and *manillensis*, though they could not determine which with any certainty.

FAMILY CUCULIDAE

Genus *CLAMATOR* Kaup

Old World genus of four or five species, one of which has been recorded in GB&I.

Great Spotted Cuckoo   *Clamator glandarius* (Linnaeus)

SM monotypic

TAXONOMY Closely related to *C. levaillantii* and *C. jacobinus* (Payne 2005). A small race ('*chlor-agium*') has been described for S Africa, but size overlaps considerably and not recognised.

DISTRIBUTION Occurs as a series of widely scattered populations across W Eurasia and Africa. Former includes Iberia, S France, Italy, Turkey, Levant, Iraq/Iran; these populations mostly winter in Africa, though a few remain in Iberia. African populations breed in a series of scattered areas S of Sahara, outside rainforests, including much of E and S Africa; those S of River Zambesi are also largely migratory, wintering N, probably into Rift Valley.

STATUS This remains a rare bird in GB&I, with only *c.* 40 records up to 2008. The first for GB&I was a first-year that was caught on Omey Is (Galway) in Mar 1842, and died a few days after capture. The first for GB was shot at Clintburn (Northumberland) in Aug 1870. Birds have been recorded Feb–May and July–Oct; although only six have been recorded in Ireland, one of the records involved two birds.

Genus *CUCULUS* Linnaeus

Old World and Australasian genus of 10–12 species, one of which breeds in GB&I.

Common Cuckoo   *Cuculus canorus* Linnaeus

MB PM *canorus* Linnaeus

TAXONOMY Morphological and molecular data show this to be part of a complex that also includes African Cuckoo *C. gularis* and 'Oriental Cuckoo' (Payne 2005), now usually split into Himalayan *C. saturatus* and Oriental *C. optatus*. Populations of Common Cuckoo vary in darkness of plumage and barring of underparts; four subspecies recognised, of which two occur in Europe.

DISTRIBUTION Occurs across Eurasia from GB&I to Kamchatka, missing only from most arid habitats and tundra regions of extreme north. Southern limits of breeding range include NW Africa, Himalayas and S China. Strongly migratory, wintering in Africa S of equator, India and Indo-China. Nominate race most widespread, replaced by *bangsi* in Iberia, N Africa, and by *subtelephonus* and *bakeri* in C & S Asia.

STATUS A bird that is easily monitored in terms of singing males, but requires more care if breeding is to be confirmed. At the time of the Second Breeding Atlas, Common Cuckoos were present in *c.* 85% of 10-km squares in GB, representing a decline of *c.* 5% on the First Breeding Atlas. In Ireland, birds were less widespread, being recorded in only 70%, a decline of 25%. The density was highest in the S & E of England, W Ireland and parts of W Scotland. The decline in Ireland was ascribed to a reduction of Meadow Pipits, the principal host in that region, which had also declined in abundance.

The CBC and BBS data for GB reveal a serious decline that has been ongoing since the 1960s. During this period, the estimated number of birds has fallen by >50%, predominantly in England, and to a lesser extent in Wales; the Scottish population appears to have held steady, or even increased slightly (Baillie *et al*. 2006). Brooke & Davies (1987) suggested that the decline in England and Wales may be due to fewer Meadow Pipits and Dunnocks as host species. Whether there are also problems on the wintering grounds in sub-Saharan Africa is unknown.

Ringing recoveries are few, but support the view that adults leave the breeding grounds early, followed by juveniles a few weeks later. Most recoveries involve exchange between GB and France, and autumn emigration seems to be SE towards Italy. Moreau (1972) suggested that Common Cuckoos overfly both the Mediterranean and the Sahara in a single stage, and the lack of ringing recoveries from N Africa supports this perceptive view. There is only one recovery of a British Common Cuckoo on the wintering grounds (in Cameroon), and a Dutch bird was found wintering in nearby Togo (Wernham *et al*. 2002).

Genus *COCCYZUS* Vieillot

New World genus of 7–13 species, two of which are vagrants to GB&I.

Black-billed Cuckoo   *Coccyzus erythropthalmus* (Wilson)

SM monotypic

TAXONOMY Molecular data suggest more closely related to Grey-capped Cuckoo *C. lansbergi* of N South America and six Caribbean species than to Yellow-billed (Payne 2005).

DISTRIBUTION More restricted than Yellow-billed Cuckoo, occurring in S Canada, and S to Maryland, Kentucky. Has declined in numbers through the 20th century (and especially since 1980), perhaps due to pesticides depleting the insects on which it feeds (Hughes 2001). Migrates to S America, but winter range little known.

STATUS This species is much rarer in GB&I than the Yellow-billed Cuckoo; a single record from Ireland and only 13 from GB. The Irish bird was the first for GB&I (and only the second for the W Palearctic), being killed at Killead (Antrim) in Sep 1871; the first for GB was found dead on Tresco (Scilly), Oct 1932; eight of the GB records are from the SW peninsula. Eleven records are from Oct; the remainder in Sep and Nov. The decline in numbers reported from N America does not seem to be reflected in occurrence in GB&I; the proportion of Black-billed to Yellow-billed during 1950–2000 is almost identical before and after 1980.

Yellow-billed Cuckoo   *Coccyzus americanus* (Linnaeus)

SM monotypic

TAXONOMY Closely related to S American Pearly-breasted Cuckoo *C. euleri*, with which it was once regarded as conspecific. Not especially close to Black-billed Cuckoo. Morphologically weakly differentiated, but west and east populations differ by four fixed mt DNA base changes, offering support for proposed western race *occidentalis* (Hughes 1999, Pruett *et al.* 2001).

DISTRIBUTION E USA, Gulf coast of Mexico and Caribbean; possibly also into S America (Hughes 1999). Strongly migratory, wintering in S America, E of Andes. Fairly common in Bahamas in Sep–Oct, suggesting not averse to sea-crossings.

STATUS There are 57 records of Yellow-billed Cuckoo from GB, nine from Ireland and one from IoM. The dates for two of the first records are rather vague: that for Ireland was killed near Youghal (Cork) in the autumn of 1825, and for GB was shot in the autumn of 1832 at Lawrenny (Pembroke). The dates of occurrence are similar, with most being recorded in Oct, but a few in Sep and Nov–Dec. As with many Nearctic vagrants, there is a bias towards the SW, but has been found from Shetland to Surrey.

FAMILY TYTONIDAE

Genus *TYTO* Billberg

Cosmopolitan genus of 10–13 species, one of which breeds in GB&I.

Barn Owl   *Tyto alba* (Scopoli)

RB *alba* (Scopoli)
SM *guttata* (C. L. Brehm)

TAXONOMY Recent molecular analyses (summarised by Wink *et al.* 2004e) show that *Tyto* is sister to, but clearly distinct from, the rest of the owls, and comprises three clades, corresponding to populations from Australasia, Europe/Africa and the Americas. Indian birds seem not to have been included in this study, but, even so, a case could be made for revising species limits to take account of prolonged isolation indicated by the genetic divergences. Barn Owl is a variable taxon, with more than 30 subspecies, some of doubtful validity and many restricted to isolated island groups, differentiated on basis of size and colour. However, these vary individually and clinally, and a thorough review is needed. Extensive analyses of plumage colour and spottiness have been undertaken in Switzerland (summarised by Roulin 2002); both characters are inherited, and there are strong differences in survival and breeding success among birds with different patterns.

Seven subspecies breed in Europe, of which two are widespread: *alba*: N, W Europe, Africa from Atlantic to Red Sea, S to Niger, Sudan; *guttata*: C & E Europe, from Balkans and Baltic states to Ukraine. Voous suggested that these evolved in glacial refugia in SW and SE Europe, respectively. Matics & Hoffmann (2002) found that the transition zone (determined from the colour of the underparts) in Hungary was clinal and up to 500 km wide. Endemic races occur in Sardinia/Corsica, Madeira, E Canaries and Cape Verde, and *erlangeri* occurs in Crete/Cyprus and perhaps E to Sinai, Iran (HBW). Claims of *guttata* in Britain complicated by occasional dark-breasted individuals among GB breeding birds (even mixed brood of white-breasted and dark-breasted individuals, French 2006).

DISTRIBUTION Very widely distributed across Europe, S of Baltic through Mediterranean to N Africa, Red Sea and much of sub-Saharan Africa; also from Indian subcontinent through SE Asia to Australia. New World populations extend from Great Lakes S to Tierra del Fuego and Falklands. Also on many isolated island groups, such as Hawaii, Bermuda.

STATUS Assessing the changes in abundance of Barn Owls is difficult because different methods of census have been used across the years and because, to some extent, the numbers vary in synchrony with cyclical populations of small mammals. Historically, it was probably the commonest owl in Britain (Holloway 1996). The number is believed to have declined from the middle of the 19th century, perhaps to 12,000 pairs in the 1930s. Further declines followed, linked with organochlorine pesticides, to <5,000 pairs in GB in the 1980s, and to *c.* 4,000 pairs in the 1990s (Toms *et al.* 2001). The last of these was based on a more systematic (and repeatable) census, which should provide greater comparability between surveys in the future. In 2000, an annual monitoring programme was initiated, which now provides data on breeding success, including site occupancy, clutch size, and hatching and fledging success (Leech *et al.* 2005). Most data come from nest box sites, of which there were *c.* 25,000 in the late 1990s, and data were obtained from *c.* 530 in 2004. Site occupancy declined from 2000 to 2004, but breeding occupancy did not; indeed, this oscillated depending on the year. Inclement winter weather may have led to the suspension of breeding, perhaps because the birds may have been in poor condition. At occupied sites, breeding success was also lower in these years.

A recovery in number of breeding birds in some regions is correlated with the provision of nest boxes. This suggests that part of the decline was due to the loss of nesting sites. A combination of the clearance of hedgerow trees and the conversion of old barns reduced the number of suitable holes.

Individual birds resembling *guttata* have been recorded on several occasions in GB&I (but see above), usually along the E coast; some of these had been ringed within the range of this race. Three of this race have also been recorded in Ireland: all in 1932.

FAMILY STRIGIDAE

Genus *OTUS* Pennant

Cosmopolitan genus of 35–40 species, one of which is a vagrant in GB&I.

Eurasian Scops Owl   *Otus scops* (Linnaeus)

SM *scops* (Linnaeus)

TAXONOMY Molecular analysis has shown that *Otus* is polyphyletic, with the New and Old World species clustering separately in both cyt-b (Wink & Heidrich 1999) and a nuclear intron tree (Wink *et al.* 2004e). Within the Old World clade, Eurasian is sister taxon to Pallid Scops *O. brucei*, as would be expected from morphology and behaviour. Six subspecies currently recognised, four or five of which occur in Europe.

DISTRIBUTION Breeds in Mediterranean broadleaved woodland and open farmland from Iberia and NW Africa to Lake Baikal, apart from the arid/desert regions of W Asia. N populations migratory, wintering S of Sahara but N of the rainforests. Populations from Iberia, Balearics, Cyprus resident or partially migrant. Nominate race breeds from France to Caucasus; replaced in E by *pulchellus*; in Iberia/N Africa by *mallorcae*; from S Greece to Jordan by *cycladum*; by *cyprius* in Cyprus; and by *turanicus* from Iraq to Pakistan.

STATUS There are less than 100 records from GB&I, with the first for GB being shot at Wetherby (W Yorkshire) in spring 1805, and the first (of 13) for Ireland being 'obtained' at Loughcrew (Meath) in July a few years previous to 1837. Since 1950, the annual totals have remained fairly constant at about two a year, with a wide distributional scatter. Despite being a summer visitor to W Europe, Scops Owls have been recorded in every month except Feb, although the peak months are Apr–June, with *c.* 57% of records. Only subspecies *scops* has been recorded in GB&I, although *mallorcae* (smaller and more heavily marked) is possible. It is notable that many recent records involve birds found injured, dead or dying; this little bird seems not to have an especially happy history in our islands.

Genus *BUBO* Dumeril

Near cosmopolitan genus of 18 species, one of which has been recorded in GB&I.

Snowy Owl   *Bubo scandiacus* (Linnaeus)

FB SM monotypic

TAXONOMY Wink *et al.* (2004e) showed that Snowy Owl is nested within *Bubo*, from both nuclear and mt DNA sequences, indicating that it should not be *Nyctea*. Combining the sequence data to increase resolving power indicated that Snowy is sister to Great Horned Owl *B. virginianus*, with which it is parapatric in N America; white plumage presumably an adaptation to an arctic existence. Low levels of genetic variation between N American, Scandinavian and E Siberian birds, suggesting high gene flow in recent past and possibly still today (Marthinsen *et al.* 2009).

DISTRIBUTION Holarctic, breeding in a relatively narrow band of arctic tundra adjacent to the northern seas. Nomadic and migratory, but movements are unpredictable (HBW), probably dependent on availability of prey. Irruptive in Nearctic, but less so in Palearctic; wanders widely, if sparsely, through N Eurasia.

STATUS Snowy Owl has been found hundreds of times in GB&I, predominantly in Scotland, where there are well over 400 records. The earliest are of birds on Unst (Shetland) in 1811 and on the S coast of Wexford in 1812. There are records from every month, with more in Apr–May than other times. Numbers increased through the 1960s to peak in the 1970s at over 15 a year, since when there has been a decline to an annual total of around four. Among these records are some remarkably long-staying individuals (especially in the northern and western isles of Scotland), culminating in breeding on Fetlar (Shetland) from 1967 to 1975, with young being successfully reared in all years except one (Robinson & Becker 1986). Permanent colonisation was not to be. The male disappeared in winter 1975–76 and was not replaced, probably because he had driven all young immigrant males from the island. Unfertilised eggs were laid in several years through the early 1980s, and gradually the population petered out. Whether this outcome is good or bad was a matter for debate (Pennington *et al.* 2004); when myxomatosis reached Fetlar in 1970 the rabbit population fell and the owls transferred their attention to wader chicks. Since these included rare breeding birds such as Red-necked Phalaropes and Whimbrels, the conservation agencies were faced with a problem that was resolved for them by the loss of the owls.

Prior to *c.* 1950 there were about 50 records from Ireland, mostly in the W & NW (Ussher & Warren 1900). This decreased considerably in the second half of the 20th century, when only around eight were observed. Since then there has been an increase, with 11 birds recorded since 2002, still predominantly in the W & NW. Some remain for extended periods and may include returning individuals. There was an unsuccessful breeding attempt in the NW in 2001.

Genus *SURNIA* Dumeril

Holarctic monospecific genus that is vagrant to GB&I.

Northern Hawk-owl   *Surnia ulula* (Linnaeus)

SM *ulula* (Linnaeus)
SM *caparoch* (P.L.S. Müller)

TAXONOMY Conventionally placed in a monotypic genus, genetic studies have, by and large, supported this, although Heidrich & Wink (1998) and Wink *et al.* (2004e) found weak support for a relationship with Old World Pygmy Owls *Glaucidium*. Three subspecies recognised by HBW: *ulula*, *caparoch*, *tianschanica* on basis of degree of contrast between darkness of plumage and whiteness of spotting.

DISTRIBUTION Holarctic, breeding in clearings of boreal forests and taiga. Nominate occurs in N Eurasia, replaced by *tianschanica* in C, E Asia, south of nominate to N China, and by *caparoch* in Nearctic. Nomadic; irruptive. Dispersing S, often showing a four-year cycle determined by availability of voles. Ringing recoveries show that when these crash, birds can be found >1,500 km S of breeding areas.

STATUS Only nine records in GB, and none from Ireland. The first was caught, exhausted, on a collier in sea area Plymouth in 1830; it was identified as *caparoch*. The first *ulula* was shot near Amesbury (Wiltshire) in 1876. There have now been four *caparoch* and three *ulula*; the remaining two are unidentified to race. Seven have been in Aug–Dec, with one in Mar.

Genus *ATHENE* Boie

Palearctic, Oriental and Ethiopian genus of three or four species, one of which has been introduced into GB&I.

Little Owl   *Athene noctua* (Scopoli)

NB *vidalii* A.E. Brehm

TAXONOMY Molecular data support morphology in placing Little Owls as sister to Pygmy Owls. Essentially sedentary nature has led to a variable taxon with >12 subspecies, five of which occur in Europe.

DISTRIBUTION Occurs across Palearctic, from *c.* 57°N S to NW Africa, Red Sea and Horn of Africa; in Asia, extends S to Himalayas and C China. Resident, with only local dispersal away from natal site. Subspecies *vidalii* breeds across much of N & W Europe, from S Baltic to Iberia; replaced in C Europe, Italy by *noctua*, and by *indigena* from Balkans, Turkey to Ukraine. Many other subspecies from N Africa across Asia to China.

STATUS Not a native member of the fauna of GB&I, and there is no conclusive evidence that birds have reached the islands naturally. Introduced to England on several occasions, mostly unsuccessfully. Lever (2005) suggests that most of today's birds are descended from birds brought from the Netherlands by Lord Lilford during 1880–90. They spread rapidly across S & C England, and by the 1920s had colonised almost every county south of the River Humber (Lever 2005), and the remaining areas of England were occupied by *c.* 1950. It is still scarce in Scotland, with very few records N of the Forth/Clyde valley. There are only four records from Ireland, with the first being killed in Kilmorony (Laois) in June 1903, having been first observed the previous Feb. None from I of Man. The main centres of abundance are in lowland England, more especially in the west and south. Since most introductions were from N & W Europe, it is unlikely that subspecies other than *vidalii* have occurred.

Although the spread across Britain shows that Little Owl has the potential for rapid dispersal, it is generally sedentary. There are a few ringing recoveries within GB, but none has left GB&I, and no overseas bird has been recovered within GB&I. Lack of records from the N Isles supports sedentary nature: there are more records of Scops Owl than Little Owl in Shetland. Génot & van Nieuwehuyse (2002) suggest that dispersal is modest, both in GB and on mainland Europe, with very few beyond 50 km. First-time breeders were usually within 10 km of their birthplace.

CBC and BBS data indicate that numbers fluctuate, with some evidence of a three- to five-year cycle, perhaps linked with small mammal numbers, but that there is evidence of an overall decline since the 1980s. This is supported by results from a dedicated survey (Toms *et al.* 2000), which reported numbers declined over a three-year period: 7,468 (1995), 6,653 (1996), 6,253 (1997). Although the confidence limits were wide, the trend agrees with recent BBS data, but the possibility of association with vole numbers cannot be ignored. These figures parallel declines elsewhere in W Europe, which have been ascribed to changes in habitat (Génot & van Nieuwehuyse 2002). In GB, the loss of hedgerow trees can lead to a reduction in nest sites. Furthermore, it is known that the main food comprises small mammals, birds and larger invertebrates; increased use of pesticides may have reduced the availability of the last of these.

Genus *STRIX* Linnaeus

Nearly cosmopolitan (mainly N Hemisphere and not Australasian) genus of 12–16 species, one of which breeds in GB&I.

Tawny Owl   *Strix aluco* Linnaeus

RB *sylvatica* Shaw

TAXONOMY Mitochonrdrial and nuclear gene sequences revealed that *Strix* is monophyletic and probably sister genus to *Bubo* (Wink *et al.* 2004e). Within *Strix*, Tawny seems to be sister to Ural Owl *S. uralensis*, although statistical support for this was weak; it is also close to Hume's Owl *S. butleri* and African Wood Owl *S. woodfordii* (Heidrich & Wink 1994). Because of its wide range and limited dispersal, there is extensive interpopulation variation in colour and size, though this is confused by extensive intrapopulation variation, including polymorphisms for plumage tone. Eleven subspecies are listed by HBW, though many others have been claimed at one time or another; five of these occur in Europe. Brito (2005, 2007) analysed control region sequences in a series of Tawny Owls from 14 populations in Europe, N Africa and Asia. These regions were markedly different in DNA sequence; indeed the genetic divergence among these three regions was almost as great as between Tawny and Ural Owls. The European birds themselves comprised three clades, although there was some mis-assortment, these broadly corresponded to Iberia and W France; Italy and S France; Greece, Austria, Fennoscandia, Denmark, N France and GB. Brito ascribes this to the existence of three glacial refugia in Iberia, Italy and the Balkans, within which the owls existed for sufficiently long to accumulate genetic differences, whose footprints are still evident today, despite Tawny Owl's expansion back across the whole of the European continent.

DISTRIBUTION Widely distributed in wooded areas across Eurasia; these might be closed or open, parkland, urban or farm woods, copses and spinneys. Subspecies *aluco* breeds in N & E Europe to Ural Mts, including Balkans; replaced by *sylvatica* in Iberia, France, GB&I; by *siberiae* E of Ural Mts; and by *mauretanica* in NW Africa. Other races range from Turkey to Taiwan. Generally resident, with only short-distance movements.

STATUS Widely distributed across GB, away from the treeless uplands and most intensively farmed areas of E England; also absent from the N Isles of Scotland, the Outer Hebrides, I of Man, and never recorded in Ireland. Evidence of a contraction of range between the two Breeding Atlases, more especially in N Scotland. A dedicated survey (Percival 1990) found that the number present in a 10-km square depended more on the area of woodland than the local geography. CBC and BBS data indicate a slight increase in abundance from 1965 to 1970, perhaps reflecting a recovery from the hard winter of 1962–63. Since then, there seems to have been a modest decline, perhaps more so since 2000. Nest failures during both egg and chick stages have decreased since 1968, the former perhaps due to a reduction in the levels of organochloride pesticides following their ban in the early 1960s; more recent rodenticides (such as difenacoum, bromodialone) are apparently fatal to Tawny Owls (Galeotti 2001).

Ringing data indicate no exchange between British and Continental populations, no doubt partly explaining the species' continuing absence from Ireland. Movement between natal site and first nesting attempt is typically < 5 km, and ringing recoveries within GB were similarly small.

Genus *ASIO* Brisson

Nearly cosmopolitan (not Australasian) genus of about six or seven species, two of which breed in GB&I.

Long-eared Owl   *Asio otus* (Linnaeus)

RB WM PM *otus* (Linnaeus)

TAXONOMY Traditionally placed close to Short-eared Owl on the basis of morphology and osteology, and molecular data support this. Genetic differentiation between them is substantial, indicating a long period of separate evolution, and they are not sister species (Wink *et al.* 2004e); Long-eared appears closer to the S American Striped Owl *A. clamator*, although statistical support for this is not strong. Four subspecies recognised: two each from Nearctic and Palearctic, although one of latter is an island endemic.

DISTRIBUTION Holarctic, breeding in dense woodland, usually close to open areas for hunting. Nominate occurs across Palearctic, from GB&I and Iberia to Sea of Okhotsk and China, and from the forests of Scandinavia to Urals and Yakutia, S to Mediterranean, N Africa; absent from treeless and desert regions. Endemic race *canariensis* occurs on Canary Is. Two races in Nearctic, ranging from British Columbia and Nova Scotia, S to California and Virginia; *tuftsi* in W and *wilsonianus* in E. Northern populations are migratory, with birds wintering S of range into Mexico (New World) and Himalayas and China (Old World).

STATUS One of the more difficult species to monitor because of its relatively low density and nocturnal habits. It was recorded in 590 10-km squares containing mature conifer and scrub woodland across GB during the First Breeding Atlas, although its occurrence was lower in Wales, and it was very scarce in the unforested parts of Scotland, including the N & W Isles. There was a contraction in range of *c.* 25% between the two Breeding Atlases, especially marked in C Wales, S & E Scotland, and parts of C & E England. Previously found in *c.* 350 10-km squares across Ireland, there was a decline of >30% here. However, Glue (in Second Breeding Atlas) points out that recording nocturnal birds in conifer woods is not without difficulty, and consequently the maps 'should be interpreted with great caution'.

Brown & Grice (2005) summarise the status in England and report that Long-eared Owls were far more widely distributed in the 19th century, coinciding with the maturity of many conifer forests. A decline set in early in the 20th century, possibly associated with increased numbers of Tawny Owls (the main competitor, Mikkola 1983), which were benefiting from a reduced level of persecution. They point out that the higher densities in Ireland, and the isles of Wight and Man (all regions lacking Tawny Owls) supports this hypothesis.

Both density and breeding success within woodland sites vary among years; in Kielder Forest (Northumberland) there is good evidence that these follow a three-year cycle, with numbers and success lagging one year behind the voles that they predate (Williams 1998). In view of the prima facie evidence of a decline, or at least a contraction in range, and the associated problems of census, this appears to be a species that would benefit from regular and dedicated national surveys.

Regularly recorded in autumn at coastal sites, the GB&I population is clearly boosted in winter by Continental immigrants. Ringing recoveries indicate that many of these are from Fennoscandia, and to a lesser extent E Europe. Long-eared Owl is irruptive, and Wernham *et al.* (2002) show that arrivals at coastal observatories fit a three- to four-year cycle of abundance, with a female bias in numbers at these sites and more widely (Wyllie *et al.* 1996). Interestingly, there are few recoveries of GB-ringed birds overseas, suggesting that the local populations are largely resident.

Short-eared Owl   *Asio flammeus* (Pontoppidan)

RB MB WM PM *flammeus* (Pontoppidan)

TAXONOMY Molecular data confirm traditional taxonomic affinity with Marsh Owl *A. capensis* (Heidrich & Wink 2004). Variable taxon, with at least 10 subspecies recognised, although several of these are island endemics (e.g. Hawaï, Galapagos, Falklands), and only one occurs in the Palearctic.

DISTRIBUTION Breeds in moorland, open areas and tundra across temperate and subarctic regions of Holarctic. Nominate occurs from Iceland, GB&I, Iberia throughout Eurasia to Bering Sea, avoiding the most extreme northern fringes of Arctic Ocean, and S to the deserts of C Asia. Northern population migratory, heading S into C Africa, India, Indo-China in winter. Superimposed on the regular migrations are irruptions following peaks in small mammal cycles, when large numbers may turn up in regions where potential prey is abundant, such as river valleys, coasts and estuaries.

STATUS The Second Breeding Atlas showed that breeding is largely restricted to upland areas of GB (chiefly heather moor), from Staffordshire northwards through the Pennines to Scotland, though here absent from the highest parts of the N & W. Breeds in the Outer Hebrides on the Uists (but not Lewis or Harris) and Orkney (but not Shetland), reflecting the need for short vegetation rich in small mammals. In Wales, breeding was confirmed on the W sides of the Cambrian mountains. There were also a few pairs breeding along the quieter coasts of England, where disturbance was minimal, and breeding was confirmed on I of Man. Although a regular winter visitor to Ireland, it is not common; there was no evidence of breeding in Ireland during the atlas fieldwork, though it has bred before and since. It has been suggested that recent breeding in the SW is associated with the arrival of bank voles (probably via the port of Limerick) some time in the 1950s

Ringing data indicate that birds move away from the breeding sites during early autumn, to winter in lowland grasslands along the major river valleys and around the coasts. Populations fluctuate with the vole cycles, breeding more successfully in years when small mammals are abundant. A sizeable proportion of the native population is migratory, moving S and W in winter, with regular records at S coast watchpoints, such as Portland Bill, and ringing recoveries as far as W France and Iberia, but also single individuals in Malta and Russia (Wernham *et al.* 2002).

Large numbers come to GB for the winter, and Short-eared Owls are regularly seen along the E coast during the autumn. Ringing recoveries indicate that many of these are from Fennoscandia, although one bird ringed in Orkney was recovered in Iceland.

Genus *AEGOLIUS* Kaup

Holarctic and Neotropical genus of four species, one of which is a vagrant to GB&I.

Boreal Owl/Tengmalm's Owl   *Aegolius funereus* (Linnaeus)

SM *funereus* (Linnaeus)

TAXONOMY Three species of *Aegolius* were included by Wink *et al.* (2004e) in their analysis of owl phylogenetics; these proved to form a compact clade, sister to a large group of little and pygmy owls *Athene* and *Glaucidium*. Although statistical support for some of these findings was weak, overall they are in agreement with traditional taxonomy. Boreal Owl has been separated into around six subspecies (HBW), largely based on colour, and to a lesser extent size.

Koopman *et al.* (2005) compared the genetic structure of populations from different subspecies using microsatellite loci. They found no difference between birds sampled in Norway and E Siberia, although the size of the latter sample was modest; nor was there much variation among birds across

North America. However, there were striking differences in allele frequency between the Nearctic and the Palearctic. This indicates a lack of gene flow between the two regions and possibly a long period of isolation. Further research is needed to determine whether species limits should be revised, and whether the two groups of populations should be treated as separate species.

DISTRIBUTION Holarctic, breeding in boreal and montane forests across N America and Eurasia. Two races in W Palearctic: *funereus* over most of range and *caucasicus* in Caucasus and (perhaps) N Turkey (HBW). Three more races occur in Asia and one in N America. Eurasian populations breed in forested areas from Scandinavia to Kamchatka, and in alpine woodlands of W Europe. In N America, breeds from Alaska to Labrador, and S through Rocky Mts. Most populations are resident, feeding on forest rodents, but irruptive when populations of these fail. Very dependent on old mature trees for nesting, but readily takes to boxes where trees are too small for suitable nest holes.

STATUS This northern owl has been recorded *c.* 55 times in GB and never in Ireland. The first was shot at Whitburn (Durham) in Oct 1848. The majority of records are from northern Britain in autumn or winter. There is clear evidence of the irruptive nature of this bird, with an average of less than one a year, but three or more records in 1860, 1861, 1872, 1881, 1884, 1901. Since 1950, there have been seven confirmed records, with three in the winter 1980–81, including a bird ringed as a nestling in Norway the previous summer.

FAMILY CAPRIMULGIDAE

Genus *CAPRIMULGUS* Linnaeus

Cosmopolitan genus of nearly 60 species; one breeds in GB&I, two others are vagrant.

European Nightjar   *Caprimulgus europaeus* Linnaeus

MB PM *europaeus* Linnaeus

TAXONOMY Reviews of the Caprimulgiformes by Barrowclough *et al.* (2006) and Larsen *et al.* (2007) show that *Caprimulgus* is not monophyletic, and a that revision of the generic placement of several species is necessary. The Old World nightjars are monophyletic and probably evolved following a single expansion out of the Neotropics. More detailed investigations of Eurasian *Caprimulgus* seem not to have been undertaken, but classical taxonomy regards European as close to Sombre Nightjar *C. fraenatus* and Rufous-cheeked Nightjar *C. rufigena* of Africa, and possibly Indian Jungle Nightjar *C. indicus* of SE Asia. Six subspecies are recognised, of which two are European: *europaeus* and *meridionalis*. Most variation is clinal.

DISTRIBUTION Breeds across W Eurasia, from *c.* 60°N to Mediterranean, Turkey and Arabian Gulf, and E to about Lake Baikal. Strongly migratory, wintering in sub-Saharan Africa, chiefly in E & S, but a proportion in the W Africa, from Senegal to Cameroon. Nominate occurs in N of Eurasian range, replaced S Eurasia, from Iberia to Caspian by *meridionalis*. Four other races across Middle East, Kazakhstan to China.

STATUS Nightjars declined in abundance during the 20th century, and this has been ascribed to loss of habitat to agriculture, building and afforestation, and also to a decrease in the large aerial insects on which they depend. The reduction in both numbers and distribution that had been recognised from regional bird club data was demonstrated during the First Breeding Atlas, when the species was located in only 656 10-km squares. A dedicated survey in 1981 (Gribble 1983) revealed that many of the traditional heathland populations were in decline and that the heartland was now in woodland, especially in clear-felled areas of plantation conifer. The Second Breeding Atlas confirmed this, revealing that the losses were especially severe in Scotland and Ireland (where 88% of squares were abandoned between the Atlases), but also parts of C & S England. There was some hope for recovery, since across much of S Scotland, England and Wales, the pattern was of a general reduction in site occupancy across the range, rather than the total loss from major areas. The potential for expansion and re-colonisation by local range expansion was available

Further dedicated surveys took place in GB during 1992 and 2004, and revealed a mixed pattern (Morris *et al.* 1994, Conway *et al.* 2007). The former showed a continued decline in some counties, but overall a substantial increase from the 1,691 churring males in 1981 to 2,864 in 1992. Most birds were found in plantation forests, but lowland heath was also utilised, especially in the S & SE of England. It was evident that continued declines were in those areas where numbers were already low, suggesting that these habitats had continued to deteriorate. Other regions showed substantial increases, and many of these were in regions where coniferous plantations had reached maturity and restocking was underway, resulting in large tracts of clear-fell, over which European Nightjars could feed and display, and within which they could nest in comparative security. Morris *et al.* (1994) also showed that some populations had benefited from a violent storm in Oct 1987, which destroyed up to 75% of standing trees, leaving similar areas for colonisation in the immediately following years. The survey in 2004 revealed 4,131 churring males in 3,264 10-km squares, which was adjusted to *c.* 4,600 when allowance was made for non-surveyed habitat. This increase masked a decline in numbers and distribution in N Wales, NW England and Scotland. In Ireland, European Nightjars have been reduced to a mere handful, with under 10 pairs known to breed.

There are several specialist ringers who concentrate on European Nightjars, ringing both nestlings and flying birds trapped using tape lures adjacent to nets. Although limited, results from these exercises have revealed that birds move relatively short distances between birthplace and subsequent breeding site (Wernham *et al.* 2002). Radio-tracking indicates that they tend to forage near the nesting site, typically over adjacent heathland or young forest, but some hunted over farm pasture several km distant. There are relatively few overseas recoveries; these indicate that birds from GB&I migrate through W France and Iberia to NW Africa. Being so strongly dependent on large flying insects, European Nightjars tend to arrive later in spring, when this prey becomes available.

Red-necked Nightjar   *Caprimulgus ruficollis* Temminck

SM *ruficollis* Temminck

TAXONOMY Two subspecies recognised by Cleere & Nurney (1998), HBW: *ruficollis, desertorum,* differing in balance of brown/sandy tones to plumage.

DISTRIBUTION Very restricted distribution in SW Palearctic; nominate breeds in Iberia, NW Morocco; *desertorum* replaces this to E, from NE Morocco to Tunisia. Both races probably winter in W Africa from Senegal, Liberia to Ivory Coast, Chad.

STATUS There is only a single record from GB&I: a bird of the race *ruficollis* shot at Killingworth (Northumberland) in Oct 1856.

Egyptian Nightjar   *Caprimulgus aegyptius* Lichtenstein

SM race undetermined

TAXONOMY Plumage varies in sandy/brown balance, and birds differ in size; variation among birds of eastern isolate is clinal. Two subspecies: *aegyptius, saharae.*

DISTRIBUTION Markedly disjunct distribution. W populations (*saharae*) breed in NW Africa from Morocco to Tunisia; E populations (*aegyptius*) from Sinai across Arabia to Iran, Kazakhstan. Both races migratory, wintering respectively in W & E of Sahel zone; Sinai populations may be resident.

STATUS Only two records from GB&I: the first was shot at Rainworth (Nottinghamshire) in June 1883; the second was seen (just over 100 years later) at Portland Bill (Dorset) in June 1984. The race of neither could be established. There are no records from Ireland.

Genus *CHORDEILES* Swainson

Nearctic and Neotropical genus of five species, one of which is a vagrant to GB&I.

Common Nighthawk   *Chordeiles minor* (J.R. Forster)

SM *minor* (J.R. Forster)

TAXONOMY Only limited molecular data, but these suggest that *Chordeiles* is sister to S American Nacunda Nighthawk *Podager nacunda*, and that these are part of a clade that also includes Eurasian *Caprimulgus* (Barrowclough *et al.* 2006). Common Nighthawk varies extensively in colouration and pattern of plumage (Poulin *et al.* 1996), with eight subspecies in the AOU Checklist.

DISTRIBUTION Breeds extensively across N America, south of tundra and boreal forests. The nominate is the most northerly race, breeding from SE Alaska, James Bay, S Labrador to Georgia. Replaced by a series of other races south to Panama. Migratory, wintering across much of S America, except Andes and S Argentina.

STATUS Twenty records from GB&I since a female was shot on Tresco (Scilly) in Sep 1927; the only Irish bird was a juv seen at Ballydonegan (Cork) in Oct 1999. Thirteen have been found in Scilly, all in Sep and Oct.

FAMILY APODIDAE

Genus *CHAETURA* Stephens

Cosmopolita genus of 9–11 species, one of which is a vagrant to GB&I.

Chimney Swift   *Chaetura pelagica* (Linnaeus)

SM monotypic

TAXONOMY As part of an investigation of swiftlet phylogenies using nuclear and mt DNA sequences, Thomassen *et al.* (2003, 2005) included several species of true swifts in their analysis. Although used essentially as an outgroup, *Chaetura* proved to be sister to *Apus* and the swiftlets. Chimney Swift is thought to be closely related to Pale-rumped Swift *C. egregia* and Vaux's Swift *C. vauxi* (Cink & Collins 2002); since Thomassen *et al.* (2005) only included the S American Grey-rumped Swift *C. cinereiventris*, this conclusion cannot as yet be tested. Generally regarded as monotypic.

DISTRIBUTION E North America, from the Great Lakes to Gulf of Mexico, and from the Missouri River to the Atlantic. Migratory, probably wintering in the Andes from Ecuador to N Chile.

STATUS The first two were seen at Porthgwarra (Cornwall) in Oct 1982. There were 20 records to 2003, including a remarkable 12 in 1999, of which the 7 in Ireland that year constituted the first records there. A second (and larger) influx occurred in Oct–Nov 2005, when *c.* 15 were found in SW Ireland, and a further 6–8 in GB. Most birds are found in the SW, but there have been birds in Yorkshire, Northumberland, Fife and Borders. All but one have been in Oct–Nov. Records of Chimney Swift in GB&I have never excluded the possibility of Vaux's; however, this species is western, less common, and a short-distance migrant that has never been recorded in E USA or NE Canada.

Genus *HIRUNDAPUS* Hodgson

Palearctic and Oriental genus of four species, one of which is a vagrant to GB&I.

White-throated Needletail   *Hirundapus caudacutus* (Latham)

SM *caudacutus* (Latham)

TAXONOMY Closely related to Silver-backed Needletail *H. cochinchinensis* (HBW). Two subspecies: *caudacutus, nudipes*; latter is generally darker, especially in the forehead and lores.

DISTRIBUTION Nominate breeds from C Siberia to Japan and Korea, wintering in E Australia. A disjunct population (*nudipes*) is resident from the Himalayas to S China.

STATUS There have probably been eight individuals, since the first at Great Horkesley (Essex) in July 1846. This includes what was probably the same bird seen successively in Kent, Staffordshire, Derbyshire and Shetland in May–June 1991. The only Irish record is from Cape Clear (Cork) in June 1964. All records have been in the period May–July.

Genus *APUS* Scopoli

Palearctic, Oriental and Ethiopian genus of 17 species; one breeds in GB&I, four others are scarce migrants or vagrants.

Common Swift   *Apus apus* (Linnaeus)

MB PM *apus* (Linnaeus)

TAXONOMY Closely related to Plain Swift *A. unicolor* and Nyanza Swift *A. niansae*, which are allopatric in W & E Africa. Three species of *Apus* (Common, Alpine, Little) formed a tight clade in analysis by Thomassen *et al.* (2005), but few taxa were included outside the swiflets. Two subspecies recognised: *apus* and *pekinensis*; latter is paler in plumage, but variation is clinal (Chantler & Driessens 1995).

DISTRIBUTION Breeds across the Palearctic, from Iberia, NW Morocco, GB&I to N China, between 69°N and 28°N. Nominate occurs in W, replaced from Iran, through Himalayas to China by *pekinensis*. Both races migratory, wintering in Africa from Zaire, Tanzania to Namibia, Mozambique; *pekinensis* may concentrate more in SW (HBW).

STATUS As an aerial insectivore, swifts are dependent on a plentiful supply of insects, especially during chick rearing. The distribution maps in the two Breeding Atlases reflect this, with the highest density of both occupied 10-km squares and bird abundance in the south and east. Occupancy and abundance both decline towards the north and west, with very few breeding records from the extreme NW of Scotland, including the N & W Isles. In Ireland, the same pattern is evident, with fewer occupied squares in the west; occupancy appeared to decline by *c.* 14% between the two Atlases, but whether this is due to changes in methodology is not clear.

Swifts were not monitored by CBC surveys, and BBS is not ideal for an essentially urban, aerial species. However, data since 1994 indicate a decline in each of England, Wales and Scotland, perhaps more markedly in the last of these. Evidence from local surveys supports this (Brown & Grice 2005). The density maps of the Second Breeding Atlas show that, in all parts of GB&I, this is locally highest in urban areas, and this reflects the need for elevated holes for nesting. Changes in building construction and improved insulation of old premises in inner cities is believed to have led to a decline in availability of nest sites, with the development of Species Action Plans in several areas. Unless there is a general movement towards the provision of nest holes in new and upgraded buildings, it seems probable that the decline will continue.

Ringing data indicate that swifts from GB&I winter across sub-Saharan Africa, especially in the Congo Basin and Malawi (Wernham *et al.* 2002). There are a few recoveries from W France, Iberia and Morocco, suggesting that birds migrate through SW Europe, and some of these indicate that movements can be very rapid. Arrival in spring tends to be towards the end of Apr, when insect food is on the wing, and departure begins soon after the young are fledged.

Pallid Swift   *Apus pallidus* (Shelley)

SM race undetermined

TAXONOMY Closely related to Bradfield's *A. bradfieldi*, African Black *A. barbatus* and Forbes-Watson's Swift *A. berliozi* of S & E Africa. Three subspecies recognised: *brehmorum, illyricus, pallidus*, differing in size and darkness of plumage.

DISTRIBUTION Generally, but not exclusively, coastal, breeding in scattered populations from Mediterranean to Persian Gulf, including mountainous parts of N Africa. Nominate race is most southerly, occurring across Sahara to E coast of Persian Gulf; *illyricus* is restricted to Adriatic coasts; *brehmorum* is most northerly, occurring along both coasts of Mediterranean, and on Madeira, Canaries. Shorter-distance migrant than Common, generally wintering in Sahel, although a few easternmost birds may winter in Pakistan.

STATUS The first record lay in a museum, labelled as Common Swift, for *c.* 75 years after being collected on in Oct 1913 at St John's Point (Down), before being correctly identified in 1990. There have been two subsequently in Ireland, in Dublin Aug 1993 and Louth Apr 1998 The first for GB was at Stodmarsh (Kent) in May 1978; since then there have been over 60 further individuals. There can be little doubt that this run of records stems from increased observer skill. There are records from Mar to Nov, with peak in Oct; whether this is due to increased vagrancy at this time or because swifts are scarce in Oct–Nov is not known. According to Finlayson (1992), Pallid Swifts are present in Gibraltar from mid-Feb–early Nov, much later than Common Swifts, so the peak in Nov is probably real.

Fork-tailed Swift/Pacific Swift   *Apus pacificus* (Latham)

SM *pacificus* (Latham)

TAXONOMY Very closely related to the extremely restricted Dark-rumped Swift *A. acuticauda* of Khasi Hills of NE India, but now regarded as separate species, leaving four races: *pacificus, kanoi, leuconyx, cooki*, varying in size and patterning, especially of rump and throat.

DISTRIBUTION Breeds in arc from Kamchatka to Lake Balkash, and S to Himalayas and coastal China, Japan. Nominate occurs across N of range, wintering from Indonesia to Australia; other three races occur S of *pacificus*, from Tibet to Taiwan and SE to Indo-China.

STATUS The first record in GB&I was of an exhausted bird caught on the Leman Bank gas platform in sea area Humber, *c.* 45 miles ENE of Happisburgh (Norfolk) in June 1991 and released later the same day at Beccles (Suffolk); it was seen nearby on the following day. Three subsequent records in May 1993 (Norfolk), July 1995 (Northamptonshire) and July 2005 (Yorkshire). No records from Ireland.

Alpine Swift   *Apus melba* (Linnaeus)

PM *melba* (Linnaeus)

TAXONOMY Molecular taxonomy, though limited (Thomassen *et al.* 2003, 2005), placed Alpine Swift in same clade as Common and Little Swifts, supporting its retention in *Apus* rather than in separate genus *Tachymarptis*. However, a more detailed analysis of the true swifts is overdue. Ten subspecies are currently recognised (HBW), of which two occur in the W Palearctic.

DISTRIBUTION Wide scatter of populations from S Europe to Lake Balkash, Himalayas, peninsular India and Sri Lanka, Mediterranean, Red Sea, Rift Valley and S Africa; many of these are isolated, and differentiated in size and balance of dark/light pigmentation, leading to an array of subspecies. Nominate breeds across S Europe to Iran; *tuneti* occurs S of this, from NW Africa, then Middle East to Kazakhstan and Pakistan (HBW).These races migratory, wintering in C Africa. Other races: five in sub-Saharan Africa, three in S Asia.

STATUS Well over 500 have been recorded in GB&I since the first authenticated bird in 1820. They have been recorded in every month except Dec–Jan, with peak in Apr–June. There is a striking difference between GB and Ireland: numbers in GB peak in May, when 23% of birds have been found; in Ireland, the peak month is Mar, with 30% of records, and only 11% in May. Presumably, this reflects a difference in origin, with Irish birds, most of which are found in the S & W, originating in Iberia, and those in GB overshooting from further east.

Little Swift   *Apus affinis* (J.E. Gray)

SM Race undetermined

TAXONOMY Sometimes regarded as conspecific with House Swift *A. nipalensis*, but more usually now treated as separate species. Six subspecies recognised, based on fragmented populations and combination of size and darkness of plumage.

DISTRIBUTION Scattered populations from Morocco, Tunisia through Middle East to Iran, Pakistan, most of peninsular India, most of sub-Saharan Africa. N African populations separated as *galilejensis*, which occurs from Morocco to Pakistan, and S to Horn of Africa (HBW); replaced further south by *aerobates* and three other races S of equator. Northern populations migratory, wintering S of breeding range into C Africa.

STATUS The first record for GB&I is also the only one from Ireland: a bird found on Cape Clear (Cork) in June 1967. There have been 21 in GB since the first at Skewjack (Cornwall) in May 1981. They have been recorded from Apr to Aug and Nov, with numbers peaking in May. Although there is a slight bias towards the S, birds have been recorded as far north as Shetland.

FAMILY ALCEDINIDAE

Genus *ALCEDO* Linnaeus

Old World and Australasian genus of about 16 species, one of which breeds in GB&I.

Common Kingfisher   *Alcedo atthis* (Linnaeus)

RB MB *ispida* Linnaeus

TAXONOMY DNA/DNA hybridisation (Sibley & Ahlquist 1990) suggested that there are three major groups within the Alcedinidae that diverged very early, and molecular anaysis supports this (Moyle 2006). The two GB&I representatives, *A. atthis* and *M. alcyon* belong in different families and, although taxon sampling was necessarily limited, it is clear that generic limits in all groups need revision. Seven subspecies of *A. atthis* are recognised by Vaurie (1965) and Fry *et al.* (1992), but there is extensive clinal variation and subspecific limits are rather arbitrary. Two races occur in the W Palearctic: *ispida, atthis*.

DISTRIBUTION Widely distributed along rivers, lakes and estuaries from Iberia to islands of the W Pacific, generally between 60°N and 10°N. Northern populations move south to avoid frozen water, and species is more frequently coastal in winter. Subspecies *ispida* breeds from Iberia to European Russia, and south to Romania; *atthis* is more southerly, breeding from NW Africa and Italy, through the Mediterranean and the Middle East to Afghanistan and Siberia.

STATUS Widely distributed across GB&I during the 19th century, south of a line from Aberdeen to Kintyre, but declined in both abundance and distribution during the latter part of the century, perhaps due to a combination of killing for feathers and specimens, river pollution and hard winters (Holloway 1996). At the time of the First Breeding Atlas, the overall range was broadly similar, but the Second Breeding Atlas revealed a gentle spread north in Scotland, with evidence of consolidation in the Forth/Clyde valley and scattered pairs breeding as far as Inverness. Other regions showed a less favourable position. Industrial pollution and agricultural run-off killing or reducing prey species were blamed for declines in several areas (Second Breeding Atlas). In Ireland, the overall range changed little, although Kingfishers were recorded in fewer squares in the Second Breeding Atlas.

The CBC survey methods are generally inappropriate for Kingfishers: the density is too low, and relatively few surveys involved suitable habitat. BBS surveying is a little better, but sample sizes are small and confidence limits generally rather broad. WeBS data are equally weak for Kingfishers; they occur at a low density across a wide range of small sites that are not included in this survey. The best data relating to abundance come from the Waterbirds Survey; this targets linear bodies such as streams and canals, which are the preferred habitat. Results indicate a decline through 1975–85, followed by a near-complete recovery by 2004. The reasons for the decline are unclear; they are very susceptible to freezing conditions, and a series of cold winters that affected other birds may have played a part. However, the fact that they can have three broods a year, with four to six young per brood, gives them a marked potential for rapid recovery.

Ringing data show extensive local movement, with juvenile birds dispersing away from the natal site, and often found in sites that are inappropriate for breeding. Mortality is high at this stage, especially in cold weather, and there is evidence of hard-weather movements (Morgan & Glue 1977). Ringing data indicate that a small proportion of the GB&I population may be migratory, with a few birds being recovered in W France and Iberia; there is also evidence of movement across the S North Sea, but again numbers are very small.

Genus *MEGACERYLE* Kaup

Old and New World genus of four species, one of which is a vagrant to GB&I.

Belted Kingfisher   *Megaceryle alcyon* (Linnaeus)

SM monotypic

TAXONOMY Their distinctive morphology and anatomical differences led earlier researchers to recognise three genera of ceryline kingfishers (*Megaceryle*, *Ceryle* and *Chloroceryle*), and the molecular data support this. A phylogenetic analysis of the kingfishers shows that *Ceryle rudis* is sister to *Chloroceryle* and is not closely related to *M. alcyon* (Moyle 2006).

DISTRIBUTION Widely distributed across N America, from S Alaska and Labrador, south to California and Gulf of Mexico. Northern populations migratory; winters through Central America and Caribbean.

STATUS Three records from GB, with the first being a female shot at Sladesbridge (Cornwall) in Nov 1908. There are also three records from Ireland. The first was a first-winter female present near Ballina (Mayo) from Dec 1978 to Feb 1979, when it was shot. Three of the four subsequent birds have also been found in late autumn or winter, although in 2005, a bird was found at Milford (Staffordshire) on 1 Apr;. Remarkably, it was relocated the following day at Howden (Yorkshire), and on the 4th it had moved on to Peterculter (Aberdeenshire), where it remained until the 8th.

FAMILY MEROPIDAE

Genus *MEROPS* Linnaeus

Old World genus of 21–22 species, two of which have been recorded in GB&I.

Blue-cheeked Bee-eater   *Merops persicus* Pallas

SM *persicus* Pallas

TAXONOMY Closely related to Olive *M. superciliosus*, with which sometimes (e.g. Vaurie 1965), but not always (e.g. Fry *et al.* 1992), treated as single species. Two subspecies, *persicus* and *chrysocercus*, differing in size and colour of upperparts.

DISTRIBUTION Strongly migratory. Subspecies *chrysocercus* breeds in NW Africa and winters in W Africa; replaced further east by *persicus*, from the Nile Delta to N India, and north to Lake Balkash, wintering in S and E Africa.

STATUS Eight have been recorded in GB, since the first was shot on St Mary's (Scilly) in July 1921. With the exception of one in Sep, all have been in a narrow window in late June–mid-July. There are no records from Ireland.

European Bee-eater   *Merops apiaster* Linnaeus

CB PM monotypic

TAXONOMY Molecular data (Ericson *et al.* 2006) suggest that the bee-eaters are sister to the todies, in a clade that includes kingfishers and rollers. Within the bee-eaters, *M. apiaster* is a member of a species group that includes Rainbow *M. ornatus*, Blue-cheeked *M. persicus,* Olive *M. superciliosus* and Blue-tailed *M. philippinus* (Marks *et al.* 2007). There is little intraspecific variation.

DISTRIBUTION Breeds around the Mediterranean, from Spain and Morocco to the Middle East, and across Arabia, north to Lake Balkash and east to Kashmir. A strong migrant, wintering in southern Africa, where a small population breeds in South Africa and Namibia (Fry *et al.* 1992).

STATUS In 1955, three pairs bred in Surrey, although only two were successful (rearing seven young). In 2002, a pair bred successfully in Durham, rearing three young, and it is possible that a second pair bred in Yorkshire in the same year (RBBP). The following year, a pair bred in Herefordshire, but the nest was predated shortly after hatching. There have been other recent breeding attempts, but none has been successful.

Bee-eaters have been recorded in every month from Apr to Nov, although there is a striking difference between GB and Ireland. In the former, there is a clear peak in numbers in May–June; in the latter, although there is also a peak in May, the season is more extended, for there are significantly more records in Apr and Nov than in GB. The number of records has increased since 1950, with parties of up to 20 being recorded in some years.

FAMILY CORACIIDAE

Genus *CORACIAS* Linnaeus

Old World genus of eight species, one of which is a vagrant to GB&I.

European Roller   *Coracias garrulus* Linnaeus

SM *garrulus* Linnaeus

TAXONOMY Closely related to Abyssinian Roller *C. abyssinicus* and Lilac-breasted Roller *C. caudatus*, with which almost parapatric. Two subspecies recognised, varying in colour and size, though latter varies clinally: *garrulus, semenowi*.

DISTRIBUTION Occurs in lightly wooded savannah, forest clearings, farmland across S, C of West Palearctic, though range contracting in W Europe (BWP). Nominate breeds from Iberia to Turkey and SW Siberia, with more isolated populations in coastal NW Africa, Levant; subspecies *semenowi* breeds from Iraq, Kazakhstan to Pakistan, W China. Migratory, wintering in sub-Saharan Africa, chiefly in E & S, crossing ocean and desert en route.

STATUS The first authenticated record seems to be one killed near Crostwick (Norfolk) in May 1664. Since then, there have been around 300 records in GB, from most counties and in all months of the year except Dec. The peak months are May–June and Sep–Oct, and numbers vary from year to year with no obvious pattern. It remains a very rare bird in Ireland, with only 17 records up to 2004; the first was at Carton (Kildare) in Sep 1831, with 10 of the records dating to before 1900.

FAMILY UPUPIDAE

Genus *UPUPA* Linnaeus

Old World monospecific genus that is a scarce visitor to GB&I.

Eurasian Hoopoe   *Upupa epops* Linnaeus

CB PM *epops* Linnaeus

TAXONOMY Varies considerably in size, black-and-white patterns of wings and tail, and rufous/buff balance to plumage. Differences sometimes treated as sufficient for species status, but variations are clinal and intermediates occur between many races. Madagascar form (*marginata*) has different call, and is treated as a separate species by some. African Hoopoe also sometimes treated as separate species *U. africana*. However, most authorities regard Eurasian Hoopoe as a single species, albeit strongly variable. Nine races recognised by HBW, but only one in Europe.

DISTRIBUTION Breeds across Africa and Eurasia, S of *c.* 60°N, missing only from rainforests and most arid deserts. Subspecies *epops* occurs from Iberia and NW Africa (including Atlantic islands) almost to Lake Baikal, also NW China and NW India. Replaced by five races in Africa and three in S, E Asia. N populations migratory; European populations of nominate winter in sub-Saharan Africa, S to Kenya (HBW), but birds in N Africa are more sedentary.

STATUS Eurasian Hoopoes breed sporadically in England: Brown & Grice (2005) list >30 instances in the 150 years since 1850, including four in 1977, but there seem to be no authenticated records from elsewhere in GB&I.

Recorded annually in small numbers across GB&I, chiefly, but by no means exclusively, in the south. In autumn, there are relatively more records along the E coast and in the N Isles. This pattern is expected in spring if the birds are overshooting migrants from further south. The more easterly bias in autumn perhaps includes birds from the breeding ground in C & E Europe. Birds may be recorded well into Nov, and there are regular records of wintering individuals. Dymond *et al.* (1989) show that there was no evidence of any change in numbers from 1958 to 1985, and this lack of trend seem to have continued through to 2003 (Fraser & Rogers 2006). It is interesting that this is one of the few scarcities that has not increased in numbers during the increase in popularity of birdwatching in the latter years of the 20th century; as Dymond *et al.* point out, this may be because the distinctive appearance of the bird means that few are missed, even by non-birders.

FAMILY PICIDAE

Genus *JYNX* Linnaeus

Palearctic and Ethiopian genus of two species, one of which has been recorded in GB&I.

Eurasian Wryneck	*Jynx torquilla* Linnaeus

CB FB PM *torquilla* Linnaeus

TAXONOMY DNA evidence revealed that the true woodpeckers (Picinae) form a single clade, sister to the piculets (Picumninae) (Webb & Moore 2005, Benz *et al.* 2006). Wryneck was sister to this assemblage, indicating an ancient separation from the rest of the woodpeckers. Four subspecies are recognised by Winkler *et al.* (1995): *torquilla, tschusii, mauretanica, himalayana*, based on size, degree of brownness and barring on the underparts.

DISTRIBUTION Breeds in forested regions of the N Palearctic, between 35°N and 64°N, from N Iberia to S Scandinavia, and across Eurasia to the Yellow Sea, with isolated populations in NW Africa, Transcaspia, Kashmir. Nominate race most widespread, with *tschusii* in Italy including Sardinia, Corsica; *mauretanica* in NW Africa; *himalayana* in Kashmir. A long-distance migrant, with the western populations wintering in sub-Saharan Africa, and eastern ones in India, S China and Indo-China.

STATUS Holloway (1996) reports breeding in most English counties south of the Humber and SE Wales during 1875–1900. By the time of the First Breeding Atlas, only in SE England, and breeding was not confirmed anywhere in England and Wales during the fieldwork for the Second Breeding Atlas. In Scotland, the pattern was slightly different: no records in the 19th century, but confirmed breeding (albeit only at one site) during the Second Breeding Atlas. Peal (1968) suggested that the decline was due to the scarcity of ants, especially *Lasius flavus*, in unmanaged grassland. The appearance of Wrynecks in Scotland during the 1960s, presumably of Scandinavian origin, led to hopes of a successful colonisation. Breeding has continued, though never more than one to five pairs were found, and it has become more sporadic in recent years. There are no breeding records for Ireland, where it is much commoner in autumn than spring.

Pennington *et al.* (2004) comment that the number of spring records in Shetland fell during 1970–2000, and also less common in spring in Dorset (Green 2004) and Norfolk (Taylor *et al.* 1999). Although the latter do not comment on changes over time, it is evident that larger spring falls occurred in earlier years. BirdLife International (2004) reports declines in most countries of N Europe, so the decline in spring numbers in GB&I may stem from a general fall in the breeding population. Autumn birds may have a different origin, and there have been slight increases in both Dorset and Shetland in recent years; it remains, however, a scarce bird. In Ireland, a rare visitor with 168 individuals to 2004, the first being at Fastnet Rock (Cork) in Sep 1898.

Genus *PICUS* Linnaeus

Palearctic and Oriental genus of 14–15 species, one of which breeds regularly in parts of GB&I.

European Green Woodpecker

Picus viridis Linnaeus

RB *viridis* Linnaeus

TAXONOMY European Green Woodpecker is closely related to Grey-headed *P. canus* (Benz *et al.* 2006, Fuchs *et al.* 2007). *Picus* itself is a poorly resolved, and possibly non-monophyletic genus (Fuchs *et al.* 2007). It is part of a large clade that includes several Asian, African and North and South American genera (Webb & Moore 2005, Benz *et al.* 2006, Fuchs *et al.* 2007). However, the resolution is weak, and further investigation is needed. Five subspecies recognised by Winkler *et al.* (1995): *viridis, karelini, innominatus, sharpei, vaillantii.* The last named is often treated as a separate species.

DISTRIBUTION Occurs over much of Europe, from C Scandinavia to the Mediterranean, coastal and mountain woodlands of NW Africa, east to the Caucasus and S Caspian; more local in Turkey, Iran, Iraq. The nominate race is most widespread, being replaced by the distinctive *vaillantii* in NW Africa; *karelini* from Italy through Turkey to N Iran; *sharpei* in Iberia and the Pyrenees; and *innominata* in the Zagros Mts of Iran.

STATUS The European Green Woodpecker is well distributed in wooded areas of GB, occupying deciduous woodland, farms with large trees, and open spaces such as parks and golf courses. It expanded its range substantially during the 20th century, colonising areas of S & E Scotland, N to Aberdeenshire, though its heartland remains the well-wooded regions of southern England and Wales. However, the story is not entirely one of success, for the Atlas surveys revealed a contraction of range in N & SW England and W Wales that may have reflected changes in farm management (Second Breeding Atlas).

CBC and BBS data reveal that there has been a continuous increase in population since 1964, apart from 1975–85, when the population declined slightly; this was probably due to a run of cold winters, for European Green Woodpecker is especially susceptible to prolonged periods of freezing conditions. Overall, however, the population just about doubled during 1964–2004, a change that was evident in both England and Wales, although more profoundly so in the former; comparative data for Scotland are lacking due to the more restrictive coverage of appropriate sites. The Repeat Woodland Survey (Amar *et al.* 2006) supported these results, and found the greatest increases to have occurred at sites where spring temperature had increased most and spring rainfall increased least. These authors also found that European Green Woodpeckers had expanded into less-wooded landscapes, perhaps with the occupation of suboptimal sites as the density of birds rose.

European Green Woodpeckers do not occur in Ireland; despite its increase in GB, there seem to have been no records since three in the 1800s.

Genus *SPHYRAPICUS* Baird

North American genus of four species, one of which has been recorded in GB&I.

Yellow-bellied Sapsucker

Sphyrapicus varius (Linnaeus)

SM monotypic

TAXONOMY Nuclear (Weibel & Moore 2002b) and mitochondrial (Weibel & Moore 2002a, Webb & Moore 2005) and combined nuclear and mitochondrial (Benz *et al.* 2006, Fuchs *et al.* 2007) DNA sequences demonstrate a close affinity between the sapsuckers and *Melanerpes*. Yellow-bellied Sapsucker is closely related to Red-naped *S. nuchalis*, with which it hybridises in SW Alberta, and Red-breasted *S. ruber* (Cicero & Johnson 1995). There is some variation in plumage colour across the range, but not sufficient for subspecific differentiation (Winkler *et al.* 1995).

DISTRIBUTION Breeds in a relatively narrow zone from British Columbia to Labrador, and south through the Appalachians to Georgia. Strongly migratory, wintering in SE North America, from Texas to Panama, including many Caribbean islands.

STATUS There is only one record from GB: an immature male present on Tresco (Scilly) in Sep–Oct 1975. The sole Irish record dates from a few years later; an immature female was present on Cape Clear (Cork) in Oct 1988.

Genus *DENDROCOPOS* Koch

Nearly cosmopolitan (not Australasian) genus of about 20 species, two of which breed in GB&I.

Great Spotted Woodpecker   *Dendrocopos major* (Linnaeus)

Endemic: RB *anglicus* Hartert
PM *major* (Linnaeus)

TAXONOMY The woodpeckers have been investigated by Weibel & Moore (2002a, b) using both nuclear and mitochondrial genes. Both data sets reveal that *Dendropicos* and *Veniliornis* are nested within the main clade, which is therefore paraphyletic. There is a need for a reassessment of generic limits in this group. Unfortunately, Syrian Woodpecker *D. syriacus* was not included, so its undoubted affinity with Great Spotted cannot be quantified. However, the data also show a sister relationship between Great Spotted and White-backed *D. leucotos*. Great Spotted is very widely distributed, with a series of isolated populations (Winkler *et al.* 1995); up to 24 subspecies are recognised, based on size, bill structure and plumage, although some are poorly defined (Michalek & Miettinen 2003). A study of three mitochondrial genes among birds from 17 sites across the Palearctic found a 3% differences between samples from Sakhalin, Primor'e, Hokkaido and the rest of the Palearctic (Zink *et al.* 2002). These birds come from subspecies *japonicus*, which may well merit specific status; the birds from the rest of the Palearctic showed no evidence of genetic subdivision, probably due to a period of recent and rapid range expansion, following the last ice age. Two subspecies from the Canary Islands (*canariensis* and *thanneri*) form a distinct but recently separated group (Garcia del Rey *et al.* 2007).

DISTRIBUTION Breeds across the Palearctic, from Canaries and Morocco to Japan and Kamchatka. Subspecies *major* breeds from Scandinavia and Poland to Siberia and N Ukraine; *anglicus*: GB (though this race is not recognised by Winkler *et al.* 1995); *pinetorum*: English Channel to Italy, and east to Rivers Don and Volga; *brevirostris*: W Siberia to Sea of Okhotsk; *hispanus*: Iberia; *harterti*: Sardinia, Corsica; *canariensis*: Tenerife; *thanneri*: Gran Canaria; *mauritanus*: Morocco; *numidus*: N Algeria, Tunisia; *poelzami*: Transcaspia. Generally sedentary, though many populations, especially those from the north, show striking irruptive movements. These are believed to be due to failure of food supply, especially pine or spruce seeds (Michalek & Miettinen 2003).

STATUS This is the most widespread woodpecker in GB and may occur anywhere that there are trees. The two breeding Atlases showed that it was found in all but the least wooded areas of Scotland, such as the mountains and the northern and western islands. It was also less abundant in the uplands of England and Wales, and the intensive arable areas of E England. There was little change in distribution between the two Breeding Atlases in England and Wales, but there was a definite reduction in the number of 10-km squares occupied in Scotland. In Ireland, it is a rare winter visitor, subject to small influxes (e.g. 1949–50 and 1968–69) that consist of *major*, but *anglicus* has also occurred. Has bred in Co. Down since 2006 and probably also in Cos Antrim and Armagh; remarkably, breeding was confirmed in Co. Wicklow in 2008 with seven successful nests in 2009 (C. Murphy and R. Coombes pers. comm.). Firm evidence that they previously bred in Ireland is lacking, although may have bred when woodland was more extensive than now; bones from two caves in Co. Clare are inconclusive.

The CBC and BBS data reveal a pattern of increase, initially through the 1970s and then again subsequent to 1990. The former is usually ascribed to Dutch Elm disease, which generated large amounts of dead trees. The second increase is less well understood, but Baillie *et al.* (2006) suggest that it might be due to a combination of the maturation of recently planted forests and the provision of food in gardens, especially in winter when natural food can be hard to come by. Smith (2005) has demonstrated that there has been a marked increase in nesting success in the last two decades at a local and national scale; this is associated with a reduction of nest site competition from Starlings, which have declined massively over the same period. Whatever the reasons for the increase, the news is not all good; Great Spotted Woodpecker can be a serious predator of nestlings, especially of hole-nesting birds such as tits.

The breeding population of GB has been assigned to an endemic race *anglicus*, which is recognised by Vaurie (1965), BWP and HBW, but not by Winkler *et al.* (1995). Birds attributable (on biometrics) to the Continental race *major* are identified in small numbers along the N & E coasts of GB during irruptions, about every 12 years. Due to clinal variation in measurements of Continental birds, possible immigrants arriving in S Britain unlikely to be identifiable (Coulson & Odin 2008).

Lesser Spotted Woodpecker   *Dendrocopos minor* (Linnaeus)

Endemic: RB *comminutus* Hartert

TAXONOMY The molecular studies of Weibel & Moore (2002a, b) showed this to be part of a strongly supported clade that also includes the American species Downy *D. pubescens*, Ladder-backed *D. scalaris* and Nuttall's *D. nuttallii*. The exact relationships among these four are not entirely clear, but Weibel & Moore (2002a) suggest two scenarios: either that the ancestor of Lesser Spotted may have colonised N America, and then radiated in the latter area; alternatively, Lesser Spotted may have spread into Eurasia from a Nearctic source. Clearly more analysis is required, ideally including more of the Old World taxa. Lesser Spotted has been separated into more than 11 subspecies, many of which are poorly defined and most of which are found in the W Palearctic.

DISTRIBUTION Breeds in a belt from Iberia, GB and Scandinavia to China and N Japan. W Palearctic races (Winkler *et al.* 1995) include: the nominate, which breeds in Scandinavia and N Europe to Urals; *hortorum*: France, Switzerland, east to Poland, Hungary; *buturlini*: S Europe from Iberia to Romania; *ledouci*: Algeria; *comminutus*: Britain; *danfordi*: Greece, Turkey; *colchicus*: Caucasus, Transcaucasia; *quadrifasciatus*: SE Transcaucasia; *morgani*: NW Iran. Most populations resident, though some northern birds may move south. Sometimes shows similar irruptive behaviour to Great Spotted, but this is little quantified.

STATUS Reclusive behaviour makes it difficult to survey accurately. However, the Atlas surveys showed that it is restricted to England and Wales (never having been recorded in Ireland), and is absent from the uplands and the treeless areas of the lowlands. The distribution decreased between the two Breeding Atlases, apart from in SE England, with an overall decline of *c.* 11% of 10-km squares. CBC data indicated a slight increase in population during 1964–80, followed by a rapid decline that is also evident in BBS counts. Indeed, it is getting so hard to locate now that it is becoming rare in BBS surveys and is no longer indexed annually. The Repeat Woodland Survey supported this decline, with numbers falling by *c.* 50% between 1980 and 2004. As with Hawfinch, Lesser-spotted Woodpecker has declined more at sites with a high density of Grey Squirrel dreys, although Great Spotted is a more likely predator, and Fuller *et al.* (2005) suggest that Great Spotted Woodpeckers may also evict Lesser Spotted from their nest cavities. Finally, Lesser Spotted forages on small-diameter dead branches, and the availability of this resource may have changed. More research is needed, especially into the interactions with Great Spotted, but the combination of a scarce and declining species that inhabits a technically difficult habitat means this will be a challenge.

PASSERINES

FAMILY TYRANNIDAE

Genus *SAYORNIS* Bonaparte

New World genus of three species, one of which has been recorded in GB&I.

Eastern Phoebe   *Sayornis phoebe* (Latham)

SM monotypic

TAXONOMY Little studied. Close to Say's *S. saya* and Black Phoebes *S. nigricans*; apparently shows ecological segregation from these, and hybridisation rare or absent (Weeks 1994).

DISTRIBUTION Breeds extensively across N America, from NW Canada, the Great Lakes and the Maritimes to SE USA. Constrained by lack of nest sites in many areas: bridges, buildings and culverts being especially favoured, but also needs closely adjacent woodlands. N populations are migratory, wintering in SE USA and C America.

STATUS There is a single record from Lundy (Devon) in Apr 1987.

Genus *EMPIDONAX* Cabanis

New World genus of about 15 species, one of which has been recorded in GB&I.

Alder Flycatcher   *Empidonax alnorum* Brewster

SM monotypic

TAXONOMY Formerly considered to be conspecific with Willow Flycatcher (*E. traillii*) under the name of Traill's Flycatcher (*E. traillii*), but now recognised as separate species on the basis of vocalisations and DNA sequence. Plumage and structural differences are not diagnostic for many individuals. Genetic analysis of *Empidonax* by Zink & Johnson (1984) using enzyme systems revealed that Alder and Willow Flycatchers are each other's nearest relatives. This result was confirmed by Johnson & Cicero (2002), who found a genetic difference of *c.* 5% based upon *c.* 3,000 base pairs of mt DNA. Playback experiments indicate that they do not respond to each other's songs, supporting their specific status.

DISTRIBUTION More northerly than Willow Flycatcher, breeding in wet woodland, especially low forest and early successional woodland, from Alaska to Newfoundland. A long-distance migrant, it generally winters south of Willow Flycatcher in W South America from Colombia to Peru.

STATUS One record: a first-winter bird at Nanjizal (Cornwall) Oct 2008 (subject to acceptance).

FAMILY VIREONIDAE

Genus *VIREO* Vieillot

A Nearctic and Neotropical genus of *c.* 30 species, three of which have been recorded in GB&I. Molecular studies suggest that vireos (Vireonidae) are closely related to orioles (Oriolidae), shrikes (Laniidae), and crow-like birds (Corvidae) (Sibley & Ahlquist 1990, Barker *et al.* 2002, 2004, Beresford *et al.* 2005, Fuchs *et al.* 2006).

Yellow-throated Vireo   *Vireo flavifrons* Vieillot

SM monotypic

TAXONOMY Molecular work (Cicero & Johnson 1998) suggested that Yellow-throated Vireo is closely related to the Blue-headed Vireo *V. solitarius* group, with which it has hybridised.

DISTRIBUTION Breeds in mixed woodland across E USA; migratory, wintering from S Mexico to N South America, with a few remaining in the W Indies.

STATUS One record: at Kenidjack (Cornwall) in Sep 1990. Not recorded in Ireland.

Philadelphia Vireo   *Vireo philadelphicus* (Cassin)

SM monotypic

TAXONOMY Behavioural, enzymic and molecular data indicate close relationship with Warbling Vireo *V. gilvus*; clinal variation in tarsus length, but insufficient for subspecific separation (Moskoff & Robinson 1996).

DISTRIBUTION Breeds in broadleaved woodland across S Canada from Newfoundland to Alberta, migrating south to winter in Central America.

STATUS Two records from GB&I. The first was at Galley Head (Cork) in Oct 1985; the second on Tresco (Scilly) in Oct 1987.

Red-eyed Vireo   *Vireo olivaceus* (Linnaeus)

SM race undetermined but likely to be *olivaceus* (Linnaeus)

TAXONOMY Enzyme data (Johnson *et al.* 1988) indicate N American *olivaceus* to be conspecific with S American *chivi* taxa, and these are here treated as two subspecies groups (Cimprich *et al.* 2000). Two weakly differentiated subspecies proposed in N America (*olivaceus* and *caniviridis*) but treated here as one; up to nine subspecies in S America.

DISTRIBUTION *Olivaceous* widespread across N, E and C USA, N into Canada as far as Hudson Bay in the E and almost up to N coast in the west. Migratory, wintering in N South America from Colombia to Brazil. There appears to be a route over the W Atlantic to S America, since birds are regular, and often common, in Bermuda, though the origin and destination of these birds is unknown (Cimprich *et al.* 2000).

STATUS One of the commonest song birds of North America, and the commonest American passerine in GB&I, with *c.* 100 records in GB and >40 in Ireland, all in Sep–Nov. The first in GB&I was at Tuskar Rock (Wexford) in Oct 1951; two on Tresco (Scilly) in Oct 1962 were the first for GB. Most are found in the SW: three-quarters of the GB records are from Scilly, Cornwall and Devon. Numbers vary among years, ranging from none to 23, with an increase after 1980, when observers started to concentrate on the SW in Sep–Oct. Perhaps unsurprisingly, there is a correlation between the number found annually in GB and Ireland. The breeding range in N America expanded NE into Newfoundland during 1920–40, which might partly explain the lack of records before the 1950s.

FAMILY ORIOLIDAE

Genus *ORIOLUS* Linnaeus

An Old World genus of *c.* 25 species, one of which breeds in GB&I.

Eurasian Golden Oriole   *Oriolus oriolus* (Linnaeus)

MB PM *oriolus* (Linnaeus)

TAXONOMY Several subspecies have been proposed; Roselaar (in BWP) and Vaurie (1959) recognise only two, *oriolus* and *kundoo*, based on differences in plumage and size. These sometimes treated as separate species but no relevant molecular studies.

DISTRIBUTION NW Africa, Iberia and across Europe to S Denmark and Baltic states; E through Russia, Turkey to River Yenesei and Himalayas. Subspecies *oriolus* breeds through Europe and most of Asia, being replaced by *kundoo* in SE, Kazakhstan, Afghanistan, N India. Nominate is migratory, wintering in sub-Saharan Africa, from Cameroon, C Kenya, south to the Cape; *kundoo* is partial migrant (BWP), wintering in S of its range and through India.

STATUS A rare, but regular, breeding species, especially in black poplar woodlands of the East Anglian fens. Brown & Grice (2005) list breeding attempts since the 19th century, but accurate censusing on a national scale is difficult because of its secretive nature. Fenland birds monitored since breeding first confirmed in 1967, but even here confidence limits are wide. There seems to have been a steady increase to a peak of *c.* 30 pairs in the early 1990s, since when numbers have declined (Milwright 1998). Active management is underway, with poplar being planted for additional nesting and foraging habitat.

Elsewhere in England and Scotland, it remains a very rare breeding bird, with only sporadic records from individual counties. More often recorded as a migrant, especially in spring, and typically from late Apr to mid-June, with Scottish records about two weeks later than those in England (BS3). Fraser & Rogers (2006) report that the average annual total rose from the 1960s to the mid-1990s, since when there has been a fall. This might be due to the declining population in France (BirdLife International 2004), from whence the spring birds are likely come. In Ireland, mainly a spring passage migrant (rarer in autumn), with 184 individuals to 2003 and no confirmed breeding.

FAMILY LANIIDAE

Genus *LANIUS* Linnaeus

A Nearctic, Palearctic, African and Oriental genus of *c.* 25 species, nine of which have been recorded in GB&I, one as an increasingly rare breeder.

Brown Shrike   *Lanius cristatus* Linnaeus

SM *cristatus* Linnaeus

TAXONOMY This and the following two species have been regarded as one, two, three or even four species at one time or another (Lefranc & Worfolk 1997). Although largely allopatric, there are areas of sympatry, but there are often ecological differences, and very few hybrids involving Brown Shrike have been found. A molecular analysis might reveal more, but meantime we follow Roselaar (in BWP) in treating them as three species. Four subspecies of Brown Shrike are recognised, though variation is clinal and intermediates exist, especially in zone of contact between *cristatus* and *confusus*.

DISTRIBUTION Breeds in E Asia, from the edge of the boreal forests to inland China and Japan. Nominate is most widespread, replaced by three other races in Manchuria (*confusus*), Japan (*supercil-iosus*) and from Korea to S China (*lucionensis*). Most populations migratory, wintering from India through the islands of SE Asia; the races may be extensively allopatric in winter (Lefranc & Worfolk 1997).

STATUS Five records from GB&I. The first was at Sumburgh (Shetland) in Sep–Oct 1985; further birds occurred on Fair Isle, Oct 2000 and Whalsay (Shetland), Sep 2004 (nominate *cristatus*), with one on Bryher (Scilly) in Sep 2001. The only record from Ireland was an adult female at Ballyferriter (Kerry) in Nov–Dec 1999.

Isabelline Shrike   *Lanius isabellinus* Hemprich & Ehrenberg

SM *isabellinus* Hemprich & Ehrenberg
SM *phoenicuroides* Schalow

TAXONOMY Largely allopatric with Red-backed and Brown Shrikes, with a paler plumage, as expected from its xeric habitat. Four subspecies are recognised, falling into two groups (*phoenicuroides* and *isabellinus*: BWP), although some (e.g. Kryukov 1995, Panov 1995) treat *phoenicuroides* as a separate species, based on morphology, behaviour and vocalisations. Hybridises with Red-backed regularly (Harris & Franklin 2000) along the zone of contact, and taxonomic re-appraisal may be necessary when more data are available.

DISTRIBUTION Breeds in C Eurasia, SW of Brown and SE of Red-backed Shrikes. Westernmost birds, *phoenicuroides* (Iran to Xinjiang), are more similar to easternmost *speculigerus* (SE Altai Mts to NC China) than to the geographically intermediate *isabellinus* (NW China) and *tsaidamensis* (Qinghai). A fifth form ('*karelini*') has also been described from the lowlands, whereas *phoenicuroides* occurs at higher altitudes; since this form is usually in zones of contact between Red-backed and Isabelline, it may be a product of hybridisation. More research is needed, especially into '*karelini*'. Isabelline Shrike is migratory, wintering S and W of the breeding range, in the dry zones of C and E Africa, through coastal Arabia to Pakistan.

STATUS Over 60 have been recorded in GB since the first on Isle of May (Fife) in Sep 1950. Found in Mar and May–Nov, but over half have been in Nov. There has been a general increase in numbers down the years, presumably reflecting increased coverage and observer skill, rather than changes in abundance. Racial identification of some birds as nominate *isabellinus* and *phoenicuroides* has recently been confirmed; it is possible that other subspecies may also occur.

Two records from Ireland: at North Slob (Wexford) in Nov–Dec 2000, and Kinsale (Cork) in Oct 2006, both believed to be *isabellinus*.

Red-backed Shrike   *Lanius collurio* Linnaeus

CB FB PM *collurio* Linnaeus

TAXONOMY The western member of the Red-backed Shrike group. Some dispute over number of subspecies (if any); Roselaar (in BWP) considers variations too limited for subspecific differentiation, as do Lefranc & Worfolk (1997) though they still list four. In detailed appraisal of subspecies, Roselaar concludes that none is reliably identifiable; very minor differences in plumage may be due either to differences in wear and abrasion among the specimens or to introgression from Isabelline Shrike.

DISTRIBUTION Breeds across Europe from S Scandinavia, Finland east to about the River Yenesei, south of about 64°N. In the south, occurs from N Spain, much of Italy and Balkans; southern Turkey and across to W side of Caspian Sea. Entirely migratory, wintering in southern Africa.

STATUS Declined progressively as a breeding species through the latter part of the 20th century, from >250 pairs in the 1960s to effective extinction in the 1990s. This was at a time when Scandinavian populations were also in decline (BirdLife International 2004), although not to the same extent as GB&I. Loss of habitat may have played some part in the decline in GB, although fairly extensive areas of heathland remain in the S & E of England, and forest clearfelling creates suitable habitat, albeit of a more temporary nature. Brown & Grice (2005) suggest that the increased use of pesticides and changes in land management may have resulted in a decline in the quantity of large insects on which this species depends, and point out that the last breeding birds were on dry heathland, which is relatively unaffected by these problems. Wintering populations may also have been adversely affected by increased locust control, although there seems to be little data linking this to shrikes. While the English population was in the process of becoming extinct, in Scotland the number of breeding records increased from 1977, presumably by birds of Scandinavian origin. The species has failed to become established in Scotland, though attempts continue, especially in years with larger than usual numbers of spring migrants. A pair bred successfully in Wales in 2005 and 2006, and possibly also in 2004.

Now largely recorded as a migrant, chiefly in May–June and Aug–Oct. Much commoner in spring in Scotland and in autumn in England; Fraser & Rogers (2006) report a progressive decline in annual totals, from *c.* 250 to *c.* 190 since 1986. This occurred at a time when the number of observers was increasing, and supports the view that numbers were declining in N Europe. Very rare in Ireland prior to 1966, now recorded in most years, mainly in autumn with a few in spring, but has never bred.

Long-tailed Shrike   *Lanius schach* Linnaeus

SM race undetermined; probably *erythronotus* (Vigors)

TAXONOMY Variable in size, extent of white on primaries, and colour of head and upperparts. Up to nine subspecies recognised, several of which are island or near-island endemics of SE Asia. The form *erythronotus* has been treated as a separate species based on a lack of interbreeding with *L. schach tricolor* (Rand & Fleming 1957), although this has been disputed (Biswas 1962).

DISTRIBUTION S & SE Asian species that breeds from S Kazakhstan through Indian subcontinent and most of S China, Malay peninsula to New Guinea. Subspecies *erythronotus* occurs from Kazakhstan to N India, replaced by *tricolor* from Nepal to Thailand, and by *schach* in E & S China. Further races in peninsular India, Malaysia to New Guinea. Northern populations of *erythronotus* are migratory, wintering across N India. Other races are largely resident or altitudinal migrants.

STATUS One record: a first-winter on S Uist (Outer Hebrides) in Nov 2000; probably ssp. *erythronotus*.

Lesser Grey Shrike   *Lanius minor* J. F. Gmelin

SM monotypic

TAXONOMY Phylogenetic affinities not clear, though evidently one of the 'grey' shrikes.

DISTRIBUTION Breeds mainly in SE Eurasia; scattered sites in SW Europe through northern Balkans, Turkey to S Caspian; extends to *c.* 55°N to about Novosibirsk (Lefranc & Worfolk 1997). Range contracted through 20th century, being lost from much of France, Germany and C Europe. Strongly migratory, moving through the E Mediterranean and Middle East to winter in southern Africa.

STATUS Another SE species; much commoner in GB than Ireland. About 150 records since the first near Christchurch (Hampshire) in Sep 1842. Apart from a singleton in Jan 1907, all have been May–Nov, with peak in May–June and Sep. Numbers have remained fairly stable since 1950, averaging between two and three per year (seven in 1977). Most have been in southern and eastern counties, but have been found in GB as far north as Shetland (where regular), and as far west as Carnarfonshire. One that was ringed at Monk's House (Northumberland) in Sep 1952 came down a chimney and died in Aberdeen (Aberdeenshire) three weeks later. There are only four records from Ireland: Great Saltee (Wexford) in May 1978 and further birds in Cork (Sep 1985), Waterford (Sep 1991) and Wexford (June 1992).

Great Grey Shrike   *Lanius excubitor* Linnaeus

WM PM *excubitor* Linnaeus

TAXONOMY Closely related to Southern Grey *L. meridionalis*, Loggerhead *L. ludovicianus*, and Chinese Grey Shrike *L. sphenocercus*. Nine races usually recognised (Vaurie 1959; BWP), differing in pallor of plumage, nature of barring on underparts, extent of white in wings and tail, but revision of species limits needed.

DISTRIBUTION Three widespread Palearctic races; four much more restricted. Nominate occurs in N Europe from France, Germany, N Scandinavia, east through N Eurasia to W Siberia, absent only from the extreme north; replaced to the S by *homeyeri* from Bulgaria through the Ural Mts into W Siberia. From C Siberia to Sea of Okhotsk, *sibiricus* occurs, with four further races (*leucopterus*, *mollis*, *bianchii*, *funereus*) occupying restricted areas of C & E Asia. Two races in Nearctic.

Movements of eastern populations little known, but presumed to be similar to western birds. Majority of northern populations vacate the breeding grounds to winter in southern parts of range or beyond, with birds found across most of Europe, except Ireland, Iberia, Italy. Further east, birds may be found well south of breeding areas, into Turkey, Trans-Caspia and N & C China.

STATUS A regular migrant and winter visitor in small numbers in GB, with main arrival generally in Oct–Nov. Autumn birds chiefly along the N & E coasts, but may winter almost anywhere with suitable habitat of open ground and bushes or small trees for perching. Annual totals vary, with some evidence of a slight long-term decline, in accord with losses in breeding populations in parts of NW Europe (BirdLife International 2004). Numbers in GB&I hard to determine because of mobility and returning birds. Occasionally remains into early summer, but there is no evidence of breeding. In Ireland, 48 individuals, mostly Oct–Mar, the most recent in 1991.

Southern Grey Shrike   *Lanius meridionalis* Temminck

SM *pallidirostris* Cassin

TAXONOMY Recently recognised as a full species, though further work needed to resolve relationships of constituent taxa. Some of these may be best treated as separate species. Analysis of mt and nuclear genes indicates paraphyly of taxa in SW Palearctic and that *koenigi+algeriensis* not conspecific with *meridionalis* (Gonzalez *et al.* 2008). Variable in size and plumage, with at least 11 races recognised (Lefranc & Worfolk 1997), nine in Europe, N Africa, Middle East, and two in Asia, India.

DISTRIBUTION Widespread, though scattered, distribution from S France, Iberia, through Africa S to Sahel, across Arabia to E side of Caspian, and E to Gobi Desert and into India. Nominate occurs in SW Europe, with *algeriensis* in NW Africa, and five further races across Africa S to Chad (*theresae*, *koenigi*, *elegans*, *leucopygos*), and three in Arabia (*aucheri*, *buryi*, *uncinatus*). Variation among these is strongly clinal (Lefranc & Worfolk 1997). The most widespread and distinctive race is *pallidirostris*, which occurs through the main Asian range, apart from E Pakistan to Himalayan foothills and S into India, where *lahtora* is found.

Most races are fairly sedentary, with comparatively local dispersal; however, N populations of *pallidirostris* are migratory, moving SW to winter from Iran through to Sudan and the Horn of Africa.

STATUS Has been found 18 times in GB and IoM, but not yet in Ireland. The first was trapped on Fair Isle (Shetland) in Sep 1956; the single Manx record spent mid-June–mid-July 2003 singing near

Ballaghennie Ayres (IoM). The frequency of records has increased in recent years, as observers have become more familiar with the key characteristics, to a maximum of five in 1994. Most have been found in Sep–Dec, but there are records from Apr and June.

Woodchat Shrike   *Lanius senator* Linnaeus

SM *senator* Linnaeus
SM *badius* Hartlaub

TAXONOMY Affinities within *Lanius* unclear; in need of detailed molecular phylogenetic analysis. Three well-defined subspecies, differentiated on extent of white in primaries and on rump.

DISTRIBUTION Predominantly European, though extends into Middle East and NW Africa, breeding in open areas in or close to Mediterranean-type woodland. Nominate occurs in Iberia, S France, Italy, coastal regions of Balkans and into W Asia Minor; also the coastal areas of Morocco, Algeria, Tunisia. Replaced in Balearics, Corsica, Sardinia by *badius*, and SE Turkey to Levant and E into Iran by *niloticus*. Range contracted during 20th century, being lost or severely reduced in the Netherlands, Poland, Hungary.

Migratory, wintering in Sahel and savannah zones of Africa, from Atlantic to Red Sea; *badius* apparently chiefly restricted to coastal areas from Liberia to W Nigeria (Lefranc & Worfolk 1997), *niloticus* from Sudan to Eritrea, and also at scattered sites through Rift Valley into Kenya.

STATUS Woodchat Shrike was removed from the list of BBRC birds in 1990, by which time there had been almost 700 records, and 63 in Ireland (to 2004).

Recorded in almost all months of the year, but there is a clear peak in May–June and a lesser one in Aug–Sep. Although numbers from Scotland are rather low (<100), they seem to be found there rather later in the year, in both spring and autumn.

There have been five records of subspecies *badius* in GB (to 2005), with the first at Sizewell (Suffolk) in June 1980. One record of this race from Ireland: an adult at Mizen Head (Cork) in June 2002.

Masked Shrike   *Lanius nubicus* Lichtenstein

SM monotypic

TAXONOMY Taxonomic affinities within *Lanius* in need of study.

DISTRIBUTION Very restricted distribution, breeding from NE Greece across W & S Turkey, through the Levant to S Israel; a separate population occurs in E Iraq and W Iran, but details are sketchy. Migratory, wintering in extreme SW of the Arabian peninsula, and from Sudan to Eritrea and W into Chad.

STATUS Two records: the first was a juvenile at Kilrenny (Fife) in Oct–Nov 2004; there was another juvenile on St Mary's (Scilly) in Nov 2006. No records from Ireland.

FAMILY CORVIDAE

Genus *PYRRHOCORAX* Tunstall

A Nearctic and marginally Oriental genus of two species, one of which breeds in GB&I.

Red-billed Chough   *Pyrrhocorax pyrrhocorax* (Linnaeus)

Endemic: RB *pyrrhocorax* (Linnaeus)

TAXONOMY Morphology and osteology suggest that choughs are closely related to *Corvus* and *Nucifraga*. However, in their analysis of intergeneric relations of corvids, both Cibois & Pasquet (1999) and Ericson *et al.* (2005) found this may not be the case. Molecular sequences indicate that Alpine and Red-billed Choughs are sister taxa, associated with a clade that included various genera of treepie, and distant from the rest of the 'black' crows. A widely distributed but relatively sedentary species, with eight subspecies generally recognised (BWP), based on size and colour of the glossiness of the plumage; four occur in Europe and the remainder in Asia.

DISTRIBUTION Scattered mountain and coastal sites in GB&I, NW Africa, and across S Europe from Iberia through the Alps, Appennines, Balkans to SW Asia and the Himalayas to W & N China. Nominate restricted to GB&I; replaced by *erythrorhamphus* across the rest of Europe to N Balkans (not recognised by Svensson 1992); by *barbarus* in NW Africa and Canary Is, and by *docilis* from N Balkans through Greece and Turkey to Afghanistan. Four other races occur in C and E Asia.

STATUS Breeds in scattered locations along the W seaboard of GB and around the coast of Ireland. Its dependence on close-cropped, insect-rich grassland near to cliff nest sites has led to its widespread decrease as this habitat has been lost. Its scarcity led to increased persecution, initially for skins and latterly by egg collectors, and an alarming contraction of range in historical times. The Second Breeding Atlas revealed a concentration of 10-km squares in the W coast of Wales, across the IoM and in the southern Hebrides. In Ireland, birds bred all round the N, W & S coasts. While this represented an increase of 12% in occupied squares in GB, and comparative stability in Ireland from the First Breeding Atlas, its status remained parlous.

Local management has combined with enhanced protection to allow the beginnings of a modest recovery. Johnstone *et al.* (2007) estimated the total population of GB as 429–497 pairs. Unringed birds have been seen at several sites in SW England in recent years, probably from colonies in NW France, culminating in successful breeding in Cornwall since 2002 (two pairs in 2006). A survey in Ireland in 2002–03 found 828 pairs, a slight decrease from the figure of 906 in 1992, although not all sites could be visited because of adverse sea conditions. These overall figures mask declines of 65% in Wexford and 53% in Galway, although the population is probably stable (Grey *et al.* 2003).

Genus *PICA* Brisson

A Holarctic genus of two or three species, one of which breeds in GB&I.

Black-billed Magpie   *Pica pica* (Linnaeus)

RB *pica* (Linnaeus)

TAXONOMY Not especially close to the *Corvus* clade; closer to the ground-jays *Podoces* (Ericson *et al.* 2005). Molecular studies suggest that the Palearctic forms (*pica*, 'Eurasian' Magpie) should be treated as separate from the North American *hudsonia* ('Black-billed' Magpie), with the latter sister to Yellow-billed Magpie *P. nuttallii*. East Asian subspecies *sericea* and *jankowskii* are sister to all other *Pica* (Lee *et al.* 2003, Haring *et al.* 2007). Species limits within *P. pica* need revising. Marked interpopulation differences in size, colour of gloss to black feathers, relative tail length, and colour of rump, although these are clinal and many adjacent subspecies intergrade. Four subspecies in Europe (*pica*, *melanotus*, *fennorum*, *bactriana* – latter two not recognised by Svensson 1992); '*galliae*' doubtfully separable (BWP); with *mauretanica* in NW Africa, *hudsonica* in N America, and at least seven in the E Palearctic (BWP).

DISTRIBUTION Breeds across Palearctic, avoiding only boreal and desert regions, from Atlantic to Yellow Sea, and S into China. Disjunct populations in NW Africa, Kamchatka and adjacent areas around Sea of Okhotsk, and N America. Nominate occurs from S Scandinavia to Mediterranean Sea, and E to Carpathian Mts; replaced by *fennorum* from N Scandinavia to W Russia, by *melanotus* from SW France into Iberia, and *bactriana* E from C Russia. Largely resident, with occasional dispersion during adverse weather.

STATUS Black-billed Magpies are almost ubiquitous in England, Wales and Ireland, although they were not recorded in Ireland until *c.* 1676 (Ussher & Warren 1900). During the 1988–91 Breeding Atlas, they were recorded in almost every 10-km square in England, Wales and Ireland. In Scotland they were more restricted, being most abundant in the NE, the Central Belt and S Scotland. There was evidence of a slight increase in distribution, although much of this was in Scotland, where earlier coverage might have been less complete. Populations are known to have declined in E England between 1950 and 1965, probably due to a combination of hedge removal and pesticides (Parslow 1973). A general increase in numbers was reported across most regions and habitats from 1964 to 1993, although numbers perhaps started to level out in the 1980s (CBC; Gregory & Marchant 1996). There was some heterogeneity within farmland types, with stronger increases on mixed and grazing farms compared with arable land. This may reflect the generalist nature of corvids in general and Black-billed Magpies in particular: they can switch between food supplies as required, consequently being better adapted to the varied ecology of a mixed farm. The increases are likely also due, at least partly, to the reduction in keepering, especially following WWII, although Gregory & Marchant (1996) suggest that the introduction of Larsen traps, especially in farmland, might also be involved with the plateau.

Black-billed Magpies are largely sedentary and most movements involve relatively modest distances.

Genus *GARRULUS* Brisson

A Palearctic and Oriental genus of three species, one of which breeds in GB&I.

Eurasian Jay   *Garrulus glandarius* (Linnaeus)

Endemic: RB *hibernicus* Witherby & Hartert

RB *rufitergum* Hartert
PM *glandarius* (Linnaeus)

TAXONOMY There are *c.* 100 species of crows around the world and only a limited number have been subject to molecular analysis. The results are broadly supportive of morphological taxonomy, especially at the generic level. Ericson *et al.* (2005) analysed two of the three *Garrulus*: Eurasian Jay and Lidth's Jay *G. lidthi* of the RyuKyu Is. These were sister taxa, in the same clade as *Corvus* and *Nucifraga*. There is extensive morphological variation among populations of Eurasian Jays, with over 30 subspecies described from across the Palearctic at one time or another, although these vary in their diagnosability. Roselaar (in BWP) recognises six to eight subspecies groups, of which the *glandarius* group occupies most of Europe apart from the Dodecanese islands. Variation within GB strongly clinal.

DISTRIBUTION Breeds in woodland, predominantly deciduous, but conifers in some regions, from Iberia, NW Africa, GB&I to Japan, and south into Levant, Iran, mountains of SE Asia. European races include: *glandarius* (Fennoscandia, east to Pechora River and south to Pyrenees); replaced by *hibernicus* (Ireland), *rufitergum* (GB, N France), *lusitanicus* (N Iberia), *fasciatus* (S & E Iberia), *corsicanus* (Corsica), *ichnusae* (Sardinia), *albipectus* (Italy, W Balkans), *jordansi* (Sicily), *graecus* (W Bulgaria, S through Balkans to Greece), *cretorum* (Crete), *ferdinandi* (SE Bulgaria, Turkey in Europe), although most adjacent subspecies show clinal connections (see Svensson 1992 for more critical assessment of valid races). Many other subspecies occur across Palearctic.

Largely resident in W Europe, though young birds disperse away from natal sites in autumn. Northern and eastern populations show eruptive migrations (e.g. 1982), probably due to failures of acorn crop (BWP). These may reach as far as GB&I in exceptional years.

STATUS Eurasian Jays are widely distributed across England and Wales. In Scotland, found N to inner Moray Firth, though strongest populations in a belt from Argyll to Perth and Kinross. Distribution in GB changed rather little between the two Breeding Atlases, but there has been a substantial contraction of *hibernicus* in Ireland. From 1964 to 1993, the average density on farmland increased by up to 50%, although figures for woodland were more stable (CBC). Density is higher in the south of Britain (Gregory & Marchant 1996). BBS figures up to 2004 indicate strong population growth since the mid-1990s. Although there is little direct evidence, it is widely believed that this stems from reduced keepering. Gregory & Marchant (1996) report a reduction in the number killed by gamekeepers during 1961–89, coinciding with a period of population growth. The data are slight and correlation is not causation. A small sample of nests suggests that failure at the egg stage declined during 1964–2004.

Most movements within GB&I are local, as would be expected from the marked regional variation in morphology. There are no ringing records of exchange with European populations, but influxes occur in autumn some years (e.g. Norfolk: Taylor *et al.* 1999) and there are a few records from Shetland (Pennington *et al* 2004); these are likely to be Scandinavian *glandarius*.

Genus *NUCIFRAGA* Brisson

A Holarctic and Oriental genus of two species, one of which has been recorded in GB&I.

Spotted Nutcracker   *Nucifraga caryocatactes* (Linnaeus)

SM *caryocatactes* (Linnaeus)
SM *macrorhynchos* C. L. Brehm

TAXONOMY Molecular analysis (Ericson *et al.* 2005) indicates that *Nucifraga* is part of the *Corvus/Garrulus* clade, but not especially close to either. Morphology suggests Spotted Nutcracker is sister to Nearctic Clark's Nutcracker *N. columbiana*, but molecular comparisons have not been made. Roselaar (in BWP) recognises eight or nine subspecies, which he separates into three groups, on basis of bill dimensions, extent of white spotting to body, and of white in tail. *Caryocatactes* group occurs in N Palearctic, including nominate, *macrorhynchos*, *rothschildi*, *japonicus*. Some authorities split the Himalayan form into a separate species.

DISTRIBUTION Breeds in boreal and upland conifers across Palearctic, from S Scandinavia to Kamchatka, and from taiga forests south into C Asia, through mountains of Japan, China to Himalayas. Isolated populations in mountain forests of C and E Europe, with small populations established in Belgium and Denmark following recent invasions. Nominate is most widespread and largely sedentary, occurring in N, C & E Europe and into S Russia; replaced by more irruptive *macrorhynchos* in Siberia, from W Urals, Iran to Sea of Okhotsk and China, by *rothschildi* in S Russia and Turkestan, and by *japonicus* in Japan.

STATUS The pattern of occurrence is heavily biased by a single huge irruption: of 363 birds recorded between 1950 and 2003, 315 were found in 1968 alone, with a further 15 apparently lingering into 1969 (Hollyer 1970). The majority of invasion birds are thought to be *macrorhynchos*, but some *caryocatactes* are also involved. Irruptions are generally driven by failure of the seed crop of Siberian stone pine *Pinus sibirica*, with four major movements in the 20th century (1911, 1933, 1954, 1968). The majority of birds involved do not survive to return to Siberia. Irruptive movements confined by waterbodies and no records from Ireland.

The first confirmed *macrorhynchos* was shot at Mostyn (Flintshire) in Oct 1753; the first *caryocatactes* was shot near Northwich (Cheshire) on an unspecified date in 1860.

Genus *CORVUS* Linnaeus

A near-cosmopolitan (not S America) genus of *c.* 45 species, five of which breed in GB&I.

Western Jackdaw   *Corvus monedula* Linnaeus

RB WM *spermologus* Vieillot
WM *monedula* Linnaeus
SM *soemmerringii* Fischer

TAXONOMY Part of the *Corvus* clade and basal in the clade identified by Ericson *et al.* (2005); only four *Corvus* included in their analysis, so this finding is of limited value in assessing relationships. Generally regarded as closely related to Daurian Jackdaw *C. dauricus*, and usually regarded as comprising four subspecies. Size, shade and extent of grey on head and neck, and presence of a white crescent behind this grey all vary clinally, and with age, sex and season (Svensson 1992; BWP).

DISTRIBUTION Widespread in Europe, except extreme north; breeds across Asia, almost to Lake Baikal, and S to Iran and Kashmir. Nominate race breeds from Denmark, Scandinavia to N Balkans and Carpathian Mts; replaced in S & W Europe and Morocco by *spermologus*, from E Europe throughout Asian range by *soemmerringii* and in Algeria by *cirtensis*. Northern populations migratory, generally heading SW so that some *monedula* and *soemmerringii* winter in range of *spermologus* (BWP).

STATUS Western Jackdaws, of the subspecies *spermologus*, have increased their distribution in GB&I over the last 100 years, probably due to changes in cultivation (Parslow 1973), although they are still absent from areas of W and NW Scotland. There was little change in distribution between the two Breeding Atlases; in some areas, there was a slight decline, perhaps being due to loss of nesting sites as old buildings were either demolished or renovated. Western Jackdaws declined in woodland during 1964–93 but increased in farmland since the mid-1970s (CBC, BBS; Gregory & Marchant 1996), but sample sizes were small. More recent BBS data indicate a fairly steady increase since about 1970. This was apparent in GB, but numbers in N Ireland have fallen by *c.* 20% since 1994. Amar *et al.* (2006) found evidence that Jackdaws have declined more at sites where there were fewer dead limbs on the trees and less grassland; a link can be made between the need for nest holes in moribund trees and open grassland for feeding.

Birds are regularly seen migrating along the E coast of GB, and ringing recoveries confirm that some of these come from Scandinavia, within the range of ssp. *monedula*. Between 1997 and 2003, 124 indeterminate individuals, referable to either *monedula* or *soemmerringii*, were found in Ireland; the latter (but not the former) regarded as occasional visitor to ROI, Northern Ireland and the Isle of Man, but no accepted records of *soemmerringii* for GB.

Rook   *Corvus frugilegus* Linnaeus

RB WM *frugilegus* Linnaeus

TAXONOMY Returned as sister taxon to Northern Raven by Ericson *et al.* (2005), but Western Jackdaw and Australian Raven *C. coronoides* were the only other *Corvus* included in their analysis. Two subspecies recognised, differing markedly in feathering of lores and chin and DNA (Haring *et al.* 2007). Clinal variation in size within both races (Roselaar in BWP).

DISTRIBUTION Breeds across Palearctic from W France, GB&I to coasts of Yellow Sea; generally absent from cold and treeless north, and arid and treeless south. Nominate race occurs from Europe to River Yenesei and NW Altai Mts; replaced further east by *pastinator*. Northern populations migratory, especially in severe winters; usually wintering within breeding range, but a proportion travels further south.

STATUS Widepread across GB&I, the Breeding Atlases revealed that it occurs in most 10-km squares away from uplands, intensive urbanisation and N & W Scotland. It was poorly recorded in the CBC fieldwork, although colonies are more frequently encountered by BBS; population trend data are not available nationally before 1994. Even here, confidence limits are wide, and there is little evidence of overall change. Colonies are relatively easy to census and there have been several long-term monitoring schemes, albeit usually at a local level. These generally indicate an increase in numbers through 1930–45, but a decline of >40% through to the 1970s (reviewed by Brenchley 1986). Current populations may reflect a considerable reduction since earlier in the 20th century. In Ireland, a small and probably fluctuating population in the NE, with occasional breeding elsewhere. In past, has bred in W & SW but never became established. Has hybridised with Hooded Crow. At sites such as Cape Clear (Cork), away from the NE breeding area, remains a rare migrant.

Studies in Scotland (e.g. Dunnet & Patterson 1968) showed there to be a critical association between soil invertebrates and both population numbers and breeding success; the intensification of pesticide applications through the 1950s and 1960s might have affected Rook numbers through this link. The modest signs of recovery since the ban on organochlorides may be a response to improving invertebrates.

Rooks in GB&I are relatively sedentary, with predominantly local movements. However, there are movements of birds from N & C Europe, perhaps best described as irregular rather than migrant. The number of recoveries of foreign birds has declined in recent years, which may relate to declines on the Continent (Wernham et al. 2002), although numbers in most European countries are stable (BirdLife International 2004).

Carrion Crow   *Corvus corone* Linnaeus

RB WM *corone* Linnaeus

TAXONOMY Closely related to Hooded Crow. Molecular data inadequate to separate these taxa, but marked assortive mating in contact zones of N Italy, NW Germany and Russia, and evidence of ecological segregation and fitness loss in hybrids in N Italy (summarised in Parkin et al. 2003). Two disjunct races (perhaps better treated as distinct species) separated by Hooded Crow, but further data required on genetics and morphology.

DISTRIBUTION W populations (*corone*) occur in England, Wales & Scotland, from Schleswig–Holstein south to Iberia, across Europe to Austria and south to Alps. E populations (*orientalis*) occur from Aral Sea and C Iran to Pacific and Japan, and from edge of taiga forest to latitudes of Kashmir. Northern populations migratory, wintering generally within breeding range and may coexist with Hooded.

STATUS Carrion Crows have increased their distribution steadily across Scotland since WWII. The western end of the hybrid zone between this species and the more northerly Hooded Crow has remained more or less anchored around Glasgow. The eastern end of the zone has, however, moved from Aberdeen towards Inverness. Much of this change has been due to Carrion Crows replacing Hooded Crows in the agricultural lowlands of NE Scotland. The hybrid zone itself has remained relatively uniform in width, indicating fairly strong selection against the hybrids.

A progressive increase in both farmland and woodland across Britain from 1964 to 1993 (CBC; Gregory & Marchant 1996), although (as with Magpie) there was evidence of a plateau being reached during the 1980s. The increases were predominantly on grazed and mixed farms; numbers on arable land remained relatively stable, possibly reflecting the species' adaptation to a generalist lifestyle.

Effectively resident. Carrion Crow occurs as a rare visitor to Ireland, mainly on the east coast in spring and autumn (where presumably from GB) and it has bred (including hybridisation with Hooded Crow).

Hooded Crow   *Corvus cornix* Linnaeus

RB WM *cornix* Linnaeus

TAXONOMY See Carrion Crow. Four subspecies, differing clinally in size and shade of grey to body.

DISTRIBUTION Nominate occurs in N Europe, from Ireland, Scotland across Scandinavia to River Yenesei, and S to N Balkans, N Caspian; replaced in Italy, Balkans and from Turkey to Altai Mts by *sharpii*, and by *pallescens* in SE Turkey, Cyprus and Middle East, except Iraq, SW Iran, where *capellanus* is found. As with Carrion Crow, northern populations migratory; also leaves high ground in winter for more sheltered lower altitudes.

STATUS Distribution has changed progressively over the past 50 years, as birds have been lost from the NE of Scotland, being replaced by Carrion Crows (see for details). The Second Breeding Atlas shows that many 10-km squares on the S side of Moray Firth had been abandoned. Studies in Italy have shown that Hooded Crow is more suited to less cultivated land, and the changes in distrbution within GB could reflect the loss of such land as agriculture becomes more intensive. A steady decline in population through the 1990s (BBS), although the Irish populations remained buoyant, with a 100% increase through the same period.

Hooded Crows are more migratory than Carrion Crows, reflecting the harsher environments in which they live. Scandinavian birds have been recovered in E England, although the number of these and of sight records has declined here over the last 25 years. Perry (1946) reported seeing parties of up to 30 birds in a day on Holy Island (Northumberland) during WWII; it is now a very scarce winter visitor to this area. Climatic amelioration has been postulated as an explanation for these changes (Holyoak 1971), although the reduced population in E Scotland could also cause a reduction of winter immigrants to England.

Northern Raven   *Corvus corax* Linnaeus

RB *corax* Linnaeus

TAXONOMY An investigation of Nearctic Ravens using mt DNA and nuclear microsatellites (Omland *et al.* 2000) revealed a deep split between populations from Idaho/California and the remainder of USA. Birds from France, Russia, Siberia, Mongolia clustered within the latter, to form a Holarctic clade, and the California clade was sister to Chihuahua Raven *C. cryptoleucus*. These results suggest that the 'Northern Raven' is a complex of (at least) two cryptic species. A subsequent analysis (Feldman & Omland 2005) using additional nuclear DNA sequences confirmed this relationship and showed Pied *Crow C. albus* to be sister to the Holarctic clade.

Extensive interpopulation differences in wing and tail lengths, also relative dimensions of bill and legs; less so in plumage (BWP). Roselaar (in BWP) recognises *c.* 11 races, although there is extensive clinal intergradation between these, so many birds cannot be assigned with any degree of confidence. The molecular data indicate that species limits within the Raven need to be revised.

DISTRIBUTION Breeds widely across Holarctic, generally away from concentrated human populations, from the northern oceans to C America, NW Africa, NW India, NW China. Three races across N Palearctic: *varius* from Faroes, Iceland; nominate from GB&I, Scandinavia, France to River Yenesei; *kamtschaticus* across E Asia to Pacific, N Japan. Across S Palearctic, *hispanus* occurs in Iberia, Balearic Is; *laurencei* (formerly '*subcorax*') from Aegean through C Asia to NW China; *canariensis* from Canary Is; *tingitanus* from NW Africa. Further races in SE Asia and two in N America that appear not to correspond with the molecular divergence, although some authorities recognise a third race ('*clarionensis*'), which coincides more closely with the Californian clade (Omland *et al.* 2000).

STATUS Once widespread across GB&I, the population was reduced by gamekeepers and farmers through the 19th century, until it remained only in the more remote regions of Wales, Cumbria and Scotland (Ratcliffe 1997). As the number of gamekeepers fell through the 20th century, and with the realisation that it was perhaps not such a villain as previously alleged, the level of persecution declined, and Northern Ravens began to expand out of these refuges. It is still illegally killed in areas of high game interest. By the time of the Breeding Atlases, they were widely distributed through Ireland and N & W Britain, though there was a contraction in range of *c.* 9% in GB, particularly in areas of conifer afforestation (Marquiss *et al.* 1978). Away from this habitat, expansion has continued, and Northern Ravens have now become established in lowland areas adjacent to the main centres of distribution and along the sea cliffs of SW England.

Ravens are relatively sedentary in GB&I, with most movements being dispersals away from the breeding sites. Although ringing recoveries show that birds will cross the Irish Sea, there is little evidence of exchange between GB&I and adjacent Continental populations (Wernham *et al.* 2002).

FAMILY REGULIDAE

Genus *REGULUS* Cuvier

Palearctic, Nearctic and Oriental genus of six species, two of which breed in GB&I.

Goldcrest   *Regulus regulus* (Linnaeus)

RB WM PM *regulus* (Linnaeus)

TAXONOMY Goldcrests show extensive interpopulation differentiation in both morphology and vocalisations, such that Vaurie (1959) recognised 12 subspecies, BWP and Baker (1997) 14. Another subspecies *ellenthalerae* was recently described by Päckert *et al.* (2006). Päckert *et al.* (2003) compared differences in vocalisations and molecular variation in mt cyt-b in nine of these: *regulus, himalayensis, yunnanensis, japonensis, tristis, inermis, azoricus, sanctaemariae* and *teneriffae*, omitting *interni* (Corsica), *buturlini* (Caucasus–Azerbaijan), *hyrcanus* (Iran), *sikkimensis* (Nepal) and *coatsi* (W Siberia) from Baker's list. Some authorities (e.g. Vaurie 1959) regard *teneriffae* as a race of Common Firecrest, while others (e.g. Sibley & Monroe 1990) treat it as a species in its own right. Sequence data from Päckert *et al.* (2003) are at variance with both of these scenarios. Firstly, they found it to sit firmly in a clade with the Goldcrests, eliminating the possibility of it being a Firecrest. Secondly, this Goldcrest clade has two parts: one including *tristis* and *japonensis*, the other *regulus, azoricus, inermis* and *sanctaemariae*. Either all these are races of a single species or Goldcrest is paraphyletic, with the Atlantic forms being part of *R. regulus sensu stricto* and *tristis/japonensis* comprising a separate species. In support of the latter hypothesis, the genetic divergence between *regulus* and *tristis/japonensis* in cyt-b is 3.6–5.0% (Martens & Päckert 2003); however, *tristis* and *japonensis* themselves differ by 3.0%, and *himalayensis* is even more divergent, differing from these three taxa by 5–6%, and basal to the remaining Goldcrest taxa. The vocal analyses of Päckert *et al.* (2003) support a phylogenetic difference between the European and Asiatic taxa. Clearly there is evidence here of cryptic species akin to those found in *Phylloscopus* warblers by Irwin *et al.* (2001a), Martens *et al.* (2004) and Olsson *et al.* (2005), but further research (including playback experiments) are needed before any changes should be made to traditional species limits.

DISTRIBUTION Widely distributed across the Palearctic, from the Atlantic islands to Urals then more discontinuously to Japan. 15 subspecies: *teneriffae* Canary Islands, Tenerife and La Gomera; *ellenthalerae* Canary Islands, El Hierro and La Palma; *inermis* Azores, except Sao Miguel, Santa Maria Azores; *azoricus* Azores, Sao Miguel; *sanctaemariae* Azores, Santa Maria; *regulus* mainland Europe to Urals and Ukraine; *interni* Corsica; *buturlini* Caucasus–Azerbaijan; *hyrcanus* Iran; *coatsi*: W Siberia to Russian Altai; *himalayensis* Afghanistan to Nepal; *sikkimensis* Nepal to S Tibet; *tristis* Tadzhikistan to N China; *yunnanensis* S China; *japonensis* Manchuria, Sakhalin to Japan.

STATUS Widely distributed, having been recorded in *c.* 85% of 10-km squares in GB&I in the First Breeding Atlas. It was absent only from locations that lack trees, but even in the generally treeless areas of Orkney, Shetland and the fens, small numbers may be found. The range in GB remained almost unchanged through the Second Breeding Atlas, although there was a decline of *c.* 7% in Ireland. The latter may, however, be due to slight changes in methodology between the two surveys. Density can be high in favoured conifer habitat, with figures of 400–600 territories per sq km being recorded in Ireland (Batten 1976).

The population increased about fivefold in 1965–75 (CBC), although this may have been from an atypically low baseline following the severe winter of 1963–64. This was followed by a period of uneven decline, although Baillie *et al.* (2006) suggest that this might stem from a series of damped oscillations as the species recovers its earlier figure. Certainly, high mortality during severe winters, combined with its potentially high reproductive rate could lead to fluctuations and instability. The more recent data from BBS have indicated a steady increase in numbers from 1994 to 2000, to a plateau since then; this is likely to be more reliable since coverage of the countryside is more comprehensive and includes conifer woodland that was generally under-represented in the CBC. Scotland has seen a steady increase since 1994; England an increase to a plateau after 2000, but numbers have apparently fallen in Wales. The Repeat Woodland Suvey identified a major increase in abundance during 1980–2004, with a doubling of the population; Amar *et al.* (2006) were unable to account for this discrepancy and postulated that differences in protocol between the monitoring schemes might be responsible.

Data from Garden Bird Surveys shows a marked seasonality in abundance. Goldcrests are commoner in Oct–Nov and Mar–Apr, periods that coincide with their main migration.

Common Firecrest   *Regulus ignicapilla* (Temminck)

RB MB WM PM *ignicapilla* (Temminck)

TAXONOMY Vaurie (1959) lists four subspecies, but *teneriffae* has been shown by Päckert *et al.* (2003) to be a Goldcrest. Baker (1997) lists three races: *ignicapilla*, *balearica* and *madeirensis*. Mt DNA sequences show that the first two of these are closely similar, and well separate from *madeirensis*; Päckert *et al.* do not give estimates of genetic divergence, but the distance seems as great as between Goldcrest and the Nearctic Golden-crowned Kinglet *R. satrapa*. Vocal analysis showed little difference in song structure between *ignicapilla* and *balearica*, but *madeirensis* had a very different format. Indeed, playback studies showed that the song of *madeirensis* did not evoke a territorial reaction from the nominate form, although the Madeiran form reacted to the song of *ignicapilla* (Päckert *et al.* 2001). In view of the diagnostic differences in morphology and vocalisations, the clear differences in DNA sequence and the unidirectional lack of response to playback, Madeira Firecrest *R. madeirensis* is treated as a separate species, so that Common Firecrest comprises two races: *ignicapilla* and *balearica*.

DISTRIBUTION Firecrest is more restricted in distribution than Goldcrest, breeding from Iberia to the Balkans, and as far north as Denmark and Poland, including Mediterranean islands east of Corsica, Sardinia, and coastal Turkey. The nominate occurs on mainland Europe; populations from Germany and Poland are migratory. *Balearica* breeds in the Balearic Islands and coastal NW Africa.

STATUS There are reports of breeding in England in 1863 and 1927 (Brown & Grice 2005), but serious colonisation did not begin until 1962, when a pair bred in the New Forest (Hampshire). Following this, populations became established at several sites across S England. The Second Breeding Atlas identified 48 10-km squares with evidence of breeding, although the populations were quite strongly aggregated; thus two woodlands held >60% of known pairs. Marchant (in Gibbons *et al.* 1993) speculates that, since many of the Second Breeding Atlas records were from known sites, more populations may remain to be discovered. Breeding sites are not all monitored every year, so RBBP Reports inevitably give an unbalanced overview of status; the picture is one of increase to *c.* 100 pairs by the 1980s, and continuing at about this level, with marked annual fluctuations, until the end of the century. Numbers rose sharply after this to *c.* 350 pairs in 2006 (RBBP), but the true figure could be still higher.

Firecrest has never bred in Ireland and was first recorded in Co Cork in Dec 1943. Since the 1960s it has become increasingly regular and annual since the 1970s. Numbers vary from single figures in a year to 30+ individuals. Most are autumn migrants, although there are a few in spring and winter. Most occur in the south, but there are records from all round the coast.

Away from the breeding areas, Firecrest remains a scarce (or even rare) passage migrant in GB; it is more frequent in autumn along the E and S coasts, often in association with foraging tit flocks. Increasingly, birds are found wintering in sheltered woodlands of S & E England, very rarely N of the River Humber. Migrants pass through coastal sites such as Scilly and Portland Bill (Dorset) in Mar–May, returning in Sep–Nov, when wintering birds are beginning to take up residence. Occasionaly significant falls in excess of 100 birds occur, especially along the English Channel coasts, but it remains rare in Scotland. Ringing data give little information concerning the movement of breeding birds; there is evidence of exchange between S England and the Netherlands, Belgium and Spain.

FAMILY REMIZIDAE

Genus *REMIZ* Jarocki

A Palearctic and Oriental genus of three or four species, one of which has been recorded in GB&I.

Eurasian Penduline Tit	*Remiz pendulinus* (Linnaeus)

SM *pendulinus* (Linnaeus)

TAXONOMY Alström *et al.* (2006) report that the penduline tits form a well-supported clade with the true tits, based on a nuclear sequence, less strong using Mt cyt-b, but compelling when the data were pooled. This supports the findings of Sheldon & Gill (1996) using DNA–DNA hybridisation. The relationships between these birds and the rest of the Sylvioidea are still unresolved. Intraspecific taxonomy is complicated. Roselaar (in BWP) recognised 11 subspecies, dividing these into three groups (Eurasian *pendulinus*, Black-headed Penduline *macronyx*, and White-crowned Penduline *coronatus*), commenting that these may merit specific status, though hybridisation occurs in places. This approach was adopted by Harrap & Quinn (1996) and followed here, while recognising the need for further analysis.

DISTRIBUTION *Pendulinus:* Europe to Ural and Caucasus Mts; *menzbieri:* C Turkey to Iran, Azerbaijan; *caspius:* SE Russia to Caspian Sea; *jaxarticus:* E of Urals to W Siberia, N Kazakhstan and Altai Mts. Northern populations are migratory, southern ones less so; some birds winter south of breeding range (e.g. Israel, Nile Valley, S Iran).

STATUS There have now been *c.* 200 in GB since the first at Spurn (Yorkshire) in Oct 1966. New birds found in every month except Feb (although long-staying birds have been present then too), with a peak in Oct, declining through to Jan, and then a small resurgence in Apr. The number of records rose in the 1990s, to a maximum of 17 in 1997, since when they seem to have fallen back. This increase led to the hope that colonisation would follow, and records of lone males building nests strengthened this to an expectation as yet unrealised. One at Pett Level (Sussex) in Oct 1988 had been ringed in early May of the same year in S Sweden. Two further birds that had been ringed in Sweden in 1997 and 2004 were retrapped in Kent and Sussex the following winters, confirming that some of our records involve dispersing birds from Scandinavia. Not recorded from Ireland. A party of four (male, female and two juveniles) were at Dingle, Suffolk in Nov 2007.

FAMILY PARIDAE

Several molecular studies have indicated that the genus *Parus* should be rearranged. The Paridae are certainly monophyletic (Gill *et al.* 2005); however, there are several clearly separate clades within the family that require a reassigning of generic limits. Gill *et al.* (2005) recommend six major genera: the grey tits and chickadees (*Poecile*), North American crested tits (*Baeolophus*), Old World crested tits (*Lophophanes*), the coal tits (*Periparus*), the 'core' tits (*Parus*), and the blue tits (*Cyanistes*). Six members of the Paridae have been recorded in GB&I, incorporating five of the six proposed genera.

Genus *CYANISTES* Kaup

Largely Palearctic genus of two to four species, one of which breeds in GB&I.

Blue Tit   *Cyanistes caeruleus* **(Linnaeus)**

Endemic: RB *obscurus* Přázák
WM *caeruleus* (Linnaeus)

TAXONOMY Fifteen subspecies have been recognised, which divide broadly into two groups: *caeruleus* (nine subspecies) from Europe and the Middle East, and *teneriffae* (six) from N Africa and the Canaries. Salzburger *et al.* (2002a) analysed seven taxa, three of Blue Tit and four of Azure Tit *C. cyanus*. The Azure Tits formed a clade that was sister to the Blue Tit ssp. *caeruleus*; Blue Tits from N Africa (*ultramarinus*) and the Canaries (*degener*) were sister to this grouping, lying outside the Azure/*caeruleus* clade and differing by *c.* 4.9%. Salzburger *et al.* (2002a) proposed that the Canary and N African populations be treated as a separate species; a view supported by vocal and plumage differences. The molecular evidence does not support the separation of Blue and Azure Tits since their sequences are very similar, suggesting 'African'and 'European' Blue Tits diverged first, and Azure Tit separated from the latter at a later date. A subsequent study of the Canary Is populations by Kvist *et al.* (2005) used a faster-evolving mitochondrial sequence (the control region). Their results supported treating these as a separate species from the rest of the European birds. A further study found genetic support for the retention of the island subspecies and recommended that birds on Gran Canaria should be recognised as *hedwigii* (Dietzen *et al.* 2008a).

While examining variation within birds from S France, Taberlet *et al.* (1992) found evidence of two distinct mitochondrial sequence groups. Kvist *et al.* (1999a, 2004) examined a transect from Barcelona (Spain) to Oulu (Finland). They found that this phenomenon was restricted to the south of the transect; from Germany northwards, they only found one of these groups. They postulated this was a consequence of a prolonged period of isolation, perhaps into glacial refugia in Iberia (*ogliastrae*) and the Balkans (*caeruleus*). Following the last ice age, the Balkan isolate expanded north and west across C & N Europe, until it came in contact, and hybridised, with the Iberian birds in S France. The results of this can be seen today through the genetic fingerprint of the eastern birds in the SW populations. In a further analysis of the mt control region, Kvist (2003) examined genetic variation among British Blue Tits (ssp. *obscurus*) and compared this with birds from elsewhere in Europe. She found slight evidence of divergence, but the sample size was small and all birds came from a single ringing station in Scotland; the study needs extending and repeating.

DISTRIBUTION Breeds across the W Palearctic, from Iberia to Kazakhstan, as far north as C Fennoscandia, and south to Lebanon. Treating the Canary Is and N African birds as a separate species *C. teneriffae* (Sangster 2006), and continuing to recognise Azure Tit, leaves nine subspecies of Blue Tit: *caeruleus* (Europe, from N Spain to Fennoscandia, and E to Ukraine), *obscurus* (GB&I, Channel Is), *balearicus* (Majorca), *ogliastrae* (S Iberia, Corsica, Sardinia), *calamensis* (S Greece to Crete, Rhodes), plus a further four from European Russia through to Iran, Kazakhstan. Generally resident in S of range, but northern populations are more migratory, generally heading S and W to avoid worst of winter food shortages. Occasionally extensively irruptive.

STATUS The resident race *obscurus* is widely distributed across GB&I, and was found in *c.* 90% of 10-km squares in the two Breeding Atlases, absent only from the highest tops in Scotland and the majority of the Outer Hebrides. The number of birds estimated from CBC and BBS showed an increase of about 35% between 1967 and 2003, but this was not uniform across the region. Scotland (up 46%), Wales (30%) and N Ireland (80%) all increased more than England (10%). The Repeat Woodland Survey (Amar *et al.* 2006) also found steeper increases away from SE England. However, numbers vary quite markedly among years, and the trends are by no means uniform over time. Blue Tit is one of the most commonly recorded birds in gardens, and the increased provision of food and nest boxes may have driven the increase in this particular habitat. However, productivity at urban sites is lower than elsewhere, especially in terms of clutch size and brood survival (Cowie & Hinsley 1987; Siriwardena *et al.*1998).

In common with many other resident species, laying date has advanced since the mid-1970s, and failure rate at the egg stage has declined. Whether these are adequate to have driven the observed increases in population has not been investigated.

The occurrence of the nominate race is difficult to quantify; brighter birds appear on the east coast in autumn in particular, and substantial irruptions occur in some years (e.g. 1957), when birds are found in numbers in the northern isles of Scotland, where they do not breed (Pennington *et al.* 2004). These are presumably all of Scandinavian origin.

Genus *PARUS* Linnaeus

Old World genus of over 20 species, one of which breeds in GB&I.

Great Tit   *Parus major* Linnaeus

Endemic: RB *newtoni* Prăzák
RB WM *major* Linnaeus

TAXONOMY The Great Tit has long been regarded as an example of a ring species around the Tibetan plateau (e.g. Mayr 1963), but (as with 'Herring' Gull) recent molecular analysis has cast doubt on this (e.g. Päckert *et al.* 2005). Over 33 subspecies have been described (Harrap & Quinn 1996), which can be divided into four principal groups (*major, minor, cinereus, bokharensis*), or three if Turkestan Tit *P. bokharensis* is treated as a separate species (Harrap & Quinn 1996). Kvist *et al.* (2003) undertook a comprehensive study of Great Tits at 23 sites from Iberia and Morocco to Japan. All four subspecies groups were included. Each proved to be monophyletic and clearly distinct, indicating long periods as independent evolutionary lineages. They found evidence of sharp differences in genetic structure, rather than progressive clinal change, as might be expected under the ring species hypothesis. A combination of diagnosable morphological differences between groups and genetic divergences of 4–5% are comparable to those between related taxa currently treated as separate species, so it might seem appropriate to treat the four lineages as distinct species: *P. major* (Great Tit), *P. minor* (Japanese Tit), *P. cinereus* (Cinereous Tit), *P. bokharensis* (Turkestan Tit). However, a more recent, and more extensive, analysis by Päckert *et al.* (2005) casts doubt on the diagnosability of *bokharensis*, finding a closer similarity with *major* in DNA. Until this difference is clarified, it seems prudent to recognise three species, retaining the *bokharensis* group within *P. major*.

Eleven subspecies are recognised in the *major* group by Harrap & Quinn (1996), of which eight occur in Europe: *major* (mainland Europe, from Fennoscandia to Lake Baikal, and south to Mediterranean, Middle East, except where replaced by following), *newtoni* (GB&I), *corsus* (S Iberia, Corsica), *mallorcae* (Balearics; not recognised by Svensson 1992), *ecki* (Sardinia), *aphrodite* (S Italy, S Greece, Cyprus), *niethammeri* (Crete; not recognised by Svensson 1992); further races occur in N Africa and the Middle East. An investigation of European populations of subspecies *major* (Kvist *et al.* 1999b) showed relatively little divergence, indicating a rapid range expansion following the last ice age, which may still be underway (Orell 1989). An investigation of the genetic status of ssp. *newtoni* (Kvist 2003) also found little divergence between British and adjacent European stocks, though (as with Blue Tit) the British sample was limited in number and came from a single site in Scotland.

DISTRIBUTION Great Tit (*sensu stricto*) occurs across the Palearctic, from the Atlantic, through Europe, Turkey, S into Iran, and across C Siberia to the Pacific. Generally resident, with most European birds existing within a few km of their birth place. However, northern birds show evidence of long-distance dispersal of juveniles (which may aid the observed limited genetic divergence across W Europe) and, as with other species that are seed-dependent in winter, populations may undertake extensive irruptive movements when the food supply fails.

STATUS Great Tit was recorded in over 85% of 10-km squares in both Breeding Atlases; it was only absent from the generally treeless areas of the N & W islands of Scotland. There was little evidence of change between the two surveys, though there was a progressive increase in numbers from 1964 to 2004 across GB&NI, with two short periods of comparative stability in the mid-1970s and mid-1980s (CBC, BBS). In common with several other woodland species, laying date has advanced since the mid-1980s, and there have been declines in both clutch size and nest failure at the egg stage over the same period.

Despite this being one of the best-studied birds in the world, the causal links between population parameters and environmental effects are still only imprecisely known. However, availability of beech mast in winter and caterpillars in summer will be critically important in some locations.

In common with Blue Tits, Great Tits are adaptable and take readily to both garden feeding and nest boxes. They are a familiar sight at bird tables, and were reported in *c.* 70% of gardens during summer and over 80% during winter. As with Blue Tit, breeding success is lower in gardens.

British Great Tits are largely resident and generally move only short distances. Formerly, the species had periodic irruptions during years with beech mast failures, with birds coming from the S Baltic and Low Countries, but there has not been an irruption for some time now, probably because of climate change and expecially winter feeding in gardens.

The endemic subspecies *newtoni* breeds across much of GB&I, intergrading with the nominate in the SE of England, although the intergrade zone has been little studied.

Genus *LOPHOPHANES* Kaup

A Palearctic and Oriental genus of two species, one of which breeds in GB&I.

European Crested Tit   *Lophophanes cristatus* (Linnaeus)

Endemic: RB *scoticus* Prǎzák
SM *cristatus* (Linnaeus)
SM *mitratus* (C. L. Brehm)

TAXONOMY Forms a sister species to Grey Crested Tit *L. dichrous* of the Himalayas. Five subspecies recognised (Harrap & Quinn 1996), based on the rufous/buff balance in the plumage; others apparently less convincing.

DISTRIBUTION N & W Europe, from Iberia to the Balkans, north to N Fennoscandia, and E to Carpathian Mts. Nominate breeds from Fennoscandia through W Russia to Bulgaria and E & S Balkans; replaced by *scoticus*: Scotland; *weigoldi*: Iberia; *abadiei*: NW France; *mitratus*: rest of W & C Europe, S to Serbia and Montenegro.

STATUS European Crested Tits are restricted to mature Scots pine forest in N Scotland, principally within the watershed of the River Spey, along the inner Moray Firth and N of the Great Glen. An apparently similar area of woodland in the valley of the River Dee is not populated, although birds occasionally cross the watershed from Strathspey. Crested Tits need open woodland with a thick understorey of heather and standing dead wood for nest sites (Summers 2000; Summers & Canham 2001). Territories are relatively large compared with other tits, and the clutch size is lower, perhaps reflecting the difficulties of living in a relatively hostile environment. During the 1990s, the winter population was estimated as 5,600–7,900 individuals.

Elsewhere in GB&I, European Crested Tits are very rare; the Scottish race has never been found south of the Scottish Central Belt. Never recorded in Ireland or Wales, the few individuals found in England that were racially identified proved to be *mitratus* (pre-1844, Yarmouth (Isle of Wight)) or *cristatus* (1872, Whitby (Yorkshire)). The remaining 11 records were all from S & E England; none was identified to race.

Genus *PERIPARUS* Selys-Longchamps

Palearctic and Oriental genus of about six species, one of which breeds in GB&I.

Coal Tit   *Periparus ater* (Linnaeus)

Endemic: RB *britannicus* (Sharpe & Dresser)
Endemic: RB *hibernicus* (W. Ingram)
CB PM *ater* (Linnaeus)

TAXONOMY Twenty subspecies have been described (e.g. Vaurie 1959), of which 13 occur in the W Palearctic. Racial differences largely based on colour saturation, but this is clinal and there is extensive intrapopulation variation (Harrap & Quinn 1996). Molecular analysis of tits by Gill *et al.* (2005) included two subspecies of Coal Tit, *aemodius* and *ater*, and found that these were not sister taxa; the former was sister to Spot-winged Tit *P. melanolophus* of the Himalayas. Martens & Eck (1995) suggest that these two races of Coal Tit are conspecific because they hybridise extensively in Nepal, in which case *melanolophus* should perhaps be included in *P. ater*. See Dickinson & Milne (in press) for author of *hibernicus*.

DISTRIBUTION Eight subspecies occur in W Europe (BWP), including *hibernicus*: Ireland except extreme NE; *britannicus*: Britain and NE Ireland; *ater*: widespread across Palearctic, from France to Sakhalin, S to a line from Ukraine to Mongolia; *vieirae*: Iberia; *sardus*: Corsica, Sardinia; *cypriotes*: Cyprus. Not all authorities recognise *hibernicus*. Two further races occur in N Africa, six from Turkey through SW Asia to Azerbaijan and six in the Far East. Birds in the S and W are largely sedentary, but northern and eastern populations are prone to irruptive behaviour, especially following failure of food crops.

STATUS Two subspecies breed regularly in GB&I: *britannicus* (GB) and *hibernicus* (Ireland). Nominate *ater* probably breeds occasionally after irruptions, e.g. on Scilly in the late 1970s, though this is based on territorial males and subsequent sightings of young birds: no nests were found (Robinson 2003). The distribution of *britannicus* changed little between the two Breeding Atlases; it was found in the majority of 10-km squares outside the N & W islands of Scotland and the fenlands of E England. Density seemed highest in areas of extensive conifer forest, and up to 100 birds per square km have been recorded in favoured sites. As with several other woodland species, numbers have varied since the 1950s, with a steep increase from 1965 to 1975, followed by an irregular plateau. This pattern is different from Blue and Great Tits, both of which show a more uniform rise in numbers from 1965 to 2005. Evidence from the Repeat Woodland Survey (Amar *et al.* 2006) suggests that the increase in Coal Tits varied across GB&I, with larger increases in the Midlands and SW regions of England. This study found that Coal Tits had increased at sites that were dominated by birch, and the authors suggest they might be spreading out of the traditional conifer habitat. Census data from gardens shows that numbers of this species peak in midwinter, but the maximum varies among years, perhaps in response to the size of the beech mast crop. In a Scottish garden, McKenzie *et al.* (2007) found higher numbers of Coal Tits in poor Sitka spruce cone years, a result not found with Great or Blue Tits, which feed preferentially in broadleaved woodland. The species is not trapped in sufficient numbers for robust data to be available on breeding success.

Little is known about the detailed status of *hibernicus*, although there is a suggestion of a contraction in range of *c.* 3% between the two Breeding Atlases. This may reflect slight changes in methodology.

Birds of the nominate race take part in irruptive movements that may be related to beech mast abundance; certainly, birds recorded as *ater* appear in autumn in Shetland and at other coastal sites, where Coal Tits are normally scarce. Ringing data confirm that some of these are of Continental origin (Wernham *et al.* 2002).

Genus *POECILE* Kaup

A Nearctic, Palearctic and marginally Oriental genus of *c.* 12 species, two of which breed in GB&I.

Willow Tit   *Poecile montana* (Conrad von Baldenstein)

Endemic: RB *kleinschmidti* (Hellmayr)
SM *borealis* (Selys-Longchamps)

TAXONOMY Conventional taxonomy has separated Willow Tit into *c.* 15 subspecies; four of these (*songara, weigoldica, affinis, stoetzneri*) have been separated on morphological grounds as Songar Tit *P. songara* by, among others, Harrap & Quinn (1996). Kvist *et al.* (2001) examined the hypervariable mt control region in 11 subspecies of Willow Tits from across the Palearctic. They found sharp differences between subspecies *affinis, songara* and an aggregate of the remaining nine, which formed a molecularly uniform group. These results support the genetic distinctness of Songar Tit, but indicate that this may, in fact, be two taxonomic units. Since the differences between these three groups are intermediate between populations and species of *Poecile*, Kvist *et al.* (2001) were unable to recommend splitting Songar Tit from Willow.

Genetic variation was also analysed by Salzburger *et al.* (2002b), who used the mt cyt-b gene to examine a series of samples from Europe, China and C Asia that spanned nine subspecies. The birds fell into four groups corresponding to subspecies *affinis, songara, weigoldica* and a fourth that included six Eurasian subspecies. These findings supported those of Kvist *et al.* (2001) but extended them by showing that *weigoldica* formed yet another grouping. Again the differences were more akin to levels of subspecific differentiation, and the authors incline towards retaining the races of 'Songar Tit' within *P. montana.*

DISTRIBUTION Continuously distributed across the Palearctic from GB to Kamchatka and the extreme E tip of Siberia; also in China. W Palearctic subspecies include *kleinschmidti* (Britain), *rhenana* (W Europe to NW Germany, N Switzerland; synonymised with next by Svensson 1992), *salicaria* (from N Germany, Poland to Austria), *borealis* (Fennoscandia, European Russia to Ukraine), *montana* (C & S Europe, to Bulgaria). W populations are generally sedentary, though N populations show occasional irruptive movements, superimposed on more regular dispersal S & W to avoid extremes of winter climate.

STATUS Does not occur in Ireland. In Scotland largely confined to SW, but formerly believed to be more widely distributed (though confusion possible with Marsh Tit). Elsewhere in GB, as with Marsh Tit, evidence of a fairly uniform decline with the Breeding Atlases; Willow Tit was lost from *c.* 10% of 10-km squares between the two Atlases. There might have been an increase in numbers after the first Atlas, with a decline back to a similar level for the second (CBC). This decline continued relentlessly through to the end of the 20th century. This was more severe in woodland and farmland than in wetter habitats (such as riverine woodland and damp scrub in or adjacent to wetland), where there was effectively no change (Siriwardena 2004). Constant-effort ringing indicated a decline in the production of juveniles during 1984–2003. The Repeat Woodland Survey confirmed the changes in that habitat and found that the declines were greatest in closed middle-aged woods.

Similar trends in wet habitats may indicate common responses to the environment, though circumstantial evidence for predation by Great-spotted Woodpeckers in farmland (Siriwardena 2004). Perrins (1979) suggested that Willow Tit is competitively inferior to other tits, but Siriwardena (2004) was unable to find any statistical evidence that increased numbers of (in particular) Blue and Great Tits impacted on Willow Tits. In general, Willow Tits were commoner at sites where potential competitors were present. Great-spotted Woodpecker is an important predator of hole-nesting birds, being able to break open the tree to extract eggs or young. Since it excavates a hole in soft, dead wood, Willow Tit might be at greater risk than species using more robust nesting trees. Siriwardena concluded that the key may lie in habitat differences in winter. Marsh and Willow Tit forage close to the ground, and such habitat may be declining due to either (or both) increased deer grazing or the closing of woodland canopy as trees mature. The enthusiasm for 'tidying up' woodland by removal of dead limbs that might otherwise cause injury to humans could also play its part by the removal of suitable nest trees. The species undertakes only limited movements within Britain.

Two English records of northern race *borealis*, (Gloucestershire, Mar 1907; Yorkshire, Feb 1975); currently under review.

Marsh Tit   *Poecile palustris* (Linnaeus)

RB *dresseri* (Stejneger)

TAXONOMY In study based on mt cyt-b, Gill *et al.* (2005) examined 12 species of *Poecile*, including two subspecies (*palustris*, *brevirostris*) of Marsh Tit. These differed by less than 1% and were sister to Willow Tit, although with weak statistical support. Populations vary in size, length of tail, and plumage, especially of upperparts and flanks; much of this is clinal. Harrap & Quinn (1996) recognise nine subspecies, of which five (the *palustris* group) occur in the W Palearctic: *dresseri*, *palustris*, *stagnatilis*, *italica*, *kabardensis*, and the remainder (the *brevirostris* and *hellmayri* groups) in the East. The genetic differences between these appear slight (Gill *et al.* 2005), although these are based on only one subspecies each.

DISTRIBUTION Two widely separate regions in E & W Palearctic (data from Harrap & Quinn 1996). Subspecies *dresseri*: NW France, GB; *palustris*: Continental Europe from S Fennoscandia to Hungary, Bulgaria, Greece: *stagnatilis*: replaces *palustris* in E Europe, W Russia, N Turkey; *italica*: French Alps, Italy; *kabardensis*: N Caucasus. Claims that birds in N England and Scotland are nominate (BWP) require confirmation. In E Palearctic, from C Siberia to Japan, and south to N China. Generally resident, though northern birds may wander in winter.

STATUS Marsh Tit breeds in England and Wales and is largely confined to SE Scotland. Absent from Ireland, apart from a single record from Bray (Wicklow) in Dec 1990. There was an unsuccessful introduction to Tipperary in the 1940s or early 1950s (P Smiddy pers. comm.).

Marsh and Willow Tits are both giving serious cause for concern. The Marsh Tit decline was already evident in the Second Breeding Atlas, where a loss of over 17% was recorded in occupied 10-km squares. Numbers fell progressively after 1964 (CBC/BBS), although there is some evidence of a plateau being reached in the 1990s. The CBC data started after the severe winter of 1962–63 and, if Marsh Tit is susceptible to cold weather, this decline may have been from an already low mark. Ringing data support the decline and subsequent plateau (Perrins 2003). Laying date has advanced by about 10 days since the mid-1980s and losses at the egg stage have increased, although not at the nestling stage (though sample sizes are small). The Repeat Woodlands Survey broadly supported these results (Amar *et al.* 2006) and found evidence that declines were greater at more elevated sites.

Causes of decline in Marsh and Willow Tits are probably not the same (Perrins 2003). Marsh Tits have declined in three major areas: wet habitats, farmland, woodland. Competition with other tits, especially for nest holes, may be important, but this is difficult to quantify. Nest predation by Great-spotted Woodpeckers may also play a part, although there is no evidence of increased nest loss (Siriwardena 2006). The increase in deer numbers in southern Britain is having adverse effects on woodland understorey and, since Marsh Tit spends most of its life in this habitat, there may be a link here. Although there is no direct evidence to test this hypothesis, Willow Tit also feeds low down and its numbers are similarly falling, unlike Great and Blue Tits, both of which generally forage higher in tree canopy and are not declining. If the selection pressure acts through habitat structure, it is presumably more significant in winter, since most coexisting tits feed on similar items during the summer. Siriwardena (2006) has commented on the extensive tree planting in lowland GB during the last decades of the 20th century. Provided this is managed appropriately, and deer numbers controlled, this could develop into mature woodland, and the Marsh Tit might recover its former status.

FAMILY PANURIDAE

Genus *PANURUS* Koch

Monospecific Palearctic genus that breeds in GB&I.

Bearded Reedling   *Panurus biarmicus* **(Linnaeus)**

RB WM PM *biarmicus* (Linnaeus)

TAXONOMY Previously regarded as a parrotbill (e.g. Vaurie 1959; Dickinson 2003). Molecular analysis shows its affinities are closer to the larks (Ericson & Johansson 2003; Alström *et al.* 2006). Data still inconclusive for inclusion in the Alaudidae (Alström *et al.* 2006), it is neither a tit nor a parrotbill. Three subspecies recognised by Roselaar (in BWP) and Svensson (1992), based largely on clinal variation in colour.

DISTRIBUTION European distribution is rather fragmented, being largely restricted to reedbeds and similar dense vegetation close to water; apparently more continuous in Asia. Subspecies *biarmicus* breeds in W Europe, from Iberia to Austria, NE Balkans, Greece; *kosswigi* bred in S Turkey (perhaps extinct: BWP); *russicus* replaces nominate from Austria, NE Balkans to S European Russia, and from Caspian to China. Largely resident, though shows patterns of irruptive dispersal, with birds being recorded outside breeding range, permitting colonisation of new reedbeds.

STATUS This inhabitant of reedbeds has suffered from a combination of habitat loss or degradation and severe winter weather, which can cause substantial mortality. Despite these setbacks, there has been a progressive expansion through the latter part of the 20th century, with colonies now established at *c.* 50 sites in GB. A national survey in 2002 recorded 504–559 pairs, an increase of *c.* 50% since 1992 (RBBP); most of these were in England, but a bird was present in suitable habitat in Wales and *c.* 20 pairs bred in Scotland. First Irish record was in Louth in 1966. The species bred in Wicklow during the 1970s and 1980s. Last recorded in Ireland in 1990.

Two factors assist rapid recovery and re-colonisation. Firstly, Bearded Reedlings have relatively large broods and an extended breeding season, so population growth can be high. Secondly, they undertake dispersal movements at the end of the breeding season, with recoveries (or recaptures) up to 400 km from the ringing site. Although many return to their natal site, the potential for rapid colonisation is evident, and the limiting factor to further expansion may be suitable habitat. Existing reedbeds need to be managed to maintain their structure; neglect can lead to drying out and the encroachment of scrub, with a consequent loss of winter food. Conservation agencies across GB&I are now creating new sites in an attempt to expand the range of Bearded Reedlings and other reed specialists, such as Great Bitterns.

FAMILY ALAUDIDAE

Genus *MELANOCORYPHA* Boie

Palearctic genus of six species, four of which have been recorded in GB&I.

Calandra Lark   *Melanocorypha calandra* **(Linnaeus)**

SM race undetermined

TAXONOMY Shows slight clinal variation in size and paleness of plumage, shape of dark centres to feathers on upperparts and extent and intensity of spotting of breast. Four subspecies recognised by HBW, three by BWP, two by Svensson (1992); last of these regards '*gaza*' and '*hebraica*' as not separable from *psammachroa*.

DISTRIBUTION Discontinuous populations along both sides of Mediterranean, from Iberia to Balkans, NW Africa to Egypt; N populations extend through Turkey, Levant to S Russia, Iran, Kazakhstan. Nominate race occurs from N Africa, Europe to Kazakhstan; replaced by *psammachroa* across Middle East to Turkmenistan. Most populations resident, though some individuals migrate S to Nile Valley, S Iran. E populations partially or completely migratory, wintering from S Russia to N Africa (HBW).

STATUS Thirteen records, all except one in GB. The first was at Portland (Dorset) in Apr 1961. Apart from this, and birds on St Kilda (Outer Hebrides) in Sep 1994 and Spurn (Yorkshire) in Oct 2004, all have been on islands between mid-Apr and mid-May (Scilly, Shetland, Farnes (Northumberland), Isle of May (Fife), Scolt Head (Norfolk)). There is a single record from I of Man in May 1987, but none from Ireland.

Bimaculated Lark   *Melanocorypha bimaculata* (Ménétriés)

SM race undetermined

TAXONOMY Slight variation in plumage tones, Svensson (1992) recognises two subspecies: nominate, NE Turkey to C Asia, and *rufescens* from S & C Asia Minor and Near East.

DISTRIBUTION More easterly than Calandra, from C, E Turkey through Iran, E Caspian to Lake Balkash. Most populations migratory, wintering S to NE Africa, parts of Middle East, SE Iran to NW India.

STATUS Three records to 2004, all from GB. The first was on Lundy (Devon) in May 1962. The others were on St Mary's (Scilly) in Oct 1975 and Fair Isle (Shetland) in June 1976. There are no records from Ireland.

White-winged Lark   *Melanocorypha leucoptera* (Pallas)

SM monotypic

DISTRIBUTION Extensively sympatric with Black Lark from SW Russia, across the steppe to Lake Balkash; bulk of population in Kazakhstan. Most populations migratory, wintering in grassland and stubble as far S & W as Black Sea.

STATUS Two records, both from GB. The first was a female caught near Brighton (Sussex) in Nov 1869; the second was seen near King's Lynn (Norfolk) in Oct 1981. There are no records from Ireland.

Black Lark   *Melanocorypha yeltoniensis* (J. R. Forster)

SM monotypic

DISTRIBUTION Breeds in a relatively narrow belt of steppe from SW Russia across much of N Kazakhstan (where most birds reside) to Lake Balkash. Most populations resident, but N populations nomadic or short-distance migrants, apparently moving S & W to Black Sea to avoid snow cover.

STATUS Two records, both from GB. The first was present at Spurn (Yorkshire) in Apr 1984, although its identity was not established for several years; the second was present at South Stack (Anglesey) in June 2003. Both were males. There are no records from Ireland.

Genus *CALANDRELLA* Kaup

Old World genus of ten species, two of which have been recorded in GB&I.

Greater Short-toed Lark   *Calandrella brachydactyla* (Leisler)

PM race(s) undetermined

TAXONOMY Formerly treated as conspecific with Red-capped Lark *C. cinerea* of Arabia and Afrotropics. Clinal variation in plumage colours, fitting broadly with Gloger's Rule: continental populations are greyer or sandier, and less streaked; those in more mesic habitats are darker and more richly coloured; however, wear and abrasion complicate these differences (HBW), and boundaries between races sometimes arbitrary (BWP). Eight subspecies recognised by BWP and HBW, but rationalisation may reduce this.

DISTRIBUTION More or less continuously distributed in dry, short-vegetation habitats across S Eurasia and coastal NW Africa from the Atlantic to C China. Absent from wooded areas and the most arid steppe and semi-desert. Most populations migratory, W populations wintering in Sahel zones of sub-Saharan Africa, those from further E in S Asia. Nominate race occurs through much of European range, replaced in Hungary, N Balkans by *hungarica* (which may not be recognisable, Svensson 1992), and by *rubiginosa* in N Africa. Five more races across Asia, becoming greyer and less intensively streaked.

STATUS The first for GB was caught near Shrewsbury (Shropshire) in Oct 1841; there were over 400 records up to 1993, when it was removed from the BBRC List. The first for Ireland was shot at Blackrock Lighthouse (Mayo) in Oct 1890; it was 61 years before the second, since when there have been over 50 individuals, with a tendency towards increasing spring records. The pattern of occurrence is broadly similar across GB&I, with records in every month except Feb, peaks of occurrence in Apr–June and Sep–Nov, and about twice as many records in the latter. Birds are geographically widely spread, although there are concentrations in the regular migrant hotspots of Shetland, Orkney, Scilly, Saltee (Wexford), Cape Clear and Dursey (both Cork).

Lesser Short-toed Lark   *Calandrella rufescens* (Vieillot)

SM race undetermined

TAXONOMY A highly variable species, with >15 subspecies described from across the range. Affinities still somewhat confused, with several forms recently split, and arguments for eastern races being treated as separate species as well. In need of a thorough, including molecular, review.

DISTRIBUTION Breeds in habitats with bare ground, grass and moderate shrub layer, from Azores, NW Africa, Iberia along S shores of Mediterranean, through Ukraine, Turkey and Middle East, across C Asia to Mongolia and NE China. Many populations sedentary, though birds from C & E of range migrate to avoid harshest of winter climate. Subspecies *apetzii* occurs in Iberia, *rufescens* and *polatzeki* in Canaries, *heinei* from Ukraine to Kazakhstan, and a series of other subspecies across N Africa and Asia.

STATUS The only record from GB was at Portland (Dorset) in May 1992. A series of records from S Ireland in 1956–58 appeared in the literature for many years. These records (42 birds, including a flock of 30) were re-examined by IRBC in 1999–2000 and found no longer to be acceptable.

Genus *GALERIDA* Boie

Old World genus of six species, one of which has been recorded in GB&I.

Crested Lark   *Galerida cristata* (Linnaeus)

SM *cristata* (Linnaeus)

TAXONOMY Closely related to Thekla Lark *G. theklae*. Its largely sedentary nature has led to extensive variation, with >35 subspecies described, based on plumage colour and streaking as well as bill and wing lengths. Variation much complicated by wear, bleaching and apparent adaptation to local soils. Populations in NW Africa form a phylogenetically distinct group (Guillaumet *et al.* 2005, 2006).

DISTRIBUTION Widespread in dry flat-lands across much of C & S Europe, N and sub-Saharan Africa, coastal regions of Arabia, and across C Asia to N China. Absent from the mountainous and desert regions of Asia. Nominate race occurs in much of European range, but replaced by *pallida* (Iberia), *meridionalis* (S Italy, Balkans and Turkey) (Svensson 1992). Many more races across Asia.

STATUS Has declined across adjacent areas of Europe (BirdLife International 2004), but still breeds regularly a relatively short distance away across the English Channel. Despite this, only about 20 birds have been recorded since the first was shot near Littlehampton (Sussex) before 1845, including two birds together in Cornwall in 1846. There are records from Mar, Apr, June and Sep–Dec, with the majority along the S coast between Kent and Cornwall. With little change in occurrence since the 1800s, it remains a seriously rare bird. There are no records from Ireland.

Genus *LULLULA* Kaup

Monospecific Palearctic genus that breeds in GB&I.

Woodlark   *Lullula arborea* (Linnaeus)

RB MB PM *arborea* (Linnaeus)

TAXONOMY Close to *Alauda* on morphology, but phylogenetics little studied. Two subspecies described, differing in paler plumage of southern form.

DISTRIBUTION Breeds in parkland and on heath and new plantations where there is open ground, across W Palearctic, from S Finland to NW Africa, and E to W Russia and Caspian Sea. Nominate breeds S to N France to N Balkans; *pallida* occurs in S & E, from NW Africa through Mediterranean, Turkey to Caspian. Generally resident, though northern populations undergo short-distance migrations to S & W.

STATUS At the time of the Second Breeding Atlas, the Woodlark population had reached a nadir, with less than 100 occupied 10-km squares across GB (it does not breed in Ireland), reflecting a decline of 60% from the First Breeding Atlas. Red Listing on the grounds of range contraction was followed by a dramatic change in fortunes. While still too scarce to be reliably monitored by CBC and/or BBS fieldwork, a dedicated survey in 1997 revealed a sixfold increase since 1986 across S & C England: changes in the number of occupied territories included <6 to >90 (Dorset), 6 to *c.* 60 (Berkshire), 1 to 30 (Nottinghamshire) and a near 10-fold increase in East Anglia. Part of the success might be due to increases in breeding success (Baillie *et al.* 2006), but the range has expanded as forestry plantations have been felled and replanted, providing the open ground with scattered bushes that it favours. The opportunistic effects of storm damage (such as the gales of Oct 1987) opening up clearings in the forests, and more sympathetic heathland management have also played their part. An estimated population of 3,064 territories was calculated following the 2006 national survey, which represented an 89% increase from 1997 (new estimate of 1,633 territories in 1997 calculated to permit a direct comparison). The range increased further to 131 10-km squares; 46% more than 1997 (G Conway pers. comm.). In Ireland, Woodlarks formerly bred in several counties in S, E and NE, but were extinct by 1900. Only two records of breeding in 20th century (most recently 1954), and only two records of singles in the 35 years up to 2002.

Largely resident, though some birds move to the near-Continent in winter. Woodlarks are now established in many parts of GB from which they had been long lost.

Genus *ALAUDA* Linnaeus

Palearctic genus of three or four species, one of which breeds in GB&I.

Eurasian Skylark   *Alauda arvensis* Linnaeus

RB WM PM *arvensis* Linnaeus

TAXONOMY Closely related to Oriental Skylark *A. gulgula* and has been treated as conspecific; molecular analysis not yet available. Populations very variable in colouration and metrics, and 12–13 subspecies are listed by Vaurie (1959), BWP, HBW, but much variation is clinal and Eurasian Skylark may be over-subdivided.

DISTRIBUTION Broadly distributed in open habitats such as grassland, farmland, parks, woodland clearings, across Palearctic from Scandinavia to NW Africa, and east to Kamchatka and Japan. Absent only from dense woodland and deserts and semi-deserts of C Asia. Nominate race widespread across N Europe to Ural Mts. Replaced in NW Iberia by *guillelmi*; C & S Iberia by *sierrae*; across S Europe and Turkey to Caucasus by *cantarella*; SE European Russia through C Asia to NW China by *dulcivox*. Other races in N Africa, C & E Eurasia. SW populations are sedentary, but N & E birds are extensively migratory, wintering in S of breeding range and beyond through N Africa, Middle East and C Asia.

STATUS More than any other bird, Eurasian Skylark epitomises the dramatic fall in farmland bird populations. Fieldwork for the Breeding Atlases showed this to be one of the most widespread species in GB&I, breeding in over 90% of 10-km squares in GB and *c.* 73% in Ireland. However, these figures masked a decline of *c.* 3% in occupied squares. The true extent of the collapse in population was revealed in the CBC/BBS data, which show a drop of *c.* 60% in numbers, mostly during 1978–2000. Perhaps more than for any species since the Whitethroat in the early 1970s, this sounded alarm bells that there was a problem with avian populations and an intensive study began into the decline. The results of these researches are reported in detail by Donald (2004), and seem to be largely due to the adverse effects of changes in agricultural practice. The change from spring to autumn sowing of cereals had the dual effect of reducing the opportunites for breeding later in the season, when the vegetation is too high, and eliminating the stubble fields that had provided essential food resources during the winter. Surprisingly, during the early years of the population collapse, breeding success increased, perhaps because of reduced intraspecific competition. Latterly, this has changed and the gains in clutch and brood size have been lost, although fluctuations in these do not necessarily translate into variations in breeding success.

Serious attention is being given to resolving the problems faced by Eurasian Skylarks. Where possible, farmers are being urged to revert to spring-sown cereals, and to leave stubble fields for winter foraging (not just for Eurasian Skylarks). A preliminary study of winter wheat by Field *et al.* (2007) found that leaving more crop residue on the surface of the ground ('conservation tillage') allowed Eurasian Skylarks to commence nesting a month earlier than conventional sowing, and potentially extend the breeding season. Where this is not feasible, they are encouraged to leave unplanted patches within the crop to provide nesting and feeding habitat in high summer; preliminary results indicate this may also be successful (Donald 2004). Key to these conservation measures (and again not just for Eurasian Skylarks) are agri-environment schemes whereby farmers are paid to maintain appropriate wildlife habitat.

Genus *EREMOPHILA* Boie

Nearctic, Palearctic and African genus of two species, one of which has been recorded in GB&I, occasionally staying to breed.

Horned Lark/Shore Lark   *Eremophila alpestris* (Linnaeus)

CB WM *flava* (J. F. Gmelin)

TAXONOMY The exact evolutionary relationships of *Eremophila* to other genera have not been subject to a molecular investigation, but morphology and plumage suggest affinities lie with *Galerida* and

Alauda larks. Shows extreme variation in size (especially wing length) and plumage, leading to >40 described subspecies, of which 27 are Nearctic). At one time, Temminck's Lark *E. bilopha* was also included, but differences in morphology and plumage led to its separation. Morphology suggests a primary separation into Nearctic and Palearctic subspecies, although three of the latter, while geographically disparate, appear closer to the former. Within these two groups, there is evidence of further associations, but the picture is confused and confusing: this taxon is in need of a modern phylogenetic analysis.

DISTRIBUTION Holarctic, breeding across much of the Americas, N of Colombia, plus the northern tundra and the southern mountains of the Palearctic. In addition to subspecies *flava*, which breeds across Eurasia from S Norway to Lake Baikal, several races exist that are potential vagrants: e.g. three that breed in NE Canada (*alpestris, hoyti, praticola*) plus *balcanica* from SE Europe. The northern populations are migratory, largely wintering within the range of more southern races; *flava* generally migrates SW to winter along the coasts of W Europe (whence its alternative English name of Shore Lark) and from E Europe eastwards into C Asia.

STATUS A very rare breeding species. Summering birds occasionally occur on mountain tops in N & W Scotland, where they sometimes breed; a fledged juvenile in 2003 was the first confirmed breeding since 1977 (RBBP).

Regular winter visitor in small numbers, especially to E coast of GB, usually on quiet areas of shingle or saltmarsh, often with Snow Buntings and Twites. Rare in Ireland with only 18 individuals, most recently in 1998; uncommon in Wales away from the River Dee mouth and Anglesey. Largely a passage bird through Scotland, though small numbers winter along the east coast. Most years, numbers are modest, but occasionally influxes occur (e.g. early 1970s and late 1990s). Colour ringing has shown that some birds show marked site fidelity, both within and between winters (Wernham *et al.* 2002). Similar studies have shown that birds migrate along the east coast, with colour-ringed individuals being seen further north through the spring.

FAMILY HIRUNDINIDAE

Genus *RIPARIA* Forster

Old and New World genus of four or five species, one of which breeds in GB&I.

Sand Martin   *Riparia riparia* (Linnaeus)

MB PM *riparia* (Linnaeus)

TAXONOMY A comprehensive study of hirundine phylogenetics has been undertaken by Sheldon *et al.* (2005), sequencing one nuclear and two mitochondrial genes, and screening 75 of the 84 species currently recognised. Three of the five *Riparia* were analysed; Banded Martin *R. cincta* was sister to *Phedina* (Mascarene and Brazza's Martins). Sand Martin was found to be sister to Brown-throated Martin *R. paludicola*, and part of a clade that included *Progne* and *Tachycineta*, the New World martins and tree-swallows, but not Banded Martin, which lay outwith the clade. This suggests that generic limits may need to be redefined.

Five subspecies are recognised by HBW, based chiefly on size, greyness of plumage and clarity of breast-band. Pale Martin *R. diluta* is now treated as a separate species.

DISTRIBUTION Breeds widely across the Holarctic and Nile Valley, usually in sand and mud banks close to water; absent from extreme north and arid regions of S Asia. Nominate is most widespread, replaced by local races in S & E Asia and Africa. All populations migratory, wintering in S America, sub-Saharan Africa outside the rainforests and deserts, and SE Asia.

STATUS Sand Martins are widely distributed across GB&I, apart from the islands of N & W Scotland, nesting in the banks of sand and gravel quarries, and in riverbanks. Since many of these are only temporarily stable, movement between sites is frequent; entire colonies may disappear between years, making accurate census very difficult. However, Sand Martins began to decline in abundance and distribution during the late 1960s, and were lost from 25% of 10-km squiares between the two Breeding Atlases. Since many winter in the same Sahel zone of Africa as Whitethroats, and since the declines were contemporary, it is likely that the drought there was similarly to blame.

Attempts to survey the populations were undertaken by a few dedicated observers (usually ringers) such as Cowley (1979), who documented a decline of 90% between 1968 and 1984, with half of the birds being lost during 1968–69. The BTO Waterbirds Survey is perhaps the most robust national data set, although it was not initiated until the late 1970s, by which time Cowley's decline was well under way. Nevertheless, this shows a continued decline until about 1985, after which there was a recovery. This seems to have been short lived and, after peaking in the mid-1990s, the population appears to be in decline again. Szép (1995) has shown that over-winter survival of Sand Martins from C Europe is negatively related to rainfall in the wintering grounds and that this affects the breeding population in the following year. Cowley & Siriwardena (2000) have shown that summer rainfall adversely affects survival, but this bird is one of a guild of Sahelian-wintering insectivores whose fate may be largely determined outwith our islands.

Ringing studies have revealed much about its movements; most birds head south through W France and Iberia to the wintering grounds in W Africa. There has also been exchange with the adjacent Continent, indicating that migrant birds may cross SE England en route to or from the breeding grounds from Norway to the Low Countries.

Genus *TACHYCINETA* Cabanis

Old World genus of nine species, one of which has been recorded in GB&I.

Tree Swallow   *Tachycineta bicolor* (Vieillot)

SM monotypic

TAXONOMY Sheldon *et al.* (2005) found a well-supported clade of New World tree-swallows.

DISTRIBUTION Breeds across N America in open wooded areas with suitable tree-holes for nesting. Northern limits are generally set by the tree-line, through it will nest beyond there if boxes are provided. Populations breed south to Texas, New Mexico. Migratory, wintering along the coasts from S USA through C America to Colombia.

STATUS There are two records, both from GB; the first was on St Mary's (Scilly) in June 1990, and the second for just a single day on Unst (Shetland) in May 2002. There are no records from Ireland.

Genus *PROGNE* Boie

New World genus of nine species, one of which has been recorded in GB&I.

Purple Martin   *Progne subis* (Linnaeus)

SM race undetermined

TAXONOMY In their analysis of swallow molecular phylogeny, Sheldon *et al.* (2005) found that *Progne* formed a well-supported clade, sister to the rough-winged swallows *Stelgidopteryx*. *P. subis* has no particularly close relatives within the genus (Moyle *et al.* 2008). Populations vary in wing and tail lengths, and also in whiteness of forehead and underparts of females, and three or four subspecies are recognised (Brown 1997).

DISTRIBUTION The nominate race breeds more or less continuously across E North America, from the Great Lakes to the Gulf of Mexico, with an extension into Alberta; there is a series of discontinuous populations elsewhere, from S tip of British Columbia to N Baja California (*arboricola*) and thence through C Mexico (*hesperia*). Strongly migratory, wintering in the lowlands of S America, apparently mostly migrating through C America, although recorded commonly in Bermuda (Brown 1997).

STATUS The only record for GB&I was on Lewis (Outer Hebrides) in Sep 2004 (the first for the Azores occurred the following day). A specimen of a female in the Natural History Museum, Dublin, was reported shot near Dun Laoghaire (then Kingstown) (Dublin) a short time previous to Mar 1840. It is included in Irish Category D1 (see Appendix 2).

Genus *PTYONOPROGNE* Reichenbach

Old World genus of three or four species, one of which has been recorded in GB&I.

Eurasian Crag Martin   *Ptyonoprogne rupestris* (Scopoli)

SM monotypic

TAXONOMY Three species of crag martin were included by Sheldon *et al.* (2005) in their analysis; they found these to form a well-supported clade, sister to, but distinct from, the barn swallows. Pale Crag Martin (*obsoleta*) is regarded as a fourth species by some (e.g. Vaurie 1959). All species have in the past been regarded as conspecific. However, the breeding ranges overlap (Turner & Rose 1989). These deliberations do not affect the British and Irish lists, since only Eurasian has been recorded here; this taxon is distinct from the Rock *P. fuligula* and Dusky Crag Martins *P. concolor*, and shows little interpopulation variation.

DISTRIBUTION Breeds in mountains and cliffs across the Palearctic from Iberia and NW Africa, through the Mediterranean, across Arabia through Kazakhstan almost to Lake Baikal, and through the Himalayas to the mountains of C China. Alpine and Asian populations are migratory, wintering in W Africa, Nile Valley and peninsular India.

STATUS The first was at Stithians Res (Cornwall) in June 1988, and the second at Beachy Head (Sussex) two weeks later in July; it is possible the same bird might have been involved. Further birds were recorded in 1989 (Caernarfonshire) and 1995 (Beachy Head). Another was seen at Swithland Res (Leicestershire) in Apr 1999, and (presumably the same bird) in Yorkshire the following day. A further bird (or possibly still the same) was seen at Finstown (Orkney) two weeks later in May, and there was a late bird in Oct 2006 in Surrey. There are no records from Ireland.

Genus *HIRUNDO* Linnaeus

Cosmopolitan genus of 14 species, one of which breeds in GB&I.

Barn Swallow   *Hirundo rustica* Linnaeus

MB PM *rustica* Linnaeus

TAXONOMY Sheldon *et al.* (2005) included 12 'barn swallows' and six 'red-rumped swallows' in their analysis, and found these to form two quite separate clades. In the past, they have been regarded as congeneric, but the molecular data indicate strongly not only that they are distinct but also that they are not especially close: the barn swallows were sister to the crag martins, and red-rumped to cliff swallows. Barn Swallow was phylogenetically close to Red-chested *H. lucida*, Ethiopian *H. aethiopica* and Angola *H. angolensis*, and they also found that New and Old World birds differed by as much as other taxa currently regarded as full species.

This study was paralleled by Zink *et al.* (2006a), who studied interpopulation phylogenetics by investigating the relationships of 13 Eurasian and six Nearctic populations using nuclear and mitochondrial gene sequences. The mt DNA data revealed three major clades that corresponded well with geography: Europe to Kazakhstan; Asia; N America. The Asian and N American clades were sister taxa; within the N American clade was a subclade of birds from the region around Lake Baikal. The nuclear data were less decisive: the pattern was broadly similar but statistical support was weaker, suggesting that the evolutionary divergence had occurred sufficiently recently that the more slowly evolving nuclear sequences had not diverged as markedly as the mitochondrial. Zink *et al.* interpret their findings as showing that Barn Swallow evolved from an ancestral Africa taxon and spread across the Holarctic; geographical isolation led first to the divergence of the W Palearctic populations, followed closely by the splitting of the E Palearctic and Nearctic populations. More surprisingly, it seems that, subsequently, birds from N America colonised the Baikal region: the genetic signatures and some plumage characteristics of these two groups are almost identical.

There is considerable interpopulation variation in morphology, which has led to a series of subspecies being described, although these only partly reflect the molecular results.

DISTRIBUTION Holarctic; breeds in farmland, usually in buildings and often associated with water; absent from tundra and taiga regions of far north. The nominate race occurs across the Palearctic from the Atlantic to C Russia and W China; four races have been described from further east, and two more from Egypt and Middle East. In N America, only *erythrogaster* is found. There is scope for a more detailed analysis of plumage differences, especially comparisons between the morphology of the Baikal and Nearctic populations. Almost entirely migratory, wintering in S, C America, sub-Saharan Africa, India, SE Asia to N Australia.

STATUS Fieldwork for the Breeding Atlases showed Barn Swallow to be extremely widespread in GB&I, breeding in *c.* 97% of 10 km squares in Ireland, and >92% in GB; the only change between the surveys being a slight expansion of range in the N & W isles of Scotland. Numbers recorded during CBC/BBS surveys have shown fluctuations, with a peak around 1980, followed by a steep drop through to 1985; there has been a subsequent recovery back towards the 1980 figure, although perhaps not as strongly in Scotland and N Ireland as in England and Wales. As with several other species, the falls in population can be linked with adverse climatic conditions on the wintering grounds (Baillie & Peach 1992). However, Mead (in Wernham *et al.* 2002) suggests that loss of nesting sites as buildings are 'improved' and the decrease in livestock and their associated insects may both play a part. In common with many other species, laying date has advanced since 1980, presumably in response to climate change.

Barn Swallows from GB&I winter in South Africa, and there are now hundreds of recoveries or recaptures indicating exchange between the two areas. There is a more limited number of recoveries of birds on active migration through W Africa, usually from coastal areas between Morocco and Nigeria, although this might indicate human predation more than a strong migration route. Spring migration may be on a broader front, since birds have been recovered as far east as Tunisia, indicating the likelihood of direct trans-Saharan flights. The mortality associated with this is high (Wernham *et al.* 2002) and increased desertification of the sub-Saharan zones will increase the length of flights where food is scarce or absent; this will select against the more eastern route, and we might predict changes in migratory pattern as a consequence of this.

Genus *DELICHON* Horsfield & Moore

Old World genus of three species, one of which breeds in GB&I.

Common House Martin   *Delichon urbicum* (Linnaeus)

MB PM *urbicum* (Linnaeus)

TAXONOMY The three species of *Delichon* were all included in their analysis by Sheldon *et al.* (2005), who found them to comprise a tight clade, with Asian *D. dasypus* and Nepal *H. nipalense* House Martins being sister taxa. Asian has been regarded as possibly conspecific with Common House Martin, but the DNA results are quite clear, with differences of >8%, compared with 4% between Asian and Nepal. Three subspecies of Common House Martin are recognised by HBW, differing in size, extent of white on the back, and depth of fork to tail, but variation largely clinal and only nominate recognised in Europe by BWP and Svensson (1992).

DISTRIBUTION Breeds across the Palearctic, nesting on buildings and coastal cliffs from Iberia, GB&I to far east of Siberia. N limit determined by tundra; in south occurs from mountains of NW Africa to Persian Gulf and across C Asia to Sea of Okhotsk. Nominate breeds from Europe and N Africa to River Yenesei; replaced by *lagopodum* in E Siberia. Migratory, wintering chiefly in E & S Africa and Bangladesh to Indo-China.

STATUS At the time of the Second Breeding Atlas, despite presence of Common House Martins in >80% of 10-km squares, there was already evidence of a contraction in range, especially in Scotland and Ireland. Surprisingly, its conspicuous habit of colonial or semi-colonial nesting on buildings makes it difficult to census; birds can move among sites between years, and this relocation between survey areas can give misleading impressions of change in abundance. However, CBC/BBS data give an indication of a decline through 1980–99, with signs of a slight recovery into the new millennium.

Ringing in GB&I has revealed a little about migration routes; as with Barn Swallows, spring migration seems to be on a broader front, with more recoveries from C France and Italy than in the autumn. There are few recoveries from Africa itself; exchange has been found involving birds from elsewhere in Europe, although the wintering areas are still only vaguely known.

Genus *CECROPIS* Boie

New World genus of seven or eight species, one of which has been recorded in GB&I.

Red-rumped Swallow   *Cecropis daurica* Laxmann

PM *rufula* (Temminck)

TAXONOMY Five species of *Cecropis* were included in their analysis by Sheldon *et al.* (2005) and these proved to be a distinct clade, sister to the cliff swallows. They also examined geographically separate populations of two taxa: substantial differences were found between Ghanaian and S Africa populations of Red-breasted Swallow *C. semirufa*, and (of more relevance to GB&I) between Red-rumped Swallows from Pakistan and Ivory Coast. These results suggest that species limits in these taxa are in need of further investigation, with the possibility of splitting both. Within the Red-rumped, most authorities separate eastern taxa as Striated Swallow *C. striolata* and some also separate the Sri Lankan race as *C. hyperythra*. Splitting of resident W African birds has also been suggested, as these differ consistently in head pattern. Clearly, further analysis is needed.

DISTRIBUTION Breeds in open habitats, usually in hills and mountains, across S Europe and Asia, S of *c.* 55°N. Absent from desert and rainforest, distribution is discontinuous south into E & W Africa, India and China. Subspecies *rufula* occurs widely from Europe through the Middle East and N India to Tien Shan; replaced by nominate race E of Kazakhstan. Eight further races occur, four each in SE Asia and sub-Saharan Africa. Northern populations are migratory, wintering in Africa and from Bangladesh through to Australia.

STATUS The first record for GB was shot on Fair Isle in June 1906, with a gap of over 25 years until the next; there had been over 500 more to 2005. In view of its regular occurrence today, the deletion of a record in the Hastings rarities purge may have been a little harsh (Fraser & Rogers 2006). A progressive increase in the number of records likely reflects a change in distribution as much as increased observer awareness: its European range extended northwards in both the SW and the SE. Most spring birds are thought to be overshooting migrants, and many are found during or following periods of southerly winds, especially in S England. There is also a pattern of more widespread arrival in mid- to late autumn. It is much less common in Ireland, with only 21 individuals to 2002.

The Fair Isle bird of 1906 was identified as subspecies *rufula*; there is no evidence of any other races being involved.

Genus *PETROCHELIDON* Cabanis

Cosmopolitan genus of eleven species, one of which has been recorded in GB&I.

American Cliff Swallow   *Petrochelidon pyrrhonota* (Vieillot)

SM race undetermined; likely to have been *pyrrhonota* (Vieillot)

TAXONOMY The cliff swallows are monophyletic, comprising a clade which is sister to the red-rumped swallow group, and well separate from the rest of the swallows (Sheldon *et al.* 2005). The American Cliff Swallow shows variation in size and colouration, and four subspecies are recognised.

DISTRIBUTION Breeds S of tundra across N America to Mexico, though less common in SE USA. Nominate is most widespread, occurring through northern part of range, three more races occur in S USA and Mexico. Strongly migratory, wintering E of Andes Mts, from S Brazil to N Argentina. Migration route predominantly overland.

STATUS The first was in Oct 1983, on St Agnes (Scilly), before moving to St Mary's for two weeks. Further birds followed in 1988, 1995 (2), 1996 and 2000 (2 or 3). The only Irish record was from Dunmore Head (Kerry) in Nov 1995.

FAMILY CETTIIDAE

Genus *CETTIA* Bonaparte

Oriental and Palearctic genus of 15–17 species, one of which breeds in GB&I.

Cetti's Warbler   *Cettia cetti* (Temminck)

RB PM *cetti* (Temminck)

TAXONOMY A molecular analysis by Alström *et al.* (2006) found that *Cettia* was part of a clade of bush-warblers that included *Tesia*, *Urosphena*, *Tickellia* and other genera. Since *Cettia* occurred in two separate subclades, it may be paraphyletic, and generic limits may need refining. Three subspecies recognised on basis of size and pigmentation, but these are both clinal (BWP, Baker 1997).

DISTRIBUTION Fairly secretive, living in thick undergrowth, usually close to water. Nominate found from S England continuously across S of continent (including Mediterranean islands) to Bulgaria, SW Russia and Crimea, and across N Africa from Morocco to Tunisia; European populations largely resident; replaced from Iran to W China by *albiventris*, which is largely migratory, wintering to S Afghanistan, Pakistan and NW India. A third race *orientalis* breeds in C & E Asia Minor but is poorly defined (Baker 1997).

STATUS The first Cetti's Warbler was found in GB&I in Mar 1961, and breeding was confirmed in 1972, with numbers increasing rapidly to over 1,100 singing males in 2004. Most are in S England, particularly in the SW and East Anglia, but also into S Midlands and Wales.

Numbers have increased markedly, especially since 1998, to over 1,400 singing males in 2006, despite productivity remaining rather constant (Robinson *et al.* 2007). The rate of population increase shows some signs of slowing down, perhaps because the birds are becoming increasingly sensitive to cold winters as they move away from the south coast. It is evident that the main centre of population has moved from the south-east, where colonisation began, to the south-west, where the climate is less severe, especially in winter.

In Ireland: the first was at Cape Clear (Cork) in Aug 1968; four records in total now: two autumn birds, a singing male at Ballymacoda (Cork) in May 2001, and a subsequent record from Five Mile Point (Wicklow) in May 2005.

FAMILY AEGITHALIDAE

Genus *AEGITHALOS* Hermann

Palearctic and Oriental genus of five to seven species, one of which breeds in GB&I.

Long-tailed Tit   *Aegithalos caudatus* (Linnaeus)

Endemic: RB *rosaceus* Mathews
SM *caudatus* (Linnaeus)
SM *europaeus* (Hermann)

TAXONOMY Variation complicated, with up to 19 subspecies recognised; these have been combined into four groups: *caudatus, europaeus, alpinus, glaucogularis*.

DISTRIBUTION Widely distributed across the Palearctic from Scandinavia, GB&I, Iberia to Kamchatka, south to Mediterranean and S Iran, roughly between *c.* 70°N and 27°N. Nine subspecies breed in W Europe *caudatus* (widespread from Fennoscandia to Kamchatka), *europaeus* (France to the Balkans), *rosaceus* (GB&I), *aremoricus* (W France, Channel Is), *taiti* (S, SW France to C Spain) (*aremoricus, taiti* not recognised by Svensson 1992), *macedonicus* (S Balkans), *irbii* (S Iberia, Corsica), *italiae* (Italy, Slovenia), *siculus* (Sicily). Northern populations partially migratory, wintering through the south of range; most populations resident, though some dispersal of juveniles occurs.

STATUS Subspecies *rosaceus* is extensively distributed across GB, though absent from the N & W islands of Scotland, and areas of the uplands that lack tree cover. There was little change in GB between the two Breeding Atlases, but there was a marked contraction in Ireland, where it was lost from *c.* 24% of 10-km squares, although this likely reflects incomplete coverage during the second survey rather than a genuine decline (P Smiddy pers. comm.).

Long-tailed Tits are especially susceptible to periods of prolonged winter cold, especially when the trees glaze over with ice. Under such conditions, mortality can be high, although the potentially high reproductive rate usually ensures a rapid recovery. The CBC and BBS data show this very clearly. The population in GB & NI was very low after the 1963–64 winter, but recovered quickly to a peak in the early 1970s. A series of cold winters led to a crash from which there has been a further recovery to a level similar to the earlier peak, with some evidence of a plateau being reached in the mid-1990s. As with several other woodland species, laying date has advanced by about two weeks since the 1970s, presumably in response to climate change. Clutch size has declined steadily since 1965, although a parallel decline in losses at the egg stage means that brood size has remained static since about 1990. For some reason, nest losses at the chick stage have risen slightly since 1964.

The Repeat Woodland Survey was in general agreement with these population changes, although it showed greater heterogeneity among regions and sites. This is probably because earlier data from the 1980s was gathered during a period of changing population, so that direct comparison is problematic (Baillie *et al.* 2006). The Garden Birds Survey indicates a slight increase in the proportion of gardens where they have been recorded since 1995, in agreement with the BBS data. In some areas, the species has taken to visiting feeders.

Two other subspecies have been recorded in GB, both as rare vagrants; on average, *caudatus* occurs less than annually; however, this masks occasional influxes, usually associated with Continental eruptions. Sometimes these remain and may even breed, their offspring presumably being incorporated into the *rosaceus* gene pool. A bird confirmed as subspecies *europeus* was taken at Dover (Kent) in 1882 and is now in the Rochester Museum.

FAMILY PHYLLOSCOPIDAE

Genus *PHYLLOSCOPUS* Boie

Old World genus of over 50 species, 14 of which have been recorded in GB&I, three as breeding birds.

Green Warbler   *Phylloscopus nitidus* (Blyth)

SM monotypic

TAXONOMY Treated as a separate species by Vaurie (1959), though more recently considered conspecific with *P. trochiloides*. There are differences in mt DNA between this and *P. t. viridanus*, although rather few individuals have been analysed (Helbig *et al.* 1995, Irwin *et al.* 2001b). Most plumage characters show slight overlap, although cumulatively diagnostic (L Svensson pers. comm.). Songs also appear to be distinct (Albrecht 1984; L Svensson pers. comm.). In view of its geographic location, it is genetically isolated from the races of Greenish Warbler, and the plumage and molecular differences indicate a long period of independent evolution.

DISTRIBUTION Breeds from N Turkey, Caucasus, to Uzbekistan, wintering in the S of peninsular India.

STATUS A single record from GB&I, on St Mary's (Scilly) in Sep–Oct 1983.

Greenish Warbler   *Phylloscopus trochiloides* (Sundevall)

PM *viridanus* Blyth
SM *plumbeitarsus* Swinhoe

TAXONOMY A member of a clade that also includes Green Warbler, Arctic Warbler and a group of Far Eastern taxa (Olsson *et al.* 2005), this species has attracted attention, from both birders and evolutionary biologists. Morphological variation and distribution across S Asia suggest it is a ring species. Five subspecies are recognised by Ticehurst (1938), Baker (1997) and Vaurie (1959): *trochiloides*, *ludlowi*, *obscuratus*, *plumbeitarsus*, *viridanus*. These form a ring around the Tibetan Plateau: sequentially, *plumbeitarsus* (NE), *viridanus* (N&W), *ludlowi* (SW), *trochiloides* (S), *obscuratus* (SE). There is a gap in the distribution between *obscuratus* and *plumbeitarsus*.

In C Siberia, *plumbeitarsus* and *viridanus* come into contact without interbreeding. They differ in plumage, structure and song; playback experiments by Irwin *et al.* (2001b) showed that sympatric males did not react to the alternate's song, indicating that they did not regard it as emanating from a conspecific. This would be strong evidence for specific status of the two forms. However, plumage, morphology and song all vary clinally (Irwin 2000), such that there is a gradual progression round the ring from *viridanus* through *ludlowi*, *trochiloides* and *obscuratus* to *plumbeitarsus*. There was strong reaction between males across the gap between *obscuratus* and *plumbeitarsus*, and little difference in morphology, suggesting that the populations were once in genetic contact and that this ecological barrier is due to habitat change (or destruction) in the relatively recent past.

Molecular studies (Irwin *et al.* 2001b, 2005) show a similar pattern of progressive change in structure round the Tibetan Plateau. The results indicate that populations at the two extremes are effectively reproductively isolated, and merit specific rank. However, the problem is 'where is the other boundary between the two species?' Evolutionarily, there is no conceptual problem with 'speciation by force of distance' (Irwin *et al.* 2005); but from a taxonomic position, deciding on a boundary is currently impossible.

DISTRIBUTION Widely distributed across the Palearctic, from the Baltic to NE China (Baker 1997). *Viridanus* S Finland and Baltic coasts across Russia and W Siberia to River Yenesei, S to Kashmir; *plumbeitarsus* River Yenesei to the Pacific, S to NE China, Mongolia, Transbaikalia; *trochiloides* Qinghai and Shaanxi, S to Nepal, Tibetan Plateau; *ludlowi* SE Afghanistan to Kashmir; *obscuratus* NE & S Qinghai to N Tibet. Some populations strongly migratory, wintering across SE Asia from peninsular India through to Thailand and Indo-China; other populations may just be altitudinal migrants.

STATUS There are *c.* 400 records from GB&I, the majority from England and Scotland. The first was shot at North Cotes (Lincolnshire) in Sep 1896, and the first Irish bird was obtained on Great Saltee (Wexford) in Aug 1952. As with many species, the number recorded annually has risen sharply, from *c.* 2 (1950–70), through *c.* 7 (1970–90) to >14 (post-1990), with occasional influxes (e.g. >30 in late Aug 2007). There has been a more modest increase in records from regularly manned bird observatories since 1950: this suggests that there has been a slight increase in frequency, but that most of the rise in numbers is due to increased observer density and skill. Birds have been recorded from May to Nov; although the main peak is Aug–Sep (>80% of records); there is a lesser peak in June. Interestingly, all nine Manx records and *c.* 60% of Welsh birds have been in May–June. Although often thought of as an early-autumn bird (certainly compared with Arctic Warbler), there are significant differences between the regions: in Scotland, 63% of autumn records were in Aug–Sep, compared with 82% in Ireland and 44% from England.

There are four records of subspecies *plumbeitarsus*, all from England; the first was on Gugh (Scilly) in Oct 1987, followed by birds in Norfolk (1996), Scilly (2003) and Yorkshire (2006).

Arctic Warbler   *Phylloscopus borealis* (Blasius)

SM *borealis* (Blasius)

TAXONOMY A member of a clade with Greenish Warbler and other Oriental taxa (Olsson *et al.* 2005), Arctic Warbler is the only *Phylloscopus* that regularly breeds in the New World. Four subspecies recognised by Ticehurst (1938), seven by Vaurie (1959), but only three by Svensson (1992), Baker (1997) and BWP: *borealis, xanthodryas* and *kennicotti.* These differ in greenness of upperparts and balance of yellow/grey below. Analysis of mt DNA suggests birds in Sakhalin and Kamchatka comprise a clade which differs by 3.8% from Palearctic/Beringian clade, and that Beringian birds form a clade within Palearctic clade, at odds with convential subspecies taxonomy (Reeves *et al.* 2008).

DISTRIBUTION Nominate breeds from N Scandinavia to Pacific, between 60°N and 75°N, wintering in SE Asia; *xanthodryas* is restricted to a small area of the Far East between Kamchatka, Sakhalin and Japan, wintering in Borneo, Philippines, Taiwan and Indonesia (E to Lesser Sundas); *kennicotti,* Alaska, migrating SW to winter principally in the Philippines (Baker 1997).

STATUS Less common than Greenish Warbler, with over 280 records from GB and a further 7 from Ireland. The first for GB was killed at Sule Skerry Lighthouse (Orkney) in Sep 1902. As for many passerines, the number recorded in GB has increased since 1950, from an average of *c.* 2 in 1950–70, through *c.* 5 (1970–90), to >7 since 1990. Birds have been recorded in all months from June to Nov, but less often in spring than Greenish. The majority of records come from Scotland, where it also occurs earlier in the autumn, with only 8% in Oct, compared with 30% from elsewhere in GB&I.

There are *c.* 65 records of Arctic Warbler from Fair Isle. Since coverage at this observatory is more consistent than at many other sites, observations from here can be used to examine trends over time, largely unconnected with observer effort. The number per decade increases slightly at Fair Isle, from *c.* 1 per year in the 1950s to *c.* 1.5 since 1990. However, this increase is significantly less than the national rate and suggests changes in GB&I are due more to an increase in the number and skill of observers than to any biological effect. Thus there may have been a slight increase in occurrence, but the likelihood that this is largely (or even entirely) due to increased observer effort is high.

Occurrences in Ireland are more sporadic; all in Sep–Oct, with the first on Tory Island (Donegal) in Sep 1960; only seven individuals in total, the most recent in Sep 2003.

Pallas's Leaf Warbler   *Phylloscopus proregulus* (Pallas)

PM monotypic

TAXONOMY The application of DNA technology and the study of vocalisations have revealed a series of closely similar taxa that exhibit good evidence of reproductive isolation (reviewed by Martens *et al.* 2004). Three of the four subspecies recognised by Ticehurst (1938; *proregulus, kansuensis, chloronotus/simlaensis*) have been elevated to specific rank (Pallas's Leaf, Gansu Leaf and Lemon-rumped Warblers), while *forresti* and *yunnanensis = sichuanensis* (Sichuan and Chinese Leaf Warblers) have also been added to the list of species. The consequence for the British and Irish Lists is that *P. proregulus* is now treated as monotypic.

DISTRIBUTION Pallas's Leaf Warbler (*sensu stricto*) breeds from SW Siberia to N Mongolia and Sakhalin, and S to NW Gansu and S Shaanxi (Baker 1997). Migrates S and SE to winter in SE China, N Thailand and N Indo-China.

STATUS The first was shot at Cley (Norfolk) in Oct 1896; the next was not seen for over 50 years. Thereafter, the number of records climbed steadily, with annual averages of 3 (1960s), 9 (1970s), 39 (1980s), 83 (1990s), >100 (since 2000). The cause of the increase is unclear, but likely due to an increase in population or westward range expansion across Siberia: Baker (1977) first drew attention to this for a range of Siberian passerines. Howey & Bell (1985) discuss the arrivals of this species in 1982, and show how there was an association with anticyclonic conditions over N Europe, leading to extensive and prolonged easterly winds across Europe and W Asia. Such conditions have also been associated with subsequent arrivals (Baker & Catley 1987).

It is much less common in the west of GB&I; the first for Ireland was on Cape Clear (Cork) in Oct 1968, and there have only been 26 to 2003, and that year was the best ever with seven records. Birds occasionally stay late through the autumn into winter, and one or two spring records are believed to be individuals that overwintered.

Yellow-browed Warbler   *Phylloscopus inornatus* (Blyth)

PM monotypic

TAXONOMY Another species whose taxonomy has been extensively revised following the application of DNA technology and analysis of vocalisations. Ticehurst (1938) and Vaurie (1959) recognised three subspecies (*inornatus, mandellii, humei*), but Irwin *et al.* (2001a) showed that *inornatus* differed markedly in DNA sequences, song and call from the other two. Furthermore, playback studies showed that *humei* and *mandellii* reacted strongly to each others' songs, but that both ignored that of *inornatus*; the latter was similarly indifferent to the songs of the other two taxa. On the strength of these studies, Yellow-browed Warbler is treated as a monotypic species, separate from the polytypic Hume's Leaf Warbler.

DISTRIBUTION Breeds S of 70°N from the Ural Mountains and River Pechora to Sea of Okhotsk; S to Lake Baikal and N Manchuria. Winters widely from Nepal and SE China to Malay peninsula.

STATUS Numbers have increased substantially since the mid-20th century. Occurrences in GB averaged *c.* 10 per year during 1950–62, when it was removed from the BBRC list. Since then numbers have continued to increase substantially, to over 300 per annum since 1980 (Fraser & Rogers 2006), mostly in autumn.

Wintering birds are found regularly, albeit in small numbers, and many of these survive through to spring. There are occasional reports of birds at coastal sites in spring; these are presumably individuals that have overwintered elsewhere in SW Europe, or further afield, since there is now at least one Dec record from Senegal (Cruse 2004). It is possible that a new migration pattern may evolve; if some of the birds that winter in SW Europe return to breeding grounds in W Asia, just as with Blackcaps, a population of westwards-migrating birds could arise.

The first Irish bird was shot on Inishtearaght (Blasket Is, Kerry) in Oct 1890. There were no further reports until one at Great Saltee (Wexford) on the early date of 6 Aug 1952. A further 15 were recorded between 1953 and 1965, and between 1966 and 1986 they occurred every year (except 1972), with influxes in 1985 (87) and 1986 (59+). Now occurs annually in varying numbers, though all in autumn and most in Sep–Oct. Most occur in the south, but records from all round the coast.

Hume's Leaf Warbler   *Phylloscopus humei* (Brooks)

SM *humei* (Brooks)

TAXONOMY Polytypic, with two subspecies *humei* and *mandellii* (see Yellow-browed Warbler).

DISTRIBUTION *Humei*: breeds S of Yellow-browed, from Altai Mountains and Xinjiang to N slopes of Himlayas, wintering widely from Iraq, Iran to Bangladesh and N India; *mandellii* breeds further S & E, from Qinghai and Shanxi S to Assam, wintering further E from Sikkim to Thailand.

STATUS The first for GB&I was at Beachy Head (Sussex) in Nov 1966. Subsequently, there have *c.* 70, mostly in Oct–Nov along the coasts of E Scotland and E & S England, although birds have also been recorded in Dec–Feb. There have been several long-staying wintering birds, including at Great Yarmouth (Norfolk) Jan–Apr 1995 and Newbiggin (Northumberland) Jan–Apr 2002, plus the only Irish records at Knockadoon Head (Cork) and Hook Head (Wexford) both in winter 2003–04. Baker & Catley (1987) note that this species tends to arrive later than Yellow-browed and is more usually associated with Pallas's Leaf Warblers.

Radde's Warbler   *Phylloscopus schwarzi* (Radde)

PM monotypic

TAXONOMY Despite the recent plethora of phylogenetic studies of *Phylloscopus* warblers, rather few have included Radde's. Johansson *et al.* (2007) found a close association with Yellow-streaked Warbler *P. armandii*.

DISTRIBUTION Breeds in Siberia, between 58°N and 38°N, from the Russian Altai to Sakhalin, S to Manchuria and N Korea. Winters from Myanmar to Indo-China.

STATUS The first Radde's Warbler was at North Cotes (Lincolnshire) in Oct 1898; the first Irish bird was at Hook Head (Wexford) in Oct 1982. There have subsequently been *c.* 250 in GB, but only 13 in Ireland to the end of 2006, with the first for NI as recently as Oct 2008 on Copeland Is (Down). It is overwhelmingly an Oct bird in GB&I, with only 30 records in Sep and Nov. The number of records per decade has increased remarkably, from none in the 1950s, through 9, 18, 67, 103 to 67 in 2000–03. The turning point seemed to come in 1987, when the average number per year rose abruptly from <2 (1960–87) to >12 (1988–2003). This coincided with the publication of a landmark paper on field identification of this and Dusky Warbler (Madge 1987), but may also relate to significant improvements in equipment and skills.

Dusky Warbler   *Phylloscopus fuscatus* (Blyth)

PM *fuscatus* (Blyth)

TAXONOMY Johansson *et al.* (2007) found a close relationship with Smoky Warbler *P. fuligiventer,* Buff-throated Warbler *P. subaffinis*, Willow Warbler and the chiffchaffs. The species shows extensive variation across its range; although Vaurie (1959) recognised four subspecies, only two were recognised by Ticehurst (1938) and Baker (1997): *fuscatus* and *weigoldi*, the latter being darker brown with a reduced superciliary streak (Vaurie 1959).

DISTRIBUTION East from the River Ob to the Sea of Okhotsk, S to N Szechwan and the Himalayas; generally between 65°N and 30°N. Nominate is more northerly, breeding as far south as Lake Baikal and Qinghai, and wintering from S China through Indo-China to Bangladesh; *weigoldi* is more restricted, from the mountains of Qinghai to Sichuan, and wintering in the E Himalayan foothills.

STATUS The first for GB was at Auskerry (Orkney) in Oct 1913, and the first for Ireland at Limerick in December 1970, having been ringed on the Calf of Man in May that year. There are now > 260 records from GB and six from Ireland. As with Radde's Warbler, there was a sharp increase in records in 1987: the annual average from 1960 to 1987 was *c.* 2; during 1987–2003, it rose to 13.4. Although numbers are similar to Radde's, there is a clear difference in the pattern of autumn occurrence, with Radde's occuring earlier: only 4% of 259 Radde's were in Nov–Dec compared with 35% of 273 Dusky Warblers. There are no records of Radde's in winter, compared with six Dusky. Records from Dec–Mar are presumably of overwintering birds: one was at Bideford (Devon) from Dec 1994–Apr 1995, another was at Bude (Cornwall) from Jan to May 1995, while a third was at Paignton (Devon) from Nov 2003 to Apr 2004. Four Dusky Warblers in May are likley to have overwintered in SW Europe.

Western Bonelli's Warbler   *Phylloscopus bonelli* (Vieillot)

SM monotypic

TAXONOMY The old 'Bonelli's Warbler' has a place in British ornithology as the first species where DNA data contributed to the splitting of the species. Molecular divergence between eastern and western populations of 'Bonelli's Warbler' is as great as that between these and Wood Warbler *P. sibilatrix* (Helbig *et al.* 1995). Clear vocal differences are supported by slight, but consistent, differences in morphology to show that the two populations are independent evolutionary lineages and merit specific status.

DISTRIBUTION Upland woods of SW Europe, from NW Balkans and N Italy, through France, Iberia to coastal Morocco, Algeria and Tunisia. Winters in Sahel zone of Africa, from Chad to the Atlantic.

STATUS The first for GB was on Skokholm (Pembrokeshire) in Aug 1948; there have been *c.* 70 subsequent records, mostly in the S & W of England. Nineteen birds have been recorded in Ireland since the first at Cape Clear (Cork) in Sep 1961. Unusually for a migrant to the W Mediterranean, there are rather few spring records: only 12 in Apr–June, compared with *c.* 60 in Aug–Nov. The number of records has remained at one or two per year since the 1960s.

There are an additional 79 records of either Western or Eastern Bonell's Warblers, mostly predating Helbig *et al.* (1995). In view of the relative abundance of the two forms since then, it seems likely that most were Western.

Eastern Bonelli's Warbler   *Phylloscopus orientalis* (C. L. Brehm)

SM monotypic

TAXONOMY Separated from *P. bonelli* on the basis of DNA, vocalisations and morphology (see Western Bonelli's Warbler).

DISTRIBUTION Allopatric to Western Bonelli's, breeding from S & E Balkans to W Turkey, and scattered areas of Levant to W Iran.

STATUS Four in GB&I: three in England and one in Scotland. The first was on St Mary's (Scilly) in Sep–Oct 1987. The others have been in Northumberland (Sep 1995), Shetland (Aug 1998) and Devon (2004). There are no records from Ireland.

Wood Warbler   *Phylloscopus sibilatrix* (Bechstein)

MB PM monotypic

TAXONOMY Closely allied to the two Bonelli's Warblers, and a member of a clade with these, Willow Warbler and the chiffchaffs (Helbig *et al.* 1995).

DISTRIBUTION Breeds across the W Palearctic, from Scandinavia, GB and France, E to the Urals, Crimea and the Caucasus, but rarely in Greece, Ireland and Iberia. Migrates to sub-Saharan Africa, wintering close to the Guinea coast, and across to Sudan and S to W Uganda and Zaire. Its winter habitat is more closely tied to tropical forests, and so the population may be less affected by climatic variations within the Sahelian savannah.

STATUS The Wood Warbler is a bird of deciduous woodland across the western side of GB; it is scarce in lowland England and NE Scotland, and almost absent from Ireland, where there is a very small breeding population mostly in the E & NE. There was a contraction in range of *c.* 22% between the two Breeding Atlases, though this may be due to less intensive surveying for the later one (Gibbons *et al.* 1993). There is little population data before 1994, but an alarming decline of *c.* 50% across the UK during 1994–2004 (BBS). The Repeat Woodland Survey revealed a similar loss of *c.* 50% since the 1980s, although the data indicated no obvious reason. The daily loss of nests at the egg stage showed a decline during 1964–2004, but sample sizes are too small for meaningful analysis.

Occurs regularly in spring and autumn at coastal sites, albeit in small numbers. BirdLife International (2004) report a decline in abundance in NW Europe, but this does not seem to be reflected in the number of migrants through GB. There was no evidence of a change in the number of bird-days at Portland and Christchurch (Dorset) over the period 1980–2000 (Green 2004); presumably these are largely birds that bred in GB. Nor is there any indication of change in Shetland (Pennington *et al.* 2004), presumably relating largely to the Scandinavian population.

Common Chiffchaff   *Phylloscopus collybita* (Vieillot)

MB WM PM *collybita* (Vieillot)
WM PM *abietinus* (Nilsson)
WM PM *fulvescens* (Severtzov)
SM *tristis* Blyth

TAXONOMY Ticehurst (1938) recognised eight subspecies: *collybita*, *abietinus*, *canariensis*, *exsul*, *lorenzii*, *tristis*, *sindianus* and *ibericus*. The last of these was incorrectly called '*brehmii*' by some authorities (Svensson 2001). Vaurie (1959) added *fulvescens* to this list, and subsequently *menzbieri* and *brevirostris* were included (BWP), to give a total of at least 11 subspecies.

The complexities of the group were partially resolved by studies in the late 1990s, including an intensive molecular analysis by Helbig *et al.* (1996), who investigated the subspecies *collybita*, *abietinus*, *caucasicus*, *brevirostris*, *tristis*, *canariensis*, *lorenzii*, *sindiacus* and *ibericus*. They analysed mt cyt-b sequences, and showed that these eight taxa fell into four groups: *collybita*, *abietinus*, *caucasicus*, *brevirostris* and *tristis*; *canariensis*; *lorenzii* and *sindianus*; and *ibericus*. Field studies showed *ibericus* and *collybita* to be different in morphology and vocalisations, with a narrow zone of overlap in SW France and N Spain. There was no evidence of prolonged mitochondrial gene flow between these two, despite occasional hybridisation. Atlantic *canariensis* is also morphologically, genetically and vocally different. *Lorenzii* and *caucasicus* are parapatric in the Caucasus Mountains, but show strong differences in DNA sequence, as well as being diagnosable in morphology and vocalisations. The grey-brown eastern form *tristis*, although differing in contact call, was only marginally different in DNA sequence from the *collybita* group; it was more divergent vocally, showing affinities with *sindianus/lorenzii*. Playback experiments showed that the well-differentiated songs of *tristis* and *collybita* were ignored by the alternate taxa. Hybridisation appears to be extensive in the area of overlap between *tristis* and *abietinus* (e.g. Marova & Leonovitch 1993), and birds in this region, which have been named *fulvescens,* are intermediate in many characters. Further investigations are needed here to establish the degree of isolation between these and the validity of *fulvescens.*

On the strength of these results, chiffchaffs are treated as four species: Iberian *P. ibericus*, Canary Is *P. canariensis* (including *exsul*), Mountain *P. sindianus* (including *lorenzii*), and Common *P. collybita* (including *abietinus*, *brevirostris*, *caucasicus* and *menzbieri*). In the area of overlap between *tristis* and *abietinus*, breeding may be assortive. Were this to be established, then *tristis* might be better regarded as a separate polytypic species (including *fulvescens;* Svensson, in Dean & Svensson 2005). In this case, the use of *fulvescens* would be restricted to those individuals from between the Urals and the Yenisey which resemble *tristis* apart from a little additional yellow and olive, and thus match the type description (Severtzov 1873). Until the relationships of *tristis* are resolved, it remains (here) a subspecies of *P. collybita.*

DISTRIBUTION Widespread across the W Palearctic from the Pyrenees to River Kolyma, and S to Black Sea and Caucasus Mts. Nominate: breeds Poland, Bulgaria to Pyrenees, including GB&I; winters in S of range and N Africa; *abietinus*: Scandinavia to Urals, and S to N Iran; winters SE Europe, NE Africa to Iran; *tristis*: E of *abietinus*, from River Pechora to River Kolyma; winters in Himalayas and into N India; *caucasicus*: Caucasus Mts; probably winters Arabia to Iran; *brevirostris*: Turkey; *menzbieri*: SW Transcaspia.

STATUS Little evidence of change in distribution between the two Breeding Atlases in GB, although there was a slight contraction in Ireland. This may be linked to variation in abundance recorded through the CBC and BBS from the mid-1960s. There was a marked decline in 1965–75, with the population at a low level until *c.* 1985. Since then, there has been a steep increase, and the population more than doubled during 1985–2004. The decline has been linked with low overwinter survival of birds in the Sahel region of Africa (Lack 1989), and the subsequent increase with climate change. Laying date is now about two weeks earlier than in 1965 (Baillie *et al.* 2006). As with other short-distance migrants, milder conditions may encourage overwintering in more northerly areas. An earlier return to breeding territories may affect potential competitors such as Willow and Wood Warblers. These two species are both showing a decline as Common Chiffchaff numbers continue to burgeon. Since the early 1990s, Common Chiffchaff productivity has been declining, largely due to a decrease in clutch size (Baillie *et al.* 2006); this may be an early indicator of density-dependent population regulation. The Repeat Woodland Survey (Amar *et al.* 2006) was unable to find any habitat changes that correlated with the changes in distribution.

Migrants resembling the N European race *abietinus* are recorded regularly at coastal sites across GB&I at both seasons, and small numbers winter in sheltered areas, particularly in the S & W. Among the more typical *abietinus* and occasional *collybita*, birds are found with significantly reduced yellow and olive compared with these two races. These are often thought to be of easterly origin and are frequently attributed to *tristis*, but, in fact, only a proportion match the characters of typical *tristis* (in the sense of matching the type). Many are much too 'grey and white' and, although some utter a plaintive and rather monosyllabic call, they lack the characteristic brown and warm ('rusty') buff of typical *tristis*. Such individuals often show yellowish elements in the bill and legs, and are sometimes equated with *fulvescens* (Dean & Svensson 2005, Dean 2007b). Although these may include intergrades, their appearance does not match the type description of *fulvescens*: birds from the breeding range of that form are not more grey and white than typical *tristis* (Dean & Svensson 2005). The distribution and plumages of eastern populations are still unclear. Problems associated with identifying wintering and migrant birds (outside their breeding range) have been discussed by Dean & Svensson (2005). They conclude that further morphological and vocal studies of Common Chiffchaffs of known provenance are required before the origins and taxonomic position of such individuals are understood. Until these are undertaken, many birds will be unidentifiable as to race and, without ringing recoveries, confirmation of provenance of individuals will continue to be fraught with difficulty. Despite these caveats, birds resembling 'classic' *tristis* are recorded in small numbers in late autumn and winter most years.

Iberian Chiffchaff   *Phylloscopus ibericus* Ticehurst

SM monotypic

TAXONOMY See Common Chiffchaff. Shown by Helbig *et al.* (1995) to be genetically distinct and apparently reproductively isolated from adjacent populations of *P. collybita*. Shows plumage differences from latter and different song. Formerly known under the name *brehmii*.

DISTRIBUTION Iberia and adjacent parts of SW France, NW Africa. Wintering grounds little known (Baker 1997).

STATUS The first was a bird singing (and recorded) at Brent Reservoir (London) in June 1972; and 14 more by 2007. Most have been in the coastal SW, although there are records from Norfolk, Oxfordshire and Kent. All have been spring males in song. Not recorded from Ireland.

Willow Warbler   *Phylloscopus trochilus* (Linnaeus)

MB PM *trochilus* (Linnaeus)
MB PM *acredula* (Linnaeus)

TAXONOMY Closely related to the chiffchaffs, forming part of same clade (Helbig *et al.* 1995). Three subspecies generally recognised (Ticehurst 1938; Vaurie 1959; Svensson 1992; BWP; Baker 1997): *trochilus*, *acredula* and *yakutensis*, though variation broadly clinal and many breeding populations variable and containing birds of various phenotypes. Genetically less diverse than Common Chiffchaff at mt loci, but nuclear genes show greater variation (Bensch *et al.* 2006). However, no inter-racial analyses have been undertaken.

DISTRIBUTION Breeds from GB&I, France across the Palearctic to River Anadyr in Siberia; S to *c.* 50°N. Nominate breeds in Europe, from GB&I, France, NE Spain S Scandinavia, and E to N Romania; winters from Guinea to Sudan, and S to Angola and the Cape; *acredula*: replaces *trochilus* from C Scandinavia to River Yenesei, and S to Volga, Urals and N Kazakhstan (Baker 1997); winters in S Africa; *yakutensis*: breeds between Rivers Yenesi and Anadyr to *c.* 72°N, and S to *c.* 50°N; also winters in southern Africa.

STATUS Willow Warbler was very widely distributed across GB&I during the two Breeding Atlas surveys, with evidence of breeding in over 90% of 10-km squares. The only region with significant absence was Shetland, although scarce in parts of Orkney and the Outer Hebrides. The population size varied from the beginning of the CBC survey (1964) until the early 1980s, at which point a serious decline set in. This was also recorded in the Repeat Woodlands Survey (Amar *et al.* 2006), which estimated a fall of *c.* 70% in the population of GB. Although losses occurred across the whole of GB, between 1986 and 1993 it was more severe south of a line from the Humber to the Mersey (47%) than north of this (7%); the former included an unprecedented run of individually significant annual declines from 1989 to 1992. In one of the first attempts to analyse changes in population size using data from a wide range of BTO surveys, Peach *et al.* (1995a) found little evidence of change in productivity, apart from a minor rise in nest failure at the chick stage in the south. The decline appeared to be largely due to a near halving of annual survival of adult birds, especially in southern Britain (45% to 24%). Productivity also declined from 1980 to 2000; whether this is due to the adults being in poorer condition, adverse climatic effects on the breeding grounds or changes in habitat is not clear. One difference that may be significant is the increase in scrub cover in the uplands of N Britain, extending the area of suitable habitat in that region at a time when the population is struggling further south (Gillings *et al.* 2000).

Breeding Willow Warblers in Scotland are dimorphic, with many showing characters resembling *acredula* or intermediates; this deserves further investigation (BS3). Otherwise, *acredula* regular as passage migrant, usually in falls in mid-May and from late August through September. Most autumn migrants in northern isles believed to be of this race, and numerous in large autumn falls in England. In Ireland, *acredula* apparently a scarce spring migrant.

FAMILY SYLVIIDAE

Genus *SYLVIA* Scopoli

Old World genus of *c.* 20–25 species, 13 of which have been recorded in GB&I, five as breeding birds.

Eurasian Blackcap   *Sylvia atricapilla* (Linnaeus)

MB WM PM *atricapilla* (Linnaeus)

TAXONOMY Sister species to Garden Warbler, though well differentiated from it; five subspecies recognised by Vaurie (1959) and Shirihai *et al.* (2001), though much variation is clinal, especially involving size and overall coloration; *atricapilla*: Europe, except far south, Levant, European Russia; *dammholzi*: Caucasus to S Caspian; *pauluccii*: C & S Italy, islands to Balearics, (possibly) Tunisia; *heineken*: Madeira, Canaries, possibly Iberia, Morocco to Tunisia; *gularis*: Cape Verde, Azores.

DISTRIBUTION Breeds across the W Palearctic from Atlantic islands to the Caucasus, and into C Asia. Typically nests in woodland (usually, but not exclusively, deciduous) with a thick understorey, but also along woodland edges, parks and gardens. Similar to Garden Warbler, and has been shown to compete for breeding territories (Garcia 1983). Most populations migratory, wintering from GB&I, S through France, Iberia, Italy to W and E Africa. In general, most northerly populations migrate farthest. Evidence is accumulating to show that migration routes and wintering areas are changing in response to natural selection, especially in relation to climate.

STATUS The Breeding Atlases revealed that Eurasian Blackcaps were widely distributed across England and Wales, S & E Scotland and through the Great Glen. In Ireland, the distribution was more scattered, though with increased densities in the east. However, this masks a marked increase between the two Breeding Atlases; more of Scotland was colonised, and distribution in Ireland expanded by almost 40%. CBC and BBS data show a steady increase from the mid-1970s, with a possible levelling out around 2000. There has also been a marked advance in laying date since the 1970s, and it is tempting to link these. As a short-distance migrant, Eurasian Blackcap is well placed to respond to climatic amelioration, combining enhanced winter survival at the northern limits of the winter range with a shorter migration (and hence reduced mortality) and an earlier arrival back on the breeding grounds. Earlier arrival allows choice of better territories and a longer breeding season, or at least an enhanced chance of multiple broods or successful repeat clutches following nest loss. These will all lead to an increasing population expanding into previously less optimal range, as revealed in the Breeding Atlas surveys.

Eurasian Blackcap is also a winter visitor. Until the 1960s, they were almost unknown at this season, but there has been a substantial increase in the winter population, mostly recorded in gardens. These birds originate from Central Europe (Berthold 1995). Autumn migration in juvenile birds is determined by course and distance, both of which are under genetic control in Eurasian Blackcaps (and presumably most other species as well). There is, however, variation about the mean orientation direction, and some of this will be genetically determined. Prior to the 1960s, juveniles from C Europe that orientated too far north of the optimal SW direction towards Gibraltar would have arrived in GB&I and presumably perished. The recent amelioration of winter climate, combined with increased availability of food through berry-bearing shrubs and garden feeding, has permitted these individuals to survive. Wintering further north and having a shorter distance to travel, they presumably return to the breeding grounds earlier, acquire better-quality territories and mate with each other. This will drive an increase in the subpopulation and the observed rise in winter numbers. Strong evidence in support of this argument has been provided from the analysis of stable isotopes in wintering birds from GB&I and breeding populations in Germany (Bearhop *et al.* 2005).

Garden Warbler   *Sylvia borin* (Boddaert)

MB PM *borin* (Boddaert)

TAXONOMY Sister species to Eurasian Blackcap, but DNA sequences are divergent. Two subspecies usually recognised based on size and coloration, though these are poorly defined and variation is clinal (Vaurie 1959; Shirihai *et al.* 2001; treated as monotypic by Svensson 1992); *borin*: Europe W of Poland, Austria, including Norway/Sweden; *woodwardi*: from C Poland through former Yugoslavia to C Siberia.

DISTRIBUTION Breeds across Palearctic from Atlantic to C Siberia. Slightly different habitat to Eurasian Blackcap, though may compete for territories (tends to be less successful in direct competition); prefers woodland with lush undergrowth and less dependent on closed canopy; rarely found in conifers. Strongly migratory, wintering in sub-Saharan Africa almost to Cape, though apparently avoiding deserts of SW.

STATUS A long-distance migrant that winters south of the Sahel zone of Africa, this species is widespread in GB south of the Grampian Mountains of Scotland; north of here, the distribution is restricted to the Great Glen and lowland woods of NE Scotland. It is scarce in the fenland of E England and is only thinly distributed in Ireland. There was little change in distribution between the two Breeding Atlases, apart from a modest spread across Scotland, but since then there have been substantial changes in abundance, with decreases reported in all areas apart from Scotland (Amar *et al.* 2006); an increase here may be a continuation of the trend evident in the two Breeding Atlases and might be a consequence of climatic amelioration.

Garden Warbler is a member of a small group of woodland species about which there is currently much concern. Numbers declined markedly through 1965–1975 (CBC), perhaps because of drought-associated problems on migration. The numbers recovered to the mid-1980s, but there has been a further slow and progressive decline. Since its inception in 1994, the BBS has recorded a decrease of 13% in England and 24% in Wales (data for the other countries are too limited for analysis). This seems to be, at least partly, due to a decline in juvenile productivity (Baillie *et al.* 2006). There has been a move towards earlier laying, but less than among short-distance migrants. It may be that this species is being forced into poorer-quality territories by the burgeoning Eurasian Blackcap population; alternatively, or perhaps additionally, there may be a decoupling of breeding season from the peak availability of nestling food availability, as has been reported for Pied Flycatchers (Both *et al.* 2006). However, there is also evidence that reduced woodland management may be leading to change in habitat, in particular closure of the canopy and consequent reduction in understorey, although increased deer grazing may also be implicated.

Barred Warbler   *Sylvia nisoria* (Bechstein)

PM *nisoria* (Bechstein)

TAXONOMY DNA sequences indicate that Barred Warbler has no close relatives among the *Sylvia* warblers (Shirihai *et al.* 2001); this and its unique plumage features indicate that it is probably one of the earliest lineages to diverge from the ancestral *Sylvia* stock. Two subspecies recognised by Shirihai *et al.* (2001) based on plumage, though these are poorly differentiated and largely clinal; *nisoria*:

Europe, including Turkey, and N Asia Minor to Ural Mountains; *merzbacheri*: C Palearctic from Tien Shan and E Kazakhstan to Mongolia.

DISTRIBUTION Breeds in open woodland with varied structure including bushes and trees, and adjacent herb layers; will nest in agricultural landscapes, often in hedgerows and field margins. In many areas, shows a close association with Red-backed/Isabelline Shrikes, apparently gaining protection from these and showing higher breeding success in such situations (Shirihai *et al.* 2001). Strongly migratory, wintering in relatively restricted area of E Africa, from S Sudan to N Tanzania.

STATUS A scarce migrant to GB&I, and overwhelmingly an autumn bird. Dymond *et al.* (1989) report <1% of records in spring, and these were largely May birds along the east coast of England and Scotland. There are few reports of singing males in spring. The first autumn records are from early Aug, extending through until late Oct; Dymond *et al.* comment that the number of records appeared to be in decline, but this seems to have been a temporary effect, for the average annual total has increased since their study (Fraser & Rogers 2006). Birds very occasionally occur in winter. Over 100 records from Ireland: all in autumn, and only one adult – though many were unaged.

Lesser Whitethroat   *Sylvia curruca* (Linnaeus)

MB PM *curruca* (Linnaeus)
PM *blythi* Ticehurst & Whistler

TAXONOMY Complex; many local forms that have been described as subspecies, some of which may merit specific rank. Most authorities (Vaurie 1969, BWP, Shirihai *et al.* 2001) recognise *curruca*, *halimodendri*, *margelanica*, *minula* and *althaea*, regarding '*blythi*' as a synonym of *curruca*. However, much variation is clinal, rendering some divisions arbitrary. Shirihai *et al.* (2001) recognise four allospecies, treating *halimodendri* as a subspecies of *curruca*, but admit that the separation of these on a combination of size and plumage is 'largely unreliable'. Unpublished DNA sequences (in Shirihai *et al.* 2001) show that *margelanica*, *minula* and *althaea* form a clade, separate from *curruca/halimodendri*, and that the first three are as mutually divergent as other taxa of *Sylvia* that are treated as species. Martens & Steil (1997) report vocal differences between *curruca*, *minula* and *althaea* that also support specific rank. However, Shirihai *et al.* (2001) stress the need for field studies in the zones of contact/overlap of several pairs of taxa. Until these have been undertaken and the DNA data have been published, it seems unwise to realign species limits in this complex.

DISTRIBUTION Breeds extensively across the Palearctic, from GB & N France to the Gobi Desert, and from the edge of the tundra zone south to Israel, S Caspian and Pakistan. Nominate race: most northerly, extending as far south as Israel, N Iran, winters NE & C Africa, Arabia, NW Indian subcontinent; *halimodendri*: breeds in steppes of C Asia from Transcaspia to W Mongolia; *margelanica* is high-altitude form, breeding in mountains of W China; *minula* breeds in dry steppe from S Caspian to W China; *althaea* breeds in the mountains from SE Kazakhstan, Afghanistan to NW Himalaya.

STATUS Extended its range between the Breeding Atlases, occupying more 10-km squares in S Scotland, Wales and NW England, but still rare in Ireland, with just a handful of breeding records in Co Wexford since 1990, and even these are mostly published as 'probable'. BBS/CBC data suggest a decline since the mid-1980s, which is supported by constant-effort ringing data. The causes are unclear; Fuller *et al.* (2005) suggest that fluctuations in productivity may have driven this, but that problems on either migration or the wintering grounds may also be involved. Shows a decided preference for hawthorn scrub, especially with thick undergrowth; intensive management of hedgerows to maximise agricultural yields may be reducing suitable nesting sites. There are a few recent records of wintering birds, in the south of both GB and Ireland.

Lesser Whitethroat is one of the few birds breeding widely in GB&I that migrates round the E Mediterranean, rather than taking the shorter route to W Africa. Ringing recoveries indicate movement in a narrow front across Europe to the Adriatic, and thence through the Balkans to the

Levant. There is a cluster of recoveries in N Italy, suggesting that this is a major refuelling area, and a further cluster of recoveries in the E Mediterranean supports evidence from fat-loads indicating flight ranges of *c.* 800 km. Problems in these areas may be adversely affecting British birds, but there seems little firm evidence. There are no recoveries of GB&I birds from the wintering grounds.

Birds resembling *blythi* occur, particularly in the N Isles in autumn, though this taxon is of doubtful validity.

Orphean Warbler   *Sylvia hortensis* (J. F. Gmelin)

SM *hortensis* (J. F. Gmelin)

TAXONOMY Molecular data reported by Shirihai *et al.* (2001) show this to be part of a clade that also includes Arabian Warbler *S. leucomelaena* and two Afro-tropical *Sylvias*. Four subspecies are recognised: *hortensis*, *crassirostris*, *balchanica* and *jerdoni*. Shirihai *et al.* (2001) recommended splitting the western *hortensis* from the rest on the basis of DNA differences (although they seem only to have examined *hortensis* and *crassirostris*), and on differences in grey/olive balance to upperparts and whiteness of underparts. There are statistical differences in size that do not attain diagnosability, and differences in song. This may be the correct conclusion, but with the overlap in all traits apart from vocalisations, and in the absence of published DNA evidence, this has not been widely adopted.

DISTRIBUTION Linearly distributed in maquis from Iberia and Morocco, along both sides of the Mediterranean, through Turkey and SC Asia to Pakistan. Subspecies *hortensis* is western, from the Atlantic to Italy, W Libya; *crassirostris* breeds through the Balkans E towards the Caspian; *balchanica* breeds from NE Iraq to SE Turkmenistan and SE Iran; *jerdoni* is most eastern, from Iran and Turkmeniya to Pakistan. It seems that *jerdoni* winters in the Indian subcontinent; the other three races in sub-Saharan Africa, from S Mauritania to Sudan and Ethiopia, and the coasts of the Arabian peninsula.

STATUS There are only five records this species from GB&I; four from England and one from Scotland; it has not been recorded in Ireland. The first was trapped at Portland Bill (Dorset) in Sep 1955; recent analysis of DNA from a retained feather is strongly indicative of the race *hortensis*. Three subsequent records were from Cornwall and NE Scotland. Surprisingly for a migrant to the Mediterranean, only one (a Cornish bird) is from the spring; the remainder were all in Sep–Oct.

Asian Desert Warbler   *Sylvia nana* (Hemprich & Ehrenberg)

SM monotypic

TAXONOMY Recently split from African Desert Warbler *S. deserti* on basis of diagnosable differences in plumage and song by Shirihai *et al.* (2001), these authors also report (unpublished) differences in DNA sequences as great as those between (e.g.) Rüppell's and Cyprus Warblers *S. melanothorax*, or Subalpine and Sardinian Warblers. There are statistical differences in size between the two desert warblers, though these are not diagnostic. The desert warblers are part of a clade that includes Common Whitethroat and several other Mediterranean warblers, though well distant from these.

DISTRIBUTION Entirely allopatric to the African Desert Warbler, whose distribution centres on Algeria. Breeds in arid habitats from the Caspian Sea, E through Kazakhstan and Uzbekistan to NW China; unlike African Desert, is largely migratory, wintering from NW India, W through the Arabian peninsula to the Red Sea.

STATUS The 11 birds found in GB&I are all from England. The first was at Portland Bill (Dorset) for three weeks in midwinter 1970–71; subsequently, there have been 10 more, mostly in the late autumn on the E & S coasts. However, one Oct bird was in Cheshire, and there have been two in May (Yorkshire and Norfolk).

Common Whitethroat   *Sylvia communis* Latham

MB PM *communis* Latham

TAXONOMY According to the molecular results reported by Shirihai *et al.* (2001), Common Whitethroat is sister to a clade of 'Mediterranean' warblers, indicating a likely early divergence. Four subspecies are recognised by Shirihai *et al.* (2001) and Baker (1997): *communis, volgensis, icterops* and *rubicola*, although the last of these was not included by Vaurie (1959). Eastern birds are paler grey and larger, though boundaries between races are imprecisely known and their relationships are in need of clarification.

DISTRIBUTION Widespread across the Palearctic, from Iberia and uplands of NW Africa to Lake Baikal, between 65°N and 32°N. According to Shirihai *et al.* (2001), nominate breeds from Atlantic to N Turkey, and NE from Crimea to Ural Mountains; *volgensis* from Crimea, E to Lake Baikal; *icterops* from S Turkey, Israel to Caspian Sea and N Iran; *rubicola* in far east of range, from Tien Shan, Tadjikistan through to Mongolia. All races winter in Africa, through the Sahel, from the Atlantic to Red Sea, and S to Zimbabwe, Mozambique.

STATUS This species showed a dramatic fall in both distribution and numbers in 1969, when *c.* 75% of the population failed to return from the wintering grounds. The migration route of Common Whitethroats from GB&I is across the Channel to N France, and thence through Iberia and NW Africa to wintering grounds in the Sahel region of sub-Saharan Africa. A few birds may fly directly to Iberia, but the majority seem to move in shorter stages. Winstanley *et al.* (1974) showed that the fall in numbers was due to excessive mortality outside GB&I, and argued that it was related to drought conditions in the Sahel. Numbers remained low for about 15 years but, since 1985, there has been a slow recovery, albeit with occasional relapses (CBC/BBS). In fact, the recovery is not uniform, with differences in both region and habitat (Baillie *et al.* 2006). Populations monitored by BBS show greater recovery in England and Scotland than in Wales, and the numbers reported by WBS surveyors increased more rapidly than those of the BBS. Irish data are more limited, with breeding surveys not established until later, but data from Cape Clear and Copeland Observatories shows a decline and recovery that parallels the picture in GB. Occasionally winters in Ireland.

The crash of 1969 led to a series of studies of Common Whitethroat ecology and population biology, which showed how breeding populations are linked to both summer habitat and winter climate. It is clear that Common Whitethroats leave areas of coppice when the canopy begins to close; coastal and heathland scrub is slower to mature and provides a more stable habitat. There are fewer of these habitats in Wales, which may account for the slower recovery there. An increase along waterways may be due to the habitat alongside rivers and canals containing more scrub and hedgerow, which are optimal for this species. Even within the recovery period, Baillie & Peach (1992) found inter-annual changes in numbers that could be related to variations in overwinter survival imposed by drought conditions on the Sahelian wintering grounds. Despite the recovery, the GB population seems still to be only *c.* 40% of the pre-1969 level.

Spectacled Warbler   *Sylvia conspicillata* Temminck

SM race undetermined; all likely to have been *conspicillata* Temminck

TAXONOMY Unpublished DNA studies (Shirihai *et al.* 2001) indicate that this is part of a clade including two other groups of Mediterranean warblers. Two subspecies are recognised (Vaurie 1959), *conspicillata* and *orbitalis*, though differences in plumage are slight (Shirihai *et al.* 2001) and differentiation is low (Dietzen *et al.* 2008b).

DISTRIBUTION Disjunct series of populations around the Mediterranean and on some Atlantic islands. Nominate breeds in Iberia, S France, Italy, coastal N Africa from Morocco to Libya; a second area includes NE Egypt, Israel, Levant and Cyprus. Western populations partially migratory, wintering S through Algeria. Subspecies *orbitalis* is resident on the Atlantic islands of Madeira, Canaries and Cape Verdes.

STATUS Recorded four times in England but not elsewhere in GB&I. The first was a male at Filey (Yorkshire) in May 1992. There have been two further birds in May and one on Tresco (Scilly) in Oct.

Dartford Warbler   *Sylvia undata* (Boddaert)

RB *dartfordiensis* Latham

TAXONOMY Blondel *et al.* (1996) and Shirihai *et al.* (2001) report that this is sister species to Tristram's *S. deserticola*, within a clade of Mediterranean warblers that also includes Marmora's *S. sarda* and Balearic Warblers *S. balearica*. Up to eight subspecies have been described, but much variation is clinal, and only three are now generally recognised: *undata*, *toni* and *dartfordiensis*.

DISTRIBUTION Relatively restricted distribution in W Europe and adjacent coastal regions of N Africa. Most widespread is *undata*, which breeds S of Bordeaux, through Iberia and into Italy, including Sicily, Corsica, Sardinia; *dartfordiensis* replaces this to the north from the extreme N of the Iberian peninsula, through NW France and extreme S of England; *toni* is restricted to S Iberia and coastal areas from Morocco to Tunisia. Nominate and *toni* both partial migrants, wintering in south of breeding area, and into C Morocco, Algeria.

STATUS Dartford Warbler suffers adversely from prolonged cold winters and extensive habitat loss, as favoured heathlands have been converted to housing and agriculture. They formerly bred as far north as Staffordshire, but a loss of *c.* 80% of lowland heath in the 150 years to 1980 severely reduced populations (Brown & Grice 2005). Severe and prolonged cold weather hits this almost exclusively insectivorous species hard. Having barely recovered from the 1946–47 winter, the winters of 1961–62 and 1962–63 almost extinguished the British population, and further severe weather in the late 1970s did little to help. At times, the number of breeding pairs may have been as low as 12.

Dartford Warblers have the potential for a high reproductive rate, and the run of mild winters since the 1980s has seen a recovery, with populations buoyant, and expansion into areas such as Suffolk, which had been lost since the 1930s (Piotrowski 2003). It now breeds in most counties along the S coast of England, and the population had risen to >1,600 pairs by 1994 (Gibbons & Wotton 1996). In 2006, the population estimate had increased by a further 70%. Distribution expanded dramatically too, from the confines of southern England in 1994, with large extensions of the range across SW England throughout coastal S Wales and into the W Midlands and East Anglia. The 2006 survey gave an estimate of 3,214 territories with a 114% increase in range over 1994 at the 10-km square level (G Conway pers. comm.). As the climate continues to change towards warmer summers and (especially) milder winters, we might expect further expansion of range.

It remains a rare bird in Ireland, where there are less than 10 records, all from the S coast. The first was a female at Tuskar Rock Lighthouse (Wexford) in Oct 1912. Eight birds have been in Aug–Oct; the others in May 1999 (Dursey Is, Cork) and Mar 2004 (Brow Head, Cork).

Marmora's Warbler   *Sylvia sarda* Temminck

SM race undetermined; likely to have been *sarda* Temminck

TAXONOMY Differences in morphology and vocalisations suggest treating the two races as separate species – Marmora's Warbler *S. sarda* and Balearic Warbler *S. balearica* (Shirihai *et al.* 2001). DNA differences support this view, both those of Blondel *et al.* (1996) and unpublished results alluded to by Shirihai *et al.* (2001), which indicate that they are sister taxa, within the same clade as Dartford. Retained as subspecies here.

DISTRIBUTION Nominate restricted to Corsica, Sardinia, and a few adjacent small islets, plus some islands off the W coast of Italy; *balearica* restricted to the Balearic Is (Shirihai *et al.* 2001).

STATUS Five records, four in May and one in June. The first was a singing male holding a territory on Midhope Moor (Yorkshire) from May to July 1982. Subsequent birds have not remained as long. Evidence strongly suggests all to have belonged to the nominate race. Not recorded from Ireland.

Rüppell's Warbler   *Sylvia rueppelli* Temminck

SM monotypic

TAXONOMY Unpublished DNA data (Shirihai *et al.* 2001) suggests that this species is close to Cyprus Warbler, as part of a clade that also includes Subalpine and Sardinian.

DISTRIBUTION Very restricted distribution, being found breeding only on maquis hillsides of Greece, some Aegean islands, and coastal Turkey to NW Syria. Migrates through Israel and Egypt to winter principally in Sudan and Chad.

STATUS Five records from GB: two each from England and Scotland, and one from Wales. The first was present at Dunrossness (Shetland) for a month in Aug–Sep 1977. Two subsequent birds have been in May; the others in autumn. Not recorded from Ireland.

Subalpine Warbler   *Sylvia cantillans* (Pallas)

PM *cantillans* (Pallas)
SM *albistriata* (C. L. Brehm)

TAXONOMY Subalpine Warbler is phylogenetically close to Sardinian and Ménétrié's Warblers *S. mystacea*, and replaces the latter and Rüppell's in the W & C Mediterranean (Shirihai *et al.* 2001). Three subspecies are recognised by Vaurie (1959) and Svensson (1992): *cantillans, inornata* and *albistriata*; Shirihai *et al.* (2001) add *moltonii* to this list. The eastern race *albistriata* is well differentiated in plumage; the others show variation in the relative amounts of orange and pink in the breast and flanks. However, *moltonii* differs in song and contact call, and this taxon at least may merit separate species status from *S. cantillans*. The oldest name for this taxon probably is *subalpina* (Baccetti *et al.* 2007).

DISTRIBUTION Mediterranean region, from Iberia to Turkey, and Morocco to Tunisia. Nominate breeds from Iberia through S France and Italy; *moltonii* breeds in Corsica, Sardinia, Balearics and recently discovered also in C Italy, sometimes in sympatry with *cantillans* (Brambilla *et al.* 2006); *inornata* is found in N Africa; *albistriata* E from the Adriatic Sea to SE Turkey. Most populations are strongly migratory, wintering in Sahel region of sub-Saharan Africa, from Senegal to N Sudan.

STATUS Commonest of the rare *Sylvia* warblers, with over 500 records since the first on St Kilda (Outer Hebrides) in June 1894, with 32 individuals in Ireland to 2002. More occurrences in spring than autumn throughout GB&I. The number recorded has increased progressively from 1 per year in the 1950s, through 2 (1960s), 5 (1970s), 17 (1980s), 20 (1990s) to >21 since 2000.

  Most (identified) records were attributed to the nominate race, but birds identified as *albistriata* have been recorded *c.* 20 times, since the first on S Uist (Outer Hebrides) in May 1993. Four others were found in 1993, two each in 2002, 2003; and four in 2007; three were found on Foula (Shetland).

Sardinian Warbler   *Sylvia melanocephala* (J. F. Gmelin)

SM *melanocephala* (J. F. Gmelin)

TAXONOMY Unpublished DNA data (Shirihai *et al.* 2001) suggest a sister relationship with Ménétriés's Warbler, as part of a clade of Mediterranean warblers. Excluding Cyprus Warbler, which at the time was regarded as conspecific with Sardinian, five subspecies were listed by Vaurie (1959), though he expressed uncertainty about two of these; only two or posibly three recognised by Svensson (1992). These three, *melanocephala, momus* and *norrisae*, are differentiated on size and plumage (Baker 1997; Shirihai *et al.* 2001). Another subspecies, *valverdei*, was recently described from Western Sahara (Cabot & Urdiales 2005).

DISTRIBUTION Widely distributed in scrub and garrigue from the Canaries and around the Mediterranean, apart from the arid coasts of Libya and Egypt. Nominate occurs throughout range, apart from the Levant and Israel (*momus*), and a restricted locality in Egypt (*norrisae*), though this last may be extinct (Goodman & Meininger 1989). Birds from N Turkey, N Balkans are migratory; the other northern populations only partially so. Winter range extends through the Nile Valley to N Sudan, around the coasts of W Africa to Senegal, and oases in the desert.

STATUS Much rarer than Subalpine, with only *c.* 70 records from GB and a further two from Ireland, both of which were in Cork in April 1993 (Cape Clear, Knockadoon Head). The first for GB was trapped on Lundy (Devon) in May 1955. Records come from Mar to Nov, with a peak in Apr–May. As befits a Mediterranean species, records from England are somewhat earlier in the year (peaking Apr) compared with Scotland (May), and often involve singing males. The number of birds found annually has increased since the 1950s, from less than one a year in 1950–80, to around four since 1990. Whether this is due to increased observer awareness or a northwards spread in response to climate change is unknown.

A male was present at Skegness (Lincolnshire) from Oct 2003 to Jan 2004; remarkably, a female was in close proximity from Nov at least into Jan.

FAMILY LOCUSTELLIDAE

Genus *LOCUSTELLA* Kaup

Predominantly Palearctic genus of eight or nine species, five of which have been recorded in GB&I, two as breeding birds.

Pallas's Grasshopper Warbler   *Locustella certhiola* (Pallas)

SM *rubescens* Blyth

TAXONOMY Molecular analysis (Drovetski *et al.* 2004a) based on the mitochondrial ND2 gene indicates that *Locustella* may need revision: this genus is closely allied to *Bradypterus*, but these may not be especially close to *Cettia*, as has previously been assumed. *Locustella* itself appears to comprise two clades, and some species of *Bradypterus* may belong in one of these, making this genus paraphyletic. Pallas' Grasshopper Warbler is a member of a clade that includes (among other eastern species) Gray's *L. fasciolata* and Middendorff's *L. ochotensis* Grasshopper Warblers and Marsh Grassbird *L. pryeri*; Drovetski *et al.* (2004a) confirm Morioka & Shigeta's (1993) placement of this in *Locustella* rather than *Megalurus*. Four subspecies of Pallas's are recognised (Vaurie 1959; BWP, Baker 1997) based on colour of upperparts, greyness of hindneck, undertail coverts and intensity of streaking, but variation seems continuous.

DISTRIBUTION Widely distributed across N and E Asia, from W Siberia to the Pacific, breeding in marshes and overgrown riverbanks. Winters in SE Asia, from India to Indonesia. Nominate *certhiola* (E Asia), *rubescens* (N Siberia from River Irtysh E to Kolyma basin), *sparsimstriata* (C Siberia from Tomsk and Novosibrisk E to Baikal, S to Altai and N Mongolia), *centralasiae* (C Asia from Kazakhstan, through NE Kyrgystan and Gobian Altai S to N China).

STATUS Most records from the Northern Isles; *c.* 38 records in GB since first at Fair Isle (Shetland) in Oct 1949; over half of subsequent records also at Fair Isle, with most of the rest elsewhere in Shetland. The first for GB&I was at Rockabill Lighthouse (Dublin) in Sep 1908. One subsequent Irish record, from Cape Clear (Cork) in Oct 1990. All recorded between 13 Sep and 19 Oct.

Lanceolated Warbler   *Locustella lanceolata* (Temminck)

SM monotypic

TAXONOMY Drovetski *et al.* (2004a) show that this is part of a clade within *Locustella* that includes Common Grasshopper, River and Savi's Warblers, plus two Far Eastern taxa, currently placed in *Bradypterus*. Lanceolated appears closer to Common Grasshopper Warbler, but statistical support within this clade is weak.

DISTRIBUTION Across European Russia and Siberia, from the Urals to the Pacific, including N Japan, between 63°N and 39°N. Winters from E Nepal, through Bangladesh, Malaysia to Indonesia and Indo-China.

STATUS Since the first at Fair Isle (Shetland) in Sep 1908, there have been *c.* 115 records in GB, of which *c.* 85 have been in Orkney and Shetland. The dates of occurrence fall between Sep and mid-Nov, with most in late Sep–mid-Oct. There are no records from Ireland.

Common Grasshopper Warbler   *Locustella naevia* (Boddaert)

MB PM *naevia* (Boddaert)

TAXONOMY Drovetski *et al.* (2004a) suggest that this is sister to Chinese Bush Warbler *Bradypterus tacsanowskius*, and perhaps rather more differentiated from the rest of *Locustella*. Four subspecies recognised (BWP; Baker 1997), on balance of grey-olive-brown of upperparts, and intensity of streaking.

DISTRIBUTION Widely distributed across the W & C Palearctic from France, GB&I, S Scandinavia to NW Mongolia, W China, slightly S of Lanceolated in the East. Winters in several separate areas, including coastal W Africa (S Morocco, Senegal), Iran, Ethiopia, Indian subcontinent. Subspecies *naevia*: Europe to Ukraine; *obscurior*: Causcasus, S to NE Turkey and Armenia; *straminea*: Urals to NW China (Tien Shan); *mongolica*: E Kazakhstan to Mongolia.

STATUS Occurs in suitable habitat across much of GB&I, though less abundant in the far north. Showed a marked contraction in occupied 10-km squares between the two Breeding Atlases; disappearing from >35% in both GB and Ireland. There was no obvious regional pattern to this rapid and profound loss. Numbers fluctuate from year to year, but there was a decline in the CBC index during the early 1970s (Glue 1990), and this was mirrored by a decline in the number both recorded at coastal bird observatories and trapped by ringers at constant-effort sites (Gibbons *et al.* 1993). More recently, numbers seem to have stabilised; BBS data indicate a slight rise in population in the mid-1990s, followed by a subsequent reversion to the previous low level. Confidence limits are wide and trends are not especially clear.

    Part of the loss since the 1970s has been ascribed to loss of habitat: maturation of conifer plantations, drainage of damp woodland, and the clearance or drying out of ditches along field margins have all been implicated. Whether this is sufficient to account for the losses seems not to be known; the possibility of factors on migration or the wintering grounds cannot be excluded. Ringing data are not especially helpful here, since the recovery rate of ringed individuals for this secretive species is very low (<0.25%: probably the lowest for any British breeding bird), although it is enhanced if the birds are also fitted with a colour ring (Wernham *et al.* 2002). The wintering areas are still little known, although birds ringed in Gambia and Senegal have been found in GB, and there have also been birds from GB&I recovered in W France. Arrival along the south coasts is usually from mid-Apr, and departure is from July to Sep.

River Warbler   *Locustella fluviatilis* (Wolf)

SM monotypic

TAXONOMY Closely related to Savi's Warbler (Drovetski *et al.* 2004a), and part of a clade that includes Lanceolated and Common Grasshopper but not Pallas's Grasshopper, which is part of a separate group.

DISTRIBUTION Breeds across W & C Palearctic, between 60°N and 44°N; S Finland and Baltic States, C & E Europe to Ukraine and W Siberia. Migrates through E Mediterrranean, Nile Valley to winter in E Africa, from Zambia to Mozambique and perhaps S Malawi.

STATUS About 30 records from GB since the first on Fair Isle (Shetland) in Sep 1961. Found in all months from May to Oct, with a peak in Sep. Several males in song in suitable breeding habitat in spring/summer, but none has attracted a mate. Records increased during 1970–90 (indeed, it was once predicted as a potential colonist), but have fallen back more recently. There are no records from Ireland.

Savi's Warbler   *Locustella luscinioides* (Savi)

CB FB SM *luscinioides* (Savi)

TAXONOMY Phylogenetically close to River Warbler, with three subspecies (Baker 1997). Plumage varies clinally from brown *luscinioides* in the west, through olive-brown *sarmatica* (European Russia) to brownish grey *fusca* in the east. Recognition of *sarmatica* is not uniform: accepted by BWP and Baker (1997), but Svensson (1992) is less convinced.

DISTRIBUTION Breeds across W & C Palearctic between 55°N and 30°N, from NW Africa, across Europe to the Ukraine, through European Russia, and from Turkey and Levant to Mongolia. The nominate race breeds in NW Africa, Europe and Ukraine; *sarmatica* in European Russia; *fusca* from Asia Minor to Mongolia. Winters in sub-Saharan Africa, in a belt from Senegal to Ethiopia.

STATUS A rare breeding bird in GB, recorded in just 27 10-km squares in the Second Breeding Atlas (with breeding confirmed in only 8), mainly in SE England and East Anglia, although this represents a doubling of range since the First Breeding Atlas. This may have been a false dawn since the number of singing males has declined from an average of *c.* 25 in the late 1970s to *c.* 5 since the late 1990s (RBBP). The preferred habitat in GB&I is reedbed with thick ground cover and scattered bushes; this is now scarce, and a combination of continued protection of existing sites and creation of new ones is essential if Savi's Warbler is to further expand its limited base.

There are >600 records from England, >80% of them singing males; elsewhere in GB&I, it is much less common, with only *c.* 25 records away from the breeding sites. In Ireland, Hutchinson (1989) reported only two records, with the first at Shannon Airport (Clare) in June 1980; he commented that breeding was likely to be imminent, and the presence of further birds, including two together and one that remained throughout the summer of 1990, supported this view. Despite singing males, breeding has not yet been confirmed, and the last record (in song) was in 1996.

FAMILY ACROCEPHALIDAE

Genus *HIPPOLAIS* Conrad von Baldenstein

Predominantly Palearctic genus of eight species, six of which have been recorded in GB&I.

Eastern Olivaceous Warbler   *Hippolais pallida* (Hemprich & Ehrenberg)

SM *elaeica* (Lindermayer)

TAXONOMY See Thick-billed Warbler (p. 297). Species limits in the Olivaceous/Booted Warbler complex have recently been clarified as a result of molecular analysis and field study, especially the use of vocalisations. Some subspecies within the traditional 'Booted' and 'Olivaceous' Warblers are more appropriately treated as full species. It is possible that further re-alignment will be necessary. Five subspecies of Olivaceous Warbler were recognised by Vaurie (1959): *pallida, elaeica, reiseri, laeneni, opaca*. Molecular and vocal/behavioural evidence (summarised in Parkin *et al.* 2004) show there are clear differences between several of these. *Elaeica* (Eastern Olivaceous) and *opaca* (Western Olivaceous) differ in their DNA sequences by as much as (e.g.) Icterine and Melodious. *Opaca* differs from some of the others in plumage and behaviour. *H. opaca* (not recorded from GB&I) is treated as a separate species from the others, which are retained for now within a polytypic *H. pallida*.

DISTRIBUTION SC Palearctic and central N Africa. Subspecies *pallida*: breeds in Egyptian Nile Valley, migrating to S Nile Valley, Sudan and Horn of Africa. *Reiseri* occurs scattered through the oases of N Africa, from Algeria to Libya and possibly SE Morocco; northern birds migrate locally to S of range. *Laeneni* breeds in C & NE Niger, Chad, N Nigeria, mostly resident, but some N populations migrate to S of range. *Elaeica* is widely distributed from S Hungary and the Balkans through Turkey to Israel, Jordan, Iran, and N to Aral Sea and Kazakhstan; possibly extreme NE Africa and Arabia (Baker 1997); migratory, wintering in Africa, from Sudan to Tanzania.

STATUS The first two records came within three days of each other; the first was trapped on the Isle of May (Fife) in Sep 1967 and killed by a Great Grey Shrike two days later; the second was trapped at Sandwich Bay (Kent). Nine subsequently: from Shetland to Scilly, and between July and Oct. Three Irish records, all from Co Cork; the first on Dursey Is in Sep 1977, and the others on Cape Clear in 1999 and 2006.

Booted Warbler   *Hippolais caligata* (M. H. C. Lichtenstein)

SM monotypic

TAXONOMY See Thick-billed Warbler (p. 297). Previously treated as conspecific with Sykes's Warbler. Molecular, morphological and vocal evidence (summarised in Parkin *et al.* 2004) indicates that these two taxa represent distinct evolutionary lineages and should be treated as separate species. They are, however, each other's closest relatives and form a clade close to, but well differentiated from, the two Olivaceous Warblers.

DISTRIBUTION W & SW Russia and Siberia, Kazakhstan E to NW China and Mongolia to *c.* 60°N. Winters in Indian subcontinent.

STATUS The first was shot on Fair Isle in Sep 1936. The next did not follow for 30 years, when one was present, again on Fair Isle, in 1966. There have now been around 100, about a third in Scotland, the rest in England and Wales. There are only three records from Ireland, all first-winter birds and all very recent: Tory Island (Donegal) Sep 2003, Ballycotton (Cork) and Annagh Head (Mayo), both in Sep 2004. About 90% of Scottish and east coast birds were found in Aug–Sep, compared with only 55% of those from the south coast. There are four spring records, all in England in June.

Sykes's Warbler   *Hippolais rama* (Sykes)

SM monotypic

TAXONOMY Closely related to Booted Warbler, but diagnosably different in DNA, plumage and vocalisations.

DISTRIBUTION Breeds in C Asia, S of Booted. Winters through Indian subcontinent.

STATUS Nine records from GB to 2006, with the first on Fair Isle (Shetland) in Aug 1959. A relatively early migrant, all GB records have been between 20 Aug and 8 Oct, apart from one at Portland (Dorset) in July 2000. The sole Irish record is from Cape Clear (Cork) in Oct 1990.

Olive-tree Warbler   *Hippolais olivetorum* (Strickland)

SM monotypic

TAXONOMY From DNA analysis Helbig & Seibold (1999) and Fregin *et al.* (2009) found this to be sister to Upcher's Warbler *H. languida*, within a clade that also included Icterine and Melodious, best retained as *Hippolais*. Evidence was strong that this quartet was phylogenetically distant from the Booted Warbler complex.

DISTRIBUTION Breeds through S Balkans, Crete, Asia Minor to N Israel and S Caspian; migratory, wintering in Africa, from Sudan to Tanzania.

STATUS One, a first-winter at Boddam (Shetland) in Aug 2006.

Icterine Warbler   *Hippolais icterina* (Vieillot)

CB PM monotypic

TAXONOMY See Olive-tree Warbler. Closely related to Melodious Warbler, though differing diagnosably in plumage and DNA (Helbig & Seibold 1999). Occasional hybridisation in the zone of contact may result in sterility among female progeny (Secondi *et al.* 2003). Songs differ in structure, but elicit bi-directional interspecific aggression among males, indicating these act as a pre-copulatory barrier to mating (Secondi *et al.* 2003).

DISTRIBUTION Breeds across C Europe, from NE France to SW Siberia, and from C Finland to N Balkans. Almost entirely allopatric to Melodious Warbler, with only narrow zone of contact, especially in E France. Winters in sub-Saharan Africa, from Zaire to South Africa.

STATUS Scarce, but regular records of singing males in spring, some of which stay for prolonged periods (e.g. Outney Common (Suffolk) in May–July 2004). A pair bred in Strathspey (Highland) in 1993, Orkney in 2002, and possibly on other occasions in Scotland (BS3), although earlier reports of breeding in England are now generally discounted (Andrew 1997).

Over 3,500 migrants recorded during 1968–2002 and, with *c.* 100 records annually, this species is about three times as common as Melodious Warbler. Icterine Warbler is predominantly an autumn migrant in GB, with 86% of records at this season, ranging along the E & S coasts, from Shetland to Scilly (Dymond *et al.* 1989). Increasingly recorded in spring, mainly along the east coasts and in the northern isles, though smaller numbers are regularly recorded further west. Only a few of the 198 records from Ireland are from this season.

Melodious Warbler   *Hippolais polyglotta* (Vieillot)

PM monotypic

TAXONOMY See Icterine Warbler.

DISTRIBUTION Replaces Icterine Warbler in SW Europe, from S Belgium, Switzerland to Iberia (including Italy and NW Balkans) and extreme NW Africa. Winters in W Africa, further north than Icterine, from Senegal to Cameroon.

STATUS Typically found in SW of GB&I, as expected from its breeding distribution (Dymond *et al.* 1989). Over 1,100 were recorded during 1968–2002, or about 30 per year. Wintering in W Africa and migrating through SW Europe, its arrival in spring is generally along the S coast:, quite distinct from the Icterine Warbler's more easterly arrivals. The autumn pattern is similarly southerly, with only a handful of records N of Yorkshire and Lancashire. The majority of the 160+ Irish records are in the autumn.

Genus *ACROCEPHALUS* Naumann & Naumann

Old World and Australasian genus of *c.* 36 species, eight of which have been recorded in GB&I, three as breeding birds.

Aquatic Warbler   *Acrocephalus paludicola* (Vieillot)

PM monotypic

TAXONOMY Molecular data place this with Sedge Warbler and Moustached Warbler *A. melanopogon*, forming a clade of 'striped' reed warblers (Leisler *et al.* 1997; Fregin *et al.* 2009).

DISTRIBUTION Restricted to E Europe, with substantial breeding populations only in Belarus, Ukraine and Poland, where *c.* 80% of world population resides, but the last of these and Lithuania are in decline. BirdLife International (2004) suggests this will continue due to habitat loss. Migrates SW through Iberia to W Africa, where wintering grounds have recently been discovered in Senegal.

STATUS Shows a striking difference in occurrence across the region: over 600 records in GB but only 13 in Ireland and none there since 1989. There are also differences in the time of occurrence: more birds are recorded in July–Aug in Scotland (75%) and Wales (80%) than in England (60%) where later in autumn, with fewer yet in Ireland (30%). The differences between Scotland/Wales, England and Ireland are statistically significant and indicate earlier arrival at more northern locations. Ringing data suggest a migration route from E Europe west to the S parts of GB, along the English Channel coast to W France and Iberia. As there are only two spring records (May in England; June in Ireland), the return routes may be different.

Sedge Warbler   *Acrocephalus schoenobaenus* (Linnaeus)

MB PM monotypic

TAXONOMY Molecular evidence places this in a clade with Aquatic and Moustached Warblers (Leisler *et al.* 1997; Fregin *et al.* 2009).

DISTRIBUTION Widely distributed across W & C Palearctic from France, GB&I and coastal Norway to River Yenesei, roughly from 71°N to Caucasus and N Iran at *c.* 30°N. W European birds migrate through Iberia to winter in W & C Africa.

STATUS Breeds across much of GB&I, apart from Shetland, and scarcer in mountainous areas. Distribution declined between the Breeding Atlases, especially in S Ireland and C & SW England. This contraction of range coincided with a fall in numbers reported by both CBC and WBS studies. Much of the decline took place between 1965 and 1985, since when the population has stabilised, or even recovered somewhat, though numbers are still >20% down on the 1960s.

Most GB&I birds winter in sub-Saharan W Africa, from Senegal to Burkino Faso, and there is good evidence for strong site fidelity between winters. Peach *et al.* (1991) showed that changes in population size in GB could be related to adult survival, and that numbers were especially low in years of poor rainfall on the wintering grounds in W Africa. Although confidence limits are wide, failure rate at the egg stage has decreased by *c.* 50% since the 1960s; this may be due to breeding birds now being concentrated in optimal habitats.

Migration in Sedge Warblers appears to be a series of long-distance flights, interspersed with periods of feeding and fattening for the next stage; protection of these sites is essential for the continued survival of this species. Arrival in GB&I tends to be from early Apr, and there is strong site fidelity between years. There seems to be a limited amount of passage through GB to NW Europe, with recoveries to and from S Scandinavia.

Paddyfield Warbler   *Acrocephalus agricola* (Jerdon)

SM monotypic

TAXONOMY Recent analyses by Leisler *et al.* (1997), Helbig & Seibold (1999) and Fregin *et al.* (2009) have gone some way to resolving the relationships of the *Acrocephalus* warblers (summarised by Parkin *et al.* 2004). Paddyfield Warbler is sister species to Manchurian Reed *A. tangorum* and Blunt-winged *A. concinens,* and this clade is sister to a clade containing the Eurasian Reed Warbler, Marsh Warbler, Blyth's Reed Warbler and others, which could collectively be treated as genus *Notiocichla*, or grouped with the striped reed warblers as *Calamodus* (Fregin *et al.* 2009). Three subspecies '*agricola*', '*septimus*' and '*capistrata*' have been recognised (e.g. Baker 1997), but the first two of these are genetically near-identical and all are morphologically non-diagnosable.

DISTRIBUTION C & SE Palearctic, from Rumania E to NW China. Migrates S to winter from S Iran across N India to Assam.

STATUS More than 60 records in GB since the first on Fair Isle (Shetland) in Sep 1925, and all coastal apart from one in Hertfordshire in Nov 1981. Half of the records come from the Northern Isles, especially Fair Isle. Most have been in Sep–Oct, but there are five records from May–June, all on the east coast. The first Irish bird was at North Slob (Wexford) on the late date of 3 Dec 1982 (dying on the 4th); there have been three others subsequently, all at more typical dates in Sep–Oct, most recently in 2000.

Blyth's Reed Warbler   *Acrocephalus dumetorum* Blyth

SM monotypic

TAXONOMY Both Leisler *et al.* (1997) and Helbig & Seibold (1999) found it lay outside the main Reed Warbler clade, despite occasional hybridisation with Marsh Warbler. Closely related to the Large-billed Reed Warbler *A. orinus* (Round *et al.* 2007). Fregin *et al.* (2009) found it to be sister to *A. orinus* and a member of the '*Notiocichla*' clade (see under Paddyfield Warbler).

DISTRIBUTION Breeds widely across Palearctic between 63°N and 35°N, from Finland across Russia, Siberia to NW Mongolia, and south to Iran, N Afghanistan. Winters in Indian subcontinent, east to Myanmar.

STATUS Over 90 records since the first on Fair Isle (Shetland) in Sep 1910, though still none in Ireland. Most have been in the autumn, with rather more in Sep than Oct, but there are also *c.* 12 records from May–June. About half have been found in the N Isles. The number of records has increased from <1 per year in 1950–80, through *c.* 2 in the 1990s to *c.* 4 per year since 2000, with 13 in 2007. Twelve of the spring birds have been since 1990. While this increase is undoubtedly partly (or largely) due to increased observer skills, it seems likely that there has been a real increase in occurrence. BirdLife International (2004) report that >2 million pairs breed in Europe, and that the populations of Russia and the Baltic republics are stable or increasing.

Marsh Warbler   *Acrocephalus palustris* (Bechstein)

CB PM monotypic

TAXONOMY Leisler *et al.* (1997) and Helbig & Seibold (1999) found that Marsh Warbler lay in the Reed Warbler clade, though its DNA was clearly different from races of Eurasian Reed.

DISTRIBUTION Breeds in the W Palearctic between 61°N and 38°N, across Europe from Britain (very local), NE France, Switzerland and S Scandinavia to the Caucasus and N Iran. Migrates through the Middle East and the Rift Valley, to winter in eastern Africa as far as the Cape.

STATUS This species has declined over the last hundred years, although it has probably never been common in GB&I (Holloway 1996). Between 1890 and 1986, it was recorded breeding in 19 counties of England and 1 in Wales (Kelsey *et al.* 1989). By the time of the First Breeding Atlas, this had reduced to 18 10-km squares and, by the Second Breeding Atlas, this had declined further to only 8. It was absent from Scotland and Ireland, never having bred in the latter. Historical strongholds were in S & W England, especially Worcestershire, where as recently as 1984 there had been 31 territory-holding males. By 1987, this had declined to only four, and it has now disappeared as a breeding bird. Kelsey *et al.* (1989) ascribe the decline to a combination of habitat loss and climate change. Some sites have lost the dense ground-layer vegetation with scattered scrub that this species prefers. However, numbers have also fallen at sites where the habitat had not changed noticeably. Average temperature in the Severn valley fell slightly during 1950–65, but the major drop in Marsh Warbler numbers occurred later than this. Kelsey *et al.* suggest that the decline may be due to both of these, but that isolation may also have played its part. Living at the western edge of the range, the Severn population may have suffered from emigration losses not being balanced by immigrants. A lack of unringed birds entering the population supports this hypothesis. Between and subsequent to the Breeding Atlases there has been a colonisation of SE England. Several new sites were occupied and small numbers continue to breed there. Elsewhere, it appears to be opportunistic, with occasional breeding well away from the S of England (e.g. Orkney 1993; Yorkshire 2001).

Marsh Warbler continues to be an uncommon migrant, with an average of *c.* 50 records each year (Fraser & Rogers 2006), although this may also be reducing. Birds are usually found along the E coasts. It is a late-spring migrant, with most individuals being reported between mid-May and June. Smaller numbers occur in Sep and Oct. It is extremely rare in Ireland, with only three confirmed records: Cork in Aug 1991 and Sep–Oct 2004, and Wexford in June 1996. Its migration route is through SE Europe into Africa via the Rift Valley. There is a recovery of a bird in Greece from Oct 1985 that had been ringed as a nestling in Worcestershire earlier the same year.

Eurasian Reed Warbler   *Acrocephalus scirpaceus* (Hermann)

MB PM *scirpaceus* (Hermann)

TAXONOMY See Paddyfield Warbler. Complex, and in need of re-analysis, using morphology, vocalisations and DNA. Two subspecies in the Palearctic based largely on plumage colour, varying clinally from warm olive-brown in the west (*scirpaceus*) to cold grey-olive in the east (*fuscus*). There is also a series of taxa in Africa, including *baeticatus, avicenniae, cinnamomeus*, that vary especially in size and wing formula, and which may or may not be separate species.

DISTRIBUTION W & C Palearctic, between 62°N and 30°N. The nominate race breeds from NW Africa and Europe, east to W Russia, Crimea and W Asia Minor; *fuscus* breeds in pockets of suitable habitat from Cyprus and Levant, through N Caspian region to SE Kazakhstan and extreme NW China, extending S to NW, C & E Iran and NW Afghanistan. Both races winter in sub-Saharan Africa as far S as Zambia, *fuscus* predominantly in the east, and *scirpaceus* in the west and centre.

STATUS As a breeding bird, largely restricted to England and Wales, with relatively small numbers in Scotland and Ireland. There was a slight decrease in distribution between the Breeding Atlases, but probably within sampling error. Baillie *et al.* (2006) suggest that CBC, BBS and WBS data (which all indicate an increase in population) are not suitable for assessing status because of the clumped distribution of this species. Bird ringers working constant-effort sites in reedbeds report a decline during 1985–92, followed by an increase, and then a further decline since 2000. Laying date has advanced since the early 1980s; during this period, clutch size, and nest failure rates, at both egg and chick stages, have declined; whether these are causally linked is not clear. This species shows confusing patterns of change.

Apart from a single pair in 1935 (Co. Down), Ireland has been colonised only since 1980. The breeding population is confined to SE coasts (between Cork and Wicklow), with only occasional breeding outside this area. The population seems to have declined since the mid-1990s (certainly in Cork: P Smiddy pers. comm.) and total breeding numbers almost certainly do not exceed 100 pairs. Otherwise a passage migrant, rare in spring and scarce in autumn.

As with other reedbed or skulking species, ringing data is heavily biased towards live-recaptures by other ringers. Consequently, within the range of the species, the data tend to represent the distribution of ringers rather than of birds. Unlike Sedge Warblers, Eurasian Reed Warblers seem to migrate through a series of shorter stages, with many leaving GB to the SE, and then moving through Biscay and Iberia towards NW Africa, generally remaining close to the coast. There are very few recoveries from the wintering grounds; these seem to be from Morocco and Gambia to Guinea-Bissau.

Great Reed Warbler   *Acrocephalus arundinaceus* (Linnaeus)

SM *arundinaceus* (Linnaeus)

TAXONOMY Included in the molecular analysis of Fregin *et al.* (2009), who found it in a clade with Oriental Reed *A. orientalis*, Clamorous Reed *A. stentoreus* and a group of mainly E Asian/Pacific taxa. Great Reed Warbler is now treated as comprising two subspecies, western *arundinaceus* and eastern *zarudnyi*.

DISTRIBUTION Breeds across the Palearctic from NW Africa and Continental Europe to Tadjikistan and Xinjiang, between 60°N and 26°N; the dividing line between *arundinaceus* and *zarudnyi* runs through Ural Mountains and N Kazakhstan, but there is extensive intergradation (Baker 1997), presumably due to hybridisation and gene flow.

STATUS There are over 220 GB records, spread from S England to the northern isles. Annual numbers have remained fairly steady at around five birds each year since the 1960s. Records come from Apr to Nov, with the majority in spring (>85% May–June), and many of these were located by their loud and strident song. Although some have remained for several weeks, there is no evidence that any have bred. There are only three records from Ireland, all from Cork, with the first dead at Castletownsend in May 1920, and two on Cape Clear in June 1964 and May 1979.

Thick-billed Warbler   *Acrocephalus aedon* (Pallas)

SM race undetermined

TAXONOMY Included in two recent molecular studies, which showed that it is a distinctive member of *Acrocephalus* (Helbig & Seibold 1999). Fregin *et al.* (2009) found it to be sister to a clade containing African *Chloropeta* along with Eastern and Western Olivaceous, Booted and Sykes's Warblers, which they collectively place in genus *Iduna*. Some (e.g. Vaurie 1959, HBW) suggest two subspecies, *aedon* and *stegmanni*, but differences are slight and may not be justified (Svensson 1992).

DISTRIBUTION Breeds across Siberia to the Pacific seaboard. The nominate is more westerly, breeding in S Siberia and N Mongolia; *stegmanni* breeds from E Siberia and NE Mongolia to Sea of Japan and NE China. Migrates S to winter in coastal E India, Malay peninsula, Indo-China.

STATUS There are only four records, all from Shetland. The first was trapped on Fair Isle in Oct 1955, where the only spring bird was also present in May 2003. The other two were on Whalsay in Sep 1971 and Out Skerries in Sep 2001. There are no records from Ireland.

FAMILY CISTICOLIDAE

Genus *CISTICOLA* Kaup

Old World genus of *c.* 50 species, one of which has been recorded in GB&I.

Zitting Cisticola   *Cisticola juncidis* (Rafinesque)

SM race undetermined

TAXONOMY A variable species, with up to 18 subspecies (Vaurie 1959; Baker 1997). Those found in W Palearctic include *juncidis* (S France, E to W Turkey and Egypt, including Mediterranean Islands E from Corsica, Sardinia); *cisticola* (W France, Balearics, Iberia, NW Africa); *neurotica* (Cyprus, Levant to Iraq and W Iran), *cursitans* (E Afghanistan, Pakistan and parts of Indian subcontinent E to Yunnan).

DISTRIBUTION Probably the most widely distributed member of the Sylviidae; certainly the most widespread *Cisticola*, being found from S Europe and Africa, across S Asia to China and through the Malay peninsula, Philippines to Australasia. Resident, with very limited powers of dispersal, accounting for the strong interpopulation variation in plumage tones, extent and intensity of streaking, patterns of moult. Breeds in open grassland and savannah, often close to fresh or saline water

STATUS First record for GB&I was on Cape Clear (Cork) in Apr 1962; one subsequently, also Cape Clear in Apr 1985: these being the only Irish occurrences. First for GB was at Cley, then Holme, (Norfolk) Aug–Sep1976. Four subsequently in GB, three in Dorset during May–June, and one in Kent (Oct 2006).

FAMILY BOMBYCILLIDAE

Genus *BOMBYCILLA* Vieillot

Holarctic genus of three species, two of which have been recorded in GB&I.

Cedar Waxwing   *Bombycilla cedrorum* Vieillot

SM monotypic

TAXONOMY *Bombycilla* forms clade with *Phainopepla*, *Phainoptila*, *Ptilogonys*, *Dulus*, *Hypocolius* and *Hylocitrea* (Spellman *et al.* 2008).

DISTRIBUTION Woodland and woodland edges across N America from SE Alaska and NE California to Newfoundland and Virginia. Northern populations migratory; winter range extends to Costa Rica and Caribbean. Numbers and distribution in winter fluctuate, largely determined in response to variable food supply.

STATUS Two records, both from GB. The first was present on Noss (Shetland) in June 1985; the second was around Nottingham (Nottinghamshire) Feb–Mar 1996. As this is a relatively popular cage bird, earlier records were considered unsafe. Not recorded in Ireland.

Bohemian Waxwing   *Bombycilla garrulus* (Linnaeus)

WM *garrulus* (Linnaeus)

TAXONOMY Three subspecies recognized, differing in plumage colouration, though one of these is doubtfully valid (Witmer 2002).

DISTRIBUTION Holarctic, breeding in northern forests and wet woodland from Scandinavia to Kamchatka; also in W North America, from Alaska to Hudson Bay, S to British Columbia. Northern populations migratory, wintering south of the breeding range across Holarctic, avoiding arid and treeless zones. Superimposed on the regular migrations is an irruptive behaviour, when large numbers may occur well south of normal range. Three subspecies; nominate: Europe to Ural Mts; *centralasiae*: Ural Mts to Kamchatka; *pallidiceps*: Nearctic.

STATUS A regular winter visitor to GB, although numbers vary markedly among years, with peak arrivals in years when breeding has been good, coupled with poor autumnal food supply on the breeding grounds. Arrivals typically late Oct–early Nov, building up rapidly in peak years, and often found in Rowan and other berry-bearing trees within urban areas. There is no obvious periodicity to invasion years, although suggestion that they might be becoming more frequent. The pattern of occurrence in Ireland is generally similar, although it does not occur every year, being scarce and local with most in the E & NE counties.

Generally, arrivals are along the E coasts, with rapid movement inland, although sometimes (as in 1995–96) birds are found early in W Ireland; such western arrivals sometimes coincide with high numbers in E Canada and may indicate birds from the Nearctic. Although there are no confirmed examples of *pallidiceps* from GB&I, the occurrence of a Cedar Waxwing in Nottingham in the winter of 1995–96 may be supportive of a Nearctic origin of at least some of the birds. Colour-ringing has shown that flocks move around GB, generally heading south and west as food suppies become depleted.

Birds may remain well into May, but there are no breeding records in GB&I.

FAMILY TICHODROMIDAE

Genus *TICHODROMA* Illiger

A monospecific Palearctic and Oriental genus that has been recorded in GB&I.

Wallcreeper   *Tichodroma muraria* (Linnaeus)

SM race undetermined, but all likely to have been *muraria* (Linnaeus)

TAXONOMY Two subspecies recognised, western *muraria* and eastern *nepalensis*, though variation clinal (Vaurie 1959).

DISTRIBUTION Scattered through mountain regions from Iberia to China. Movements little known, but seems to be partly an altitudinal migrant with some birds transferring to lower-level ranges in winter. Occasionally disperses far from native range.

STATUS Ten records from GB. The first was at Stratton Strawless (Norfolk) in Oct 1792. Most have been in S England, between Apr–June and Sep–Nov. Several of these have wintered in GB, including Dorset Nov 1969–Apr 1970, Somerset Nov 1976–Apr 1977 and at the same site, presumably the same bird, from Nov 1977 to Apr 1978. Not recorded in Ireland.

FAMILY SITTIDAE

Genus *SITTA* Linnaeus

A Holarctic and Oriental genus of *c.* 24 species, one of which breeds in GB&I and another has occurred as a vagrant.

Red-breasted Nuthatch   *Sitta canadensis* Linnaeus

SM monotypic

TAXONOMY Member of a group of small Nuthatches that includes Krüper's *S. krueperi*, Corsican *S. whiteheadi*, and Algerian *S. ledanti* from the W Palearctic (Pasquet 1998).

DISTRIBUTION Boreal forests from S Alaska to Newfoundland, breeds regularly south to California in west, and to N Carolina in east. Northern populations irruptive migrants, depending on seed crop; birds may winter as far south as Gulf of Mexico.

STATUS A single record from GB&I: a first-year male in Holkham Wood (Norfolk) Oct 1989–May 1990.

Eurasian Nuthatch   *Sitta europaea* Linnaeus

RB *caesia* Wolf

TAXONOMY Very complex. Seventeen subspecies listed by Harrap & Quinn (1996), which can be partitioned into three groups on basis of geography and breast colour: *caesia* (7 sspp.), *europaea* (9 sspp.), *sinensis* (1 ssp.). Within the groups, variation is poorly marked (Harrap & Quinn 1996) and clinal; where the groups meet, populations are very variable. This suggests secondary contact and consequent disruption of locally adapted genomes through hybridisation. In view of the generally sedentary nature of the species, the apparent breadth of these hybrid zones implies less profound differentiation than (e.g.) in Carrion/Hooded Crows, but more detailed study is needed to determine the existence or otherwise of cryptic species *sensu* Irwin *et al.* (2001a). A molecular study (Zink *et al.* 2006b) found a very deep split between the taxon *arctica* and the other subspecies of *S. europaea* but did not revise species limits. Red'kin and Konovalova(2006) proposed specific status for *arctica* based on morphological differences and sympatric breeding with *S. europaea* in NE Siberia.

DISTRIBUTION Breeds in a belt of predominantly deciduous woodland from Iberia and S Scandinavia to Kamchatka, and S into E China. The southern populations of W Europe, Balkans, Turkey to Caucasus form *caesia* group, including *hispaniensis* (S Iberia, Morocco), *cisalpina* (S Switzerland, Italy, NE shores of Adriatic), *caesia* (W & C Europe to Carpathian Mts, and S to Greece) plus four more from Turkey and Middle East; central populations from Scandinavia to Pacific comprise *europaea* group, including *europaea* (Scandinavia, European Russia to Ukraine) plus eight more subspecies across Siberia to Japan.

STATUS Does not occur in Ireland. Eurasian Nuthatch is expanding its GB range northwards on both sides of the Pennines and it now occurs across S Scotland. Numbers increased steadily since the 1970s, and by 2003 the population was more than double that in 1967 (CBC/BBS). This general statement masks variable patterns of abundance, with lower densities away from S & W England and Wales. As with many other woodland birds, laying date has advanced since the 1980s, presumably in response to climate change. There was also a marked increase in brood size through to the 1990s, with a slight decline subsequently. This does not seem to be associated with clutch size, which has remained fairly uniform since 1968, although sample sizes are smaller than for brood size.

Dispersal is low in Eurasian Nuthatches, though several widely scattered records in Scotland N to Caithness. Isolation is an important constraint on woodland occupancy. Bellamy *et al.* (1998)

examined a series of small, apparently suitable woods in Cambridgeshire that lacked Eurasian Nuthatches. Many of these were effectively isolated and received insufficient immigrants for successful colonisation. Improving the quality of individual woodlands would permit more small woods to be colonised, but the extinction risk would still be high. More valuable would be increasing the overall area of woodland to create larger units capable of sustaining a population.

FAMILY CERTHIIDAE

Genus *CERTHIA* Linnaeus

An Holarctic and Oriental genus of six to nine species, two of which have been recorded in GB&I, one as a breeding bird.

Eurasian Treecreeper   *Certhia familiaris* Linnaeus

Endemic: RB *britannica* Ridgway
PM *familiaris* Linnaeus

TAXONOMY About 13 subspecies (Vaurie 1959; Harrap & Quinn 1996) have been described, falling into two groups based on plumage and geography. Northern group (*familiaris*) shows clinal variation in colour saturation; Himalayan/Chinese birds of second group are much darker. Phylogenetic and vocal analyses (Tietze *et al.* 2006) indicate the need for a revision of species limits within *C. familiaris*. Molecular data identified two major clades, each of which included two subclades. One of these comprised the Nearctic forms of Brown Creeper *C. americana* as sister taxa to the W Palearctic Short-toed Treecreeper. The second clade showed that the races from Eurasia and N China form a clearly distict subclade to those from the Sino-Himalayan region. Vocal data support the separation of the latter as *C. hodgsoni* Hodgson's Treecreeper. Tietze *et al.* (2006) draw attention to the molecular divide between the European and N African races of *C. familiaris*, suggesting that further cryptic species may exist in this complex.

DISTRIBUTION More or less continuous in woodland from GB&I to Japan. Four of the nine *familiaris* subspecies occur in W Europe. These are: *britannica* (GB&I), *macrodactyla* (W Europe, from France to River Oder, and S to C Hungary and NW Balkans), *corsa* (Corsica), *familiaris* (Fennoscandia, Poland, S to E Hungary, Bulgaria, Greece to W Siberia and River Yenesei); the other five range from Turkey to Japan. The southern group breed from the Himalayas to C China. Most W & S populations are resident, but individuals may wander well beyond their subspecies borders; eastern birds may show altitudinal movement to avoid worst of winter conditions.

STATUS Between the Breeding Atlases there was an apparent decline in distribution in Ireland, with *c.* 20% fewer 10-km squares showing evidence of breeding; this may be partly due to a lower coverage in the second Atlas. In general, this is a bird of mixed or deciduous woodland, with mature conifers an important secondary habitat. There was a strong increase in the population during 1965–72, followed by a decline to 1980, since when the numbers have been generally stable (CBC). The first increase may have been a recovery from the severe winter of 1962–63, and the subsequent decline in response to cold winters in the early 1980s. There may have been a further decline during 1995–2004 in England, but not in Wales; data from elsewhere are too limited for interpretation (BBS). The Repeat Woodland Survey of 1980–2002 (Amar *et al.* 2006) also shows evidence of a decline, supporting the BBS results.

Numbers and survival are adversely affected by wet winters, as well as prolonged periods of cold (Peach *et al.* 1995b). Despite the few data, there is evidence that failures at the egg stage have declined since the 1960s, and as with many other woodland species, there has been a strong move towards earlier laying.

Eurasian Treecreepers are largely sedentary with few movements of any great distance. The breeding birds belong to the race *britannica*. Eurasian Treecreepers occur at east coast sites, mainly in autumn, and some are likely to be *familiaris* from the Continent. Many, if not most, appearing in the N Isles have been identified to this race, and some have overwintered (BS3).

Short-toed Treecreeper   *Certhia brachydactyla* **C. L. Brehm**

SM race undetermined

TAXONOMY Believed closest to Brown Creeper *C. americana* of N America based on similarities in songs and calls (Baptista & Krebs 2000) and phylogenetic analyses (Tietze *et al.* 2006). Five subspecies recognised by Vaurie (1959), Harrap & Quinn (1996), based on clinal variation in size and colour.

DISTRIBUTION More restricted than Eurasian Treecreeper, being confined to W Palearctic. Breeds from Iberia and NW Africa to S Denmark, Poland to Balkans and coastal Turkey. Isolated populations around Black Sea. Subspecies: *megarhyncha* (N Iberia, across N France to W Germany), *brachydactyla* (S & E of *megarhyncha*, from C Iberia, S France, Denmark, E to Balkans), *mauritanica* (NW Africa), *dorothea* (Cyprus), *harterti* (Turkey to Caucasus). Resident, with small amount of post-breeding dispersal outside of natural range.

STATUS The first for GB was at Dungeness (Kent) in Sep 1969; subsequently, there have been 25 records, of which 19 have been in Kent (12 at Dungeness). Others have been in Essex, Yorkshire and Dorset. Records come from most months of the year, but there is a peak in Sep–Oct. Although it breeds as close as the Channel Islands, there are no breeding records from GB&I. Not recorded from Ireland.

FAMILY TROGLODYTIDAE

Genus *TROGLODYTES* Vieillot

Predominantly New World genus of 9–14 species (one Holarctic); one of which breeds in GB&I.

Winter Wren   *Troglodytes troglodytes* **(Linnaeus)**

Endemic: RB *fridariensis* Williamson
Endemic: RB *hebridensis* Meinertzhagen
Endemic: RB *hirtensis* Seebohm
Endemic: RB *indigenus* Clancey
Endemic: RB *zetlandicus* Hartert
WM PM *troglodytes* (Linnaeus)

TAXONOMY Molecular analysis of mitochondrial and nuclear DNA (Barker 2004) confirmed that *Troglodytes* is monophyletic within the Troglodytidae, and that Treecreepers and Gnatcatchers *Polioptila* are close allies. A more detailed analysis of *T. troglodytes* phylogeography by Drovetski *et al.* (2004b) did not include the island races from the Atlantic coasts of Europe, but found six clades that fitted well with geography. Although not all of these were equally supported statistically, European populations were monophyletic and sister to those from the Caucasus. Two clades from E Palearctic were similarly monophyletic, and Palearctic birds themselves were also monophyletic. They identified two clades in the Nearctic: E & W; the former were sister to the Palearctic birds. The historical scenario they propose is that W Nearctic wrens split from the remainder first; a second divergence was between E Nearctic and Palearctic birds, and that the latter subsequently split into E & W isolates. Divergence occurred within both of these last to yield the groupings that are evident today: E Asia and Nepal, Caucasus and Europe. Drovetski *et al.* (2004b) were able to estimate timescales for the divergences, and these fitted with known glaciations.

Brewer (2001) lists 27 races of Winter Wren in the Palearctic and a further 12 in the Nearctic; HBW records 44 (30 in the Palearctic). Of these, six occur in GB&I, five as resident endemics, the last as a migrant. These vary in size, colour and vocalizations. Differentiation presumably rapid, for many of the races occupy locations that are likely to have been uninhabitable during the last ice age.

DISTRIBUTION Holarctic; breeds in a wide range of habitats from boreal forests to tree-less islands. In the Nearctic, from S Alaska across James Bay to Newfoundland; W populations extend through the Rocky Mts to California; E populations through New England and S through the Appalachian Mts. In the Palearctic, occurs across much of Europe to the Caucasus, Himalayas to China and discontinuously N to Kamchatka. Subspecies endemic to GB&I are *zetlandicus*: Shetland Is; *fridariensis*: Fair Isle; *hirtensis*: St Kilda; *hebridensis*: rest of Outer Hebrides; *indigenus*: rest of GB&I. These differ from nominate *troglodytes* (mainland Europe to Ural Mts) in size and general plumage colours, especially in rufous/grey balance (refs in McGowan *et al.* 2003). Across Holarctic, many N populations migratory; being exceptionally susceptible to prolonged cold and ice, birds usually move S to avoid extremes of climate, to winter in temperate regions of both continents, though other populations (usually on small islands) quite sedentary.

STATUS During both Breeding Atlases, Winter Wren was one of the most widely distributed species in GB&I, with evidence of breeding in over 97% of 10-km squares, and no sign of any change in distribution between the two survey periods. Numbers fluctuate markedly, perhaps more so than for any other breeding bird. CBC and BBS fieldwork has shown that numbers were very low following the 1962–63 winter, but recovered to reach a peak in the early 1970s. Numbers dipped during the cold winters of the early 1980s, but have recovered again to their previous level, although Baillie *et al.* (2006) indicate that the increases since 1994 have been limited to Scotland (87%) and N Ireland (63%) rather than England (0%) and Wales (6%). The Repeat Woodland Surveys confirm the overall increase since the 1980s but, because of the time gap between the fieldwork, the intervening peaks and troughs were unrecorded. Records of Winter Wrens in gardens have only been recorded systematically since 1995; these peak in winter and showed a dip through 1997–98 as numbers struggled to recover from poor breeding during the cold, wet spring of 1995.

Sharp declines in cold winters are due to dependency on a ready supply of invertebrate food, and the adverse effects of snow cover or the glazing of trees, stone walls and other feeding areas by ice, which interrupt feeding, with fatal results. High reproductive rates ensure that recovery can be rapid. There has been a slight increase in brood size since 1964, and productivity (measured by the proportion of juveniles to adults at constant-effort ringing sites) tends to be higher following severe winters. Laying date has advanced by about a week since 1968.

The status of the individual subspecies is known with varying degrees of accuracy. Pennington *et al.* (2004) report that the populations of *zetlandicus* and *fridariensis* are 1,500–3,000 and 25–40 singing males respectively; the small size of the latter putting it at particular risk from chance extinction. Boyd & Boyd (1990) suggest that there may be *c.* 225 pairs of *hirtensis* on the main islands of St Kilda. The population size of *hebridensis* is broadly estimated at 5,000–10,000 pairs (BS3). The number of nominate birds migrating to or through GB&I is also unknown, although there are counts of 30–40 birds at sites such as Out Skerries, Sumburgh and Fair Isle (Shetland), usually during Oct.

FAMILY MIMIDAE

Genus *MIMUS* Boie

New World genus of nine species, one of which has been recorded in GB&I.

Northern Mockingbird　　　　　　　　　　　　　*Mimus polyglottos* (Linnaeus)

SM race undetermined

TAXONOMY Molecular phylogenetic studies indicate that mockingbirds (Mimidae) and starlings (Sturnidae) are closely related. Mimidae and Sturnidae are both monophyletic if the Philippine 'creepers' *Rhabdornis* are included in the latter (Lovette & Rubenstein 2007; see also Cibois & Cracraft 2004). Mimidae comprises two subclades: *Mimus* and *Toxostoma* in one, and *Dumetella* in the other, along with some C American and Caribbean genera. Three subspecies of Northern Mockingbird are recognized by HBW.

DISTRIBUTION Much of N America from S British Columbia, Great Lakes, Maritimes, S to Mexico and Caribbean islands. The nominate race breeds in E N America, and is the most likely to have occurred; other races occur in the W and the Caribbean. Largely resident; most northerly populations move south; apparently also dispersive in search of suitable food outside breeding season.

STATUS Two accepted records for GB&I. The first was at Saltash (Cornwall) in Aug 1982; the second at Horsey Is (Essex) in May 1988. Earlier records from Norfolk (1971) and Gower (1978) were regarded as likely to be of captive origin and ship-assisted, respectively. Not recorded in Ireland.

Genus *TOXOSTOMA* Wagler

New World genus of 10 species, one of which has been recorded in GB&I.

Brown Thrasher　　　　　　　　　　　　　　*Toxostoma rufum* (Linnaeus)

SM *rufum* (Linnaeus)

TAXONOMY Phylogenetically close to mockingbirds *Mimus* (Cibois & Cracraft 2004; Lovette & Rubenstein 2007), and usually treated as comprising two subspecies.

DISTRIBUTION Breeds from central N America to Canadian Maritimes, S to Florida and Gulf Coast. W populations assigned to *longicauda*, E to *rufum*. N populations are mostly short-distance migrants; those further south are resident.

STATUS The only record was at Durlston Head (Dorset) from Nov 1966 to Feb 1967. This bird may have been ship-assisted, although it did arrive shortly after the transit of Atlantic Hurricane Lois (Andrew Harrop pers. com.). Not recorded in Ireland.

Genus *DUMETELLA* Wood

Monospecific New World genus that has been recorded in GB&I.

Grey Catbird　　　　　　　　　　　　　　*Dumetella carolinensis* (Linnaeus)

SM monotypic

TAXONOMY DNA sequences confirm that this is part of the Mimidae (Cibois & Cracraft 2004; Lovette & Rubenstein 2007).

DISTRIBUTION Generally breeds in thick secondary woodland from British Columbia to Great Lakes and Nova Scotia, and S to Gulf of Mexico. Populations along Atlantic coast of USA are resident; others are migratory, wintering in Caribbean islands and along the coasts the Gulf of Mexico.

STATUS Two records; the first was on Cape Clear (Cork) in Nov 1986; the second at South Stack (Anglesey) in Oct 2001. There is also a record of a bird on board the *QEII* when it docked at Southampton (Hampshire) in Oct 1998.

FAMILY STURNIDAE

Genus *STURNUS* Linnaeus

A Palearctic and Oriental genus of *c.* 15 species, one of which breeds in GB&I.

Common Starling   *Sturnus vulgaris* Linnaeus

Endemic: RB *zetlandicus* Hartert
RB WM PM *vulgaris* Linnaeus

TAXONOMY Sister species to Spotless Starling *S. unicolor*, with which it is largely allopatric in the breeding season, and these two basal to a large clade of Eurasian starlings (Lovette *et al.* 2008). A variable species, with 12 races recognised by Feare & Craig (1998), which differ in glossing of head and body; underwing colour, size, bill dimensions, juvenile plumage.

DISTRIBUTION Breeds across Palearctic, from Pyrenees, GB&I, Iceland through Europe (except S Italy, S Balkans); N populations extend to E of Lake Baikal; more scattered in S, through Turkey to S Caspian, and into Kazakhstan. Introduced populations in N America, S Africa, Australia, New Zealand. Nominate occurs from Iceland, most of GB&I, Pyrenees S to Italy, N Balkans, E to European Russia, Ukraine. Replaced by *faroensis* in Faroe Is; by *zetlandicus* in Shetland (and possibly the Outer Hebrides); by *granti* in Azores; and by *poltaratskyi* in E European Russia. Other races across Asia (BWP).

N & E populations are migratory. Recent years have seen a decrease in migration, with more birds overwintering in or near breeding grounds, especially in warmer urban areas. Many birds winter in SW Europe and N Africa; ringing data indicate these originate in N Scandinavia and NW Russia. The island races *zetlandicus* and *faroensis* are largely resident.

STATUS There are three components to the population of Starlings in GB&I: the subspecies *zetlandicus*; the resident breeding birds and those that come to these islands to winter, chiefly from Scandinavia and the Baltic. There have been few attempts to determine either of the latter populations, though the Second Breeding Atlas suggested a population of >6.6 million birds (Gibbons *et al.* 1993). BBS data for 1994–2000 indicated a figure of 8.5 million (Robinson *et al.* 2005), although this includes non-breeding individuals. Densities were 10-fold higher in urban/suburban habitats than in the wider countryside, and perhaps 60% of the population now breeds in these habitats. Despite the limited data on population size, there is no doubt that breeding abundance has declined since the 1960s, with a decrease in woodland of >90% in the period 1965–2000 (CBC). Declines in farmland are less, though still substantial (*c.* 66%, 1962–2000), with significant differences between farm type. Declines have been more severe on livestock than arable farms, perhaps because of changes in food availability (Robinson *et al.* 2005). It is now included in the 'Red List', and research is underway to identify the driving forces behind its decline. In a multivariate analysis of its demography, Freeman *et al.* (2007) found evidence that the national decline might be due to changes in the survival of first-year birds. This appeared to vary regionally, being less marked in E England, where adult survival (possibly in the breeding season) seemed to be more involved.

Assessing winter abundance is equally problematic. Fewer birds are using gardens in winter (refs in Robinson *et al.* 2005), and those using farmland in Oxfordshire fell by 50% between 1975and 1996. It is surprising that so little quantitative data exist about this important member of our fauna.

Common Starlings that breed in GB&I are largely resident, with very few overseas recoveries (Wernham *et al.* 2002). The population is boosted in winter by the arrival of large numbers, chiefly from the Low Countries, S Scandinavia, the Baltic States, and Poland. These birds are declining as well (Feare 1994). BirdLife International (2004) reports severe declines in the populations of Germany, Poland and Russia, and moderate declines in most other N European states. Declining overseas populations combined with reduced migration ('short stopping') due to climatic amelioration will inevitably reduce winter numbers in GB&I.

Subspecies *zetlandicus* confined to Shetland, with intermediate birds in Fair Isle (Shetland) and Outer Hebrides (BS3). Little population information. Adults apparently show only local movements, though first-year birds from Fair Isle have been recovered in N & E Scotland.

Genus *PASTOR* Temminck

A monotypic Palearctic genus, of which the only species has occurred in GB&I.

Rosy Starling *Pastor roseus* (Linnaeus)

PM monotypic

TAXONOMY A recent phylogenetic study (Lovette *et al.* 2008) supports placement of Rosy Starling in monotypic genus *Pastor* (e.g. Feare & Craig 1998).

DISTRIBUTION Breeds from Turkey through Iran to Afghanistan and extreme W China; in the north, occurs as far as S Ukraine, N Caspian and Aral Seas. Breeding range is determined by availability of grasshoppers and locusts. Migratory, wintering in peninsular India; in some years, may spread W into Europe, and this may be followed by breeding in new locations.

STATUS The first was killed near Norwood (Middlesex) in 1742; the first for Ireland was at Roxton (Clare) in about 1808. Subsequently, hundreds have been recorded in GB&I. There has been a dramatic change since the late 1970s; the annual means (recorded in the Scarce Migrant Report) increased progressively from 3 in the 1960s through 6 (1970s), 8 (1980s), 18 (1990s) to 92 since 2000. Numbers from the bird observatories showed a parallel trend, suggesting that there has been a real increase in occurrence. The three years with the highest number recorded have all been since 2000. Found in every month, but the majority are in June–Oct. In some years (e.g. 2003), more birds are recorded in the SW of England than elsewhere, although East Anglia is also a favoured area. Whether this indicates arrival from a more southerly direction is unknown.

The nearest breeding populations are in Ukraine, Bulgaria, Romania and European Russia, with smaller numbers in the Balkans, including Greece. BirdLife International (2004) suggests that these are generally stable but fluctuating; increasing records from GB&I likely stem from expanding populations.

FAMILY CINCLIDAE

Genus *CINCLUS* Borkhausen

Holarctic genus of five species, one of which breeds in GB&I.

White-throated Dipper *Cinclus cinclus* (Linnaeus)

Endemic: RB *gularis* (Latham)
Endemic: RB *hibernicus* Hartert
PM *cinclus* (Linnaeus)

TAXONOMY A molecular analysis of the genus by Voelker (2002a) found *C. cinclus* to be sister to Brown Dipper *C. pallasii* of E Asia; these two were themselves sister to a clade including the three New World species. Twelve subspecies are recognized by HBW, of which four are European. A study of these (Lauga *et al.* 2005), using a shorter DNA sequence than Voelker and limited by small sample sizes, found no statistical difference between *hibernicus* and either *aquaticus* or *cinclus*, although the latter two differed from each other, and all three from *gularis*. Hourlay *et al.* (2008) recovered five distinct genetic lineages in the W Palearctic, and suggested the validity of some of the subspecies would benefit from a reassessment.

DISTRIBUTION Largely resident on clear, fast-flowing waterways across the W Palearctic, from NW Africa, GB&I, Scandinavia to Mongolia, though absent from many polluted or lowland rivers across E Europe and W Asia. Local movements away from ice-bound rivers in winter.

Nominate subspecies breeds across much of W Europe, E to River Pechora. Replaced in W Scotland, Hebrides and Ireland by *hibernicus*; in remainder of GB by *gularis*; from Belgium through France to parts of Iberia, and SE to Italy, Greece by *aquaticus*. Eight further races breed in NW Africa and E from Turkey through the Ural Mts to W China.

STATUS Breeds along clean and fast-flowing streams and rivers across GB&I, being notably absent from the low-lying agricultural land of England and C Ireland. Between the two Breeding Atlases, there was evidence that White-throated Dippers had disappeared from those parts of GB where acidification of the watercourses had increased. White-throated Dippers are not especially well covered by CBC and BBS fieldwork or even the Waterways Survey. There was a slight decline in population during 1975–2003, although there are fluctuations within this period and the confidence limits are wide. An apparent decline in Ireland was apparently due to reduced coverage. In Cork and Waterford there has been no change in White-throated Dipper (*hibernicus*) populations over the last 25 years (refs in Smiddy & O'Halloran 2004).

White-throated Dippers are a good monitor of water quality, especially acidity, and several local studies have monitored population dynamics. These have shown that breeding performance has improved since the 1960s; in particular, nest failure at the egg stage and brood size have improved markedly. Since there was evidence of organochlorine, PCB and even DDT contamination of eggs in the 1980s (see Ormerod & Tyler 1992 for detailed refs), these improvements in breeding performance may be due to the gradual reduction of pollutants in the riverine environments following the bans on their use.

Ringing studies have shown that there is very little exchange of birds between regions, and this lack of gene flow has no doubt accelerated the divergence of the two endemic subspecies. Movement tends to be along waterways, with birds descending to lower altitudes in hard weather.

Nominate race *cinclus* is a rare visitor to GB&I, with about 70–80 records to 2003; the vast majority of these are from Shetland, and there is a generally easterly bias to the remainder. The peak months are Mar–Apr and Oct–Nov. There is a single record from Ireland; a bird on the River Tolka (Dublin) in Jan–Feb 1956.

FAMILY MUSCICAPIDAE, SUB-FAMILY TURDINAE

Genus *ZOOTHERA* Vigors

Oriental, Ethiopian, Australasian and Palearctic genus of *c.* 35 species, two of which have been recorded in GB&I.

Scaly Thrush/White's Thrush   *Zoothera dauma* (Latham)

SM *aurea* (Holandre)

TAXONOMY Klicka *et al.* (2005) examined the molecular phylogeny of 54 species of thrushes using mt DNA sequences and found strong evidence that *Zoothera* is polyphyletic. They noted that Scaly and Siberian Thrushes lie in separate clades and that these are not closely related (see under Siberian Thrush). The clade that includes Scaly is sister to most of the species that they included. Siberian *aurea* (Scaly Thrush) and Himalayan *dauma* (White's Thrush) show marked differences in song and differences in morphology (Martens & Eck 1995), indicating that they may be better treated as separate species. Interpopulation variation in size and plumage of Scaly Thrush is largely clinal. Six races recognised by HBW, two further 'doubtful' by Vaurie (1959), but Clement *et al.* (2000) recognise only four, splitting them largely on the grounds of distribution and isolation.

DISTRIBUTION Rather disjunct distribution, with populations assigned to *aurea* across C & E Siberia from S Urals to Korea, generally between 64°N and 43°N; nominate *dauma* breeds from Himalayas to Indo-China. Further races or species occur in Far East and peninsular India. Subspecies *dauma* is largely resident or an altitudinal migrant, but *aurea* migrates extensively to winter from NE India to coastal China.

STATUS Over 60 records, with the first for GB being shot near Christchurch (Hampshire) in Jan 1828. The first (of four) in Ireland was also shot, near Brandon (Cork) in early Dec 1842. There are records for most months between Sep and May, with peaks in Oct and Jan; the former presumably correlating with migration to the wintering grounds and the latter perhaps with midwinter hard-weather movements. Of 24 Scottish records, 21 were in Sep–Nov and only 3 in Jan–Feb; the figures for England are 11 in each period. Of the three Scottish winter records, two were in Perthshire and Lanarkshire, well away from the N & E coasts, where almost all of the rest originate, further supporting the view that autumn and winter birds have different histories.

Siberian Thrush   *Zoothera sibirica* (Pallas)

SM race undetermined

TAXONOMY Recent molecular analyses have shown that *Zoothera* as currently recognised comprises two distinct clades, *Zoothera* and *Geokichla*. Siberian Thrush is sister to the other species in the *Geokichla* clade and best included under that name (Klicka *et al.* 2005, Voelker & Klicka 2008, Voelker & Outlaw 2008). Two subspecies are recognised (Vaurie 1959, Svensson 1992, Clement *et al.* 2000), chiefly on basis of blackness of plumage.

DISTRIBUTION A bird of C & E Palearctic between 69°N and 35°N, from River Yenesei to Sea of Okhotsk, and south to Japan. Nominate across most of range; *davisoni* restricted to region between S Sakhalin and S Japan. Migratory, wintering from E India to Indo-China and Indonesia.

STATUS Only nine records in GB&I to 2007. The first for GB was on the Isle of May (Fife) in Oct 1954; the first of two Irish records was at Cape Clear (Cork) in Oct 1985. All records are in the period Sep–Dec.

Genus *IXOREUS* Bonaparte

Monotypic Nearctic genus that has been recorded once in GB&I.

Varied Thrush   *Ixoreus naevius* (J.F. Gmelin)

SM race undetermined

TAXONOMY Previously placed in both *Zoothera* and *Turdus*. The molecular results of Klicka *et al.* (2005) indicate that neither of these is supported, but that it is probably best placed in its own genus *Ixoreus*, in a clade with, but phylogenetically distinct from, *Catharus*, *Hylocichla* and Aztec Thrush *Ridgwayia pinicola*. Two subspecies are recognised, separated on characteristics of female plumage, although their validity is still unclear (George 2000).

DISTRIBUTION The nominate race breeds on the coastal side of the mountains from SE Alaska to N California; *meluroides* breeds across much of the rest of Alaska, west to Mackenzie River, and south through British Columbia to the extreme NW USA. Both races are migratory, wintering south through the coastal states of USA to N Baja California.

STATUS One only, at Nanquidno (Cornwall) in Nov 1982. The bird was not only out of range but also showed an aberrant plumage (lacking the buff/orange pigment) that has apparently been recorded only once before (Law 1931).

Genus *HYLOCICHLA* Baird

Monospecific Nearctic genus that has been recorded in GB&I.

Wood Thrush   *Hylocichla mustelina* (J. F. Gmelin)

SM monotypic

TAXONOMY Clement *et al.* (2000) place this in *Catharus* with the other Nearctic thrushes, but Roth *et al.* (1996), BWP and HBW retain it in its own monotypic genus. Molecular evidence places this in clade with *Ridgwayia* and *Catharus* (Voelker *et al.* 2007).

DISTRIBUTION Breeds in mixed and deciduous woodland in E North America, from the Canadian Maritimes to the Gulf of Mexico. Migratory, wintering in the forests of Central America.

STATUS One: a first-winter bird on St Agnes (Scilly) in Oct 1987. Not recorded in Ireland.

Genus *CATHARUS* Bonaparte

Neotropical, Nearctic and marginally Palearctic genus of about 12 species, four of which have been recorded in GB&I.

Hermit Thrush   *Catharus guttatus* (Pallas)

SM race undetermined

TAXONOMY Molecular data (Klicka *et al.* 2002; Outlaw *et al.* 2003; Winker & Pruett 2006) suggest this to be part of a *Catharus* clade and sister taxon to Russet Nightingale-Thrush *C. occidentalis*. Shows extensive variation in both size and plumage across its range. Up to 13 subspecies have been described (Jones & Donovan 1996), which form three groups: Pacific coast, interior mountains of the west, and eastern.

DISTRIBUTION Widespread across N America, including most of Canada and both NW & NE USA. Pacific populations are resident or altitudinal migrants; remaining birds migrate south, to winter in south of USA, and Mexico to N Central America.

STATUS Seven records for GB&I since the first on Fair Isle (Shetland) in June 1975; apart from this and a bird at the end of Apr, all records were found in the second half of Oct. There are only two Irish records, both in Oct: the first was a bird at Galley Head (Cork) in 1998, a second on Cape Clear (Cork) in 2006.

Swainson's Thrush   *Catharus ustulatus* (Nuttall)

SM *swainsoni* (Tschudi)

TAXONOMY Molecular data (Klicka *et al.* 2005) support treatment as a member of *Catharus*. Most recent authors (e.g. Evans Mack & Yong 2000; HBW) recognise six subspecies, though Clement *et al.* (2000) only accept four. These separate into two groups: western russet-backed *ustulatus* and eastern olive-backed *swainsoni*, which differ in size, plumage, habitat, winter range, DNA and vocalisations (Ruegg 2007).

DISTRIBUTION Three western subspecies (*ustulatus, phillipsi, oedicus*) breed from SE Alaska to California, wintering generally in Central America. The eastern races (*swainsoni, incanus, appalachiensis*) breed in the boreal forests of Alaska and Canada, and in the E USA south to Virginia, and winter from Panama to Peru. Western populations of these apparently fly east towards Atlantic before turning south and crossing Gulf of Mexico.

STATUS The first for GB&I was found dead at Blackrock Lighthouse (Mayo) in May 1956; there have been three more in Ireland, all in Oct in County Cork (1968, 1990, 1999). The first for GB was on Skokholm (Pembrokeshire) in Oct 1967, since when there have been 24 more. In total, 24 of the 29 records have been in Oct, and only one has been away from the northern and western fringes. The dates fit with migration patterns in N America: >95% of birds trapped at a site in Pennsylvania were between 1 Sep and 15 Oct, and 70% in Alabama between last week of Sep and first half of Oct (Evans Mack & Yong 2000). The only GB&I record allocated to subspecies was the first, in Ireland, which was determined to be *swainsoni*.

Grey-cheeked Thrush   *Catharus minimus* (Lafresnaye)

SM race undetermined

TAXONOMY Using mt DNA sequences, Outlaw *et al.* (2003), Klicka *et al.* (2005) and Winker & Pruett (2006) identified a subclade of *Catharus* thrushes that comprised Veery, Grey-cheeked and Bicknell's Thrushes; these three differ among each other by c. 2.2% at the mt control region (Lowther *et al.* 2001). Although these figures are not high, there are apparently fixed differences among them, supporting the case for treating the last of these as a distinct species *C. bicknelli* (Ouellet 1993). Grey-cheeked has two weakly differentiated races, and some authorities treat the species as monotypic.

DISTRIBUTION Breeds in the taiga forests and tundra shrubs of the Nearctic and adjacent NE Siberia. Subspecies *minimus* restricted to Newfoundland and 'possibly N Quebec' (Lowther *et al.* 2001); elsewhere subspecies *aliciae*. Strongly migratory, wintering in Caribbean and N South America.

STATUS The first record from GB&I was a first-winter bird trapped on Fair Isle (Shetland) in Oct 1953, with a further 45 subsequently. The first for Ireland was on Cape Clear (Cork) in Oct 1982; there have been three more since, all in Oct. Only three of the GB&I records to 2003 were not in Oct, and only four of the GB records were away from the northern and western fringes. Grey-cheeked Thrushes arrive in GB&I significantly later than Swainson's Thrushes. Two-thirds of the autumn records of Grey-cheeked were after 16 Oct, compared with only one-third of Swainson's.

After a review of the specimens available from GB, Knox (1996) concluded that birds from Lossiemouth (Moray) Nov 1965, St Mary's (Scilly) Oct 1986, and probably St Kilda (W Isles) Oct 1965 seemed to be closer to *aliciae*, and one from Bardsey (Gwynedd) Oct 1971 appeared closer to *minimus*. However, in view of the uncertainty over the validity of the races themselves, it seems safest to record all birds in GB&I as 'race undetermined'.

Veery   *Catharus fuscescens* (Stephens)

SM race undetermined

TAXONOMY Closely related to Grey-cheeked and Bicknell's Thrushes (Outlaw *et al.* 2003; Klicka *et al.* 2005; Winker & Pruett 2006). Up to six subspecies have been described, but several of these are doubtfully diagnosable (Bevier *et al.* 2004).

DISTRIBUTION Breeds in damp woodland across S Canada from British Columbia to Newfoundland, and S through USA at increasing altitudes to the mountains of C Colorado. Subspecies *fuscescens* breeds in SE Canada and E USA; replaced from Newfoundland to Quebec by *fuliginosus*; other races further S & W. Migratory, flying overland through USA, C America to wintering grounds in Brazil.

STATUS The first was a first-winter bird trapped at Porthgwarra (Cornwall) in Oct 1970. There were six further records to 2006, all in the northern isles or SW. There are no records from Ireland. The predominantly inland migration route presumably accounts for its rarity in GB&I relative to other *Catharus* thrushes.

Genus *TURDUS* Linnaeus

Near-cosmopolitan genus of over 70 species, 12 of which have been recorded in GB&I, six as breeding birds.

Ring Ouzel   *Turdus torquatus* Linnaeus

MB PM *torquatus* Linnaeus
SM *alpestris* (C. L. Brehm) or *amicorum* Hartert

TAXONOMY Klicka *et al.* (2005) used mitochondrial sequences to show that *Turdus* forms a well-supported clade, though this includes four species not currently in *Turdus* that might merit inclusion within the genus. Voelker *et al.* (2007) examined 60 of the 65 species within *Turdus* in more detail; they identified a Eurasian clade, within which Ring Ouzel lay as sister taxon to a subclade including Dark-throated and Naumann's Thrushes as sister species. The superficially similar plumage and ecology with Common Blackbird would thus seem to be convergent. Three subspecies are recognised based on size of pectoral band, extent of the pale edges to the wing panel, and degree of scalloping on the underparts. The degree of differentiation among these needs clarification.

DISTRIBUTION Restricted to W Palearctic, breeding from GB&I NE through Scandinavia to White Sea (*torquatus*), and discontinuously across the montane regions of S Europe from Iberia to W Turkey (*alpestris*) and from here to Iran (*amicorum*). Most populations are migratory (excepting only those in the extreme S), wintering in S Spain and NW Africa, Balkans, Turkey and Iran.

STATUS Largely restricted to the uplands and not well covered by most breeding bird surveys. There was clear evidence of a contraction of range between the two Breeding Atlases, with birds being lost from *c.* 27% of 10-km squares between 1968–72 and 1988–91. Ring Ouzel has always been less common in Ireland; it bred in most counties at the beginning of the 20th century but, by the 1960s, had decreased substantially in range, so that only the most mountainous areas were occupied. The decline continued through to the Second Breeding Atlas, though this might be partly a result of poorer coverage in upland habitats (Buchanan *et al.* 2003).

These contractions caused alarm, and dedicated surveys were undertaken in selected regions, which revealed a population decline of >50% through the 1990s. Burfield & Brooke (2005) showed that numbers recorded at western bird observatories during 1970–98 fell by a similar amount, although this was not matched by observatories on the eastern side of GB&I.

Birds breeding in GB&I appear to depart in Sep and winter from SW France to NW Africa. There is a secondary peak of birds at coastal observatories during Oct–Nov; these are migrants originating in Fennoscandia. The differences between observatories on the two sides of GB&I presumably reflect the different origins and destinations of their birds. Those in the west are predominatly birds that are part of the declining population within our islands; those along the eastern side are migrants on their way to or from Fennoscandia, where the population is stable (BirdLife International 2004). The reasons for the decline in GB&I are not fully understood, although recent work by the RSPB (Sim *et al.* 2007b) found that the extent of heather cover was important in determining site occupancy. Furthermore, population declines were greater two years after high spring rainfall in Morocco (Beale *et al.* 2006), a factor that is known to reduce juniper pollination and consequently the abundance 18 months later of the berries that form an important food for wintering ouzels.

Birds in GB from May and Sep–Oct are likely to have been *alpestris* or (less likely) *amicorum* but these require review.

Common Blackbird   *Turdus merula* Linnaeus

RB MB WM PM *merula* Linnaeus

TAXONOMY Vœlker *et al.* (2007) found Common Blackbird to be basal to most of the Eurasian *Turdus* and not especially close to any other taxa that they analysed. The species has been divided into many subspecies, from 9 (HBW) or 10 (Vaurie 1959) to 15 (Clement *et al.* 2000), though Collar (in HBW) suggests that several of these merit specific rank. A more detailed molecular analysis of the Common Blackbird complex is required.

DISTRIBUTION Widely distributed through the Old World, from the Atlantic islands and Iberia, GB&I, Scandinavia across C Europe through the Balkans, Turkey to Afghanistan and C & E China. Northerly populations are migratory, wintering through the range, and beyond into Iran, Iraq and S China. Nominate breeds across Europe, replaced by island races in Atlantic; further races in NW Africa, SE Europe, S Aegean, and across S & E Asia.

STATUS One of our most widespread species, being recorded in over 95% of 10-km squares in both Breeding Atlases; there was a modest contraction in distribution of 1.9% in GB and 0.8% in Ireland. CBC and BBS data indicate that there was a steady decline in numbers from 1965 to 1995, with a recovery through to 2004. The earlier decline led to Common Blackbird being added to the Amber List, but it has now reverted to the Green List. Siriwardena *et al.* (1998) suggest that the decline might have been driven by reduced adult survival. Laying date appears to have been advancing since the 1980s, and nest failure at the chick stage is also declining; these may be partly due to the run of warmer, earlier springs.

The pattern is not uniform across the archipelago; numbers are increasing more in the N and in W England than elsewhere, and the recovery is less evident in woodland (Amar *et al.* 2006). Hatchwell *et al.* (1996) compared the breeding biology in woodland, farmland and woodland edges; they found evidence that the recruitment rate to maintain a farmland population was approximately twice that required for stability in woodland. They concluded that farmland showed characteristics of suboptimal habitat for Common Blackbirds. The lower recovery in this habitat may stem from this suboptimality.

Common Blackbirds in GB&I are generally sedentary, although a small number migrate south and west in winter. Birds from northern Britain move further than those from the south, with many more wintering in Ireland. Very large numbers of Common Blackbirds come to Britain in late autumn, when they can be seen arriving off the sea along the east coasts of Scotland and England. These are predominantly from Scandinavia, but birds also come from the Netherlands across to the Baltic states. It has been suggested (Wernham *et al.* 2002) that the proportion of birds originating in northern Scandinavia declined towards the end of the 20th century, perhaps due to amelioration of the winters in those areas.

Eyebrowed Thrush   *Turdus obscurus* J. F. Gmelin

SM monotypic

TAXONOMY With Pale *T. pallidus* and Brown-headed *T. chrysolaus* Thrushes of SE Asia in a clade that is sister to Island Thrush *T. poliocephalus*; has been regarded as conspecific with former two, but molecular data lend little support to this.

DISTRIBUTION Breeds in C & E Siberia, from River Ob to Sea of Okhotsk, with isolated populations in Kamchatka. Migratory, wintering in S China, through to Sumatra.

STATUS The first was at Oundle (Northamptonshire) in Oct 1964, followed by another two that year. Since then, there have been 15 further records, although only one since 2000 and none from Ireland. There is an easterly bias among these records, with most in Sep–Oct.

Dusky Thrush   *Turdus eunomus* Temminck

SM monotypic

TAXONOMY Dusky *T. eunomus* and Naumann's Thrushes *T. naumanni* are very closely related taxa and treatment as separate species is borderline. Molecular data published to date (e.g. Voelker *et al.* 2007) do not include both subspecies, and decisions are based on morphological and behavioural data. The plumages are diagnosably distinct, and reports of birds with intermediate phenotypes appear to be overstated (L Svensson pers. comm.); songs, although little studied, are reported to be quite different. There are slight, but apparently consistent, size differences, and these are counter-intuitive. The northern taxon (Dusky) has a longer migration, but shorter wings, than Naumann's. These three differences lend support to the taxa being separate and independent evolutionary lineages, and they are now recognised as specifically distinct (e.g. HBW). The degree of parapatry is little known, and there seem to be no data relating to pair composition in the zone of contact.

DISTRIBUTION Breeds in a narrow band of N Siberia from River Yenesei to Kamchatka, replaced further south by *T. naumanni*, though with a long common border and an unknown degree of sympatry. The more northerly taxon, Dusky, is recorded more often in montane habitats (HBW). Both species are migratory, wintering from NE India (*eunomus*) through China (*eunonus* in S, *T. naumanni* in N) to Japan (where both occur).

STATUS The first for GB was shot at Gunthorpe (Nottinghamshire) in Oct 1905. There have been seven subsequent records, two of which remained for several weeks: Hartlepool (Cleveland) Dec 1959– Feb 1960; Major's Green (W Midlands) Feb–Mar 1979. Not recorded in Ireland.

Naumann's Thrush   *Turdus naumanni* Temminck

SM monotypic

TAXONOMY See Dusky Thrush.

DISTRIBUTION Breeds further south than Dusky Thrush.

STATUS Two individuals: the first was at Woodford Green (London) Jan–Mar 1990; the second at nearby S Woodford (London) in Jan 1997. Not recorded in Ireland.

Black-throated Thrush   *Turdus atrogularis* Jarocki

SM monotypic

TAXONOMY Red- and Black-throated Thrushes are very closely related and species status is debatable. They are sister to Dusky/Naumann's Thrushes. Voelker *et al.* (2007) examined two males, one from inside the range of Black-throated, the other from just into the range of Red-throated; they found no difference in sequence. Although hybridisation is known to occur, it seems improbable that they had the misfortune to examine a recent backcross combining the phenotype of Red with the mitochondrion of Black. More likely, the genetic divergence between the two taxa is negligible, supporting a recent common ancestry and also that (as with Dusky/Naumann's) decisions about specific status should be based on non-molecular attributes. The phenotypes are diagnosably distinct (although intergrades occur), principally in the colouration, which could be a single gene difference. There are striking vocal differences, but the effectiveness of these as an isolating mechanism has not been tested.

Occasional mixed pairs in zone of contact (summarised by Clement *et al.* 2000), and occasional individuals with hybrid appearance. This implies some gene flow between the taxa. In the absence of data on frequency and relative fitnesses of pure and mixed pairs in the contact zone, it is difficult to reach a conclusion relating to their genetic difference. Strong diagnosability and different vocalisations support these being treated as independent evolutionary lineages, and they are treated here as separate species.

DISTRIBUTION Breeds in a range of woodland habitats in C Siberia from Ural Mts to Lake Baikal, and S to Mongolia. *T. atrogularis* is more extensive and western, *T. ruficollis* more towards the S & E. Zone of contact is extensive, from Lake Baikal to Russian Altai Mts, within which mixed pairs occur. Migratory, wintering from the Red Sea and S Caspian across N India (*atrogularis*) and NE India to SW China (*ruficollis*).

STATUS The first was shot near Lewes (Sussex) in Dec 1868; there have been over 60 subsequently, although none in Ireland. The number of records has increased since 1950: yearly averages per decade being 0.1 (1950s), 0.0 (1960s), 0.7 (1970s), 1.0 (1980s), 2.5 (1990s), 1.75 (2000s). Whether this is due to increased observer awareness or to changes in status is not clear.

Dividing the records into 'autumn' (Sep–Nov) and 'winter' (Jan–Mar) reveals a similar difference between regions to that found with Scaly Thrush. Almost all of the Scottish records are autumnal, compared with half of English. This suggests a difference in origin; presumably, the autumn birds are freshly arrived from the breeding grounds, whereas those in winter either arrived earlier in the autumn and spread gradually (and unnoticed) southwards or are freshly arrived but from the near-Continent, where they spent time before moving to GB as the winter progressed.

Red-throated Thrush   *Turdus ruficollis* Pallas

SM monotypic

TAXONOMY See Black-throated Thrush.

DISTRIBUTION See Black-throated Thrush.

STATUS Only one: a first-winter female at the Naze (Essex) in Sep–Oct 1994. There are no records from Ireland.

Fieldfare   *Turdus pilaris* Linnaeus

CB WM PM monotypic

TAXONOMY Molecular data (Voelker *et al.* 2007) show this to be in a weakly supported clade with Kessler's Thrush *T. kessleri* of the Tibetan Plateau.

DISTRIBUTION Breeds in parks, farmland and clearings in pine and birch forests from Scandinavia, C Europe, E through central latitudes of Siberia to Tien Shan. A small population exists in Iceland, but that in Greenland may be extinct (HBW). Migratory, wintering generally to S & E of breeding range, through most of W & C Europe and SW Asia.

STATUS Very small numbers breed, typically one to two pairs per year in GB, mostly in Scotland. For several years through the 1970s and 1980s, birds bred almost annually in N England, but this has declined since 2000, and breeding is now very seldom reported outside Scotland.

A regular passage migrant and winter visitor to GB&I, with many tens of thousands of birds arriving through the autumn. Most birds move rapidly inland, feeding mainly on soil invertebrates on farmland and pasture, though also seeking fruit and berries. As with Redwing, when the food stocks become depleted, Fieldfares move on. This process is enhanced by harsh weather, when flocks can be seen steadily heading S & W towards the milder conditions in SW England, Ireland and N France.

Most Fieldfares that winter in GB&I originate in Fennoscandia, with a few from further east into Russia. There appear to be differences in the wintering patterns of birds from Norway and Finland/Sweden; the former winter widely across GB&I, but juveniles of the latter winter in SE England. In subsequent years, these birds seem to winter further east into S France and N Italy (Wernham *et al.* 2002).

Song Thrush

Turdus philomelos C.L. Brehm

Endemic: RB MB *hebridensis* Clarke
RB MB WM PM *clarkei* Hartert
WM PM *philomelos* C. L. Brehm

TAXONOMY Sister to all other Eurasian *Turdus*, apart from Mistle Thrush (Klicka *et al.* 2005, Voelker *et al.* 2007). The distinctiveness of these two, from each other as well as from the rest of the Eurasian *Turdus*, led Voelker *et al.* (2007) to speculate that they should be placed in their own genera. However, *Turdus* is a natural group and splitting it would add complexity. Song Thrush is a unique and distinctive taxon, closely related only to Chinese Thrush *T. mupinensis* in Voelker's phylogeny (although the DNA sequence they obtained for this taxon was limited).

Four subspecies are recognised on the basis of size, darkness of plumage, uniformity of rump and back, and nature and extent of spotting: nominate, *hebridensis*, *clarkei*, *nataliae*.

DISTRIBUTION Breeds across Europe and W Asia from Atlantic to Lake Baikal, and from N Scandinavia south to Mediterranean. Nominate race is most widespread. Replaced by: *hebridensis* in Skye and Outer Hebrides; birds approaching *hebridensis* occur in W Ireland; *clarkei* in GB&I (except where *hebridensis* occurs), also C Netherlands to NW France; *nataliae* is little known, but probably breeds in W & S Siberia (HBW). Birds in W, C Europe from Denmark, C Germany to N Balkans are intermediate between *clarkei* and *philomelos*. Partially migrant; birds from N & E populations move into Middle East, S & W Europe to avoid harshest winter conditions.

STATUS Subspecies *clarkei* is widespread in lowland Britain, occurring in farmland, woodland and human habitats such as parks and gardens. Consequently, it is the best studied, not least because of an alarming decline in numbers through 1970–95 (Baillie *et al.* 2006). The patterns of change appeared to differ among habitats since CBC data between 1968 and 1999 indicated that the population had declined by 69% in farmland but by only 46% in woodland. However, since CBC plots were biased towards SE Britain, the changes may not be geographically uniform. Indeed, the Repeat Woodland Survey and BBS data (which are scattered more randomly across GB&NI) suggested there were increases in all regions during the 1990s, but less so in England than elsewhere. Nevertheless, numbers appeared to have fallen markedly from the levels of the 1960s.

A preliminary study (Thomson *et al.* 1997) showed that the average survival of first-years fell from *c.* 48% to *c.* 40% at about the time the decline began (1975). R.A. Robinson *et al.* (2004) followed this up by analysing a longer series of summer ringing recoveries, and confirmed a lower daily survival for the post-fledging period compared with both the remainder of the first year and among adults; this may be a common pattern among passerines, but the data are not available to seek it in most species (G Siriwardena pers. comm.). They also found significant variation in survival across the years; survival of first-years was negatively correlated with long periods of frost, whereas adults survived less well during summer drought. Adverse weather alone was insufficient to explain all of the changes in abundance; other, as yet unidentified, effects must also be involved. A variety of factors has been suggested, including changes in agricultural practice, land drainage, pesticides and predators, but none has been proven. Peach *et al.* (2004) make a strong case for the increased drainage of agricultural land and the loss of grassland in E Britain reducing the populations of earthworms, which are a vital food resource for Song Thrushes. Unusually among resident passerines, laying date does not appear to be advancing.

Less is known about population levels in Ireland, or changes over the years since the 1960s (no CBC or equivalent until recently). Nevertheless, it appears that the Song Thrush may not have declined to the same extent as in the UK. Recently, numbers seem to be stable (Kelleher & O'Halloran 2006). Also, a small sample showed relatively low levels of contaminants in eggs and tissues; mercury and the organochlorine HEOD were found but no PCBs (P Smiddy pers. comm.).

Little is known about *hebridensis*, but ringing data show that *clarkei* is a partial migrant, with many British birds wintering in Ireland, W France and Iberia. Movement between hatching and first breeding appears to be modest, with >90% remaining within 20 km. Scandinavian birds (*philomelos*) arrive in large numbers in the autumn; the relative proportions of these that remain in GB&I or continue to winter in Iberia is unknown.

Redwing   *Turdus iliacus* Linnaeus

RB or MB WM PM *iliacus* Linnaeus
MB WM PM *coburni* Sharpe

TAXONOMY Recent molecular data returned Redwings in a clade that consisted predominantly of Central American–Caribbean thrushes. The implications of this are unclear, but might be due to west to east colonisation by an ancestral taxon (G Voelker pers. comm.). Two subspecies are recognised: *coburni*, *iliacus*, differing in size, darkness of upperparts and strength of streaking below; many birds cannot be racially recognised.

DISTRIBUTION Breeds in a band of woodland interspersed with open country from Iceland to River Kolyma; southern limits are Ukraine to Lake Baikal. Nominate race is most widespread, replaced by *coburni* in Iceland, Faroes. Both races migratory, except in W Norway and parts of Iceland, where resident. Most populations of *iliacus* winter well south of breeding range, and predominantly from Europe to Caspian Sea; *coburni* winters in SW Europe.

STATUS Small numbers breed most years. The majority are *iliacus* in N & W Scotland, though very occasional pairs of *coburni* also breed in N & W isles (BS3); there are sporadic nesting records from England (Brown & Grice 2005), but apparently none from Wales or Ireland.

A regular passage migrant and winter visitor to GB&I, with hundreds of thousands of birds passing through, or remaining in, GB&I. Peak arrivals are through Oct–Nov, when, given appropriate wind and weather conditions, thousands of birds may arrive in the N & E of GB&I. Large numbers may accumulate on isolated islands (e.g. Fair Isle, Cape Clear), while many of those arriving at coastal sites typically move quickly inland in search of food. Feeding extensively on berries, birds move progressively S & W as these become depleted; hard-weather movements also occur, with flocks of birds recorded across GB&I heading towards milder conditions to the SW.

Both subspecies winter in GB&I. Many *iliacus* are trapped at N & E coastal sites in autumn, and these move on south to winter through lowland Britain, W France and Iberia. There is evidence that birds may winter in widely different ragions between years. Recoveries involving Icelandic birds are more limited, but it seems that there is a tendency for individuals from E & W Iceland to winter in Ireland and Scotland, respectively.

Mistle Thrush   *Turdus viscivorus* Linnaeus

RB MB WM PM *viscivorus* Linnaeus

TAXONOMY Voelker *et al.* (2007) confirmed the findings of Klicka *et al.* (2005) that showed Mistle Thrush to be sister to the rest of *Turdus*. Its morphological similarity to Song Thrush is not phylogenetically significant since they are as distant from each other as they are to the rest of the turdines, and must be either parallel evolution or the retention of the ancestral phenotype. Three subspecies are recognised (HBW, Clement *et al.* 2000), differing in size and paleness of plumage: Mediterranean resident *deichleri* is smaller; strongly migratory *bonapartei* is larger than the nominate.

DISTRIBUTION Palearctic, present in mixed and open woodland from NW Africa, Iberia, GB&I E through Turkey, European Russia to Nepal, Altai Mts. Absent from northern fringes and most arid parts of Transcaspia, Kazakhstan. Nominate is most widespread, replaced in NW Africa, Corsica, Sardinia by *deichleri*, and from Turkmenistan to Altai by *bonapartei*. Northern and eastern populations are migratory, wintering in Mediterranean, Black Sea areas, and as far SE as Kazakhstan.

STATUS Very widely distributed across GB&I, being recorded in c. 85% of 10-km squares, only absent from Orkney, Shetland and Outer Hebrides and parts of the highlands. These distributional figures mask evidence of a decline between the Breeding Atlases of *c.* 2% in GB and (perhaps more concerning) of *c.* 9% in Ireland. CBC/BBS data revealed a more substantial decrease in population of c. 35%, mostly during 1975–95. BBS data indicate some evidence of a recovery in Scotland, Wales and N Ireland since then, although numbers are continuing to fall in England. Nest record data reveal an increase in clutch size since the 1960s, but latterly brood size may be declining. Mistle Thrush is one of the few species that does not seem to be advancing its laying date. Siriwardena *et al.* (1998) suggest that the decline in numbers is driven by reduced annual survival, similar to Common Blackbird and Song Thrush.

Mistle Thrushes in the southern parts of GB&I are largely resident, though Scottish birds (especially young birds) leave the breeding grounds in autumn; a few British-ringed birds have been recovered in France and Belgium, but most are found within a few km of their ringing site. Small numbers are recorded most autumns at coastal observatories, but whether they remain in GB or move on to Ireland or France is not known (Wernham *et al.* 2002). A handful of birds trapped at coastal sites on passage are subsequently recovered further south, even into SW France.

American Robin   *Turdus migratorius* Linnaeus

SM *migratorius* Linnaeus

TAXONOMY Voelker *et al.* (2007) found this to be a member of a clade of Central America and Caribbean species, and sister to Rufous-collared Thrush *T. rufitorques* of C America; this last has been suggested before on the basis of behaviour and vocalisation (see Sallabanks & James 1999). A variable taxon, in both size and plumage details, with seven subspecies, though species limits are in need of investigation.

DISTRIBUTION The most widespread N American thrush, breeding across the entire region, apart from the extreme N of Canada and Alaska, and the arid or mountainous regions of C America. The nominate race breeds across much of Canada and N USA, as far S as Kansas; the very dark-plumaged *nigredeus* breeds in Labrador, Newfoundland; other races occur elsewhere across the continent. Migratory, with the Canadian populations of *migratorius* wintering S as far as Mexico; *nigredeus* is a more short-distance migrant, wintering into E USA, but is a possible vagrant here. More southerly birds are resident.

STATUS The first for GB was seen to fly in off the sea at Dover (Kent) in Apr or May 1876; there have now been over 20. The first for Ireland was shot at Shankhill (Dublin) in May 1891, since when there have been nine further records. Most records have been in the winter months Oct–Feb, but also three in May–June. There have been several long-staying individuals: e.g. Jan–Mar 1966 (Dorset), Feb–Mar 1966 (Surrey), Dec 2003–Feb 2004 (Cornwall). A bird summered in County Offaly in 1983, being present from June to 'end' July.

FAMILY MUSCICAPIDAE, SUB-FAMILY MUSCICAPINAE

Genus *MUSCICAPA* Brisson

African, Palearctic and Oriental genus of over 20 species, one of which breeds in GB&I and one has occurred as a vagrant.

Asian Brown Flycatcher   *Muscicapa dauurica* (Pallas)

SM *dauurica* (Pallas)

TAXONOMY Presumed closely related to Spotted Flycatcher, but molecular analyses have not included both species. Five races recognised by HBW, but none by Svensson (1992). Relationships within and between species of *Muscicapa* need further study.

DISTRIBUTION Nominate race breeds from E Russia through C Siberia and Mongolia to Sakhalin, N Korea and Japan; winters from S China to Philippines. Other races breed in Indian subcontinent and through Indo-China to Borneo.

STATUS Three: single birds on Fair Isle (Shetland), July 1992; Flamborough (Yorkshire), Oct 2007; Fair Isle, Sep 2008. [There are two records from Ireland, but neither has been assigned to any list.]

Spotted Flycatcher   *Muscicapa striata* (Pallas)

MB PM *striata* (Pallas)

TAXONOMY Vœlker & Spellman (2004) examined a series of passerines, including Spotted and Red-breasted Flycatchers. Using mt DNA sequences, they found evidence that these two genera are phylogenetically not especially close; more detailed investigation is needed. BWP and Svensson (1992) recognise five subspecies in the W Palearctic, with two more in Asia. Not all are well defined.

DISTRIBUTION Breeds from NW Africa, Iberia to Fennoscandia, E to Black and Caspian Seas, and across Siberia to Lake Baikal. Subspecies include: *striata* (NW Africa, mainland Europe plus GB&I, east to River Irtysh), *neumanni* (east of nominate to Baikal), *inexpectata* (Crimea), *balearica* (Balearics), *tyrrhenica* (Corsica, Sardinia); and there are further races in Asia. A long-distance migrant, mostly wintering in Africa south of Equator.

STATUS This species has shown a massive decline across Europe since the 1960s (Tucker & Heath 1994) and numbers have decreased by more than 50% in GB&I. The decline had begun by the 1980s in Ireland, with a contraction of over 18% between the two Breeding Atlases in the number of 10-km squares with evidence of breeding. This accelerated from the late 1980s, and occurred in both woodland and farmland, suggesting broad-scale causes. Results from the Repeat Woodland Survey indicated that numbers in SW England were increasing (Baillie *et al.* 2006). Freeman & Crick (2003) attempted to relate the decline from 1965 to 1996 (measured using results from the CBC) to data from ringing and nest records. They found no evidence of a change in clutch or brood size; indeed these may even have increased slightly – perhaps as a response to lowered density. They could predict the falling numbers on CBC plots by assuming that post-fledging survival declined over time. The factors causing this were unclear, but could involve enhanced mortality within GB immediately post-fledging, perhaps due to a reduction in the quantity of large aerial insects at a critical time for fledglings. Stevens *et al.* (2007) found that birds nesting in gardens are significantly more successful than those in woods and farmland, with avian predation (chiefly by Jays) being a major cause of failure in the latter habitats. This finding is supported by Stoate & Szczur (2006), who found increased nest survival in a population when predator control was in place. Further evidence came from a remarkable series of studies (Stevens *et al.* 2008) using digital nest cameras in two areas of England. Of 65 monitored nests, 20 were fully or partially predated, three by domestic cats and the remainder by birds. Of the latter, 12 were taken by Jays, two by Great-spotted Woodpeckers, and one each by Buzzard, Sparrowhawk and Jackdaw. Interestingly, Grey Squirrels were recorded at nests during construction and after fledging, but were not seen to predate either eggs or young. It seems unlikely that predation is the main cause of population decline, with a strong possibility that conditions may be worsening on migration or on the wintering grounds.

Genus *CERCOTRICHAS* Boie

Predominantly African genus of about 10 species, one of which has been recorded in GB&I.

Rufous-tailed Scrub Robin   *Cercotrichas galactotes* (Temminck)

SM *galactotes* (Temminck)
SM *syriacus* (Hemprich & Ehrenberg) or *familiaris* (Ménétriés)

TAXONOMY Sometimes placed in genus *Erythropygia*. Five subspecies recognised by Vaurie (1959), HBW, and traditionally divided into two groups on basis of colour, wing formula, tail length (Roselaar in BWP). Western (rufous) forms include *galactotes*, *minor*, *hamertoni*; eastern (grey-brown) forms *syriacus*, *familiaris*.

DISTRIBUTION Breeds in scattered locations in a band of dry scrub and open woodland, from Spain and Morocco to Lake Balkash. Nominate *galactotes*: Iberia, Africa N of Sahara, from Morocco to Nile Valley, Israel, S Syria; *syriaca*: Balkans, E & S Turkey, N Syria; *familiaris*: SE Turkey and Caucasus, E to Kazakhstan and W Pakistan; *minor*: resident in belt of N Africa, S of nominate, from Gambia to N Somalia; *hamertoni*: resident in E Somalia. Northern populations migratory, wintering in arid acacia and scrub woodlands of W African Sahel and from Sudan to Kenya.

STATUS The first for GB was identified as *galactotes*, shot near Brighton (Sussex) in Sep 1854, and the first for Ireland (also *galactotes*) met a similar fate at Old Head of Kinsale (Cork) in Sep 1876. There have been nine further records: one in Apr, the rest in Aug–Oct. The four records from the 1800s were all *galactotes*; subsequently none has been assigned to race, apart from a bird on Great Saltee (Wexford) in Sep–Oct 1968 that was identified as *syriacus* or *familiaris*. Despite the increase and expertise of birders, this has remained a rare bird, with only seven records in the 20th century, and none since 1980. The steep declines in the western strongholds of Iberia and Turkey (BirdLife International 2004) lend little hope of any change in occurrence here.

Genus *ERITHACUS* Cuvier

Palearctic genus of one to three species, one of which breeds in GB&I.

European Robin   *Erithacus rubecula* (Linnaeus)

Endemic: RB MB *melophilus* Hartert

WM PM *rubecula* (Linnaeus)

TAXONOMY Rather variable with a series of races of varying merit; eight now generally recognised (eg Roselaar in BWP; HBW). Races *rubecula* and *melophilus* differ in orange of breast, size and olive-grey balance to mantle.

DISTRIBUTION Breeds across Europe, from the Atlantic Is to W Siberia, and S into NW Africa in the west and Iran in the east. Nominate *rubecula*: most of Continental Europe to Ural Mts, including NW Morocco, Azores, Madeira, W Canaries; *melophilus*: GB&I; *superbus*: Tenerife, Gran Canaria; *witherbyi*: N Algeria, Tunisia. Four more races in E Europe and W Asia, including *tataricus* from the Ural Mts and SW Siberia. N populations migratory, wintering within range of S populations, but also along Mediterranean coasts of N Africa.

STATUS Britain's national bird with an appropriately wide distribution; European Robins were recorded in *c.* 94% of 10-km squares during the two Breeding Atlases, absent only from parts of Orkney and Shetland. Numbers declined from 1965 to 1985, probably due to a series of cold winters (Baillie *et al.* 2006), but there has been a marked increase across GB&I since the mid-1980s. The main increase was in England, with less in Scotland and Wales, and the smallest increase in NI. The Repeat

Woodlands Survey (Amar *et al.* 2006) confirmed this for all regions of GB. Statistical analysis of habitat characteristics indicated that numbers had increased most at sites with less grass and where vegetation cover at 2–4 m had increased, presumably taking advantage of areas with more bare earth and leaf litter, and increased cover for nesting.

Failures at the egg stage have declined significantly since *c.* 1980, probably in response to the increasingly milder winters, which have also led to an advance in the laying date of *c.* 6 days over the same period. A familiar bird around houses, the European Robin is regularly recorded in the Garden BirdWatch survey. Here minimum numbers are recorded in late summer, but the proportion of gardens that host European Robins at this season seems to be rising.

British-bred European Robins (*melophilus*) tend to be fairly sedentary, usually breeding within a few km of their birthplace. There are a few examples of long-distance movement, with recoveries of birds in SW France and Spain, but these seem to be exceptional. Many individuals depart from the breeding site, particularly in the N and at higher altitudes, in late summer, returning the following spring. For such a common bird, movements outside the breeding season are surprisingly little known. Large numbers come to GB&I in autumn, with occasional massive falls of *rubecula* along the east coast, e.g. the 'great Robin rush' of Oct 1951, when thousands of birds appeared along the coasts of Lincolnshire and Norfolk (see Williamson 1965). Most of these individuals seem to come from Scandinavia and the Baltic states to Poland. The status of *rubecula* in Ireland is unclear. It undoubtedly occurs, and there are three records from Great Saltee (Wexford) in May and Oct; *c.* 50 were present at Cape Clear (Cork) in Oct 1959, and two trapped in Tipperary in Jan had the characters of this race.

Genus *LUSCINIA* Forster

Palearctic, Oriental and marginally Nearctic genus of 11 species, six of which have been recorded in GB&I, two as a breeding birds, one regularly, the other sporadic.

Rufous-tailed Robin   *Luscinia sibilans* (Swinhoe)

SM monotypic

TAXONOMY Precise placement of this species still unclear; sometimes put in *Erithacus*, sometimes (e.g. Vaurie 1959) in its own genus *Pseudoaedon*. Vaurie gave it the splendid name of Swinhoe's Pseudorobin, and also recognised two subspecies; HBW treats it as monotypic.

DISTRIBUTION Breeds in damp lowland forest with dense undergrowth from C & E Siberia to NE China; migratory, wintering in S China into Indo-China.

STATUS The only record from GB&I is of a first-winter bird on Fair Isle (Shetland) in Oct 2004.

Thrush Nightingale   *Luscinia luscinia* (Linnaeus)

SM monotypic

TAXONOMY Closely related to Common Nightingale, with which it occasionally hybridises.

DISTRIBUTION Breeds N & C Europe, from S Fennoscandia south to Romania, and E in narrow band to N Kazakhstan and Caucasus. Migratory, wintering in E Africa, from S Sudan to S Africa.

STATUS Over 150 records from GB&I, of which almost two-thirds have been in Scotland. There are only three records from Ireland: the first on Cape Clear (Cork) in Oct 1989, and two further individuals in County Cork in Oct 1990, 1999. It is rather more of a spring bird in Scotland, with 81% of records in May–July, compared with only 68% elsewhere in GB&I. This suggests the birds are overshooting migrants heading for N Europe and arriving from the SE. In the autumn, most birds that have been aged are young on their first migration.

Common Nightingale   *Luscinia megarhynchos* C.L. Brehm

MB PM *megarhynchos* C.L. Brehm
SM *golzii* Cabanis

TAXONOMY Variation is clinal, from darker birds with short wings and tail in the west to paler, longer winged and tailed in east; more data from the intervening regions would be valuable in determining the nature of the linking populations. Up to eight races have been proposed, but only three recognised by Roselaar (in BWP), Svensson (1992) and HBW.

DISTRIBUTION More southerly than Thrush Nightingale, breeding from Iberia, NW Africa to SW Mongolia. Nominate *megarhynchos* occurs NW Africa, Europe to C Turkey and Levant; *africana*: E Turkey, Caucasus to N & W Iran; *golzii* (formerly called '*hafizi*'): C Asia, especially Kazakhstan. All populations migratory, wintering in forest edges and riverine woodland of sub-Saharan Africa, from Gambia to Kenya.

STATUS Restricted as a breeding bird to SE England, and distribution declined dramatically between the two Breeding Atlases; almost 30% of 10-km squartes were abandoned, leaving few sites N of a line from River Severn to the Wash. Even here, distribution was not continuous, with few sites in Cornwall and Devon or in the region N & W of London. A dedicated survey in 1999 revealed a further contraction in range. There is little data from CBC and/or BBS fieldwork: the species is too scarce to be represented in most survey sites. There is a little evidence from constant-effort ringing that suggests the decline may have been stemmed, but sample sizes are low and confidence limits correspondingly wide. It has never bred in Ireland and the 23 records are all of spring or autumn passage birds, mostly in Cos Wexford and Cork.

Reasons for the population decline may include loss of breeding habitat in SE England and perhaps a decline in the quality of that which remains (Fuller *et al.* 2005). There are no ringing recoveries from the wintering grounds, which are presumed to be in W Africa south of the Sahara. Here, habitat degradation and increased desertification may be reducing the area suitable for wintering, and the increasing width of the Sahara will add pressures to migrants. It is of concern that we have little idea why the species is in such decline.

There are three records of subspecies *golzii*, all in Oct. The first was found dead on Fair Isle (Shetland) in 1971; subsequent birds were at St Agnes (Scilly) in 1987 and Spurn (Yorkshire) in 1991.

Siberian Rubythroat   *Luscinia calliope* (Pallas)

SM monotypic

TAXONOMY Closely similar in morphology to White-tailed Rubythroat *L. pectoralis*. Shows slight interpopulation variation, but this largely clinal and considered insufficient for racial separation (BWP, Svensson 1992, HBW). However, the clines between Siberian and White-tailed are sufficiently continuous for the possibility of them being conspecific still to need investigation.

DISTRIBUTION Breeds from Ural Mts to Kamchatka, and S to N Mongolia, N Japan; isolated populations in Qinhai, Gansu. Migratory, wintering SE Asia to Philippines.

STATUS Seven records, all in Oct. Five in Shetland, including the first on Fair Isle (Shetland) in Oct 1975. There are no records from Ireland.

Bluethroat   *Luscinia svecica* (Linnaeus)

CB PM *svecica* (Linnaeus)
CB PM *cyanecula* (Meisner)

TAXONOMY Variable, especially in intensity of blue breast, and presence, colour and extent of breast spot. Ten subspecies recognised by HBW, fewer by Vaurie (1959), Roselaar (in BWP). Much taxonomic differentiation based on colour of breast spot. Molecular analysis by Questiau *et al.* (1998) indicates very little genetic divergence between the small, white-spotted race *namnetum* and the larger red-spotted Scandinavian *svecica*, although there appeared to be one consistent base difference. A more comprehensive analysis showed that this difference is not consistent and that phylogeographic structure is weak (Zink *et al.* 2003). Another study using microsatellites found a well-differentiated southern group of subspecies with white or no throat spots and a less-differentiated northern group of chestnut-spotted populations. Phylogenetic analyses indicated that the southern all-blue and white-spotted forms are ancestral to the chestnut-spotted subspecies (Johnsen *et al.* 2006).

DISTRIBUTION Widespread across N Eurasia, from Iberia, Scandinavia to Bering Straits and into NW Alaska; south into Iran, C Asia, Himalayas, Manchuria. Nominate *svecica* is most northerly and most widespread, breeding from Scandinavia across N Asia to Alaska; *namnetum*: restricted to C, SW France; *cyanecula*: rest of N Iberia, N, E France, Netherlands to Ukraine, Belarus; *volgae*: Ukraine, European Russia; *magna*: E Turkey, Caucasus to Iran; with five more subspecies from Asia.

STATUS There are a few breeding records from GB. A nest containing six eggs was found in Scotland in 1968; no male was seen, so the race could not be established, and the nest was subsequently predated. Further breeding attempts occurred in 1985 and 1995; these were successful and involved red-spotted *svecica* males. Two pairs of white-spotted *cyanecula* birds bred successfully at Thorne (Yorkshire) in 1996, with males rerturning to the area in several successive years. Permanent colonisation seems possible, but has not yet materialised.

Two subspecies are regularly recorded, but because of the difficulty of racial identification, many are simply recorded as 'Bluethroats'. White-spotted *cyanecula* is much the rarer with only *c.* 10 confirmed records annually, though only males in spring can confidently be identified. White-spotted birds typically arrive a month earlier than *svecica*, with many in Mar–Apr, whereas the latter is more usually a May bird. Birds arrive anywhere along the coasts, but are more often found in the N & E, from Shetland to East Anglia. In autumn, Bluethroats are much commoner, but again show a northern bias, with most records coming from the northern isles. Much rarer in Ireland, with only 34 individuals recorded to 2004; of these, five were identifiable as *svecica* and one as *cyanecula*.

The Scarce Migrant report for 2003 gives annual averages per decade of 98 (1970s), 186 (1980s), 116 (1990) and 85 (post-2000). This is at variance with data from BirdLife International (2004), which suggest that populations across Europe are stable. There may be other reasons for the apparent decline in GB&I, or perhaps the breeding data are less accurate.

Siberian Blue Robin   *Luscinia cyane* (Pallas)

SM race undetermined

TAXONOMY Two subspecies traditionally recognised (e.g. HBW), differing in darkness of blue upperparts and, slightly, in size. Birds from Sakhalin, S Kurils and Japan differ in size and wing formula from elsewhere in Asia and have recently been separated as a third (Red'kin 2006).

DISTRIBUTION Breeds across E Palearctic, from Altai Mts to Kamchatka, and south through E China, Japan to Korea. Precise range of subspecies unclear. Nominate *cyane* breeds from Altai to N Mongolia (HBW), to Sea of Okhotsk and Korea (BWP); *bochaiensis* is more eastern, Sakhalin, Ussuriland, Japan – perhaps N Korea (HBW). Migratory, majority wintering from Indo-China to Borneo, Sumatra.

STATUS Two records; first at Minsmere (Suffolk) in Oct 2000; second on N Ronaldsay (Orkney) in Oct 2001. No records from Ireland.

Genus *TARSIGER* Hodgson

Palearctic and Oriental genus of five or six species, one of which has been recorded in GB&I.

Red-flanked Bluetail   *Tarsiger cyanurus* (Pallas)

SM monotypic

TAXONOMY Two subspecies formerly recognised. Geographically widely separate; these differ in size, plumage and vocalisations (Martens & Eck 1995), and ongoing research (L Svensson, P Alström pers. comm.) suggests these should be regarded as separate monotypic species: Red-flanked Bluetail *T. cyanurus* and Himalayan Bluetail *T. rufilatus*.

DISTRIBUTION Red-flanked Bluetail extended its range in 1940s to colonise Finland, though numbers there remain small. Breeds from here across northern Eurasia to Kamchatka; in east, breeds across a wider latitudinal range, occurring as far south as N Mongolia, NE China, Japan. *T. rufilatus* breeds through Himalayas and into C & S China. *T. cyanurus* is strongly migratory, wintering through China into Indo-China.

STATUS Forty records to 2006, all in England and Scotland, with the first being an adult male at N Cotes (Lincolnshire) mid-Sep 1903. There are four spring records, the rest are from Sep to Nov. The pattern of occurrence shows an increase, from about one each decade before 1970, to about one every couple of years through 1980–2000, and more than one a year since then. For such a distinctive species, this must reflect changes in abundance, with increased records resulting from a westwards spread of the breeding range.

Genus *IRANIA* Filippi

Monospecific Ethiopian, Palearctic and marginally Oriental genus that has been recorded in GB&I.

White-throated Robin   *Irania gutturalis* (Guérin-Méneville)

SM monotypic

TAXONOMY Not included in any molecular analyses, so taxonomic placement relies on morphological assessments. Conventionally placed close to *Tarsiger*, the evolutionary relationships of these near-chats would repay investigation.

DISTRIBUTION Restricted to Middle East (Turkey, Levant to Iran) and C Asia (S Kazakhstan, Afghanistan, Tadjikistan). Migratory, wintering in E Africa, from Kenya to Tanzania.

STATUS Two records, both in spring. The first was a male at Calf of Man (Isle of Man) in June 1983; the first for GB was a first-summer female, on Skokholm (Pembrokeshire) in May 1990. Not recorded in Ireland.

Genus *PHOENICURUS* Forster

Palearctic and marginally Oriental genus of 11 species, three of which have been recorded in GB&I, two as breeding birds the other as a vagrant.

Black Redstart   *Phoenicurus ochruros* (S. G. Gmelin)

RB MB WM PM *gibraltariensis* (J.F. Gmelin)

TAXONOMY A molecular analysis of the mt cyt-b gene (Ertan 2006) showed that Common and Black Redstarts are not especially closely related. Black is a member of a (weakly supported) clade that also includes Hodgson's *P. hodgsoni*, Güldenstädt's *P. erythrogastrus* and Daurian Redstart *P. auroreus*. Common is sister to this group. Subspecies of Black Redstart are traditionally based on size and relative extent of red and black on underparts, and five are recognised by HBW. Two further races are recognised by Vaurie (1959) and BWP; however, HBW (followed here) merges *aterrimus* with *gibraltariensis* and regards *xerophilus* as intermediate between *phoenicuroides* and *rufiventris*. Ertan (2006) investigated the relationships of these five races; he showed that the western races *gibraltariensis*, *ochrurus* and *semirufus* form a monophyletic group, the races *phoenicuroides* and *rufiventris* were more distant. He explains this in terms of evolution within and between successive range contractions and expansions. His data give no molecular justification for revising species limits in the Black Redstart.

DISTRIBUTION Widespread across S Europe, Middle East, and from Caucasus through Iran to E Tibet. Subspecies *gibraltariensis*: Europe, NW Africa; *ochrurus*: Turkey, Caucasus, Iran; *semirufus*: Syria to Israel; *phoenicuroides*: scattered sites from S Russia to W Mongolia, and south to Kazakhstan and Pakistan; *rufiventris*: Turkmenistan, through Himalayas to C China.

STATUS Has long been a scarce breeding bird in England, with the number of pairs rising steadily from 1940 to 1990, reaching a peak at *c.* 120 pairs. Dedicated surveys are undertaken occasionally; that of 1977 reported 104 territory-holding males with 61 pairs proved to breed (Morgan & Glue 1981). Many of these were in inner cities, where they nest in crevices in buildings and forage in open spaces, preferring sparse vegetation to lawns and parkland. Large industrial complexes such as power stations are also popular as breeding sites. Many of these sites are prone to disturbance, and several areas have established Black Redstart Action Plans in an attempt to improve the nesting environment. Despite these efforts, numbers are now fairly stable and further spread and colonisation seems limited. It has never bred in Ireland.

Regularly recorded in small numbers at coastal sites on passage in both spring and autumn. Sites in S & W England in particular, such as Scilly and Portland (Dorset), regularly hold birds through the winter, and some remain on or close to breeding sites in inner cities.

Black and Common Redstarts occasionally hybridise, and the offspring resemble the orange-bellied eastern races of Black Redstart: *phoenicuroides*, *semirufus*, *ochruros*. Several records of eastern Black Redstarts have been reported, but in each case it was not possible to exlude hybrid origin. Although *phoenicuroides* is a possible vagrant, until criteria are established to separate these from hybrids, safe identification seems unlikely without DNA support.

Common Redstart   *Phoenicurus phoenicurus* (Linnaeus)

MB PM *phoenicurus* (Linnaeus)
SM *samamisicus* (Hablizl)

TAXONOMY Two subspecies recognised, based chiefly on the colour of the edges to the secondaries and tertials; Ertan (2006) included these in his analysis of redstart phylogeny and found them to be sister taxa, well separated from the rest of the forms that he examined.

DISTRIBUTION Forested areas of Europe, from the Mediterranean to *c.* 70°N; and from the Atlantic to C Asia. Nominate subspecies occurs across the range, except in SE, where replaced by *samamisicus* from Turkey, Crimea to Turkmenistan, and Iran; possible zone of intergradation in Greece and S Balkans, but this needs reviewing. Migratory, with most birds wintering in Sahel zones of Africa and some in S Arabian peninsula.

STATUS The range of the Common Redstart contracted by *c.* 20% between the two Breeding Atlases, especially in S & C England. CBC data confirm a collapse in population during 1965–75, which is thought to have resulted from the drought conditions in the Sahel zone of Africa that also severely affected the Common Whitethroat. The following 20 years saw a near doubling of the population, although to a level that is still below the peak of the 1960s. These national figures mask regional differences. The population continued to rise in England to the early 1990s, but has fallen back slightly since then; in Wales, the decline has been more uniform. Repeat Woodland Surveys in Wales were more extensive than BBS surveys, and confirmed a decline of over 50% since the 1980s. This lack of consistency probably reflects the relatively small number of monitored sites. While the driving force behind the decline might have been Sahel drought (Marchant *et al.* 1990), the recovery seems to have been aided by a general rise in both clutch and brood sizes, and declines in the daily losses of eggs and nestlings. In common with several other relatively short-distance migrants, laying date has also advanced since the 1980s. In Ireland, has always been a scarce and sporadic breeder, with rarely more than a pair or two recorded and not in every year. Otherwise it is a scarce passage migrant in spring and autumn.

Rather few Common Redstarts ringed in the nest in GB&I have been recovered as adults in subsequent years, so the dispersal distance in poorly known. However, there are many recoveries overseas, and these indicate migration south through Iberia, but few records from the wintering grounds. Apart from in the northern isles, the majority of birds trapped at coastal observatories in spring are native to GB&I. In the autumn, there may be strong movements of Continental birds; ringing recoveries indicate that most of these originate in Germany and Fennoscandia, suggesting a westerly dispersal away from the breeding grounds prior to departure.

Four birds identified as subspecies *samamisicus* have been recorded, with the first at Heacham (Norfolk) in Oct 1975.

Moussier's Redstart   *Phoenicurus moussieri* (Olphe-Galliard)

SM monotypic

DISTRIBUTION Restricted to NW Africa, from Morocco to Tunisia. Upland populations are altitudinal migrants.

STATUS One male at Dinas Head (Pembrokeshire) in Apr 1988. The timing is in agreement with vagrancy to Malta (Mar–May). No records from Ireland.

Genus *SAXICOLA* Bechstein

Palearctic, Ethiopian and Oriental genus of about 12 species, two of which breed in GB&I.

Whinchat   *Saxicola rubetra* (Linnaeus)

MB PM monotypic

TAXONOMY Slight clinal variation, but proposed subspecies not recognised by Roselaar (in BWP), Svensson (1992) or HBW.

DISTRIBUTION Predominantly European, but extends east through Caucasus into W Siberia and NW Iran. A trans-Saharan migrant, wintering in moist savannah and forest clearings from Atlantic to Red Sea, and south through Kenya, Tanzania.

STATUS The period between the two Breeding Atlases saw a dramatic contraction in range, with a loss of birds from >16% of 10-km squares in GB, and an even more concerning 35% in Ireland. The losses were especially severe in the lowland areas of England and C Ireland. Callion (in Gibbons *et al.* 1993) speculates that agricultural intensification may have reduced the uncultivated grassland that Whinchats favour. Simultaneously, conifer planting in the uplands led to new habitat, but this was lost again as trees matured. There is no recent information on its breeding status in Ireland, where it is otherwise a rather scarce spring and autumn passage migrant.

The largely upland distribution of Whinchats in GB means that few were included in CBC sites. It was not until 1994 and the start of the BBS, with its more random and inclusive array of sites, that population estimates became available. Rather limited evidence from this suggests that the GB population of Whinchats has declined overall by *c.* 15% since the mid-1990s, and by >30% in England. There is little data relating to population dynamics, although there is some evidence of an increase in nest failures.

Whinchats from GB&I migrate through W France and Iberia into NW Africa, presumably skirting the Sahara, to wintering grounds in the savannahs from Senegal through N Nigeria. Speculatively, increased desertification in these regions may be adversely affecting birds through either a reduction in suitable wintering habitat or the loss of fattening sites prior to the trans-Saharan flight.

Eurasian Stonechat   *Saxicola torquatus* (Linnaeus)

RB MB *hibernans* (Hartert)
SM *maurus* (Pallas) and *maurus/stejnegeri* (Parrot)
SM *variegatus* (S.G.Gmelin)

TAXONOMY In need of revision. HBW list 24 subspecies, of which five breed in Europe. Molecular analyses by Wink *et al.* (2002b) and Illera *et* al. (2008) supported earlier (more limited) analysis in showing that 'European' *rubicola*, 'Siberian' *maurus* and 'African' *torquatus* have diverged substantially, with differences of 3–5%. This compares with *c.* 6.7% between Stonechats and Whinchat and would seem to support specific differentiation. Zink *et al.* (2009) showed that European *rubicola*, central Palearctic *maura* and eastern Palearctic *stejnegeri* are all evolving independently and that *maura* and *stejnegeri* are not each other's closest relatives, suggesting that all three may be better treated as separate species.

DISTRIBUTION Occurs widely across Africa and Eurasia. Distribution in Africa is fragmented and has led to population divergence and the recognition of a plethora of subspecies. In Eurasia, the distribution is more continuous, from NW Africa, Iberia to the far east of Siberia, generally south of 70°N. European subspecies include *hibernans*: W Iberia, W France, GB&I; *rubicola*: NW Africa, Europe south of *maurus*, Turkey to Caucasus; *variegatus*: E of Caucasus to Iran; *armenicus*: SE Turkey to SW Iran; *maurus*: E Finland to Mongolia and Pakistan. Birds from W & S of Eurasia generally resident; those from N & E more migratory, wintering in N Africa, Arabia and Indian subcontinent; *stejnegeri*: E Siberia to Korea, Japan, wintering south to Malay peninsula.

STATUS There was a severe reduction in distribution between the two Breeding Atlas surveys, with birds being lost from c. 15% of 10-km squares in GB and 28% in Ireland; partly due to a run of cold winters through the 1980s (Callion, in Gibbons *et al.* 1993). There was a marked loss of birds from the east coast of Scotland and many lowland sites across GB&I, so the the species showed a strong westerly distribution. As with the other chats, CBC fieldwork involved few sites with this species, and abundance data prior to the start of BBS are scant. There has been a significant increase in numbers since the mid-1990s, with population estimates more than doubling in GB and expanding in Ireland.

Some Eurasian Stonechats leave GB&I to winter in SW Europe, from Iberia to S France, the Balearics and the N coast of Morocco and Algeria. Many remain on their breeding territories, and warmer springs have allowed an advancement of laying by almost a week since the mid-1980s, with a consequent movement towards three broods a year. Nest record data suggest an increase in clutch size, plus a reduction in losses at both egg and chick stage. Although the sample sizes are small, these changes would be sufficient to assist in driving up productivity.

The lack of a confirmed record of *rubicola* is doubtless due to the absence of a ringing recovery from the appropriate region. Birds with paler rumps than the 'typical' breeding stock are regularly found in winter, and may even breed in SE England (Kehoe 2006).

Several other races have been recorded as vagrants; *maurus* was first recorded when a male was collected on Isle of May (Fife) in Oct 1913; an adult male that appeared to be closer to *stejnegeri* was found at Cley (Norfolk) in May 1972; two *variegatus* have been recorded: the first at Porthgwarra (Cornwall) in Oct 1985. Many other records, plus six in Ireland, have been recorded as *maurus* or *stejnegeri*. In view of the distribution and migration route of the latter, the latter subspecies seems much less likely to occur; also the problems of separating some of these from brightly marked *rubicola* add to the complexity. These records are in need of review.

Genus *OENANTHE* Vieillot

Palearctic, Ethiopian and Oriental genus of 20–23 species, six of which have been recorded in GB&I, one as a breeding bird.

Isabelline Wheatear   *Oenanthe isabellina* (Temminck)

SM monotypic

TAXONOMY A recent morphological analysis of the wheatears by Kaboli *et al.* (2007) has shown that differences in morphology, especially wing length and primary emargination, are related to foraging and migratory behaviour. A concurrent molecular study by essentially the same team (Aliabadian *et al.* 2007) confirmed the monophyly of the wheatears, and the sister relationship of Isabelline and Northern Wheatears, within a subclade of *Oenanthe* that also includes Desert and Pied Wheatears.

DISTRIBUTION Breeds from E Greece, across Turkey and Ukraine to Inner Mongolia and NE China. Migratory, wintering in N & E Africa, Arabia to Pakistan.

STATUS First for GB was first-year female, shot at Allonby (Cumbria) in Nov 1887. Apart from one in Norfolk in May 1977, the 26 records to 2006 cluster between mid-Sep and mid-Nov, when birds would normally be arriving in the N African wintering grounds (HBW). The only Irish record is from Mizen Head (Cork) in Oct 1992.

Northern Wheatear   *Oenanthe oenanthe* (Linnaeus)

MB PM *oenanthe* (Linnaeus)
PM *leucorhoa* (J. F. Gmelin)

TAXONOMY Four subspecies recognised by Roselaar (in BWP), Svensson (1992) and HBW. Nominate smaller than *leucorhoa* and less rich buff below; *libanotica* has longer bill, paler plumage and narrow tail tip (Roselaar); male *seebohmi* has striking black throat and is sometimes regarded as a separate species. However, Aliabadian *et al.* (2007) found no molecular divergence between this and *libanotica*. Only these two races were included in their study, so this casts no light on the relationships of *oenanthe* and/or *leucorhoa*.

DISTRIBUTION Widely distributed across N, C Palearctic, also Greenland, NE Canada, Alaska. Nominate breeds from GB&I, Iberia to far east of Siberia and into Alaska, south as far as Himalayas, Lake Baikal; *leucorhoa* replaces nominate from Canada to Faroe Is; *libanotica* occurs south of nominate, from S Europe across Turkey to Tien Shan, Mongolia and N China; *seebohmi* is restricted to NW Africa. All populations are migratory, wintering in sub-Saharan Africa, and south through E Africa to Zimbabwe; a small number of *libanotica* winter in Iran/Iraq.

STATUS This species showed a withdrawal from lowland sites between the two Breeding Atlases, with >6% 10-km squares being abandoned in GB and *c.* 19% in Ireland. It remains a bird of the uplands, with small populations on Dartmoor and Exmoor, on the chalk downland of S England, and around the coasts of East Anglia. It continues to occur widely on the hills of Wales and N England and across most of Scotland except the agricultural lowlands. In Ireland, it breeds extensively in the W & N, with smaller pockets along the Irish Sea coast. The changes are likely due to the loss of rabbit-grazed pastures to arable in the lowlands and to afforestation of upland areas.

BBS data indicate that the breeding population (subspecies *oenanthe*) is fluctuating, perhaps with evidence of a modest decline since the mid-1990s. Sample sizes are very small. There is slight evidence that brood size is declining, and (in common with many other small passerines) nest failure at the egg stage is falling. Southward migration is through W France and Iberia to NW Africa; there are few recoveries beyond this, but presumably from here, Northern Wheatears cross the Sahara to wintering grounds in the dry savannahs from Senegal to Nigeria.

Subspecies *leucorhoa* occurs as a passage migrant, moving through GB&I from late Apr to mid-May, appreciably later than the nominate (Wernham *et al.* 2002); it is similarly later in the autumn, with peak passage occurring from mid-Sep; in Shetland, they usually pass through the islands after the local birds have left (Pennington *et al.* 2004).

Pied Wheatear   *Oenanthe pleschanka* (Lepechin)

SM monotypic

TAXONOMY Previously regarded as conspecific with Cyprus Wheatear *O. cypriaca* (Vaurie 1959; Roselaar in BWP), but now generally treated as separate species, on basis of differences in plumage, morphology and vocalisations. Hybridises extensively with Black-eared Wheatear in regions E, W & S of Caspian Sea, where pair formation between the taxa appears random (Panov 2005). Alibadian *et al.* (2007) did not include Cyprus Wheatear in their analysis, but found that Pied and Desert Wheateats were sister taxa, in the same clade as Northern and Isabelline.

DISTRIBUTION Breeds discontinuously from SE Europe through C Asia to N Himalayas and NW China. Migratory, wintering in E Africa, from Sudan to Tanzania, and adjacent Arabian peninsula.

STATUS First record for GB was a male (originally reported as a female) shot on the Isle of May (Fife) in Oct 1909; the first for Ireland was at Knockadoon Head (Cork) in Nov 1980. Three spring records from GB, and *c.* 50 in the autumn, over 70% in Oct. The three Irish birds were all in Nov, significantly later than their GB counterparts.

Black-eared Wheatear   *Oenanthe hispanica* (Linnaeus)

SM *hispanica* (Linnaeus)
SM *melanoleuca* (Güldenstädt)

TAXONOMY Previous authorities such as Dementiev & Gladkov (1954) and Portenko (1954) treated this as conspecific with Pied Wheatear, but more recently they have been regarded as separate. Two subspecies of Black-eared Wheatear recognised by Vaurie (1959), Svensson (1992), BWP and HBW; western form (*hispanica*) has sandier plumage than eastern (*melanoleuca*). White- and black-throated forms occur in both subspecies, with a progressive increase in the latter from W to E. Not included in study by Aliabadian *et al.* (2007), although one bird identified as a hybrid with Pied was found to be genetically indistinguishable from Pied; however, as the study was based on maternally inherited mt DNA, this would be expected if the mother had been a Pied Wheatear.

DISTRIBUTION Nominate breeds from Iberia, NW Africa east along Mediterranean to Croatia; *melanoleuca* in rest of Balkans, through Turkey to Caspian Sea. Both races migratory, wintering in W & C and E African Sahel, respectively.

STATUS The first for GB was a male *melanoleuca* collected at Radcliffe Res (Lancashire) in May 1875; the first *hispanica* was a male at Spurn (Yorkshire) in Sep 1892. The first of four Irish records (all in May) was at Tuskar Rock (Wexford) in 1916; two of these (1916, 1992) were *hispanica*; the other two were not assigned to race. There have now been *c.* 60 in GB&I, recorded in all months between Mar and Oct, but with distinct peaks in May and Sep. Of those listed by Naylor that were identified to subspecies up to 2003, 11 of the 21 *hispanica* were in spring, and all of the 12 *melanoleuca*, but there seems to be no difference in regional distribution within GB&I.

Desert Wheatear   *Oenanthe deserti* (Temminck)

SM *atrogularis* (Blyth)
SM *deserti* (Temminck)
SM *homochroa* (Tristram)

TAXONOMY Shows considerable clinal variation among populations; four subspecies recognised by Vaurie (1959) and Roselaar (in BWP), but only three by Svensson (1992) and HBW, who merge *atrogularis* with *deserti*, despite former's larger size. Until a thorough review, we retain these as separate. Two subspecies were included in study by Aliabadian *et al.* (2007): *homochroa* and *deserti*; they were indistinguishable in DNA sequence.

DISTRIBUTION Nominate *deserti*: NE Egypt to NW Arabia; *homochroa*: N Africa from Morocco to NW Egypt; *atrogularis*: south Asia, from S Caucasus to Mongolia; *oreophila*: south of *atrogularis*, from Himalayas to Inner Mongolia. *Homochroa* is relatively short-distance migrant; remaining populations migrate south and west to winter in N Africa, Arabia, NW India, Pakistan.

STATUS About 100 individuals have been recorded in GB to 2007, with a further four in Ireland. The first for GB was a male at Alloa (Clackmannanshire) in Nov 1880; the first for Ireland was in Wexford in Mar 1990. Only six individuals have been assigned to race: the first of four *homochroa* was shot at Spurn (Yorkshire) in Oct 1926; the only *deserti* met a similar fate on Fair Isle (Shetland) in Oct 1926, as did the sole *atrogularis* on Pentland Skerries (Orkney) in June 1906.

White-crowned Wheatear   *Oenanthe leucopyga* (C. L. Brehm)

SM race undetermined

TAXONOMY Aliabadian *et al.* (2007) found this to be a member of a weakly supported clade of wheatears that included Mourning *O. lugens*, Variable *O. picata*, Finsch's *O. finschii* and Hume's *O. albonigra*. Two subspecies, differing in size and glossiness of black plumage.

DISTRIBUTION Breeds across N Africa (*leucopyga*) into Sinai, Arabia (*ernesti*). Generally resident, though some local movement (HBW).

STATUS One, at Kessingland (Suffolk) in early June 1982. A bird of this species, or Black Wheatear *O. leucura*, was at Portnoo (Donegal) in June 1964.

Genus *MONTICOLA* Boie

Palearctic, Ethiopian and Oriental genus of about 10 species, two of which have been recorded in GB&I.

Rufous-tailed Rock Thrush   *Monticola saxatilis* (Linnaeus)

SM monotypic

TAXONOMY Rock thrushes are usually regarded as thrushes (Turdinae), but recent molecular results (Voelker & Spellman 2004) indicate a closer affiliation with the chats (Saxicolini). This species is closely related to Blue Rock Thrush (Outlaw *et al.* 2007). According to Svensson (1992), BWP and Clement *et al.* (2000), limited interpopulation variation in the Rufous-tailed Rock Thrush is insufficient for subspecific differentiation.

DISTRIBUTION Widespread, but only locally common, in stony habitats from NW Africa and Iberia across the Mediterranean, Asia Minor, through the Caucasus and Afghanistan to W & C China, between 56°N and 28°N. Migratory, wintering in sub-Saharan Africa from the Atlantic to the Indian Ocean and south into Tanzania.

STATUS About 25 records from GB and a further two from Ireland, scattered through the year from Feb to Nov, but with a clear peak in May. The first was shot at Therfield (Hertfordshire) in May 1843, and the first for Ireland was at Clogher Head (Louth) in May 1974. It seems likely, though unproven, that spring birds are largely European, whereas those in the autumn might be of more eastern origin.

Blue Rock Thrush   *Monticola solitarius* (Linnaeus)

SM race(s) undetermined

TAXONOMY Variation among populations is largely clinal, except for eastern *philippensis*, which is orange from lower breast to under tail. Clement *et al.* (2000) report this intergrades with *pandoo* so specific separation seems unlikely.

DISTRIBUTION Breeds on mountain plateaux, cliffs, rocky outcrops from NW Africa and Iberia to China and Japan. Nominate race is westernmost, breeding from Atlantic to N Turkey and S Caucasus; replaced by *longirostris* from C Turkey to Kashmir. Three further races outside W Palearctic (*pandoo*, *philippensis*, *madoci*). Mostly partial or altitudinal migrant in Europe; eastern populations more truly migratory, wintering from E Africa and S Arabia through Indian subcontinent to New Guinea.

STATUS Six birds have been recorded, between April and Oct, since a first-summer male at Skerryvore Lighthouse (Argyll) in June 1985. Despite having died and being preserved at the Natural History Museum, Tring, clinal variation means neither its race, nor that of any other, has been established. Where there is sufficient detail, it seems that all have belonged to one of of the blue-bellied W Palearctic races. There are no records from Ireland.

Genus *FICEDULA* Brisson

Predominantly Palearctic and Oriental genus of over 30 species, four of which have been recorded in GB&I, one as a breeding bird.

Red-breasted Flycatcher   *Ficedula parva* (Bechstein)

PM monotypic

TAXONOMY Recently split from Taiga Flycatcher on basis of plumage, bill colour, vocalisations, moult (summarised by Svensson *et al.* 2005). Analysis of the Mt cyt-b gene (Li & Zhang 2004) indicates a sequence divergence of *c.* 6.5%, more than the differences between Pied, Collared and Semicollared *F. semitorquata* (Saetre *et al.* 2001a).

DISTRIBUTION Chiefly from C Europe to Ural Mts, with isolated pockets in Baltic, Balkans, Germany. Migratory, wintering in India, principally in W & C.

STATUS A regular visitor in small numbers, chiefly in May–June and Sep–Oct, through birds are regularly found outside these periods. Commoner in autumn, when may occur widely within falls of migrants, especially along the east coasts of England and Scotland. The annual average for GB has remained relatively constant at *c.* 100 individuals (Fraser & Rogers 2006). In view of the increase in observers, this may mask a decline. European populations appear stable (BirdLife International 2004).

There have been 203 individuals in Ireland to 2004, where it is largely an autumn migrant with only one record in spring; most have occurred in Cork and Wexford.

Taiga Flycatcher   *Ficedula albicilla* (Pallas)

SM monotypic

TAXONOMY Recently separated from Red-breasted Flycatcher *(q.v.)*.

DISTRIBUTION Eastern sister species to Red-breasted Flycatcher, breeding from Ural Mts to Kamchatka. Zone of contact imprecisely known, as is pair composition and degree of isolation in sympatry, if indeed they are sympatric. The degree of genetic divergence (see Red-breasted Flycatcher) indicates a prolonged period of isolation. Migratory, wintering east of Red-breasted Flycatcher to Indo-China, SE China, though with some overlap in Himalayas, N India.

STATUS Two records from GB&I. The first was a first-summer male at Flamborough (Yorkshire) in Apr 2003, shortly followed by a first-winter at Sandgarth (Shetland) in Oct 2003. There are no records from Ireland.

Collared Flycatcher   *Ficedula albicollis* (Temminck)

SM monotypic

TAXONOMY Closely related to the other Eurasian black-and-white flycatchers, but genetically distinct (Saetre *et al.* 2001a,b). Sympatric with Pied Flycatcher in several parts of C & E Europe and Baltic islands, with some hybridisation. However, progeny are less fit and female hybrids are usually infertile (Alatalo *et al.* 1990).

DISTRIBUTION Principally from Poland to European Russia, but scattered populations further west in Italy, France, Germany, Sweden (BWP). Trans-Saharan migrant, wintering areas not entirely clear, but probably from Tanzania to Zimbabwe.

STATUS There are *c.* 28 records of this species (including 25 males). All but one have been in spring, and 21 of these in May or the first few days of June. The majority have been along the east coast and in the northern isles, but there are single records from Carnarfonshire, Cumbria and Sussex. There are no records from Ireland.

European Pied Flycatcher   *Ficedula hypoleuca* (Pallas)

MB PM *hypoleuca* (Pallas)

TAXONOMY Mitochondrial analysis (e.g. Saetre *et al.* 2001a,b) suggests that European Pied Flycatcher is sister to Collared Flycatcher, with Semicollared Flycatcher *F. semitorquata* sister to these; nuclear sequences are not congruent with this (Saetre *et al.* 2003). Until recently, four subspecies recognised, but DNA data indicate that the distinctive form from the Atlas Mts is as differentiated from European Pied as this is from Collared and Semicollared; now treated as a separate species – Atlas Pied Flycatcher *F. speculigera*. Three remaining subspecies of European Pied: nominate, *iberiae*, *sibirica*. Despite its relative isolation from other European Pied Flycatchers, and its proximity to NW Africa, the Iberian race is genetically close to the nominate.

DISTRIBUTION The most widespread of the Eurasian black-and-white flycatchers, breeding from Iberia across N & C Europe to the upper River Yenesei. Nominate breeds from France to the Ural Mts, including GB&I; *iberiae* is restricted to the Iberian penisula; *sibirica* replaces *hypoleuca* E of the Ural Mts. A trans-Saharan migrant, wintering in W Africa, but precise distribution unclear because of confusion with other three species.

STATUS European Pied Flycatcher expanded its range quite markedly between the two Atlases, being recorded in 34% more 10-km squares in the Second Breeding Atlas. It also reached Ireland, with breeding proved at sites in Antrim and Wicklow; breeding has not been confirmed subsequently, although birds present in suitable habitat. The majority of the GB range expansion was in the N & W heartland, but breeding also recorded in the SE. There were few CBC sites that held European Pied Flycatchers, so changes in numbers before 1990 are difficult to assess. However, recent data from BBS are more extensive, and indicate a fairly steep decline of *c.* 35% since 1994.

Many European Pied Flycatcher populations have adapted to warmer springs by advancing the date of breeding. There is now evidence suggesting that, despite their earlier breeding, this does not always correspond with the peak emergence of their insect food (which has advanced more rapidly: Both *et al.* 2006). The consequences are that some populations are increasingly out of synchrony with their food supply. Furthermore, the accelerated development of invertebrates may be leading to a shorter period of availability. This is not the case in all of Europe; birds breeding in Finland, for example, have advanced their arrival, perhaps because they benefit from earlier emergence of invertebrate food on their migration, allowing them to gain weight more rapidly.

FAMILY PRUNELLIDAE

Genus *PRUNELLA* Vieillot

Predominantly Palearctic genus of 13 species, two of which have been recorded in GB&I, one as a breeding bird the other as a vagrant.

Dunnock	***Prunella modularis* (Linnaeus)**

Endemic: RB *hebridium* Meinertzhagen
RB *occidentalis* (Hartert)
WM PM *modularis* (Linnaeus)

TAXONOMY The systematic position of the accentors has long been unresolved, and various relationships have been postulated using skeletal and plumage characters. Several studies have shown that the accentors (Prunellidae) are closely related to, or part of, a large assemblage that also includes the pipits, wagtails, sparrows, finches, buntings, grosbeaks, New World warblers and New World blackbirds (Sorenson & Payne 2001, Ericson & Johansson 2003, Barker *et al.* 2004, Beresford *et al.* 2005). The closest relative of the accentors may be the Olive Warbler *Peucedramus* (Ericson & Johansson 2003).

Eight (HBW) or nine (BWP) subspecies recognised on basis of plumage colouration; four in W Europe.

DISTRIBUTION Breeds in a diverse range of habitats, from woodland edges and scrub through to montane sites beyond the tree-line (Davies 1992). Found across W Palearctic, from Iberia, GB&I, Scandinavia to Ural and Caucasus Mts, though absent from steppe lowlands of European Russia, Ukraine, Kazakhstan. Nominate race breeds in N, C Europe to Ural Mts; replaced in Hebrides, Ireland by *hebridium*; in GB and W France by *occidentalis*; Iberia to Greece by *mabbottii*; other races occur from Balkans to Iran. N populations migratory, wintering in S of range (beyond in E) to Mediterranean coasts and Iran.

STATUS A common and familiar bird, present across GB&I, apart from Shetland and the extreme highlands of C & N Scotland. The range contracted slightly between the two Breeding Atlases, with birds deserting some of the most severe 10-km squares in N Scotland. CBC/BBS fieldwork revealed a substantial decline in the decade following 1975, and the beginnings of a recovery only since the mid-1990s. It is not clear what caused this decline, nor indeed the reasons for the limited recovery. Fuller *et al.* (2005) suggest that this may be one of the woodland species that has been adversely affected by the elevated numbers of deer; increased grazing may have reduced the quantity and quality of ground and scrub cover, which are essential for feeding and breeding Dunnocks, and an increase in nest failure at the egg stage may stem from this. Alternatively, farmland populations could have been affected by lack of winter food: they are generalists but eat a lot of seed in winter, and experiments suggest populations can be stabilized by feeding in winter (Siriwardena *et al.* 2006). On the other hand, clutch size has increased since the 1970s, which may be due to the gradual erosion of pesticide levels in the environment. At first sight, these results seem contradictory; the population fell at a time when clutch size was increasing and has begun to recover at a time when nest failures are rising. However, as with other species, increases in aspects of breeding success are more likely to be consequences of population change (e.g. reduced competition) than causes.

Little study has been undertaken of the two native subspecies; most research combines data from across GB&I. Dunnocks of the race *modularis* are regular passage migrants through the northern isles (Pennington *et al.* 2004). Evidence of migration is supported by ringing data that show immigration from Scandinavia during the autumn and that some GB birds move to the near-Continent in winter. However, the extent of these movements is generally unknown, although birds sometimes recorded in unusually large numbers at Portland and Christchurch (both Dorset) in both spring and autumn (Green 2004).

Alpine Accentor   *Prunella collaris* (Scopoli)

SM *collaris* (Scopoli)

TAXONOMY Morphologically similar to Altai Accentor *P. himalayana*, and these sometimes placed in separate genus to rest of accentors. Number of recognised races varies, generally 9–10, differing in plumage tones, colour and extent of streaking, and general body colour.

DISTRIBUTION Discontinuously distributed above tree-line on mountains of S Eurasia, from Pyrenees to Japan, though occurs at sea level in east. Two races in Europe: nominate from NW Africa, through SW, C Europe to Carpathian Mts; replaced in SE Europe by *subalpina*. Other races occur from Turkey through the Himalayas and Altai Mts to NE China.

STATUS 45 accepted records up to 2006, all from GB. The first was killed in Epping Forest (Essex) in Aug 1817, and birds have now been recorded in every month except Sep, with rather more in Apr–May and Nov–Jan. As might be expected from its European distribution, the majority occur in southern England, with only two each from Scotland and Wales, and none from Ireland. All records probably relate to European races.

FAMILY PASSERIDAE

Genus *PETRONIA* Kaup

A Palearctic, African and marginally Oriental genus of one to five species, one of which has been recorded in GB&I.

Rock Sparrow   *Petronia petronia* (Linnaeus)

SM race undetermined

TAXONOMY Mt DNA sequences indicate that *Petronia* is associated with a *Passer* clade (Allende *et al.* 2001), but with weak statistical support. Using both nuclear and mitochondrial sequences, Ericson & Johansson (2003) showed petronias to be sister to snow finches, and these to be in a strongly supported clade with *Passer*. A discontinuous distribution combines with relatively sedentary life style to generate a series of subspecies across the range, but the differences among these are finely drawn and some races are unlikely to be recognisable in the field (Clement *et al.* 1993).

DISTRIBUTION Breeds on barren or rocky hillsides across the Palearctic, from Iberia to W China, with further populations on Atlantic islands and in N Africa. Nominate race occurs through European range; replaced by *barbara* in N Africa and *madeirensis* in Madeira group and Canary Is. Further races in Turkey, Middle East and C Asia. Most populations sedentary, though some altitudinal migration, and eastern populations may disperse away from the breeding areas.

STATUS One record: at Cley (Norfolk) in June 1981. There are no records from Ireland.

Genus *PASSER* Brisson

A Palearctic, African and Oriental genus of about 20 species, three of which have been recorded in GB&I, two as breeding birds.

House Sparrow   *Passer domesticus* (Linnaeus)

RB *domesticus* (Linnaeus)

TAXONOMY House and Spanish Sparrows are closely related and formed a well-defined clade when analysed using mt cyt-b (Allende *et al.* 2001). House Sparrows and Italian Sparrows *italiae* were returned as sister taxa, with Spanish being slightly more distant. This finding weakens the case for Italian being a race of Spanish, but see Töpfer (2006) for contrary view; the taxonomic position of *italiae* remains unresolved (see also Roselaar in BWP).

Varies in colour, especially in facial pattern; also in general plumage tones. Two subspecies groups recognised: grey-cheeked *domesticus* and white-cheeked *indicus*.

DISTRIBUTION Natural range probably extends from NW Africa, GB&I, Scandinavia across Europe and Asia to Sea of Okhotsk and extreme E of Siberia; S to Mediterranean, Arabia, Indian subcontinent including Sri Lanka, and into Myanmar. Reputedly expanded its range E across Siberia following the construction of the Trans-Siberian railway. Nominate race occurs across N Eurasia; replaced by *balearoibericus* from Iberia through S France, Balkans to C Turkey, and by *tingitanus* in N Africa, and other races through Middle East into C Asia. SE populations generally belong to *indicus* group, including *indicus* itself from S Israel, S Arabia to India, and other races through into the Himalayas.

Has been introduced extensively around the world, including N & S America, S & W Africa, Australia and New Zealand. Most introductions from European stock, though those in S Africa include birds believed to be predominantly, if not exclusively, *indicus*.

STATUS There is no doubt that this species has declined markedly over the last 30 years, although precise breeding data are poor for the earlier years. It was not included in many CBC studies, and the long-term data from that source relate mostly to lowland farmland, where there has been a decline of *c.* 50% since 1976 (Crick *et al.* 2002). The decline is not uniform: it has been most severe in E & SE England (90%, 65%), whereas populations in the N & SW have been comparatively stable. The BTO's Garden BirdWatch survey allows a more precise dissection of the changes, revealing declines that started in the early 1980s (rather later than in farmland), and steeper losses in suburban and urban gardens than in rural ones. These results support the view that the decline began earlier in rural areas than urban sites and may have different causes. The seasonality of garden use is also diverging. Numbers are relatively uniform through the winter at rural sites, but in suburban gardens there has been a shift from a peak in midwinter in the 1970s to Oct in the 1990s. This shift is more pronounced in pastoral than arable areas, suggesting that seed availability may be a critical determinant in the former.

Analysis of a series of data (refs in Crick *et al.* 2002) indicates that the changes in population size are likely to have different causes among regions and/or habitats. However, the decline does not appear to be due to failures at the breeding stage; indeed, if anything, breeding performance in all regions of GB has increased over the past 40 years. This increase has been smaller in the SE (where the decline has been most severe) than in the N & W (where some populations have increased). Ringing data indicate that the survival of first-year birds declined during the 1970s, but that of adults remained constant: this effect could be largely responsible for the population decline (Siriwardena *et al.* 1999). Crick *et al.* (2002) note that adult survival might be expected to rise as the population levels fall; there is no evidence of this, and a failure of adult survival to improve may be partially responsible for the slow recovery of the populations.

A variety of factors have been suggested as drivers of the decline in House Sparrows, and it is likely that some or all of these play their part. Churcher & Lawton (1987) reported that at least 30% of House Sparrow mortality in a Bedfordshire village was due to cats, and this might represent a sizeable proportion of juvenile mortality. Sparrowhawk predation may also have an effect on the mortality of inexperienced juveniles, but, while it is tempting to note the association between the decline in sparrows and the increase in this predator, there is no direct evidence of cause and effect. Indeed, many other species included in Sparrowhawk diet, such as Common Chaffinch and Blue Tit, have not declined. A more likely cause of the decline, especially in juvenile survival, may be a decrease in autumn and winter seed supplies. Agriculture has become more 'efficient': less grain is spilled, increased herbicides result in less weed seed, and crops are stored in more secure buildings. Availability of nest sites may have declined due to the conversion of farm buildings to housing and the increase in house insulation; although this will not affect juvenile survival, it might reduce the overall productivity. Summers-Smith (2003) has proposed the decline is linked to unleaded petrol. This contains chemical additives whose by-products might kill invertebrates that comprise an important food for nestlings. Although sparrow populations in (e.g.) London are declining more rapidly than in nearby suburban sites, the same effects have not been found in other urban centres, nor have other birds that also feed on insects been affected. There has been a loss of overgrown and weedy corners in towns, villages and farmyards that generate lots of seed; such 'untidy' places often attract parties of feeding sparrows. This merits further investigation, as does the possibility of disease transmission through feeding stations.

Spanish Sparrow   *Passer hispaniolensis* (Temminck)

SM race undetermined, but unlikely to have been other than *hispaniolensis* (Temminck)

TAXONOMY Sister to the House/Italian clade (Allende *et al.* 2001), or conspecific with latter, in which case Spanish Sparrow would become *P. italiae hispaniolensis* (Töpfer 2006); two subspecies. Races differ in extent of rufous pink on fringes of mantle and upperwing feathering (BWP).

DISTRIBUTION Nominate race occurs as scattered populations from Canaries across N Africa to Libya, Israel, Jordan Valley; N of Mediterranean, from Iberia, Sardinia, Balkans, Bulgaria, Asia Minor to NW Iran. Replaced further east through Caspian, Kazakhstan to Afghanistan by *transcaspicus*. Hybridises with House Sparrow on islands in C Mediterranean (e.g. Malta, Sicily, Crete). Italian Sparrows suggested to be stable hybrid population of *domesticus* and *hispaniolensis*, formed following southwards spread of former into range of latter in C Italy. Populations in south of Italian peninsula show progressively more *hispaniolensis* characteristics (Roselaar in BWP).

   N populations are partially migratory, wintering within or beyond breeding range. S and island populations generally resident.

STATUS Seven records from GB, with the first on Lundy (Devon) in June 1966. Four of the subsequent six were in the SW, with one each in Orkney and Cumbria. There is no pattern to the date of arrival: birds have been found from May to Nov. A bird at Waterside (Cumbria) remained from July 1996 to Dec 1998. Not recorded in Ireland.

Eurasian Tree Sparrow   *Passer montanus* (Linnaeus)

RB *montanus* (Linnaeus)

TAXONOMY The position of Tree Sparrow within the *Passer* phylogeny could not be resolved conclusively by Allende *et al.* (2001), who found it to be in different parts of the tree depending on the analytical methodology, but very weakly supported in either instance. While clearly a *Passer*, it is not especially close to House and/or Spanish, despite its similar ecology in the Orient. Over 30 subspecies have been described, but many (if not most) of dubious validity and intermediate in morphology between other races (BWP, Clement *et al.* 1993). Roselaar (in BWP) recognises 10, though these differ from Vaurie (1959), and shows that size varies clinally from north (large) to south (small), with darker birds in areas of high humidity, fitting with the classical zoogeographical rules of Bergmann and Gloger.

DISTRIBUTION Breeds in woodland, farmland and forest edges across Eurasia from Iberia to Japan, absent only from the far north. In the west, generally absent from N Scandinavia, and further east, from India and the desert regions of C Asia. Nominate race is most widely distributed, replaced by *transcaucasicus* from Turkey to N Iran, and a series of other races through Asia (see BWP). Most populations are sedentary, although irruptions occur sporadically (usually of juveniles: BWP), especially from northern regions. Numbers fluctuate, with sites being occupied and abandoned (often for no obvious reason).

Introduced to N America, Australia and several island groups in SE Asia.

STATUS The species has shown alarming changes in distribution in the last 25 years. Although occupancy of 10-km squares rose slightly in Ireland between the two Breeding Atlases, in GB it fell by *c.* 20%. Abandoned squares were generally around the margins of the range, possibly indicating the desertion of less optimal habitats. The Breeding Atlases did not reveal the full extent of the problem: CBC data showed a decline in population of *c.* 97% during 1975–90. Although BBS data indicate the beginnings of a recovery, with a rise of 30–50% across GB, this is from such a low base that the population is still only *c.* 3% of that in the late 1960s. This does not appear to be due to failures in reproduction; clutch and brood sizes have both increased since 1968, and nest failures at both egg and chick stages have decreased. As in GB, Irish populations have shown periods of decline and recovery; the exact status here is difficult to judge at the present time.

The species is relatively sedentary, with very few recent movements of more than a few km. Data are too limited for detailed analysis of adult survival, but since breeding failure does not seem the driving force behind the population crash (and it can be called little else), adult survival seems to be the most likely cause. Tree Sparrows are granivorous and need a reliable seed supply for the winter. In the past, this was available from stubble and livestock feeding stations. Changes in agriculture have reduced this resource; in particular, the change from sowing cereals in the spring to the autumn has removed overwinter stubbles, and more efficient harvesting has resulted in less grain spillage. Changes in livestock management may also be implicated, as may the loss of wet patches in farmland, thereby reducing insect food (G Siriwardena pers. comm.).

In some regions, conservation agencies have begun initiatives such as providing nest boxes and ensuring that food is available throughout the winter; this seems to be having an effect, at least locally. However, Tree Sparrows have a long way to go to regain their former abundance.

FAMILY MOTACILLIDAE

Genus *MOTACILLA* Linnaeus

Old World genus of at least 11 species, four of which have been recorded in GB&I, three as breeding birds.

Yellow Wagtail	*Motacilla flava* Linnaeus

MB PM *flavissima* (Blyth)
CB PM *flava* Linnaeus
CB PM *thunbergi* Billberg
CB SM *cinereocapilla* Savi
SM *beema* (Sykes)
SM *feldegg* Michahelles
SM *simillima* Hartert

TAXONOMY The taxonomic relations of the various taxa of Yellow Wagtail have long been confused. Attempts to disentangle inter-relationships based on plumage, morphology, vocalisations, distribution and observed hybridisation have revealed little. Recent mt DNA studies have clarified some relationships and, although there is still uncertainty over where species boundaries should be drawn, a few conclusions are emerging. The most comprehensive review is that of Alström *et al.* (2003), who list 18 taxa in which they have varying degrees of confidence. These are differentiated largely on the basis of male plumage and geographic distribution.

Molecular data (Ödeen & Alström 2001; Alström & Ödeen 2002; Voelker 2002b; Pavlova *et al.* 2003; Ödeen & Björklund 2003) support the separation of Yellow Wagtail into two species, 'Western Yellow Wagtail' *M. flava*, and 'Eastern Yellow Wagtail' *M. tschutschensis*. This is a major split. Four subspecies that occur largely east of Lake Baikal *(plexa, macronyx, taivana, tschutschensis*, plus '*simillima*', which Alström & Mild regard as a synonym of *tschutschensis*) are well separated from the rest in both mitochondrial and nuclear genes, although *leucocephala* from NW Mongolia has not been analysed. In both nuclear and mitochondrial data sets, Grey Wagtail is sister to 'Eastern Yellow Wagtail' – albeit without strong statistical support. A similar pattern in nuclear and mt genes is hard to explain other than by invoking striking parallel evolution. Within each of the two groups of Yellow Wagtail, the various forms are genetically very close; there is no evidence that any subspecies is diagnosable on the basis of DNA sequence. Apart from *pygmaea* (Egypt), which forms a tight cluster, the subspecies are scattered through the phylogeny. Two northerly taxa (*thunbergi* and *plexa*), which Alström *et al.* (2003) considered inseparable in plumage, belong in the separate groups: 'Western' and 'Eastern' Yellow Wagtail, respectively.

The Yellow Wagtail has most likely been subject to repeated fragmentation, probably driven by glaciations. An early one led to partition into eastern and western refuges, and during this period of isolation the DNA sequences diverged (and Grey Wagtail became separate from the eastern Yellow Wagtail group). Climatic amelioration led to range expansion, with or without geographical contact; the two populations of Yellow Wagtail diverged sufficiently to function as separate evolutionary lineages: the *flava* and *tschutschensis* groups. During one or more subsequent ice ages, populations again became fragmented, and within the two taxa, the various subspecies diverged, probably driven by sexual selection by females for particular male patterns. Again, climatic amelioration was followed by range expansion, leading to the patterns of secondary contact that we see today. Some taxa are more striking, but the DNA data do not support any further specific separation. Some of the hybrid zones are steep, and in other areas non-random mating has been reported (e.g. Bril 1998). Both of these may suggest that the fitness of hybrids is reduced, but strong female preference for the parental phenotype will produce a similar pattern.

DISTRIBUTION The 'Western Yellow Wagtail' group breeds across the Palearctic from the Atlantic and N Africa to C Siberia (see Alström *et al.* 2003). Subspecies reported from GB&I include: *flavissima*: GB&I, N coast of France and along southern N Sea coasts to S Norway; *flava*: W & C Europe from Pyrenees to S Sweden, E to Urals; *cinereocapilla:* Italy; *beema*: Russia, central latitudes of W Siberia; *thunbergi*: C Fennoscandia, across Siberia N of *c.* 60°N, merging into *plexa*; *feldegg*: Balkans through C Asia to extreme W China. Almost entirely migratory, wintering across most of sub-Saharan Africa and the Indian subcontinent. A degree of sub-structuring of subspecies on the wintering grounds, but more than one taxon frequently occurs at any location.

Four subspecies in the 'Eastern Yellow Wagtail' group: nominate (including '*simillima*' (Kamchatka), according to Alström *et al.* (2003), breeds C & E Siberia to Alaska; *plexa,* N Siberia, from Taimyr to Kolyma; *macronyx*, E Mongolia, NE China, SE Russia; *taivana,* E Siberia (Amurland), Sakhalin, N Japan; species winters in SE China, SE Asia through to Indonesia, and the Philippines, or India (*plexa*).

STATUS See Citrine Wagtail in relation to possible hybrid pairing with that species. Most of the world population of the distinctive subspecies *flavissima* breeds in GB&I, though largely absent from Ireland and most of Scotland. In GB, the range contracted between the Breeding Atlas surveys, with almost 10% of 10-km squares being vacated between 1968–72 and 1988–91. There was a parallel decline in abundance, as reflected in the CBC/BBS data; the population estimates for England fell by 65% between 1967 and 2003, mostly since 1978 (there are too few monitored sites elsewhere for meaningful analysis). The Waterbirds Survey shows an even greater decline, with the estimated population falling by *c.* 90% since 1975. Clearly, Western Yellow Wagtails are in serious trouble in GB&I. There is relatively little information concerning breeding performance, but the limited nest record data indicate a decline in brood size of *c.* 10% since 1968. The Breeding Atlas data indicate a contraction in range towards a core area of C England, with breeding sites being lost from the N, W & S (Gibbons *et al.* 1993). This has coincided with major changes in farming practice: in particular, drainage of wet areas, loss of grassland to arable cultivation, the consequent decline in livestock and associated insect food, and the transition from spring to autumn sowing of cereals. These will combine to reduce both nesting and foraging habitats (Nelson *et al.* 2003). *Flava* has occasionally bred in GB as far N as Scotland.

The picture in Ireland is rather different (P Smiddy pers. comm.). The race *flavissima* ceased to breed in the west of the Republic in the 1920s, and in NI by 1944. Since the 1950s, there has been a scattering of breeding records in different counties, but nowhere has this been sustained, and it always involved just a handful of pairs. In several years since the 1990s, a few pairs have been suspected or proved to breed in County Wicklow. Otherwise, this race is a scarce spring and autumn passage migrant. The race *flava* bred in 1963, 1965, 1983, 1997 and 2002, and was suspected of breeding on other occasions. Mixed pairs have also occurred.

Ringing data indicate that *flavissima* moves through Iberia into NW Africa. Little is known of the extent of the wintering range in W Africa.

Racial identification of the races occurring in GB&I is confused by variation within races and the complexities of hybrisation. Nominate *flava* is recorded regularly on passage in GB, especially in spring, and it is a rare migrant in spring and autumn in Ireland (about 53 individuals). *Thunbergi* is a scarce migrant in GB (and much rarer in Ireland, where only six recorded) and breeds occasionally (BS3). There have been *c.* 10 records of *feldegg* since a male on Fair Isle (Shetland) in May 1970. Birds somewhat paler than typical *flava* have occasionally bred; these are sometimes referred to as 'Channel Wagtails', and are probably the result of hybridisation between *flava* and *flavissima*. They can look superficially like C Asian *beema* (and make identification of that race very difficult as a result), while the palest examples resemble Mongolian *leucocephala*, with the consequence that that race would also be hard to confirm as a vagrant (although it does not seem a likely candidate).

Remarkably, two pairs of *cinereocapilla* bred in Belfast in 1956, and two other individuals have been recorded: one on Rathlin Island (Antrim) in May 1985; another trapped at Copeland Bird Observatory (Down) in May 1998 was seen in Belfast the following day.

Two specimens of *simillima* from Fair Isle (Shetland) in Oct 1909 and Sep 1912. Genetic material from these individuals is currently being analysed to confirm (or otherwise) their taxonomic affinities. These are the only records relating to the 'Eastern Yellow Wagtail' group.

Citrine Wagtail   *Motacilla citreola* Pallas

HB SM race undetermined

TAXONOMY Analysis of wagtail mt DNA (Ödeen & Alström 2001, Voelker 2002b, Pavlova *et al.* 2003) has revealed that Citrine Wagtail comprises two mitochondrial clades, closer to 'Eastern' and 'Western' Yellow Wagtail respectively than they are to each other. At first sight, this would make Citrine Wagtail paraphyletic. However, results from analysis of nuclear genes (Ödeen & Alström 2001, Ödeen & Björklund 2003) do not support this result. The latter authors in particular find the subspecies of Citrine to be very similar, and sister to 'Western' Yellow Wagtail. It seems likely that, following the recognition of E & W Yellow Wagtail lineages, one of these diverged into Grey and 'Eastern' Yellow Wagtails, and the other into Citrine and 'Western' Yellow. The presence of the 'wrong' mt DNA in subspecies *citreola* is then explained by the eastward expansion of Citrine and the subsequent hybridisation between male Citrine and female 'Eastern' Yellow Wagtails, with the transfer of the maternally inherited mt DNA across the species barrier. In any event, the finding that Citrine is sister to 'Western' Yellow Wagtail and Grey to 'Eastern' Yellow Wagtail supports the splitting of the Yellow Wagtail into two species.

DISTRIBUTION Breeds in Russia, from the Kola Peninsula to River Khatanga, and south to the Himalayas. Nominate covers most of the range, being replaced by *calcarata* south from the Tien Shan Mts. Migratory, wintering from S China through N Indo-China, across the N of the Indian subcontinent and along the shores of the Arabian Gulf. A third race *werae* is often recognized (e.g. BWP, HBW, Svensson 1992); after examining a long series of specimens, Alström *et al.* (2003) concluded that the differences were inconsistent and they were unable to sustain this taxon.

STATUS One breeding instance involving a hybrid pairing: in July 1976, a male was present at Walton-on-the-Naze (Essex), feeding four young. Although no female was seen in attendance at the nest, from the plumage of the juveniles, it is believed that this represented a mixed pair of a male Citrine and a female *M. flava flavissima* (Cox & Inskipp 1978).

Citrine Wagtail shows a marked difference in occurrence across GB&I. The first was present on Fair Isle (Shetland) in Oct 1954 and, of the next 150 records, more than 50% were in Shetland. There were 13 Irish records to the end of 2006, of which the first was at Ballycotton (Cork) in Oct 1968 and the first for NI at Mullagh (Londonderry) in Sep 2008. Overall, there has been a steady increase in frequency, from less than 1 each year in the 1950s to *c.* 12 each year since 1990; data from Fair Isle Bird Observatory, where coverage is fairly consistent across the years, suggest that the increase may be a consequence of the gradual westward spread of the species (BWP). The majority of records (>90%) are from Aug to Oct, but there have been 16 in spring, 13 of which are from England. It is likely that some of these were newly arrived from the east, rather than birds surviving from the previous autumn.

Grey Wagtail   *Motacilla cinerea* Tunstall

RB PM *cinerea* Tunstall

TAXONOMY DNA sequences indicate a sister relationship with 'Eastern' Yellow Wagtail. Although the statistical support is not strong, two independent data sets (nuclear and mitochondrial) agree on this relationship. Up to five or six (Vaurie 1959, Svensson 1992) subspecies have been recognised, but Alström *et al.* (2003) accept only three: one widespread and the other two restricted to Atlantic islands.

DISTRIBUTION The nominate race is widely distributed from the Canaries and the Atlas Mts to the hills of Norway, across W & C Europe to the Balkans, N Turkey and the Caucasus Mts. There is a gap in distribution in W Russia, but the species is again widespread across Siberia to Kamchatka, Korea and N China ('*robusta*' and '*melanope*'; BWP, HBW, Svensson 1992). Northern populations are migratory, wintering along coastal regions of N Africa and Arabia, the Rift Valley, the Indian subcontinent, Indo-China, S China and the islands from Malaysia to New Guinea. Distinct subspecies occur on Madeira (*schmitzi*) and the Azores (*patriciae*).

STATUS Grey Wagtail is widespread in GB&I away from the arable lowlands of England, and the N & W isles of Scotland. It is more closely associated with water than Pied, especially clean, fast-flowing upland streams and rivers. There was evidence of an increase in distribution of *c.* 7% between the two Breeding Atlases in GB; a decline of 8% in Ireland is probably due to reduced coverage. It is not well covered by CBC/BBS fieldwork, and data on population trends comes mostly from the Waterbirds Survey. The latter indicated a steep decline in population of *c.* 40% through the decade following 1974, followed by a period of rather low numbers to the late 1990s, with the beginnings of a recovery since then. Overall, numbers in 2003 were still about 20% down on the early 1970s.

The initial decrease in population revealed by the Waterbird Survey coincided with the fieldwork for the Second Breeding Atlas, which showed an increase in distribution, and studies in Wales that showed a marked increase (reviewed in Tyler & Ormerod 1991). As with Pied Wagtail, there may be different ecological forces at work here; Waterbird Survey sites tend to be in lowland areas, which Tyler & Ormerod regard as suboptimal, compared with the buoyant populations in the uplands.

Ringing recoveries show that there is a general movement southwards in winter, with a few birds crossing the English Channel into France or Iberia. There are also local movements towards lower land in response to harsh winter weather. Some Continental birds winter in GB&I, indicated by a small number of birds ringed on the near-Continent (Wernham *et al.* 2002) and a regular passage through Shetland in spring and autumn (Pennington *et al.* 2004).

White Wagtail/Pied Wagtail   *Motacillla alba* Linnaeus

RB MB *yarrellii* Gould
CB PM *alba* Linnaeus
SM *leucopsis* Gould

TAXONOMY Extensive morphological variation has led to the recognition of many subspecies; Alström *et al.* (2003) list nine; more are included by Vaurie (1959) – although he regards African Pied Wagtail *M. aguimp* as a race of *M. alba*. White Wagtails were included in molecular analyses by Odeen & Alström (2001), Alström & Odeen (2002), Voelker (2002b) and Pavlova *et al.* (2003). Differentiation among races was slight, indicating a recent and rapid divergence. There was evidence that the S Asian races *personata, alboides, leucopsis* formed a separate clade to the remaining subspecies, but nuclear sequences did not support this. There was no evidence of genetic differentiation between *alba* and *yarrellii*.

DISTRIBUTION Breeds across the Palearctic, from SE Greenland, Iceland to W Alaska, and as far south as the Himalayas. The nominate occurs across the entire W Palearctic, apart from GB&I (*yarrellii*) and NW Africa (*sub-personata*). In the E Palearctic, the most widespread race is *ocularis,* replaced by *lugens* from Kamchatka through Japan to E China, and by *personata, baicalensis, alboides* and *leucopsis* in the south of the region.

STATUS Most of the world population of the distinctive subspecies *yarrellii* (Pied Wagtail) breeds in GB&I.

Nominate race birds occasionally breed, especially in the northern isles of Scotland. There is evidence of non-random mating in Shetland, with a deficiency of Pied x White pairs. Whether this is due to differences in mate choice or because of pair formation in migrating flocks of single-taxon wagtails is not known. Nevertheless, this will serve to maintain genetic differences between the two subspecies, even in areas of apparent sympatry.

There was little change in distribution across GB&I between the Breeding Atlases, with >95% of 10-km squares occupied by breeding birds, apart from a slight expansion northwards into Orkney and Shetland. CBC/BBS data show an increase in population into the 1970s, perhaps as the species recovered from the severe winter of 1962–63. Numbers fell in the decade following 1975, but have recently shown a progressive rise back towards the maximum of the early 1970s. Data from the Waterbirds Survey conflicts with this, with evidence of a decline of numbers in this habitat during 1975–2003, suggesting different ecological forces at work. Nest record data show a progressive decrease in clutch size since 1968 and, although nest failures are also declining, brood size shows a downturn since the early 1980s. Whether this is driving down the riverine populations requires investigation.

Ringing recoveries show that many British Pied Wagtails leave the islands to winter in W France and Iberia, with southern populations wintering further south (Irish data are too few for comparison). The nominate race occurs widely as a passage migrant through much of Britain and there is evidence that Icelandic birds pass through northern parts of GB&I in autumn to winter further south. In Ireland, the nominate race is commoner in autumn than spring.

Subspecies *leucopsis* ('Amur Wagtail'): one, a male at Seaham (Durham), April 2005.

Genus *ANTHUS* Bechstein

Near-cosmopolitan genus of over 40 species, 11 of which have been recorded in GB&I, three as breeding birds.

Richard's Pipit
Anthus richardi Vieillot

PM monotypic

TAXONOMY Five groups are traditionally identified: *richardi* (Siberia), *rufulus* (India), *cinnamomeus* (Africa), *australis* (Australia) and *novaeseelandiae* (New Zealand). Four of these were included in a molecular study (Voelker 1999) that suggested that the Richard's Pipit complex is not monophyletic. The statistical support for this was weak. Alström *et al.* (2003) recommend that four species should be recognised, combining the two Australasian taxa into *A. novaeseelandiae*. There is slight morphological variation within Richard's Pipit (*sensu stricto*) that has led some (e.g. HBW) to recommend several subspecies; however, Alström *et al.* (2003) suggest that Richard's Pipit is best treated as monotypic.

DISTRIBUTION Breeds in the E Palearctic, from about Omsk to China, including Mongolia, S of *c.* 65°N. Migratory, with northern populations wintering in S & E India. Small numbers winter in the Middle East, and regularly straggle W into Europe and N Africa, with increasing numbers being recorded in winter – perhaps as observers become more confident about its identification, or increased overwinter survival.

STATUS The first record for GB&I was a bird caught alive in Copenhagen Fields (Middlesex) in Oct 1812, and there have been over 3,500 birds since 1958. The majority of records are in the autumn, and (in GB) birds are especially found in Shetland, Norfolk and Scilly. This bias is unlikely to be real and is likely due to the distribution of observers. There are 78 Irish records (to 2004), of which the first was a bird taken at Lucan (Dublin) in Nov 1907. The number of records has increased somewhat down the years, although annually there are rarely more than 1 or 2; however, in 1994, an exceptional 13 birds were found. Over half of the Irish records have been in County Cork, again more likely due to density of observers than abundance of birds.

The overall increase in records has not been mirrored at the bird observatories. Numbers at Fair Isle (Shetland) and Portland Bill (Dorset) have been relatively stable or even declined slightly (R Riddington, M Cade, pers. comm.), suggesting that increasing numbers elsewhere are due to improved coverage. The recent increase in winter and inland records is presumably not a new phenomenon, but simply due to birders now finding birds that were always there.

Blyth's Pipit
Anthus godlewskii (Taczanowski)

SM monotypic

TAXONOMY Formerly regarded as a subspecies of both Richard's and Tawny, but sympatric with both of these in parts of its range. Voelker (1999) examined mt cyt-b sequences and found it to comprise a separate lineage, although not especially well resolved. Alström *et al.* (2003) argue that the streaked plumage of Blyth's and Richard's Pipits may be a product of convergent evolution as a result of the more vegetated habitat compared with the more open habitat of Tawny Pipit, or the retention of a common ancestral character.

DISTRIBUTION Apparently restricted to a relatively small area of the E Palearctic (Alström *et al.* 2003), south of Lake Baikal, from the Altai Mts eastwards to NE China, with core area in Mongolia. Precise distribution still unclear, but Alström & Mild doubt that it breeds on the Tibetan Plateau. Migratory, wintering locally across the Indian subcontinent to Sri Lanka.

STATUS The first record for GB&I was collected at Brighton (Sussex) in Oct 1882. It was over 100 years before the next was found on Fair Isle (Shetland) in Oct 1988. There have now been 21 (to 2007), including two at Portland Bill (Dorset) in Nov 1998. All have been during Sep–Dec, and there is one inland record at Gringley Carr (Nottinghamshire) in Dec 2002. There are no records from Ireland.

Tawny Pipit   *Anthus campestris* **(Linnaeus)**

PM *campestris* (Linnaeus)

TAXONOMY Voelker (1999) found Tawny Pipit to be sister to Berthelot's Pipit *A. berthelotii*, unsurprisingly in view of its range and plumage similarities, and supporting molecular and morphological work by Arctander *et al.* (1996) and Alström *et al.* (2003), respectively. Intraspecific variation largely clinal; regard as monotypic by Alström *et al.* (2003), or with two subspecies (Svensson 1992) or three (Roselaar in BWP).

DISTRIBUTION Breeds across W Palearctic, from NW Africa and Iberia to S Sweden (excluding GB&I, NW France), and E to C Mongolia, between 60°N and 30°N, and in parts of the Middle East. Migratory, with western populations wintering in the Sahel from Senegambia to Ethiopia; more eastern populations winter in Arabia and NW Indian subcontinent. Subspecies (following BWP) – *campestris*: W Palearctic, NW Kazakhstan, W Siberia to about Omsk; *griseus:* S Caspian to Tien Shan Mts; *kastchenkoi:* Omsk and Altai Mts, E to Mongolia.

STATUS The first was a bird obtained at Shoreham-by-Sea (Sussex) in Aug 1858; many of the subsequent occurrences have also been along the S coast of England. There were over 500 records from GB by 1982, when the species was removed from the BBRC list. Numbers had risen steadily over the preceeding decade following an identification paper by Grant (1972). The main months of occurrence are Aug–Oct, but there is also a small peak in Apr–May. By 2002, the total exceeded 1,100, but the average had fallen back to *c.* 10 per year (Fraser & Rogers 2006).

Tawny Pipit remains a rare bird in Ireland, with only 33 records, since the first at Great Saltee (Wexford) in May 1953. The same patterns hold here as for GB, with peaks in spring and autumn, and >80% of birds being found along the south coast in Wexford or Cork.

Olive-backed Pipit   *Anthus hodgsoni* **Richmond**

SM *yunnanensis* Uchida & Kuroda

TAXONOMY Voelker (1999) found Olive-backed Pipit to be sister to Tree Pipit, in a subclade of the small pipits that also included Pechora Pipit. There are slight differences in choice of breeding habitat between Olive-backed and Tree Pipits where they coexist (Alström *et al.* 2003). Two geographically separate subspecies are recognised (*hodgsoni* and *yunnanensis*), based on plumage and size.

DISTRIBUTION Occurs in two geographically separate areas. Subspecies *yunnanensis* is the more widely distributed, from N Ural Mts across Palearctic, S of Lake Baikal to Kamchatka, and through China, possibly into N Korea. Conversely, *hodgsoni* is restricted to an area from the Himalayas to Gansu, and possibly the N islands of Japan (Alström *et al.* 2003). Migratory, wintering from India (*hodgsoni*), east through Indo-China to S China, and the Philippines (*yunnanensis*).

STATUS The first for GB was trapped on Skokholm (Pembrokeshire) in Apr 1948, but its identification was not confirmed for many years (Conder 1979). As observers gained familiarity with rare and scarce pipits, more followed, so that the total stood at over 300 by 2007. It remains rare in Ireland, with just seven records, all in County Cork apart from the first on Great Saltee (Wexford) in Oct 1978. There have been none in Ireland since 1993. Records across GB&I fall into three clusters, of which by far the largest is Sep–Nov (95%); nine birds have been recorded in spring, and there are also six from midwinter, including birds at Pitsea (Essex) Jan–Apr 1994 and Brixham (Devon) Jan–Apr 1997.

Slightly more have been recorded in Scotland than England. Numbers in the two regions are closely similar for Oct and Nov, but far more are found in Scotland in Sep: there are no records from Scilly in Sep, yet 20% of Shetland birds are found in this month.

Tree Pipit   *Anthus trivialis* (Linnaeus)

MB PM *trivialis* (Linnaeus)

TAXONOMY Sister taxon to Olive-backed Pipit. Two subspecies recognised (Vaurie 1959) though variation slight.

DISTRIBUTION Breeds (*trivialis*) through Europe (not S Iberia, Iceland, Ireland), across the boreal forests of the Palearctic to Verkhoyansk Mts and S to Lake Baikal. Also breeds around the Black Sea to the S Caspian. Subspecies *haringtoni* more restricted, breeding in Turkestan, Tien Shan Mts and W Himalayas (Alström *et al.* 2003). Breeds in the western Himalayas, from Kashmir, Russian Turkestan, and W Pamirs. Strongly migratory, wintering across sub-Saharan Africa, from Senegambia to Ethiopia, S to Zimbabwe except rainforests; also Indian subcontinent.

STATUS Tree Pipits are widely distributed in GB, avoiding the coastal fringes and intensively arable land. The number of occupied squares declined by *c.* 15% between the two Breeding Atlases, predominantly in lowland England. The decline in S England had earlier been recorded by Parslow (1973), and is paralleled by a reduction in numbers shown by CBC data. The population declined slowly from 1965 to 1980, and then fell rapidly through to *c.* 1995. More recently, there are signs of a modest overall recovery, largely driven by the Scottish population. The Repeat Woodland Survey confirmed severe declines identified from CBC data. Fuller *et al.* (2005) have suggested that the changes in abundance reflect the changing pattern of planting and maturing of woodland since WWII. Brood size has increased since 1964, as has nest failure at the egg stage, especially in recently clear-felled forest, perhaps contributing to local extinctions (Burton 2009). There is also a modest advance in laying date, although these results are all based on a small sample of nests.

There is no fully satisfactory and proved breeding record for Ireland (P Smiddy pers. comm., *contra* the Second Breeding Atlas). Individuals have been heard singing in Ireland in habitats suitable for breeding, but sometimes also when clearly on passage. It is a scarce passage migrant, occurring in spring (Apr–May) and autumn (Aug–Oct).

Pechora Pipit   *Anthus gustavi* Swinhoe

SM *gustavi* Swinhoe

TAXONOMY Voelker (1999) found that the Pechora Pipit is sister to the species pair of Tree and Olive-backed Pipits, with these three forming a clade separate from the rest of the small pipits, although its statistical support is weak. Pechora Pipit is ecologically the northern replacement of these two, being largely allopatric and breeding in wet and marshy tundra with only scattered trees. Two subspecies are recognised, one of which is very restricted and geographically isolated; from its distinct song it may be a separate species (Leonovich *et al.* 1997).

DISTRIBUTION Nominate *gustavi* breeds across the Palearctic tundra, from the Pechora River to the Bering Sea, between about 72°N and 45°N, and part of the Kamchatka Peninsula. Subspecies *menzbieri* is restricted to NE China. Migratory: winter range incompletely known, but records from Philippines, Borneo, Wallacea (Alström *et al.* 2003).

STATUS The first was shot on Fair Isle (Shetland) in Sep 1925 and this site has hosted the majority of birds since then; 41 of the 75 British examples to 2007 were found here, with 25 of the remainder elsewhere in Shetland. Only one spring record, at Calf of Man (I of Man) in May 1991; the rest have been in Sep–Oct, and one in Pembrokeshire in Nov 2007. Two have occurred in Ireland: one at Garinish (Cork) in Sep 1990 and one at Tory Island (Donegal) in Sep 2001.

Meadow Pipit   *Anthus pratensis* (Linnaeus)

Endemic: RB or MB *whistleri* Clancey
RB MB WM PM *pratensis* (Linnaeus)

TAXONOMY Meadow Pipit is part of a clade that also includes Red-throated, Rosy *A. roseatus*, Buff-bellied, Eurasian Rock and Water Pipits (Alström & Odeen, in Alström *et al.* 2003). This is supported by similarities in vocalisations and territorial behaviour. According to Voelker (1999), the closest relatives of Meadow Pipit are Water and Eurasian Rock Pipits, while Buff-bellied Pipit is sister to these three. Although this hypothesis is only weakly supported by his data, it receives strong support from additional mt DNA sequences (Alström & Ödeen, unpublished). These results are unexpected and need to be corroborated by independent data. Vaurie (1959) only recognised two subspecies, one of which (*'theresae'*, W Ireland) he regarded as only moderately well differentiated. Roselaar (in BWP) and Svensson (1992) recognise only nominate and *whistleri*. Alström *et al.* (2003) are equivocal, regarding variation as clinal.

DISTRIBUTION Breeds across the W Palearctic north of *c.* 42°N, from France to Romania, and east to the River Ob (though precise geographic limits unclear); also E Greenland, Iceland, Faroes. Nominate occurs across most of the range; *whistleri* breeds in Ireland and W Scotland, but not Iceland, Greenland (BWP). Northern populations migratory, wintering as far south as coastal N Africa, S Europe, Middle East.

STATUS Meadow Pipits contracted their range between the two Breeding Atlases. Although still widespread in Scotland and Wales, breeding in almost every single 10-km square, there were serious losses from wide areas of lowland England and Ireland. This contraction in range is parallelled by a decline of 40% in the English population during 1979–2003. BBS data are sufficient to permit partitioning of the regions, and these show a continued decline in England (and possibly Scotland) since 1994, but signs of a recovery elsewhere. The decline has been linked to deterioration of conditions on the Iberian wintering grounds, but it is also likely due to the loss of marginal land on farmland and the conversion of grassland to arable. In common with many other species, laying date has advanced since 1980, and there is evidence of a simultaneous decline in nest failure.

Ringing has shown that many birds from GB winter in Iberia; there are fewer data from Ireland for this species. Birds from Iceland and Scandinavia pass through Britain in the autumn.

Red-throated Pipit   *Anthus cervinus* (Pallas)

PM monotypic

TAXONOMY According to Voelker (1999), Red-throated Pipit is sister to a clade that includes the 'Water Pipit' complex as well as Meadow and Rosy Pipits. Alström *et al.* (2003) argue that this is generally in agreement with observations of bill colour and seasonal variation. However, the clade is not supported very strongly by the published sequence data, and the precise position of Red-throated Pipit remains uncertain.

DISTRIBUTION Breeds on wet and scrubby tundra across the Palearctic, from N Norway to the Bering Straits, and into W Alaska, south into Kamchatka. Strongly migratory, wintering in sub-Saharan Africa, from Mauritania to Ethiopia, north of the rainforests; also in SE Asia. Small numbers winter along the coasts of the Mediterranean and Red Seas, and in the Nile Valley.

STATUS The first was heard, seen and shot on Fair Isle (Shetland) in Oct 1908; the first for Ireland was on Great Saltee (Wexford) in May 1955. The number recorded in GB&I increased to a maximum of *c.* 18 per year in the 1990s, since when there may have been a small decline. There are now over 450 records from GB&I, in every month from Apr to Dec, but with clear peaks in May and Oct.

There are marked differences in occurrence between England and Scotland, in both spring and autumn. In both regions, the peak spring month is May, but arrivals are earlier in England and later in Scotland: 10% of spring birds are found in Apr in England, compared with only 1% in Scotland;

the comparable figures for June are 5% and 22%. In autumn, more than 60% of Scottish birds are in Aug–Sep, compared with 25% in England. Clearly, birds are arriving earlier in England in spring and in autumn in Scotland, and more occur everywhere following easterly winds in May. Numbers in Ireland (28 individuals to 2004) and Wales are too few for a similar analysis, but pattern apparently corresponds more to England.

Eurasian Rock Pipit   *Anthus petrosus* (Montagu)

RB *petrosus* (Montagu)
WM PM *littoralis* C.L. Brehm

TAXONOMY The relationships of the Eurasian Rock/Water Pipit complex have become clearer in recent years, following re-assessments of morphology and zoogeography (summarised by Knox 1988a, Alström *et al.* 2003) and molecular analyses of mt DNA sequences (Voelker 1999). On the strength of the morphological and ecological similarities and differences, three species were recognised: *A. spinoletta* (Water Pipit), *A. petrosus* (Eurasian Rock Pipit) and *A. rubescens* (Buff-bellied Pipit). The molecular data show that the complex is not monophyletic; Meadow Pipit is closer to Eurasian Rock and Water Pipits than these are to Buff-bellied Pipit, supporting the view that the latter is a separate species to Eurasian Rock and Water Pipits. Mt DNA data were adequate to distinguish conclusively between these latter two; the genetic difference is less than 2% (Arctander *et al.* 1996), but they are diagnosably distinct in fresh plumage, show apparently consistent differences in moult, and striking differences in habitat, ecology and song, and Alström *et al.* (2003) concur that they should be retained as separate species. Two subspecies are recognised by Alström *et al.* (2003): *petrosus* and *littoralis*. A number of subspecies have been described in the past, including *kleinschmidti* (Faroes, Shetland, Fair Isle), *meinerzhageni* (Outer Hebrides), *hesperianus* (Arran), *immutabilis* (Ouessant), but Alström & Mild and Svensson (1992) are inclined to include all of these in *petrosus*.

DISTRIBUTION Breeds around the coasts of NW Europe: *littoralis* from the Kola Peninsula to the Skaggerak, and along the Baltic coasts of Finland and Sweden; *petrosus* around the rocky coasts of GB&I and NW France. Fennoscandian populations are wholly or partially migrant, wintering along the coasts of SW Europe to Morocco.

STATUS The nominate race breeds around the coasts of GB&I, essentially being absent only from the sandy and shingle shorelines of E Scotland, Lancashire and SE England. Also breeds away from the shore on rocky islands around Scotland. Rarely included in CBC/BBS surveys, so estimates of the population are not easily available; census data are based on a combination of density along the shoreline and length of suitable coastal habitat.

Ringing has shown that most *petrosus* are resident within GB&I, with only a few short-distance movements, usually along the coast. The Scandinavian race *littoralis* occurs as a passage migrant and winter visitor, with birds being recorded regularly at Scottish sites such as Fair Isle (Shetland) and Isle of May (Fife), but less commonly in Wales, and only 19 individuals had been recorded in Ireland to 2004. There are ringing recoveries of birds from Scandinavia, predominantly in S & SE England, but also one of the Irish birds had been ringed in Norway. It has been suggested (C Kehoe pers. comm.) that most *petrosus* tend to remain along the splash zone throughout the year, while the winter immigrant *littoralis* are more usually in alternative shoreline habitats such as saltmarshes.

Water Pipit   *Anthus spinoletta* (Linnaeus)

WM PM *spinoletta* (Linnaeus)

TAXONOMY Three subspecies recognised by Alström *et al.* (2003) differ quite markedly in plumage. The eastern form (*blakistoni*) breeds in sympatry with the *japonicus* race of Buff-bellied Pipit, and shows apparently consistent ecological separation, supporting their specific distinctness.

DISTRIBUTION Widely distributed in mountainous regions of the Palearctic, from S & C Europe (*spinoletta*), Turkey, to the Caucasus and Iran (*coutellii*), and from C Asia through Mongolia to N and C China (*blakistoni*). Western populations winter on the lowlands, from Denmark to N Africa. W Asian birds winter on the plains from the Middle East to NW India; those from further east in S China.

STATUS Generally a winter visitor to GB&I, though rare in Scotland and Ireland. Most records are from S & E England, where they may arrive as early as mid-Aug, but more typically are found from Oct to Apr. Migrants are not usually found in saline environments, preferring fresh or mildly brackish marshes; some remain in such sites close to the North Sea perhaps with freshwater flushes, but many move further S & W to winter in small numbers on watercress beds and other wetlands, or beside rivers and gravel pits. In Ireland, 33 individuals recorded to 2004, with a noticeable increase since 2000, possibly as observers become more familiar with the species.

Buff-bellied Pipit   *Anthus rubescens* (Tunstall)

SM *rubescens* (Tunstall)

TAXONOMY Three subspecies recognised: *rubescens, alticola, japonicus*, based on plumage differences in summer or winter plumage. Zink *et al.* (1995) and Voelker (1999) both found differences in mt DNA between *japonicus* and *rubescens* that were characteristic of species-level divergence. Since these taxa also differ in plumage and vocalisations, they appear to be separate evolutionary lineages, and a case might be made for separating these (summarised by Alström *et al.* 2003).

DISTRIBUTION Nominate *rubescens* breeds extensively across the tundra of Alaska, N & W Canada to W Greenland, with isolated pockets in NE USA. Replaced in W USA by *alticola*. Subspecies *japonicus* breeds in E Palearctic, from Lake Baikal to the Bering Straits, and through the Kamchatka Peninsula to the Kuril Is. This taxon winters in Japan, Korea and China, but also is regularly seen in the Middle East, leading Alström *et al.* (2003) to speculate that it may breed much further west than is currently believed.

STATUS The first for GB was caught on St Kilda (Outer Hebrides) in Sep 1910; further birds have been found on Fair Isle (Shetland) in 1953, Scilly in 1988, 1996, and Lincolnshire in 2005, and six in 2007. There are two Irish records; the first on Great Saltee (Wexford) in Oct 1951 and the second near Newcastle (Wicklow) in Oct 1967. The first for NI was at Mullagh (Londonderry) in Dec 2008. Those that have been identified to subspecies were the American race *rubescens*.

FAMILY FRINGILLIDAE

Genus *FRINGILLA* Linnaeus

A largely Palearctic genus of three species, two of which have been recorded in GB&I, one as a breedng bird.

Common Chaffinch   *Fringilla coelebs* Linnaeus

Endemic: RB *gengleri* Kleinschmidt
WM PM *coelebs* Linnaeus

TAXONOMY The taxonomic affinities of the finches have been examined by several researchers. Using cyt-b sequences, Groth (1998) identified a clade that included *Pinicola, Carpodacus* and *Carduelis*, plus other non-Palearctic genera, although this was not strongly supported statistically. Van der Meij *et al.* (2005), using a combination of nuclear and mitochondrial sequences, confirmed this 'finch' clade, and noted that *Fringilla* was basal but only weakly associated with the remainder. Common Chaffinch is variable in size and plumage, and up to 14 subspecies have been recognised (Clement *et al.* 1993). Eight are restricted to the Atlantic islands, and the relationships of these have been analysed by Baker *et al.* (1990) and Marshall & Baker (1999), using polymorphic enzyme systems and mt DNA sequences, respectively. The latter found *africana* (Morocco) and *coelebs* (Iberia) to be sister taxa, well separate from *spodiogenys* (Tunisia), and that the island forms are monophyletic; they conclude that the Azores, Madeira, Canaries were probably colonised by a single wave of immigrants from Morocco.

DISTRIBUTION Breeds across the W & C Palearctic, from the Atlantic to C Siberia, and from Fennoscandia to N Africa. Subspecies closest to GB&I include *gengleri* (GB&I); *coelebs* (Europe to Lebanon, Kazakhstan, excluding Crete, Sardinia); *africana* (NW Africa to Tunisia); *spodiogenys* (Tunisia to Libya). Northern populations are strongly migratory, moving S & W to avoid freezing conditions of winter. More eastern populations winter in Afghanistan Kashmir, Nepal.

STATUS The Common Chaffinch (subspecies *gengleri*) is one of the most widespread and abundant birds in GB&I; the species was recorded in *c.* 93% of 10-km squares during the two Breeding Atlases, being scarce only in Shetland and the Outer Hebrides. Numbers increased steadily during the period 1970–85, perhaps from an artificial low following the exceptional winter of 1963–64. There is evidence of a levelling out in the 1990s, though numbers may be increasing again in some regions since 2000 (Baillie *et al*. 2006). Clutch and brood sizes appear to have declined since the 1980s, perhaps a density-dependent effect, but there is no evidence of any other changes in breeding performance. This is a species that has shown a marked change in laying date, with initiation now *c.* 10 days earlier than in the 1980s. Changes in abundance were mirrored in the Repeat Woodland Survey (Amar *et al.* 2006): the largest increases were where spring temperature and Apr rainfall had increased but May rainfall had declined. This may be due to previously suboptimal habitats becoming more productive as a result of changes in climate.

Common Chaffinches occur in just about every habitat where there are trees. During the breeding season, they take invertebrates as protein for the nestlings, but for the rest of the year the diet is predominantly seeds, taken almost exclusively from the ground. The Garden BirdWatch survey shows a peak in attendance from Jan to Apr, presumably due to the shortage of natural food at this time.

The nominate race occurs as a migrant and winters in GB&I; there is extensive immigration from Scandinavia.

Brambling   *Fringilla montifringilla* Linnaeus

CB WM PM monotypic

TAXONOMY Van der Meij *et al.* (2005) and Arnaiz-Villena *et al.* (2007) found this to be sister to the chaffinches, in agreement with conventional taxonomy.

DISTRIBUTION Breeds in woodland and scrub across the north of Eurasia, from Scandinavia to Kamchatka, between 70°N and 50°N. Almost entirely migratory, with only a few of the most southerly breeders remaining through the winter. Winters as far south as N Africa, Middle East, N India, C China, Japan (Clement *et al.* 1993).

STATUS Breeds sporadically in Scotland, with the earliest records in 1899, 1920 and 1979, since when it has bred erratically without establishing a permanent presence (RBBP; BS3). Has never bred in Ireland.

A winter visitor to GB&I. Not included in most of the national surveys, so determining trends is not easy. There is a strong winter dependency on beech mast, although alternative seed supplies will be utilised. To some extent, the winter distribution reflects the occurrence of beech trees, with birds moving rapidly to these habitats after arrival in early autumn. Numbers fluctuate annually, depending on breeding success and food availability on the nesting grounds. Large concentrations sometimes occur at rich food sources. In Ireland, winter numbers also vary widely, but in most years it is rather scarce.

Winter birds come from Scandinavia and the Baltic states, sometimes moving on through Britain to areas further south. Data from successive years indicates that Bramblings may winter in widely separate areas from one year to the next, with little apparent fidelity to the wintering site.

Genus *SERINUS* Koch

A Palearctic and Ethiopian genus of *c.* 30–40 species, one of which is a rare breeding bird in GB&I.

European Serin   *Serinus serinus* (Linnaeus)

CB PM monotypic

TAXONOMY Using mt DNA sequences, *Carduelis*, *Serinus* and *Carpodacus* were found to be para- or polyphyletic (Arnaiz-Villena *et al.* 2007; Nguembock *et al.* 2008). From their work, the European Serin is closest to several other Asian–African canaries (including *S. canaria*) and part of *Carduelis*, but not to other clades of African canaries. The family is in need of a thorough revision.

DISTRIBUTION Breeds in W Europe, from Iberia to Ukraine, and from the southern Baltic to the Mediterranean coast of Africa and around the shores of Turkey. Scarce or absent from Channel and North Sea coasts of Europe. Northern populations migratory, wintering in southern part of the breeding range and into SE Mediterrranean.

STATUS Breeding was first confirmed in Dorset in 1967 and, through to 1980, further reports came from Devon, Hampshire and Sussex (detailed in Brown & Grice 2005). The number of nests varied among years, with the maximum apparently in 1988, with nine pairs. Colonisation has not been sustained, although with the current trend towards warmer summers, the establishment of a permanent population is perhaps again more likely. There are no reports of breeding from Ireland.

Otherwise, a scarce visitor to GB&I. The number recorded in GB increased steadily through the period 1958–95 (Fraser & Rogers 2006), with slight evidence of a decline since that date. Eight individuals in Ireland to 2000. There are two peaks in occurrence: Apr–May and Oct–Nov, with the former much larger. At both seasons, the majority of records are from the S coasts of GB&I. The increase in records may be partly due to greater observer cover, but parallels data from the near-Continent, where numbers rose through 1970–90; the population in France is reported to have fallen, perhaps by as much as a third, through 1990–2000 (BirdLife International 2004). The apparent decrease in GB&I could be related to this.

Genus *CARDUELIS* Brisson

Near-cosmopolitan (not Australasian) genus of over 30 species, eight of which have been recorded in GB&I, seven as breeding birds.

European Greenfinch   *Carduelis chloris* (Linnaeus)

RB WM *chloris* (Linnaeus)

TAXONOMY Molecular data shows that the greenfinches form a clade within the cardueline finches, close to part of *Serinus* (Arnaiz-Villena *et al.* 1998, 2007; Nguembock *et al.* 2008). The cardueline finches are in need of thorough revision. A detailed investigation of W European populations by Merila *et al.* (1997), based on the hypervariable mt control region, found levels of genetic variation that were generally low for a passerine, and decreased northwards from Iberia and S Italy to Finland. They postulate that this is due to a population bottleneck in southern Europe (probably Iberia) some 3,000–5,000 years ago, followed by a progressive expansion northwards – possibly with further, lesser bottlenecks. This date, of course, fits with the end of the last Ice Age. Variation clinal and minor; several subspecies have been described, but only four are recognised by Clement *et al.* (1993).

DISTRIBUTION Breeds across Europe, S of Fennoscandia to N Africa, and east to Kazakhstan. Nominate *chloris* (N Europe, from GB&I to S Fennoscandia and E to Switzerland, Austria, including Corsica, Sardinia); *aurantiiventris* (Europe, south of nominate, to N Africa); *chlorotica* (Middle East); *turkestanicus* (Crimea, N Iran to S Kazakhstan). Northern populations are migratory, wintering within, or south of, the breeding range.

STATUS Widespread across GB&I during the Breeding Atlases, though an apparent decline of *c.* 12% in Ireland might have been due to more than just reduced coverage. Absent from the highlands of Scotland and scarce in the N & W isles. There is evidence of a change in distribution within the regions: a decline in farmland, perhaps due to the reduced levels of stubbles and spilt grain, offset by a movement towards gardens and parks with their increased amount of berry-bearing bushes and feeding stations. A pattern of relative stability through 1965–75, with perhaps a slight decline over the next decade, and then a sustained increase through 1985–2005 (CBC/BBS). These changes seem to be fairly similar across GB&I, although the recent increases in NI have been rather higher than elsewhere in GB. There is little change in breeding parameters that might explain the recent increase; indeed, clutch and brood size have both declined, and nest losses at the egg stage are reducing. Population growth may be due to increased adult survival, though this analysis has not yet been undertaken (G Siriwardena pers. comm.). Laying date has advanced by *c.* 14 days since 1965, which may allow an additional brood to be reared.

European Greenfinches from GB tend to move S & W in winter, with recoveries from Ireland and N France, returning to the natal areas the following spring. Large numbers are sometimes recorded in autumn, passing through S coast watchpoints, such as Portland Bill (Green 2004). There is also a substantial immigration into the region in autumn, primarily of birds from S Scandinavia. The species is a scarce, but regular, passage migrant through Orkney and Shetland in both spring and autumn.

Citril Finch   *Carduelis citrinella* (Pallas)

SM monotypic

TAXONOMY Closely related to Corsican Finch *C. corsicana*; both are sister to European Goldfinch within *Carduelis* (see below). Recent molecular analyses (Förschler *et al.* 2009) found clear differences in mt DNA between Citril and Corsican, but not in the more slowly evolving nuclear sequences. Vocalisations are diagnosably distinct.

DISTRIBUTION Occurs at high altitudes in the mountains of S Europe, from Spain to W Austria, including Italian peninsula.

STATUS One, Fair Isle, June 2008 (subject to acceptance).

European Goldfinch   *Carduelis carduelis* (Linnaeus)

MB RB *britannica* (Hartert)

TAXONOMY *Carduelis* is a polyphyletic genus in need of revision. Arnaiz-Villena *et al.* (1998, 2007; Nguembock *et al.* 2008) examined many *Carduelis* using mt cyt-b. They identified a clade that included the Citril Finch *C. citrinella*, with European Goldfinch being sister to the latter. Other researchers included fewer species of carduelines, so their results cannot be compared directly. Morphological taxonomy recognises at least 12 subspecies (Clement *et al.* 1993), which fall into two groups: black-headed *carduelis* and grey-headed *caniceps*.

DISTRIBUTION Breeds from Iberia, GB&I and Fennoscandia, east across Europe and Asia to Mongolia and the Himalayas, between 64°N and 27°N. Nine black-headed subspecies occur in the W Palearctic, including the nominate, which breeds from N Spain to the Black Sea and the Ural Mountains, and north to S Scandinavia; subspecies *britannica* breeds in GB&I, the Channel Islands and the Netherlands. Three grey-headed races breed from S Iran to Tibet, to about 60°N. Northern populations are migratory, usually wintering within the southern parts of the range. Southern populations are generally resident.

STATUS A widespread species that is absent only from the highlands and islands of Scotland. There was evidence of a change in distribution between the Breeding Atlases, with a contraction in Ireland, and a recolonisation of the Moray Firth region of Scotland following historic losses. Because of its widespread distribution and conspicuous behaviour, Goldfinches were well represented in CBC fieldwork and this has continued into BBS studies; consequently, the changes in abundance are well documented. Numbers increased through the early 1970s, before falling back dramatically to 1985; since then there has been a continuous increase back towards the peak levels of the 1970s. Baillie *et al.* (2006) suggest that these changes are largely due to variations in adult survival, and found no evidence of change in any reproductive parameters such as clutch size or nest survival. Agricultural intensification through the 1970s and 1980s led to a reduction of thistle and other weed seeds across the countryside and, after struggling for a while, the species seems to have adapted to changed circumstances by moving into gardens and taking advantage of feeding stations. Results from the BTO Garden BirdWatch show that there has been a progressive increase in the proportion of gardens that record European Goldfinches at all times of the year.

A large part of the population moves S & W in autumn to avoid the extremes of winter weather; most ringing recoveries involve birds marked in the S of England and the situation further north is less clear. Recoveries come from Belgium through N & W France to Iberia. Observational data from watchpoints on the coast of S England, including Dungeness (Kent) and Portland Bill (Dorset), record peak movements in Oct or early Nov, and return passage through Apr. Siriwardena *et al.* (1999) believe that lower winter mortality, including reduced hunting on the Continent, may be key to the population increases in recent years.

Immigrants and passage birds of the nominate race from the Continent suspected amongst east coast migrants and wintering flocks, but as yet no conclusive proof.

Eurasian Siskin   *Carduelis spinus* (Linnaeus)

RB WM PM monotypic

TAXONOMY Arnaiz-Villena *et al.* (2007) showed Eurasian Siskin to be sister to N American Pine Siskin *C. pinus* and Antillean Siskin *C. dominicensis*, in a larger clade with an array of Central and South American siskins.

DISTRIBUTION Occurs in two widely separate regions in the W & E Palearctic. In the west, from N France, GB&I east through Scandinavia and across W Siberia to *c.* 70°E. South and east of this, the populations become increasingly fragmented and discontinuous, to the Mediterranean, Asia Minor and S Caspian. In the E Palearctic, breeds from Lake Baikal to the Sea of Okhotsk and Japan. Northern populations are strongly migratory, wintering as far south as NW Africa, Iraq, and (in the east) S China. In some years, when food is short, large numbers may irrupt into the winter range (Clement *et al.* 1993).

STATUS The Eurasian Siskin underwent a dramatic expansion in both range and numbers during the 1970s and 1980s. Between the two Breeding Atlases, the number of 10-km squares in which there was evidence of breeding increased by 85% in GB and *c.* 20% in Ireland. Siskins are a bird of coniferous forest and the area of mature trees doubled between the Breeding Atlases. Data on population sizes and trends are not available for the period of main increase because conifer woodland was not well represented in the CBC. The enforced randomisation of BBS sites has resulted in better coverage of this difficult habitat, and Eurasian Siskins are now monitored more closely. The picture has reversed since the mid-1990s, and Eurasian Siskins have declined by 35–40% in England and Scotland. The Repeat Woodland Survey recorded increased numbers of Eurasian Siskins, but since it was primarily directed towards deciduous woodland, the sample sizes are small, and the data may be inadequate.

The Garden Birdwatch data show that Eurasian Siskins generally visit gardens from Jan to May, and the proportion of gardens where they are recorded varies markedly among years. This likely reflects availability of natural food, but there is no evidence of any consistent trend over time, although McKenzie *et al.* (2007) report that numbers in a Scottish garden were lower in years with a high crop of Sitka spruce cones.

Eurasian Siskins are very mobile, and ringing data reflect this. Both marking and recapture is biased towards birds trapped during the winter. Eurasian Siskins leave the conifer forests when the food begins to decline, and many birds move south in autumn, with a few birds being recorded as far south as Gibraltar and Morocco. There is extensive movement between GB&I and regions to the NW, with many birds crossing the North Sea and exchanges between most countries from Belgium to Finland, though individuals from the Low Countries are likely still on passage.

Common Linnet   *Carduelis cannabina* (Linnaeus)

Endemic: RB MB *autochthona* (Clancey)
RB MB WM PM *cannabina* (Linnaeus)

TAXONOMY Mt cyt-b shows this to be a member of the larger *Carduelis* clade (i.e. excluding *C. carduelis, C. flammea, C. hornemanni, Loxia*) (Arnaiz-Villena *et al.* 1998, 2007; Nguembock *et al.* 2008), and sister species to Twite. Variation in the brown of the upperparts and the extent of white on flanks and rump has led to the recognition of six subspecies (Vaurie 1959, BWP), though Svensson (1992) and Clement *et al.* (1993) do not recognise *autochthona*, the validity of which requires attention.

DISTRIBUTION Breeds across the W & C Palearctic, from C Scandinavia and the Baltic states to W China, and south to the Sahara, S Israel and Afghanistan. Nominate subspecies occurs from the Atlantic to C Kazakhstan, W Turkey and Crimea, except in Scotland, where replaced by *autochthona*; *bella* replaces nominate in the east, from C Turkey to China. Three further races are restricted to Madeira and the Canaries. As with many small granivores, northern populations are migratory, generally wintering within the range of resident southern birds, but also into Arabia, Egypt.

STATUS A widespread, though declining, species in GB&I. During the fieldwork for the Second Breeding Atlas, Common Linnets were recorded in most of GB, apart form the Highlands of Scotland, Shetland and large parts of the Hebrides (perhaps being replaced here by Twite). Less widely distributed across Ireland, it was lost from >15% of 10-km squares, and appeared to have withdrawn from central areas to become slightly more coastal. The decline is also apparent from CBC/BBS and constant-effort ringing data. During 1965–85, there was a numeric decrease of >60% across GB, although this appears to have levelled out since the 1990s (Baillie *et al.* 2006). The decline was paralleled in other seedeaters. In contrast, there has been a general increase in S & E Europe, and in France, Germany and Denmark, the populations were stable during 1970–90 (when the most severe decline was underway in GB&I), but declined subsequently (BirdLIfe International 2004). Similar changes were also reported from Switzerland, the Low Countries and Sweden, although to a lesser extent.

Nest record data indicate that nest failure at both egg and chick stages has increased, a result supported by the finding that the juvenile/adult ratio has fallen in recent years. Statistical analysis indicates that these losses are sufficient to explain the population decline (Siriwardena *et al.* 2000); Baillie *et al.* (2006) suggest that changes in hedgerow management may have resulted in nests becoming less well concealed and more prone to predation. Whether this could also be the cause of the declines across NW Europe is unclear. The picture is confused and further research is needed across the European range to determine what is happening to this species. It is becoming clear that oilseed rape and dandelion are increasingly important in nestling diet as agricultural intensification reduces the quantity of traditional weed seeds in the environment (Moorcroft *et al.* 2006).

While many Common Linnets remain around the breeding sites throughout the year, some undertake extensive movements to winter as far south as Gibraltar. There is little evidence of exchange between Britain and NW Europe, although small numbers of presumed nominate race birds are recorded in both spring and autumn in Shetland, where Common Linnets breed only rarely.

Twite   *Carduelis flavirostris* (Linnaeus)

Endemic: RB MB *pipilans* (Latham)
WM *flavirostris* (Linnaeus)

TAXONOMY Molecular data indicate a sister relationship with Linnet (Arnaiz-Villena *et al.* 1998, 2007). Extensive, largely clinal, variation in warmth and paleness of plumage has led to eight (Vaurie 1959, Clement *et al.* 1993) or nine (BWP) subspecies being recognised, falling into two geographical groups.

DISTRIBUTION One group of populations in NW Europe, including subspecies *flavirostris* (coastal Scandinavia from Kola Peninsula to C Sweden), and *pipilans* (GB&I). The other extends from E Turkey to W & C China, and south to the northern Himalayas. Birds from northern Scandinavia are migratory, wintering along the coast of W Europe from Poland and the Baltic states to NW France. Eastern populations are nomadic and partially migratory, wintering at lower altitudes within the breeding range or further south.

STATUS In GB&I, Twite breed on more exposed areas of moorland and low-intensity agricultural land, feeding in adjacent farmland. It is thus restricted to the uplands of N England, Scotland N of the Forth–Clyde valley and the outer isles. There is evidence of a strong decline in distribution across GB during the last 100 years. In Ireland, it is principally a bird of the W coast, and there was a serious decline in distribution between the Breeding Atlases, with >50% of Irish 10-km squares being deserted. The reasons for this are unknown, but may be due to deterioration of habitat through either agriculture or climate change. In the rest of GB&I, there was also a contraction in distribution between the Breeding Atlases, but this was less severe and might indeed have been due to sampling error.

A national survey in 1999 (Langston *et al.* 2006) indicated *c.* 10,000 pairs in GB&NI, of which *c.* 6% in England and <1% in Wales and N Ireland. It was also evident that the decline between the Breeding Atlases was real, and that it had continued (or even accelerated), with serious losses in the highlands and islands of Scotland and the S Pennines of England. There is evidence from the latter of declining brood size and indications that the loss of seed and insect-rich hay meadows to pasture and silage may have been partly, or even largely, responsible (Brown *et al.* 1995). After breeding, both adults and juveniles disperse widely in search of feeding sites (Raine *et al.* 2006b). Most Twite show natal philoptry, returning to within a few km of their birthplace. Raine *et al.* (2006b) suggest that even this limited dispersal would allow the species to expand its range back over the uplands were the ecological conditions more suitable for breeding.

Many coastal Twite remain on the breeding grounds throughout the year, especially in the northern isles and NE Scotland (Lack 1986). Substantial flocks winter on the saltmarshes of coastal England, especially in the SE. Many of the latter are *pipilans* from the S Pennines wintering from the Wash to Kent, and to a lesser extent into the Low Countries. Only a few of the Pennines birds move to Lancashire and S Cumbria, where the wintering flocks seem to originate from the Hebrides (Raine *et al.* 2006a). Although there is some interchange between wintering grounds, most birds appear to be site-faithful between winters.

Twite sometimes appear along the east coast in autumn in the company of migrant species from Scandinavia, but there have been no ringing recoveries to support a movement across the North Sea. Flocks wintering in SE England are assumed, on scant evidence, to contain some birds of Continental origin. The nominate race has not been recorded in Scotland, Ireland or the Isle of Man.

Lesser Redpoll   *Carduelis cabaret* (P. L. S. Müller)

RB MB monotypic

TAXONOMY Mt DNA analysis indicates that the redpolls are close to European Goldfinch, Citril Finch and crossbills (Arnaiz-Villena *et al.* 2007; Nguembock *et al.* 2008). A number of redpoll taxa have been described, based primarily on morphology and plumage; these include *cabaret, flammea,* *rostrata, islandica, exilipes, hornemanni.* Enzyme and molecular studies have revealed no significant differences among these, indicating only recent divergence. Morphological differences among various taxa point to periods of isolation during which genetic drift or selection (natural or sexual) have overcome any residual gene flow to drive the phenotypes apart. At least some are separate evolutionary lineages and comprise different species.

Exilipes and *flammea* are both circumpolar in the Low Arctic; *hornemanni* and *rostrata* occur predominantly at higher latitudes in the Canadian Arctic and Greenland. Phenotypic divergence within the sympatric pairs indicates restricted gene flow and *exilipes* and *flammea* should be treated as separate species, as should *hornemanni* and *rostrata*. It would be possible to treat this quartet as two (*exilipes–hornemanni* and *flammea–rostrata*) or four species; there seems no good reason for the latter, and we follow conventional taxonomy in regarding them as two species: *C. flammea* (including *rostrata*) and *C. hornemanni* (including *exilipes*).

Lesser Redpolls in Iceland show extensive variation, with some as pale as *exilipes* and others as dark as *rostrata*: the latter have been described as ssp. *islandica*. Amongst other hypotheses, an endemic subspecies (*islandica*) may have evolved in Iceland; alternatively, the country may have been populated by *rostrata* from Greenland and subsequently colonised by *exilipes*, so that the Icelandic population now includes representatives or derivitaves of both. We follow conventional taxonomy in regarding *islandica* as a subspecies of *C. flammea*, although further research is needed into its affinities.

Knox (1988b) and Herremans (1990) examined the small, brown redpolls described as *cabaret* and concluded that they are diagnosably distinct from other redpolls. Herremans (1989) also identified distinct differences in vocalisations. In Sweden, *cabaret* spread into the range of *flammea* (Götmark 1978), and apparently occupied different habitats (coastal/heathland pines and inland spruce, respectively). In 1994, a site in Norway hosted 11 pairs of redpolls – six *cabaret* and five *flammea* – with no mixed pairs (Lifjeld & Bjerke 1996). On the strength of these findings, C. *cabaret* was elevated to specific level (Knox *et al.* 2001). There are anecdotal reports of weakness in this analysis, but no refutation has been published, and we retain C. *cabaret* as a third species of redpoll. Evidence is emerging of *flammea* (and/or *islandica* or *rostrata*) breeding in NW Scotland, especially in the Outer Hebrides and Shetland (BS3; RBBP).

DISTRIBUTION Lesser Redpoll breeds in GB&I and adjacent areas of Continental Europe, from France to Norway, extending into C Europe. Until recently, it was isolated from other European redpolls during the breeding season, although it has now spread into S Norway, where it is parapatric or even sympatric with *flammea*. Most populations are resident, though many birds make short-distance movements, sometimes as far as the Mediterranean (Clement *et al.* 1993).

STATUS Lesser Redpolls were already showing evidence of a serious decline by the 1990s. Between the two Breeding Atlases, they had withdrawn from over 11% of 10-km squares in GB and c. 38% in Ireland. Contractions were especially severe in the S & E of Ireland and NE Scotland. CBC data suggest that the population increased during 1964–74, since when there has been a sustained decline. The peak number occurred between the two Breeding Atlases. The decline has continued through to about 1990, since when it has remained around this very low level (BBS). There are few correlates with breeding performance (perhaps due to small sample sizes), although there was a slight indication that productivity had fallen. The Repeat Woodland Survey also revealed a massive loss, with declines well in excess of 50% between the 1980s and 2002–05.

Reasons for the decline are still hard to come by. Fuller *et al.* (2005) suggested that there might be a geographical bias in the monitoring programmes, but the recent Repeat Woodland Surveys were designed to be more extensive, and support the census data. Lesser Redpolls are ecologically associated with birch trees, and this species has declined in many woodlands as these have matured. There is evidence that the survival rates of both adults and juveniles were higher during the period of population growth but fell during the subsequent decline. Whether this is cause and effect is unknown. It seems likely that changes in forest structure may be involved; birch is often associated with conifer planting, and the move towards increased broadleaved forest may be partly responsible for the decline.

Common Redpoll   *Carduelis flammea* (Linnaeus)

CB WM PM *flammea* (Linnaeus)
CB PM *rostrata* (Coues)
PM *islandica* (Hantzsch)

TAXONOMY See Lesser Redpoll.

DISTRIBUTION Holarctic, breeding in deciduous and coniferous woodlands across subarctic regions of Canada, Alaska, Iceland and Eurasia as far north as the tundra edges. Subspecies are largely allopatric: *flammea* (Low Arctic circumpolar), *rostrata* (Baffin Island, Labrador (Canada); Greenland), *islandica* (Iceland). Latter two winter on or close to the breeding areas, making only short-distance movements (Knox & Lowther 2000a), although have been found as far south as New Jersey, Iceland, Scotland, Ireland. Northern populations of nominate are more migratory, wintering to S of breeding range across Palearctic. Occasionally, major irruptive movements occur, probably in response to failing food supplies or harsh weather, when larger numbers of birds may be found hundreds of km beyond normal winter range.

STATUS Since the 1970s, Common Redpoll has been found breeding in small numbers in the Highland region of Scotland, in the northern isles and the Outer Hebrides (BS3; RBBP). It is likely that most are *flammea*, but the possibility of *islandica* and/or *rostrata* cannot currently be eliminated. Both Lesser and Common Redpoll breed in the Outer Hebrides, and the opportunity exists to investigate the extent of assortative pairing and/or habitat segreggation. There is an older record (1959) of *rostrata* breeding successfully in Strathspey (Highland).

Common Redpoll is a regular winter visitor in small numbers, with substantial irruptions in some years. During an irruption in 1995–96, the largest numbers of *flammea* were counted along the east coast, but the influx continued inland, and small numbers were reported as far west as Wales and Scilly as the winter progressed (Riddington *et al.* 2000). The irruption was also reported from Scandinavian sites, with exceptionally high numbers in Nov 1995 at Utsira (Norway), Falsterbo (Sweden) and Heligoland (Germany); the arrival in GB extended into Dec, and this was associated with very cold weather. Birds appeared to return through a different route; arrivals were generally along the northern parts of the E coast, whereas departing birds were noted significantly further south.

Subspecies *rostrata* and (less often) *islandica* occur in NW Scotland in most years, with irruptions of the former in earlier years (1925, 1955, 1959: Wernham *et al.* 2002). There are no ringing recoveries involving either race.

In Ireland, *flammea* is much less common than elsewhere in GB&I, with only 73 individuals recorded to 2003, mainly during Sep–Feb. There have also been 34 indeterminate individuals, referable to either *rostrata* or *islandica*.

Arctic Redpoll   *Carduelis hornemanni* (Holboell)

PM *exilipes* (Coues)
SM *hornemanni* (Holboell)

TAXONOMY Two subspecies; see Lesser Redpoll.

DISTRIBUTION Holarctic, breeding in shrub and dwarf birch, across tundra of N America and Eurasia, north of *c.* 57°N. Nominate is more restricted, replacing *exilipes* from Ellesmere to Baffin Islands (Canada), and around N Greenland. Both winter largely in or close to breeding areas, though northern populations of *exilipes* migrate S to avoid winter harshness, wintering on occasion as far south as GB&I, C Europe, N China, Mongolia, and N USA (Knox & Lowther 2000b). Sometimes irrupts in substantial numbers, when flocks may occur hundreds of km south of typical range. Subspecies *hornemanni* appears to be more sedentary, rarely showing irruptive movements of the magnitude of *exilipes*.

STATUS Both subspecies have been recorded in GB&I; nominate *hornemanni* is very rare, with about 30 records. The first *hornemanni* was obtained in Apr 1855 near Whitburn (Durham), and the first *exilipes* near Easington (Yorkshire) in Feb 1894. There are only three Irish records of this species; the first on Dursey Is (Cork) in Oct 1999, and two on Tory Is (Donegal) in Sep 2000 and 2001.

The number reported in GB since the 1950s varies considerably among years, from none (e.g. 1974, 1983) to winter 1995–96, when 431 records were accepted by BBRC (Riddington *et al.* 2000). Probably all of these were *exilipes*, and most were associated with an irruption of Mealy Redpolls. Birds that winter were found across GB. Peak arrival was in Nov–Dec, but birds continued to be found throughout the winter, and some were recorded on the east coast into May 1996, presumably on their way back to the breeding grounds. More typically, *hornemanni* (which is much less common: *c.* 40 records) arrives in the N Isles from Sep to Dec, occurring south as far as Dungeness (Kent), with occasional records in the spring; *exilipes* is typically a little later, rarely arriving before mid-Oct and much less often in the N Isles.

Genus *LOXIA* Linnaeus

Essentially Holarctic genus of four or five (or perhaps many more) species, four of which have been recorded in GB&I, three as breeding birds.

Two-barred Crossbill   *Loxia leucoptera* J. F. Gmelin

SM *bifasciata* (C. L. Brehm)

TAXONOMY Arnaiz-Villena *et al.* (2001, 2007; Nguembock *et al.* 2008) found this to be sister to Common Crossbill and Parrot Crossbill, and close to the redpolls, but more detailed analyses have not been undertaken. Traditionally regarded as three subspecies, though Benkman (1992) reported differences in bill size between *megaplaga* and *bifasciata* (large) and *leucoptera* (small) likely to reflect significant differences in feeding strategies and justify specific separation. Subsequently, *megaplaga* and *leucoptera* shown to differ in vocalisations, and the specific status of Hispaniolan Crossbill *L. megaplaga* now generally accepted. *Bifasciata* and *leucoptera* differ diagnostically in vocalisations (Elmberg 1993, Constantine *et al.* 2006), biometrics, plumage and bill structure (Harrop *et al.* 2007), and DNA (Parchman *et al.* 2007), and are clearly separate evolutionary lineages that merit specific status.

DISTRIBUTION Breeds in a narrow belt of northern conifers, especially larch and cedars (Clement *et al.* 1993) from Finland to the mountains of Lake Baikal, and perhaps Altai. Usually sedentary, or short-distance migrant, but occasionally irrupts (approximately 7-year periodicity) as far SW as France, Italy, and SE to Sakhalin, Japan.

STATUS The first for GB&I was shot at Belfast in Jan 1802, one of only four records from Ireland, all from the north; subsequent birds were in Antrim (about 1867, 1927) and Fermanagh (1895). The first for GB was shot near Brampton (Cumbria) in Nov 1845, during an invasion that lasted into 1846. Since then, there have been *c.* 200 records, with birds found in every month, and peak in Aug in Scotland and Dec–Mar in England and Wales. Annual numbers vary; there can be periods when none are seen (e.g. 1954–58, 1992–96) and years with significant influxes, especially to the northern isles, e.g. 22 in Aug 1987; 26 in July 1990–Apr 1991; 18 in Aug 2002–Mar 2003. These correlate well with data from Scandinavia, where high numbers were recorded in 1987, 1990, 1997–98, 2002–04 (T Tyrberg pers. comm.). The exception seems to be 1997–98, when only four recorded in GB but good breeding population and high numbers in the north in Scandinavia; rather than immigrants from further east, the latter are likely to be the source of birds found in GB.

Red Crossbill/Common Crossbill
Loxia curvirostra Linnaeus

RB MB WM PM *curvirostra* Linnaeus

TAXONOMY Recent molecular analysis has shown that *curvirostra* is closely related to *scotica* and *pytyopsittacus* and that DNA sequences cannot separate this trio; *curvirostra* and *pytyopsittacus* sister to the two-barred crossbills and close to the redpolls (Arnaiz-Villena *et al.* 2007; Nguembock *et al.* 2008); see Scottish Crossbill. About 20 subspecies listed by Clement *et al.* (1993); many of these have restricted range; subspecies generally differ in size and the balance of orange/red in male plumage; recent work has identified apparently diagnostic vocal differences, both within and between currently recognised races (Robb 2000). Vocal types correlate with bill morphology (related to diet; Benkman 1992) in apparently assortatively mating populations, some of which may be better treated as separate species (Edelaar 2008).

DISTRIBUTION Holarctic, breeding in larch, spruce or pine forests across the New and Old Worlds, with populations SE to Philippines. Nominate subspecies breeds in W & C Europe, as far E as Sea of Okhotsk. In W Palearctic, other races occur in Crimea, Cyprus, Corsica, Balearics, Atlas Mts; others across Asia, N & C America. Many populations resident (generally pine-feeding forms), but northern birds may be nomadic, searching for food when cone crops fail or are exhausted, sometimes taking part in large, continent-wide irruptions.

STATUS Breeds in conifer woodlands across GB, although the picture in Scotland is confused because of the difficulty separating this from Scottish Crossbill. The number and distribution of breeding birds often changes markedly between years, but increased across GB between the two Breeding Atlases as areas of planted conifers approached maturity. In Ireland, it is an irruptive species, breeding (often widely) in years following irruptions, then populations petering out until next irruption.

Common Crossbills can be found nesting in almost any patch of conifer, although they are most regular in the larger woods of Scotland, N England, East Anglia, W Wales and W Ireland. Irruptions occur in most decades, and birds frequently remain to breed, staying in the same place or relocating each year as the food supply dictates. During irruptive arrivals in summer and autumn, birds may be found at coastal sites from Shetland to Kent, often in an emaciated state, and feeding (or attempting to feed) on inappropriate food plants such as thistles or dock. Arrivals may approach Britain from Scandinavia, or along the S coasts of the Baltic and N Sea; source areas are generally unknown but some irruptions at least are likely to be from much further east.

Scottish Crossbill
Loxia scotica Hartert

Endemic: RB monotypic

TAXONOMY Britain's only endemic species, with a bill intermediate in size between Parrot and Common Crossbill. Long believed to be the only crossbill breeding in the Scottish Highlands, until Knox (1975, 1990) showed Common Crossbill often bred sympatrically. More recently, Parrot Crossbill also shown to breed in old pinewoods (Marquiss & Rae 2002, Summers *et al.* 2002, 2007). As a generalisation, Parrot Crossbills feed in Scots pine; Common Crossbills feed on larch and spruce; Scottish Crossbills feed on a more mixed diet. There is marked assortive mating among bill sizes and apparently diagnostic vocal types (Knox 1990, Summers *et al.* 2002, 2007).

Despite differences not being absolute, the recent findings are encouraging. Absence of molecular differentiation within or between crossbills supports a recent origin for the three taxa; indeed, the highlands of Scotland were uninhabitable during the last glaciation, which ended only 10,000 years ago. The parallels with Darwin's Finches *Geospiza* spp. are striking (Grant 1986); precise bill morphology is critical for extracting seeds efficiently and, for crossbills, different bills are best adapted to feed on the seeds of differing conifers. Bill size is heritable in crossbills, so selection for larger or smaller bills will have a rapid effect. Matings between birds of differing bill dimensions will produce young with intermediate bill structures unsuitable for the available food options. Vocal differences will assist in flock (and hence taxon) integrity, and (whether inherited or learned) be selectively advantageous in mate recognition.

The recent results from Scotland confirm the conclusions reached by Knox (1975, 1976) that Scottish Crossbill is a separate evolutionary lineage and should be regarded as a distinct species.

DISTRIBUTION Restricted to the older conifer forests of highland Scotland. Largely resident, with few records away from the core breeding areas.

STATUS Knox (1975, 1990) and, more recently, Marquiss & Rae (2002) and Summers *et al.* (2002, 2007) showed that previous distributional data may be unreliable, and estimates of population numbers correspondingly uncertain. Development of new methodology for surveying all the crossbills is under way using vocalisations, and a detailed survey began in winter 2007. There have been no confirmed records of Scottish Crossbill outside of Scotland.

Parrot Crossbill   *Loxia pytyopsittacus* Borkhausen

RB SM monotypic

TAXONOMY See Common Crossbill, Scottish Crossbill.

DISTRIBUTION Restricted to W Palearctic, breeding in Scots pine forests from Scandinavia through Finland to NW Russia. Nomadic rather than migratory, but occasionally irrupts S & W from core areas, and may remain to breed for several years.

STATUS Following the 1982–83 invasion, breeding was confirmed at several sites in East Anglia, and may have occurred more widely. In recent years, it has become apparent that Parrot Crossbills are breeding alongside both Common and Scottish Crossbills in the forests of Strathspey and Deeside in particular (BS3). The population of Scottish breeding birds may be in the region of 100 pairs (BS3). The breeding birds are largely sedentary, though movements of up to 42 km have been recorded.

Parrot Crossbill otherwise shows an irruptive pattern of occurrence in GB, though is much rarer than Common Crossbill and there may be long periods between arrivals. The first was collected at Blythburgh (Suffolk) in 1818, and there have now been *c.* 500 records, excluding the breeding population. More recent, larger movements include 1962–63 (77 birds), 1982–83 (109), 1990–91 (264). Birds typically arrive first in the N Isles in late Sept–Oct or along the E coasts of England and Scotland, moving inland and being recorded in conifer woods across GB. Not recorded in Ireland.

Genus *BUCANETES* Cabanis

Palearctic genus of one or two species, one of which has been recorded in GB&I.

Trumpeter Finch   *Bucanetes githagineus* (M. H. C. Lichtenstein)

SM race undetermined

TAXONOMY Placed in a clade with *Eremopsaltria mongolica* in mt DNA analysis by Zamora *et al.* (2006), though branch length suggests best placed in separate genera. Four subspecies recognised by Vaurie (1959), Svensson (1992), Clement *et al.* (1993), based generally on the intensity of pink in the plumage.

DISTRIBUTION Breeds in desert or semi-desert regions of N Africa, parts of Arabia E to NW India. Subspecies *zedlitzi* occurs in S Spain, NW Sahara to Tunisia; replaced in Egypt by *githagineus;* by *amantum* in Canary Is, and by *crassirostris* E of Sinai. E populations migratory; W birds more dispersive, even nomadic, wandering in search of food.

STATUS This species has been recorded 11 times in GB. The first was at Minsmere (Suffolk) May–June 1971 and a second on Handa Is (Highland), also in June 1971. Subsequent birds have been found from

May to Sep, and from Sussex to Orkney. There have been multiple arrivals on more than one occasion: e.g. four in late May–early June 2005, when there were also birds in France, Sweden and Switzerland. The movements from late May correspond with post-breeding dispersal, possibly from the increasing Spanish and NW African population. Not recorded in Ireland.

Genus *CARPODACUS* Kaup

Holarctic genus of *c.* 21 species, one of which has been recorded in GB&I.

Common Rosefinch   *Carpodacus erythrinus* (Pallas)

CB PM *erythrinus* (Pallas)

TAXONOMY Arnaiz-Villena *et al.* (2007) identified two groups of *Carpodacus*: Asian species formed a clade with *Uragus* and *Haematospiza*; American rosefinches were only distantly related. Yang *et al.* (2006) confirmed that *Carpodacus* is not monophyletic and identified four lineages. Generic realignment is clearly necessary. Five subspecies of Common Rosefinch were listed by Vaurie (1959) and four of these were included in a molecular analysis by Pavlova *et al.* (2005). They screened birds from Moscow to Kamchatka, almost the entire longitudinal range of the species, and found evidence of three groups corresponding to birds in the SW (*kubanensis*), the NE (*grebnitskii*) and the NW (*erythrinus*, *ferghanensis*). These were not fully differentiated, indicating relatively recent separation and a degree of continuing gene flow.

DISTRIBUTION Breeds in scrub and bushes along woodland edges, farmland, river valleys, across the Palearctic from S Scandinavia to Kamchatka, south from edges of taiga to N Turkey, S Caspian and Himalayas. Nominate race occurs N & C Europe to Altai Mts, replaced by *grebnitskii* in E Palearctic, and by three other races from Turkey through Himalayas to S China. Northern populations entirely migratory, wintering SW, S or SE of breeding range. Southern populations less so, or even resident.

STATUS This is a rare breeding bird, though singing males in spring are now annual in occurrence. The first breeding record was in Ross and Cromarty in 1982, and this was followed by further nesting, especially in the 1990s, with a maximum of 20 possible pairs in 13 localities during 1992. The latter followed an unusually large number of coastal migrants that spring. The momentum of this colonisation was not sustained and the last confirmed breeding was in 2001 in Yorkshire.

Thousands of Common Rosefinches have been recorded in GB&I since the first was trapped near Brighton (Sussex) in Sep 1869: unlike many prized vagrants at that time, this one survived in an aviary until June 1876. Common Rosefinch was removed from the BBRC list in 1982, and records from GB are now collated in the Scarce Migrants Reports. Birds may be found between May and Nov, with clear peaks in late May–early June and mid-Aug–end Oct; numbers increased markedly in the late 1980s, and since 1990, *c.* 150 have been recorded annually. The increase in records in GB&I correlates with the breeding range spreading west through NW Europe through most of the 20th century. BirdLife International (2004) report the European population is very large (>3 million pairs, mostly in European Russia), and generally stable since the 1970s, apart from Finland, where there has been a serious decline.

In Ireland, the first record was a bird on Tory Island (Donegal) in Sep 1954; since then there have been a further 122 records (to 2004); mostly in autumn, and a few in May–June. It is now found annually in small numbers, with occasional singing males, although it has never bred.

Genus *PINICOLA* Vieillot

Holarctic genus of one or two species, one of which has been recorded in GB&I.

Pine Grosbeak   *Pinicola enucleator* (Linnaeus)

SM *enucleator* (Linnaeus)

TAXONOMY Arnaiz-Villena *et al.* (2001, 2007) included Pine Grosbeak in their analysis of finch phylogeny and found it to be sister to the bullfinches; these conclusions appear to be robust and are at variance with classical taxonomy, which places *Pinicola* close to crossbills and rosefinches. There are up to 10 subspecies (three from the Nearctic), based on bill proportions, size and intensity of pigmentation.

DISTRIBUTION Breeds in boreal forests, both coniferous and deciduous, across the Holarctic. A series of races occur in N America, from Alaska down the Rocky Mts and E to Labrador and the Canadian maritimes. Nominate race occurs from N Fennoscandia through W Siberia to the River Yenesei; replaced in C & E Siberia by *pacata* and *kamtschatkensis*, respectively. Most populations are resident or partially migrant, but irruptive dispersals occur when food supplies fail in breeding grounds. At these times, birds can be found hundreds of km S of typical range.

STATUS The first for GB was shot at Bill Quay (Durham) in 1831; since then there have been a further 10. Found in Mar–May and Oct–Nov, and from Shetland to Kent. Recent birds have been in Nov 1954, Apr 1955, Nov 1957, May 1971, May 1975, Mar 1992, Nov 2000, Nov 2004. Irruption winters in Scandinavia were 1954–55, 1956–57, 1974–75, 1976–77, 1978–79, 1998–99, 2002–03 (T Tyrberg, pers. comm.), so the agreement is less than for Two-barred Crossbill. Not recorded in Ireland.

Genus *PYRRHULA* Brisson

Palearctic and Oriental genus of six species, one of which breeds in GB&I.

Eurasian Bullfinch   *Pyrrhula pyrrhula* (Linnaeus)

Endemic: RB *pileata* MacGillivray
PM *pyrrhula* (Linnaeus)

TAXONOMY Molecular analysis has shown this to be close to Pine Grosbeak (Arnaiz-Villena *et al.* 2007). Restricted to the Palearctic, more than 10 subspecies have been described (though all currently recognised: Clement *et al.* 1993), based on size and the intensity of red pigment in the male; this variation is generally clinal, with larger and brighter birds in the north.

DISTRIBUTION Breeds in mixed and open woodland across the Palearctic, from the Atlantic to Kamchatka and Japan; and from the Arctic Circle south to Italy, Balkans, Caucasus Mts, S Siberia. Nominate race occurs across most of N & W Palearctic, replaced in W by *pileata* (GB&I), *europoea* (NW Spain to Denmark), *iberiae* (Pyrenees, above *europoea*), *murina* (Azores), and five more S of nominate, from Turkey to Japan. The Azores taxon is sometimes treated as a full species *P. murina*.

Most populations resident, but those in N leave the breeding grounds in winter. Irruptions occur irregularly, probably in response to failure of food supply; at these times, large numbers of the N & E subspecies may disperse far south of their usual range.

STATUS Widespread throughout GB&I, apart from Caithness, N & W Isles and westernmost fringes of Ireland. Parslow (1973) reported a general and widespread increase in Eurasian Bullfinches, especially since the mid-1950s, and Newton (1967) suggested the decline in Sparrowhawks allowed Eurasian Bullfinches to spread into more open habitats. The range of the Eurasian Bullfinch decreased by *c.* 6.5% between the Breeding Atlases. A decline began in the mid-1970s, and was more marked in farmland (65%) than woodland (28%) (CBC, BBS). Annual population estimates were correlated in the two habitats, suggesting that the causes of decline were generally similar, if not the same (Siriwardena *et al.* 2001). Examining various aspects of breeding biology was rather inconclusive: the decline could best be explained by nest failure at the egg stage; however, nestling loss, juvenile and adult survival could equally be important, but untestable due to limited sample sizes. None of these could be linked with environmental change. Predation, orchard management and pest control in fruit areas seemed not to be involved either. However, Marquiss (2007) found Eurasian Bullfinches to be vulnerable to predation when they are compelled to forage far from cover, concluding that, especially in winter, their populations may be (at least partly) limited by food availability close to cover.

Large bright Northern Bullfinches of the nominate race are regularly recorded along the N & E coasts of GB, only rarely extending far west: there is just one record from Ireland prior to 2000 (Ballinasloe, Galway in Feb 1965). The majority are found in autumn, but there are also spring records, presumably relating to returning birds. In some years, the number of immigrants can be substantial, although there seems to be no evidence of periodicity. Recent irruptions have occurred in 1968, 1988, 1994, 1999, 2001, with more than 1,000 being recorded in GB&I during 1994. There does not appear to be a strong correlation between the years of irruption into GB&I and elsewhere in NW Europe, though this may be due to caution in identifying birds to race (Pennington & Meek 2006). The most recent major influx was in autumn 2004, with over 1,000 birds recorded in Orkney and Shetland alone (Pennington & Meek 2006). Some gave a 'trumpet' call unfamiliar to many observers and unlike the typical call of *pyrrhula* and *pileata*. Recordings suggested that these birds might have originated from European Russia, rather than Scandinavia. Fox (2006) found that large individuals (identifiable from colour rings) consistently gave only one or other call. He suggested that this indicated the irrupting birds were from a wide range (perhaps extending from Scandinavia into E Russia), a conclusion supported by stable isotope analysis (Newton *et al.* 2006). The trumpet call occurs amongst breeding birds in Finland (Fox 2006), although this may be of recent origin, perhaps due to colonisation following an earlier irruption.

Genus *COCCOTHRAUSTES* Brisson

Monospecific Palearctic genus that breeds in GB&I.

Hawfinch   *Coccothraustes coccothraustes* (Linnaeus)

RB PM *coccothraustes* (Linnaeus)

TAXONOMY Analysis of mt DNA found the Hawfinch to be in a clade with the grosbeaks *Eophona* and *Mycerobas* (Arnaiz-Villena *et al.* 2007) and near the root of the cardueline clade (Nguembock *et al.* 2008). Five (Clement *et al.* 1993) or six (BWP) subspecies are recognised, only one of which occurs in W Europe.

DISTRIBUTION Breeds across the Holarctic in suitable woodlands, especially cherry, beech and hornbeam. Nominate is most widespread, from France, GB&I, S Fennoscandia, across Eurasia to Lake Baikal and Altai Mts. Replaced by *buvryi* in NW Africa; *nigricans* in Crimea to N Iran; and by *humii* across Siberia, through N Afghanistan; and *japonicus*, Sakhalin to Japan. Northern populations are migratory, moving southwards towards the Mediterranean, C Asia, S China. Occasional larger influxes occur, though perhaps not substantial enough to be termed 'irruptions'.

STATUS Remarkably little known, largely due to its secretive habits when the trees are in leaf. It disappeared from five British counties during the first half of the 20th century (Holloway 1996), and there was a further decline of over 30% in the number of occupied 10-km squares between the two Breeding Atlases. Langston *et al.* (2002) found evidence of a reduction of about 40% after 1985. This was supported by the Repeat Woodland Survey (Amar *et al.* 2006), which found similar losses through to 2005. These were more severe at sites with high densities of Grey Squirrel dreys, suggesting that nest predation might be contributing to the decline (e.g. Fuller *et al.* 2005).

In Ireland, >125 individuals recorded 1950–2004, mostly in Oct–Mar, though precise numbers are difficult to establish because of occasionally nomadic behaviour. Large flocks have occurred, such as 25 in Phoenix Park (Dublin) sometime in the 1880s (Ussher & Warren 1900). During the autumn of

1988 there was an influx to Currachase Forest Park (Limerick). Initially, about 35 birds were noted, increasing to about 95 by Feb 1989. Two adults were observed tending a juvenile at Ballyvaughan (Clare) in Sep 1991, suggestive of breeding by survivors from the Currachase flock. Birds continue to be observed at Currachase Park from time to time but there is no further evidence of breeding; the park has large numbers of hornbeam, which may explain their occurrence there.

Genus *HESPERIPHONA* Bonaparte

New World genus of two or three species, one of which has been recorded in GB&I.

Evening Grosbeak	*Hesperiphona vespertina* (W. Cooper)

SM race undetermined, but unlikely to be other than *vespertina* (W. Cooper)

TAXONOMY Gillihan & Byers (2001) suggest that this is closely related to Hooded Grosbeak *H. albeillei*; detailed molecular analyses not yet undertaken. *Hesperiphona* sometimes subsumed within *Coccothraustes*. Three subspecies are recognised, based on bill metrics, facial pattern and vocalisations.

DISTRIBUTION Breeds in a relatively narrow band of coniferous forest from British Columbia to Newfoundland and the Canadian maritimes; also S through W USA as far as N Mexico. Nominate occurs from Alberta to Atlantic, with *brooksi* and *montanus* replacing this in N & S parts of W range. Largely resident, though shows substantial irruptions south when seed crops fail. Gillihan & Byers (2001) report that these (which occurred on an approximately two-year cycle) have become less frequent since the mid-1980s.

STATUS There are two records from Scotland. The first was on St Kilda (Outer Hebrides) in Mar 1969; the second at Nethybridge (Highland) in Mar 1980. Not recorded in Ireland.

FAMILY EMBERIZIDAE, TRIBE CALCARIINI

Genus *PLECTROPHENAX* Stejneger

Holarctic (circumpolar) genus of two species, one of which breeds thinly in GB&I.

Snow Bunting	*Plectrophenax nivalis* (Linnaeus)

RB WM PM *nivalis* (Linnaeus)
RB WM PM *insulae* Salomonsen

TAXONOMY See Lapland Longspur. Molecular analysis (Klicka *et al.* 2003) confirms close affinity with McKay's Bunting *P. hyperboreus* of Bering Straits.

DISTRIBUTION Breeds further north than any other passerine, occurring in rocky places on tundra, mountain sides and cliffs around the Arctic. Four subspecies recognised (Byers *et al.* 1995), based on pattern of black in wings and rump, and colour of fresh feather fringes; however, much of this is clinal, with many intermediates (BWP). Nominate is most widespread, occurring across Nearctic from Alaska to Greenland, and from N Scandinavia to White Sea; replaced in remainder of N Eurasia by *vlasowae*; in Iceland, Scotland by *insulae* (some *nivalis* in latter), and by *townsendi* from Kamchatka to Aleutian Is. Strongly migratory, though routes are complex (Byers *et al.* 1995), wintering on the coasts of NW Europe, and across C Asia to China, Japan; Nearctic birds winter around the coasts of N North America, and inland across the prairies to the Great Lakes.

STATUS Small, but variable, numbers breed on several of the highest Scottish mountains. Numbers decreased between 1920s and 1960s, and then increased through to 1990s. In 1991 the estimate was 70–100 breeding pairs, but a count in the Cairngorms in 2005 suggested numbers were down by a third and that the total Scottish breeing population was perhaps 50 pairs (BS3). Most Scottish breeding birds are *insulae*, with less than 10% of the Cairngorm breeders *nivalis* (BS3).

The breeding birds are relatively sedentary, moving to lower ground when the weather deteriorates and rarely wintering more than 100 km away.Otherwise, a relatively common passage migrant through N GB&I, remaining to winter in sand dunes and saltmarshes close to the sea, often associating with Shore Larks and Twites. Most flocks are along the North Sea coast, but birds also winter in the northern and western islands of Scotland and saltmarshes of Ireland. Most ringing recoveries involve birds from Iceland (*insulae*), and a small number from Scandinavia, Greenland and Newfoundland (*nivalis*). In Scotland, most wintering birds are *insulae*, but in Norfolk both races present in approximate equal numbers. In Ireland its status is similar to GB, but less common.

Genus *CALCARIUS* Bechstein

Holarctic genus of four species, one of which has been recorded in GB&I.

Lapland Longspur   *Calcarius lapponicus* (Linnaeus)

CB WM PM *lapponicus* (Linnaeus)

TAXONOMY A review of longspurs and snow buntings using the complete cyt-b sequence revealed that these form a well-supported clade, not especially close to either the New World sparrows or the true buntings *Emberiza* (Klicka *et al.* 2003), and placed near the base of the nine-primaried oscine assemblage. Lapland Longspur is part of a clade that includes Chestnut-collared *C. ornatus* and Smith's *C. pictus* (the 'collared longspurs'). McCown's Longspur *C. mccownii* is not part of this clade and is sister to the snow buntings, suggesting that generic placements may need revising. Five subspecies are recognised, of which four are Palearctic.

DISTRIBUTION Circumpolar, breeding on the tundra; only significant absences are from Iceland and Svalbard. Differentiation of three races, *subcalcaratus* (Mackenzie River to Greenland), *lapponicus* (Norway to E Siberia), *alascensis* (Chukotski peninsula, Bering Sea, Alaska to Mackenzie River), is slight and clinal. Two further races in Sea of Okhotsk and Komander Is (Byers *et al.* 1995). Migratory, although the wintering grounds are incompletely known, especially in Asia; relatively small numbers winter in Europe.

STATUS Normally a passage or winter visitor, breeding was first confirmed in 1977, when at least 2, and possibly as many as 16, pairs bred in the Grampian and Cairngorm hills of Scotland. Two (to 6) pairs bred again the following year, and a remarkable 11 (to 14) pairs in 1979, but only 1 in 1980. Although this was the last proven breeding, there continue to be occasional records of birds in the breeding season in the hills (BS3).

A regular migrant and winter visitor to coastal areas, with occasional records inland, although numbers vary markedly among years. Passage birds may be found almost anywhere along the N, W & E coasts from Sep onwards, with small parties remaining through the winter in grassland near to the sea, chiefly along the E coast of England, but occasionally elsewhere in GB&I (Lack 1986). Measurements of trapped birds suggest that birds wintering on the E coast of England come from Norway (*lapponicus*), whereas dates, arrival patterns, moult status, biometrics and associated birds indicate that those in Ireland and the N Isles are from Greenland (*subcalcaratus*). Highest numbers arrive in the Northern Isles in late Sep–early Oct, though peak numbers in SE Scotland not until Nov (BS3). In Ireland its status is similar to GB, but in much smaller numbers; recorded chiefly from Sep to Mar.

FAMILY EMBERIZIDAE, TRIBE CARDINALINI

Genus *PIRANGA* Vieillot

New World genus of about 10 species, two of which have been recorded in GB&I.

Summer Tanager   *Piranga rubra* (Linnaeus)

SM race undetermined, but likely to have been *rubra* (Linnaeus)

TAXONOMY Klicka *et al.* (2007) redefined the limits of the Cardinalini (cardinal-grosbeaks, *sensu* Sibley & Monroe 1990; Cardinalinae *sensu* Voous 1977) and found three tanager genera, including *Piranga*, belonged within the Cardinalini. Within their scheme, the Cardinalini are more closely related to the Thraupini (tanagers) than the Emberizini (buntings/sparrows). Two subspecies recognised; *rubra*: E USA to C Texas; *cooperi*: W Texas to S California is paler red in males, greyer in females.

DISTRIBUTION A bird of S & E USA and Mexico, breeding in mixed woodlands. Winters from S Mexico through to Bolivia and Brazil, and almost all birds leave breeding grounds by Sep–Oct. Robinson (1996) suggests that E populations may cross the Gulf of Mexico.

STATUS One record: a first-winter male on Bardsey (Caernarfonshire) in Sep 1957. Not recorded in Ireland.

Scarlet Tanager   *Piranga olivacea* (J.F. Gmelin)

SM monotypic

TAXONOMY See Summer Tanager.

DISTRIBUTION More northerly distribution than Summer Tanager, breeding from Canadian border through E USA, S to Missouri and Virginia. Winters from lowlands of Panama to SE Bolivia. Migrates later and further E than Summer Tanager, being recorded on Atlantic coast more frequently, and with peak passage at Cape May in mid-Sep, and Florida mid-Oct.

STATUS Seven records, most in Oct. The first of three from Ireland was a female on Copeland Island (Down) in Oct 1963; two further birds were found at Firkeel (Cork) in Oct 1985. The first of four in GB was a first-winter male on St Mary's (Scilly) in Oct 1970.

Genus *PHEUCTICUS* Reichenbach

Nearctic and Neotropical genus of six species, one of which has been recorded in GB&I.

Rose-breasted Grosbeak   *Pheucticus ludovicianus* (Linnaeus)

SM monotypic

TAXONOMY Sister to Black-headed Grosbeak *P. melanocephalus* (Klicka *et al.* 2007), with which it has hybridised.

DISTRIBUTION Breeds in mixed and deciduous woodland across N America, from N British Columbia through the Great Lakes to the Canadian Maritimes and S Newfoundland. To E of Rockies, breeds as far S as Kansas and the Dakotas, and through the Appalachians. Migratory, wintering in coastal Mexico and through C America to Venezuela, Colombia. Recorded from Bermuda from mid-Sep, and crosses the Caribbean, where some birds winter.

STATUS The first for GB&I was on Cape Clear (Cork) in Oct 1962; the first of over 20 records for GB was on St Agnes (Scilly) in Oct 1966. Over 75% in GB and seven of the eight Irish records have been in Oct, which is the peak of migration down the E coast of North America. More than half the total have been on Cape Clear or in Scilly.

Genus *PASSERINA* Vieillot

Nearctic and Neotropical genus of six or seven species, one of which has been recorded in GB&I.

Indigo Bunting   *Passerina cyanea* (Linnaeus)

SM monotypic

TAXONOMY Traditionally regarded as close to Lazuli Bunting *P. amoena*, with which it hybridises, and sometimes regarded as conspecific. Molecular analysis by Klicka *et al.* (2001) showed that these two are not sister taxa. They identified two sister clades comprising 'blue' (Lazuli Bunting, Blue Grosbeak *P. caerulea*) and 'painted' (Painted *P. ciris*, Rose-bellied *P. rositae*, etc.), and showed that Indigo is sister to a clade containing all other *Passerina* species.

DISTRIBUTION Widespread in shrubby habitats and in early succession of fields back to shrubland across North America, from S Canada to N Mexico, from Great Plains to Atlantic. Migratory, moving S & W from breeding grounds to winter in C America and Caribbean (Payne 2006).

STATUS Two records for GB&I. The first was an immature at Cape Clear (Cork) in Oct 1985; the other, a first-winter male on Ramsey Is (Pembroke) in Oct 1996. [Further records deemed unacceptable; the pattern of their occurrence is atypical for Nearctic vagrants – none from SW England, and three of five from the E coast: Lark Sparrow shows a similar pattern.]

FAMILY EMBERIZIDAE, TRIBE EMBERIZINI

Genus *PIPILO* Vieillot

New World genus of seven or eight species, one of which has been recorded in GB&I.

Eastern Towhee   *Pipilo erythrophthalmus* (Linnaeus)

SM race undetermined, likely to have been *erythrophthalmus* (Linnaeus).

TAXONOMY *Pipilo* and *Aimophila* close and polyphyletic; *Pipilo* (*sensu stricto*, including *P. erythrophthalmus*) sister to tropical *Atlapetes* (DaCosta *et al.* 2009). Until recently, treated as conspecific with Spotted Towhee *P. maculatus*, but early molecular analysis (Ball & Avise 1992) found appreciable genetic divergence. Four subspecies recognised, but Greenlaw (1996) suggests much variation in plumage and size within Eastern is clinal, and subspecies limits need revisiting. Florida race (*alleni*) may be specifically distinct.

DISTRIBUTION Largely restricted to the eastern side of USA; meets Spotted Towhee in W of range, and occasionally hybridises. N populations migratory, and winters from New England to Gulf of Mexico.

STATUS One record, an adult female trapped on Lundy (Devon) in June 1966. There are no records from Ireland.

Genus *CHONDESTES* Swainson

Monospecific New World genus that has been recorded in GB&I.

Lark Sparrow   *Chondestes grammacus* (Say)

SM *grammacus* (Say)

TAXONOMY Mt DNA analyses of Carson & Spicer (2003) and Klicka & Spellman (2007) place the Lark Sparrow in a clade with Lark Bunting *Calamospiza melanocorys* and Black-throated Sparrow *Amphispiza bilineata* (but not Sage Sparrow *A. belli*). Two subspecies are recognised (*grammacus*, *strigatus*), but differences in colour and back pattern are slight.

DISTRIBUTION Breeds in grass and farmland across W, C USA, south to Mexico. A short-distance or partial migrant, depending on range, wintering from S USA to C America. E populations are *grammacus*, W are *strigatus*, but apparently merge clinally in zone of contact.

STATUS Two records, both from East Anglia. The first was at Landguard Point (Suffolk) June–July 1981; the second at Waxham (Norfolk) in May 1991. There are no records from Ireland. This is an unlikely vagrant, and these birds may have travelled on board ship from the Gulf of Mexico, although this does not prevent them being admitted to Category A. The species was present in captivity in Belgium and the Netherlands in the 1980s, with successful breeding reported by one keeper (Clement & Gantlett 1993).

Genus *PASSERCULUS* Bonaparte

Monospecific New World genus that has been recorded in GB&I.

Savannah Sparrow   *Passerculus sandwichensis* (J.F. Gmelin)

SM *oblitus* (Peters & Griscom) or *labradorius* (Howe)
SM *princeps* (Maynard)

TAXONOMY In mt DNA analyses of Carson & Spicer (2003), Klicka *et al.* (2001), Klicka & Spellman (2007) *Passerculus* is in a clade with *Xenospiza, Melospiza* and part of a non-monophyletic *Ammodramus*, in a group clearly needing revision. Considerable variation in size, colouration, length of supercilium, etc. Byers *et al.* (1995) list 22 subspecies, although they do suggest that this might be overspilt; Wheelwright & Rising (1993) are more conservative, restricting this to 17. Seven of the latter are generally resident from SE USA through Mexico to C America; Ipswich Sparrow *princeps* is larger and paler than the other northern races. The remaining nine are poorly differentiated, and variation is largely clinal; identification of single individuals to race is often difficult. Zink *et al.* (2005) examined mt DNA variation within the species and suggested saltmarsh populations in the SW of the range should be treated as a separate species, *P. rostratus*. They found no phylogeographic structure elsewhere.

DISTRIBUTION One of the most widespread N American sparrows, breeding across much of continental N America, north of Arizona; extending further S on W coast, with populations along the seaboard as far as California and Mexico, and isolated populations in C Mexico (Wheelwright & Rising 1993). Northern populations migratory, wintering south to Guatemala. Race *princeps* globally rare and breeds only on Sable Is, wintering along the Atlantic coast to Georgia. A series of nine races breed across Canada and USA from Alaska (*sandwichensis*), through N centre (*oblitus*) to Labrador (*labradorius*).

STATUS Three records, two from Fair Isle. The first (*princeps*) at Portland (Dorset) in Apr 1982. The second on Fair Isle (Shetland) Sep–Oct 1987 was either *oblitus* or *labradorius*. Not recorded from Ireland.

Genus *PASSERELLA* Swainson

New World genus with one or more species; recorded once in GB&I.

Fox Sparrow   *Passerella iliaca* (Merrem)

SM *iliaca* (Merrem) group

TAXONOMY Molecular analysis of New World sparrows (Carson & Spicer 2003) found Fox Sparrow to be sister to American Tree Sparrow *Spizella arborea*, and the latter in a separate clade from other *Spizella*, suggesting that generic limits may need revision. Zink and colleagues have published extensively on this species (summarised in Zink & Weckstein 2003, Zink 2008), finding that mt DNA sequences support the separation of the many subspecies of Fox Sparrow into four broadly allopatric groups. On the basis of their molecular results, they recommend that these be treated as four distinct species: *P. iliaca*, *P. unalaschcensis*, *P. megarhyncha*, *P. schistacea*.

DISTRIBUTION Fox Sparrow (*sensu lato*) breeds across northern Canada and down the W side of North America from Alaska to Baja California. The reddish group *iliaca* is most northerly, and is replaced by the slate-coloured *schistacea* group in the Rocky Mts, by the thick-billed group *megarhyncha* in the uplands from S Oregon to California, and by the sooty-plumaged *unalaschcensis* group along the Pacific coast from the Aleutians to Vancouver Island. Most populations are migratory, wintering across the southern USA and N Mexico.

STATUS Recorded once (one of the reddish *iliaca* group) in Ireland on Copeland Island (Co. Down) in June 1961. Not recorded in rest of GB&I.

Genus *MELOSPIZA* Baird

New World genus of three species, one of which has been recorded in GB&I.

Song Sparrow   *Melospiza melodia* (Wilson)

SM race undetermined

TAXONOMY See Savannah Sparrow. Song Sparrow is in a clade with Swamp *M. georgiana* and Lincoln's Sparrows *M. lincolnii* (Carson & Spicer 2003). A widespread and variable species; Arcese *et al.* (2002) recognise 24 races, although up to 52 have been proposed at one time or another, based on size and plumage colour. Molecular analysis (Fry & Zink 1998) suggests that these bear little relationship to underlying molecular structure, indicating perhaps that plumage and morphology have evolved rapidly since the last post-glacial expansion in range. The subspecies fall into around five groups based on size and plumage, which correlate well with geography.

DISTRIBUTION Breeds across N America, from Aleutian Is, to S Hudson Bay and Newfoundland; S to Baja California in the west, but only to Georgia in the east. Northern populations migrant or partially migrant; extent of migration declines further south. Winters to N Mexico.

STATUS The first of seven records from GB was on Fair Isle (Shetland) Apr–May 1959. Of these, six were in a narrow window between mid-Apr and mid-May, coinciding with the peak of spring migration along the eastern seaboard of N America (peaks through New York in Mar–Apr: Arcese *et al.* 2002). The only autumn record was in the docks at Seaforth (Lancashire) in Oct. Not recorded from Ireland.

Genus *ZONOTRICHIA* Swainson

New World genus of five species, two of which have been recorded in GB&I.

White-crowned Sparrow   *Zonotrichia leucophrys* (J. R. Forster)

SM race undetermined

TAXONOMY *Zonotrichia* is sister to *Junco* in the mt DNA analysis of Carson & Spicer (2003), and this species very close to Golden-crowned Sparrow *Z. atricapilla*. Weckstein *et al.* (2001) also found the latter affinity, and suggest a period of relatively recent hybridisation – perhaps during sympatry enforced by a glacial maximum. Five subspecies recognised on colour of bill, lores, flanks and back (Chilton *et al.* 1995).

DISTRIBUTION Breeds across boreal Canada from N Newfoundland to Alaska, and down W side of N America, in the mountains and uplands through British Columbia, Alberta to California, New Mexico. Most northern populations are totally migratory; those from W USA less so; winters across S USA through Mexico, but scarce in extreme SE USA. Nominate race occurs S & E of Hudson Bay, intergrading clinally into other races to S & W. Migration routes apparently overland, with spring arrivals in Nova Scotia and Maine during mid-May (Chilton *et al.* 1995).

STATUS The first of four for GB was on Fair Isle (Shetland) in May 1977; subsequent birds were at Hornsea Mere (Yorkshire) a week later, in Seaforth Docks (Lancashire) in Oct 1995 and at Cley (Norfolk) through Jan–Mar 2008. The first (and only) for Ireland was at Dursey Sound (Cork) in May 2003.

    This Nearctic granivore is often associated with shipping. The first one recorded in GB&I was on board *SS Nova Scotia* in sea area Shannon off SW Ireland in June 1948, and another was present on *Queen Elizabeth II* in Southampton Water (Hampshire) in Sep 1988. Being ship-assisted, these birds were relegated to Irish Category D (1948, see Appendix 2) and British Category E (1988).

White-throated Sparrow   *Zonotrichia albicollis* (J. F. Gmelin)

SM monotypic

TAXONOMY See White-crowned Sparrow. Molecular data (Zink *et al.* 1991) indicate that White-throated Sparrow is closely allied with White-crowned *Z. leucophrys* and Golden-crowned *Z. atricapilla*. No morphological variation among populations, though there is polymorphism within (see Falls & Kopachena 1994); two forms exist, differing in colour of head stripes: white and tan. There are also differences in behaviour; white-striped males sing more, show more aggression and less parental care than tan-stripe males, while white-striped females show more territorial defence than tan-striped, which in turn show greater parental care. These parameters appear to be associated with a chromosomal inversion. Mating is non-random, with a significant excess of disassortive pairs within the populations and, in view of the chromosomal inversion, this will maintain the polymorphisms. Indeed, a mixed pair would provide the optimal combination: enhanced territorial defence from one partner and heightened parental care from the other.

DISTRIBUTION Breeds in conifer forest across Canada and NE USA, from Rocky Mts to Newfoundland, Maritimes and Appalachian Mts. Migratory, wintering along Pacific coast, and through much of eastern USA to NE Mexico. Migration routes not well known, but passage through Ontario is late Apr–mid-May, with some even later.

STATUS The first for GB was on shot at Eilan Mor (Outer Hebrides) in May 1909. There have subsequently been 29 more records, of which 24 have been in the spring, and the rest in Sep–Dec. There are two records from Ireland: the first on Cape Clear (Cork) in Apr 1967, and another in Belfast Dec 1984–May 1985. Perhaps more than any other bird, this species turns up on ships in GB&I waters. There are several records of birds either at sea or on liners when they docked – including four on a Cunard liner at Southampton (Hampshire) in autumn 1958, and one seen in a city park in Oct 1962, two days after the *Mauretania* arrived at the same place with two White-throated Sparrows on board. More recently, birds have been seen in 2001 and 2002 on North Sea oil rigs.

Genus *JUNCO* Wagler

Nearctic and very marginally Neotropical genus of three or four species, one of which has been recorded in GB&I.

Dark-eyed Junco   *Junco hyemalis* (Linnaeus)

SM race undetermined, but likely to have been *hyemalis* (Linnaeus)

TAXONOMY See White-crowned Sparrow. There is still some confusion over the precise relationships among the Juncos; until 1973, those with dark eyes were split into five separate species, three of which comprised two subspecies each. These were then lumped into a single species *J. hyemalis* with 15 subspecies, although the past splits continue to be acknowledged in the arrangement of the races into five groups (BWP). A recent molecular analysis found little differentiation among North American subspecies of *J. hyemalis* (Milá *et al.* 2007a).

DISTRIBUTION Breeds across Alaska and Canada south of the tree-line, to S Hudson Bay, Labrador, Newfoundland and S through Appalachian Mts; in the west of N America, breeds through most states to mountains of S California. Most northern populations migratory; in winter, occurs across entire USA to C Mexico. Autumn migration may extend into Dec, while northward migration is chiefly mid-Mar–mid-Apr, though some birds linger on wintering grounds in NE USA until early June.

STATUS The first record for GB&I (and one of only three for Ireland) was at Loop Head Lighthouse (Clare) in May 1905; the other Irish birds were at Ballygannon (Wicklow) in Aug 2000 and Whitehead (Antrim) in May 2004. There have been *c.* 30 records in GB since a male was trapped at Denge Marsh (Kent) in May 1960. Birds have been seen in every month from Nov to May, with a concentration in spring, and 21 of 26 in Mar–May. Birds were recorded flying from the *Mauretania* in Southampton Water (Hampshire) in Oct 1962, another on *Shell Bravo* gas rig in sea area Thames in May 1980, and a third on *Maersk Curlew* (sea area Dogger) in May 2000.

Genus *EMBERIZA* Linnaeus

Palearctic, Oriental and Ethiopian genus of about 40 species, 16 of which have been recorded in GB&I, four as breeding birds.

Black-faced Bunting   *Emberiza spodocephala* Pallas

SM *spodocephala* Pallas

TAXONOMY Three well-defined subspecies, differing in colour of head, throat, and underparts; also extent of white in tail. Relationships to other species not well known. A recent molecular study placed it in an unresolved polytomy that also included Rustic, Little, Yellow-breasted, Yellow-browed and several eastern Palearctic buntings.

DISTRIBUTION Breeds in damp undergrowth, often beside watercourses, in E Asia, from about Krasnoyarsk to coast of China, N Japan, and S as far as treeless areas along Mongolian border. Nominate is most northerly, and most migratory, replaced from Sakhalin to Japan by *personata*. In C China, geographically isolated race *sordida*. N populations migratory, wintering from NE India to S China and in N Indo-China.

STATUS As with many eastern buntings, the possibility of captive origin is hard to eliminate, but five individuals in GB seem to be acceptable. The first was at Pennington Flash (Lancashire) in Mar–Apr 1994. Four subsequently, all in Oct: Newbiggin (Northumberland) in 1999; Lundy (Devon) in 2001; Fair Isle (Shetland) in 2002; Flamborough (Yorkshire) 2004. There are no records from Ireland.

Pine Bunting   *Emberiza leucocephalos* S. G. Gmelin

SM *leucocephalos* S. G. Gmelin

TAXONOMY Closely related to Yellowhammer (Alström *et al.* 2008), with which it hybridises (e.g. Panov *et al.* 2003); these have been regarded as conspecific in the past, but coexist in Siberia (BWP). Two subspecies (Vaurie 1959), differing in width of black band across forehead, and intensity of chestnut markings.

DISTRIBUTION E Palearctic, widely distributed from Ural Mts to Sea of Okhotsk in broken woodland and wooded steppe. Nominate race is replaced in W China by the geographically isolated (and apparently resident) *fronto*. Most populations migrate S to winter at scattered sites across S Eurasia, including Afghanistan to NW India, and N China (Byers *et al.* 1995), but isolated sites occur in Israel, Iran, S Caspian, and probably elsewhere: e.g. wintering flocks have been found in N Italy and to a lesser extent S France.

STATUS About 45 in GB&I since the first on Fair Isle (Shetland) in Oct 1911; first of the two Irish records more recent, at N Slob (Wexford) Jan–Feb 1995, second in Dublin in Mar 1996. Recorded in most months of the year, with a peak in Oct. Of the 22 in Scotland to 2004, 20 were in autumn (Aug–Dec), compared with only 5 of 21 elsewhere in GB&I.

Yellowhammer   *Emberiza citrinella* Linnaeus

Endemic: RB *caliginosa* Clancey
RB WM PM *citrinella* Linnaeus

TAXONOMY Sister species to Pine Bunting (Alström *et al.* 2008), which it replaces in the W Palearctic. Three subspecies; differences in size and intensity of streaking are clinal (Vaurie 1959), although birds from GB&I are smaller (BWP). Other subspecies have been described, but Roselaar (in BWP) casts doubt on most of these. Hybridises regularly with Pine Bunting in E Kazakhstan.

DISTRIBUTION Widespread across Europe, and E to C Siberia, Caucasus and Iran; wide altitudinal range, from sea level in the north to uplands in S (e.g. Appenine Mts of Italy, mountain regions of Balkans). Nominate occurs across most of European range, intergrading through Baltic states to N Balkans, Ukraine, to be replaced by *erythrogenys* through into Siberia. Race *caliginosa* breeds in N & W of GB&I, merging into nominate in S & E England (Vaurie 1959). European populations are largely resident; northern populations are partial migrants to avoid harsh winter conditions. Further east, migration is more complete, with birds wintering from Caucasus through Iran, Iraq to Turkmenistan.

STATUS Yellowhammer was not included by Fuller *et al.* (1995) in their list of seriously declining farmland birds. However, in GB, there was a modest contraction in range of *c.* 8% between the Breeding Atlases, chiefly in the N & W. In Ireland, the decline was much more pronounced; >35% of 10-km squares had been abandoned, especially from those in the N & W with low levels of tilled land (Gibbons *et al.* 1993). Its decline in GB seems to have started in the late 1980s, and by the end of the 1990s numbers were falling by *c.* 10% per year. Siriwardena *et al.* (1998) suggested that it was resistant to the factors that affected other species that began their decline earlier, without being able to identify what the cause of the decline might be. Bradbury *et al.* (2000) were unable to identify any aspect of reproductive performance that might drive the decline, and suggested that overwinter survival might be important. Now largely absent from N & W Scotland.

Scottish breeding birds largely sedentary. Small numbers of birds, mainly in N Isles and on E coasts in spring and autumn presumed migrants (nominate) from Continent.

Cirl Bunting   *Emberiza cirlus* Linnaeus

RB monotypic

TAXONOMY Sister to Yellowhammer, Pine Bunting and White-capped Bunting *E. stewarti* (Alström *et al.* 2008). Three subspecies recognised by Byers *et al.* (1995), only two by Vaurie (1959); Svensson (1992) and BWP report that even these are not separable; treated here as monotypic.

DISTRIBUTION Small population in GB; main range extends S from Normandy, SW through most of Iberia into NW Africa, and SW through Italy and Balkans, Romania into N & W Turkey. Resident over most of range, with some winter dispersal in search of food.

STATUS Cirl Bunting underwent a collapse in distribution during the latter part of the 20th century. In the 1930s, it bred as far north as N Wales and through much of the S Midlands of England (Wotton *et al.* 2004). As recently as the First Breeding Atlas (1968–72) it occurred widely across S England, especially south of a line between the Severn and Thames. By the Second Breeding Atlas, over 80% of the occupied range had been lost, leaving a relict population in S Devon, isolated pairs elsewhere in the SW peninsula and fewer than 150 pairs by 1990. The decline coincided with agricultural change. The ploughing of stubble for autumn-planted cereals brought a loss of winter foraging for many species, including Cirl Buntings. Increases in silage production led to changes in fertiliser and insecticide use, and the structure of hay fields, and fewer invertebrates for growing chicks.

Bradbury *et al.* (2008) compared the use of low- and high-input pesticide (especially herbicide) regimes on the winter availability of seed in cereal stubble and on Cirl Bunting numbers. They found higher seed density on low-input regimes, generally due to more broadleaved weeds, and similarly high numbers of Cirl (and other) Buntings here. There was also evidence of a preference for smaller fields, perhaps because of shelter and proximity to hedges to escape predators. Highest numbers of winter Cirl Buntings were found on fields with elevated numbers of territories the previous year, indicating that the low dispersal of the species requires local management in the early stages of any recovery plans.

Some areas of S Devon have seen a return to more traditional field management, through the introduction of set-aside, the actions of sympathetic landowners such as the National Trust, or following inducements for less intensive farming. This has resulted in a population recovery, and the number of territories in Devon had increased to *c.* 700 by 2003 (Wotton *et al.* 2004). Birds are now recorded regularly in Cornwall, and a re-introduction programme is underway.

Away from the breeding areas, this remains a rare bird. In Ireland one was present at Mizen Head (Cork) in May 2006.

Rock Bunting   *Emberiza cia* Linnaeus

SM race undetermined, but all likely to have been *cia* Linnaeus

TAXONOMY Forms a natural group with Meadow *E. cioides*, Jankowski's *E. jankowskii* and Godlewski's Buntings *E. godlewskii*, and closest to the last of these (Alström *et al.* 2008); indeed, Vaurie (1959) treated these two buntings as conspecific, though Mauersberger (1972) did not. Within *cia*, five subspecies recognised by Vaurie (1959) and BWP, but only four by Byers *et al.* (1995); differences based on size and intensity of plumage colouration, but mostly clinal.

DISTRIBUTION Inhabits dry rocky hills and valleys with limited vegetation, generally avoiding woodlands. Breeds from Iberia, NW Africa through S Europe to Turkey, Caucasus, Iran to S Altai Mts and W Himalayas. Nominate race occurs in Europe, Africa, replaced by other races from NE Turkey, Caucasus Mts. Most populations are resident, though northern ones may move south in winter, and all may descend to lower altitudes during most severe weather.

STATUS Four records (five individuals), all from England and Wales. The first comprises two birds caught near Shoreham (Sussex) at the end of Oct 1902. Subsequent records from Feb to June; none since 1967. Not recorded in Ireland.

Ortolan Bunting   *Emberiza hortulana* Linnaeus

PM monotypic

TAXONOMY Sister to Cretschmar's Bunting (Alström *et al.* 2008).

DISTRIBUTION Breeds widely across W Eurasia (though rarely near the sea, except Baltic), from Fennoscandia to Altai Mts in the north, and from Iberia through N Mediterranean, Turkey to S Caspian in the south. Strongly migratory, wintering S of Sahara in Sierra Leone, Guinea, E Sudan and Ethiopia; apparently rarely recorded elsewhere.

STATUS Scarce migrant, with average of 50–70 per year since 1968 (Fraser & Rogers 2006). Most are in the autumn, and chiefly along the E coast of GB and in SW England. In Ireland, it is much less common, with only *c.* 100 individuals to 2004, again mostly in Sep–Oct, and very rare in spring. BirdLife International (2004) documents a contraction in population across much of W Europe, especially in France and Fennoscandia.

Cretzschmar's Bunting   *Emberiza caesia* Cretzschmar

SM monotypic

TAXONOMY Closely similar to Ortolan Bunting, which it replaces in E Mediterranean.

DISTRIBUTION Breeds in more arid sites than Ortolan, on hillsides in S Greece, Aegean Islands, W & S Turkey, Cyprus and Levant. Migratory, wintering in Nile Valley of Sudan, and adjacent SW coast of Red Sea.

STATUS Three records from GB&I, all from May–June in the northern isles. The first two were found on Fair Isle (Shetland) in June 1967 and June 1979; the third was on Stronsay (Orkney) in May 1998. There are no records from Ireland.

Chestnut-eared Bunting   *Emberiza fucata* Pallas

SM *fucata* Pallas

TAXONOMY The E Palearctic buntings have not been included in any molecular analyses, and their taxonomy is based on traditional morphology, biogeography and behaviour. Vaurie (1959) calls this 'Grey-hooded Bunting' and places it with Tristram's *E. tristrami* and Little, on the basis of morphology and plumage. Three subspecies recognised, based on breadth of breast-band and colour intensity: *fucata, kuatunensis, arcuata*.

DISTRIBUTION Occurs discontinuously from N Japan to Himalaya. Nominate in NE China and Japan; *kuatunensis* in SE China; *arcuata* in India and Yunnan. The northern populations (*fucata*) are migratory, wintering S into China; the other races move less far; those in Himalaya are more altitudinal migrants.

STATUS One record from GB&I; a first-winter male on Fair Isle (Shetland) in Oct 2004.

Yellow-browed Bunting   *Emberiza chrysophrys* Pallas

SM monotypic

TAXONOMY Morphology, vocalisations and recent phylogenetic study indicate close relationship with Tristram's Bunting (Alström *et al.* 2008).

DISTRIBUTION A little-known species from C Siberia (Byers *et al.* 1995), found in woodland clearings and clearances from Irkutsk to Yakutsk. Migratory, wintering in C, SE China.

STATUS Five records, all from GB. The first was at Holkham Meals (Norfolk) in Oct 1975. This was followed by birds at Fair Isle (Shetland) in Oct 1980; Orkney (Sep 1992, May 1998), and Scilly (Oct 1994). There are no records from Ireland.

Rustic Bunting   *Emberiza rustica* Pallas

PM *rustica* Pallas

TAXONOMY A recent molecular phylogeny placed Rustic Bunting in a group that also includes Little, Yellow-breasted and Black-faced Buntings along with several eastern Palearctic buntings (Alström *et al.* 2008). Two subspecies recognised: *rustica* and *latifascia*, differing in darkness of crown and intensity of orange/red streaking, but these are very slight (BWP).

DISTRIBUTION Northern Palearctic, breeding in damp taiga forest from tree-line south to *c.* 59°N. Nominate occurs from Scandinavia to about Lake Baikal; replaced by *latifascia* from here to Kamchatka, N Japan. Migratory, wintering chiefly in China, Japan, Korea.

STATUS There are hundreds of records from GB, but only 17 from Ireland, where the first was on Cape Clear (Cork) in Oct 1959. The number of birds recorded in GB has increased in recent years, from an average of less than 1 each year in the 1950s to over 20 in the 1990s. There has also been a 10-fold increase at the bird observatories, especially Fair Isle. The population in Fennoscandia, where many of these birds will have come from, grew during 20th century (BirdLife International 2004).

There is a striking preponderance of records from the N & W islands of Scotland: of 141 Scottish records during autumn 1990–2003, 115 were found in Orkney, Shetland and, to a lesser extent, the Outer Hebrides. A significantly higher proportion of Scottish records are in spring (48%) than elsewhere (31%) and, in Scotland, 82 of 141 autumn records (54%) have been in Sep; the comparable figures for the rest of GB&I are 29 out of 129 (22%). The main migration routes from the western breeding areas are eastwards then turning south in E Siberia (Byers *et al.* 1995), with a reversal of this route in spring. Arrivals in GB&I in spring could be of birds overshooting the Scandinavian breeding grounds; those in autumn, which are predominantly juveniles, presumably represent post-breeding dispersal or those orienting to the extreme SW. Of the 19 Irish records, 17 were in County Cork.

A female trapped on Fair Isle (Shetland) in June 1963 was recovered at Khios (Greece) in Oct 1963.

Little Bunting   *Emberiza pusilla* Pallas

PM monotypic

TAXONOMY See Rustic Bunting.

DISTRIBUTION Broadly similar distribution to Rustic, though slightly further north; less widespread in Scandinavia and far east of Siberia. Migratory, with W populations following similar route to Rustic, first heading E then turning S in eastern Siberia. Winters further south than Rustic, from NE India to C China. Has spread westwards since the 1930s.

STATUS A scarce bunting that shows striking differences in pattern of occurrence between regions within GB&I. Numbers increased in GB from an average of <5 per year in 1950–70, through *c.* 10 in the 1970s to >25 when it was removed from the BBRC List in the mid-1990s. It has remained rare in Ireland; the first was found dead at Rockabill Lighthouse (Dublin) in Oct 1908, the next at Rineanna (Clare) in Sep 1949. Since then there have been a further 26 records to 2004. The number recorded at bird observatories (especially Fair Isle) increased at a similar rate to the GB&I figures, suggesting that there has been a real increase in the pattern of occurrence, and this might be related to the westward spread during the second half of 20th century.

As with Pine and Rustic Buntings, many more occur in the N & W islands of Scotland, and they arrive in Scotland earlier than elsewhere in GB&I: of birds that are found in the autumn, 41% of Scottish birds arrive in Sep, compared with only 19% elsewhere. Little Bunting is occurring more frequently as a winter resident, with increasing numbers being found in finch and bunting flocks. Again there is a difference between regions; data from Naylor (2005) show that, although numbers are small, 4 of 388 birds found in Scotland came from Dec to Feb, compared with 22 of 308 birds from elsewhere ($P < 0.01$). Origins probably similar: overshooting migrants in spring, juvenile dispersal and extreme SW orientation of migrants in autumn. This may be a species (like Yellow-browed Warbler) that is gaining a small winter foothold in the W Palearctic.

Yellow-breasted Bunting   *Emberiza aureola* Pallas

SM *aureola* Pallas

TAXONOMY Morphology and molecular phylogenetics suggest close to Chestnut Bunting *E. rutila* (BWP; Alström *et al.* 2008). Two subspecies are recognised, though variation in intensity of chestnut upperparts and yellow underparts is clinal, and intermediates probably not recognisable (Byers *et al.* 1995).

DISTRIBUTION Breeds on meadowland south of Taiga (Byers *et al.* 1995) from NE Europe to Kamchatka, and S to Lake Baikal and into Manchuria, N Japan. Colonised Finland since 1920s, and has continued to extend west during second half of 20th century (BWP). Nominate occurs across north of range; *ornata* breeds in S parts of eastern range: Manchuria, Sakhalin to Japan. Strongly migratory, wintering in SE Asia

STATUS Well over 200 records from GB&I; mostly in Sep; 75% from Scotland, especially the northern isles (and particularly Fair Isle). The first two were at Cley (Norfolk) in Sep 1905. Only four in Ireland, since the first on Tory Island (Donegal) in 1959, and all in a tight window in mid-Sep. The number recorded across GB&I increased from around one per year through 1950–70, to around six subsequently, but has declined since 2000. The increase may have been due to observer awareness, although the number recorded at bird observatories also increased markedly to 1990, suggesting the increase to be real and related to westward expansion.The recent decline may be associated with a fall in the breeding population through the 1990s in the core area of Russia (BirdLife International 2004).

Common Reed Bunting   *Emberiza schoeniclus* (Linnaeus)

RB WM PM *schoeniclus* (Linnaeus)

TAXONOMY One of three reed buntings (with Pallas's Reed and Japanese Reed Bunting *E. yessoensis*) that are broadly allopatric across the Palearctic. Extensive interpopulation variation has led to a long series of (up to 30) subspecies in Common Reed Bunting. Vaurie (1959) lists 15 in three groups; Byers *et al.* (1995) give 20 in four groups; Roselaar (in BWP) prefers two groups, but suggests there are subgroups within these. This complexity is based on size (wing length), bill size (especially depth, which varies latitudinally), and colour differences, which Byers *et al.* describe as 'subtle' (which vary with longitude). In general, northern birds tend to be smaller with thin bills, southern ones larger with thick bills; there are intermediates, both geographically and anatomically. Grapputo *et al.* (1998) examined molecular differences among a series of populations in N Italy that straddled the border between two races: thin-billed *schoeniclus* and thick-billed *intermedia*. There was no difference in mitochondrial sequence between the races, indicating a relatively recent origin; more rapidly evolving microsatellites showed a larger degree of differentiation, implying that bill morphology has diverged recently and is probably maintained by differential selection imposed through food size and hardness (Grapputo *et al.* 1998).

DISTRIBUTION Widespread across the Palearctic, breeding in damp scrub from C France, GB&I across Scandinavia and C Europe, with scattered isolated populations in Iberia, Mediterranean islands, NW Africa, Balkans. Distribution across C Asia incompletely known (Byers *et al.* 1995), but breeds as far east as Kamchatka, Mongolia, N Japan. According to Roselaar (BWP), nominate race (one of the thin-billed group) occurs through most of N Europe, being replaced by other thin-billed races across Siberia, through the Carpathian Mts to the Altai Mts, and discontinuously through N Asia to Japan. Thick-billed races occur from Iberia through Mediterranean and S Europe and Balkans, through Turkey, Iran, Kazakhstan to C China (BWP; Byers *et al.* 1995). N populations almost entirely migratory, many wintering sympatrically with southern birds of different bill type, where ecological differences may be evident (Grapputo *et al.* 1998).

STATUS Numbers of this widespread species about doubled during the decade following 1965 (CBC), apparently as it adapted to drier farmland habitats, where it could be found alongside Yellowhammers and Corn Buntings. In common with these species (and other farmland granivores) a steep decline began in the mid-1970s, which was also seen in the Waterways Bird Survey and, in little over 10 years, numbers were so low that Common Reed Bunting was Red Listed. The change to winter cereal, and loss of winter stubble with its spilt grain and weed seeds, has likely played its part in this decline. The Breeding Atlases more or less coincide with the decade of decline; across GB&I, *c.* 12% of 10-km squares were abandoned between the two Breeding Atlases, but especially in Scotland, Wales, SW England and S Ireland.

Intensive research (e.g. Brickle & Peach 2004) demonstrated the importance of even small patches of wetland, such as ditches, for successful breeding on farmland, providing both shelter for nests and invertebrate food for nestlings, and confirmed the importance of oilseed rape for nesting and summer food in some regions (Gruar *et al.* 2006). A demographic analysis of nest record and ringing data (Peach *et al.* 1999) indicated that the decline was probably driven by reduced adult survival, and revealed that nest losses have increased since the 1980s, perhaps as damp patches have been removed in the interests of agricultural intensification.

Common Reed Buntings are partial migrants within GB&I; most are resident, but some move S & W in autumn (Wernham *et al.* 2002), even crossing the English Channel into N France, with females showing a greater dispersal distance. Good numbers are regularly recorded at coastal sites such as Portland Bill (Dorset) and Dungeness (Kent) in autumn. There is also exchange between GB and S Scandinavia; some of these are birds that come for the winter, others are passage migrants, on their way to France. Peak numbers are recorded in Shetland during May and Oct; these are undoubtedly migrants since Common Reed Buntings are rare breeding birds there.

Pallas's Reed Bunting *Emberiza pallasi* (Cabanis)

SM race undetermined

TAXONOMY Closely related to Common Reed and Japanese Reed Buntings (Alström *et al.* 2008). Varies in size and paleness, leading to four (BWP) to six (Byers *et al.* 1995) subspecies. But integrity of these is unclear: one (*minor*) may be a race of Eurasian Reed (Byers *et al.* 1995) and another (*lydiae*) may be distinct species (BWP).

DISTRIBUTION Breeds in Asia, in two, apparently disjunct, areas. Nominate race occurs in mountains of C Asia; *polaris* replaces this in N & E, from NE European Russia to Chukotski peninsula. Further races occur in SW Mongolia and from Lake Baikal to E Mongolia, but these all still unclear.

STATUS Three records: the first two were both on Fair Isle (Shetland) in Sep–Oct 1976 and Sep 1981. The third was trapped at Icklesham (Sussex) in Oct 1990. Since these arrive slightly earlier than other eastern vagrants, they may originate from the west of the species range. There are no records from Ireland.

Black-headed Bunting *Emberiza melanocephala* Scopoli

SM monotypic

TAXONOMY Closely related to Red-headed Bunting *E. bruniceps* (Alström *et al.* 2008), with which it hybridises in an area of sympatry near the Caspian Sea (BWP).

DISTRIBUTION Breeds in wooded steppe, agricultural woodlands from the Balkan coasts across Turkey, NE Black Sea to Caspian Sea, and S through Iran to Persian Gulf. Migratory, wintering in NW of Indian subcontinent.

STATUS There are *c.* 190 records from GB&I, though only seven of these are from Ireland, with the first being on Great Saltee (Wexford) in May 1950. Perhaps unsurprisingly for a species that arrives in its breeding range in SE Europe from the SE, many of these records are in the spring, presumably overshooting migrants. Birds arrive later in Scotland, where only 18% of spring records are in Apr–May, compared with 63% for the rest of GB&I. As with many other species, the average number per year has increased since 1950, from around one (1951–70), through around three (1971–90) to over six (since 1991). It is also evident that far more males have been recorded; this may reflect differences in migration pattern between the sexes, or the difficulty of separating female Black- and Red-headed Buntings.

Corn Bunting   *Emberiza calandra* **(Linnaeus)**

RB PM *calandra* (Linnaeus)

TAXONOMY Corn Bunting is possibly sister to Chestnut-eared Bunting *E. fucata* (Alström *et al.* 2008) and lies within the genus *Emberiza* as currently defined (Lee *et al.* 2001). Varies clinally, especially in plumage tones, also some size differences among more isolated populations. Five subspecies recognised by Byers *et al.* (1995), but only three by Roselaar (in BWP), of which '*clanceyi*' endemic to GB&I not recognised by Svensson (1992) or here.

DISTRIBUTION Breeds in open habitats, especially grasslands, cereals with suitable song posts; may vary in density across apparently uniform habitat and disappear from locations where breeding has been successful to appear in previously uninhabited sites. Occurs around Mediterranean from Morocco, Tunisia, Iberia, France, GB across C Europe to Poland, Ukraine, to Caspian Sea, and from Turkey, Lebanon to Iran. A disjunct population breeds from S Kazakhstan to W China. Nominate occurs across most of European range; replaced by two further races in W Mediterranean, Canaries (though these not included by BWP), and a fourth from Turkey to Kazakhstan (Byers *et al.* 1995). Some populations (especially in W Europe) are resident or disperse locally; others are more migratory, wintering into Middle East.

STATUS Along with several other birds of farmland, Corn Buntings have declined markedly since the mid-1970s, not just in Britain but across Europe (Hagemeijer & Blair 1997). Extinct as a breeding species in Ireland in recent years, although in 1900 it bred in 30 of the 32 Irish counties. Within GB&I, its range contracted by *c.* 35% between the two Breeding Atlases, and the population declined by *c.* 75% during 1968–1991 (CBC). It is one of seven severely affected species (all farmland birds) sharing a predominantly granivorous diet, although insects are also taken, particularly during the breeding season. The decline has been ascribed to changing agricultural practices, including a trend towards autumn sowing of cereals, the consequent loss of winter stubble, and also to agricultural intensification with increased use of pesticides and inorganic fertiliser (Fuller *et al.* 1995). Brickle *et al.* (2000) report on the interactions between farm practice and breeding success. They found that, while provisioning young, adults foraged in field margins, unimproved grassland and set-aside, avoiding winter wheat and intensive grassland, both habitats that are treated with insecticide. Nest losses were due to predation or farming practice, and the former was more likely when invertebrate food was not immediately adjacent. As with other passerine species, demographic evidence shows increasing breeding success during the decline, perhaps due to lowered density, but there is not enough ringing data to analyse survival of either adults or juveniles post-fledging. An ongoing study by the RSPB in E Scotland (Perkins *et al.* 2008) supports the view that agricultural practices are heavily implicated in the decline; the retention of overwinter stubbles combined with spring-sown cereal ('Farmland Bird Lifeline') reduced the rate of decline compared with control sites.The population in the Outer Hebrides has declined by *c.* 60% since 1995, especially in areas where cereals are harvested as arable silage. This contains fewer grains, and more of them unripe, than crops harvested in the traditional reaper/binder method. Reporting this, Wilson *et al.* (2007) suggest these practices result in less grain being available for Corn Buntings (and other granivores) through the winter months.

Comparatively few birds have been ringed, and recoveries are correspondingly modest. One bird has been recovered in France; otherwise most movements have been only a few km.

FAMILY EMBERIZIDAE, TRIBE ICTERINI

Genus *DOLICHONYX* Swainson

Nearctic monospecific genus that has been recorded in GB&I.

Bobolink   *Dolichonyx oryzivorus* (Linnaeus)

SM monotypic

TAXONOMY Cyt-b analysis of Lanyon & Omland (1999) placed this species in a clade with *Xanthocephalus*, *Leistes* and *Sturnella*, one of five major lineages within the Icteridae, though the relationship between *Dolichonyx* and the other taxa was distant and statistical support for this grouping was not strong.

DISTRIBUTION Breeds in hay and mixed grassland across S Canada and N USA, from British Columbia to Newfoundland, and S to Virginia. Strongly migratory, flying up to 10,000 km to wintering grounds in the pampas of C South America. Not averse to crossing oceans, with records from Bermuda and Galapagos.

STATUS 26 records from GB&I, including the first on St Agnes (Scilly) in Sep 1962, and three from Ireland, with the first at Hook Head (Wexford) in Oct 1971 and the others at Cape Clear (Cork) in Sep 1982 and Oct 2003. As with many American passerines, the majority were in the SW (including 13 in Scilly) and all in Sep–Oct.

Genus *MOLOTHRUS* Swainson

Nearctic and Neotropical genus of four to six species, one of which has been recorded in GB&I.

Brown-headed Cowbird   *Molothrus ater* (Boddaert)

SM *ater* (Boddaert) or *artemisiae* Grinnell

TAXONOMY *Molothrus* paraphyletic in cyt-b analysis of Lanyon & Omland (1999), where *M. ater* placed in a clade (in the 'grackles and allies' lineage) with Bronzed *M. aeneus*, Shiny *M. bonariensis*, Giant *Scaphidura oryzivora* and Screaming Cowbirds *M. rufoaxillaris*. Three subspecies recognised (eastern, central and western), predominantly on size and colour.

DISTRIBUTION Widely distributed across N America, and expanding as forests are opened up for agriculture. Breeds in scattered woodland, forest edges, parks and (especially) farmland, from central Canada south to Mexico. Northern populations are short-distance migrants, wintering S into Mexico. The spread into new regions has exposed species unadapted to brood parasitism to serious threat, and species such as Kirtland's Warbler *Dendroica kirtlandii* are at particular risk.

STATUS One, on Islay (Argyll) in Apr 1988. Not recorded from Ireland.

Genus *ICTERUS* Brisson

New World, predominantly Neotropical, genus of 25 to 30 species, one of which has been recorded in GB&I.

Baltimore Oriole

Icterus galbula (Linnaeus)

SM monotypic

TAXONOMY Hybridises with Bullock's Oriole *I. bullockii* in central USA and, on the strength of this, previously treated as conspecific. Within the hybrid zones, both parental forms occur, and the zone itself is relatively narrow and stable, suggesting a lack of gene flow and perhaps reduced fitness of the hybrids. Molecular analysis (Omland *et al.* 1999) indicated that these are not each other's closest relatives: Baltimore Oriole is sister to Black-backed Oriole *I. abeillei* (Omland & Kondo 2006); Streak-backed Oriole *I. pustulatus* is sister to Bullock's.

DISTRIBUTION Widespread in E North America, from Great Lakes to Gulf of Mexico, but more restricted in west, where replaced by Bullock's. Migratory, wintering predominantly in C America to Colombia, Venezuela, but also Florida and W Caribbean islands. Most migration is overland, but some cross Caribbean.

STATUS The first for GB was caught alive on Unst (Shetland) in Sep 1890. The second was on Lundy (Devon) in Oct 1958, since when there have been *c.* 20 more. Most have been in the SW, but there are inland records from Warwickshire and Oxfordshire. Some have stayed for several weeks, including Jan–Apr 1989 (Pembrokeshire) and Dec 2003–Jan 2004 (Oxfordshire). The first record for Ireland was (appropriately) at Baltimore (Cork) in Oct 2001; a second was reported from Cape Clear (Cork) in Oct 2006. There have been three spring records, most recently at John O'Groats (Caithness) in May 2007.

FAMILY EMBERIZIDAE, TRIBE PARULINI

Genus *MNIOTILTA* Vieillot

Monospecific New World genus that has been recorded in GB&I.

Black-and-white Warbler

Mniotilta varia (Linnaeus)

SM monotypic

TAXONOMY Mt DNA study by Klein *et al.* (2004) places *Mniotilta* in a clade with *Basileuterus*, *Wilsonia* (part), *Cardellina* and *Myiobiorus*, although statistical support is not strong.

DISTRIBUTION Widespread across forested areas from the taiga of NE British Columbia, S through C & E Canada, through E USA to Mexico. Winters in Florida, Caribbean and C America, into N South America.

STATUS 15 records from GB, mostly Sep–Oct, but 1 each Mar, Nov, Dec; first was found dead at Scalloway (Shetland) in Oct 1936. First for Ireland bird was on Cape Clear (Cork) in Oct 1978; subsequently, one in Londonderry, Sep–Oct 1984.

Genus *VERMIVORA* Swainson

New World genus of about 10 species, three of which have been recorded in GB&I.

Golden-winged Warbler   *Vermivora chrysoptera* (Linnaeus)

SM monotypic

TAXONOMY Mt DNA studies by Klein *et al.* (2004) and Lovette & Hochachka (2006) showed *Vermivora* to be polyphyletic and comprising two clades. One of these clades contains the Golden-winged Warbler and the closely related Blue-winged Warbler *V. pinus* (with which it hybridises). Previously believed to be genetically almost indistinguishable (Gill 1987); later molecular studies indicated wider difference than had been thought (Dabrowski *et al.* 2005); most recent analysis, using a combination of genetic markers, found no evidence of the private alleles that would indicate prolonged periods of isolation (Vallender *et al.* 2007). Many birds in the zone of contact contained alleles more frequently found in the other species, indicating ongoing hybridisation. Sangster (2008a) split *Vermivora* into three genera based on molecular analyses, retaining *chrysoptera* and *pinus* in *Vermivora*, and placing six 'plain' species (including *peregrina*) in a new genus *Leiothlypis*.

DISTRIBUTION Restricted breeding range in E North America, principally from S Ontario through Pennsylvania, New York to Maryland and Virginia. Winters from Guatemala and Honduras to Venezuela and Colombia. Status declining rapidly due to changes in habitat and northward expansion of Blue-winged Warbler. The extensive introgression of Blue-winged genes into phenotypically 'pure' Golden-winged Warbler reported by Vallender *et al.* (2007) must lead to concerns for the future of this declining species.

STATUS One record: Maidstone (Kent) Jan–Apr 1989.

Blue-winged Warbler   *Vermivora pinus* (Linnaeus)

SM monotypic

TAXONOMY Phylogenetically close to Golden-winged Warbler (which see), with which it hybridises, and which it replaces to the south.

DISTRIBUTION More southerly than Golden-winged, breeding in eastern USA from Minnesota and Arkansas to New England. Winters in Central America, from S Mexico to Panama.

STATUS The only record for GB&I is of a first-year male at Cape Clear (Cork) in Oct 2000.

Tennessee Warbler   *Vermivora peregrina* (Wilson)

SM monotypic

TAXONOMY See under Golden-winged Warbler. Mt DNA studies by Klein *et al.* (2004) and Lovette & Hochachka (2006) showed *Vermivora* to be polyphyletic and comprising two clades. Tennessee Warbler belongs to the larger, rather plainer clade of species inhabiting S, W & N N America. Closely related to Orange-crowned *V. celata* and Nashville Warblers *V. ruficapilla*, and has probably hybridised with latter.

DISTRIBUTION Breeds across the Canadian taiga forest, from Atlantic to Pacific, relatively uncommon in N USA. Winters from S Mexico to Ecuador. Relatively early departures (July–Aug) from James Bay could be due to breeding failure; more typically departs Canada late Aug.

STATUS Four records, all in Sep from the N & W Isles of Scotland. The first was a first-winter on Fair Isle, (Shetland) in Sep 1975. Not recorded from Ireland.

Genus *PARULA* Bonaparte

New World genus of four species, one of which has been recorded in GB&I.

Northern Parula   *Parula americana* (Linnaeus)

SM monotypic

TAXONOMY Mt DNA sequences showed *Parula* to be polyphyletic and to comprise two clades. *P. americana* is closely related to Tropical Parula *P. pitiayumi* (Klein *et al.* 2004; Lovette & Hochachka 2006), and these two species should probably be placed in *Dendroica*. Variation in body size and colour of adult males, but much overlap in these characters. Moldenhauer (1992) recognised two song types (eastern and western) and recommended they should be treated as separate races: *americana* and *ludoviciana*, respectively. Further investigation is required.

DISTRIBUTION Breeds across the E half of N America from the Great Lakes to the Gulf of Mexico. Winters occasionally in S USA, but chiefly in Caribbean, S Mexico and Central America. Peak migration late Aug to late Oct, with many recorded through Florida and along Atlantic coast.

STATUS 16 records from GB&I, all Sep–Nov, with 11 in Oct. First for GB was a male on Tresco (Scilly) in Oct 1966. First (of three) in Ireland was a first-winter male at Firkeel (Cork) in Oct 1983.

Genus *DENDROICA* Gray

New World genus of about 30 species, eight of which have been recorded in GB&I.

American Yellow Warbler   *Dendroica petechia* (Linnaeus)

SM *aestiva* (J. F. Gmelin)

TAXONOMY Mt DNA studies by Klein *et al.* (2004) and Lovette & Hochachka (2006) showed *Dendroica* comprises at least three major clades. American Yellow Warbler is in clade with Chestnut-sided Warbler, Blackpoll Warbler, Bay-breasted Warbler and others. About 43 subspecies recognised, in three subspecies groups, or even species (Lowther *et al.* 1999), of which N American *aestiva* group (nine subspecies, in two lineages – eastern and western; Boulet & Gibbs 2006) are migratory, and yellow-headed Golden (*petechia* group) and Mangrove (*erithachorides* group) are largely resident forms from Caribbean and coastal N South America, respectively.

DISTRIBUTION Widespread across N America, from the tundra edge of N Canada to N Mexico. Winters from coastal Mexico S to W South America as far as N Bolivia. One of the earliest autumn migrant wood warblers, with departures often under way by mid-July in many eastern populations, and largely over by end September (Lowther *et al.* 1999).

STATUS Six records, all in Aug–Nov. First of five in GB was on Bardsey (Caernarfonshire) in Aug 1964, identified as subspecies *aestiva*, it died the following day; other four all in N & W Isles. The first in Ireland was a first-winter male at Brownstown Head (Waterford) in Oct 1995 and the second, a day later, at Kilbaha (Clare).

Chestnut-sided Warbler   *Dendroica pensylvanica* (Linnaeus)

SM monotypic

TAXONOMY See American Yellow Warbler.

DISTRIBUTION Breeds in mixed and hardwood forests in S Canada, from Alberta to Newfoundland, S into NE USA, extending down the Appalachian Mountains. Restricted in winter to Central America, from SE Mexico to Panama. Begins migration mid-Aug, on broad front, apparently avoiding Atlantic coast (Richardson & Brauning 1995)

STATUS Two records: the first was a first-year on Fetlar (Shetland) in Sep 1985. The second was in Devon in Oct 1995. Not recorded in Ireland.

Blackburnian Warbler   *Dendroica fusca* (P. L. S. Müller)

SM monotypic

TAXONOMY Lovette & Bermingham (1999) and Lovette & Hochachka (2006) found evidence that this was closest to Bay-breasted Warbler in mt DNA sequence, and represents part of an evolution of *Denrdoica* warblers that has seen a rapid divergence in molecular structure but relatively little morphological differentiation.

DISTRIBUTION Breeds primarily in conifer forests from W Alberta and Saskatchewan though E Canada and S along the Appalachian Mountains. Winters in NW South America, and an early migrant with chief passage from Aug to Sep.

STATUS Two records from GB: Skomer (Pembrokeshire) in Oct 1961; Fair Isle (Shetland) in Oct 1988. Not recorded from Ireland. This scarcity perhaps reflects its predominantly overland migration through the Americas.

Cape May Warbler   *Dendroica tigrina* (J. F. Gmelin)

SM monotypic

TAXONOMY Mt DNA analysis by Lovette & Hochachka (2006) found this to be close to Northern Parula (which properly belongs within *Dendroica*; Klein *et al*. 2004).

DISTRIBUTION Breeds across the taiga forests of Canada, and just S into USA close to Great Lakes. Winters in Caribbean and coast from Yucatan to Panama. Migrates through coastal Atlantic states, leaving breeding grounds by end of Sep, and arriving Coast Rica by end of Nov; apparently northward migration is further west.

STATUS Only one record from GB&I; a singing male in Paisley Glen (Renfrewshire) in June 1977. The more westerly spring migration in N America (Baltz & Latta 1998) suggests that this bird might have arrived the previous autumn and overwintered somewhere in SW Europe or NW Africa. Not recorded in Ireland.

Magnolia Warbler   *Dendroica magnolia* (Wilson)

SM monotypic

TAXONOMY Relationships unclear, though mt DNA analysis by Klein *et al*. (2004) places this close to Bay-breasted and Blackpoll Warblers, though statistical support is not strong.

DISTRIBUTION Breeds widely in the taiga forests of Canada, from British Columbia S & E into NE USA as far as Maryland and Virginia. Winters in Caribbean and Central America, leaving breeding grounds in mid-Aug. Many cross the Gulf of Mexico directly, but others follow coast.

STATUS Only one record from GB&I: a bird on St Agnes (Scilly) in Sep 1981. No records from Ireland.

Yellow-rumped Warbler   *Dendroica coronata* (Linnaeus)

SM race undetermined, but likely to have been *coronata* (Linnaeus)

TAXONOMY Has been regarded as two species, Myrtle *coronata* and the western Audubon's *auduboni*, but presently treated as a single species, with four subspecies in two groups: the more northerly and migratory *coronata* and *auduboni*, and the largely sedentary, C American *nigrifrons* and *goldmani*, which diverged much earlier (Milá *et al.* 2007b). Mt DNA analysis by Lovette & Hochachka (2006) found this to be in a clade with Yellow-throated Warbler *D. dominica* and Pine Warbler *D. pinus*.

DISTRIBUTION Yellow-rumped breeds from Alaska to Newfoundland in the taiga forests, just extending into NE USA. Winters along Pacific seabaords of USA, across S & E USA, and much of Mexico to Panama. Migratory behaviour variable, depending on population, but eastern birds are long-distance migrants. One of the latest species to migrate, with peak southwards movements in Sep–Oct. Spring migration also starts early, with many departing Mexico by late Apr (Hunt & Flaspohler 1998).

STATUS One of the commoner parulids in GB&I, with 27 records, 1 in winter; 3 in May–June; 21 in Oct–Nov. First of 17 for GB was a male near Exeter (Devon) Jan–Feb 1955. First of 11 Irish birds was on Cape Clear (Cork) in Oct 1976.

Blackpoll Warbler   *Dendroica striata* (J. R. Forster)

SM monotypic

TAXONOMY See American Yellow Warbler.

DISTRIBUTION Breeds across coniferous forests (especially Black Spruce *Picea mariana*) of N & C Canada from Alaska to Newfoundland. Winters in South America, E of the Andes to the Amazon. Leaves breeding grounds in Newfoundland by early Oct, arriving South America from end Sep, depending on breeding area. Many appear to cross the W Atlantic directly, which probably accounts for the relative abundance of this species in our islands.

STATUS One of the commoner *Dendroica* warblers in GB&I, with *c.* 40 records, 1 each June and Dec, the rest heavily concentrated in Oct. First of >30 in GB was on St Agnes (Scilly) in Oct 1968. The first of six in Ireland was on Cape Clear (Cork) in Oct 1976; the most recent being in 2000.

Bay-breasted Warbler   *Dendroica castanea* (Wilson)

SM monotypic

TAXONOMY See American Yellow Warbler.

DISTRIBUTION Breeds across Canada in the spruce forest belt from Alberta to Newfoundland, marginal in northern USA. Winters Panama and NW South America. Depart from breeding grounds in Aug–Sep, arriving in winter range by mid-Oct.

STATUS Only one record from GB&I; a first-winter male at Lands End (Cornwall) in Oct 1995. Not recorded in Ireland.

Genus *SETOPHAGA* Swainson

Monospecific New World genus that has been recorded in GB&I.

American Redstart   *Setophaga ruticilla* (Linnaeus)

SM monotypic

TAXONOMY Mt DNA study by Klein *et al.* (2004) found *Dendroica* to be paraphyletic, and that the species currently in *Dendroica*, Northern and Tropical Parulas, Hooded Warbler and American Redstart should all be placed in the same genus. Results from Lovette & Hochachka (2006) broadly the same. Morphology and ecology of American Redstart due to flycatching adaptations.

DISTRIBUTION Very widely distributed across a variety of open wooded habitats in Canada and USA; chiefly absent from northern tundra and SW USA. Winters from coastal California and Caribbean, S though Central America to N Brazil and Ecuador. Broad range of migration routes and dates of movement, depending on population.

STATUS Eight records, all in Oct–Nov. First of six in GB was a first-winter male at Porthgwarra (Cornwall) in Oct 1967. The first of two in Ireland was a male on Cape Clear (Cork) in Oct 1968; second was at Galley Head (Cork) in Oct 1985.

Genus *SEIURUS* Swainson

New World genus of three species, two of which have been recorded in GB&I.

Ovenbird   *Seiurus aurocapilla* (Linnaeus)

SM *aurocapilla* (Linnaeus)

TAXONOMY Mt DNA studies by Klein *et al.* (2004) and Lovette & Hochachka (2006) revealed *Seiurus* not to be monophyletic. The Ovenbird is placed near the root of the parulid tree in both analyses, and distant to the waterthrushes. Sangster (2008b) retains the Ovenbird in a monotypic *Seiurus* and places the waterthrushes in new genus *Parkesia*. Three subspecies recognised, based on colour of back; *aurocapilla* is most olive and breeds across much of N and E America; *cinereus* is greyer and breeds S from Alberta to Nebraska; *furvior* is darker than either of the other races and breeds only in Newfoundland (van Horn & Donovan 1994).

DISTRIBUTION Breeds in deciduous and mixed woodland across C & E Canada and USA S to Missouri and Virginia. Winters chiefly in Florida, Caribbean and Central America to N Colombia. Evidence suggests they use Atlantic coastal or Mississippi flyways, arriving in winter range by end Sep.

STATUS The first evidence that this species occurs in GB&I was a wing found on the tideline at Formby Point (Lancashire) in Jan 1969. Five subsequent records comprise three in Sep–Oct, two in Dec. First of three for GB: in Oct 1973 on Out Skerries (Shetland). First for Ireland was found dead (ssp. *aurocapilla*) at Lough Carra Forest (Mayo) in Dec 1977; a second was on Dursey Island (Cork) in Sep 1990.

Northern Waterthrush   *Seiurus noveboracensis* (J. F. Gmelin)

SM monotypic

TAXONOMY See Ovenbird. Clinal variation in size and colour, especially extent of yellow or white on breast; also wings longer among northern birds. Treated as monotypic here, but three subspecies recognised by Eaton (1995), based on size and colour of upper- and underparts; '*noveboracensis*' plain olive above and yellowish below, breeds in E of range; '*notabilis*' larger, with white underparts, breeds in NW Canada; '*limnaeus*' darker above, with intermediate underparts, breeds in British Columbia.

DISTRIBUTION Widespread across woodland ponds and streams from Alaska to New England. Winters from S Mexico and Caribbean, through Central America to Colombia and N Brazil. An early migrant, with most southerly populations (earliest breeders) leaving by end of July, peak movements in mid-Sep.

STATUS Eight records in GB&I, from Aug to Oct. First of six in GB was a first-winter bird on St Agnes (Scilly) in Sep–Oct 1958. Four have been in Scilly; also at Gibraltar Point (Lincolnshire) in 1988 and Portland Bill (Dorset) in 1996. The only Irish record is from Cape Clear (Cork) in Sep 1983.

Genus *GEOTHLYPIS* Cabanis

New World genus of 9–14 species, one of which has been recorded in GB&I.

Common Yellowthroat   *Geothlypis trichas* (Linnaeus)

SM race undetermined

TAXONOMY *Geothlypis* comprised a well-defined clade with *Opornis* in mt DNA analyses by Klein *et al.* (2004) and Lovette & Hochachka (2006). Intraspecific variation confused by extensive overlap, but 13 subspecies recognised by Guzy & Ritchison (1999). A limited study based on polymorphic enzyme systems (Zink & Klicka 1990) indicated that three subspecies from Texas were more similar to each other than to geographically more distant subspecies from Minnesota, but the genetic distances were broadly in line with other avian subspecies.

DISTRIBUTION Breeds in marshy habitats across most of Canada and USA, apart from far north and arid south. Southern populations more resident; northern populations winter S to Caribbean and Panama. Departs from N breeding grounds by mid-Sep, more southern populations earlier.

STATUS Nine records from GB&I, with birds being found in Jan, May–June and Sep–Nov. The first was a first-winter male on Lundy (Devon) in Nov 1954. Subsequent birds have been in Shetland, Scilly and Carnarfonshire; there is also a record of a first-winter male that remained at Sittingbourne (Kent) Jan–Apr 1989. A single bird in Ireland: first-winter male at Loop Head Lighthouse (Clare) in Oct 2003.

Genus *WILSONIA* Bonaparte

New World genus of three or four species, three of which have been recorded in GB&I.

Hooded Warbler   *Wilsonia citrina* (Boddaert)

SM monotypic

TAXONOMY In mt DNA studies of Klein *et al.* (2004) and Lovette & Hochachka (2006), the three species of *Wilsonia* did not form a monophyletic group. Hooded Warbler was close to *Dendroica*, while Canada and Wilson's Warblers form a monophyletic group with *Myioborus* and *Cardellina*.

DISTRIBUTION Breeds in dense woodland in E USA, from Great Lakes to Texas. Winters from C Caribbean down E coasts of Central America. Migrations starts late July, arriving wintering grounds by mid-Oct.

STATUS Two records, both in Sep. First was a female on St Agnes (Scilly) in Sep 1970; the second on Hirta, St Kilda (Outer Hebrides) in Sep 1992. Not recorded from Ireland.

Wilson's Warbler   *Wilsonia pusilla* (Wilson)

SM race undetermined

TAXONOMY See Hooded Warbler. Clinal variation, with Pacific birds brightest in colour. Three subspecies; the Alaskan, W Central race *pileolata* is slightly larger.

DISTRIBUTION Breeds, often close to water, in the taiga forests across Canada, from Alaska to Newfoundland; S down Rocky Mountains. Winters from C Mexico to Panama. Migrates during Aug–Oct; recorded Bermuda early Sep to mid-Nov.

STATUS Only one record from GB&I: a male at Rame Head (Cornwall) in Oct 1985. Not recorded in Ireland.

Canada Warbler   *Wilsonia canadensis* (Linnaeus)

SM monotypic

TAXONOMY See Hooded Warbler.

DISTRIBUTION Breeds across N America, from E British Columbia to Labrador and the Canadian maritimes, and south through the Great Lakes to Georgia. Migratory, wintering in the Andes from Colombia to Peru.

STATUS A first-winter female was at Kilbaha (Clare) in Oct 2006. Not recorded elsewhere in GB&I.

SPECIES LISTS

The table which follows shows the species and subspecies recorded in each of the main political areas of GB&I, with details of their respective categorisation.

CATEGORISATION

After the identification of a new species for Britain has been accepted by British Bords Rarities Committee and British Ornithologists' Union Records Committee, it is assigned by BOURC to one of a series of categories. Originally, these were intended to give an indication of the nature of the record, but over time the categories have evolved to enhance their value, especially to the conservation agencies. Within the UK, the protection given to a species is in part determined by its origin, and policies regarding possible re-introduction also depend upon this. Consequently, in 1997, categorisation was revised to assist bird protection under national wildlife legislation (Holmes *et al.* 1998), especially of naturalised species (Category C). More recently, this category has been expanded to allow species with different histories to be distinguished, again to assist the conservation agencies with their work. At the same time, Category D (colloquially: 'doubtful') was reduced in scope, and Category E ('escapes') was formalised to enable local and national recorders to monitor known escaped species.

The categories currently used are as follows:

GREAT BRITAIN

A   Species that have been recorded in an apparently natural state at least once since 1 January 1950.

B   Species that were recorded in an apparently natural state at least once between 1 January 1800 and 31 December 1949, but have not been recorded subsequently.

C   Species that, although introduced, now derive from the resulting self-sustaining populations.

C1   *Naturalized introduced species* – species that have occurred *only* as a result of introduction, e.g. Egyptian Goose *Alopochen aegyptiacus*

C2   *Naturalized established species* – species with established populations resulting from introduction by man, but which also occur in an apparently natural state, e.g. Greylag Goose *Anser anser*

C3   *Naturalized re-established species* – species with populations successfully re-established by man in areas of former occurrence, e.g. Red Kite *Milvus milvus*

C4   *Naturalized feral species* – domesticated species with populations established in the wild, e.g. Common Pigeon *Columba livia*.

C5   *Vagrant naturalized species* – species from established naturalized populations abroad, e.g. possibly some Ruddy Shelducks *Tadorna ferruginea* occurring in Britain. There are currently no species in Category C5.

C6   *Former naturalized species* – species formerly placed in C1 whose naturalized populations are either no longer self-sustaining or are considered extinct, e.g. Lady Amherst's Pheasant *Chrysolophus amherstiae*

D   Species that would otherwise appear in Category A except that there is reasonable doubt that they have ever occurred in a natural state. Species that only occur in Category D form no part of the British List.

E   Species that have been recorded as introductions, human-assisted transportees or escapees from captivity, and whose breeding populations (if any) are thought not to be self-sustaining. Species in Category E that have bred in the wild in Britain are designated as E*. Category E species form no part of the British List (unless already included within Categories A, B or C).

F   In GB, a subcommittee of the BOU has recently been established to produce a list of species to be assigned to a new Category F ('fossils' – though many are more recent than this). The terms of reference are to provide a historical record of bird species in the British Isles from 1800 back to 700,000 BP (*Ibis* 149: 652–654). Full details of the subcategories can also be found on the BOU website.

REPUBLIC OF IRELAND, NORTHERN IRELAND AND THE ISLE OF MAN

Northern Ireland, the Republic of Ireland and the Isle of Man also use sets of categories, but these differ slightly from Britain. As with Britain, the Manx and Northern Irish lists are used by governmental and non-governmental agencies to inform their conservation policies.

REPUBLIC OF IRELAND AND NORTHERN IRELAND

A   Species that have been recorded in an apparently natural state at least once since 1st January 1950.
B   Species that have been recorded in an apparently natural state at least once up to 31st December 1949, but have not been recorded subsequently.
C1   Species that, although originally introduced by man, have established feral breeding populations which apparently maintain themselves without necessary recourse to further introduction.
C2   Species that have occurred but are considered to have originated from established naturalised populations overseas.
D1   Species that would otherwise appear in Category A or B except that there is a reasonable doubt that they have ever occurred in a natural state.
D2   Species that have arrived through ship or other human assistance.
D3   Species that have only ever been found dead on the tideline.
D4   Species that would otherwise appear in Category C1 except that their feral populations may or may not be self-supporting.
E   Species that have been recorded as introductions, transportees or escapes from captivity.

ISLE OF MAN

A   Species that have been recorded on the Isle of Man in an apparently natural state since 1 January 1950.
B   Species that would have otherwise been in Category A, but have not been recorded since 31 December 1949.
B2   As B but extinct.
C   Species that have established breeding populations derived from introduced stock and which maintain themselves on the Isle of Man without necessary recourse to further introduction.
C*   Species fulfilling the above requirements in Britain but do not breed on the Isle of Man and have occurred there naturally as visitors.
D   Species that would otherwise appear in categories A or B except that there is reasonable doubt that they have ever occurred in a natural state.
E   Species that have been recorded as introductions, transportees or escapes from captivity and whose breeding populations (if any) are thought not to be self-sustaining.

Categories A and B are essentially the same in all parts of Britain and Ireland. Category C has more subsets in Britain, but Category D has fewer. In Britain, Category E has a subclass that allows an indication of whether the species has ever bred within the region following release, transportation or introduction. A more significant difference is that, in Ireland, birds found dead on the tideline are placed in D3. In Britain, an assessment is made as to the likelihood of their having arrived in British waters under their own power; if this is deemed probable, then the species is admitted into Category A or B, alternatively it may be relegated to Category D or E. In all regions, species in Categories D and E form no part of the national lists.

It is possible for a species to be placed in more than one category. For example, wild Barnacle Geese occur regularly as winter visitors and are placed in Category A. There are also apparently self-sustaining breeding populations of captive origin (in East Anglia and Lincolnshire); these individuals belong in Category C2. Finally, in some parts of Britain, there are casual escapes that survive and breed but are not self-sustaining; since they are escapes, they go into E and since some have bred following their escape, E* is the appropriate category. Thus, Barnacle Goose is categorised as A C2 E*.

The taxonomy adopted by the regions differs slightly, and attempting to incorporate these (albeit minor) differences would have complicated the layout. We have therefore standardised on the current BOU taxonomy and apologise to colleagues in Ireland and the Isle of Man; no disrespect is intended, but considerations of space and legibility have imposed this upon us. Similarly, we have standardised our English names, using the list published for the International Ornithological Congress (Gill & Wright 2006, updated to version 2.1 at www.worldbirdnames.org) and adopted by the BOU. No decision has yet been reached by the other three regions over the adoption of these names.

We have given brief details of those species in Category D in Appendix 2; some have been recorded as Category A or B in one part of GB&I but in D elsewhere.

We have not listed those species solely in Category E in GB; details of these can be found on the BOU website. Approximately 300 species fall into this category, ranging from the more obvious and expected ornamental waterfowl and escaped cage birds to such bizarre and improbable species as Common Ostrich *Struthio camelus*, Burrowing Parrot *Cyanoliseus patagonus* and Helmeted Guineafowl *Numida meleagris* (the last of which bred successfully in Norfolk in 2001). Included in the GB Category E list are a number of other species currently breeding in the wild, some of which would be candidates for addition to the British List, should they become established. Amongst these is Red-winged Laughingthrush *Garrulax formosus*, which has bred successfully for several years in the Isle of Man; if this became established it would be the first naturalized passerine to gain admission to Category C in any part of GB&I.

In the lists below, records of taxa currently under consideration are shown as **P** (pending); monotypic species are indicated by **M** and race undetermined by **U**.

			GB	RoI	NI	IoM
Mute Swan	*Cygnus olor*	M	A C2	A C1	A	A C
Tundra Swan	*Cygnus columbianus*		E			
	bewickii		A	A	A	A
	columbianus		A	A		
Whooper Swan	*Cygnus cygnus*	M	A E*	A	A	A
Bean Goose	*Anser fabalis*		E			A
	fabalis		A	A	A	
	rossicus		A	A	A	
Pink-footed Goose	*Anser brachyrhynchus*	M	A E*	A	A	A
Greater White-fronted Goose	*Anser albifrons*		E*			
	flavirostris		A	A	A	A
	albifrons		A	A	A	A
Lesser White-fronted Goose	*Anser erythropus*	M	A E*	A		
Greylag Goose	*Anser anser*		E*			
	anser		A C2 C4	A C1	A C1	A C

			GB	RoI	NI	IoM
Snow Goose	*Anser caerulescens*		E*			
	caerulescens		A C2	A	A	
	atlanticus		C2	A C1		
Canada Goose/Greater Canada Goose	*Branta canadensis*		E*			
	canadensis		C2	A	A C1	A C
Cackling Goose/Lesser Canada Goose	*Branta hutchinsii*					
	hutchinsii			A C1		
Barnacle Goose	*Branta leucopsis*	M	A C2 E*	A C1	A C1	A
Brant Goose/Brent Goose	*Branta bernicla*		E			
	hrota		A	A	A	A
	bernicla		A	A	A	A
	nigricans		A	A	A	
Red-breasted Goose	*Branta ruficollis*	M	A E*	D1		
Egyptian Goose	*Alopochen aegyptiaca*	M	C1 E*			
Ruddy Shelduck	*Tadorna ferruginea*	M	B D E*	B	B	
Common Shelduck	*Tadorna tadorna*	M	A	A	A	A
Mandarin Duck	*Aix galericulata*	M	C1 E*	C1	C1	
Eurasian Wigeon	*Anas penelope*	M	A E*	A	A	A
American Wigeon	*Anas americana*	M	A E	A	A	A
Gadwall	*Anas strepera*	M	A C2	A	A	A C
Eurasian Teal	*Anas crecca*					
	crecca		A	A	A	A
Green-winged Teal	*Anas carolinensis*	M	A	A	A	A
Mallard	*Anas platyrhynchos*		E*			
	platyrhynchos		A C2 C4	A	A	A
American Black Duck	*Anas rubripes*	M	A	A		
Northern Pintail	*Anas acuta*		E			
	acuta		A	A	A	A
Garganey	*Anas querquedula*	M	A	A	A	A
Blue-winged Teal	*Anas discors*	M	A E	A	A	
Northern Shoveler	*Anas clypeata*	M	A	A	A	A
Red-crested Pochard	*Netta rufina*	M	A C2 E*	A	A	A
Canvasback	*Aythya valisineria*	M	A E			
Common Pochard	*Aythya ferina*	M	A E*	A	A	A
Redhead	*Aythya americana*	M	A E	A		
Ring-necked Duck	*Aythya collaris*	M	A E	A	A	A
Ferruginous Duck	*Aythya nyroca*	M	A E	A	A	A
Tufted Duck	*Aythya fuligula*	M	A	A	A	A
Greater Scaup	*Aythya marila*					
	marila		A	A	A	A

			GB	RoI	NI	IoM
Lesser Scaup	*Aythya affinis*	M	A	A	A	
Common Eider	*Somateria mollissima*					
	mollissima		A	A	A	A
King Eider	*Somateria spectabilis*	M	A	A	A	A
Steller's Eider	*Polysticta stelleri*	M	A			
Harlequin Duck	*Histrionicus histrionicus*	M	A			
Long-tailed Duck	*Clangula hyemalis*	M	A	A	A	A
Black Scoter/Common Scoter	*Melanitta nigra*	M	A	A	A	A
American Scoter/Black Scoter	*Melanitta americana*	M	A			
Surf Scoter	*Melanitta perspicillata*	M	A	A	A	A
Velvet Scoter	*Melanitta fusca*	M	A	A	A	A
Bufflehead	*Bucephala albeola*	M	A	A		
Barrow's Goldeneye	*Bucephala islandica*	M	A E			
Common Goldeneye	*Bucephala clangula*		E			
	clangula		A	A	A	A
Hooded Merganser	*Lophodytes cucullatus*	M	A	A	A	
Smew	*Mergellus albellus*	M	A	A	A	A
Red-breasted Merganser	*Mergus serrator*	M	A	A	A	A
Common Merganser/Goosander	*Mergus merganser*					
	merganser		A	A	A	A
Ruddy Duck	*Oxyura jamaicensis*		E*			
	jamaicensis		C1	C1 C2	C1/2	C*
Willow Ptarmigan/Red Grouse	*Lagopus lagopus*					
	scotica		A	A	A	A
Rock Ptarmigan	*Lagopus muta*					
	millaisi		A			
Black Grouse	*Tetrao tetrix*		E			
	britannicus		A			
Western Capercaillie	*Tetrao urogallus*					
	urogallus		B C3	B		
Red-legged Partridge	*Alectoris rufa*					
	rufa		C1 E*	D4	D4	C
Grey Partridge	*Perdix perdix*		E*			
	perdix		A C2	A C1	A C1	A C
Common Quail	*Coturnix coturnix*		E*			
	coturnix		A	A	A	A
Common Pheasant	*Phasianus colchicus*		E*			
	colchicus		C1	C1	C1	C
	torquatus		C1	C1	C1	
	principalis		C1			
	mongolicus		C1			
	pallasi		C1			
	satschuensis		C1			

			GB	RoI	NI	IoM
Golden Pheasant	*Chrysolophus pictus*	M	C1 E*			
Lady Amherst's Pheasant	*Chrysolophus amherstiae*	M	C6 E*			
Red-throated Loon/ Red-throated Diver	*Gavia stellata*	M	A	A	A	A
Black-throated Loon/Black-throated Diver	*Gavia arctica* *arctica*		A	A	A	A
Pacific Loon/Pacific Diver	*Gavia pacifica*	M	A			
Great Northern Loon/ Great Northern Diver	*Gavia immer*	M	A	A	A	A
Yellow-billed Loon/ White-billed Diver	*Gavia adamsii*	M	A	A	A	
Black-browed Albatross	*Thalassarche melanophris* *melanophris*		A	A		
Atlantic Yellow-nosed Albatross	*Thalassarche chlororhynchos*	M	P			
Northern Fulmar	*Fulmarus glacialis* *glacialis*		A	A	A	A
Fea's Petrel	*Pterodroma feae*	U	A		A	
Fea's Petrel or Zino's Petrel				A		
Black-capped Petrel/ Capped Petrel	*Pterodroma hasitata* *hasitata*		A			
Bulwer's Petrel	*Bulweria bulwerii*	M		A		
Cory's Shearwater	*Calonectris diomedea* *diomedea* *borealis*		A A	P A	A	A
Great Shearwater	*Puffinus gravis*	M	A	A	A	A
Sooty Shearwater	*Puffinus griseus*	M	A	A	A	A
Manx Shearwater	*Puffinus puffinus*	M	A	A	A	A
Balearic Shearwater	*Puffinus mauretanicus*	M	A	A	A	A
Macaronesian Shearwater	*Puffinus baroli* *baroli*		A	A		A
Wilson's Storm Petrel	*Oceanites oceanicus* *exasperatus*		A	A	B	
White-faced Storm Petrel	*Pelagodroma marina* *hypoleuca*		B			
European Storm Petrel	*Hydrobates pelagicus*	M	A	A	A	A
Leach's Storm Petrel	*Oceanodroma leucorhoa* *leucorhoa*		A	A	A	A
Swinhoe's Storm Petrel	*Oceanodroma monorhis*	M	A	A		
Band-rumped Storm Petrel/ Madeiran Storm Petrel	*Oceanodroma castro*	M	B	B		
Red-billed Tropicbird	*Phaethon aethereus*	U	A			
Northern Gannet	*Morus bassanus*	M	A	A	A	A

			GB	RoI	NI	IoM
Great Cormorant	*Phalacrocorax carbo*					
		carbo	A	A	A	A
		sinensis	A	A	A	A
Double-crested Cormorant	*Phalacrocorax auritus*	U	A	A		
European Shag	*Phalacrocorax aristotelis*					
		aristotelis	A	A	A	A
Ascension Frigatebird	*Fregata aquila*	M	A			
Magnificent Frigatebird	*Fregata magnificens*	M	A			A
Frigatebird sp.				A		
Eurasian Bittern	*Botaurus stellaris*					
		stellaris	A	A	A	A
American Bittern	*Botaurus lentiginosus*	M	A	A	B	
Little Bittern	*Ixobrychus minutus*					
		minutus	A	A	A	A
Black-crowned Night Heron	*Nycticorax nycticorax*		E*			
		nycticorax	A	A	A	A
Green Heron	*Butorides virescens*	M	A	P		
Squacco Heron	*Ardeola ralloides*	M	A	A	B	
Cattle Egret	*Bubulcus ibis*		E			
		ibis	A	A		
Snowy Egret	*Egretta thula*	U	A			
Little Egret	*Egretta garzetta*					
		garzetta	A	A	A	A
Great Egret	*Ardea alba*					
		alba	A	A	A	A
Grey Heron	*Ardea cinerea*					
		cinerea	A	A	A	A
Great Blue Heron	*Ardea herodias*	U	A			
Purple Heron	*Ardea purpurea*					
		purpurea	A	A		A
Black Stork	*Ciconia nigra*	M	A E	A		
White Stork	*Ciconia ciconia*		E			
		ciconia	A	A	A	A
Glossy Ibis	*Plegadis falcinellus*		E			
		falcinellus	A	A	B	
Eurasian Spoonbill	*Platalea leucorodia*		E			
		leucorodia	A	A	A	A
Pied-billed Grebe	*Podilymbus podiceps*	U	A	A		
Little Grebe	*Tachybaptus ruficollis*					
		ruficollis	A	A	A	A
Great Crested Grebe	*Podiceps cristatus*					
		cristatus	A	A	A	A
Red-necked Grebe	*Podiceps grisegena*					
		grisegena	A	A	A	A
		holboellii	B			

			GB	RoI	NI	IoM
Horned Grebe/Slavonian Grebe	*Podiceps auritus*					
	auritus		A	A	A	A
Black-necked Grebe	*Podiceps nigricollis*					
	nigricollis		A	A	A	A
European Honey Buzzard	*Pernis apivorus*	M	A	A	A	A
Black Kite	*Milvus migrans*		E			
	migrans		A	A	A	A
Red Kite	*Milvus milvus*		E*			
	milvus		A C3	A C2	A C2	A C*
White-tailed Eagle	*Haliaeetus albicilla*	M	A C3 E*	A C1	A C2	B
Bald Eagle	*Haliaeetus leucocephalus*					
	washingtonensis		E	A	A	
Egyptian Vulture	*Neophron percnopterus*		E			
	percnopterus		B			
Griffon Vulture	*Gyps fulvus*	U	B			
Short-toed Snake Eagle/ Short-toed Eagle	*Circaetus gallicus*	M	A			
Western Marsh Harrier	*Circus aeruginosus*					
	aeruginosus		A	A	A	A
Northern Harrier/Hen Harrier	*Circus cyaneus*					
	cyaneus		A	A	A	A
	hudsonius		A			
Pallid Harrier	*Circus macrourus*	M	A			
Montagu's Harrier	*Circus pygargus*	M	A	A		
Northern Goshawk	*Accipiter gentilis*		E*			
	gentilis		A C3	A	A	A C*
	atricapillus		B	A	A	
Eurasian Sparrowhawk	*Accipiter nisus*					
	nisus		A	A	A	A
Common Buzzard	*Buteo buteo*		E*			
	buteo		A	A	A	A
Rough-legged Buzzard	*Buteo lagopus*					
	lagopus		A	A	A	A
	sanctijohannis			A		
Greater Spotted Eagle	*Aquila clanga*	M	B	B		
Golden Eagle	*Aquila chrysaetos*		E			
	chrysaetos		A	A C1	A	A
Osprey	*Pandion haliaetus*		E*			
	haliaetus		A	A	A	A
Lesser Kestrel	*Falco naumanni*	M	A	B		
Common Kestrel	*Falco tinnunculus*					
	tinnunculus		A	A	A	A
American Kestrel	*Falco sparverius*					
	sparverius		A			

			GB	RoI	NI	IoM
Red-footed Falcon	*Falco vespertinus*	M	A	A	A	A
Merlin	*Falco columbarius*					
	columbarius			P		
	aesalon		A	A	A	A
	subaesalon		A	A	A	A
Eurasian Hobby	*Falco subbuteo*					
	subbuteo		A	A	A	A
Eleonora's Falcon	*Falco eleonorae*	M	A			
Gyrfalcon	*Falco rusticolus*	M	A E	A	A	B
Peregrine Falcon	*Falco peregrinus*		E			
	peregrinus		A	A	A	A
Water Rail	*Rallus aquaticus*					
	aquaticus		A	A	A	A
Spotted Crake	*Porzana porzana*	M	A	A	A	B
Sora	*Porzana carolina*	M	A	A		
Little Crake	*Porzana parva*	M	A	B		B
Baillon's Crake	*Porzana pusilla*					
	intermedia		A	B		B
Corn Crake	*Crex crex*	M	A	A	A	A
Common Moorhen	*Gallinula chloropus*					
	chloropus		A	A	A	A
Allen's Gallinule	*Porphyrio alleni*	M	A			
Purple Gallinule	*Porphyrio martinica*	M	A			
Eurasian Coot	*Fulica atra*					
	atra		A	A	A	A
American Coot	*Fulica americana*	U	A	A		
Common Crane	*Grus grus*					
	grus		A	A	A	A
Sandhill Crane	*Grus canadensis*		A			
	canadensis			B		
Little Bustard	*Tetrax tetrax*	M	A	B		
Macqueen's Bustard	*Chlamydotis macqueenii*	M	A			
Great Bustard	*Otis tarda*					
	tarda		A	B		
Eurasian Oystercatcher	*Haematopus ostralegus*					
	ostralegus		A	A	A	A
Black-winged Stilt	*Himantopus himantopus*					
	himantopus		A	A	A	
Pied Avocet	*Recurvirostra avosetta*	M	A	A	A	A
Eurasian Stone-curlew	*Burhinus oedicnemus*					
	oedicnemus		A	A	B	A
Cream-coloured Courser	*Cursorius cursor*					
	cursor		A	A		

			GB	RoI	NI	IoM
Collared Pratincole	*Glareola pratincola*					
	pratincola		A		A	
Oriental Pratincole	*Glareola maldivarum*	M	A			
Black-winged Pratincole	*Glareola nordmanni*	M	A	A	A	
Little Ringed Plover	*Charadrius dubius*					
	curonicus		A	A	A	A
Common Ringed Plover	*Charadrius hiaticula*					
	hiaticula		A	A	A	A
	tundrae		A	A	A	A
Semipalmated Plover	*Charadrius semipalmatus*	M	A	A		
Killdeer	*Charadrius vociferus*					
	vociferus		A	A	A	
Kentish Plover	*Charadrius alexandrinus*					
	alexandrinus		A	A	A	A
Lesser Sand Plover	*Charadrius mongolus*					
atrifrons group *atrifrons, pamirensis, schaeferi*			A			
mongolus group *mongolus, stegmanni*			A			
Greater Sand Plover	*Charadrius leschenaultii*	U	A			
Caspian Plover	*Charadrius asiaticus*	M	A			
Eurasian Dotterel	*Charadrius morinellus*	M	A	A	A	A
American Golden Plover	*Pluvialis dominica*	M	A	A	A	
Pacific Golden Plover	*Pluvialis fulva*	M	A	A		
European Golden Plover	*Pluvialis apricaria*	M	A	A	A	A
Grey Plover	*Pluvialis squatarola*	M	A	A	A	A
Sociable Lapwing	*Vanellus gregarius*	M	A	A		
White-tailed Lapwing	*Vanellus leucurus*	M	A			
Northern Lapwing	*Vanellus vanellus*	M	A	A	A	A
Great Knot	*Calidris tenuirostris*	M	A	A		
Red Knot	*Calidris canutus*					A
	canutus		A	A		
	islandica		A	A	A	
Sanderling	*Calidris alba*	M	A	A	A	IoM
Semipalmated Sandpiper	*Calidris pusilla*	M	A	A	A	
Western Sandpiper	*Calidris mauri*	M	A	A		
Red-necked Stint	*Calidris ruficollis*	M	A	A		
Little Stint	*Calidris minuta*	M	A	A	A	A
Temminck's Stint	*Calidris temminckii*	M	A	A	A	A
Long-toed Stint	*Calidris subminuta*	M	A	A		
Least Sandpiper	*Calidris minutilla*	M	A	A		
White-rumped Sandpiper	*Calidris fuscicollis*	M	A	A	A	A
Baird's Sandpiper	*Calidris bairdii*	M	A	A	A	

			GB	RoI	NI	IoM
Pectoral Sandpiper	*Calidris melanotos*	M	A	A	A	A
Sharp-tailed Sandpiper	*Calidris acuminata*	M	A	A		
Curlew Sandpiper	*Calidris ferruginea*	M	A	A	A	A
Stilt Sandpiper	*Calidris himantopus*	M	A	A	A	
Purple Sandpiper	*Calidris maritima*	M	A	A	A	A
Dunlin	*Calidris alpina*					
	schinzii		A	A	A	A
	alpina		A	A	A	A
	arctica		A	A	A	A
Broad-billed Sandpiper	*Limicola falcinellus*					
	falcinellus		A	A	A	
Buff-breasted Sandpiper	*Tryngites subruficollis*	M	A	A	A	A
Ruff	*Philomachus pugnax*	M	A	A	A	A
Jack Snipe	*Lymnocryptes minimus*	M	A	A	A	A
Common Snipe	*Gallinago gallinago*					
	gallinago		A	A	A	A
	faeroeensis		A	A	A	
Wilson's Snipe	*Gallinago delicata*	M	A		A	
Great Snipe	*Gallinago media*	M	A	A	A	B
Short-billed Dowitcher	*Limnodromus griseus*	U	A	A		
Long-billed Dowitcher	*Limnodromus scolopaceus*	M	A	A	A	A
Eurasian Woodcock	*Scolopax rusticola*	M	A	A	A	A
Black-tailed Godwit	*Limosa limosa*					
	limosa		A	A		
	islandica		A	A	A	A
Hudsonian Godwit	*Limosa haemastica*	M	A			
Bar-tailed Godwit	*Limosa lapponica*					
	lapponica		A	A	A	A
Little Curlew	*Numenius minutus*	M	A			
Eskimo Curlew	*Numenius borealis*	M	B	B		
Whimbrel	*Numenius phaeopus*				A	
	phaeopus		A	A		
	hudsonicus		A	A		
Slender-billed Curlew	*Numenius tenuirostris*	M	A			
Eurasian Curlew	*Numenius arquata*					
	arquata		A	A	A	A
Upland Sandpiper	*Bartramia longicauda*	M	A	A	A	
Terek Sandpiper	*Xenus cinereus*	M	A	A		
Common Sandpiper	*Actitis hypoleucos*	M	A	A	A	A
Spotted Sandpiper	*Actitis macularius*	M	A	A	A	
Green Sandpiper	*Tringa ochropus*	M	A	A	A	A
Solitary Sandpiper	*Tringa solitaria*	U	A	A		

			GB	RoI	NI	IoM
Grey-tailed Tattler	*Tringa brevipes*	M	A			
Spotted Redshank	*Tringa erythropus*	M	A	A	A	A
Greater Yellowlegs	*Tringa melanoleuca*	M	A	A	A	
Common Greenshank	*Tringa nebularia*	M	A	A	A	A
Lesser Yellowlegs	*Tringa flavipes*	M	A	A	A	B
Marsh Sandpiper	*Tringa stagnatilis*	M	A	A		
Wood Sandpiper	*Tringa glareola*	M	A	A	A	A
Common Redshank	*Tringa totanus*					
	totanus		A	A	A	A
	robusta		A	A	A	
Ruddy Turnstone	*Arenaria interpres*					
	interpres		A	A	A	A
Wilson's Phalarope	*Phalaropus tricolor*	M	A	A	A	A
Red-necked Phalarope	*Phalaropus lobatus*	M	A	A	A	A
Red Phalarope/Grey Phalarope	*Phalaropus fulicarius*	M	A	A	A	A
Pomarine Skua	*Stercorarius pomarinus*	M	A	A	A	A
Parasitic Jaeger/Arctic Skua	*Stercorarius parasiticus*	M	A	A	A	A
Long-tailed Jaeger/ Long-tailed Skua	*Stercorarius longicaudus*					
	longicaudus		A	A	A	A
	pallescens		A			
Great Skua	*Stercorarius skua*					
	skua		A	A	A	A
Ivory Gull	*Pagophila eburnea*	M	A	A	A	
Sabine's Gull	*Xema sabini*	M	A	A	A	A
Black-legged Kittiwake	*Rissa tridactyla*					
	tridactyla		A	A	A	A
Slender-billed Gull	*Chroicocephalus genei*	M	A			
Bonaparte's Gull	*Chroicocephalus philadelphia*	M	A	A	A	
Black-headed Gull	*Chroicocephalus ridibundus*	M	A	A	A	A
Little Gull	*Hydrocoloeus minutus*	M	A	A	A	IoM
Ross's Gull	*Rhodostethia rosea*	M	A	A	A	
Laughing Gull	*Larus atricilla*	M	A	A	A	
Franklin's Gull	*Larus pipixcan*	M	A	A	A	
Mediterranean Gull	*Larus melanocephalus*	M	A	A	A	A
Audouin's Gull	*Larus audouinii*	M	A			
Pallas's Gull/Great Black-headed Gull	*Larus ichthyaetus*	M	B			
Mew Gull/Common Gull	*Larus canus*					
	canus		A	A	A	A
	heinei		A			

			GB	RoI	NI	IoM
Ring-billed Gull	*Larus delawarensis*	M	A	A	A	A
Lesser Black-backed Gull	*Larus fuscus*					
	graellsii		A	A	A	A
	intermedius		A	A	A	A
	fuscus		A	A	A	A
European Herring Gull	*Larus argentatus*					
	argenteus		A	A	A	A
	argentatus		A	A	A	A
Yellow-legged Gull	*Larus michahellis*					
	michahellis		A	A	A	A
Caspian Gull	*Larus cachinnans*	M	A	A	A	
American Herring Gull	*Larus smithsonianus*					
	smithsonianus		A	A	A	A
Iceland Gull	*Larus glaucoides*					
	glaucoides		A	A	A	A
	kumlieni		A	A	A	
	thayeri			A	A	
Glaucous-winged Gull	*Larus glaucescens*	M	A			
Glaucous Gull	*Larus hyperboreus*					
	hyperboreus		A	A	A	A
Great Black-backed Gull	*Larus marinus*	M	A	A	A	A
Aleutian Tern	*Onychoprion aleuticus*	M	A			
Sooty Tern	*Onychoprion fuscatus*					
	fuscatus		A	A	A	
Bridled Tern	*Onychoprion anaethetus*			D3		A
	antarcticus		B			
Little Tern	*Sternula albifrons*					
	albifrons		A	A	A	A
	antillarum / athalassos / browni		A			
Gull-billed Tern	*Gelochelidon nilotica*					
	nilotica		A	A	A	
Caspian Tern	*Hydroprogne caspia*	M	A	A		
Whiskered Tern	*Chlidonias hybrida*					
	hybrida		A	A		
Black Tern	*Chlidonias niger*					
	niger		A	A	A	A
	surinamensis		A	A		
White-winged Tern	*Chlidonias leucopterus*	M	A	A	A	
Elegant Tern	*Sterna elegans*	M		A	A	
Sandwich Tern	*Sterna sandvicensis*					
	sandvicensis		A	A	A	A
	acuflavida		A			
Royal Tern	*Sterna maxima*					
	maxima		A	D3		
Lesser Crested Tern	*Sterna bengalensis*					
	torresii		A	A		

			GB	RoI	NI	IoM
Forster's Tern	*Sterna forsteri*	M	A	A	A	
Common Tern	*Sterna hirundo*					
	hirundo		A	A	A	A
Roseate Tern	*Sterna dougallii*					
	dougallii		A	A	A	A
Arctic Tern	*Sterna paradisaea*	M	A	A	A	A
Common Murre/ Common Guillemot	*Uria aalge*					
	aalge		A	A	A	A
	albionis		A	A	A	A
Thick-billed Murre/ Brünnich's Guillemot	*Uria lomvia*	U	A	A		
Razorbill	*Alca torda*					
	islandica		A	A	A	A
	torda		A	A		
Great Auk	*Pinguinus impennis*	M	B	B	B	B2
Black Guillemot	*Cepphus grylle*					
	arcticus		A	A	A	A
Long-billed Murrelet	*Brachyramphus perdix*	M	A			
Ancient Murrelet	*Synthliboramphus antiquus*	M	A			
Little Auk	*Alle alle*					
	alle		A	A	A	A
Atlantic Puffin	*Fratercula arctica*	M	A	A	A	A
Pallas's Sandgrouse	*Syrrhaptes paradoxus*	M	A	B	B	B
Common Pigeon/Rock Dove	*Columba livia*		E*			
	livia		A C4	A	A	A
Stock Dove	*Columba oenas*					
	oenas		A	A	A	A
Common Wood Pigeon	*Columba palumbus*					
	palumbus		A	A	A	A
Eurasian Collared Dove	*Streptopelia decaocto*					
	decaocto		A	A	A	A
European Turtle Dove	*Streptopelia turtur*					
	turtur		A	A	A	A
Oriental Turtle Dove	*Streptopelia orientalis*					
	orientalis		B			
	meena		A			
Mourning Dove	*Zenaida macroura*	U	A			A
Rose-ringed Parakeet	*Psittacula krameri*	U	C1 E*			
Great Spotted Cuckoo	*Clamator glandarius*	M	A	A	A	A
Common Cuckoo	*Cuculus canorus*					
	canorus		A	A	A	A
Black-billed Cuckoo	*Coccyzus erythropthalmus*	M	A	B	B	

			GB	RoI	NI	IoM
Yellow-billed Cuckoo	*Coccyzus americanus*	M	A	A	A	A
Barn Owl	*Tyto alba*		E*			
	alba		A	A	A	A
	guttata		A	A		
Eurasian Scops Owl	*Otus scops*					
	scops		A	A	A	A
Snowy Owl	*Bubo scandiacus*	M	A E	A	A	B
Northern Hawk-owl	*Surnia ulula*					
	ulula		A			
	caparoch		B			
Little Owl	*Athene noctua*					
	vidalii		C1	A	C2	C*
Tawny Owl	*Strix aluco*					
	sylvatica		A			A
Long-eared Owl	*Asio otus*					
	otus		A	A	A	A
Short-eared Owl	*Asio flammeus*					
	flammeus		A	A	A	A
Boreal Owl/Tengmalm's Owl	*Aegolius funereus*					
	funereus		A			
European Nightjar	*Caprimulgus europaeus*					
	europaeus		A	A	A	A
Red-necked Nightjar	*Caprimulgus ruficollis*					
	ruficollis		B			
Egyptian Nightjar	*Caprimulgus aegyptius*	U	A			
Common Nighthawk	*Chordeiles minor*					
	minor		A	A		
Chimney Swift	*Chaetura pelagica*	M	A	A		
White-throated Needletail	*Hirundapus caudacutus*					
	caudacutus		A	A		
Common Swift	*Apus apus*					
	apus		A	A	A	A
Pallid Swift	*Apus pallidus*	U	A	A	B	
Fork-tailed Swift/Pacific Swift	*Apus pacificus*					
	pacificus		A			
Alpine Swift	*Apus melba*					
	melba		A	A	A	A
Little Swift	*Apus affinis*	U	A	A		
Common Kingfisher	*Alcedo atthis*					
	ispida		A	A	A	A
Belted Kingfisher	*Megaceryle alcyon*	M	A	A	A	
Blue-cheeked Bee-eater	*Merops persicus*					
	persicus		A			
European Bee-eater	*Merops apiaster*	M	A	A	A	A

			GB	RoI	NI	IoM
European Roller	*Coracias garrulus*					
	garrulus		A	A	A	A
Eurasian Hoopoe	*Upupa epops*		E			
	epops		A	A	A	A
Eurasian Wryneck	*Jynx torquilla*					
	torquilla		A	A	A	A
European Green Woodpecker	*Picus viridis*					
	viridis		A	B		A
Yellow-bellied Sapsucker	*Sphyrapicus varius*	M	A	A		
Great Spotted Woodpecker	*Dendrocopos major*					
	anglicus		A		A	A
	major		A	A	A	A
Lesser Spotted Woodpecker	*Dendrocopos minor*					
	comminutus		A			
Eastern Phoebe	*Sayornis phoebe*	M	A			
Alder Flycatcher	*Empidonax alnorum*	M	P			
Yellow-throated Vireo	*Vireo flavifrons*	M	A			
Philadelphia Vireo	*Vireo philadelphicus*	M	A	A		
Red-eyed Vireo	*Vireo olivaceus*	M	A	A		
Eurasian Golden Oriole	*Oriolus oriolus*					
	oriolus		A	A	A	A
Brown Shrike	*Lanius cristatus*	U	A	A		
Isabelline Shrike	*Lanius isabellinus*			A		
	isabellinus		A			
	phoenicuroides		A			
Red-backed Shrike	*Lanius collurio*					
	collurio		A	A	A	A
Long-tailed Shrike	*Lanius schach*	U	A			
Lesser Grey Shrike	*Lanius minor*	M	A	A		
Great Grey Shrike	*Lanius excubitor*					
	excubitor		A	A	A	A
Southern Grey Shrike	*Lanius meridionalis*		E			
	pallidirostris		A			A
Woodchat Shrike	*Lanius senator*					
	senator		A	A	A	A
	badius		A	A		
Masked Shrike	*Lanius nubicus*	M	A			
Red-billed Chough	*Pyrrhocorax pyrrhocorax*					
	pyrrhocorax		A E	A	A	A
Black-billed Magpie	*Pica pica*					
	pica		A	A	A	A
Eurasian Jay	*Garrulus glandarius*					
	rufitergum		A			A
	hibernicus			A	A	
	glandarius		A			

		GB	RoI	NI	IoM
Spotted Nutcracker	*Nucifraga caryocatactes*				
	caryocatactes	B			
	macrorhynchos	A			
Western Jackdaw	*Corvus monedula*				
	spermologus	A	A	A	A
	monedula	A			
	soemmerringii		A	A	A
Rook	*Corvus frugilegus*				
	frugilegus	A	A	A	A
Carrion Crow	*Corvus corone*				
	corone	A	A	A	A
Hooded Crow	*Corvus cornix*				
	cornix	A	A	A	A
Northern Raven	*Corvus corax*				
	corax	A	A	A	A
Goldcrest	*Regulus regulus*				
	regulus	A	A	A	A
Common Firecrest	*Regulus ignicapilla*				
	ignicapilla	A	A	A	A
Eurasian Penduline Tit	*Remiz pendulinus*				
	pendulinus	A			
Blue Tit	*Cyanistes caeruleus*				
	obscurus	A	A	A	A
	caeruleus	A			
Great Tit	*Parus major*				
	newtoni	A	A	A	A
	major	A			A
European Crested Tit	*Lophophanes cristatus*				
	scoticus	A			
	cristatus	B			
	mitratus	B			
Coal Tit	*Periparus ater*				
	britannicus	A			A
	hibernicus		A	A	
	ater	A	A		
Willow Tit	*Poecile montana*				
	kleinschmidti	A			A
	borealis	B			
Marsh Tit	*Poecile palustris*				
	dresseri	A	A		A
Bearded Reedling	*Panurus biarmicus*				
	biarmicus	A	A		A
Calandra Lark	*Melanocorypha calandra* U	A			A
Bimaculated Lark	*Melanocorypha bimaculata* U	A			
White-winged Lark	*Melanocorypha leucoptera* M	A			
Black Lark	*Melanocorypha yeltoniensis* M	A			

			GB	RoI	NI	IoM
Greater Short-toed Lark	*Calandrella brachydactyla*	U	A	A	A	
Lesser Short-toed Lark	*Calandrella rufescens*	U	A			
Crested Lark	*Galerida cristata*					
	cristata		A E			
Woodlark	*Lullula arborea*					
	arborea		A	A	B	A
Eurasian Skylark	*Alauda arvensis*					
	arvensis		A	A	A	A
Horned Lark/Shore Lark	*Eremophila alpestris*					
	flava		A	A	A	A
Sand Martin	*Riparia riparia*					
	riparia		A	A	A	A
Tree Swallow	*Tachycineta bicolor*	M	A			
Purple Martin	*Progne subis*	U	A	D3		
Eurasian Crag Martin	*Ptyonoprogne rupestris*	M	A			
Barn Swallow	*Hirundo rustica*					
	rustica		A E	A	A	A
Common House Martin	*Delichon urbicum*					
	urbicum		A	A	A	A
Red-rumped Swallow	*Cecropis daurica*					
	rufula		A	A	A	A
American Cliff Swallow	*Petrochelidon pyrrhonota*	U	A	A		
Cetti's Warbler	*Cettia cetti*					
	cetti		A	A		
Long-tailed Tit	*Aegithalos caudatus*					
	rosaceus		A	A	A	A
	caudatus		A			A
	europaeus		B			
Green Warbler	*Phylloscopus nitidus*	M	A			
Greenish Warbler	*Phylloscopus trochiloides*					
	viridanus		A	A		A
	plumbeitarsus		A			
Arctic Warbler	*Phylloscopus borealis*					
	borealis		A	A		
Pallas's Leaf Warbler	*Phylloscopus proregulus*	M	A	A		A
Yellow-browed Warbler	*Phylloscopus inornatus*	M	A	A	A	A
Hume's Leaf Warbler	*Phylloscopus humei*					
	humei		A	A		
Radde's Warbler	*Phylloscopus schwarzi*	M	A	A		
Dusky Warbler	*Phylloscopus fuscatus*					
	fuscatus		A	A		A
Western Bonelli's Warbler	*Phylloscopus bonelli*	M	A	A		A
Eastern Bonelli's Warbler	*Phylloscopus orientalis*	M	A			

			GB	**RoI**	**NI**	**IoM**
Wood Warbler	*Phylloscopus sibilatrix*	M	A	A	A	A
Common Chiffchaff	*Phylloscopus collybita*					
	collybita		A	A	A	A
	abietinus		A	A	A	A
	fulvescens		A			
	tristis		A	A	A	A
Iberian Chiffchaff	*Phylloscopus ibericus*	M	A			
Willow Warbler	*Phylloscopus trochilus*					
	trochilus		A	A	A	A
	acredula		A	A	A	A
Eurasian Blackcap	*Sylvia atricapilla*					
	atricapilla		A	A	A	A
Garden Warbler	*Sylvia borin*					
	borin		A	A	A	A
Barred Warbler	*Sylvia nisoria*					
	nisoria		A	A	A	A
Lesser Whitethroat	*Sylvia curruca*					
	curruca		A	A	A	A
	blythi		A	A		
Orphean Warbler	*Sylvia hortensis*					
	hortensis		A			
Asian Desert Warbler	*Sylvia nana*	M	A			
Common Whitethroat	*Sylvia communis*					
	communis		A	A	A	A
Spectacled Warbler	*Sylvia conspicillata*	U	A			
Dartford Warbler	*Sylvia undata*					
	dartfordiensis		A	A		
Marmora's Warbler	*Sylvia sarda*	U	A			
Rüppell's Warbler	*Sylvia rueppelli*	M	A			
Subalpine Warbler	*Sylvia cantillans*					
	cantillans		A	A	A	A
	albistriata		A			A
Sardinian Warbler	*Sylvia melanocephala*					
	melanocephala		A	A		A
Pallas's Grasshopper Warbler	*Locustella certhiola*					
	rubescens		A	A		
Lanceolated Warbler	*Locustella lanceolata*	M	A			
Common Grasshopper Warbler	*Locustella naevia*					
	naevia		A	A	A	A
River Warbler	*Locustella fluviatilis*	M	A			
Savi's Warbler	*Locustella luscinioides*					
	luscinioides		A	A		
Eastern Olivaceous Warbler	*Hippolais pallida*					
	elaeica		A	A		

			GB	RoI	NI	IoM
Booted Warbler	*Hippolais caligata*	M	A	A		
Sykes's Warbler	*Hippolais rama*	M	A	A		
Olive-tree Warbler	*Hippolais olivetorum*	M	A			
Icterine Warbler	*Hippolais icterina*	M	A	A	A	A
Melodious Warbler	*Hippolais polyglotta*	M	A	A	A	A
Aquatic Warbler	*Acrocephalus paludicola*	M	A	A		A
Sedge Warbler	*Acrocephalus schoenobaenus* M	A	A	A	A	
Paddyfield Warbler	*Acrocephalus agricola*	M	A	A		
Blyth's Reed Warbler	*Acrocephalus dumetorum*	M	A			
Marsh Warbler	*Acrocephalus palustris*	M	A	A		A
Eurasian Reed Warbler	*Acrocephalus scirpaceus*					
	scirpaceus		A	A	A	A
Great Reed Warbler	*Acrocephalus arundinaceus*					
	arundinaceus		A	A		
Thick-billed Warbler	*Acrocephalus aedon*	U	A			
Zitting Cisticola	*Cisticola juncidis*	U	A	A		
Cedar Waxwing	*Bombycilla cedrorum*	M	A			
Bohemian Waxwing	*Bombycilla garrulus*					
	garrulus		A E	A	A	A
Wallcreeper	*Tichodroma muraria*	U	A			
Red-breasted Nuthatch	*Sitta canadensis*	M	A			
Eurasian Nuthatch	*Sitta europaea*					
	caesia		A			A
Eurasian Treecreeper	*Certhia familiaris*					
	britannica		A	A	A	A
	familiaris		A			
Short-toed Treecreeper	*Certhia brachydactyla*	U	A			
Winter Wren	*Troglodytes troglodytes*					
	troglodytes		A			
	zetlandicus		A			
	fridariensis		A			
	hebridensis		A			
	hirtensis		A			
	indigenus		A	A	A	A
Northern Mockingbird	*Mimus polyglottos*	U	A E			
Brown Thrasher	*Toxostoma rufum*					
	rufum		A			
Grey Catbird	*Dumetella carolinensis*	M	A	A		
Common Starling	*Sturnus vulgaris*					
	vulgaris		A	A	A	A
	zetlandicus		A			
Rosy Starling	*Pastor roseus*	M	A E	A	A	A

			GB	RoI	NI	IoM
White-throated Dipper	*Cinclus cinclus*					
		cinclus	A	A		A
		hibernicus	A	A	A	A
		gularis	A			
Scaly/White's Thrush	*Zoothera dauma*					
		aurea	A	A	A	
Siberian Thrush	*Zoothera sibirica*	U	A E	A		
Varied Thrush	*Ixoreus naevius*	U	A			
Wood Thrush	*Hylocichla mustelina*	M	A			
Hermit Thrush	*Catharus guttatus*	U	A	A		
Swainson's Thrush	*Catharus ustulatus*		A			
		swainsonii		A		
Grey-cheeked Thrush	*Catharus minimus*	U	A	A		
Veery	*Catharus fuscescens*	U	A	A		
Ring Ouzel	*Turdus torquatus*					
		torquatus	A	A	A	A
		alpestris or *amicorum*	A			
Common Blackbird	*Turdus merula*					
		merula	A	A	A	A
Eyebrowed Thrush	*Turdus obscurus*	M	A			
Dusky Thrush	*Turdus eunomus*	M	A			
Naumann's Thrush	*Turdus naumanni*	M	A			
Black-throated Thrush	*Turdus atrogularis*	M	A			
Red-throated Thrush	*Turdus ruficollis*	M	A			
Fieldfare	*Turdus pilaris*	M	A	A	A	A
Song Thrush	*Turdus philomelos*					
		philomelos	A			A
		clarkei	A	A	A	A
		hebridensis	A	A		A
Redwing	*Turdus iliacus*					
		iliacus	A	A	A	A
		coburni	A	A	A	A
Mistle Thrush	*Turdus viscivorus*					
		viscivorus	A	A	A	A
American Robin	*Turdus migratorius*					
		migratorius	A E	A	A	
Asian Brown Flycatcher	*Muscicapa dauurica*	U	A			
Spotted Flycatcher	*Muscicapa striata*					
		striata	A	A	A	A
Rufous-tailed Scrub Robin	*Cercotrichas galactotes*					
		galactotes	A	A		
		syriacus or *familiaris*		A		
European Robin	*Erithacus rubecula*					
		melophilus	A	A	A	A
		rubecula	A	A	A	A

			GB	RoI	NI	IoM
Rufous-tailed Robin	*Luscinia sibilans*	M	A			
Thrush Nightingale	*Luscinia luscinia*	M	A	A		A
Common Nightingale	*Luscinia megarhynchos*					
	megarhynchos		A	A	A	A
	golzii		A			
Siberian Rubythroat	*Luscinia calliope*	M	A			
Bluethroat	*Luscinia svecica*					
	svecica		A	A	A	A
	cyanecula		A	A		
Siberian Blue Robin	*Luscinia cyane*	U	A			
Red-flanked Bluetail	*Tarsiger cyanurus*	M	A E			
White-throated Robin	*Irania gutturalis*	M	A			A
Black Redstart	*Phoenicurus ochruros*					
	gibraltariensis		A	A	A	A
Common Redstart	*Phoenicurus phoenicurus*					
	phoenicurus		A	A	A	A
	samamisicus		A			
Moussier's Redstart	*Phoenicurus moussieri*	M	A			
Whinchat	*Saxicola rubetra*	M	A	A	A	A
Eurasian Stonechat	*Saxicola torquatus*					
	hibernans		A	A	A	A
	maurus		A			A
	maurus/stejnegeri		A	A		
	variegatus		A			
Isabelline Wheatear	*Oenanthe isabellina*	M	A	A		
Northern Wheatear	*Oenanthe oenanthe*					
	oenanthe		A	A	A	A
	leucorhoa		A	A	A	A
Pied Wheatear	*Oenanthe pleschanka*	M	A	A		
Black-eared Wheatear	*Oenanthe hispanica*					A
	hispanica		A	A		
	melanoleuca		A			
Desert Wheatear	*Oenanthe deserti*			A		A
	deserti		B			
	homochroa		A			
	atrogularis		B			
White-crowned Wheatear	*Oenanthe leucopyga*	U	A			
'Black' Wheatear (unidentified)	*Oenanthe leucopyga/ leucura*			A		
Rufous-tailed Rock Thrush	*Monticola saxatilis*	M	A	A		
Blue Rock Thrush	*Monticola solitarius*	U	A E			
Red-breasted Flycatcher	*Ficedula parva*	M	A	A	A	A
Taiga Flycatcher	*Ficedula albicilla*	M	A			
Collared Flycatcher	*Ficedula albicollis*	M	A			

			GB	**RoI**	**NI**	**IoM**
European Pied Flycatcher	*Ficedula hypoleuca*					
	hypoleuca		A	A	A	A
Dunnock	*Prunella modularis*					
	hebridium		A	A	A	
	occidentalis		A			A
	modularis		A			
Alpine Accentor	*Prunella collaris*					
	collaris		A			
Rock Sparrow	*Petronia petronia*	U	A			
House Sparrow	*Passer domesticus*					
	domesticus		A	A	A	A
Spanish Sparrow	*Passer hispaniolensis*	U	A			
Eurasian Tree Sparrow	*Passer montanus*					
	montanus		A	A	A	A
Yellow Wagtail	*Motacilla flava*					
	flava		A	A	A	A
	flavissima		A	A	A	A
	cinereocapilla		A		A	
	thunbergi		A	A		A
	beema		B			
	feldegg		A			
	simillima		A			
Citrine Wagtail	*Motacilla citreola*	U	A	A		
Grey Wagtail	*Motacilla cinerea*					
	cinerea		A	A	A	A
White Wagtail/Pied Wagtail	*Motacillla alba*					
	yarrellii		A	A	A	A
	alba		A	A	A	A
	leucopsis		A			
Richard's Pipit	*Anthus richardi*	M	A	A	A	A
Blyth's Pipit	*Anthus godlewskii*	M	A			
Tawny Pipit	*Anthus campestris*					
	campestris		A	A	A	A
Olive-backed Pipit	*Anthus hodgsoni*					
	yunnanensis		A	A		A
Tree Pipit	*Anthus trivialis*					
	trivialis		A	A	A	A
Pechora Pipit	*Anthus gustavi*					
	gustavi		A	A		A
Meadow Pipit	*Anthus pratensis*					
	whistleri		A	A	A	
	pratensis		A	A	A	A
Red-throated Pipit	*Anthus cervinus*	M	A	A	A	A
Eurasian Rock Pipit	*Anthus petrosus*					
	petrosus		A	A	A	A
	littoralis		A	A	A	A
Water Pipit	*Anthus spinoletta*					
	spinoletta		A	A	A	A

			GB	RoI	NI	IoM
Buff-bellied Pipit	*Anthus rubescens*					
	rubescens		A	A		
Common Chaffinch	*Fringilla coelebs*		E			
	gengleri		A	A	A	A
	coelebs		A	A	A	A
Brambling	*Fringilla montifringilla*	M	A	A	A	A
European Serin	*Serinus serinus*	M	A	A	A	A
European Greenfinch	*Carduelis chloris*					
	chloris		A E	A	A	A
Citril Finch	*Carduelis citrinella*	M	P			
European Goldfinch	*Carduelis carduelis*					
	britannica		A	A	A	A
Eurasian Siskin	*Carduelis spinus*	M	A	A	A	A
Common Linnet	*Carduelis cannabina*					
	autochthona		A			
	cannabina		A	A	A	A
Twite	*Carduelis flavirostris*					
	pipilans		A	A	A	A
	flavirostris		A			
Lesser Redpoll	*Carduelis cabaret*	M	A	A	A	A
Common Redpoll	*Carduelis flammea*					
	flammea		A	A	A	A
	rostrata		A	A	A	
	islandica		A			
Arctic Redpoll	*Carduelis hornemanni*					
	hornemanni		A	A		
	exilipes		A	P		
Two-barred Crossbill	*Loxia leucoptera*					
	bifasciata		A	A	B	
Red Crossbill/Common Crossbill	*Loxia curvirostra*					
	curvirostra		A	A	A	A
Scottish Crossbill	*Loxia scotica*	M	A			
Parrot Crossbill	*Loxia pytyopsittacus*	M	A			
Trumpeter Finch	*Bucanetes githagineus*	U	A E			
Common Rosefinch	*Carpodacus erythrinus*					
	erythrinus		A	A	A	A
Pine Grosbeak	*Pinicola enucleator*					
	enucleator		A			
Eurasian Bullfinch	*Pyrrhula pyrrhula*					
	pileata		A	A	A	A
	pyrrhula		A	A	A	
Hawfinch	*Coccothraustes coccothraustes*					
	coccothraustes		A	A	A	A
Evening Grosbeak	*Hesperiphona vespertina*	U	A			

			GB	RoI	NI	IoM
Snow Bunting	*Plectrophenax nivalis*					
	nivalis		A	A	A	A
	insulae		A			
Lapland Longspur	*Calcarius lapponicus*					
	lapponicus		A	A	A	A
Summer Tanager	*Piranga rubra*	U	A			
Scarlet Tanager	*Piranga olivacea*	M	A	A	A	
Rose-breasted Grosbeak	*Pheucticus ludovicianus*	M	A	A		
Indigo Bunting	*Passerina cyanea*	M	A E	A		
Eastern Towhee	*Pipilo erythrophthalmus*	U	A			
Lark Sparrow	*Chondestes grammacus*					
	grammacus		A			
Savannah Sparrow	*Passerculus sandwichensis*					
	princeps		A			
	oblitus or *labradorius*		A			
Fox Sparrow	*Passerella iliaca*					
	iliaca				A	
Song Sparrow	*Melospiza melodia*	U	A E			A
White-crowned Sparrow	*Zonotrichia leucophrys*	U	A E	A		
White-throated Sparrow	*Zonotrichia albicollis*	M	A E	A	A	A
Dark-eyed Junco	*Junco hyemalis*	U	A E	A	A	
Black-faced Bunting	*Emberiza spodocephala*					
	spodocephala		A E			
Pine Bunting	*Emberiza leucocephalos*					
	leucocephalos		A	A		
Yellowhammer	*Emberiza citrinella*					
	caliginosa		A	A	A	A
	citrinella		A			
Cirl Bunting	*Emberiza cirlus*	M	A	P		A
Rock Bunting	*Emberiza cia*	U	A			
Ortolan Bunting	*Emberiza hortulana*	M	A E	A	A	A
Cretzschmar's Bunting	*Emberiza caesia*	M	A			
Chestnut-eared Bunting	*Emberiza fucata*					
	fucata		A			
Yellow-browed Bunting	*Emberiza chrysophrys*	M	A			
Rustic Bunting	*Emberiza rustica*					
	rustica		A	A	A	A
Little Bunting	*Emberiza pusilla*	M	A	A	A	A
Yellow-breasted Bunting	*Emberiza aureola*					
	aureola		A E	A		
Common Reed Bunting	*Emberiza schoeniclus*					
	schoeniclus		A	A	A	A
Pallas's Reed Bunting	*Emberiza pallasi*	U	A			

			GB	RoI	NI	IoM
Black-headed Bunting	*Emberiza melanocephala*	M	A E	A	A	A
Corn Bunting	*Emberiza calandra* *calandra*		A	A	A	A
Bobolink	*Dolichonyx oryzivorus*	M	A	A		
Brown-headed Cowbird	*Molothrus ater* *ater* or *artemisiae*		A			
Baltimore Oriole	*Icterus galbula*	M	A E	A		A
Black-and-white Warbler	*Mniotilta varia*	M	A	A	A	
Golden-winged Warbler	*Vermivora chrysoptera*	M	A			
Blue-winged Warbler	*Vermivora pinus*	M		A		
Tennessee Warbler	*Vermivora peregrina*	M	A			
Northern Parula	*Parula americana*	M	A E	A		
American Yellow Warbler	*Dendroica petechia* *aestiva*		A	A		
Chestnut-sided Warbler	*Dendroica pensylvanica*	M	A			
Blackburnian Warbler	*Dendroica fusca*	M	A			
Cape May Warbler	*Dendroica tigrina*	M	A			
Magnolia Warbler	*Dendroica magnolia*	M	A E			
Yellow-rumped Warbler	*Dendroica coronata*	U	A	A		A
Blackpoll Warbler	*Dendroica striata*	M	A E	A		
Bay-breasted Warbler	Dendroica castanea	M	A			
American Redstart	*Setophaga ruticilla*	M	A E	A		
Ovenbird	*Seiurus aurocapilla* *aurocapilla*		A	A		
Northern Waterthrush	*Seiurus noveboracensis*	M	A	A		
Common Yellowthroat	*Geothlypis trichas*	U	A	A		
Hooded Warbler	*Wilsonia citrina*	M	A			
Wilson's Warbler	*Wilsonia pusilla*	U	A			
Canada Warbler	*Wilsonia canadensis*	M		A		

CATEGORY D

The species that follow have been assigned to Category D by the British and Irish committees. Species in this category do not form part of the list of GB, NI, IoM or RoI. Category D is reserved for species where there is reasonable doubt that they have occurred naturally. In the light of recent records or international developments (e.g. Baikal Teal in Denmark), it is expected that all British Category D records will be reviewed by BOURC during 2008–10. Where species have occurred and are known *not* to be of natural origin, they are included in Category E (not listed in this book – see www.bou.org.uk for details).

Ross's Goose   *Anser rossii* Cassin

MONOTYPIC Closely related to Snow Goose, with which it is has hybridised to produce fertile offspring. Typically white, although the downy young are polymorphic, and a blue morph occurs very rarely among adults. Breeds in small numbers along the S & W shores of Hudson Bay, and north to the main colonies along the Perry River. Migratory, wintering in coastal California and across to the Gulf of Mexico. Population increasing in Canada.

Relatively common in captivity in GB&I, birds are known to escape and join up with flocks of feral or wild geese, sometimes surviving for several years. This makes assessment of potential vagrants difficult. As with Snow Goose, no wild-ringed bird has been found in GB&I, though individuals are increasingly found with migrant geese from Greenland and Iceland. Genuine vagrants may occur, though none has yet been formally accepted on either side of the Irish Sea.

Red-breasted Goose   *Branta ruficollis* (Pallas)

MONOTYPIC see main text. Birds in the Republic of Ireland have been assigned to Category D1 on the grounds that there is reasonable doubt that they occurred naturally.

Falcated Duck   *Anas falcata* Georgi

MONOTYPIC Part of a clade that also includes Gadwall and Chiloe, American and Eurasian Wigeons. A migratory species that breeds from Lake Baikal across China to N Japan, and winters south to Vietnam.

A popular bird in wildfowl collections, which makes assessment of vagrancy difficult. A likely vagrant, with a scattering of records across W Asia. Birds occur regularly in Europe, often in association with Eurasian Wigeon. Several records from GB, but BOURC considers that none has yet occurred in circumstances that confirm vagrancy beyond reasonable doubt.

Baikal Teal   *Anas formosa* (Georgi)

MONOTYPIC A very distinctive taxon. A molecular study, based on two mtDNA genes (Johnson & Sorenson 1998, 1999) found that the affinities of Baikal Teal were unclear, except that it was well differentiated from all other *Anas* species. This contrasted with analyses of courtship and post-copulatory displays, both of which indicated a loose affinity with a group including Garganey (Johnson *et al.* 2000). Breeds in E Asia, migrating south to winter in Japan and China.

Baikal Teal has, in the past, been regarded as an unlikely vagrant to GB&I. Recent observations have indicated that it is much commoner in the wild than was previously thought and, despite its popularity in captivity, the likelihood of natural vagrancy cannot be ignored. This was substantiated when a first-winter bird was shot in Denmark in Nov 2005; stable isotope analysis indicated that its old, unmoulted feathers had in all probability grown in the Far East (Fox *et al.* 2007). It was thus unlikely to have been captive-bred in Europe – although the remote chance that it had been transported to the west as an immature and subsequently escaped could not formally be excluded. All existing records of Baikal Teal in GB&I have been placed in Category D on both sides of the Irish Sea but, in view of the Danish bird, GB records are now being re-assessed, especially in relation to age, sex, date of arrival, companions and location.

Marbled Duck   *Marmaronetta angustirostris* (Ménétries)

MONOTYPIC Taxonomy little studied at a molecular level, but Johnson & Sorenson (1999) found it to be basal to, and genetically divergent from, most other ducks. Has a fragmented distribution across the W Palearctic, from Iberia and Morocco through the Mediterranean to the borders of China; some populations migratory, though much movement is more dispersive, and nomadic during droughts (HBW). The Turkish population declined through the 1990s, following loss of habitat associated with drainage and excessive water extraction, and it is now regarded as vulnerable. Populations elsewhere seem to be more stable, but the W European population numbers a maximum of only about 200 pairs (BirdLife International 2004).

Individuals are recorded almost annually in England, often in late summer, when post-breeding dispersal might be expected; this is also the time when captive-bred stock might escape, as either newly fledged juveniles or previously clipped adults that have grown new flight feathers. Consequently, no bird has yet been sufficiently authenticated, and the British records have all been assigned to Category D. There are no records from Ireland.

White-headed Duck   *Oxyura leucocephala* (Scopoli)

MONOTYPIC A detailed analysis of the stiff-tails by McCracken & Sorenson (2005) showed that the Australian Blue-billed *O. australis*, African Maccoa *O. maccoa* and Palearctic White-headed Ducks are sister species, and that these were closer to Ruddy Duck than to the Argentine Lake Duck *O. vittata*. Despite analysing >8,000 base pairs of sequence data, they were unable to resolve the finer relationships of the three Old World taxa. White-headed Duck has a fragmented distribution across the Palearctic, from Iberia to W China; eastern populations migratory, wintering south from Caspian Sea to Pakistan. The Turkish population declined massively during the 1990s (BirdLife International 2004), but that in Iberia increased by *c.* 80% through the same decade. The Iberian population went through a major bottleneck in the 1970s and 1980s, and Muñoz-Fuentes *et al* (2005) found that the contemporary population shows a significant loss of genetic variation compared with museum specimens taken prior to the 1970s. Breeds well in captivity, and stocks have been increased to assist re-introduction programmes in the W Mediterranean.

Recorded annually in Britain, but no bird has yet made its way up from Category D. The increased research in Iberia has resulted in more birds being ringed there, and finding one of these in GB&I would help confirm a natural origin.

## Red-legged Partridge															*Alectoris rufa* (Linnaeus)

POLYTYPIC see main text. Small, feral populations exist in Ireland, but these are believed not to be self-sustaining, so the species is assigned to the Irish Category D4.

## Great White Pelican													*Pelecanus onocrotalus* Linnaeus

MONOTYPIC Breeds at scattered wetland sites across the Palearctic from E Europe to Mongolia, and in sub-Saharan Africa. Palearctic populations migratory, wintering from Iraq to N India (HBW). Wintering grounds of European birds unclear, but birds certainly migrate though SE Turkey and Israel into the Nile Valley.

Birds regularly escape from captivity, and recapture proves difficult or impossible. Many are of known provenance and some survive for months in the wild. No bird has yet been found that can justify movement from Category D.

## Greater Flamingo												*Phoenicopterus roseus* Pallas

MONOTYPIC Morphology suggests that *ruber* and *roseus* should be treated as separate species. Breeds in saline lagoons around the Mediterranean Sea and east to Kazakhstan; further populations breed through the Nile Valley into E & S Africa, and also through Middle East to Indian subcontinent. N populations are migratory, eastern birds disperse south to wetlands of Arabian Sea, Indian Ocean and into Rift Valley, where they may winter alongside Mediterranean birds, though many of these migrate SW into W Africa.

Relatively common in captivity, whence birds escape regularly – even breeding occasionally (e.g. in Netherlands/Germany). However, many thousands have been ringed in S France and Iberia and, despite dispersal northwards in adverse weather, there are almost no sightings of colour-ringed birds in N France. No bird has ever been seen in GB&I bearing French or Spanish rings and, until such an individual arrives, it seems likely that the species will remain in Category D in both GB and Ireland.

## Bald Eagle													*Haliaeetus leucocephalus* (Linnaeus)

POLYTYPIC see main text. Birds escape from captivity from time to time, and because of their size and high profile, usually achieve national media coverage. Although there are acceptable records from Ireland, and natural arrival remains a possibility in Britain, no bird there has yet been judged to be other than of captive origin.

## Booted Eagle													*Aquila pennata* (J F. Gmelin)

MONOTYPIC related to Bonelli's Eagle and the African Hawk-eagles. Polymorphic, with dark and pale forms. Genetics unclear. Breeds in a narrow belt of open woodland from Iberia and NW Africa through the Black and Caspian Seas to Lake Baikal. Strongly migratory, wintering in sub-Saharan Africa and the Indian subcontinent. Dislikes crossing open water, and main migration routes out of Europe via Suez and Gibraltar.

A pale-phase bird was seen at a wide range of locations across Ireland, England and Scotland in Mar 1999– Jun 2000. Recognisable by feather damage where photographs were available, it was evident that only a single bird was involved. Its damaged plumage, the species' dislike of crossing oceans, and an arrival in March (before migration is really underway in Gibraltar) all raised suspicions about its origins. The relevant committees in the Republic, NI and GB independently decided that the bird was probably of captive origin and placed it in their respective Category D.

Saker Falcon   *Falco cherrug* J.E. Gray

POLYTYPIC closely related to Lanner *F. biarmicus* and Laggar Falcons, *F. jugger*. Breeds from C Europe (nominate) across S Russia to Altai, replaced by race *milvipes* east to N China. Northern populations migratory, wintering from SE Europe to E Africa (nominate) and from Arabia to Himalayas and central China (*milvipes*).

Popular in falconry, many captive birds are lost and then 'found' by birders. As powerful fliers, both this and Lanner Falcon must be potential vagrants, but the confusion over falconers' escapes means that none has so far escaped this taint, and the species has been firmly rooted in Category D.

Bridled Tern   *Onychoprion anaethetus* (Scopoli)

POLYTYPIC see main text. A bird of the race *melanopterus* found dead at North Bull Island (Dublin) in Nov 1953 is placed in the Irish Category D3 as a tideline corpse.

Royal Tern   *Sterna maxima* Boddaert

POLYTYPIC see main text. A bird found dead at North Bull Island (Dublin) in Mar 1954 is placed in the Irish Category D3 as a tideline corpse.

Northern Flicker   *Colaptes auratus* (Linnaeus)

POLYTYPIC *Colaptes* forms a clade that also includes the New World *Piculus* (refs in Webb & Moore 2005), suggesting that generic limits may need redefining.

A bird flew ashore from an ocean liner in Cork Harbour in Oct 1962, having landed on the ship during a crossing of the North Atlantic. As a 'ship assisted' bird, it is placed in the Irish Category D2. Photographs establish that it was of the nominate *aureus* ('Northern Flicker') group.

House Crow   *Corvus splendens* Vieillot

POLYTYPIC The molecular phylogenetics of crows have attracted little attention in recent years. Native populations breed from Iran across the Indian subcontinent to Burma. Largely sedentary, but has spread to many coastal areas of Africa and Middle East, especially sea-ports, through deliberate introduction and casual transportation on shipping.

A bird was found in Dunmore East (Waterford) in Nov 1974, having presumably arrived by ship. It survived for almost six years, being last seen in Oct 1980. As a ship-assisted vagrant, it was placed in the Irish Category D2.

Purple Martin   *Progne subis* (Linnaeus)

POLYTYPIC see main text. A bird of this species was shot near Kingstown (now Dun Laoghaire, Dublin) in autumn 1839. The skin is still available, and though the identification is not questioned, there is some doubt as to the authenticity of the specimen and the Irish Rare Birds Committee has placed the bird in the Irish Category D1, indicating uncertainty over its natural arrival.

Daurian Starling   *Agropsar sturninus* (Pallas)

MONOTYPIC Included in the molecular study by Lovette *et al.* (2008), who found this and Chestnut-cheeked Starling *S. philippensis* best placed in separate genus *Agropsar*, or perhaps into *Gracupica* with Black-collared Starling *S. nigricollis* and Pied Myna *S. contra*. Migratory, breeding from E Mongolia to N China, and wintering from S Burma through peninsula Malaysia to Java.

One on Fair Isle (Shetland) in May 1985 was originally admitted to Category A. A review of the record revealed no evidence of westward vagrancy. At the time, the species occurred in captivity and in trade in W Europe, and this led BOURC to move the record into Category D. This bird and a subsequent one at Durness (Sutherland) in 1998 are currently under review.

Mugimaki Flycatcher   *Ficedula mugimaki* (Temminck)

MONOTYPIC In a preliminary study of *Ficedula*, Outlaw & Voelker (2006) found this species to be part of a clade of predominantly Asiatic flycatchers; it appeared to be basal within this, but statistical support was weak. They concluded that 'classical' taxonomic relationships within the genus were not supported by mtDNA data, and suggested that certain aspects of morphology were probably due to ecological and behavioural factors rather than phylogeny.

A bird was present at Sunk Island (E Yorkshire) in Nov 1991. Despite this being a long-distance migrant, there is no pattern of westwards vagrancy, and the species was regularly available in trade at the time. On the strength of these arguments, BOURC felt that there was a possibility of captive origin, and the record was assigned to Category D.

American Goldfinch   *Carduelis tristis* (Linnaeus)

POLYTYPIC Molecular analysis (Arnaiz-Villena *et al.* 1998) found this to be in a clade with the New World taxa Lawrence's Goldfinch *C. lawrencei* and Lesser Goldfinch *C. psaltria*. The affinity of this trio to the rest of *Carduelis* was unclear and differed depending upon the mode of analysis. Breeds across much of N America, between Hudson Bay and N Mexico. Central and northern populations migratory, wintering S to Gulf of Mexico.

A bird shot on Achill Island (Mayo) in Sep 1894 has been assigned to the Irish Category D1 on the grounds that there is reasonable doubt that it arrived naturally.

Red-headed Bunting   *Emberiza bruniceps* Brandt

MONOTYPIC Sister species to Black-headed Bunting (Alström *et al.* 2008), with which, in the past, it was regarded as conspecific. Nested among the *Emberiza* in a clade that also includes Crested Bunting *Melophus lathami*, leading Alström *et al.* to suggest that generic realignment is necessary. Breeds in C Asia, from the Caspian Sea to W Mongolia and NW China. Migratory, wintering in N & W of Indian subcontinent.

For several decades this species occurred quite frequently, often adult males in summer, and in the N & W Isles of Scotland. In view of its popularity in captivity, all records were deemed to be suspect and the species has remained firmly in both British and Irish Category D. Since traffic in this species reduced in the 1980s, the number occurring has fallen (especially of adult males), supporting the view that many were previously of captive origin. There remain several records at migration localities, at appropriate times of year, and sometimes with a supporting cast of Siberian vagrants (Vinicombe 2007). A review of recent British records is under way, which may result in a change of status.

Yellow-headed Blackbird   *Xanthocephalus xanthocephalus* (Bonaparte)

MONOTYPIC Lanyon & Omland (1999) found this to be sister to Bobolink, within a clade that also included the meadowlarks *Sturnella*. Breeds from the Canadian prairies south across the Great Plains of N America to N California. A medium- to long-distance migrant, wintering from S USA and Mexico, to Costa Rica.

There have been several records in GB, but the species was regular in trade in the latter part of the 20th century. Regarded as an unlikely vagrant, BOURC decided that they could not rule out the possibility of captive origin, and the species was added to Category D. There has been a fall in the number in trade in recent years and in the number of records in GB&I.

REFERENCES

To locate the sources of BTO survey results quoted in the texts, see www.bto.org. A few key references are used particularly frequently in the text and are referred to in an abbreviated format:

BS3	Forrester, R., Andrews, I., McInerny, C., Murray, R., McGowan, R., Zonfrillo, B., Betts, M., Jardine, D. and Grundy, D. 2007. *The Birds of Scotland.* 2 vols. SOC, Edinburgh.
BWP	Cramp, S. and others. 1977–1994. *Birds of the Western Palearctic.* 9 vols. OUP, Oxford.
HBW	del Hoyo, J. and others. 1992– *Handbook of the Birds of the World.* Lynx Publications, Barcelona.
RBBP	The reports of the Rare Breeding Birds Panel, as published (usually annually) in *British Birds.*
Seabird 2000	Lloyd, C. S., Tasker, M. L. and Partridge, K. 1991. *The Status of Seabirds in Britain and Ireland.* T and AD Poyser, Calton.
Seafarer	Cramp, S., Bourne, W. R. P. and Saunders, D. 1974. *The Seabirds of Britain and Ireland.* Collins, London.
First Breeding Atlas	Sharrock, J. T. R. 1976. *The Atlas of Breeding Birds in Britain and Ireland.* T and AD Poyser, Berkhamsted.
Second Breeding Atlas	Gibbons, D. W., Reid, J. B. and Chapman, R. A. 1993. *The New Atlas of Breeding Birds in Britain and Ireland: 1988–1991.* T and AD Poyser, London.

Ainslie, J. A. and Atkinson, R. 1937. On the breeding habits of Leach's Fork-tailed Petrel. *British Birds* 30: 234–248.

Alatalo, R. V., Ericsson, D. P., Gustafsson, L. and Sundberg, A. 1990. Hybridisation between Pied and Collared Flycatchers – sexual selection and speciation theory. *Journal of Evolutionary Biology.* 3: 375–389.

Albrecht, J. S. M. 1984. Some notes on the identification, song and habitat of the Green Warbler in the western Black Sea coast of Turkey. *Sandgrouse* 6: 69–75.

Alderman, R., Double, M. C., Valencia, J. and Gales, R. P. 2005. Genetic affinities of newly sampled populations of Wandering and Black-browed Albatross. *Emu* 105: 169–179.

Aliabadian, M., Kaboli, M., Prodon, R., Nijman, V. and Vences, M. 2007. Phylogeny of Palearctic wheatears (genus *Oenanthe*): congruence between morphometric and molecular data. *Molecular Phylogenetics and Evolution* 42: 665–675.

Allan, R. C. 1962. The Madeiran Storm Petrel (*Oceanodroma castro*). *Ibis* 105b: 274–295.

Allen, D., Mellon, C., Enlander, I. and Watson, G. 2004. Lough Neagh diving ducks: recent changes in wintering populations. *Irish Birds* 7: 327–336.

Allende, L. M., Rubio, I., Ruiz del Valle, V., Guillén, J., Martinez-Laso, J., Lowy, E., Varela, P., Zamora, J. and Arnaiz-Villena, A. 2001. The old-world sparrows (genus *Passer*): phylogeography and their relative abundance of nuclear mtDNA pseudogenes. *Journal of Molecular Evolution* 53: 144–154.

Alström, P. 1985. Artsbestämning av storskarv *Phalacrocorax carbo* och toppskarv *P. aristotelis*. *Vår Fågelvärld* 44: 325–350.

Alström, P., Ericson, P. G. P., Olsson, U. and Sundberg, P. 2006. Phylogeny and classification of the avian superfamily *Sylvioidea*, based on nuclear and mitochondrial sequence data. *Molecular Phylogenetics and Evolution* 38: 381–397.

Alström, P., Mild, K. and Zetterström, B. 2003. *Pipits and Wagtails of Europe, Asia and North America.* Christopher Helm, London.

Alström, P. and Odeen, M. 2002. Incongruence between mitochondrial DNA, nuclear DNA and non-molecular data in the avian genus *Motacilla*: implications for estimates of species phylogenies. In P. Alström (ed.),

Species Limits and Systematics in Some Passerine Birds. Acta Universitatis Upsaliensis, Uppsala.

Alström, P. and Olsson, U. 1995. A new species of *Phylloscopus* warbler from Sichuan Province, China. *Ibis* 137: 459–468.

Alström, P., Olsson, U., Lei, F., Wang, H.-T., Gao, W. and Sundberg, P. 2008. Phylogeny and classification of the Old World Emberizini (Aves, Passeriformes). *Molecular Phylogenetics and Evolution* 47: 960–973.

Amadon, D. and Short, L. L. 1992. Taxonomy of lower categories – suggested guidelines. *Bulletin of the British Ornithologists' Club Centenary Suppl.* 112A: 11–38.

Amar, A., Hewson, C. M., Smith, K. W., Fuller, R. J., Lindsell, J. A., Conway, G., Butler, S. and MacDonald, M. A. 2006. *What's Happening to our Woodland Birds? Long-term Changes in the Populations of Woodland Birds.* BTO Research Report No. 169, RSPB Research Report No. 19. BTO, Thetford.

Andersson, M. 1973. Behaviour of the Pomarine Skua *Stercorarius pomarinus* Temm. with comparative remarks on Stercorariinae. *Ornis Scandinavica* 4: 1–16.

Andersson, M. 1999a. Hybridisation and skua phylogeny. *Proceedings of the Royal Society of London* 266: 1579–1585.

Andersson, M. 1999b. Phylogeny, behaviour, plumage evolution and neoteny in skuas Stercorariidae. *Journal of Avian Biology* 30: 205–215.

Andrew, D. G. 1997. The earlier breeding records of Icterine Warbler in England. *British Birds* 90: 187–189.

Arcese, P., Sogge, M. K., Marr, A. B. and Patten, M. A. 2002. Song Sparrow (*Melospiza melodia*). In A. Poole and F. Gill (eds), *The Birds of North America, No. 704.* The Birds of North America, Inc., Philadelphia, PA.

Arctander, P., Folmer, O. and Fjeldså, J. 1996. The phylogenetic relationships of Berthelot's Pipit (*Anthus berthelotti*) illustrated by DNA sequence data, with remarks on the genetic distance between Rock and Water Pipits (*A. spinoletta*). *Ibis* 138: 263–272.

Armstrong, I. H., Coulson, J. C., Hawkey, P. and Hudson, M. J. 1978. Further mass seabird deaths from paralytic shellfish poisoning. *British Birds* 71: 58–68.

Arnaiz-Villena, A., Álvarez-Tejado, M., Ruíz-

del-Valle, V., García-de-la-Torre, C., Varela, P., Recio, M. J., Ferre, S. and Martinez-Laso, J. 1998. Phylogeny and rapid northern and southern hemisphere speciation of goldfinches during the Miocene and Pliocene epochs. *Cellular and Molecular Life Sciences* 54: 1031–1041.

Arnaiz-Villena, A., Guillén, J., Ruíz-del-Valle, V., Lowy, E., Zamora, J., Varela, P., Stefani, D. and Allende, L. M. 2001. Phylogeography of crossbills, bullfinches, grosbeaks, and rosefinches. *Cellular and Molecular Life Sciences* 58: 1159–1166.

Arnaiz-Villena, A., Moscoso, J., Ruiz-del-Valle, V., Gonzalez, J., Reguera, R., Wink, M. I. and Serrano-Vda, J. 2007. Bayesian phylogeny of Fringillinae birds: status of the singular African oriole finch *Linurgus olivaceus* and evolution and heterogeneity of the genus *Carpodacus. Acta Zoologica Sinica* 53: 826–834.

Atkinson-Willes, G. L. and Matthews, G. V. T. 1960. The past status of the Brent Goose. *British Birds* 53: 352–357.

Austin, G. E., Rehfisch, M. M., Allan, J. R. and Holloway, S. J. 2007. Population size and differential population growth of introduced Greater Canada *Branta canadensis* and re-established Greylag Goose *Anser anser* across habitats in Great Britain in the year 2000. *Bird Study* 54: 343–352.

Austin, J. E., Custer, C. M. and Afton, A. D. 1998. Lesser Scaup (*Aythya affinis*). In A. Poole and F. Gill (eds), *The Birds of North America, No. 338.* The Birds of North America, Inc., Philadelphia, PA.

Austin, J. J. 1996. Molecular phylogenetics of *Puffinus* shearwaters: preliminary evidence from mitochondrial cytochrome b gene sequences. *Molecular Phylogenetics and Evolution* 6: 77–88.

Austin, J. J., Bretagnolle, V. and Pasquet, E. 2004. A global molecular phylogeny of the small *Puffinus* shearwaters and implications for systematics of the Little–Audubon's Shearwater complex. *Auk* 121: 847–864.

Avise, J. C., Ankney, C. D. and Nelson, W. S. 1990. Mitochondrial gene trees and the evolutionary relationship of Mallard and Black Ducks. *Evolution* 44: 1109–1119.

Avise, J. C., Nelson, W. S., Bowen, B. W. and Walker, D. 2000. Phylogeography of colonially nesting seabirds, with special reference to global matrilineal patterns in the sooty tern (*Sterna fuscata*). *Molecular Ecology* 9: 1783–1792.

Avise, J. C. and Zink, R. M. 1988. Molecular genetic divergence between avian sibling species: King and Clapper Rails, Long-billed and Short-billed Dowitchers, Boat-tailed and Great-tailed Grackles, and Tufted and Black-crested Titmice. *Auk* 105: 516–528.

Baccetti, N., Massa, B. and Violani, C. 2007. Proposed synonymy of *Sylvia cantillans moltonii* Orlando, 1937, with *Sylvia cantillans subalpina* Temminck, 1820. *Bulletin of the British Ornithologists' Club* 127: 107–110.

Bagenal, T. B. and Baird, D. E. 1958. The birds of North Rona in 1958, with notes on Sula Sgeir. *Bird Study* 6: 153–174.

Baillie, S. R., Marchant, J. H., Crick, H. Q. P., Noble, D. G., Balmer, D. E., Coombes, R. H., Downie, I. S., Freeman, S. N., Joys, A. C., Leech, D. I., Raven, M. J., Robinson, R. A. and Thewlis, R. M. 2006. *Breeding Birds in the Wider Countryside: their Conservation Status 2005*. BTO Research Report No. 435. BTO, Thetford.

Baillie, S. R. and Peach, W. J. 1992. Population limitation in Palaearctic–African migrant passerines. *Ibis* 134 Suppl. 1: 120–132.

Baker, A. J., Dennison, M. D., Lynch, A. and Le Grand, G. 1990. Genetic divergence in peripherally isolated populations of Chaffinches in the Atlantic islands. *Evolution* 44: 981–999.

Baker, A. J., Pereira, S. L. and Paton, T. A. 2007. Phylogenetic relationships and divergence times of Charadriiformes genera: multigene evidence for the Cretaceous origin of at least 14 clades of shorebirds. *Biology Letters* 3: 205–209.

Baker, H., Stroud, D. A., Aebischer, N. J., Cranswick, P. A., Gregory, R. D., McSorley, C. A., Noble, D. G. and Rehfisch, M. M. 2006. Population estimates of birds in Great Britain and the United Kingdom. *British Birds* 99: 25–44.

Baker, J. K. and Catley, G. P. 1987. Yellow-browed Warblers in Britain and Ireland, 1968–85. *British Birds* 80: 93–109.

Baker, K. 1977. Westward vagrancy of Siberian passerines in autumn 1975. *Bird Study* 24: 233–242.

Baker, K. 1997. *Warblers of Europe, Asia and North Africa*. Christopher Helm, London.

Ball, R. M., Jr and Avise, J. C. 1992. Mitochondrial DNA phylogeographic differentiation among avian populations and the evolutionary significance of subspecies. *Auk* 109: 626–636.

Baltz, M. E. and Latta, S. C. 1998. Cape May Warbler (*Dendroica tigrina*). In A. Poole and F. Gill (eds), *The Birds of North America, No. 332*. The Birds of North America, Inc., Philadelphia, PA.

Banks, A., Collier, M., Austin, G., Hearn, R. and Musgrove, A. 2006. *Waterbirds in the UK 2004/5: the Wetlands Bird Survey*. BTO/WWT/RSPB/JNCC, Thetford.

Banks, A. N., Burton, N. H. K., Calladine, J. R. and Austin, G. E. 2007. *Winter Gulls in the UK: Population Estimates from the 2003/04–2005/06 Winter Gull Roost Survey*. BTO Research Report No. 456 to English Nature, the Countryside Council for Wales, Scottish Natural Heritage, the Environment and Heritage Service (Northern Ireland), the Joint Nature Conservation Committee and Northumbrian Water Ltd. BTO, Thetford.

Banks, R. C., Chesser, R. T., Cicero, C., Dunn, J. L., Kratter, A. W., Lovette, I. J., Rasmussen, P. C., Remsen, J. V., Rising, J. D. and Stotz, D. F. 2007. Forty-eighth supplement to the American Ornithologists' Union Check-list of North American Birds. *Auk* 124: 1109–1115.

Banks, R. C., Cicero, C., Dunn, J. L., Kratter, A. W., Rasmussen, P. C., Remsen, J. V., Rising, J. D. and Stotz, D. F. 2003. Forty-fourth supplement to the American Ornithologists' Union Check-list of North American Birds. *Auk* 120: 923–931.

Banks, A. N., Coombes, R. H. and Crick, H. Q. P. 2003. *The Peregrine Falcon Breeding Population of the UK and Isle of Man in 2002*. Research Report 330. BTO, Thetford.

Baptista, L. F. and Krebs, R. 2000. Vocalizations and relationships of Brown Creepers *Certhia americana*: a taxonomic mystery. *Ibis* 142: 457–465.

Barker, F. K. 2004. Monophyly and relationships of wrens (Aves: Troglodytidae): a congruence analysis of heterogeneous mitochondrial and nuclear DNA sequence data. *Molecular Phylogenetics and Evolution* 31: 486–504.

Barker, F. K., Barrowclough, G. and Groth, J. G. 2002. A phylogenetic hypothesis for passerine birds: taxonomic and biogeographic implications of an analysis of nuclear DNA sequence data. *Proceedings of the Royal Society of London B* 269: 295–308.

Barker, F. K., Cibois, A., Schikler, P., Feinstein, J. and Cracraft, J. 2004. Phylogeny and diversification of the largest avian radiation. *Proceedings of the National Academy of Sciences of the USA* 101: 11040–11045.

Barrowclough, G. F., Groth, J. G. and Mertz, L. A. 2006. The RAG-1 exon in the avian order Caprimulgiformes: phylogeny, heterozygosity and base composition. *Molecular Phylogenetics and Evolution* 41: 238–248.

Batten, L. A. 1976. Bird communities of some Killarney woodlands. *Proceedings of the Royal Irish Academy* 76: 285–313.

Baxter, E. V. and Rintoul, L. J. 1953 *The Birds of Scotland*. Oliver and Boyd, Edinburgh.

Beale, C. M., Burfield, I. J., Sim, I. M. W., Rebecca, G. W., Pearce-Higgins, J. W. and Grant, M. C. 2006. Climate change may account for the decline in British Ring Ouzels *Turdus torquatus*. *Journal of Animal Ecology* 75: 826–835.

Bearhop, S., Fiedler, W., Furness, R. W., Votier, S. C., Waldron, S., Newton, J., Bowen, G. J., Berthold, P. and Farnsworth, K. 2005. Assortative mating as a mechanism for rapid evolution of a migratory divide. *Science* 310: 502–504.

Beasley, I., Robertson, K. M. and Arnold, P. 2005. Description of a new dolphin, the Australian snubfin dolphin *Orcaella heinsohni* sp. n. (Cetacea, Delphinidae). *Marine Mammal Science* 21: 365–400.

Belik, V. P. 2005. The Sociable Lapwing in Eurasia: what does the future hold? *British Birds* 98: 476–485.

Bellamy, P. E., Brown, N. J., Enoksson, B., Firbank, L. G., Fuller, R. J., Hinsley, S. A. and Schotman, A. G. M. 1998. The influences of habitat, landscape structure and climate on local distribution patterns of the nuthatch (*Sitta europaea* L.). *Oecologia* 115: 127–136.

Benkman C. W. 1992. White-winged Crossbill. In A. Poole, P. Stettenheim and F. Gill (eds), *The Birds of North America, No. 27*. The Academy of Natural Sciences, Philadelphia, and The American Ornithologists' Union, Washington, DC.

Bensch, S., Irwin, D. E., Irwin, J. H., Kvist, L. and Åkesson, S. 2006. Conflicting patterns of mitochondrial and nuclear DNA diversity in *Phylloscopus* warblers. *Molecular Ecology* 15: 161–171.

Benz, B. W., Robbins, M. B. and Peterson, A. T. 2006. Evolutionary history of woodpeckers and allies (Aves: Picidae): Placing key taxa on the phylogenetic tree. *Molecular Phylogenetics and Evolution* 40: 389–399.

Beresford, P., Barker, F. K., Ryan, P. G. and Crowe, T. M. 2005. African endemics span the tree of songbirds (Passeri): molecular systematics of several evolutionary 'enigmas'. *Proceedings of the Royal Society of London B* 272: 849–858.

Berthold, P. 1995. Microevolution of migratory behaviour illustrated by the Blackcap *Sylvia atricapilla*: 1993 Witherby Lecture. *Bird Study* 42: 89–100.

Berthold, P. 2001. *Bird Migration: a General Survey*. 2nd Edition. Oxford University Press, Oxford.

Bevier, L., Poole, A. F. and Moskoff, W. 2004. Veery (*Catharus fuscescens*). *The Birds of North America Online*. (Poole, A., ed.) Ithaca: Cornell Laboratory of Ornithology; Retrieved from The Birds of North American Online database.

Bibby, C. J. and Etheridge, B. 1993. Status of the Hen Harrier *Circus cyaneus* in Scotland in 1988–89. *Bird Study* 40: 1–11.

Bibby, C. J. and Nattrass, M. 1986. Breeding status of the Merlin in Britain. *British Birds* 79: 170–185.

Bildstein, K. L., Bechard, M. J., Farmer, C. and Newcomb, L. 2009. Narrow sea crossings present major obstacles to migrating Griffon Vultures *Gyps fulvus*. *Ibis* 151: 382–391.

BirdLife International 2004. *Birds in Europe: Population Estimates, Trends and Conservation Status*. BirdLife Conservation Series No. 12. BirdLife International, Cambridge UK.

Biswas, B. 1962. Further notes on the shrikes *Lanius tephronotus* and *Lanius schach*. *Ibis* 104: 112–115.

Björklund, M. 1994. Phylogenetic relationships among Charadriiformes: reanalysis of previous data. *Auk* 111: 825–832.

Blondel, J., Catzeflis, F. and Perret, P. 1996. Molecular phylogeny and the historical biogeography of the warblers of the genus *Sylvia* (Aves). *Journal of Evolutionary Biology* 9: 871–891.

Bocheński, Z. M. 1994. The comparative osteology of grebes (Aves: Podicipediformes) and its systematic implications. *Acta Zoologica Cracoviensia* 37: 191–346.

Boere, G. C., Galbraith, C. A. and Stroud, D. A. (eds) 2006. *Waterbirds around the World*. The Stationery Office, Edinburgh, UK.

Bolton, M., Smith, A. L., Gómez-Diaz, E., Friesen, V. L., Medeiros, R., Bried, J., Roscales, J. L. and Furness, R. W. 2008. Monteiro's Storm-petrel *Oceanodroma monteiroi*: a new species from the Azores. *Ibis* 150: 717–727.

Borowik, O. A. and McLennan, D. A. 1999. Phylogenetic patterns of parental care in Calidrine sandpipers. *Auk* 116: 1107–1117.

Both, C., Bouwhuis, S., Lessells, C. M. and Visser, M. E. 2006. Climate change and population declines in a long-distance migratory bird. *Nature* 441: 81–83.

Boudewijn, T. J. and Dirksen, S. 1995. Impact of contaminants on the breeding success of the Cormorant *Phalacrocorax carbo sinensis* in the Netherlands. *Ardea* 83: 325–338.

Boulet, M. and Gibbs, H. L. 2006. Lineage origin and expansion of a Neotropical migrant songbird after recent glaciation events. *Molecular Ecology* 15: 2505–2525.

Boulet, M., Potvin, C., Shaffer, F., Breault, A. and Bernatchez, L. 2005. Conservation genetics of the threatened horned grebe (*Podiceps auritus* L.) population of the Magdalen Islands, Québec. *Conservation Genetics* 6: 539–550.

Bourne, W. R. P. 1983. The Soft-plumaged Petrel, the Gon-gon and the Freira *Pterodroma mollis*, *P. feae* and *P. madeira*. *Bulletin of the British Ornithologists' Club* 103: 52–58.

Bourne, W. R. P. 1992. Leach's Storm-petrels visiting ships at sea. *British Birds* 85: 556–557.

Bourne, W. R. P. 2008. Storks and other possible past British breeding birds. *British Birds* 101: 214.

Boyd, J. M. and Boyd, I. L. 1990. *The Hebrides*. Collins, London.

Bradbury, R. B., Bailey, C. M., Wright, D. and Evans, A. D. 2008. Wintering Cirl Buntings *Emberiza cirlus* in southwest England select cereal stubbles that follow a low-input herbicide regime. *Bird Study* 55: 23–31.

Bradbury, R. B., Kyrkos, A., Morris, A. J., Clack, S. C., Perkins, A. J. and Wilson, J. D. 2000. Habitat associations and breeding success of Yellowhammers on lowland farmland. *Journal of Applied Ecology* 37: 789–805.

Brambilla, M., Catzeflis, F. and Perret, P. 2006. Geographical distribution of Subalpine Warbler *Sylvia cantillans* subspecies in mainland Italy. *Ibis* 148: 568–571.

Braun, M. J. and Brumfield, R. T. 1998. Enigmatic phylogeny of skuas: an alternative hypothesis. *Proceedings of the Royal Society of London* 265: 995–999.

Brazil, M. 2003. *The Whooper Swan*. T and AD Poyser, London.

Brazil, M. A. 1991. *The Birds of Japan*. Smithsonian Institution Press, Washington DC.

Brenchley, A. 1986. The breeding distribution and abundance of the Rook *Corvus frugilegus* L. in Great Britain since the 1920s. *Journal of the Zoological Society of London A* 210: 261–278.

Bretagnolle, V. and Lequette, B. 1990. Structural variation in the call of the Cory's Shearwater (*Calonectris diomedea*, Aves, Procellariidae). *Ethology* 85: 313–323.

Brewer, D. 2001. *Wrens, Dippers and Thrashers*. Christopher Helm, London.

Brickle, N. W., Harper, D. G. C., Aebischer, N. J. and Cockayne, S. H. 2000. Effects of agricultural intensification on the breeding success of Corn Buntings *Milaria calandra*. *Journal of Applied Ecology* 37: 742–755.

Brickle, N. W. and Peach, W. J. 2004. The breeding ecology of Reed Buntings *Emberiza schoeniclus* in farmland and wetland habitats in lowland Britain. *Ibis* 146 (Suppl. 2): 69–77.

Bridge, E. S., Jones, A. W. and Baker, A. J. 2005. A phylogenetic framework for the terns (Sternini) inferred from mtDNA sequences: implications for taxonomy and plumage evolution. *Molecular Phylogenetics and Evolution* 35: 459–469.

Bril, B. 1998. Les sous-espèces de Bergeronnette Printanière la type *Motacilla f. flava* et la Flavéole *M. f. flavissima* dans le nord-oeust de la région Nord – Pas de Calais: distribution, effectifs. *Heron* 31: 81–96.

Brindley, E., Mudge, G., Dymond, N., Lodge, C., Ribbands, B., Steele, D., Ellis, P. M., Meek, E., Suddaby, D. and Ratcliffe, N. 1999. The status of Arctic Terns *Sterna paradisea* at Shetland and Orkney in 1994. *Atlantic Seabirds* 1: 135–143.

Brisbin, I. L., Jr, Pratt, H. D. and Mowbray, T. B. 2002. American Coot (*Fulica americana*) and Hawaiian Coot (*Fulica alai*). In A. Poole and F. Gill (eds), *The Birds of North America, No. 697*. The Birds of North America, Inc., Philadelphia, PA.

British Ornithologists' Union (BOU) 1883. *A List of British Birds*. BOU, London. 229pp.

British Ornithologists' Union (BOU) 1915. *A List of British Birds*. BOU, London. 430pp.

British Ornithologists' Union (BOU) 1923. *A List of British Birds*. BOU, London. 33pp.

British Ornithologists' Union (BOU) 1952. *Check-list of the Birds of Great Britain and Ireland*. BOU, London. 106pp.

British Ornithologists' Union (BOU) 1971. *The Status of Birds in Britain and Ireland*. Blackwell, Oxford. 333pp.

British Ornithologists' Union (BOU) 1992. *Checklist of Birds of Britain and Ireland*. Sixth edn. BOU, Tring. 50pp.

British Ornithologists' Union (BOU) 1994. Records Committee: Twentieth Report. *Ibis* 136: 253–256.

British Ornithologists' Union (BOU) 2006. The British List: a Checklist of Birds of Britain (7th edition). *Ibis* 148: 526–563.

Brito, P. 2005. The influence of Pleistocene glacial refugia on Tawny Owl genetic diversity and phylogeography in Western Europe. *Molecular Ecology* 14: 3077–3094.

Brito, P. H. 2007. Contrasting patterns of mitochondrial and microsatellite genetic structure among Western European populations of tawny owls (*Strix aluco*). *Molecular Ecology* 16: 3423–3437.

Brodeur, S., Savard, J.-P. L., Robert, M., Laporte, P., Lamothe, P., Titman, R. D., Marchand, S., Gilliland, S. and Fitzgerald, G. 2002. Harlequin Duck *Histrionicus histrionicus* population structure in eastern Nearctic. *Journal of Avian Biology* 33: 127–137.

Brooke, M. de L. 1990. *The Manx Shearwater*. T and AD Poyser, London.

Brooke, M. de L. and Davies, N. B. 1987. Recent changes in host usage by cuckoos *Cuculus canorus* in Britain. *Journal of Animal Ecology* 56: 873–883.

Brown, A. and Grice, P. 2005. *Birds in England*. T and AD Poyser, London.

Brown, A. F., Crick, H. P. Q. and Stillman, R. A. 1995. The distribution, numbers and breeding ecology of Twite *Acanthis flavirostris* in the south Pennines of England. *Bird Study* 42: 107–121.

Brown, C. R. 1997. Purple Martin (*Progne subis*). In A. Poole and F. Gill (eds), *The Birds of North America, No. 287*. The Academy of Natural Sciences, Philadelphia, PA, and The American Ornithologists' Union, Washington, DC.

Brown, J. W., Van Coeverden de Groot, P. J., Birt, T. P., Seutin, G., Boag, P. T. and Friesen, V. L. 2007. Appraisal of the consequences of the DDT-induced bottleneck on the level and geographic distribution of neutral genetic variation in Canadian peregrine falcons, *Falco peregrinus*. *Molecular Ecology* 16: 327–343.

Brown, P. and Waterston, G. 1962. *The Return of the Osprey*. Collins, London.

Browne, S. and Aebischer, N. 2005. Studies of West Palearctic birds: Turtle Dove. *British Birds* 98: 58–72.

Brua, R. B. 2001. Ruddy Duck (*Oxyura jamaicensis*). In A. Poole and F. Gill (eds), *The Birds of North America, No. 696*. The Birds of North America, Inc., Philadelphia, PA.

Buchanan, G. M., Pearce-Higgins, J. W., Wotton, S. R., Grant, M. C. and Whitfield, D. P. 2003. Correlates of the change in Ring Ouzel *Turdus torquatus* abundance in Scotland from 1988–91 to 1999. *Bird Study* 50: 97–105.

Budworth, D., Canham, M., Clark, H., Hughes, B. and Sellers, R. M. 2000. Status, productivity, movements and mortality of Great Cormorants *Phalacrocorax carbo* breeding in Caithness, Scotland: a study of a declining population. *Atlantic Seabirds* 2: 165–180.

Buehler, D. M. and Baker, A. J. 2005. Population divergence times and historical demography in Red Knots and Dunlins. *Condor* 107: 497–513.

Burfield, I. J. and Brooke, M. de L. 2005. The decline of the Ring Ouzel *Turdus torquatus* in Britain: evidence from bird observatory data. *Ringing and Migration* 22: 199–204.

Burg, T. M. and Croxall, J. P. 2001. Global relationships among black-browed albatrosses: analysis of population structure using mtDNA and microsatellites. *Molecular Ecology* 10: 2647–2660.

Burg, T. M., Lomax, J., Almond, R., Brooke, M. de L. and Amos, W. 2003. Unravelling dispersal patterns in an expanding population of a highly mobile seabird, the Northern Fulmar. *Proceedings of the Royal Society of London B* 270: 979–984.

Burger, J. and Gochfeld, M. 2002. Bonaparte's Gull (*Larus philadelphia*). In A. Poole and F. Gill (eds), *The Birds of North America, No. 634*. The Birds of North America, Inc., Philadelphia, PA.

Burton, N. H. K. 2009. Reproductive success of Tree Pipits *Anthus trivialis* in relation to habitat selection in conifer plantations. *Ibis* 151: 361–372.

Bush, K. L. and Strobeck, C. 2003. Phylogenetic relationships of the Phasianidae reveals possible non-pheasant taxa. *Journal of Heredity* 94: 472–489.

Byers, C., Olsson, U. and Curson, J. 1995. *Buntings and Sparrows: a Guide to the Buntings and North American Sparrows*. Pica Press, Sussex.

Byrkjedal, I. and Thompson, D. B. A. 1998. *Tundra Plovers: the Eurasian, Pacific and American Golden Plover and Grey Plover*. T and AD Poyser, London.

Cabot, J. and Urdiales, C. 2005. The subspecific status of Sardinian Warblers *Sylvia melanocephala* in the Canary Islands with the description of a new subspecies from Western Sahara. *Bulletin of the British Ornithologists' Club* 125: 230–240.

Cadbury, C. J., Hill, D. J. and Sorenson, J. 1989. The history of the Avocet population and its management in England since re-colonisation. *RSPB Conservation Review* 3: 9–13.

Cadiou, B., Riffaut, L., McCoy, K. D., Cabelguen, J., Fortin, M., Gelinaud, G., Le Roch, A., Tirard, C. and Boulinier, T. 2004. Ecological impact of the 'Erika' oil spill: determination of the geographic origin of the affected common guillemots. *Aquatic Living Resources* 17: 369–377.

Cagnon, C., Lauga, B., Hémery, G. and Mouchés, C. 2004. Phylogeographic differentiation of Storm Petrels (*Hydrobates pelagicus*) based on cytochrome *b* mitochondrial DNA variation. *Marine Biology* 145: 1257–1264.

Caizergues, A., Bernard-Laurent, A., Brenot, J.-F., Ellison, L. and Rasplus, J. Y. 2003. Population genetic structure of rock ptarmigan *Lagopus mutus* in Northern and Western Europe. *Molecular Ecology* 12: 2267–2274.

Calladine, J. 2004. Lesser Black-backed Gull *Larus fuscus*. In P. I. Mitchell, S. F. Newton, N. Ratcliffe and T. E. Dunn (eds), *Seabird Populations of Britain and Ireland*, pp. 226–241. T and AD Poyser, London.

Campbell, B. 1960. The Mute Swan census in England and Wales. *Bird Study* 7: 208–223.

Campbell, L. H. 1984. The impact of changes in sewage treatment on seaducks wintering in the Firth of Forth, Scotland. *Biological Conservation* 14: 111–124.

Camphuysen, C. J. 2005. Seabirds at sea in summer in the northwest North Sea. *British Birds* 98: 2–19.

Carson, C. A., Cornford, G. A. and Thomas, G. J. 1977. Little Gulls nesting on the Ouse Washes. *British Birds* 70: 331–332.

Carson, R. J. and Spicer, G. S. 2003. A phylogenetic analysis of the emberizid sparrows based on three mitochondrial genes. *Molecular Phylogenetics and Evolution* 29: 43–57.

Carss, D. and Marquiss, M. 1992. Avian

predation at farmed and natural fisheries. *Proceedings of the 22nd Institute of Fisheries Management, Annual Study Course*: 179–196.

Carter, H. R., Wilson, U. W., Lowe, R. W., Rodway, M. S., Manuwai, D. A., Takekawa, J. E. and Yee, J. L. 2001. Population trends of the Common Murre (*Uria aalge californica*). In D. L. Orthomeyer (ed.), *Biology and Conservation of the Common Murre in California, Oregon, Washington and British Columbia*. Vol. 1 US Geological Survey, Washington DC.

Carter, I. and Grice, P. 2000. Studies of re-established Red Kites in England. *British Birds* 93: 304–322.

Casey, C. 1998. Distribution and conservation of the Corncrake in Ireland 1993–1998. *Irish Birds* 6: 159–176.

Casey, S., Moore, N., Ryan, L., Merne, O. J., Coveney, J. A. and del-Nevo, A. 1995. The Roseate Tern conservation project on Rockabill, Co. Dublin: a six year review 1989–1994. *Irish Birds* 5: 251–264.

Chabrzyk, K. G. and Coulson, J. C. 1976. Survival and recruitment in the Herring Gull *Larus argentatus*. *Journal of Animal Ecology* 51: 187–203.

Chamberlain, D. E. and Crick, H. Q. P. 2003. Temporal and spatial associations in aspects of reproductive performance of Lapwings *Vanellus vanellus* in the United Kingdom, 1962–1999. *Ardea* 91: 183–196.

Chandler, R. 2003. Rose-ringed Parakeets – how long have they been around? *British Birds* 96: 407–408.

Chandler, R. J. 1981. Influxes into Britain and Ireland of Red-necked Grebes and other waterbirds during winter 1978/78. *British Birds* 74: 55–81.

Chandler, R. J. 1998. Dowitcher identification and ageing: a photographic review. *British Birds* 91: 93–106.

Chantler, P. and Driessens, G. 1995. *Swifts: a Guide to the Swifts and Treeswifts of the World*. Pica Press, Sussex.

Chilton, G., Baker, M. C., Barrentine, C. D. and Cunningham M. A. 1995. White-crowned Sparrow (*Zonotrichia leucophrys*). In A. Poole and F. Gill (eds), *The Birds of North America, No. 183* The Academy of Natural Sciences, Philadelphia, and The American Ornithologists' Union, Washington, DC.

Chisholm, K. 2007. History of the Wood Sandpiper as a breeding bird in Britain. *British Birds* 100: 112–121.

Christian, N. and Hancock, M. H. 2009. A 25-year study of breeding Greenshanks: territory occupancy, breeding success and the effects of new woodland. *British Birds* 102: 203–210.

Chu, P. C. 1995. Phylogenetic reanalysis of Strauch's osteological data set for the Charadriiformes. *Condor* 97: 174–196.

Chu, P.C. 1998. A phylogeny of the gulls (Aves: Larini) inferred from osteological and integumentary characters. *Cladistics* 14: 1–43.

Churcher, P. B. and Lawton, J. H. 1987. Predation by domestic cats in an English village. *Journal of Zoology (Lond.)* 212: 439–455.

Cibois, A. and Cracraft, J. 2004. Assessing the passerine 'tapestry': phylogenetic relationships of the Muscicapoidea inferred from nuclear DNA sequences. *Molecular Phylogenetics and Evolution* 32: 264–273.

Cibois, A. and Pasquet, E. 1999. Molecular analysis of the phylogeny of 11 genera of the Corvidae. *Ibis* 141: 297–306.

Cicero, C. and Johnson, N. K. 1995. Speciation in sapsuckers (*Sphyrapicus*): III. Mitochondrial-DNA sequence divergence at the cytochrome-B locus. *Auk* 112: 547–563.

Cicero, C. and Johnson, N. K. 1998. Molecular phylogeny and ecological diversification in a

clade of New World songbirds (genus *Vireo*). *Molecular Ecology* 7: 1359–1370.

Cimprich D. A., Moore, F. R. and Guilfoyle, M. P. 2000. Red-eyed Vireo (*Vireo olivaceus*). In A. Poole and F. Gill (eds), *The Birds of North America, No. 527*. The Birds of North America, Inc., Philadelphia, PA.

Cink, C. L. and Collins, C. T. 2002. Chimney Swift (*Chaetura pelagica*). In A. Poole and F. Gill (eds), *The Birds of North America, No. 646*. The Birds of North America, Inc., Philadelphia, PA.

Cleere, N. and Nurney, D. 1998. *Nightjars: a Guide to Nightjars and Related Nightbirds*. Pica Press, Sussex.

Clement, P. and Gantlett, S. 1993. The origin of species. *Birding World* 6: 206–213.

Clement, P., Harris, A. and Davis J. 1993. *Finches and Sparrows: an Identification Guide*. Christopher Helm, London.

Clement, P., Hathway, R., Byers, C. and Wilczur, J. 2000. *Thrushes*. Christopher Helm, London.

Clements, R. 2001. The Hobby in Britain: a new population estimate. *British Birds* 94: 402–408.

Clements, R. 2002. The Common Buzzard in Britain: a new population estimate. *British Birds* 95: 377–383.

Clements, R. 2005. Honey Buzzards in Britain. *British Birds* 98: 153–155.

Cohen, B. L., Baker, A. J., Blechschmidt, K., Dittmann, D. L., Furness, R. W., Gerwin, J. A., Helbig, A. J., de Korte, J., Marshall, H. D., Palma, R. L., Peter, H.-U., Ramli, R., Siebold, I., Willcox, M. S., Wilson, R. H. and Zink, R. M. 1997. Enigmatic phylogeny of skuas (Aves: Stercorariidae). *Proceedings Of the Royal Society of London* 264: 181–190.

Collier, M. B., Banks, A. N., Austin, G. E., Girling, T., Hearn, R. D. and Musgrove, A. J. 2005. *The Wetland Bird Survey 2003/04: Wildfowl and Wader Counts*. BTO/WWT/RSPB/JNCC, Thetford.

Collinson, J. M., Parkin, D. T., Knox, A. G., Sangster, G. and Svensson, L. 2008. Species boundaries in the Herring Gull and Lesser Black-backed Gull complex. *British Birds* 101: 340–363.

Collinson, M., Parkin, D. T., Knox, A. G., Sangster, G. and Helbig, A. J. 2006. Species limits within the genus *Melanitta*, the scoters. *British Birds* 99: 183–201.

Colwell, M. A. and Jehl, J. R. 1994. Wilson's Phalarope (*Phalaropus tricolor*). In A. Poole and F. Gill (eds), *The Birds of North America, No. 83*. The American Ornithologists' Union, Washington, DC.

Conder, P. 1979. Britain's first Olive-backed Pipit. *British Birds* 72: 2–4.

Connors, P. G. 1983. Taxonomy, distribution and evolution of Golden Plovers (*Pluvialis dominica* and *Pluvialis fulva*). *Auk* 100: 607–620.

Connors, P. G., McCaffery, B. J. and Maron, J. L. 1993. Speciation in Golden Plovers, *Pluvialis dominica* and *Pluvialis fulva*: evidence from their breeding grounds. *Auk* 110: 9–20.

Constantine, M. and The Sound Approach 2006. *The Sound Approach to Birding*. The Sound Approach, Dorset.

Conway, G., Wotton, S., Henderson, I., Langston, R., Drewitt, A. and Currie, F. 2007. The status and distribution of European Nightjars *Caprimulgus europaeus* in the UK in 2004. *Bird Study* 54: 98–111.

Cooke, F., Rockwell, R. F. and Lank, D. B. 1995. *The Snow Geese of La Pérouse Bay: Natural Selection in the Wild*. OUP, Oxford.

Coulson, J. and Odin, N. 2008. Continental Great Spotted Woodpeckers in mainland Britain – fact or fiction? *Ringing and Migration* 23: 217–222.

Coulson, J. C. and Coulson, B. A. 2008. Measuring immigration and philopatry in seabirds: recruitment to Black-legged Kittiwake colonies. *Ibis* 150: 288–299.

Coulson, J. C. and Strowger, J. 1999. The annual mortality rates of Black-legged Kittiwakes in NE England from 1954 to 1998, and a recent exceptionally high mortality. *Waterbirds* 22: 3–13.

Cowie, R. J. and Hinsley, S. A. 1987. Breeding success of Blue Tits and Great Tits in urban gardens. *Ardea* 75: 81–90.

Cowley, E. 1979. Sand Martin population trends in Britain, 1965–1978. *Bird Study* 26: 113–116.

Cowley, E. and Siriwardena, G. 2000. Long term variation in the survival of Sand Martins *Riparia riparia*: dependence on breeding ground and wintering ground weather, age and sex, and their population consequences. *Bird Study* 52: 237–251.

Cox, S. and Inskipp, T. 1978. Male Citrine Wagtail feeding young wagtails in Essex. *British Birds* 71: 209–213.

Coyne, J. A. 1985. The genetic basis of Haldane's Rule. *Nature* 314: 736–738.

Craik, J. C. A. 1998. Long-term effects of North American Mink *Mustela vison* on seabirds in western Scotland. *Bird Study* 44: 303–309.

Craik, J. C. A. 2002. Results of the mink–seabird project 2001. Unpublished Report to JNCC.

Cramp, S., Bourne, W. R. P. and Saunders, D. 1974. *The Seabirds of Britain and Ireland*. Collins, London.

Cranswick, P. A., Hall, C. and Smith, L. 2004. *All Wales Common Scoter Survey: Report on 2002/3 Work Programme*. WWT Wetlands Advisory Service report to Countryside Council for Wales. CCW Contract Science Report No. 615.

Cranswick, P., Worden, J., Ward, R., Rowell, H., Hall, C., Musgrove, A., Hearn, R., Holloway, S., Banks, A., Austin, G., Griffin, L., Hughes, B., Kershaw, M., O'Connell, M., Pollitt, M., Rees, E. and Smith, L. 2005. *The Wetland Bird Survey 2001–03: Wildfowl and Wader Counts*. BTO/WWT/RSPB/JNCC,Thetford.

Crick, H. Q. P., Robinson, R. A. and Siriwardena, G. M. 2002. Causes of the population declines: summary and recommendations. In H. Q. P. Crick, R. A. Robinson, G. F. Appleton, N. A. Clark and A. D. Rickard (eds), *Investigation into the Causes of the Decline of Starlings and House Sparrows in Great Britain*, pp. 262–292. BTO Research Report No. 290, BTO, Thetford.

Crochet, P.-A., Bonhomme, F. and Lebreton, J.-D. 2000. Molecular phylogeny and plumage evolution in gulls (Larini). *Journal of Evolutionary Biology* 13: 47–57.

Crowe, O. 2005. *Ireland's Wetlands and their Waterbirds: Status and Distribution*. Birdwatch Ireland, Newcastle, Co. Wicklow.

Crowe, O., Austin, G. E., Colhoun, K., Cranswick, P. A., Kershaw, M. and Musgrove, A. J. 2008. Estimates and trends of waterbird numbers wintering in Ireland, 1994/95 to 2003/04. *Bird Study* 55: 66–77.

Cruse, R. 2004. Yellow-browed Warbler *Phylloscopus inornatus* in Senegal in December 2003. *Bulletin of African Bird Club* 11: 147–148.

Cubitt, M. 1995. Swinhoe's Storm-petrels at Tynemouth: new to Britain and Ireland. *British Birds* 88: 342–348.

Dabrowski, A., Fraser, R., Confer, J. L. and Lovette, I. J. 2005. Geographic variability in mitochondrial introgression among hybridizing populations of Golden-winged (*Vermivora chrysoptera*) and Blue-winged (*V. pinus*) Warblers. *Conservation Genetics* 6: 843–853.

DaCosta, J. M., Spellman, G. M., Escalante, P. and Klicka, J. 2009. A molecular systematic revision of two historically problematic songbird clades: *Aimophila* and *Pipilo*. *Journal of Avian Biology* 40: 206–216.

Davies, A. K. 1988. The distribution and status of the Mandarin Duck *Aix galericulata* in Britain. *Bird Study* 35: 203–208.

Davies, N. B. 1992. *Dunnock Behaviour and Social Evolution*. OUP, Oxford.

Davis, P. 1993. The Red Kite in Wales: setting the record straight. *British Birds* 86: 295–298.

Davis, W. E. and Kushlan, J. A. 1994. Green Heron (*Butorides virescens*). In A. Poole and F. Gill (eds), *The Birds of North America, No. 129*. The Academy of Natural Sciences, Philadelphia, and The American Ornithologists' Union, Washington, DC.

Day, R. H., de Gange, A. R., Divoky, G. J. and Troy, D. M. 1988. Distribution and subspecies of Dovekie in Alaska. *Condor* 90: 712–714.

de Korte, J. 1972. Birds, observed and collected by 'De Nederlandse Spitsbergen Expeditie' in west and east Spitsbergen, 1967 and 1968–'69; first part. *Beaufortia* 19: 113–150, 197–232.

de León, A., Minguez, E., Harvey, P., Meek, E., Crane, J. E. and Furness, R. W. 2006. Factors affecting breeding distribution of Storm-petrels *Hydrobates pelagicus* in Orkney and Shetland. *Bird Study* 53: 64–72.

de Queiroz, K. 1998. The general lineage concept of species, species criteria and the process of speciation. In D. J. Howard and S. H. Berlocher (eds) *Endless Forms: Species and Speciation*. OUP, Oxford.

Dean, A. R. 2007a. The British Birds Rarities Committee: a review of its history, publications and procedures. *British Birds* 100: 149–176.

Dean, A. R. 2007b. Enigmatic grey-and-white Common Chiffchaffs. *British Birds* 100: 497–499.

Dean, A. R. and Svensson, L. 2005. Siberian Chiffchaff revisited. *British Birds* 98: 396–410.

Dean, B. J., Webb, A., McSorley, S. A., Schofield, R. A. and Reid, J. A. 2004. *Surveillance of Wintering Seaducks, Divers and Ggrebes in UK Inshore Areas: Aerial Surveys and Shore-based Counts 2003/04*. JNCC Report 357, Peterborough.

Dementiev, G. P. and Gladkov, N. A. 1954. *The Birds of the Soviet Union*. Vol. 6. Soviet Nauka Press, Moscow.

Dennis, R. H. 1991. *Ospreys*. Colin Baxter, Lanark.

Dennis, R. 2007. *A Life of Ospreys*. Whittles, Dunbeath, Scotland.

Denny, M. J. H., Clausen, P., Percival, S. M., Anderson, G. Q. A., Koffijberg, K. and Robinson, J. A. 2004. *Light-bellied Goose Branta bernicla hrota in Svalbard, Greenland, Franz Josef Land, Norway, Denmark, The Netherlands and Britain 1960/1–1999/2000*. Waterbird Review Series, WWT/JNCC, Slimbridge.

Dickinson, E. C. 2003. *The Howard and Moore Complete Checklist of the Birds of the World*. 3rd edn. Christopher Helm, London.

Dickinson, E. C. and Milne, P. In press. The authorship of *Parus ater hibernicus*. *Bulletin of the British Ornithogists' Club*.

Dietzen, C, Garcia del Rey, E, Castro, G. D. and Wink, M. 2008a. Phylogeography of the Blue Tit (*Parus teneriffae*-group) on the Canary Islands based on mitochondrial DNA and morphometrics. *Journal of Ornithology* 149: 1–12.

Dietzen, C., Garcia del Rey, E., Castro, G. D. and Wink, M. 2008b. Phylogenetic differentiation of *Sylvia* species (Aves Passeriformes) of the Atlantic islands (Macaronesia) based on mitochondrial DNA sequence data and morphometrics. *Biological Journal of the Linnean Society* 95: 157–174.

Dillon, I. A., Smith, T. D., Williams, S. J.,

Haysom, S. and Eaton, M. A. 2009. Status of Red-throated Divers *Gavia stellata* in Britain in 2006. *Bird Study* 56: 147–157.

Dittmann, D. L. and Zink, R. M. 1991. Mitochondrial DNA variation among phalaropes and allies. *Auk* 108: 771–779.

Dixey, A. E., Ferguson, A., Heywood, R. and Taylor, A. R. 1981 Aleutian Tern: new to the western Palearctic. *British Birds* 74: 411–416.

Donald, P. F. 2004. *The Skylark*. T and AD Poyser, London.

Donázar, J. A., Negro, J. J., Palacios, C. J., Gangoso, L., Godoy, J.A., Ceballos, O., Hiraldo, F. and Capote, N. 2002. Description of a new subspecies of the Egyptian Vulture (Accipitridae: *Neophron percnopterus*) from the Canary Islands. *Journal of Raptor Research* 36: 17–23.

Donne-Goussé, C., Laudet, V. and. Hanni, C. 2002. A molecular phylogeny of Anseriformes based on mitochondrial DNA analysis. *Molecular Phylogenetics and Evolution* 23: 339–356.

Dougall, T. W., Holland, P. K. and Yalden, D. W. 2004. A revised estimate of the breeding population of Common Sandpipers *Actitis hypoleucos* in Great Britain and Ireland. *Wader Study Group Bulletin* 105: 42–49.

Drovetski, S. V. 2002. Molecular phylogeny of grouse: individual and combined performance of W-linked, autosomal, and mitochondrial loci. *Systematic Biology* 51: 930–945.

Drovetski, S., Zink, R. M., Fadeev, I. V., Nesterov, E. V., Koblik, E. A., Red'kin, Y. A. and Rohwer, S. 2004a. Mitochondrial phylogeny of *Locustella* and related genera. *Journal of Avian Biology* 35: 105–110.

Drovetski, S., Zink, R. M., Rohwer, S., Fadeev, I. V., Nesterov, E. V., Karagodin, I., Koblik, E. A. and Red'kin, Y. A. 2004b. Complex biogeographic history of a Holarctic passerine. *Proceedings of the Royal Society of London B* 271: 545–551.

Dudley, S. P. 2005. Changes to Category C of the British List. *Ibis* 147: 803–820.

Dunn, T. E. 2004. Black-headed Gull *Larus ridibundus*. In P. I. Mitchell, S. F. Newton, N. Ratcliffe and T. E. Dunn (eds), *Seabird Populations of Britain and Ireland*, pp. 196–213. T and AD Poyser, London.

Dunnet, G. M., Ollason, J. C. and Anderson, A. 1979. A 28-year study of breeding Fulmars (*Fulmarus glacialis*) in Orkney. *Ibis* 121: 293–300.

Dunnett, G. M. and Patterson, I. J. 1968. The Rook problem in northeast Scotland. In R. K. Murton and E. N. Wright (eds), *The Problem of Birds as Pests*. Academic Press, London.

Dunstone, N. 1993. *The Mink*. T and AD Poyser, London.

Durant, J. M., Anker-Nilsen, T. and Stenseth, N. C. 2003. Trophic interactions and climate change: the Atlantic Puffin as an example. *Proceedings of the Royal Society of London* 270: 1461–1466.

Dwight, J. 1925. The Gulls (Laridae) of the world: their plumages, moults, variations, relationships and distribution. *Bulletin of the American Museum of Natural History* 52: 63–401.

Dyke, G. J., Gulas, B. E. and Crowe, T. M. 2003. Suprageneric relationshiops of galliform birds (Aves, Galliformes): a cladistic analysis of morphological characters. *Zoological Journal of the Linnean Society* 137: 227–244.

Dymond, J. N. 1991. *The Birds of Fair Isle*. Ritchie, Edinburgh.

Dymond, J. N., Fraser, P. A. and Gantlett, S. J. M. 1989. *Rare Birds in Britain and Ireland*. T and AD Poyser, Calton.

Eaton, M. A., Brown, A. F., Noble, D. G., Musgrove, A. J., Hearn, R. D., Aebischer, N. J., Gibbons, D. W., Evans, A. and Gregory, R.

D. 2009. Birds of conservation concern 3: the population status of birds in the United Kingdom, Channel islands and Isle of Man. *British Birds* 102: 296–341.

Eaton, M. A., Dillon, I. A., Stirling-Aird, P. and Whitfield, D. P. 2007a. Status of Golden Eagle *Aquila chrysaetos* in Britain in 2003. *Bird Study* 54: 212–220.

Eaton, M. A., Marshall, K. B. and Gregory, R. D. 2007b. Status of Capercaillie *Tetrao urogallus* in Scotland during 2003/4. *Bird Study* 54: 145–153.

Eaton, S. W. 1995. Northern Waterthrush (*Seiurus noveboracensis*). In A. Poole and F. Gill (eds) *The Birds of North America No. 182*. The American Ornithologists' Union, Washington, DC.

Edelaar, P. 2008. Assortative mating also indicates that common crossbill *Loxia curvirostra* vocal types are species. *Journal of Avian Biology* 39: 9–12.

Efe, M. A., Tavares, E. S., Baker, A. J. and Bonatto, S. 2009. Multigene phylogeny and DNA barcoding indicate that the Sandwich Tern complex (*Thalasseus sandvicensis*, Laridae, Sternini) comprises two species. *Molecular Phylogenetics and Evolution* 52: 263–267.

Elmberg, J. 1993. Song differences between North American and European White-winged Crossbills. *Auk* 110: 385.

Elphick, C. S and Klima, J. 2002. Hudsonian Godwit (*Limosa haemastica*). In A. Poole and F. Gill (eds), *The Birds of North America, No. 629*. The Birds of North America Inc., Philadelphia, PA.

Ely, C. R., Fox, A. D., Alisauskas, R. T., Andreev, A., Bromlay, R. G., Degtyarev, A. G., Ebbinge, B., Gurtovaya, E. N., Kerbes, R., Kondratyev, A. V., Kostin, I., Krechmar, A. V., Litvin, K. E., Miyabayashi, Y., Mooij, J. H., Oates, R. M., Orthmeyer, D. L., Sabano, Y., Simpson, S. G., Solovieva, D. V., Spindler, M. A., Syroechkovsky, Y. V., Takekawa, J. Y. and Walsh, A. 2005. Circumpolar variation in morphological characteristics of Greater White-fronted Geese *Anser albifrons*. *Bird Study* 52: 104–119.

Engelmoer, M. and Roselaar, C. 1998. *Geographical Variation in Waders*. Kluwer, Dordrecht.

Eraud, C., Boutin, J.-M., Riviere, M., Brun, J., Barbraud, C. and Lormee, H. 2009. Survival of Turtle Doves *Streptopelia turtur* in relation to western Africa environmental conditions. *Ibis* 151: 186–190.

Ericson. P. G. P., Anderson, C. L., Britton, T., Elzanowski, A., Johansson, U. S., Källersjö, M., Ohlson, J. I., Parsons, T. J., Zuccon, D. and Mayr, G. 2006. Diversification of Neoaves integration of molecular sequence data and fossils. *Biology Letters* 2: 543–547.

Ericson, P. G. P., Envall, I., Irestedt, M. and Norman, J. A. 2003. Inter-familial relationships of the shorebirds (Aves: Charadriiformes) based on nuclear DNA sequence data. *BMC Evolutionary Biology* 3: 16.

Ericson, P. G. P., Jansén, A.-L., Johansson, U. S. and Ekman, J. 2005. Inter-generic relationships of the crows, jays, magpies and allied groups (Aves: Corvidae) based on nucleotide sequence data. *Journal of Avian Biology* 26: 222–234.

Ericson, P. G. P. and Johansson, U. S. 2003. Phylogeny of Passerida (Aves: Passeriformes) based on nuclear and mitochondrial sequence data. *Molecular Phylogenetics and Evolution* 29: 126–138.

Ertan, K. T. 2006. The evolutionary history of Eurasian redstarts, *Phoenicurus*. *Acta Zoologica* 52 (Supp.): 310–313.

Evans Mack, D. and Yong, W. 2000. Swainson's

Thrush (*Catharus ustulatus*). In A. Poole and F. Gill (eds), *The Birds of North America, No. 540*. The Birds of North America, Inc., Philadelphia, PA.

Evans, I. M., Dennis, R. H., Orr-Ewing, D. C., Kjellén, N., Andersson, P.-O., Sylvén, M., Senosiain, A., Carbo, F. C. 1997. The re-establishment of Red Kite breeding populations in Scotland and England. *British Birds* 90: 123–138.

Evans, M. E. and Sladen, W. J. 1980. A comparative analysis of the bill markings of Whistling and Bewick's Swans and out-of-range occurrences of the two taxa. *Auk* 97: 697–703.

Evans, R. J., Wilson, J. D., Amar, A., Douse, A., MacLennan, A., Ratcliffe, N. and Whitfield, D. P. 2009. Growth and demography of a re-introduced population of White-tailed Eagles *Haliaeetus albicilla*. *Ibis* 151: 244–254.

Ewins, P. 1985. Colony attendance and censusing of Black Guillemots *Cepphus grylle* in Shetland. *Bird Study* 32: 176–185.

Falls, J. B. and Kopachena, J. G. 1994. White-throated Sparrow (*Zonotrichia albicollis*). In A. Poole and F. Gill (eds), *The Birds of North America, No. 128*. The Academy of Natural Sciences, Philadelphia, and The American Ornithologists' Union, Washington, DC.

Fasola, M., Sanchez, J. M. and Roselaar, C. S. 2002. *Sterna albifrons*: Little Tern. *BWP Update* 4: 89–114.

Feare, C. J. 1994. Changes in numbers of Starlings and farming practice in Lincolnshire. *British Birds* 87: 200–204.

Feare, C. and Craig, A. 1998. *Starlings and Mynahs*. Christopher Helm, London.

Fefelov, I. V. 2001. Comparative breeding ecology and hybridization of Eastern and Western Marsh Harriers *Circus spilonotus* and *C. aeruginosus* in the Baikal region of Eastern Siberia. *Ibis* 143: 587–592.

Feldman, C. R. and Omland, K. E. 2005. Phylogenetics of the Common Raven complex (*Corvus*: Corvidae) and the utility of ND4, COI and intron 7 of the b-fibrinogen gene in avian molecular systematics. *Zoologica Scripta* 34: 145–156.

Field, R. H., Kirby, W. B. and Bradbury, R. B. 2007. Conservation tillage encourages early breeding by Skylarks *Alauda arvensis*. *Bird Study* 54: 137–141.

Finlayson, C. 1992. *Birds of the Strait of Gibraltar*. T and AD Poyser, London.

Fisher, J. 1952. *The Fulmar*. Collins, London.

Förschler, M. I., Senar, J. C., Perret, P. and Björklund, M. 2009. The species status of the Corsican Finch *Carduelis corsicana* assessed by three genetic markers with different rates of evolution. *Molecular Phylogenetics and Evolution* 52: 234–240.

Forsyth, I. 1980. A breeding census of Mute Swans in 1978. *Irish Birds* 1: 492–501.

Fox, A. D. 1991. History of the Pochard breeding in Britain. *British Birds* 84: 83–88.

Fox, A. D. 2006. Calls of 'Northern Bullfinches'. *British Birds* 99: 370–371.

Fox, A. D. and Aspinall, S. J. 1987. Pomarine Skuas in Britain and Ireland in autumn 1985. *British Birds* 80: 401–421.

Fox, A. D., Christensen, T. K., Bearhop, S. and Newton, J. 2007. Using stable isotope analysis of multiple feather tracts to identify moulting provenance of vagrant birds: a case study of Baikal Teal *Anas formosa* in Denmark. *Ibis* 149: 622–625.

Fox, A. D., Stroud, D., Walsh, A., Wilson, J., Norris, D. and Francis, I. 2006. The rise and fall of the Greenland White-fronted Goose: a case study in international conservation. *British Birds* 99: 242–261.

Fraser, P. A. and Rogers, M. J. 2005. Report on scarce migrant birds in Britain in 2002. *British Birds* 98: 73–88.

Fraser, P. A. and Rogers, M. J. 2006. Report on scarce migrant birds in Britain in 2003. *British Birds* 99: 129–147.

Frederiksen, M., Wanless, S., Harris, M. P., Rothery, P. and Wilson, L. J. 2004. The role of industrial fisheries and oceanographic change in the decline of North Sea black-legged Kittiwakes. *Journal of Applied Ecology* 41: 1129–1139.

Freeland, J., Allen, D. and Anderson, S. 2006. *DNA Analysis of Red Grouse: an Analysis of Taxonomy and Genetic Diversity*. Environment and Heritage Service, Northern Ireland.

Freeman, S. N. and Crick, H. Q. P. 2003. The decline of the Spotted Flycatcher *Muscicapa striata* in the UK: an integrated population model. *Ibis* 145: 400–412.

Freeman, S. N., Robinson, R. A., Clark, J. A., Griffin, B. M. and Adams, S. Y. 2007. Changing demography and population decline in the Common Starling *Sturnus vulgaris*: a multisite approach to Integrated Population Monitoring. *Ibis* 149: 587–596.

Fregin, S., Haase, M., Olsson, U. and Alström, P. 2009. Multi-locus phylogeny of the family Acrocephalidae (Aves: Passeriformes) – the traditional taxonomy overthrown. *Molecular Phylogenetics and Evolution* 52: 866–878.

French, P. R. 2006. Dark-breasted Barn Owl in Devon. *British Birds* 99: 210–211.

Friesen, V. L. and Anderson, D. J. 1997. Phylogeny and evolution of the Sulidae (Aves: Pelecaniformes): a test of alternative modes of speciation. *Molecular Phylogenetics and Evolution* 7: 252–260.

Friesen, V. L., Baker, A. J. and Piatt, J. F. 1996a. Phylogenetic relationships within the Alcidae (Charadriiformes: Aves) inferred from total molecular evidence. *Molecular Biology and Evolution* 13: 359–367.

Friesen, V. L., Piatt, J. F. and Baker, A. J. 1996b. Evidence from cytochrome b sequences and allozymes for a 'new' species of alcid: the Long-billed Murrelet (*Brachyramphus perdix*). *Condor* 98: 681–690.

Friesen, V. L., Smith, A. L., Gómez-Díaz, E., Bolton, M., Furness, R. W., González-Solís, J. and Monteiro, L. R. 2007. Sympatric speciation by allochrony in a seabird. *Proceedings of the National Academy of Sciences of the USA* 104: 18589–18594.

Fry, A. J. and Zink, R. M. 1998. Geographic analysis of nucleotide diversity and Song Sparrow (Aves: Emberizidae) population history. *Molecular Ecology* 7: 1303–1313.

Fry, C. H., Fry, K. and Harris, A. 1992. *Kingfishers, Bee-eaters and Rollers*. Christopher Helm, London.

Fuchs, J., Fjeldså, J., Bowie, R. C. K., Voelker, G. and Pasquet, E. 2006. The African warbler genus *Hyliota* as a lost lineage in the Oscine songbird tree: molecular support for an African origin of the Passerida. *Molecular Phylogenetics and Evolution* 39: 186–197.

Fuchs, J., Ohlson, J. I., Ericson, P. G. P. and Pasquet, E. 2007. Synchronous intercontinental splits between assemblages of woodpeckers suggested by molecular data. *Zoologica Scripta* 36: 11–25.

Fuller, R. J., Gregory, R. D., Gibbons, D. W., Marchant, J. H., Wilson, J. D., Baillie, S. R. and Carter, N. 1995. Population declines and range contractions among lowland farm birds in Britain. *Conservation Biology* 9: 1425–1441.

Fuller, R. J., Noble, D. G., Smith, K. W. and Vanhinsbergh, D. 2005. Recent declines in populations of woodland birds in Britain: a review of possible causes. *British Birds* 98: 116–143.

Furness, R. M. and Ratcliffe, N. 2004. Great Skua *Stercorarius skua*. In P. I. Mitchell, S. F. Newton, N. Ratcliffe and T. E. Dunn (eds), *Seabird Populations of Britain and Ireland*, pp. 173–186. T and AD Poyser, London.

Furness, R. W. 1987. *The Skuas*. T and AD Poyser, Calton.

Furness, R. W. 1997. A 1997 survey of the Rum Manx Shearwater population. *Scottish Natural Heritage Research, Survey and Monitoring Report* No. 73.

Galeotti, P. 2001. *Strix aluco*: Tawny Owl. *BWP Update* 3: 43–77.

Gamauf, A. and Haring, E. 2004. Molecular phylogeny and biogeography of Honey-buzzards (genus *Pernis* and *Henicopernis*). *Journal of Zoological Systematics and Evolutionary Research* 42: 145–153.

Gantlett, S. and Pym, T. 2007. The Yellow-nosed Albatross from Somerset to Lincolnshire: a new British Bird. *Birding World* 20: 279–295.

Garcia, E. F. J. 1983. An experimental test of competition for space between Blackcaps *Sylvia atricapilla* and Garden Warblers *Sylvia borin* in the breeding season. *Journal of Animal Ecology* 52: 795–805.

Garcia del Rey, E., Delgado, G., Gonzalez, J. and Wink, M. 2007. Canary Island Great Spotted Woodpecker (*Dendrocopos major*) has distinct mtDNA. *Journal of Ornithology* 148: 531–536.

Gauthier, G. 1993. Bufflehead (*Bucephala albeola*). In A. Poole and F. Gill (eds), *The Birds of North America, No. 67*. The Academy of Natural Sciences, Philadelphia, and The American Ornithologists' Union, Washington, DC.

Gay, L., Bell, D. A. and Crochet, P.-A. 2005. Additional data on mitochondrial DNA of North American large gull taxa. *Auk* 122: 684–688.

Gay, L., Defos du Rau, P., Mondain-Monval, J.-Y. and Crochet, P.-A. 2004. Phylogeography of a game species: the red-crested pochard (*Netta rufina*) and consequences for its management. *Molecular Ecology* 13: 1035–1045.

Génot, J.-C. and van Nieuwehuyse, D. 2002. *Athene noctua*: Little Owl. *BWP Update* 4: 35–63.

George, T. L. 2000. Varied Thrush (*Ixoreus naevius*). In A. Poole and F. Gill (eds), *The Birds of North America, No. 541*. The Birds of North America, Inc., Philadelphia, PA.

Gibbons, D. W., Amar, A., Anderson, G. Q. A., Bolton, M., Bradbury, R. B., Eaton, M. A., Evans, A. D., Grant, M. C., Gregory, R. D., Hirons, G. J. M., Hughes, J., Johnstone, I., Newbery, P., Peach, W. J., Ratcliffe, N., Smith, K. W., Summers, R. W., Walton, P. and Wilson, J. D. 2007. *The Predation of Wild Birds in the UK: a Review of its Conservation Impact and Management*. RSPB Research Report No. 23. RSPB, Sandy.

Gibbons, D. W., Bainbridge, I. P., Mudge, G. P. and Thane, P. M. 1997. The status and distribution of the Red-throated Diver *Gavia stellata* in Britain in 1994. *Bird Study* 44: 194–205.

Gibbons, D. W., Reid, J. B. and Chapman, R. A. 1993. *The New Atlas of Breeding Birds in Britain and Ireland: 1988–1991*. T and AD Poyser, London.

Gibbons, D. W. and Wotton, N. 1996. The Dartford Warbler in the United Kingdom in 1994. *British Birds* 89: 203–212.

Gibbs, D., Barnes, E. and Cox, J. 2001. *Pigeons and Doves: a Guide to the Pigeons and Doves of the World*. Pica Press, Sussex.

Gibbs, J. P., Melvin, S. and Reid, F. A. 1992. American Bittern (*Botaurus lentiginosus*). In A. Poole, P. Stettenheim and F. Gill (eds), *The Birds of North America, No. 18*. The Academy of Natural Sciences, Philadelphia, and The American Ornithologists' Union, Washington, DC.

Gilbert, G. 2002. The status and habitat of Spotted Crake *Porzana porzana* in Britain in 1999. *Bird Study* 49: 79–86.

Gill, F. and Wright, M. 2006. *Birds of the World: Recommended English Names.* Princeton UP, Princeton.

Gill, F. B. 1987. Allozymes and genetic similarity of Blue-winged and Golden-winged warblers. *Auk* 104: 444–449.

Gill, F. B., Slikas, B. and Sheldon, F. H. 2005. Phylogeny of titmice (Paridae): II. Species relationships based on sequences of the mitochondrial cytochrome b gene. *Auk* 122: 121–143.

Gill, J. A., Norriss, K., Potts, P. M., Gunnrasson, T. G., Atkinson, P. W. and Sutherland, W. J. 2001. The buffer effect and large-scale population in migratory birds. *Nature* 412: 436–438.

Gillihan, S. W. and Byers, B. 2001. Evening Grosbeak (*Coccothraustes vespertinus*). In A. Poole and F. Gill (eds), *The Birds of North America, No. 599.* The Birds of North America, Inc., Philadelphia, PA.

Gillings, S. 2005a. International workshop on passage and wintering Eurasian Golden Plovers: workshop summary. *Wader Study Group Bulletin* No 108: 5–12.

Gillings, S. 2005b. Trends in winter abundance and distribution of European Golden Plovers and Lapwings in Britain between 1970 and 2003. *Wader Study Group Bulletin* No. 108: 8.

Gillings, S., Austin, G. E., Fuller, R. J. and Sutherland, W. J. 2006. Distribution shifts in wintering Golden Plover *Pluvialis apricaria* and Lapwing *Vanellus vanellus* in Britain. *Bird Study* 53: 274–284.

Gillings, S. and Fuller, R. J. 2009. How many Eurasian Golden Plovers *Pluvialis apricarius* and Northern Lapwings *Vanellus vanellus* winter in Great Britain? Results from a large-scale survey in 2006/07. *Wader Study Group Bull.* 116: 21–28.

Gillings, S., Fuller, R. J. and Balmer, D. E. 2000. Breeding birds in scrub in the Scottish Highlands: variation in community composition between scrub type and successional stage. *Scottish Forestry* 54: 73–85.

Given, A. D., Mills, J. A. and Baker, A. J. 2005. Molecular evidence for recent radiation in southern hemisphere masked gulls. *Auk* 122: 268–279.

Glue, D.E. 1990. Breeding biology of the Grasshopper Warbler in Britain. *Brit. Birds* 83: 131–145.

Glutz von Blotzheim, U. N., Bauer, K. M. and Bezzel, E. 1975. *Handbuch der Vögel Mitteleuropas. Band 6. Charadriiformes (Teil 1).* Akademische, Wiesbaden.

Gollop, J. B., Barry, T. W. and Iverson, E. H. 1986. *Eskimo Curlew: a Vanishing Species?* Special Publication No. 17. Saskatchewan NHS, Saskatchewan.

Gomersall, C. H., Morton, J. S. and Wynde, R. M. 1984. Status of breeding Red-throated Divers in Shetland, 1983. *Bird Study* 31: 223–229.

Gómez-Díaz, E., González-Solís, J., Peinado, M. A. and Page, R. D. M. 2006. Phylogeography of the *Calonectris* shearwaters using molecular and morphometric data. *Molecular Phylogenetics and Evolution* 41: 322–332.

Gonzalez, J., Wink, M., Garcia del Rey, E. and Castro, G. D. 2008. Evidence from DNA nucleotide sequences and ISSR profiles indicates paraphyly in the Southern Grey Shrike (*Lanius meridionalis*). *Journal of Ornithology* 149: 495–506.

Goodman. S. M. and Meininger, P. L. 1989. *The Birds of Egypt.* OUP, Oxford.

Goostrey, A., Carss, D. N., Noble, L. R. and Piertney, S. B. 1998. Population introgression and differentiation in the Great Cormorant *Phalacrocorax carbo* in Europe. *Molecular Ecology* 7: 329–338.

Götmark, F. 1978. Gråsiskans *Carduelis flammea* förekomst i södra Sverige under sydhäckningsåret 1975. *Vår Vågelvärld* 41: 315–322.

Grant, P. J. 1972. Field identification of Richard's and Tawny Pipits. *British Birds* 65: 287–290.

Grant, P. R. 1986. *Ecology and Evolution of Darwin's Finches.* Princeton University Press, Princeton.

Grapputo, A., Pilastro, A. and Marin, G. 1998. Genetic variation and bill size dimorphism in a passerine bird, the Reed Bunting *Emberiza schoeniclus. Molecular Ecology* 7: 1173–1182.

Green, A. J. and Hughes, B. 1996. Action plan for the White-headed Duck *Oxyura leucocephala.* In B. Heredia, L. Rose and M. Painter (eds), *Globally Threatened Birds in Europe,* pp. 119–146. Council of Europe Publishing, Strasbourg.

Green, A. J. and Hughes, B. 2001. *Oxyura leucocephala* White-headed Duck. *BWP Update* 3: 79–90.

Green, G. 2004. *The Birds of Dorset.* Christopher Helm, London.

Green, R. 1996. Factors affecting the population density of the Corncrake *Crex crex* in Britain and Ireland. *Journal of Applied Ecology* 33: 237–248.

Greenlaw, J. S. 1996. Eastern Towhee (*Pipilo erythrophthalmus*). In A. Poole and F. Gill (eds), *The Birds of North America, No. 262.* The Academy of Natural Sciences, Philadelphia, PA, and The American Ornithologists' Union, Washington, DC.

Greenwood, J. G. 1986. Geographical variation and taxonomy of the Dunlin *Calidris alpina. Bulletin of the British Ornithologists' Club* 106: 43–56.

Greenwood, J. G. 1988. The Northern Ireland Black Guillemot survey 1987. *Irish Naturalists' Journal* 22: 490–491.

Greenwood, J. J. D., Delany, S. and Kirby, S. 1994. A method for estimating Mute Swan breeding populations in Great Britain in three years. In E. J. M. Hagemeijer and T. J. Verstrael (eds), *Bird Numbers 1992. Distribution, Monitoring and Ecological Aspects,* pp. 533–542. Proceedings of the 12th International Conference of IBCC and EOAC. Noordwijkerhout, The Netherlands.

Gregory, R. D. and Marchant, J. H. 1996. Population trends of Jays, Magpies, Jackdaws and Carrion Crows in the United Kingdom. *Bird Study* 43: 28–37.

Gregory, R. D., Wilkinson, N. I., Noble, D. G., Robinson, J. A., Brown, A. F., Hughes, J., Procter, D, Gibbons, D. W. and Galbraith, C. A. 2002. The population status of birds in the United Kingdom, Channel Islands, and Isle of Man. *British Birds* 95: 410–448.

Gretton A., Yurlov A. K. and Boere G. C. 2002. Where does the Slender-billed Curlew nest and what future does it have? *British Birds* 95: 334–344.

Grey, N., Thomas, G., Trewby, M. and Newton, S. 2003. The status and distribution of Choughs *Pyrrhocorax pyrrhocorax* in the Republic of Ireland 2002/3. *Irish Birds* 7: 147–156.

Gribble, F. C. 1983. Nightjars in Britain and Ireland in 1981. *Bird Study* 30: 165–176.

Griffiths, C. S., Barrowclough, G. F., Groth, J. G. and Mertz, L. 2007. Phylogeny, diversity, and classification of the Accipitridae based on DNA sequences of the RAG-1 exon. *Journal of Avian Biology* 38: 587–602.

Groth, J. 1998. Molecular phylogenetics of finches and sparrows: consequences of character state removal in cytochrome b sequences. *Molecular Phylogenetics and Evolution* 10: 377–390.

Gruar, D., Barritt, D. and Peach, W. J. 2006. Summer utilization of oilseed rape by Reed Buntings *Emberiza schoeniclus* and other farmland birds. *Bird Study* 53: 47–54.

Guilford, T. C., Meade, J., Freeman, R., Biro, D., Evans, T., Bonadonna, F., Boyle, D., Roberts, S. and Perrins, C. M. 2008. GPS tracking of the foraging movements of Manx Shearwaters *Puffinus puffinus* breeding on Skomer Island, Wales. *Ibis* 150: 462–473.

Guillaumet, A., Crochet, P.-A. and Godelle, B. 2005. Phenotypic variation in *Galerida* larks in Morocco: the role of history and natural selection. *Molecular Ecology* 14: 3809–3821.

Guillaumet, A., Pons, J.-M., Godelle, B. and Crochet, P.-A. 2006. History of the Crested Lark in the Mediterranean area as revealed by mtDNA sequences and morphology. *Molecular Phylogenetics and Evolution* 39: 645–656.

Gunnarsson, T. G., Gill, J. A., Petersen, A., Appleton, G. F. and Sutherland, W. J. 2005. A double buffer effect in a migratory shorebird population. *Journal of Animal Ecology* 74: 965–971.

Guzy, M. J. and Ritchison, G. 1999. Common Yellowthroat (*Geothlypis trichas*). In A. Poole and F. Gill (eds), *The Birds of North America, No. 448.* The Birds of North America, Inc., Philadelphia, PA.

Haag-Wackernagel, D., Heeb, P. and Leiss, A. 2006. Phenotype-dependent selection of juvenile urban Feral Pigeons *Columba livia. Bird Study* 53: 163–170.

Hafner, H., Fasola, M., Voisin, C. and Kaiser, Y. 2002. *Egretta garzetta* Little Egret. *BWP Update* 1: 1–19.

Haftorn, S. 1971. *Norges fugler.* Univeirsitatsforlaget, Oslo.

Hagemeijer, E. J. M. and Blair, M. J. 1997. *The EBCC Atlas of European Breeding Birds: their Distribution and Abundance.* T and A D. Poyser, London.

Haig, S., Gratto-Trevor, C. L., Mullins, T. D. and Colwell, M. A. 1997. Population identification of Western Hemisphere shorebirds throughout the annual cycle. *Molecular Ecology* 6: 413–427.

Hall, C., Smith, L. E., Trinder, M. N. and Cranswick, P. A. 2005. *All Wales Common Scoter Survey: Report on 2003/04 Work Programme.* WWT Wetlands Advisory Service report to Countryside Council for Wales, CCW Contract Science Report No. 709.

Hamer, K. C. 2001. *Catharacta skua* Great Skua. *BWP Update* 3: 91–110.

Hamer, K. C. 2003. *Puffinus puffinus* Manx Shearwater. *BWP Update* 5: 203–214.

Hancock, J. and Kushlan, J. 1984. *The Herons Handbook.* Croom Helm, Beckenham.

Hancock, M. H. 2000. Artificial floating islands for Black-throated Divers *Gavia arctica* in Scotland: their construction, use and effect on breeding success. *Bird Study* 47: 165–175.

Haring, E., Gamauf, A. and Kryukov, A. 2007. Phylogeographic patterns in widespread corvid birds. *Molecular Phylogenetics and Evolution* 45: 840–862.

Harrap, S. and Quinn, D. 1996. *Tits, Nuthatches and Treecreepers.* Christopher Helm, London.

Harrington, B. A. and Morrison, R. I. G. 1979. Semipalmated Sandpiper migration in North America. *Studies in Avian Biology* 2: 83–100.

Harris, M. P. 1969. The biology of storm petrels in the Galapagos Islands. *Proceedings of the California Academy of Sciences* 37: 95–165.

Harris, M. P. 1979. Measurements and weights of British Puffins. *Bird Study* 26: 179–186.

Harris, M. P. 2004. Atlantic Puffin *Fratercula arctica.* In P. I. Mitchell, S. F. Newton, N. Ratcliffe and T. E. Dunn (eds), *Seabird Populations of Britain and Ireland,* pp. 392–406. T and AD Poyser, London.

Harris, M. P., Frederiksen, M. and Wanless, S. 2007. Within- and between-year variation in the juvenile survival of Common Guillemots *Uria aalge*. *Ibis* 149: 472–481.

Harris, M. P., Heubeck, M., Shaw, D. N. and Okill, J. D. 2006. Dramatic changes in the return date of Guillemots *Uria aalge* to colonies in Shetland, 1962–2005. *Bird Study* 53: 247–252.

Harris, M. P., Newell, M., Daunt, F., Speakman, J. R. and Wanless, S. 2008. Snake Pipefish *Entelurus aequoreus* are poor food for seabirds. *Ibis* 150: 413–415.

Harris, M. P., Rothery, P. and Wanless, S. 2003. Increase in frequency of the bridled morph of the Common Guillemot *Uria aalge* on the Isle of May, 1946–2000: a return to former levels. *Ibis* 145: 22–29.

Harris, M. P. and Wanless, S. 1996. Differential responses of Guillemot *Uria aalge* and Shag *Phalacrocorax aristotelis* to a late winter wreck. *Bird Study* 43: 220–230.

Harris, M. P. and Wanless, S. 2004. Common Guillemot *Uria aalge*. In P. I. Mitchell, S. F. Newton, N. Ratcliffe and T. E. Dunn (eds), *Seabird Populations of Britain and Ireland*, pp. 350–363. T and AD Poyser, London.

Harris, M. P., Wanless, S. and Rothery, P. 2000. Adult survival rates of Shag (*Phalacrocorax aristotelis*), Common Guillemot (*Uria aalge*), Razorbill (*Alca torda*), Puffin (*Fratercula arctica*) and Kittiwake (*Rissa tridactyla*) on the Isle of May 1986–1996. *Atlantic Seabirds* 2: 133–150.

Harris, T. and Franklin, K. 2000. *Shrikes and Bush-shrikes*. Christopher Helm, London.

Harrop, A. H. J. 2002. The Ruddy Shelduck in Britain: a review. *Brit. Birds* 95: 123–128.

Harrop, A. H. J., Knox, A. G. and McGowan, R. Y. 2007. Britain's first Two-barred Crossbill. *British Birds* 100: 650–657.

Hartley, C. 2004. Little Gulls at sea off Yorkshire in autumn 2003. *British Birds* 97: 448–455.

Harvey, N. G. 2000. A hierarchical genetic analysis of swan relationships. *Swan Specialist News Letter* 9: 11.

Hatchwell, B. J., Chamberlain, D. E. and Perrins, C. M. 1996. The demography of Blackbirds *Turdus merula* in rural habitats: is farmland a sub-optimal habitat? *Journal of Applied Ecology* 33: 1114–1124.

Hayman, P., Marchant, J. and Prater, T. 1986. *Shorebirds: an Identification Guide to the Waders of the World*. Croom Helm, London.

Hazevoet, C. J. 1995. *The Birds of the Cape Verde Islands*. BOU, Tring.

Heaney, V., Ratcliffe, N., Brown, A., Robinson, P and Lock, L. 2002. The status and distribution of European Storm-petrels *Hydrobates pelagicus* and Manx Shearwaters *Puffinus puffinus* on the Isles of Scilly. *Atlantic Seabirds* 4: 1–16.

Hearn, R. D. 2004a. Bean Goose *Anser fabalis* in *Britain and Ireland 1960/1–1999/2000*. Waterbird Review Series, WWT/JNCC, Slimbridge.

Hearn, R. D. 2004b. *Greater White-fronted Goose* Anser albifrons albifrons (*Baltic/North Sea population*) *in Britain and Ireland 1960/1–1999/2000*. Waterbird Review Series, WWT/JNCC, Slimbridge.

Hebert, P. D. N., Stoeckle, M. Y., Zemlak, T. S. and Francis, C. M. 2004. Identification of birds through DNA Barcodes. *PloS Biology* 2: 10 e312.

Heidrich, P. and Wink, M. 1994. Tawny Owl (*Strix aluco*) and Hume's Tawny Owl (*Strix butleri*) are distinct species: evidence from nucleotide sequences of the cytochrome b gene. *Zeit. Für Naturforschung* 49c: 230–234.

Heidrich, P. and Wink, M. 1998. Phylogenetic relationships in holarctic owls (Order

Strigiformes): evidence from nucleotide sequences of the mitochondrial cytochrome b gene. In R. D. Chancellor, B.-U. Meyburg and J. J. Ferrero (eds) *Holarctic Birds of Prey*. Adenex and WWGBP, Mérida and Berlin.

Helbig, A., Knox, A. G., Parkin, D. T., Sangster, G. and Collinson, M. 2002. Guidelines for assigning species rank. *Ibis* 144: 518–525.

Helbig, A. J., Kocum, A., Seibold, I. and Braun, M. J. 2005a. A multi-gene phylogeny of aquiline eagles (Aves: Accipitriformes) reveals extensive paraphyly at the genus level. *Molecular Phylogenetics and Evolution* 35: 147–164.

Helbig, A. J., Martens, J., Seibold, I., Henning, F., Schottler, B. and Wink, M. 1996. Phylogeny and species limits in the Palearctic Chiffchaff *Phylloscopus collybita* complex: mitochondrial genetic differentiation and bioacoustic evidence. *Ibis* 138: 650–666.

Helbig, A. J. and Seibold, I. 1999. Molecular phylogeny of Palearctic–African *Acrocephalus* and *Hippolais* warblers (Aves: Sylviidae). *Molecular Phylogenetics and Evolution* 11: 246–260.

Helbig, A. J., Seibold, I., Kocum, A., Liebers, D., Irwin, J., Bergmanis, U., Meyburg, B. U., Scheller, W., Stubbe, M. and Bensch, S. 2005b. Genetic differentiation and hybridisation between Greater and Lesser Spotted Eagles (Accipitriformes: *Aquila clanga*, *A. pomarina*). *Journal of Ornithology* 146: 226–234.

Helbig, A. J., Seibold, I., Martens, J. and Wink, M. 1995. Genetic differentiation and phylogenetic relationships of Bonelli's Warbler *Phylloscopus bonelli* and Green Warbler *P. nitidus*. *Journal of Avian Biology* 26: 139–153.

Hepp, G. R., Novak, J. M., Scribner, K. T. and Stangel, P. W. 1988. Genetic distance and hybridization of Black Ducks and Mallards: a morph or a different color? *Auk* 105: 804–807.

Herremans, M. 1989. Vocalisations of Common, Lesser and Arctic Redpolls. *Dutch Birding* 11: 9–15.

Herremans, M. 1990. Taxonomy and evolution in redpolls *Carduelis flammea–hornemanni*: a multivariate study of their biometry. *Ardea* 78: 441–458.

Heubeck, M. 1993. Moult flock surveys indicate a continued decline in the Shetland Eider population. *Scottish Birds* 14: 146–152.

Heubeck, M. 1997. The direct effect of the Braer Oil Spill on seabird populations and an assessment of the role of the Wildlife Response Centre. In J. M. Davies and G. Topping (eds), *The Impact of an Oil Spill on Turbulent Waters: the Braer*. The Stationery Office, Edinburgh.

Heubeck, M. 2004. Black-legged Kittiwake *Rissa tridactyla*. In P. I. Mitchell, S. F. Newton, N. Ratcliffe and T. E. Dunn (eds), *Seabird Populations of Britain and Ireland*, pp. 277–290. T and AD Poyser, London.

Heubeck, M., Mellor, R. M. and Harvey, P. V. 1997. Changes in the breeding distribution and numbers of Kittiwakes *Rissa tridactyla* around Unst, Shetland, and the presumed role of predation by Great Skuas *Stercorarius skua*. *Seabird* 19: 12–21.

Heubeck, M. and Richardson, M. G. 1980. Bird mortality following the *Esso Bernica* oil spill, December 1978. *Scottish Birds* 11: 97–108.

Hill, D. and Robertson, P. 1988. *The Pheasant: Ecology, Management and Conservation*. BSP Professional Books, Oxford.

Hillcoat, B., Keijl, G. O. and Wallace, D. I. M. 1997. *Calonectris edwardsii* Cape Verde Shearwater. *BWP Update* 1: 128–130.

Hille, S., Nesje, M. and Segelbacher, G. 2003. Genetic structure of Kestrel populations and colonization of the Cape Verde archipelago. *Molecular Ecology* 12: 2145–2151.

Hockey, P. A. R. 1996. *Haematopus ostralegus* in

perspective: comparisons with other Oystercatchers. In J. D. Goss-Custard (ed.), *The Oystercatcher: from Individual to Populations*, pp. 251–285. Oxford University Press, Oxford.

Höglund, J., Johansson, T., Beintema, A. and Schekkerman, H. 2009. Phylogeography of the Black-tailed Godwit *Limosa limosa*: substructuring revealed by mtDNA control region sequences. *Journal of Ornithology* 150: 45–53.

Holder, K., Montgomerie, R. and Friesen, V. L. 1999. A test of the glacial refugium hypothesis using patterns of mitochondrial and nuclear DNA sequence variation in Rock Ptarmigan (*Lagopus mutus*). *Evolution* 53: 1936–1950.

Holder, K., Montgomerie, R. and Friesen, V. L. 2000. Glacial vicariance and historical biogeography of Rock Ptarmigan (*Lagopus mutus*) in the Bering region. *Molecular Ecology* 9: 1265–1278.

Holloway, S. 1996. *The Historical Atlas of Breeding Birds in Britain and Ireland: 1875–1900*. T and AD Poyser, London.

Hollyer, J. N. 1970. The invasion of Nutcrackers in autumn 1968. *British Birds* 63: 353–373.

Holmes, J., Marchant, J., Buckland, N., Stroud, D. and Parkin, D. 1998. The British List: new categories and their relevance to conservation. *British Birds* 91: 2–11.

Holyoak, D. 1971. Movements and mortality of Corvidae. *Bird Study* 18: 97–106.

Hoodless, A. N., Lang, D., Aebischer, N. J., Fuller, R. J. and Ewald, J. A. 2009. Densities and population estimates of breeding Eurasian Woodcock *Scolopax rusticola* in Britain in 2003. *Bird Study* 56: 15–25.

Hourlay, F., Libois, R., D'Amico, F., Sarà, M., O'Halloran, J. and Michaux, J. R. 2008. Evidence of a highly complex phylogeographic structure on a specialist river bird species, the dipper (*Cinclus cinclus*). *Molecular Phylogenetics and Evolution* 49: 435–444.

Howey, D. H. and Bell, M. 1985. Pallas's Warblers and other migrants in Britain and Ireland in October 1982. *British Birds* 78: 381–392.

Hudson, A. V., Stowe, T. J. and Aspinall, S. J. 1990. Status and distribution of Corncrakes in Britain in 1988. *British Birds* 83: 173–187.

Hudson, P. J. 1992. *Grouse in Space and Time. The Population Biology of a Managed Gamebird*. Game Conservancy, Fordingbridge.

Hudson, P. J., Dobson, A. P. and Newborn, D. 1999. Prevention of population cycles by parasite removal. *Science* 282: 2256–2258.

Hudson, R. 1965. The spread of the Collared Dove in Britain and Ireland. *British Birds* 58: 105–139.

Hudson, R. 1976. Ruddy Ducks in Britain. *British Birds* 69: 132–142.

Hughes, B. 1998. *Oxyura jamaicensis* Ruddy Duck. *BWP Update* 2: 159–171.

Hughes, J. M. 1999. Yellow-billed Cuckoo (*Coccyzus americanus*). In A. Poole and F. Gill (eds), *The Birds of North America, No. 418*. The Birds of North America, Inc., Philadelphia, PA.

Hughes, J. M. 2001. Black-billed Cuckoo (*Coccyzus erythropthalmus*). In A. Poole and F. Gill (eds), *The Birds of North America, No. 587*. The Birds of North America, Inc., Philadelphia, PA.

Hughes, R., Robinson, J., Green, A., Li, D. and Mundkur, T. 2005. *International Single Species Action Plan for the White-headed Duck* Oxyura leucocephala. Wetlands International, Slimbridge.

Hughes, S. W. M., Bacon, P. and Flegg, J. J. M. 1979. The 1975 census of the Great Crested Grebe in Britain. *Bird Study* 26: 213–226.

Hunt, P. D. and Flaspohler, D. J. 1998. Yellow-rumped Warbler (*Dendroica coronata*). In A. Poole and F. Gill (eds), *The Birds of North America, No. 376*. The Birds of North America, Inc., Philadelphia, PA.

Hutchinson, C. D. 1989. *Birds in Ireland*. T and AD Poyser, Berkhamsted.

Hutchinson, C. D. and Neath, B. 1978. Little Gulls in Britain and Ireland. *British Birds* 71: 563–582.

Illera, J. C., Richardson, D. S., Helm, B., Atienza, J. C. and Emerson, B. C. 2008. Phylogenetic relationships, biogeography and speciation in the avian genus *Saxicola*. *Molecular Phylogenetics and Evolution* 48: 1145–1154.

Ilyashenko, V.Yu. 2008. [The new form of crane from Trans-Caucasus.] *Russian Journal of Ornithology* 17: 559–562.

Inglis, I. R., Isaacson, A. J., Thearle, R. J. P. and Westwood, N. J. 1990. The effects of changing agricultural practice on Woodpigeon *Columba palumbus* numbers. *Ibis* 132: 262–272.

Irwin, D. E. 2000. Song variation in an avian ring species. *Evolution* 54: 998–1010.

Irwin, D. E., Alström, P., Olsson, U. and Benowitz-Fredericks, Z. M. 2001a. Cryptic species in the genus *Phylloscopus* (Old World leaf warblers). *Ibis* 143: 233–247.

Irwin, D. E., Bensch, S., Irwin, J. H. and Price, T. D. 2005. Speciation by distance in a ring species. *Science* 307: 414–416.

Irwin, D. E., Bensch, S. and Price, T. D. 2001b. Speciation in a ring. *Nature* 409: 333–337.

Isenmann, P. 1976. Contribution à l'étude de la biologie de la reproduction et de l'étho-écologie du Goéland Railleur (*Larus genei*). *Ardea* 64: 48–61.

Jackson, B. J. S and Jackson, J. A. 2000. Killdeer (*Charadrius vociferus*). In A. Poole and F. Gill (eds), *The Birds of North America, No. 517*. The Birds of North America, Inc., Philadelphia, PA.

James, P. C. 1986. Little Shearwaters in Britain and Ireland. *British Birds* 79: 28–33.

Jefferies, D. J. and Parslow, J. L. F. 1976. The genetics of bridling in Guillemots from a study of hand-reared birds. *Journal of Zoology, London* 179: 411–420.

Jehl, J. R., Klima, J. and Harris, R. E. 2001. Short-billed Dowitcher (*Limnodromus griseus*). In A. Poole and F. Gill (eds), *The Birds of North America, No. 564*. The Birds of North America, Inc., Philadelphia, PA.

Jesus, J., Menezes, D. L., Gomes, S., Oliveira, P., Nogales, M. and Brehm, A. N. In press. Phylogenetic relationships of gadfly petrels *Pterodroma* spp. from the northeastern Atlantic Ocean: molecular evidence for specific status of Bugio and Cape Verde petrels and implications for conservation. *Bird Conservation International*.

Johansson, U. S., Alström, P., Olsson, U., Ericson, P.G.P., Sundberg, P. and Price, T.D. 2007. Build-up of the Himalayan avifauna through immigration: a biogeographical analysis of the *Phylloscopus* and *Seicercus* warblers. *Evolution* 61: 324–333.

Johnsen, A., Andersson, S., Fernandez, J. G., Kempenaers, B., Pavel, V., Questiau, S., Raess, M., Rindal, E. and Lifjeld, J. T. 2006. Molecular and phenotypic divergence in the Bluethroat (*Luscinia svecica*) complex. *Molecular Ecology* 15: 4033–4047.

Johnsgard, P. A. 1965. *Handbook of Waterfowl Behavior*. Cornell University Press, Ithaca.

Johnsgard, P.A. 1988. *The Quails, Partridges and Francolins of the World*. Oxford University Press, Oxford.

Johnsgard, P. A. 1993. *Cormorants, Darters and Pelicans of the World*. Smithsonian Institute Press, Washington, DC.

Johnson, J. A., Lerner, H. R. L., Rasmussen, P. C. and Mindell, D. P. 2006. Systematics within

Gyps vultures: a clade at risk. *BMC Evolutionary Biology* 6: 65–76.

Johnson, J. A., Watson, R. T. and Mindell, D. P. 2005. Prioritizing species conservation: does the Cape Verde Kite exist? *Proceedings of the Royal Society of London B* 272: 1365–1371.

Johnson, K. P. and Clayton, D. H. 2000. A molecular phylogeny of the dove genus *Zenaida*: mitochondrial and nuclear DNA sequences. *Condor* 102: 864–870.

Johnson, K. P., de Kort, S., Dinwoodey, K., Mateman, A. C., ten Cate, C., Lessells, C. M. and Clayton, D. H. 2001. A molecular phylogeny of the dove genera *Streptopelia* and *Columba*. *Auk* 118: 874–887.

Johnson, K. P., McKinney, F., Wilson, R. and Sorenson, M. D. 2000. The evolution of postcopulatory displays in dabbling ducks (Anatini): a phylogenetic perspective. *Animal Behaviour* 59: 953–963.

Johnson, K. P. and Sorenson, M. D. 1998. Comparing molecular evolution in two mitochondrial protein coding genes (cytochrome *b* and ND2) in the dabbling ducks (Tribe Anatini). *Molecular Phylogenetics and Evolution* 10: 82–94.

Johnson, K. P. and Sorenson, M. D. 1999. Phylogeny and biogeography of dabbling ducks (genus *Anas*): a comparison of molecular and morphological evidence. *Auk* 116: 792–805.

Johnson, N. K. and Cicero, C. 2002. The role of ecologic diversification in sibling speciation of *Empidonax* flycatchers (Tyrannidae): multigene evidence from mtDNA. *Molecular Ecology* 11: 2065–2081.

Johnson, N. K., Zink, R. M. and Marten, J. A. 1988. Genetic evidence for relationships in the avian family Vireonidae. *Condor* 90: 428–445.

Johnson, O. W. and Connors, P. G. 1996. American Golden-Plover (*Pluvialis dominica*), Pacific Golden-Plover (*Pluvialis fulva*). In A. Poole and F. Gill (eds), *The Birds of North America, No. 201–202*. The American Ornithologists' Union, Washington, DC.

Johnson, O. W., Morton, M. L., Bruner, P. L. and Johnson, P. M. 1989. Fat cyclicity, predicted migratory flight ranges, and features of wintering behavior in Pacific Golden-Plovers. *Condor* 91: 156–177.

Johnstone, I., Thorpe, R., Moore, A. and Finney, S. 2007. Breeding status of Choughs *Pyrrhocorax pyrrhocorax* in the UK and Isle of Man in 2002. *Bird Study* 54: 23–34.

Jones, K. L., Krapu, G. L., Brandt, D. A. and Ashley, M. V. 2005. Population genetic structure in migratory sandhill cranes and the role of Pleistocene glaciations. *Molecular Ecology* 14: 2645–2657.

Jones, P. W. and Donovan, T. M. 1996. Hermit Thrush (*Catharus guttatus*). In A. Poole and F. Gill (eds), *The Birds of North America, No. 261*. The Academy of Natural Sciences, Philadelphia, PA, and The American Ornithologists' Union, Washington, DC.

Joseph, L., Lessa, E. P. and Christidis, L. 1999. Phylogeography and biogeography in the evolution of migration: shorebirds of the *Charadrius* complex. *Journal of Biogeography* 26: 329–342.

Jukema, J., Piersma, T., Hulscher, J. B., Bunskoeke, E. J., Koolhaas, A. and Veenstra, A. 2001. *Golden Plovers and Wilsternetters: a Deeply Rooted Fascination with Migrating Birds*. KNNV Uitgeverij, Utrecht.

Juniper, T. and Parr, M. 1998. *Parrots: a Guide to the Parrots of the World*. Pica Press, Sussex.

Kaboli, M., Aliabadian, M., Guillaumet, A., Roselaar, C. S. and Prodon, R. 2007. Ecomorphology of the wheatears (genus *Oenanthe*). *Ibis* 126: 410–415.

Kampe-Persson, H. 2002. *Anser anser* Greylag Goose. *BWP Update* 4: 181–216.

Kehoe, C. 2006. Racial identification and assessment in Britain: a report from the RIACT subcommittee. *British Birds* 99: 619–645.

Kelleher, K. and O'Halloran, J. 2006. The breeding biology of the Song Thrush, *Turdus philomelus*, in an island population. *Bird Study* 53: 142–155.

Kelsey, M. G., Green, G. H., Garnett, M. C. and Hayman, P. V. 1989. Marsh Warblers in Britain. *British Birds* 82: 239–256.

Kennedy, M., Gray, R. D. and Spencer, H. G. 2000. The phylogenetic relationships of the shags and cormorants: can sequence data resolve a disagreement between behaviour and morphology? *Molecular Phylogenetics and Evolution* 17: 345–359.

Kennedy, M. and Spencer, H. G. 2003. Phylogenies of the frigatebirds (Fregatidae) and tropicbirds (Phaethonidae), two divergent groups of the traditional order Pelecaniformes, inferred from mitochondrial DNA sequences. *Molecular Phylogenetics and Evolution* 31: 31–38.

Kershaw, M. and Cranswick P. A. 2003. Numbers of wintering waterbirds in Great Britain, 1994/95–1998/99: I. wildfowl and selected waterbirds. *Biological Conservation* 111: 91–104.

Kidd, M. G. and Friesen, V. L. 1998a. Sequence variation in the guillemot (Alcidae: *Cepphus*) mitochondrial control region and its nuclear homolog. *Molecular Biology and Evolution* 15: 61–70.

Kidd, M. G. and Friesen, V. L. 1998b. Analysis of mechanisms of microevolutionary change in *Cepphus* guillemots using patterns of control region variation. *Evolution* 52: 1158–1168.

Kimball, R. T., Braun, E. L., Zwartjes, P. W., Crowe, T. M. and Ligon, J. D. 1999. A molecular phylogeny of the pheasants and partridges suggests that these lineages are not monophyletic. *Molecular Phylogenetics and Evolution* 11: 38–54.

Kirby, J. S., Kirby, K. K. and Woolfall, S. J. 1989. Curlew Sandpipers in Britain and Ireland in autumn 1988. *British Birds* 82: 399–409.

Klein, N. K., Burns, K. J., Hackett, S. J. and Griffiths, C. S. 2004. Molecular phylogenetic relationships among the wood warblers (Parulidae) and historical biogeography in the Caribbean basin. *Journal of Caribbean Ornithology* 17: 3–17.

Klicka, J., Burns, K. and Spellman, G. M. 2007. Defining a monophyletic Cardinalini: a molecular perspective. *Molecular Phylogenetics and Evolution* 45: 1014–1032.

Klicka, J., Fry, A. J., Zink, R. M. and Thompson, C. W. 2001. A cytochrome-b perspective on *Passerina* bunting relationships. *Auk* 118: 611–623.

Klicka, J. and Spellman, G. M. 2007. A molecular evaluation of the North American 'Grassland' sparrow clade. *Auk* 124: 537–551.

Klicka, J., Voelker, G. and Spellman, G. M. 2005. A molecular phylogenetic analysis of the 'true thrushes' (Aves: Turdinae). *Molecular Phylogenetics and Evolution* 34: 486–500.

Klicka, J., Zink, R. M. and Winker, K. 2003. Longspurs and snow buntings: phylogeny and biogeography of a high-latitude clade. *Molecular Phylogenetics and Evolution* 26: 165–175.

Knox, A. 1996. Grey-cheeked and Bicknell's Thrushes: taxonomy, identification and the British and Irish records. *British Birds* 89: 1–9.

Knox, A. G. 1975. Crossbill Taxonomy In D. Nethersole-Thompson (ed.) *Pine Crossbills*, pp. 191–201. Poyser, Berkhamsted.

Knox, A. G. 1976. The taxonomic status of the

Scottish Crossbill. *Bulletin of the British Ornithologists' Club* 96: 15–19.

Knox, A. G. 1987. Taxonomic status of 'Lesser Golden Plovers'. *British Birds* 80: 482–487.

Knox, A. G. 1988a. Taxonomy of the Rock/Water Pipit superspecies *Anthus petrosus, spinoletta, rubescens. British Birds* 81: 206–211.

Knox, A. G. 1988b. The taxonomy of redpolls. *Ardea* 76: 1–26.

Knox, A. G. 1990. The sympatric breeding of Common and Scottish Crossbills *Loxia curvirostra* and *L. scotica*, and the evolution of crossbills. *Ibis* 132: 454–466.

Knox, A. G. 1993. Richard Meinertzhagen – a case of fraud examined. *Ibis* 135: 320–325.

Knox, A. G. 2007. Order or chaos? Taxonomy and the British List over the last 100 years. *British Birds* 100: 609–623.

Knox, A. G., Helbig, A. J., Parkin, D. T. and Sangster, G. 2001. The taxonomic status of Lesser Redpoll. *British Birds* 94: 260–267.

Knox, A. G. and Lowther, P. E. 2000a. Common Redpoll (*Carduelis flammea*). In A. Poole and F. Gill (eds), *The Birds of North America, No. 543*. The Birds of North America, Inc., Philadelphia, PA.

Knox, A. G. and Lowther, P. E. 2000b. Hoary Redpoll (*Carduelis hornemanni*). In A. Poole and F. Gill (eds), *The Birds of North America, No. 544*. The Birds of North America, Inc., Philadelphia, PA.

Koop, B. 2003. *Podiceps nigricollis* Black-necked Grebe. *BWP Update* 5: 185–202.

Koopman, M. E., McDonald, D. B., Hayward, G. D., Eldegarde, K., Sonerud, G. A. and Sermach, S. G. 2005. Genetic similarity among Eurasian subspecies of boreal owls *Aegolius funereus. Journal of Avian Biology* 36: 179–183.

Krajewski, C. and King, D. G. 1996. Molecular divergence and phylogeny: rates and patterns of cytochrome *b* evolution in cranes. *Molecular Biology and Evolution* 13: 21–30.

Kretzmann, M. B., Capote, N., Gautschi, B., Godoy, J. A., Donázar, J. A. and Negro, J. J. 2003. Genetically distinct island populations of the Egyptian Vulture (*Neophron percnopterus*). *Conservation Genetics* 4: 697–706.

Kruckenhauser, L., Haring, E., Pinsker, W., Riesing, M. J., Winkler, H., Wink, M. and Gamauf, A. 2004. Genetic vs. morphological differentiation of Old World buzzards (genus *Buteo*, Accipitridae). *Zoologica Scripta* 33: 197–211.

Kryukov, A. P. 1995. Systematics of small Palearctic shrikes of the 'cristatus group'. In R. Yosef and F. E. Lohrer (eds), *Shrikes (Laniidae) of the World: Biology and Conservation. Proceedings of the Western Foundation of Vertebrate Zoology* 6: 22–25.

Kulikova, I. V., Drovetski, S. V., Gibson, D. D., Harrigan, R. J., Rohwer, S., Sorenson, M. D., Winker, K., Zhuravlev, Y. N. and McCracken, K. G. 2005. Phylogeography of the Mallard (*Anas platyrhynchos*): hybridization, dispersal, and lineage sorting contribute to complex geographic structure. *Auk* 122: 949–956.

Kulikova, I. V., Zhuravlev, Y. N. and McCracken, K. G. 2004. Asymmetric hybridisation and sex-biased gene flow between Eastern Spot-billed Ducks (*Anas zonorhyncha*) and Mallards (*Anas platyrhynchos*) in the Russian Far East. *Auk* 121: 930–949.

Kvist, L. 2003. On the phylogenetic status of the British Great Tit *Parus major newtoni* and Blue Tit *Parus caeruleus obscurus. Avian Science* 3: 31–35.

Kvist, L., Broggi, J., Illera, J. C. and Koivula, K. 2005. Colonisation and diversification of the blue tits (*Parus caeruleus teneriffae*-group) in

the Canary Islands. *Molecular Phylogenetics and Evolution* 34: 501–511.

Kvist, L., Martens, J., Ahola, A. and Orell, M. 2001. Phylogeography of a Palearctic sedentary passerine, the Willow Tit (*Parus montanus*). *Journal of Evolutionary Biology* 14: 930–941.

Kvist, L., Martens, J., Higuchi, H., Nazarenko, A. A., Valchuk, O. P and Orell, M. 2003. Evolution and genetic structure of the Great Tit (*Parus major*) complex. *Proceedings of the Royal Society of London B* 270: 1447–1454.

Kvist, L., Ruokonen, M., Lumme, J. and Orell, M. 1999a. Different population structures in northern and southern populations of the European Blue Tit (*Parus caeruleus*). *Journal of Evolutionary Biology* 12: 798–805.

Kvist, L., Ruokonen, M., Lumme, J. and Orell, M. 1999b. The colonisation history and present-day population structure of the European Great Tit (*Parus major major*). *Heredity* 82: 495–502.

Kvist, L., Vliri, K., Dias, P. C., Rytkönen, S. and Orell, M. 2004. Glacial history and colonisation of Europe by the Blue Tit *Parus caeruleus. Journal of Avian Biology* 35: 352–399.

Lack, P. C. 1986. *The Atlas of Wintering Birds in Britain and Ireland*. T and AD Poyser, Calton.

Lack, P. C. 1989. Overall and regional trends in warbler population of British farmland over 25 years. *Annales Zoologici Fennici* 26: 219–225.

Lalanne, Y., Hémery, G., Cagnon, C., d'Ebbée, E. and Mouchés, C. 2001. Discrimination morphologique des sous-especes d'océanites tempête: nouveaux résultants pour deux populations méditerranéenes. *Alauda* 69: 475–482.

Lanctot, R., Goatcher, B., Scribner, K., Talbot, S., Pierson, B., Esler, D. and Zwiefelhofer, D. 1999. Harlequin recovery from the Exxon Valdez oil spill: a population genetics perspective. *Auk* 116: 781–791.

Langston, R., Gregory, R. and Adams, R. 2002. The status of the Hawfinch in the UK 1975–1999. *British Birds* 95: 166–173.

Langston, R. H. W., Smith, T., Brown, A. F. and Gregory, R. D. 2006. Status of breeding Twite *Carduelis flavirostris* in the UK. *Bird Study* 53: 55–63.

Lanyon, S. M. and Omland, K. E. 1999. A molecular phylogeny of the blackbirds (Icteridae): five lineages revealed by cytochrome-*B* sequence data. *Auk* 116: 629–639.

Larsen, C., Speed, M., Harvey, N. and Noyes, H. A. 2007. A molecular phylogeny of the Nightjars Aves: Caprimulgidae suggests extensive conservation of primitive morphological traits across multiple lineages. *Molecular Phylogenetics and Evolution* 42: 789–796.

Lashko, A. 2004. *Population Genetic Relationships in the Roseate Tern: Globally, Regionally and Locally*. Unpublished PhD thesis, James Cook University, Townsville.

Lauga, B., Cagnon, C., D'Amico, F., Karama, S. and Mouchès, C. 2005. Phylogeography of the white-throated dipper *Cinctus cinclus* in Europe. *Journal of Ornithology* 146: 257–262.

Law, J. R. 1931. An orangeless mutant of the Varied Thrush and its bearing on sex color differences. *Condor* 33: 151–153.

Lebreton J.-D, Hines, J. E., Pradel, D., Nicholls, J. D. and Spendelow, J. A. 2003. Estimation by capture–recapture of recruitment and dispersal over several sites. *Oikos* 101: 253–264.

Lee, P. L. M., Richardson, L.. J. and Bradbury, R. B. 2001. The phylogenetic status of the Corn Bunting *Miliaria calandra* based on

mitochondrial control-region NDA sequences. *Ibis* 143: 299–303.

Lee, S.-I., Parr, C. S., Hwang, Y., Mindell, D. P. and Choe, J. C. 2003. Phylogeny of magpies (genus *Pica*) inferred from mtDNA data. *Molecular Phylogenetics and Evolution* 29: 250–257.

Leech, D. I., Crick, H. Q. P. and Shawyer, C. R. 2005. *The BTO Barn Owl Monitoring Programme: Fifth Year 2004*. BTO Research Report 424. BTO, Thetford.

Lefranc, N. and Worfolk, T. 1997. *Shrikes: a Guide to the Shrikes of the World*. Pica Press, Sussex.

Leisler, B., Heidrich, P., Schultze-Hagen, K. and Wink, M. 1997. Taxonomy and phylogeny of reed warblers (genus *Acrocephalus*) based on mtDNA sequences and morphology. *Journal of Ornithology* 138: 469–496.

Leonovich, V. V., Deminia, G. V. and Veprintseva, O. D. 1997. On the taxonomy and phylogeny of pipits (Genus *Anthus*, Motacillidae, Aves) in Eurasia. *Otdel biologicheskii*. 102: 14–22. [In Russian.]

Lerner, H. R. L. and Mindell, D. P. 2005. Phylogeny of eagles, Old World vultures, and other Accipitridae based on nuclear and mitochondrial DNA. *Molecular Phylogenetics and Evolution* 37: 327–346.

Lever, C. 2005. *Naturalised Birds of the World*. Christopher Helm, London.

Li, G., Jones, G., Rossiter, S. J., Chen, S. F., Parsons, S. and Zhang, S. Y. 2006. Phylogenetics of small horseshoe bats from east Asia based on mitochondrial DNA sequence variation. *Journal of Mammalogy* 87: 1234–1240.

Li, W. and Zhang, Y.-Y. 2004. Subspecific taxonomy of *Ficedula parva* based on sequences of mitochondrial cytochrome b gene. *Zoological Research* 25: 127–131.

Liebers, D., de Knijff, P. and Helbig, A. J. 2004. The Herring Gull complex is not a ring species. *Proceedings of the Royal Society of London B* 271: 893–901.

Liebers, D. and Helbig, A. J. 2002. Phylogeography and colonization history of Lesser Black-backed Gulls (*Larus fuscus*) as revealed by mtDNA sequences. *Journal of Evolutionary Biology* 15: 1021–1033.

Liebers, D., Helbig, A. J. and de Knijff, P. 2001. Genetic differentiation and phylogeography of gulls in the *Larus cachinnans–fuscus* group (Aves: Charadriiformes). *Molecular Ecology* 10: 2447–2462.

Lifjeld, J. T. and Bjerke, B. A. 1996. Evidence for assortative pairing by the *cabaret* and *flammea* subspecies of the Common Redpoll *Carduelis flammea* in SE Norway. *Cinclus* 19: 1–8.

Lifjeld, J. T., Bjornstad, G., Steen, O. F. and Nesje, M. 2002. Reduced genetic variation in Norwegian Peregrine Falcons *Falco peregrinus* indicated by minisatellite DNA fingerprinting. *Ibis* 144: E19–E26.

Limpert, R. J. and Earnst, S. L. 1994. Tundra Swan (*Cygnus columbianus*). In A. Poole and F. Gill (eds), *The Birds of North America, No. 89*. The Academy of Natural Sciences, Philadelphia.

Lindström, J., Rintamäki, P. T. and Storch, I. 1998. *Tetrao tetrix* Black Grouse. *BWP Update* 2: 173–191.

Little, B. and Furness, R. W. 1985. Long distance moult migration by British Goosanders *Mergus merganser. Ringing and Migration* 6: 77–82.

Livezey, B. C. 1986. A phylogenetic analysis of recent Anseriform genera using morphological characters. *Auk* 103: 737–754.

Livezey, B. 1995. Phylogeny and evolutionary ecology of modern seaducks (Anatidae: Mergini). *Condor* 97: 233–255.

Livezey, B. C. 1996. A phylogenetic analysis of geese and swans (Anseriformes: Anserinae), including selected fossil species. *Systematic Biology* 45: 415–450.

Lloyd, C. S., Tasker, M. L. and Partridge, K. 1991. *The Status of Seabirds in Britain and Ireland.* T and AD Poyser, Calton.

Long, J. L. 1981. *Introduced Birds of the World.* David and Charles, London.

Longcore, J. R., McAuley, D. G., Hepp, G. R. and Rhymer, J. M. 2000. American Black Duck (*Anas rubripes*). In A. Poole and F. Gill (eds), *The Birds of North America, No. 481.* The Birds of North America, Inc., Philadelphia, PA.

Love, J. A. 1978. Leach's and Storm-petrels on North Rona 1971–1974. *Ringing and Migration* 2: 15–19.

Love, J. A. 1983. *The Return of the Sea Eagle.* Cambridge University Press, Cambridge.

Love, J. A. 2003. A history of the White-tailed Eagle in Scotland. In B. Helander, M. Marquiss and W. Bowerman (eds), *Sea Eagle 2000,* pp. 39–51. Swedish Society for Nature Conservation, Stockholm.

Lovette, I. J. and Bermingham, E. 1999. Explosive speciation in the New World *Dendroica* warblers. *Proceedings of the Royal Society of London B* 266: 1629–1636.

Lovette, I. J. and Hochachka, W. M. 2006. Simultaneous effects of phylogenetic niche conservatism and competition on avian community structure. *Ecology* 87: S14–S28.

Lovette, I. J., McCleery, B. V., Talaba, A. L. and Rubenstein, D. R. 2008. A complete species-level molecular phylogeny for the 'Eurasian' starlings (Sturnidae: *Sturnus, Acridotheres,* and allies): recent diversification in a highly social and dispersive avian group. *Molecular Phylogenetics and Evolution* 47: 251–260.

Lovette, I. J. and Rubenstein, D. R. 2007. A comprehensive molecular phylogeny of the starlings (Aves: Sturnidae) and mockingbirds (Aves: Mimidae): congruent mtDNA and nuclear trees for a cosmopolitan avian radiation. *Molecular Phylogenetics and Evolution* 44: 1031–1056.

Lowther, P. E., Celada, C., Klein, N. K., Rimmer, C. C. and Spector, D. A. 1999. Yellow Warbler (*Dendroica petechia*). In A. Poole and F. Gill (eds), *The Birds of North America, No. 454.* The Birds of North America, Inc., Philadelphia, PA.

Lowther, P. E., Rimmer, C. C., Kessel, B., Johnson, B. S. L. and Ellison, W. G. 2001. Gray-cheeked Thrush (*Catharus minimus*). In A. Poole and F. Gill (eds), *The Birds of North America, No. 591.* The Birds of North America, Inc., Philadelphia, PA.

Lucchini, V., Hoglund, J., Klaus, S., Swenson, J. and Randi, E. 2001. Historical biogeography and a mitochondrial DNA phylogeny of grouse and ptarmigan. *Molecular Phylogenetics and Evolution* 20: 149–162.

Luikkonen-Anttila, T., Uimaniemi, L., Orell, M. and Lumme, J. 2002. Mitochondrial DNA variation and the phylogeography of the Grey Partridge (*Perdix perdix*) in Europe: from Pleistocene history to present day populations. *Journal of Evolutionary Biology* 15: 971–982.

Madden, B. and Newton, S. 2004 Herring Gull *Larus argentatus.* In P. I. Mitchell, S. F. Newton, N. Ratcliffe and T. E. Dunn (eds), *Seabird Populations of Britain and Ireland,* pp. 242–262. T and AD Poyser, London.

Madge, S. 1987. Field identification of Radde's and Dusky Warblers. *British Birds* 80: 595–603.

Madge, S. and Burn, H. 1988. *Waterfowl: An Identification Guide to the Ducks, Geese and Swans of. the world.* Christopher Helm, London.

Madge, S. and McGowan, P. 2002. *Pheasants, Partridges and Grouse.* Christopher Helm, London.

Madsen, J., Cracknell, G. and Fox, A. D. (eds) 1999. *Goose Populations of the Western Palearctic: a Review of Status and Distribution.* Wetlands International Publ. No. 48. Wetlands International, Wageningen, The Netherlands. National Environmental Research Institute, Rönde, Denmark.

Mank, J. E., Carlson, J. E. and Brittingham, M. C. 2004. A century of hybridization: decreasing genetic distance between American Black Ducks and Mallards. *Conservation Genetics* 5: 395–403.

Marchant, J. H., Hudson, R., Carter, S. P. and Whittington, P. A. 1990. *Population Trends in British Breeding Birds.* BTO, Tring.

Markkola, J. and Tynjälä, M. 1993. The Finnish Lesser White-fronted Goose project. *The Ring* 15: 390–392.

Marks, B. D., Weckstein, J. D. and Moyle, R. G. 2007. Molecular phylogenetics of the bee-eaters (Aves: Meropidae) based on nuclear and mitochondrial DNA sequence data. *Molecular Phylogenetics and Evolution* 45: 23–32.

Marova, I. M. and Leonovitch, V. V. 1993. Hybridisation between Siberian (*Phylloscopus collybita tristis*) and East European (*P. collybita abietinus*) chiffchaffs in the area of sympatry. In O.L. Rossolimo (ed.), *Hybridisation and the problem of species in vertebrates. Archives of the Zoological Museum of the Moscow State University* 30: 147–163.

Marquiss, M. 1981. The Goshawk in Britain: its provenance and current status. In R. E. Kenwood and M. Lindsay (eds). *Understanding the Goshawk,* pp 43–55. International Association for Falconry, Oxford.

Marquiss, M. 2007. Seasonal pattern in hawk predation on Common Bullfinch *Pyrrhula pyrrhula*: evidence of an interaction with habitat affecting food availability. *Bird Study* 54: 1–11.

Marquiss, M. and Newton, I. 1982. The Goshawk in Britain. *British Birds* 75: 243–260.

Marquiss, M., Newton, I. and Ratcliffe, D. A. 1978. The decline of the Raven *Corvus corax* in relation to afforestation in southern Scotland and northern England. *Journal of Applied Ecology* 15:129–144.

Marquiss, M. and Rae, R. 2002. Ecological differentiation in relation to bill size amongst sympatryic, genetically undifferentiated cross-bills. *Ibis* 144: 494–508.

Marr, B. A. E. 1993. The BOU Records Committee: through a newcomer's eyes. *British Birds* 86: 423–429.

Marshall, H. D. and Baker, A. J. 1999. Colonization history of Atlantic Island Common Chaffinches (*Fringilla coelebs*) revealed by mitochondrial DNA. *Molecular Phylogenetics and Evolution* 11: 201–212.

Martens, J. and Eck, S. 1995. Towards an Ornithology of the Himalayas. Systematics, ecology and vocalizations of Nepal birds. *Bonner Zoological Monographs* 38: 1–445.

Martens, J. and Päckert, M. 2003. Disclosure of songbird diversity in the Palearctic/Oriental transition zone. *Proceedings of the 18th International Congress of. Zoology*: 551–558.

Martens, J. and Steil, B. 1997. Territorial songs and species differentiation in the lesser whitethroat superspecies *Sylvia (curruca). Journal of Ornithology* 138: 1–23.

Martens, J., Tieze, D. T., Eck, S. and Veith, M. 2004. Radiation and species limits in the Asian Pallas's Warbler complex (*Phylloscopus proregulus* s.l.). *Journal of Ornithology* 145: 206–222.

Marthinsen, G., Wennerberg, L., Solheim, R. and Lifjeld, J. T. 2009. No phylogeographic structure in the circumpolar Snowy Owl (*Bubo scandiacus*). *Conservation Genetics* 10: 923–933.

Mason, J. R. and Clark, L. 1990. Sarcosporidiosis observed more frequently in hybrids of Mallards and American Black Ducks. *Wilson Bulletin* 102: 160–162.

Masuda, R., Noro, M., Kurose, N., Nishida-Umehara, C., Takechi, H., Yamazaki, T., Kosuge, M. and Yochida, M. C. 1998. Genetic characteristics of endangered Japanese Golden Eagles (*Aquila chrysaeetus japonica*) based on mitochondrial DNA D-loop sequences and karyotypes. *Zoo Biology* 17: 111–121.

Mather, J. R. 1986. *The Birds of Yorkshire.* Croom Helm, London.

Matics, R. and Hoffmann, G. 2002. Location of the transition zone of the Barn Owl subspecies *Tyto alba alba* and *Tyto alba guttata* (Strigiformews: Tytonidae). *Acta Zoologica Cracoviensia* 45: 245–250.

Mauersberger, G. 1972. Über den taxonomischen Rang von *Emberiza godlewskii* Taczanowski. *Journal of Ornithology* 113: 53–59.

Mavor, R. A., Pickerell, G., Heubeck, M. and Mitchell, P. I. 2002. *Seabird Numbers and Breeding Success in Britain and Ireland 2000.* Joint Nature Conservation Committee, Peterborough.

Maynard Smith, J. 1966. Sympatric speciation. *American Naturalist* 100: 637–650.

Mayr, E. 1942. *Systematics and the Origin of Species.* Columbia University Press, New York.

Mayr, E. 1951. Speciation in birds. *Proceedings of the X International Ornithological Congress, Uppsala*: 91–131.

Mayr, E. 1963. *Animal Species and Evolution.* Harvard University Press, Cambridge, Mass.

McCoy, K. D., Boulinier, T. and Tirard, C. 2005. Comparative host-parasite population structures: disentangling prospecting and dispersal in the Black-legged Kittiwake *Rissa tridactyla. Molecular Ecology* 14: 2825–2838.

McCracken, K. G., Johnson, W. P. and Sheldon, F. H. 2001. Molecular population genetics, phylogeography, and conservation biology of the mottled duck (*Anas fulvigula*). *Conservation Genetics* 2: 87–102.

McCracken, K. G. and Sheldon, F. H. 1998. Molecular and osteological heron phylogenies: sources of incongruence. *Auk* 115: 127–141.

McCracken, K. G. and Sorenson, M. D. 2005. Is homoplasy or lineage sorting the source of incongruent mtDNA and nuclear gene trees in the stiff-tailed ducks (*Nomomyx-Oxyura*)? *Systematic Biology* 54: 35–55.

McGowan, R. Y. 2006. Comment on 'Holboell's Red-necked Grebe' in Wester Ross in 1925. *British Birds* 99: 481.

McGowan, R. Y., Clugston, D. L. and Forrester, R. W. 2003. Scotland's endemic subspecies. *Scottish Birds* 24: 18–35.

McGowan, R. Y. and Weir, D. N. 2002. Racial identification of Fair Isle Solitary Sandpiper. *British Birds* 95: 313–314.

McKenzie, A. J., Petty, S. J., Toms, M. P. and Furness, R. W. 2007. Importance of Sitka Spruce *Picea sitchensis* seed and garden bird-feeders for Siskins *Carduelis spinus* and Coal Tits *Periparus ater. Bird Study* 54: 236–247.

McKinney, F. 1961. An analysis of the displays of the European Eider *Somateria mollissima mollissima* (Linnaeus) and the Pacific Eider *Somateria mollissima v. nigra* Bonaparte. *Behaviour* Suppl. 7.

Mead, C. J. and Hudson, R. 1986. Report on bird-ringing for 1985. *Ringing and Migration* 7: 139–188.

Merila, J., Björklund, M. and Baker, A. J. 1997. Historical demography and present day population structure of the Greenfinch, *Carduelis chloris* – an analysis of mtDNA control-region sequences. *Evolution* 51: 946–956.

Merne, O. J. and Mitchell, P. I. 2004. Razorbill *Alca torda*. In P. I. Mitchell, S. F. Newton, N. Ratcliffe and T. E. Dunn (eds), *Seabird Populations of Britain and Ireland*, pp. 364–376. T and AD Poyser, London.

Messenger, D. 2001. Adult Little Gulls summering in Britain, 1975–1997 with comments on the likelihood of an expansion of the breeding range. *British Birds* 94: 310–314.

Michalek, K. G. and Miettinen, J. 2003. *Dendrocops major* Great Spotted Woodpecker. *BWP Update* 5: 101–184.

Mikkola, H. 1983. *Owls of Europe*. T and AD Poyser, Calton.

Milá, B., McCormack, J. E., Castaneda, G., Wayne, R. K. and Smith, T. B. 2007a. Recent postglacial range expansion drives the rapid diversification of a songbird lineage in the genus *Junco*. *Proceedings of the Royal Society of London B* 274: 2653–2660.

Milá, B., Smith, T. B. and Wayne, R. K. 2007b. Speciation and rapid phenotypic differentiation in the yellow-rumped warbler *Dendroica coronata* complex. *Molecular Ecology* 16: 159–173.

Miller, E. H., Gunn, W. W. H. and Harris, R. 1983. Geographic variation in aerial song of the Short-billed Dowitcher (Aves, Scolopacidae). *Canadian Journal of Zoology* 61: 2191–2198.

Milne, H., and Robertson, F. W. 1965. Polymorphisms in egg albumen protein and behaviour in the Eider Duck. *Nature* 205: 367–369.

Milwright, R. D. P. 1998. Breeding biology of the Golden Oriole *Oriolus oriolus* in the fenland basin of eastern England. *Bird Study* 45: 320–330.

Mirarchi, R. E. and Baskett, T. S. 1994. Mourning Dove (*Zenaida macroura*). In A. Poole and F. Gill (eds), *The Birds of North America, No. 117*. The Academy of Natural Sciences, Philadelphia, and The American Ornithologists' Union, Washington, DC.

Mitchell, C. and Hearn, R. D. 2004. Pink-footed Goose Anser brachyrhynchus (*Greenland/Iceland Population*) *in Britain and Ireland 1960/1–1999/2000*. Waterbird Review Series, WWT/JNCC, Slimbridge.

Mitchell, C., MacDonald, R. and Boyer, P. 1995. *Greylag Geese on the Uists*. WWT Report to JNCC.

Mitchell, P. I. 2004a. European Storm-petrel *Hydrobates pelagicus*. In P. I. Mitchell, S. F. Newton, N. Ratcliffe and T. E. Dunn (eds), *Seabird Populations of Britain and Ireland*, pp. 81–100. T and AD Poyser, London.

Mitchell, P.I. 2004b. Leach's Storm-petrel *Oceanodroma leucorhoa*. In P. I. Mitchell, S. F. Newton, N. Ratcliffe and T. E. Dunn (eds), *Seabird Populations of Britain and Ireland*, pp. 101–114. T and AD Poyser, London.

Mitchell, P.I. 2004c. Black Guillemot *Cepphus grylle*. In P. I. Mitchell, S. F. Newton, N. Ratcliffe and T. E. Dunn (eds), *Seabird Populations of Britain and Ireland*, pp. 377–391. T and AD Poyser, London.

Mitchell, P. I., Newton, S. F., Ratcliffe, N. and Dunn, T. E. 2004. *Seabird Populations of Britain and Ireland*. T and AD Poyser, London.

Moen, S. M. 1991. Morphologic and genetic variation among breeding colonies of the Atlantic Puffin (*Fratercula arctica*). *Auk* 108: 755–763.

Moldenhauer, R. R. 1992 Two song populations of the Northern Parula. *Auk* 109: 215–222.

Monaghan, P., Uttley, J. D. and Burns, M.D. 1992. Effects of changes in food availability on reproductive effort in Arctic Terns. *Ardea* 80: 71–81.

Monroe, B. L. and Browning, M. R. 1992. A reanalysis of *Butorides*. *Bulletin of the British Ornithologists'. Club* 112: 81–85.

Monteiro, L. R. and Furness, R. W. 1998. Speciation through temporal segregation of Madieran Storm-petrel (*Oceanodroma castro*) population in the Azores? *Proceedings of the Royal Society of London B* 353: 945–953.

Moon, S. J. 1983. The eventual identification of a Royal Tern in Mid-Glamorgan. *British Birds* 76: 335–339.

Moorcroft, D., Wilson, J. D. and Bradbury, R. B. 2006. Diet of nestling Linnets *Carduelis cannabina* on lowland farmland before and after agricultural intensification. *Bird Study* 53: 156–162.

Moorman, T. E. and Gray, P. N. 1994. Mottled Duck (*Anas fulvigula*). In A. Poole and F. Gill (eds),*The Birds of North America, No. 81*. The Academy of Natural Sciences, Philadelphia, and The American Ornithologists' Union, Washington, DC.

Moreau, R. E. 1972. *The Palearctic–Africa Bird Migration Systems*. Academic Press, London.

Morgan, D. H. W. 1993. Feral Rose-ringed Parakeets in Britain. *British Birds* 86: 561–564.

Morgan, R. and Glue, D. 1977. Breeding, mortality and movements of Kingfishers. *Bird Study* 24: 15–24.

Morgan, R. A. and Glue, D. 1981. Breeding survey of Black Redstarts in Britain, 1977. *Bird Study* 28: 163–168.

Morioka, H. and Shigeta, Y. 1993. Generic allocation of the Japanese Marsh Warbler *Megalurus pryeri* (Aves: Sylviidae). *Bulletin of the National Science Museum, Tokyo, (A)* 19: 37–43.

Morris, A., Burges, D., Fuller, R. J., Evans, A. D. and Smith, K. W. 1994. The status and distribution of Nightjars *Caprimulgus europaeus* in Britain in 1992. *Bird Study* 41: 181–191.

Morrison, R. I. G. 1984. Migration system of some New World shorebirds. In J. Burger and B. L. Olla (eds), *Behaviour of Marine Animals*, Vol. 6. Plenum Press, New York.

Moskoff, W. and Robinson, S. K. 1996. Philadelphia Vireo (*Vireo philadelphicus*). In A. Poole and F. Gill (eds),*The Birds of North America, No. 214*. The Academy of Natural Sciences, Philadelphia, and The American Ornithologists' Union, Washington, DC.

Moss, R., Oswald, J. and Baines, D. 2001. Climate change and breeding success: decline of the Capercaillie in Scotland. *Journal of Animal Ecology* 70: 47–61.

Moss, R., Wanless, S. and Harris, M. P. 2002. How small Northern Gannet colonies grow faster than big ones. *Waterbirds* 25: 442–448.

Moss, R., Watson, A. and Parr, R. 1996. Experimental prevention of a population cycle in Red Grouse. *Ecology* 77: 1512–1530.

Moum, T., Arnason, U. and Arnason, E. 2002. Mitochondrial DNA sequence evolution and phylogeny of the Atlantic Alcidae, Including the extinct Great Auk (*Pinguinus impennis*). *Molecular Biology and Evolution* 19: 1434–1439.

Mowbray, T. B. 2002. Canvasback (*Aythya valisineria*). In A. Poole and F. Gill (eds), *The Birds of North America, No. 659*. The Birds of North America, Inc., Philadelphia, PA.

Moyle, R. G. 2006. A molecular phylogeny of Kingfishers (Alcedinidae) with insights into early biogeographic history. *Auk* 123: 487–499.

Moyle, R. G., Slikas, B., Whittingham, L. A., Winkler, D. W. and Sheldon, F. H. 2008. DNA sequence assessment of phylogenetic relation-

ships among New World martins (Hirundinidae: *Progne*). *The Wilson Journal of Ornithology* 120: 683–691.

Moynihan, M. 1959. A revision of the family Laridae (Aves). *American Museum Novitates* 1928: 1–42.

Mulard, H., Aubin, T., White, J. F., Wagner, R. H. and Danchin. E. 2009. Voice variance may signify ongoing divergence among Black-legged Kittiwake populations. *Biological Journal of the Linnean Society* 97: 289–297.

Muñoz-Fuentes, V., Green, A. J., Negro, J.J. and Sorenson, M. D. 2005. Population structure and loss of genetic diversity in the endangered White-headed Duck *Oxyura leucocephala*. *Conservation Genetics* 6: 999–1015.

Muñoz -Fuentes, V., Green, A. J., Sorenson, M. D. Negro, J. J. and Vila, C. 2006. The Ruddy Duck *Oxyura jamaicensis* in Europe: natural colonization or human introduction? *Molecular Ecology* 15: 1441–1453.

Murton, R. K., Westwood, N. J. and Thearle, R. J. P. 1973. Polymorphism and the evolution of a continuous breeding season in the pigeon, *Columba livia*. *Journal of Reproduction and Fertility, Suppl.* 19: 563–577.

Musgrove, A. J. 2002. The non-breeding status of the Little Egret in Britain. *British Birds* 95: 62–80.

Nankinov, D., Dalakchieva, S., Popov, K. and Kirilov, S. 2003. Records of the Slender-billed Curlew *Numenius tenuirostris* in Bulgaria during the last ten years (1993–2002). *Wader Study Group Bulletin* 26: 101–102.

Naylor, K. 2005. *Rare Birds of Great Britain and Ireland*. Privately published CD.

Nelson, J. B. 1978. *The Sulidae: Gannets and Boobies*. OUP, Oxford.

Nelson, J. B. 2002. *The Atlantic Gannet*. Fenix Books, Norfolk.

Nelson, S. H., Court, I., Vickery, J. A., Watts, P. N. and Bradbury, R. B. 2003. The status and ecology of the Yellow Wagtail in Britain. *British Wildlife* 14: 270–274.

Nesje, M., Røed, K. H., Bell, D. A., Lindberg, P. and Lifjeld, J. T. 2000. Microsatellite analysis of population structure and genetic variability in Peregrine Falcons (*Falco peregrinus*). *Animal Conservation* 3: 267–275.

Nettleship, D. N. and Birkhead, T. R. 1985. *The Atlantic Alcidae*. Academic Press, London.

Newbery, P., Kitchin, C., Grice, P. and Ellis, J. 2004. The Corn Crake returns to England. *British Birds* 97: 548–549.

Newson, S. E., Hughes, B., Russell, I. C., Ekins, G. R. and Sellers, R. M. 2004. Sub-specific differentiation and distribution of Great Cormorants *Phalacrocorax carbo* in Europe. *Ardea* 92: 3–10.

Newton, I. 1967. The feeding ecology of the Bullfinch (*Pyrrhula pyrrhula* L.) in southern England. *Journal of Animal Ecology* 36: 721–744.

Newton, I. 1973. Egg breakage and breeding failure in British Merlins. *Bird Study* 20: 241–244.

Newton, I. 1974. Changes attributed to pesticides in the nesting success of the Sparrowhawk in Britain. *Journal of Applied Ecology* 11: 95–102.

Newton, I. 1979. *Population Ecology of Raptors*. T and AD Poyser, Berkhamsted.

Newton, I. and Haas, M. B. 1988. Pollutants in Merlin eggs and their effects on breeding. *British Birds* 81: 258–269.

Newton, I., Haas, M. B. and Freestone, P. 1990. Trends in organochlorine and mercury levels in Gannet eggs. *Environmental Pollution* 63: 1–12.

Newton, I., Hobson, K. A., Fox, A. D. and Marquiss, M. 2006. An investigation into the provenance of Northern Bullfinches *Pyrrhula p. pyrrhula* found in winter in Scotland and

Denmark. *Journal of Avian Biology* 37: 431–435.

Newton, S. F. 2004. Roseate Tern *Sterna dougallii*. In P. I. Mitchell, S. F. Newton, N. Ratcliffe and T. E. Dunn (eds), *Seabird Populations of Britain and Ireland*, pp. 302–314. T and AD Poyser, London.

Newton, S. F., Thompson, K. and Mitchell, P. I. 2004. Manx Shearwater *Puffinus puffinus*. In P. I. Mitchell, S. F. Newton, N. Ratcliffe and T. E. Dunn (eds), *Seabird Populations of Britain and Ireland*, pp. 63–80. T and AD Poyser, London.

Nguembock, B., Fjeldså, J., Couloux, A. and Pasquet, E. 2008. Molecular phylogeny of Carduelinae (Aves, Passeriformes, Fringillidae) proves polyphyletic origin of the genera *Serinus* and *Carduelis* and suggests redefined generic limits. *Molecular Phylogenetics and Evolution* 51: 169–181.

Nightingale, B. 2005. The status of Lady Amherst's Pheasant in Britain. *British Birds* 98: 20–25.

Nittinger, F., Gamauf, A., Pinsker, W., Wink, M. and Haring, E. 2007. Phylogeography and population structure of the Saker Falcon (*Falco cherrug*) and the influence of hybridization: mitochondrial and microsatellite data. *Molecular Ecology* 16: 1497–1517.

Nittinger, F., Haring, E., Pinsker, W., Wink, M. and Gamauf, A. 2005. Out of Africa? Phylogenetic relationships between *Falco biarmicus* and the other hierofalcons (Aves: Falconidae). *Journal of Zoological Systematics and Evolutionary Research* 43: 321–331.

Nol, E. and Blanken, M. S. 1999. Semipalmated Plover (*Charadrius semipalmatus*). In A. Poole and F. Gill (eds), *The Birds of North America, No. 444*. The Birds of North America Inc., Philadelphia, PA.

Nunes, M. 2000. Madeiran Storm-petrel (*Oceanodroma castro*) in the Desertas Islands (Madeira Archipelago): a new case of two distinct populations breeding annually? *Arquipelago*. Life and Marine Sciences Supplement 2 (Part A): 175–179. Ponta Delgada, Madeira.

Nunn, G. B., Cooper, J., Jouventin, P., Robertson, C. J. R. and Robertson G. G. 1996. Evolutionary relationships among extant albatrosses (Procellariiformes: Diomedeidae) established from complete cytochrome-*b* gene sequences. *Auk* 113: 784–801.

Nygård, T., Frantzen, B. and Švažas, S. 1995. Steller's Eider *Polysticta stelleri* wintering in Europe: numbers, distribution and origin. *Wildfowl* 46: 140–155.

O'Brien, M., Green, R. E. and Wilson, J. 2006. Partial recovery of the population of Corncrakes *Crex crex* in Britain, 1993–2004. *Bird Study* 53: 213–224.

O'Brien, S. H., Wilson, L. J., Webb, A. and Cranswick, P. A. 2008. Revised estimate of wintering Red-throated Divers *Gavia stellata* in Great Britain. *Bird Study* 55: 152–160.

O'Connor, R. J. and Shrubb, M. 1986. *Farming and Birds*. Cambridge University Press, Cambridge.

O'Donald, P. 1983. *The Arctic Skua: a Study of the Ecology and Evolution of a Seabird*. Cambridge University Press, Cambridge.

Odeen, A. and Alström, P. 2001. Evolution of secondary sexual traits in wagtails (genus *Motacilla*). In A. Odeen, *Effects of Postglacial Range Expansion and Population Bottlenecks on Species Richness*. PhD Thesis, Uppsala University.

Odeen, A. and Björklund, M. 2003. Dynamics in the evolution of sexual traits: losses and gains, radiation and convergence in yellow wagtails (*Motacilla flava*). *Molecular Ecology* 12: 2113–2130.

O'Flynn, W. J. 1983. Population changes of the Hen Harrier in Ireland. *Irish Birds* 2: 337–343.

Olsen, K. M. and Larsson, H. 1995. *Terns of Europe and North America*. Helm, London.

Olsen, K. M. and Larsson, H. 1997. *Skuas and Jaegers*. Pica Press, Sussex.

Olsen, K. M. and Larsson, H. 2003. *Gulls of Europe, Asia and North America*. Helm, London.

Olsson, U., Alström, P., Ericson, P. G. P. and Sundberg, P. 2005. Non-monophyletic taxa and cryptic species – evidence from a molecular phylogeny of leaf-warblers (*Phylloscopus*, Aves). *Molecular Phylogenetics and Evolution* 36: 261–276.

Omland, K. E. and Kondo, B. 2006. Phylogenetic studies of plumage evolution and speciation in New World orioles (*Icterus*). *Acta Zoologica Sinica* 52 (Suppl.): 320–326.

Omland, K. E., Lanyon, S. M. and Fritz, S. J. 1999. A molecular phylogeny of the New World Orioles (*Icterus*): the importance of dense taxon sampling. *Molecular Phylogenetics and Evolution* 12: 224–239.

Omland, K. E., Tarr, C. L., Boarman, W. I., Marzluff, J. M. and Fleischer, R. C. 2000. Cryptic genetic variation and paraphyly in ravens. *Proceedings of the Royal Society of London B* 267: 2475–2482.

Orell, M. 1989. Population fluctuations and survival of Great Tits *Parus major* dependent on food supplied by man in winter. *Ibis* 131: 112–27.

Ormerod, S. J. and Tyler, S. 1992. Patterns of contamination by organochlorines and mercury in the eggs of two river passerines in Britain and Ireland, with reference to individual PCB congeners. *Environmental Pollution* 76: 223–243.

Ottvall, R., Höglund, J., Bensch, S. and Larsson, K. 2005. Population differentiation in the Redshank (*Tringa totanus*) as revealed by mitochondrial DNA and amplified fragment length polymorphism markers. *Conservation Genetics* 6: 321–331.

Ouellet, H. 1993. Bicknell's Thrush: taxonomic status and distribution. *Wilson Bulletin* 105: 545–572.

Outlaw, D. C. and Voelker, G. 2006. Systematics of *Ficedula* flycatchers (Muscicapidae): a molecular reassessment of a taxonomic enigma. *Molecular Phylogenetics and Evolution* 41: 118–126.

Outlaw, D. C., Voelker, G., Milá, B. and Girman, D. J. 2003. Evolution of long-distance migration in and historical biogeography of *Catharus* thrushes: a molecular phylogenetic approach. *Auk* 120: 299–310.

Outlaw, R. K., Voelker, G. and Outlaw, D. C. 2007. Molecular systematics and historical biogeography of *Monticola* Rock-Thrushes. *Auk* 124: 561–577.

Päckert, M., Dietzen, C., Martens, J., Wink, M. and Kvist, L. 2006. Radiation of Atlantic goldcrests *Regulus regulus* spp.: evidence of a new taxon from the Canary Islands. *Journal of Avian Biology* 37: 364–380.

Päckert, M., Martens, J. and Hofmeister, T. 2001. Lautäusserungen der Sommergoldhähnchen von den Inseln Madeira und Mallorca (*Regulus ignicapilla madeirensis*, *R. i. Balearicus*). *Journal of Ornithology* 142: 16–29.

Päckert, M., Martens, J., Kosuch, J., Nazarenko, A. A. and Veith, M. 2003. Phylogenetic signal in the song of crests and kinglets (Aves: *Regulus*). *Evolution* 57: 616–629.

Päckert, M., Martens, J., Nazarenko, A. A., Vakhuk, O., Petri, B. and Veith, M. 2005. The great tit (*Parus major*) – a misclassified ring species. *Biological Journal of the Linnean Society*. 86: 153–174.

Pálsson, S., Vigfúsdóttir, F. and Ingólfsson, A.

2009. Morphological and genetic patterns of hybridization of Herring Gulls (*Larus argentatus*) and Glaucous Gulls (*L. hyperboreus*) in Iceland. *Auk* 126: 376–382.

Panov, E. N. 1995. Superspecies of shrikes in the former USSR. In R. Yosef and F. E. Lohrer (eds), *Shrikes (Laniidae) of the World: Biology and Conservation. Proceedings of the Western Foundation of Vertebrate Zoology* 6: 26–33.

Panov, E. N. 2005. *Wheatears of Palearctic: Ecology, Behaviour and Evolution of the Genus Oenanthe*. Pensoft, Sofia-Moscow.

Panov, E. N., Roubtsov, A. S. and Monzikov, D. G. 2003. Hybridisation between Yellowhammer and Pine Bunting in Russia. *Dutch Birding* 25: 17–31.

Parchman, T. L., Benkman, C. W. and Mezquida, E. T. 2007. Coevolution between Hispaniolan Crossbills and pine: does more time allow for greater phenotypic escalation at lower latitude? *Evolution* 61: 2142–2153.

Parkin, D. T. 1980. The ecological genetics of plumage polymorphism in birds. In R. I. C. Spearman and P. A. Riley (eds), *The Skin of Vertebrates*, pp. 219–234. Linnean Society Symposium Series No. 9.

Parkin, D. T., Collinson, M., Helbig, A. J., Knox, A. G. and Sangster, G. 2003. The taxonomic status of Carrion and Hooded Crows. *British Birds* 96: 274–290.

Parkin, D. T., Collinson, M., Helbig, A. J., Knox, A. G. and Sangster, G. 2006. Developing guidelines to assist in defining species limits. *Acta Zooogica Sinica* 52 (Suppl.): 435–438.

Parkin, D. T., Collinson, J. M., Helbig, A. J., Knox, A. G., Sangster, G. and Svensson, L. 2004. Species limits in *Acrocephalus* and *Hippolais* warblers from the Western Palearctic. *British Birds* 97: 276–299.

Parrot, D. and McKay, H. V. 2001. Mute Swan grazing on winter crops: estimation of yield loss in oilseed rape and wheat. *Crop Protection* 20: 913–929.

Parslow, J. L. F. 1973. *Breeding Birds of Britain and Ireland*. Poyser, Berkhamsted.

Parslow-Otsu, M. 1991. Bean Geese in the Yare valley, Norfolk. *British Birds* 84: 161–170.

Parsons, K. C. and Master, T. L. 2000. Snowy Egret (*Egretta thula*). In A. Poole and F. Gill (eds), *The Birds of North America, No. 489*. The Birds of North America Inc., Philadelphia, PA.

Pasquet, E. 1998. Phylogeny of the nuthatches of the *Sitta canadensis* group and its evolutionary and biogeographic implications. *Ibis* 140: 150–156.

Paton, T. A., Baker, A. J., Groth, J. G. and Barrowclough, G. F. 2003. RAG-1 sequences resolve phylogenetic relationships within Chardriiform birds. *Molecular Phylogenetics and Evolution* 29: 268–278.

Pavlova, A., Zink, R. M., Drovetski, S. V., Red'kin, Y. and Rohwer, S. 2003. Phylogeographic patterns in *Motacilla flava* and *Motacilla citreola*: species limits and population history. *Auk* 120: 744–758.

Pavlova, A., Zink, R. M. and Rohwer, S. 2005. Evolutionary history, population genetics and gene flow in the Common Rosefinch (*Carpodacus erythrinus*). *Molecular Phylogenetics and Evolution* 36: 669–681.

Paxinos, E. E., James, H. F., Olson, S. L., Sorenson, M. D., Jackson, J. and Fleischer, R. C. 2002. MtDNA from fossils reveals a radiation of Hawaiian geese recently derived from the Canada Goose (*Branta canadensis*). *Proceedings of the National Academy of Sciences of the USA* 99: 1399–1404.

Payne, R. B. 2005. A molecular genetic analysis of cuckoo phylogeny. In R. B. Payne, *Bird Families of the World: Cuckoos*. Oxford University Press, Oxford.

Payne, R. B. 2006. Indigo Bunting (*Passerina cyanea*). In A. Poole (ed.), The Birds of North America Online Cornell Laboratory of Ornithology, Ithaca. Retrieved from The Birds of North America Online database: http://bna.birds.cornell.edu/BNA/account/Indigo_Bunting/.

Peach, W. J., Baillie, S. R. and Underhill, L. 1991. Survival of British Sedge Warblers *Acrocephalus schoenobaenus* in relation to West African rainfall. *Ibis* 133: 300–305.

Peach, W.J., Crick, H. Q. P. and Marchant, J. H. 1995a. The demography of the decline in the British Willow Warbler population. *Journal of Applied Statistics* 22: 905–922.

Peach, W.J., du Feu, C. and McMeeking, J. 1995b Site tenacity and survival rates of Wrens *Troglodytes troglodytes* and Treecreepers *Certhia familiaris* in a Nottinghamshire wood. *Ibis* 137: 497–507.

Peach, W.J., Robinson, R. A. and Murray, K. A. 2004. Demographic and environmental causes of the decline of rural Song Thrushes *Turdus philomelos* in lowland Britain. In J. A. Vickery, A. D. Evand, P. V. Grice, N. J. Aebischer and R. Brand-Hardy (eds), *Ecology and conservation of lowland farmland birds II: the road to recovery. Ibis* 146 (Suppl. 2): 50–59.

Peach, W. J., Siriwardena, G. M. and Gregory, R. D. 1999. Long-term changes in the abundance and demography of British Reed Buntings *Emberiza schoeniclus. Journal of Applied Ecology* 36: 798–811.

Peach, W.J., Thompson, P.S. and Coulson, J. C. 1994. Annual and long-term variation in the survival rates of British Lapwings *Vanellus vanellus. Journal of Animal Ecology* 63: 60–70.

Peal, R. E. F. 1968. The distribution of the Wryneck in the British Isles, 1964–66. *Bird Study* 15: 111–126.

Pearce, J. M., Talbot, S. L., Petersen, M. R. and Rearick, J. R. 2005. Limited genetic differentiation among breeding, molting, and wintering groups of the threatened Steller's eider: the role of historic and contemporary factors. *Conservation Genetics* 6: 743–757.

Pearson, D. J. and Ash, J. S. 1996. The taxonomic position of the Somali courser *Cursorius (cursor) somalensis. Bulletin of the British Ornithologists' Club* 116(4): 225–229.

Penhallurick, J. and Wink, M. 2004. Analysis of the taxonomy and nomenclature of the Procellariformes based on complete nucleotide sequences of the mitochondrial cytochrome b gene. *Emu* 104: 125–147.

Pennington, M. G. and Meek, E. R. 2006. The 'northern' Bullfinch invasion of autumn 2004. *British Birds* 99: 2– 24.

Pennington, M., Osborn, K., Harvey, P., Riddington, R., Okill, D., Ellis, P. and Heubeck, M. 2004. *The Birds of Shetland.* Helm, London.

Percival, S. 1990. *Population Trends in British Barn Owls* Tyto alba *and Tawny Owls* Strix aluco *in Relation to Environmental Change.* Research Report 57. BTO, Tring.

Pereira, S. L. and Baker, A. J. 2005. Multiple gene evidence for parallel evolution and retention of ancestral morphological states in the shanks (Charadriiformes: Scolopacidae). *Condor* 107: 514–526.

Pereira, S. L. and Baker, A. J. 2006. A mitogenomic timescale for birds detects variable phylogenetic rates of molecular evolution and refute the standard molecular clock. *Molecular Biology and Evolution* 9: 1731–1740.

Pereira, S. L., Johnson, K. P., Clayton, D. H. and Baker, A. J. 2007. Mitochondrial and nuclear DNA sequences support a Cretaceous origin of Columbiformes and a dispersal-driven radiation in the Paleogene. *Systematic Biology* 56: 656–672.

Perkins, A. J., Maggs, H. E., Wilson, J. D., Watson, A. and Smout, C. 2008. Targeted management intervention reduces rate of population decline of Corn Buntings *Emberiza calandra* in eastern Scotland. *Bird Study* 55: 52–58.

Perrins, C. 1961. The 'lesser scaup' problem. *British Birds* 54: 49–54.

Perrins, C. M. 1979. *British Tits.* Collins, London.

Perrins, C. 2003. The status of Marsh and Willow Tits in the UK. *British Birds* 96: 418–426.

Perry, R. 1946. *A Naturalist on Lindisfarne.* Lindsay Drummond, London.

Peters, J. L., McCracken, K. G., Zhuravlev, Y. N., Wilson, R. E., Johnson, K. P. and Omland, K. E. 2005. Phylogenetics of wigeons and allies (Anatidae: Aves): the importance of sampling multiple loci and multiple individuals. *Molecular Phylogenetics and Evolution* 35: 209–224.

Peters, J. L., Zhuravlev, Y., Fefelov, I., Logie, A. and Omland, K. E. 2007. Nuclear loci and coalescent methods support ancient hybridization as cause of mitochondrial paraphyly between Gadwall and Falcated Duck (*Anas* spp.). *Evolution* 61: 1992–2006.

Petersen, A. 1976. Size variable in Puffins *Fratercula arctica* from Iceland and bill features as criteria of age. *Ornis Scandinavica* 7: 185–192.

Pethon, P. 1967. The systematic position of the Norwegian Common Murre (*Uria aalge*) and Puffin (*Fratercula arctica*). *Nytt Magasin for Zoologi* 14: 84–95.

Pettifor, R. A., Fox, A. D. and Rowcliffe, J. M. 1999. *Greenland White-fronted Goose (Anser albifrons flavirostris) – the Collation and Statistical Analysis of Data and Population Viability Analyses.* Scottish Natural Heritage. Surveying and Monitoring Report No. 140.

Pettifor, R. A., Rowcliffe, J. M. and Mudge, G. P. 1997. *Population Viability Analysis of Icelandic/Greenlandic Pink-footed Geese.* Scottish Natural Heritage Research and Survey Monitoring Report No. 107.

Phillips, R. A. and Furness, R. W. 1998. Polymorphism, mating preferences and sexual selection in the Arctic Skua. *Journal of Zoology* 245: 245–252.

Phillips, R. A., Thompson, D. R. and Hamer, K. C. 1999. The impact of Great Skua predation on seabird populations at St Kilda: a bioenergetic model. *Journal of Applied Ecology* 36: 218–232.

Pickerell, G. 2004. Little Tern *Sterna albifrons*. In P. I. Mitchell, S. F. Newton, N. Ratcliffe and T. E. Dunn (eds), *Seabird Populations of Britain and Ireland*, pp. 339–349. T and AD Poyser, London.

Pielowski, Z. 1959. Studies on the relationship: predator (Goshawk)–prey (Pigeon). *Bulletin of the Polish Academy of Sciences* 7: 401–403.

Piersma, T. and Davidson, N. C. 1992. The migrations and annual cycles of five subspecies of Knots in perspective. *Wader Study group Bulletin 64 (Suppl.):* 187–197.

Piotrowski, S. 2003. *The Birds of Suffolk.* Christopher Helm, London.

Pithon, J. A. and Dytham, C. 2001. Determination of the origin of British feral Rose-ringed Parakeets. *British Birds* 94: 74–79.

Pitra, C., Lieckfeldt, D. and Alonso, J. C. 2000. Population subdivision in Europe's Great Bustard inferred from mitochondrial and nuclear DNA sequence variation. *Molecular Ecology* 9: 1165–1170.

Pitra, C., Lieckfeldt, D., Frahnert, S. and Fickel, J. 2002. Phylogenetic relationships and ancestral areas of the bustards (Gruiformes: Otididae), inferred from mitochondrial DNA

and nuclear intron sequences. *Molecular Phylogenetics and Evolution* 23: 63–74.

Pollitt, M. S., Hall, C., Holloway, S. J, Hearn, P. E., Musgrove, A. J., Robinson, J. A. and Cranswick, P. A. 2003. *The Wetland Bird Survey 2000–2001: Wildfowl and Wader Counts.* BTO/WWT/RSPB/JNCC, Slimbridge.

Pons, J.-M., Hassanin, A. and Crochet, P.-A. 2005. Phylogenetic relationships within the *Laridae* (Charadriiformes: Aves) inferred from mitochondrial markers. *Molecular Phylogenetics and Evolution* 37: 686–699.

Portenko, L. A. 1954. *The Birds of the USSR.* Zoological Institute Academy of Sciences USSR, Moscow–Leningrad.

Potapov, E. and Sale, R. 2004. *The Gyrfalcon.* T and AD Poyser, London.

Potts, G. R. 1969 . The influence of eruptive movements, age, population size and other factors on the survival of the Shag (*Phalacrocorax aristotelis*). *Journal of Animal Ecology* 38: 53–102.

Potts, G. R. 1986. *The Partridge: Pesticides, Predation and Conservation.* Collins, London.

Poulin, R. G., Grindal, S. D. and Brigham, R. M. 1996. Common Nighthawk (*Chordeiles minor*). In A. Poole and F. Gill (eds), *The Birds of North America, No. 213.* The Academy of Natural Sciences, Philadelphia, and The American Ornithologists' Union, Washington, DC.

Prater, A. J. 1989. Ringed Plover *Charadrius hiaticula* breeding in the United Kingdom in 1984. *Bird Study* 36: 154–159.

Prestt, I. 1965. An enquiry into the recent breeding status of some of the smaller birds of prey and crows in Britain. *Bird Study* 12: 196–221.

Pruett, C. L., Gibson, D. D. and Winker, K. 2001. Molecular 'cuckoo clock' suggests listing of western Yellow-billed Cuckoos may be warranted. *Wilson Bulletin* 113: 228–231.

Puigcerver, M., Rodríguez-Teijero, J. D. and Gallego, S. 2001. The problem of the subspecies in *Coturnix coturnix* quail. *Game and Wildlife Science* 18: 561–572.

Questiau, S., Eybert, M.-C., Gaginskaya, A. R., Gielly, L. and Taberlet, P. 1998. Recent divergence between two morphologically differentiated subspecies of Bluethroat (Aves: Muscicapidae: *Luscinia svecica*) inferred from mitochondrial DNA sequence variation. *Molecular Ecology* 7: 239–245.

Raine, A. F., Sowter, D. J, Brown, A. F. and Sutherland, W. J. 2006a. Migration patterns of two populations of Twite *Carduelis flavirostris* in Britain. *Ringing and Migration* 23: 45–52.

Raine, A. F., Sowter, D. J, Brown, A. F. and Sutherland, W. J. 2006b. Natal philopatry and local movement patterns of Twite *Carduelis flavirostris. Ringing and Migration* 23: 89–94.

Rand, A. L. and Fleming, R. L. 1957. Birds from Nepal. *Fieldiana Zoology* 41: 1–218.

Randi, E. 1996. A mitochondrial cytochrome b phylogeny of the *Alectoris* partridges. *Molecular Phylogenetics and Evolution* 6: 214–227.

Randi, E. and Bernard-Laurent, A. 1999. Population genetics of a hybrid zone between the Red-legged Partridge and Rock Partridge. *Auk*: 116: 324–337.

Randi, E. and Lucchini, V. 1998. Organisation and evolution of the mitochondrial DNA control region in the avian genus *Alectoris. Journal of Molecular Evolution* 47: 449–462.

Randi, E., Meriggi, A., Lorenzini, R., Fusco, G. and Alkon, P.-U. 1992. Biochemical analysis of relationships of Mediterranean *Alectoris* partridges. *Auk* 109: 358–367.

Ransome, A. 1947. *Great Northern?* Jonathan Cape, London.

Ratcliffe, D. 1970. Changes attributable to pesticides in egg breakage frequency and eggshell thickness in some British birds. *Journal of Applied Ecology* 7: 67–115.

Ratcliffe, D. 1980. *The Peregrine Falcon*. T and AD Poyser, Calton.

Ratcliffe, D. 1993. *The Peregrine Falcon* (2nd Edition). T and AD Poyser, London.

Ratcliffe, D. 1997. *The Raven*. T and AD Poyser, London.

Ratcliffe, N. 2004a. Sandwich Tern *Sterna sandvicensis*. In P. I. Mitchell, S. F. Newton, N. Ratcliffe and T. E. Dunn (eds), *Seabird Populations of Britain and Ireland*, pp. 291–301. T and AD Poyser, London.

Ratcliffe, N. 2004b. Common Tern *Sterna hirundo*. In P. I. Mitchell, S. F. Newton, N. Ratcliffe and T. E. Dunn (eds), *Seabird Populations of Britain and Ireland*, pp. 315–327. T and AD Poyser, London.

Ratcliffe, N. 2004c. Arctic Tern *Sterna paradisea*. In P. I. Mitchell, S. F. Newton, N. Ratcliffe and T. E. Dunn (eds), *Seabird Populations of Britain and Ireland*, pp. 328–338. T and AD Poyser, London.

Ratcliffe, N., Nisbet, I. C. T. and Newton, S. 2004. *Sterna dougallii* Roseate Tern. *BWP Update*: 6: 77–90.

Ratcliffe, N., Pickerell, G. and Brindley, E. 2000. Population trends of Little and Sandwich Terns *S. albifrons* and *S. sandvicensis* in Britain and Ireland from 1969 to 1998. *Atlantic Seabirds* 2: 211–226.

Rebecca, G. W. and Bainbridge, I. P. 1998. The breeding status of the Merlin *Falco columbarius* in Britain in 1993–94. *Bird Study* 45: 172–187.

Red'kin, Y. A. 2006. A new subspecies of the Siberian Blue Robin, *Luscinia* (*Larivora*) *cyane nechaevi* (Turdidae). *Zoologichesky Zhurnal* 85: 614–620.

Red'kin, Y. and Konovalova, M. 2006. Systematic notes on Asian birds. 63. The eastern Asiatic races of *Sitta europaea* Linnaeus, 1758. *Zoologische Mededelingen* (*Leiden*) 80: 241–261.

Redpath, S. and Thirgood, S. 1997. *Birds of Prey and Red Grouse*. HMSO, London.

Rees, E. C., Bowler, J. M. and Beekman, J. H. 1997. *Cygnus columbianus* Bewick's Swan and Whistling Swan. *BWP Update* 1: 63–74.

Reeves, A. B., Drovetski, S. V. and Fadeev, I. V. 2008. Mitochondrial DNA data imply a stepping-stone colonization of Beringia by Arctic Warbler *Phylloscopus borealis*. *Journal of Avian Biology* 39: 567–575.

Rehfisch, M. M., Austin, G. E., Armitage, M. J. S., Atkinson, P. W., Holloway, S. J., Musgrove, A. J. and Pollitt, M. S. 2003a. Numbers of wintering waterbirds in Great Britain and the Isle of Man (1994/95–1998/99): II. Coastal waders (Charadrii). *Biological Conservation* 112, 329–341.

Rehfisch, M. M. and Crick, H. Q. P. 2003. Predicting the impact of climate change on Arctic-breeding waders. *Wader Study Group Bulletin* 100: 86–95.

Rehfisch, M. M., Holloway, S. J. and Austin, G. E. 2003b. Population estimates of waders on the non-estuarine coasts of the UK and the Isle of Man during the winter of 1997/98. *Bird Study* 50: 22–32.

Rehfisch, M. M., Wernham, C. V. and Marchant, J. H. (eds) 1999. *Population, Distribution, Movements and Survival of Fish-eating Birds in Great Britain*. DETR, London.

Reid, J. B. 2004. Great Black-backed Gull *Larus marinus*. In P. I. Mitchell, S. F. Newton, N. Ratcliffe and T. E. Dunn (eds), *Seabird Populations of Britain and Ireland*, pp. 263–276. T and AD Poyser, London.

Reynolds, N. and Harradine, J. 1996. Grey Goose *Shooting Kill and Duck Recruitment, 1995/96*. BASC Report to WWT.

Rheindt, F. E. and Austin, J. J. 2005. Major analytical and conceptual shortcomings in a recent taxonomic revision of the Procellariiformes: a reply to Penhallurick and Wink (2004). *Emu* 105: 181–186.

Rhymer, J. M., Williams, M. J. and Braun, M. J. 1994. Mitochondrial analysis of gene flow between New Zealand Mallards (*Anas platyrhynchos*) and Grey Ducks (*A. superciliosa*). *Auk* 111: 970–978.

Richardson, M. and Brauning, D. W. 1995. Chestnut-sided Warbler (*Dendroica pensylvanica*). In A. Poole and F. Gill (eds), *The Birds of North America, No. 20*. American Ornithologists' Union, Washington, DC.

Riddington, R., Votier, S. C. and Steele, J. 2000. The influx of redpolls into Western Europe, 1995/96. *British Birds* 93: 59–67.

Riffaut, L., McCoy, K. D., Tirard, C., Friesen, V. L. and Boulinier, T. 2005. Population genetics of the common guillemot *Uria aalge* in the North Atlantic: geographic impact of oil spills *Marine Ecology Progress Series* 291: 263–273.

Ritz, M. S., Hahn, S., Janicke, T. and Peter, H.-U. 2006. Hybridisation between South Polar Skua (*Catharacta maccormicki*) and Brown Skua (*C. antarctica lonnbergi*) in the Antarctic peninsula region. *Polar Biology* 29: 153–159.

Ritz, M. S., Millar, C., Miller, G. D., Phillips, R. A., Ryan, P., Sternkopf, V., Liebers-Helbig, D. and Peter, H.-U. 2008. Phylogeography of the southern skua complex – rapid colonisation of the southern hemisphere during a glacial period and reticulate evolution. *Molecular Phylogenetics and Evolution* 49: 292–303.

Robb, M. 2000. Introduction to vocalizations of crossbills in north-western Europe. *Dutch Birding* 22: 61–107.

Robertson, C. J. R. 2002. The scientific name of the Indian yellow-nosed albatross *Thalassarche carteri*. *Marine Ornithology* 30: 48–49.

Robertson, C. J. R. and Nunn, G. B. 1998. Towards a new taxonomy for albatrosses. In G. Robertson and R. Gales (eds), *Albatross Biology and Conservation*, pp. 13–19. Surrey Beatty, Sydney.

Robinson, J. A., Colhoun, K., Gudmundsson, G., Boertmann, D., Merne, O. J., O'Briain, M., Portig, A., Mackie, K. and Boyd, H. 2004b. *Light-bellied Goose* Branta bernicla hrota *in Canada, Ireland, Iceland, France, Greenland, Scotland, Wales, England, the Channel Islands and Spain 1960/1–1999/2000*. Waterbird Review Series, WWT/JNCC, Slimbridge.

Robinson, J. A., Colhoun, K., McElwaine, J. G. and Rees, E. C. 2004a. *Whooper Swan* Cygnus cygnus (Iceland Population) in Britain and Ireland 1960/1–1999/2000*. Waterbird Review Series, WWT/JNCC, Slimbridge.

Robinson, M. and Becker, C. D. 1986. Snowy Owls on Fetlar. *British Birds* 79: 228–242.

Robinson, P. 2003. *The Birds of the Isles of Scilly*. Christopher Helm, London.

Robinson, R. A., Freeman, S. N., Balmer, D. E. and Grantham, M. J. 2007. Cetti's Warbler: anatomy of an expanding population. *Bird Study* 54: 230–235.

Robinson, R. A., Green, R. E., Baillie, S. R., Peach, W. J. and Thomson, D. L. 2004. Demographic mechanisms of the population decline of the Song Thrush *Turdus philomelos* in Britain. *Journal of Animal Ecology* 73: 670–682.

Robinson, R. A., Learmonth, J. A., Hutson, A. M., Macleod, C. D., Sparks, T. M., Leech, D. I., Pierce, G. J., Rehfisch, M. M. and Crick, H. Q. P. 2005. *Climate Change and Migratory Species*. BTO Research Report No. 414. BTO, Thetford.

Robinson, W. D. 1996. Summer Tanager (*Piranga rubra*). In A. Poole and F. Gill (eds), *The Birds of North America, No. 248*. American Ornithologists' Union, Washington, DC.

Robson, M. J. H. 1968. The breeding birds of North Rona. *Scottish Birds* 5: 126–156.

Rodgers, J. A., Jr., and Smith, H. T. 1995. Little Blue Heron (*Egretta caerulea*). In A. Poole and F. Gill (eds), *The Birds of North America, No. 145*. The Academy of Natural Sciences, Philadelphia, PA, and American Ornithologists' Union, Washington, DC.

Rogers, M. J. and the Rarities Committee. 2004. Report on rare birds in Great Britain in 2003. *British Birds* 97: 558–625.

Roselaar, C. S. 1979. Fluctuaties in aantallen Krombekstrandlopers *Calidris ferruginea*. *Watervogels* 4: 203–210.

Roth, R. R., Johnson, M. S. and Underwood, T. J. 1996. Wood Thrush (*Hylocichla mustelina*). In A. Poole and F. Gill (eds), *The Birds of North America, No. 246*. The Academy of Natural Sciences, Philadelphia, PA, and American Ornithologists' Union, Washington, DC.

Roulin, A. 2002. *Tyto alba* Barn Owl. *BWP Update* 4: 115–138.

Round, P. D. 1996. Long-toed Stint in Cornwall: the first record for the Western Palearctic. *British Birds* 89: 12–24.

Round, P. D., Hansson, B., Pearson, D. J., Kennerley, P. R. and Bensch, S. 2007. Lost and found: the enigmatic Large-billed Reed Warbler *Acrocephalus orinus* rediscovered after 139 years, *Journal of Avian Biology* 38: 133–138.

Rowell, H. E. 2005. *The 2004 Icelandic-breeding Goose Census*. Wildfowl and Wetlands Trust/Joint Nature Conservation Committee, Slimbridge.

Ruegg, K. 2007. Divergence between subspecies groups of Swainson's Thrush (*Catharus ustulatus ustulatus* and *C. u. swainsoni*). *Ornithological Monographs* 63: 67–77.

Ruokonen, M., Aarvak, T. and Madsen, J. 2005. Colonization history of the Pink-footed Goose *Anser brachyrhynchus*. *Molecular Ecology* 14: 171–178.

Ruokonen, M., Kvist, L. and Lumme, J. 2000. Close relatedness between mitochondrial DNA from seven *Anser* goose species. *Journal of Evolutionary Biology* 13: 532–540.

Ruokonen, M., Kvist, L., Aarvak, T., Markkola, J., Morozov, V. V., Oien, I. J., Syroechkovsky E. E., Jr, Tolvanen, P. and Lumme, J. 2004. Population genetic structure and conservation of the Lesser White-fronted Goose *Anser erythropus*. *Conservation Genetics* 5: 501–512.

Ruttledge, R. F. 1975 *A List of the Birds of Ireland*. The Stationery Office, Dublin.

Sæther, T. 1989. A new taxonomic approach to the Norwegian island Willow Grouse *Lagopus lagopus variegatus*. *Cinclus* 12: 79–99.

Saetre, G. P., Borge, T., Lindell, J., Moum, T., Primmer, C. R., Sheldon, B. C., Haavie, J., Johnsen, A. and Ellegren, H. 2001b. Speciation, introgressive hybridisation and nonlinear rate of molecular evolution in flycatchers. *Molecular Ecology* 10: 737–749.

Saetre, G.-P., Borge, T., Lindroos, K., Haavie, J., Sheldon, B. C., Primmer, C. and Syvänen, A.-C. 2003. Sex chromosome evolution and speciation in *Ficedula* flycatchers. *Proceedings of the Royal Society of. London B* 270: 53–60.

Saetre, G. P., Borge, T. and Moum, T. 2001a. A new bird species? The taxonomic status of 'the Atlas Flycatcher' assessed from DNA sequence analysis. *Ibis* 143: 494–497.

Sallabanks, R. and James, F. C. 1999. American Robin (*Turdus migratorius*). In A. Poole and F. Gill (eds), *The Birds of North America, No.*

462. The Birds of North America, Inc., Philadelphia, PA.

Salzburger, W., Martens, J., Nazarenko, A. A., Yua-Hue Sun, Dallinger, R. and Sturmbauer, C. 2002b. Phylogeography of the Eurasian Willow Tit (*Parus montanus*) based on DNA sequences of the mitochondrial cytochrome b gene. *Molecular Phylogenetics and Evolution* 24: 26–34.

Salzburger, W., Martens, J. and Sturmbauer, C. 2002a. Paraphyly of the Blue Tit (*Parus caeruleus*) suggested from cytochrome b sequences. *Molecular Phylogenetics and Evolution* 24: 19–25.

Sangster, G. 2006. The taxonomic status of 'phylogroups' in the *Parus teneriffae* complex (Aves): comments on the paper by Kvist *et al.* (2005). *Molecular Phylogeneics and Evolution* 38: 288–289.

Sangster, G. 2008a. A revision of *Vermivora* (Parulidae), with the description of a new genus. *Buletin of the British Ornithologists' Club* 128: 207–211.

Sangster, G. 2008b. A new genus for the waterthrushes (Parulidae). *Bulletin of the British Ornithologists' Club* 128: 212–215.

Sangster, G., Collinson, J. M., Helbig, A. J., Knox, A. G. and Parkin, D. T. 2002a. The generic status of Black-browed Albatross and other albatrosses. *British Birds* 95: 383–385.

Sangster, G., Collinson, J. M., Helbig, A. J., Knox, A. G. and Parkin, D. T. 2002b. The specific status of Balearic and Yelkouan Shearwaters. *British Birds* 95: 636–639.

Sangster, G., Collinson, J. M., Helbig, A. J., Knox, A. G. and Parkin, D. T. 2004a. The taxonomic status of Macqueen's Bustard. *British Birds* 97: 60–67.

Sangster, G., Collinson, J. M., Helbig, A. J., Knox, A. G. and Parkin, D. T. 2004b. Taxonomic recommendations for British birds: second report. *Ibis* 146: 153–157.

Sangster, G., Collinson, J. M., Helbig, A. J., Knox, A. G. and Parkin, D. T. 2005. Taxonomic recommendations for British Birds: third report. *Ibis* 147: 821–826.

Sangster, G., Collinson, J. M., Helbig, A. J., Knox, A. G., Parkin, D. T. and Prater, T. 2001. The taxonomic status of Green-winged Teal *Anas carolinensis*. *British Birds* 94: 218–226.

Sangster, G. , Collinson, J. M., Knox, A. G., Parkin, D. T. and Svensson, L. In press. Taxonomic recommendations for British birds: Sixth report. *Ibis* 152:

Sangster, G. and Oreel, G. J. 1996. Progress in taxonomy of Taiga and Tundra Bean Geese. *Dutch Birding* 18: 310–316.

Schnell, G. D. 1970a. A phenetic study of the sub-order Lari (Aves). I. Methods and results of principal components analyses. *Systematic Zoology* 19: 35–57.

Schnell, G. D. 1970b. A phenetic study of the sub-order Lari (Aves). II. Phenograms, discussion, and conclusions. *Systematic Zoology* 19: 264–302.

Scott, D. A. and Rose, P. M. 1996. *Atlas of Anatidae Populations in Africa and Western Eurasia*. Wetlands International Publication 41, Wageningen.

Scribner, K. T., Malecki, R. A., Batt, B. D. J., Inman, R. L., Libants, S. and Prince, H. H. 2003b. Identification of source population for Greenland Canada Geese: genetic assessment of a recent colonization. *Condor* 105: 771–782.

Scribner, K. T., Talbot, S. L., Pearce, J. M., Pierson, B. J., Bollinger, K. S. and Derksen, D. V. 2003a. Phylogeography of Canada geese (*Branta canadensis*) in western North America. *Auk* 120: 889–907.

Secondi, J., Bretagnolle, V., Compagnon, C. and Faivre, B. 2003. Specific-song convergence in a moving hybrid zone between two passerines.

Biological Journal of the Linnean Society 80: 507–517.

Segelbacher, G., Höglund, J. and Storch, I. 2003. From connectivity to isolation: genetic consequences of population fragmentation in Capercaillie across Europe. *Molecular Ecology* 12: 1773–1780.

Segelbacher, G. and Piertney, S. 2008. Phylogeography of the European Capercaillie (*Tetrao urogallus*) and its implications for conservation. *Journal of Ornithology* 148: 269–274.

Seibold, I, and Helbig, A. J. 1995. Systematic position of the Osprey (*Pandion haliaetus*) according to mitochondrial DNA sequences. *Vogelwelt* 116: 209–217.

Seibold, I. and Helbig, A. J. 1996. Phylogenetic relationships of the sea eagles (genus *Haliaeetus*): reconstructions based on morphology, allozymes and mitochondrial DNA sequences. *Journal of Zoological Systematics and Evolutionary Research* 34: 103–112.

Sellers, R. 2004. Great Cormorant *Phalacrocorax carbo*. In P. I. Mitchell, S. F. Newton, N. Ratcliffe and T. E. Dunn (eds), *Seabird Populations of Britain and Ireland*, pp. 128–145. T and AD Poyser, London.

Severtzov, N. A. 1873. *Vertical and Horizontal Distribution of Turkestan Wildlife*. Turkestan. (In Russian.)

Sharrock, J. T. R. 1974. *Scarce Migrant Birds in Britain and Ireland*. T and AD Poyser, Berkhamsted.

Sharrock, J. T. R. and Grant, P. J. 1982. *Birds New to Britain and Ireland*. T and AD Poyser, Calton.

Sheldon, F. H. 1987. Phylogeny of herons estimated from DNA–DNA hybridization data. *Auk* 104: 97–108.

Sheldon, F. H. and Gill, F. B. 1996. A reconsideration of songbird phylogeny, with emphasis on titmice and their sylvioid relatives. *Systematic Biology* 45: 473–495.

Sheldon, F. H., Jones, C. E. and McCracken, K. G. 2000. Relative patterns and rates of evolution in heron nuclear and mitochondrial DNA. *Molecular Biology and Evolution* 17: 437–450.

Sheldon, F. H., Whittingham, L. A., Moyle, R. G., Slikas, B. and Winkler, D. W. 2005. Phylogeny of swallows (Aves: Hirundinidae) estimated from nuclear and mitochondrial genes. *Molecular Phylogenetics and Evolution* 35: 253–270.

Sheldon, R. D., Chaney, K. and Tyler, G. A. 2007. Factors affecting nest survival of Northern Lapwings *Vanellus vanellus* in arable farmland: an agri-environment scheme prescription can enhance nest survival. *Bird Study* 54: 168–175.

Shirihai, H., Gargallo, G. and Helbig, A. J. 2001. *Sylvia Warblers: Identification, Taxonomy and Phylogeny of the Genus Sylvia*. Christopher Helm, London.

Shrubb, M. 2003a. Farming and birds: an historic perspective. *British Birds* 96: 158–177.

Shrubb, M. 2003b. *Birds, Scythes and Combines: a History of Birds and Agricultural Change*. Cambridge University Press, Cambridge.

Sibley, C. G. and Ahlquist, J. E. 1990. *Phylogeny and Classification of Birds: a Study in Molecular Evolution*. Yale University Press, New Haven.

Sibley, C. G. and Monroe, B. L., Jr 1990. *Distribution and Taxonomy of Birds of the World*. Yale University Press, New Haven.

Siegel-Causey, D. 1988. Phylogeny of the Phalacrocoracidae. *Condor* 90: 885–905.

Silbernagl, H. P. 1982. Seasonal and spatial distribution of the American Purple Gallinule in South Africa. *Ostrich* 53: 236–240.

Sim, I. M. W., Burfield, I. J., Grant, M. C.,

Pearce-Higgins, J. W. and Brooke, M.de L. 2007b. The role of habitat composition in determining breeding site occupancy in a declining Ring Ouzel *Turdus torquatus* population. *Ibis* 149: 374–385.

Sim, I. M. W., Dillon, I. A., Eaton, M. A., Etheridge, B., Lindley, P., Riley, H., Saunders, R., Sharpe, C. and Tickner, M. 2007a. Status of the Hen Harrier *Circus cyaneus* in the UK and Isle of Man in 2004, and a comparison with the 1988/89 and 1998 surveys. *Bird Study* 54: 256–267.

Sim, I. M. W., Gibbons, D. W., Bainbridge, I. P. and Mattingley, W. A. 2001. Status of the Hen Harrier *Circus cyaneus* in the UK and the Isle of Man in 1998. *Bird Study* 48: 341–353.

Simmons, R. E. and Simmons, J. R. 2000. *Harriers of the World: Their Behaviour and Ecology*. OUP, Oxford.

Simpson, G. G. 1951. The species concept. *Evolution* 5: 285–298.

Sinclair, J. C., Brooke, R. K. and Randall, R. M. 1982. Races and records of the Little Shearwater *Puffinus assimilis* in South African waters. *Cormorant* 10: 19–26.

Siriwardena, G. M. 2004. Possible roles of habitat, competition and avian nest predation in the decline of the Willow Tit *Parus montanus* in Britain. *Bird Study* 51: 193–202.

Siriwardena, G. M. 2006. Avian nest predation, competition and the decline of British Marsh Tits *Parus palustris*. *Ibis* 148: 255–265.

Siriwardena, G. M., Baillie, S. R., Crick, H. Q. P. and Wilson, J. D. 2000. The importance of variation in the breeding performance of seed-eating birds for their population trends on farmland. *Journal of Applied Ecology* 37: 1–22.

Siriwardena, G. M., Baillie, S. R. and Wilson, J. D. 1998. Variation in the survival rates of some British passerines with respect to their population trends on farmland. *Bird Study* 45: 276–292.

Siriwardena, G. M., Baillie, S. R. and Wilson, J. D. 1999. Temporal variation in the annual survival rates of six granivorous birds with contrasting population trends. *Ibis* 141: 621–636.

Siriwardena, G. M., Cakbrade, N. A., Vickery, J. A. and Sutherland, W. J. 2006. The effect of the spatial distribution of winter seed food resources on their use by farmland birds. *Journal of Applied Ecology* 43: 628–639.

Siriwardena, G. M., Freeman, S. N. and Crick, H. Q. P. 2001. The decline of the Bullfinch *Pyrrhula pyrrhula* in Britain: is the mechanism known? *Acta Ornithologica* 36: 143–152.

Smallwood, J. A., and Bird. D. M. 2002. American Kestrel (*Falco sparverius*). In A. Poole and F. Gill (eds), *The Birds of North America, No. 602*. The Birds of North America, Inc., Philadelphia, PA.

Smiddy, P. and Duffy, B. 1997. Little Egret *Egretta garzetta*: a new breeding bird for Ireland. *Irish Birds* 6: 55–56.

Smiddy, P. and O'Halloran, J. 2004. The ecology of river bridges: their use by birds and mammals. In J. Davenport and J. L. Davenport (eds), *The Effects of Human Transport on Ecosystems: Cars and Planes, Boats and Trains*, pp. 83–97 Royal Irish Academy, Dublin.

Smith, K. W. 2005. Has the reduction in nest-site competition from Starlings *Sturnus vulgaris* been a factor in the recent increase of Great Spotted Woodpecker *Dendrocops major* in Britain? *Bird Study* 52: 307–313.

Smith, K. W., Reed, J. M. and Trevis, B. E. 1999. Nocturnal and diurnal activity patterns and roosting sites of Green Sandpipers *Tringa ochropus* wintering in southern England. *Ringing and Migration* 19: 315–322.

Snow, D. W. and Snow, B. K. 1966. The breeding season of the Madieran Storm-petrel *Oceanodroma castro* in the Galapagos. *Ibis* 108: 283–284.

Solovieva, D. V., Pihl, S., Fox, A.. D. and Bustnes, J.-O. 1998. *Polysticta stelleri* Steller's Eider. *BWP Update* 2: 145–158.

Sorensen, M. D. and Fleischer, R. C. 1996. Multiple independent transpositions of mitochondrial DNA control region sequences to the nucleus. *Proceedings of the National Academy of Sciences of the USA* 93: 15239–15243.

Sorenson, M. D. and Payne, R. B. 2001. A single ancient origin of brood parasitism in African finches: implications for host–parasite coevolution. *Evolution* 55: 2550–2567.

Southern, H. N. 1962. Survey of bridled Guillemots, 1959–1960. *Proceedings of the Zoological Society of London* 138: 455–472.

Southern, H. N. 1966. Distribution of bridled Guillemots in east Scotland over eight years. *Journal of Animal Ecology* 35: 1–11.

Spellman, G. M., Cibois, A., Moyle, R. G., Winker, K. and Barker, F. K. 2008. Clarifying the systematics of an enigmatic avian lineage: What is a bombycillid? *Molecular Phylogenetics and Evolution* 49: 1036–1040.

Squires, J. R. and Reynolds, R. T. 1997. Northern Goshawk (*Accipiter gentilis*). In A. Poole and F. Gill (eds), *The Birds of North America, No. 298*. The Academy of Natural Sciences, Philadelphia, PA, and The American Ornithologists' Union, Washington, DC.

Staton, J. 1945. The breeding of Black-winged Stilts in Nottinghamshire in 1945. *British Birds* 38: 322–328.

Steele, J. and Vangeluwe, D. 2002. From the Rarity Committee's files: the Slender-billed Curlew at Druridge Bay, Northumberland, in 1998. *British Birds* 95: 279–299.

Stempniewicz, L. 2001. *Alle alle* Little Auk. *BWP Update* 3: 175–201.

Stevens, D. K., Anderson, G. O. A., Grice, P. V. and Norris, K. 2007. Breeding success of Spotted Flycatchers *Muscicapa striata* in southern England: is woodland a good habitat for the species? *Ibis* 149: 214–223.

Stevens, D. K., Anderson, G. Q. A., Grice, P. V., Norris, K. and Butcher, N. 2008. Predators of Spotted Flycatcher *Muscicapa striata* nests in southern England as determined by digital nest cameras. *Bird Study* 55: 179–187.

Stoate, C. and Szczur, J. 2006. Potential influence of habitat and predation on local breeding success and population in Spotted Flycatchers *Muscicapa striata*. *Bird Study* 53: 328–330.

Storch, I. 2001. *Tetrao urogallus* Capercaillie. *BWP Update* 3: 1–24.

Strauch, J. G., Jr 1978. The phylogeny of the Charadriiformes (Aves): an estimate using the method of character compatibility analysis. *Transactions of the Zoological Society of London* 34: 263–345.

Stroud, D. A., Chambers, D., Cook, S., Buxton, N., Fraser, B., Clement, P., Lewis, P., McLean, I., Baker, H. and Whitehead, S. 2001. *The UK SPA Network: its Scope and Content*. Joint Nature Conservation Committee, Peterborough.

Stroud, D. A., Davidson, N. C., West, R., Scott, D. A., Haanstra, L., Thorup, O., Ganter, B. and Delany, S. (compilers) on behalf of the International Wader Study Group. 2004. Status of migratory wader populations in Africa and Western Eurasia in the 1990s. *International Wader Studies* 15: 1–259.

Suddaby, D., Shaw, K. D., Ellis, P. M. and Brockie, K. 1994. King Eiders in Britain and Ireland in 1958–90: occurrences and ageing. *British Birds* 87: 418–430.

Summers, R. W. 2000. The habitat requirements of the Crested Tit *Parus cristatus* in Scotland. *Scottish Forestry* 54: 197–201.

Summers, R. W. and Canham, M. 2001. The distribution of Crested Tits in Scotland duing the 1990s. *Scottish Birds* 22: 20–27.

Summers, R. W., Dawson, R. J. G. and Phillips, R. E. 2007. Assortative mating and patterns of inheritance indicate that the three crossbill taxa in Scotland are species. *Journal of Avian Biology* 38: 153–162.

Summers, R. W., Jardine, D. C., Marquiss, M. and Rae, R. 2002. The distribution and habitats of crossbills *Loxia* spp. in Britain, with special reference to the Scottish Crossbill *Loxia scotica*. *Ibis* 144: 393–410.

Summers-Smith, J. D. 2003. The decline of the House Sparrow: a review. *British Birds* 96: 439–446.

Sutherland, W. J. and Allport, G. A. 1991. The distribution and ecology of naturalised Egyptian Geese *Alopochen aegyptiacus* in Britain. *Bird Study* 38: 128–134.

Svensson, L. 1992. *Identification Guide to European Passerines*. Privately published, Stockholm.

Svensson, L. 2001.The correct name of the Iberian Chiffchaff *Phylloscopus ibericus* Ticehurst 1937, its identification and new evidence of its winter grounds. *Bulletin of the British Ornithologists' Club* 121: 281–296.

Svensson, L., Collinson, M., Knox, A. G., Parkin, D. T. and Sangster, G. 2005. Species limits in the Red-breasted Flycatcher. *British Birds* 98: 538–541.

Syroechkovski, E. E., Jr 2002. Distribution and Population Estimates for Swans in the Siberian Arctic in the 1990s. *Waterbirds* 25: 100–113.

Szczys, P., Hughes, C. R. and Kesseli, R. V. 2005. Novel microsatellite markers used to determine the population genetic structure of the endangered Roseate Tern, *Sterna dougallii*, in Northwest Atlantic and Western Australia. *Conservation Genetics* 6: 461–466.

Szép, T. 1995. Relationship between West African rainfall and the survival of central European Sand Martins *Riparia riparia*. *Ibis* 137: 162–168.

Taberlet, P., Meyer, A. and Bouvet, J. 1992. Unusual mitochondrial DNA polymorphism in two local populations of Blue Tit *Parus caeruleus*. *Molecular Ecology* 1: 27–36.

Tasker, M. L. 2004a. Northern Fulmar *Fulmarius glacialis*. In P. I. Mitchell, S. F. Newton, N. Ratcliffe and T. E. Dunn (eds), *Seabird Populations of Britain and Ireland*, pp. 49–62. T and AD Poyser, London.

Tasker, M. L. 2004b. Common Gull *Larus canus*. In P. I. Mitchell, S. F. Newton, N. Ratcliffe and T. E. Dunn (eds), *Seabird Populations of Britain and Ireland*, pp. 214–225. T and AD Poyser, London.

Taverner. J. H. 1970. Mediterranean Gulls nesting in Hampshire. *British Birds* 63: 67–79.

Taylor, B. and van Perlo, B. 1998. *Rails: a Guide to the Rails, Crakes, Gallinules and Coots of the World*. Pica Press, Sussex.

Taylor, M., Seago, M., Allard, P. and Dorling, D. 1999. *The Birds of Norfolk*. Pica Press, Sussex.

Telfair, R. C. 1994. Cattle Egret (*Bubulcus ibis*). In A. Poole and F. Gill (eds), *The Birds of North America, No. 113*. The Academy of Natural Sciences, Philadelphia, PA, and The American Ornithologists' Union, Washington, DC.

Thévenot, M., Vernon, J. D. R. and Bergier, P. 2003. *The Birds of Morocco: an Annotated Checklist*. BOU, Tring.

Thom, V. 1986. *Birds in Scotland*. T and AD Poyser, Calton.

Thomas, G. H., Wills, M. A. and Székely, T. 2004a. Phylogeny of shorebirds, gulls and alcids (Aves: Charadrii) from the cytochrome-b gene: parsimony, Bayesian inference, minimum evolution and quartet puzzling. *Molecular Phylogenetics and Evolution* 30: 516–526.

Thomas, G. H., Wills, M. A. and Székely, T. 2004b. A supertree approach to shorebird phylogeny. *BMC Evolutionary Biology* 4: 28.

Thomassen, H. A., den Tex, R.-J., de Bakker, M. A. G. and Povel, D. E. 2005. Phylogenetic relationships among swifts and swiftlets: a multi-locus approach. *Molecular Phylogenetics and Evolution* 37: 264–277.

Thomassen, H. A., Wiersema, A. T., de Bakker, M. A. G., de Knijff, P., Heterbrij, E. and Povel, D. E. 2003. A new phylogeny of swiftlets (Aves: Apodidae) based upon cytochrome-b DNA. *Molecular Phylogenetics and Evolution* 29: 86–93.

Thompson, B. C., Jackson, J. A., Burger, J., Hill, A., Kirsch, E. M. and Attwood, J. L. 1997. Least Tern (*Sterna antillarum*). In A. Poole and F. Gill (eds), *The Birds of North America, No. 290*. The Academy of Natural Sciences, Philadelphia, PA, and The American Ornithologists' Union, Washington, DC.

Thompson, D. B. A. and Brown, A. 1992. Biodiversity in montane Britain: habitat variation, vegetation diversity and objectives for conservation. *Biodiversity Conservation* 1: 179–208.

Thompson, D. B. A., Stroud, D. A. and Pienkowski, M. W. 1988. Effects of afforestation on upland birds: consequences for population ecology. In M.B. Usher and D.B.A. Thompson (eds), *Ecological Change in the Uplands*. Blackwell, Oxford.

Thomson, D. L., Baillie, S. R. and Peach, W. J. 1997. The demography and age-specific annual survival of Song Thrushes during periods of population stability and decline. *Journal of Animal Ecology* 66: 414–424.

Thorup, O. 2006. *Breeding Waders in Europe 2000*. International Wader Studies 14. International Wader Study Group, UK.

Ticehurst, C. B. 1938. *A Systematic Review of the Genus Phylloscopus*. British Museum, London.

Tiedemann, R., Paulus, K. B., Scheer, M., von Kistowski, K. G., Skirnisson, K., Bloch, D. and Dam, M. 2004. Mitochondrial DNA and microsatellite variation in the eider duck (*Somateria molissima*) indicate stepwise postglacial colonization of Europe and limited current long-distance dispersal. *Molecular Ecology* 13: 1481–1494.

Tietze, D. T., Martens, J. and Sun, Y.-H. 2006. Molecular phylogeny of treecreepers (*Certhia*) detects hidden diversity. *Ibis* 148: 477–488.

Tomkovich, P. 1986. Geographic variation of the Dunlin in the Far East. (In Russian.) *Biol. Mosk.* 91: 3–15.

Tomkovich, P. S. 2008. [A new subspecies of the Whimbrel (*Numenius phaeopus*) from central Russia.] *Zoologichesky Zhurnal* 87: 1092–1099.

Toms, M. P., Crick, H. Q. P. and Shawyer, C. R. 2000. *Project Barn Owl Final Report*. Research Report 197. BTO, Tring.

Toms, M. P., Crick, H. Q. P. and Shawyer, C. R. 2001. The status of breeding Barn Owls *Tyto alba* in the United Kingdom 1995–97. *Bird Study* 48: 23–37.

Töpfer, T. 2006. The taxonomic status of the Italian Sparrow – *Passer italiae* (Vieillot 1817): Speciation by stabilised hybridisation? A critical analysis. *Zootaxa* 1325: 117–145.

Trinder, M., Rowdiffe, M., Pettifor, R., Rees, E., Griffin, L., Ogilvie, M. and Percival, S. 2005. *Status and Population Viability Analyses of Geese in Scotland*. Scottish Natural Heritage Report No. 107, Edinburgh.

Tucker, G. M. and Heath, M. F. 1994. *Birds in

Europe: their Conservation Status. BirdLife International, Cambridge.

Turner, A. and Rose, C. 1989. *Swallows and Martins of the World.* Christopher Helm, London.

Tyler, S. J. and Ormerod, S. J. 1991. The influence of stream acidification and riparian land use on the breeding biology of Grey wagtails *Motacilla cinerea* in Wales. *Ibis* 133: 286–292.

Underhill, L. G. 1987. Changes in the age structure of Curlew Sandpiper populations at Langebaan Lagoon, South Africa, in relation to lemming cycles in Siberia. *Transactions of the Royal Society of South Africa* 46: 209–214.

Underhill, M. C., Gittings, T., Callaghan, D. A., Kirby, J. S., Hughes, B. and Delaney, S. 1998. Pre-breeding status and distribution of the Common Scoter *Melanitta fusca* in Britain and Ireland 1995. *Bird Study* 45: 146–156.

Underhill-Day, J. 1998. Breeding Marsh Harriers in the United. Kingdom, 1983–95. *British Birds* 91: 210–218.

Ussher, R. and Warren, R.1900. *The Birds of Ireland.* Gurney and Jackson, London.

Vallender, R., Robertson, R. J., Friesen, V. L. and Lovette, I. J.2007. Complex hybridization dynamics between golden-winged and blue-winged warblers (*Vermivora chrysoptera* and *Vermivora pinus*) revealed by AFLP, microsatellite, intron and mtDNA markers. *Molecular Ecology* 16: 2017–2029.

van der Meij, M. A. A., Bakker, M. A. G. and Bout, R. G. 2005. Phylogenetic relationships of finches and allies based on nuclear and mitochondrial DNA. *Molecular Phylogenetics and Evolution* 34: 97–105.

van Horn, M. A. and Donovan, T. M. 1994. Ovenbird (*Seiurus aurocapilla*). In A. Poole and F. Gill (eds), *The Birds of North America, No. 277.* American Ornithologists' Union, Washington, DC.

van Roomen, M. W. J., Hustings, F. and Koffijberg, K. 2003. *Handleiding monitoringproject watervogels.* SOVON Vogelonderzoek Nederland, Beek-Ubbergen.

van Roomen, M., van Winden, E., Hustings, F., Koffijberg, K., Kleefstra, R., SOVON Ganzen- en zwanenwerkgroep and Soldaat L. 2005. *Watervogels in Nederland in 2003/2004.* SOVON-monitoringrapport 2005/03, RIZA-rapport BM05.15, SOVON Vogelonderzoek Nederland, Beek-Ubbergen.

van Roomen, M., van Winden, E., Koffijberg, K., Boele, A., Hustings, F., Kleefstra, R., Schoppers, J., van Turnhout, C., SOVON Ganzen- en Zwanenwerkgroep and Soldaat, L. 2004. *Watervogels in Nederland in 2002/2003.* SOVON Vogelonderzoek Nederland, Beek-Ubbergen.

van Tets, G. F. 1976. Australasia and the origin of shags and cormorants. *Proc. XVI Int. Ornithol. Congr.*: 121–124.

Vaurie, C. 1959. *Birds of the Palearctic Fauna: Passerines.* Witherby, London.

Vaurie, C. 1965. *Birds of the Palearctic Fauna: Non-passerines.* Witherby, London.

Vinicombe, K. E. 2002. Ruddy Shelducks in Britain. *British Birds* 95: 398–399.

Vinicombe, K. 2007. The status of Red-headed Bunting in Britain. *British Birds* 100: 540–551.

Vinicombe, K. E. and Chandler, R. J. 1982. Movements of Ruddy Ducks during the hard winter of 1978–79. *British Birds* 75: 1–11.

Vinicombe, K. E. and Harrop, A. H. J. 1999. Shelducks in Britain and Ireland, 1986–94. *British Birds* 92: 225–255.

Vlug, J. J. 2002. *Podiceps grisegena* Red-necked Grebe. *BWP Update* 4: 139–179.

Voelker, G. 1999. Molecular evolutionary relationships in the avian genus *Anthus* (Pipits: Motacillidae). *Molecular Phylogenetics and Evolution* 11: 84–94.

Voelker, G. 2002a. Molecular phylogenetics and the historical biogeography of dippers (*Cinclus*). *Ibis* 144: 577–584.

Voelker, G. 2002b. Systematics and historical biogeography of wagtails (Aves: *Motacilla*): dispersal versus vicariance revisited. *Condor* 104: 725–739.

Voelker, G. and Klicka, J. 2008. Systematics of *Zoothera* thrushes, and a synthesis of true thrush molecular systematic relationships. *Molecular Phylogenetics and Evolution* 49: 377–381.

Voelker, G. and Outlaw, R. K. 2008. Establishing a perimeter position: thrush speciation around the Indian Ocean Basin. *Journal of Evolutionary Biology* 21: 1779–1788.

Voelker, G., Rohwer, S., Bowie, R. C. K. and Outlaw, D. C. 2007. Molecular systematics of a speciose, cosmopolitan songbird genus: defining the limits of, and relationships among, the *Turdus* thrushes. *Molecular Phylogenetics and Evolution* 42: 422–434.

Voelker, G. and Spellman, G. M. 2004. Nuclear and mitochondrial evidence of polyphyly in the avian superfamily Muscicapoidea. *Molecular Phylogenetics and Evolution* 30: 386–394.

Voous, K. H. 1960. *Atlas of European Birds.* Nelson, London.

Voous, K. H. 1977. *List of Recent Holarctic Bird Species.* Revised reprint. BOU.

Votier, S. C., Harrop, A. H. J. and Denny, M. 2003. A review of the status and identification of American Wigeon in Britain and Ireland. *British Birds* 96: 2–22.

Votier, S. C., Kennedy, M., Bearhop, S., Newell, R. G., Griffiths, K., Whitaker, H., Ritz, M. S. and Furness, R. W. 2007. Supplementary DNA evidence fails to confirm presence of Brown Skuas *Stercorarius antarctica* in Europe: a retraction of Votier *et al.* (2004). *Ibis* 149: 619–621.

Walbridge, G., Small, B. and McGowan, R. Y. 2003. From the Rarities Committee's files: Ascension Frigatebird on Tiree – new to the Western Palearctic. *British Birds* 96: 58–73.

Wallace, D. I. M. 1970. Identification of Spotted Sandpipers out of breeding plumage. *British Birds* 63: 168–173.

Walter, H. 1979. *Eleonora's Falcon. Adaptations to Prey and Habitat in a Social Raptor.* University of Chicago, Chicago.

Wanless, S., Frederiksen, M., Harris, M. P. and Freeman, S. N. 2006. Survival of gannets *Morus bassana* in Britain and Ireland 1959–2002. *Bird Study* 53: 79–85.

Wanless, S. and Harris, M. P. 2004a. Northern Gannet *Morus bassana.* In P. I. Mitchell, S. F. Newton, N. Ratcliffe and T. E. Dunn (eds), *Seabird Populations of Britain and Ireland,* pp. 115–127. T and AD Poyser, London.

Wanless, S. and Harris, M. P. 2004b. European Shag *Phalacrocorax aristotelis.* In P. I. Mitchell, S. F. Newton, N. Ratcliffe and T. E. Dunn (eds), *Seabird Populations of Britain and Ireland,* pp. 146–159. T and AD Poyser, London.

Wanless, S., Harris, M., Murray, S. and Wilson, L. J. 2003. Status of the Atlantic Puffin *Fratercula arctica* on the Isle of May National Nature Reserve, Craigleith and Fidra, Forth Islands Special Protection Area. Report to Scottish Natural Heritage, Cupar.

Ward, R., Evans, P. and O'Connell, M. 2003. *Study of Long Term Changes in Bird Usage of the Tees Estuary.* WWT Research Dept. Report to English Nature, Slimbridge.

Ward, R. M. 2004. *Dark-bellied Goose* Branta bernicla hrota *in Britain 1960/1–1999/2000.* Waterbird Review Series, WWT/JNCC, Slimbridge.

Ward, R. M., Cranswick, P. A., Kershaw, M., Austin, G., Brown, A. W., Brown, L. M.,

Coleman, J. T., Chrisholm, H. and Spray, C. 2007. *National Mute Swan Census 2002.* WWT, Slimbridge.

Warham, J. 1990. *The Petrels: their Ecology and Breeding Systems.* Academic Press, London.

Warham, J. 1996. *The Behaviour, Population Biology and Physiology of the Petrels.* Academic Press, London.

Warren, P. and Baines, D. 2004. Black Grouse in northern England: stemming the decline. *British Birds* 97: 183–189.

Warren, P. and Baines, D. 2008a. Abundance of male Black Grouse *Tetrao tetrix* in Britain in 2005, and change since 1995–96. *Bird Study* 55: 304–313.

Warren, P. and Baines, D. 2008b. Current status and recent trends in numbers and distribution of Black Grouse *Tetrao tetrix* in northern England. *Bird Study* 55: 94–99.

Watson, A., Moss, R. and Rae, S. 1998. Population dynamics of Scottish Rock Ptarmigan cycles. *Ecology* 79: 1174–1192.

Watson, A., Moss, R. and Rothery, P. 2000. Weather and synchrony in 10-year population cycles of Rock Ptarmigan and Red Grouse in Scotland. *Ecology* 81: 2126–2136.

Watson, A. and Rae, R. 1987. Dotterel numbers, habitat and breeding success in Scotland. *Scottish Birds* 149: 191–198.

Watson, J. 1997. *The Golden Eagle.* T and AD Poyser, London.

Watson, J. and Dennis, R. H. 1992. Nest site selection by Golden Eagles *Aquila chrysaetos* in Scotland. *British Birds* 85: 469–481.

Webb, D. M. and Moore, W.S. 2005. A phylogenetic analysis of woodpeckers and their allies using 12S, Cyt b, and COI nucleotide sequences (class Aves; order Piciformes). *Molecular Phylogenetics and Evolution* 36: 233–248.

Weckstein, J. D., Afton, A. D., Zink, R. M. and Alisauskas, R. T. 2002. Hybridization and population subdivision within and between Ross's Geese and Lesser Snow Geese: a molecular perspective. *Condor* 104: 432–436.

Weckstein, J. D., Zink, R. M., Blackwell-Rago, R. C. and Nelson, D. A. 2001. Anomalous variation in mitochondrial genomes of White-crowned (*Zonotrichia leucophrys*) and Golden-crowned (*Zonotrichia atricapilla*) Sparrows: pseudogenes, hybridisation or incomplete lineage sorting? *Auk* 118: 231–236.

Weeks, H. P., Jr 1994. Eastern Phoebe (*Sayornis phoebe*). In A. Poole and F. Gill (eds), *The Birds of North America, No. 94.* The Birds of North America, Inc., Philadelphia, PA.

Weibel, A. C. and Moore, W.S. 2002a. A test of a mitochondrial gene-based phylogeny of woodpeckers (Genus *Picoides*) using an independent nuclear gene, ß-fibrinogen intron 7. *Molecular Phylogenetics and Evolution* 22: 247–257.

Weibel, A. C. and Moore, W. S. 2002b. Molecular phylogeny of a cosmopolitan group of woodpeckers (Genus *Picoides*) based on COI and cyt b mitochondrial gene sequences. *Molecular Phylogenetics and Evolution* 22: 65–75.

Weir, D. N., Kitchener, A. C. and McGowan, R. Y. 2000. Hybridization and changes in the distribution of Iceland Gulls (*Larus glaucoides/ kumlienel thayeri*). *J. Zool. Lond.* 252: 517–530.

Wenink, P.W. and Baker, A.J. 1996. Mitochondrial DNA lineages in composite flocks of migratory and wintering Dunlins (*Calidris alpina*). *Auk* 113: 744–756.

Wenink, P. W., Baker, A. J., Rösner, H. U. and Tilanus, M. G. L. 1996. Global mitochondrial DNA phylogeography of Holarctic breeding Dunlins (*Calidris alpina*). *Evolution* 50: 318–330.

Wenink, P. W., Baker, A. J. and Tilanus, M. G. L.

1993. Hypervariable control-region sequences reveal global population structuring in a long-distance migrant shorebirds, the Dunlin (*Calidris alpina*). *Proceedings of the National Academy of Sciences of the USA* 90: 94–98.

Wenink, P. W., Baker, A. J. and Tilanus, M. G. L. 1994. Mitochondrial control-region sequences in two shorebird species, the Turnstone and the Dunlin, and their utility in population genetic studies. *Molecular Biology and Evolution* 11: 22–31.

Wennerberg, L. 2001. Breeding origin and migration pattern of Dunlin (*Calidris alpina*) revealed by mitochondrial DNA analysis. *Molecular Ecology* 10: 1111–1120.

Wennerberg, L., Klaassen, M. and Lindström, Å. 2002. Geographical variation and population structure in the White-rumped Sandpiper *Calidris fuscicollis* as shown by morphology, mitochondrial DNA and carbon isotope ratios. *Oecologia* 131: 380–390.

Wernham, C., Toms, M., Marchant, J., Clark, J., Siriwardena, G. and Baillie, S. (eds) 2002. *The Migration Atlas: Movements of the Birds of Britain and Ireland.* T and AD Poyser, London.

Wetlands International 2002. *Waterbird Population Estimates – Third Edition.* Global Series No. 10. Wetlands International, Wageningen.

Wetlands International 2006. *Waterbird Population Estimates – Fourth Edition.* Wetlands International, Wageningen, The Netherlands.

Wheelwright, N. T. and Rising, J. D. 1993. Savannah Sparrow (*Passerculus sandwichensis*). In A. Poole and F. Gill (eds), *The Birds of North America, No. 45.* The Academy of Natural Sciences, Philadelphia , and the American Ornithologists' Union, Washington, DC.

Whilde, A. 1979. Auks trapped in salmon drift-nets. *Irish Birds* 1: 370–376.

White, C. M. and Boyce, D. A. 1988. An overview of Peregrine Falcon subspecies. In T. J. Cade, J. H. Enderson, C. G. Thelander and C. M. White (eds), *Peregrine Falcon Populations. Their Management and Recovery.* The Peregrine Fund Inc., Boise, Idaho.

Whitfield, D. P. 2002. Status of breeding Dotterel *Charadrius morinellus* in Britain in 1999. *Bird Study* 49: 237–249.

Whitfield, D. P., Fielding, A. H., McLeod, D. R. A., Morton, K., Stirling-Aird, P. and Eaton, M. A. 2007. Factors constraining the distribution of Golden Eagles *Aquila chrysaetos* in Scotland. *Bird Study* 54: 199–211.

Whittier, J. B., Leslie, D. M., Jr and van dev Bussche, R. A. 2006. Genetic variation among subspecies of Least Tern (*Sterna antillarum*): implications for conservation. *Waterbirds* 29: 176–184.

Wieloch, M., Włodarczyk, R. and Czapulak, A. 2004. *Cygnus olor* Mute Swan. *BWP Update* 6: 1–38.

Wiley, E. O. 1978. The evolutionary species concept reconsidered. *Systematic Zoology* 39: 399–413.

Williams, C. L., Brust, R. C., Fendley, T. T., Tiller, G. R., Jr and Rhodes, O. E. 2005. A comparison of hybridisation between Mottled Ducks (*Anas fulvigula*) and Mallards (*A. platyrhynchos*) in Florida and South Carolina using microsatellite DNA analysis. *Conservation Genetics* 6: 445–453.

Williams, R. S. R. 1998. Unequal sex ratio of the Long-eared Owl *Asio otus* in Northern India. *Journal of the Bombay Natural History Society* 95: 343–344.

Williams, T. C. and Williams, J. M. 1990. The orientation of trans-oceanic migrants. In E. Gwinner (ed.) *Bird Migration: Physiology and Ecophysiology.* Springer-Verlag, Berlin.

Williamson, K. 1960. Juvenile and winter plumages of the marsh terns. *British Birds* 53: 243–252.

Williamson, K. 1965. *Fair Isle and its Birds.* Oliver and Boyd, Edinburgh.

Wilson, G. E. 1976. Spotted Sandpipers nesting in Scotland. *British Birds* 69: 288–292.

Wilson, J. D., Boyle, J., Jackson, D. B., Lowe, B. and Wilkinson, N. I. 2007. Effect of cereal harvesting method on a recent population decline of Corn Buntings *Emberiza calandra* on the Western Isles of Scotland. *Bird Study* 54: 362–370.

Wink, M., Clouet, M., Goar, J.-L. and Barrau, C. 2004a. Sequence variation in the cytochrome b gene of the subspecies of Golden eagles *Aquila chrysaetos*. In R. D. Chancellor and B.-U. Meyburg (eds) *Raptors Worldwide.* WWGBP, Berlin.

Wink, M. and Heidrich, P. 1999. Molecular evolution and systematics of the owls (Strigiformes). In Konig *et al.* (eds) *Owls: a Guide to the Owls of the World.* Pica Press, Robertsbridge, E Sussex.

Wink, M., Kuhn, M., Sauer-Gürth, H. and Witt, H.-H. 2002a. Ein Eistaucher (*Gavia immer*) bei Düren – Fundgeschichte und erste genetische Herkunftsuntersuchungen. *Charadrius* 38: 239–245.

Wink, M., Sauer-Gürth, H., Ellis, D. and Kenward, R. 2004d. Phylogenetic relationships in the *Hierofalco* complex (Saker, Gyr, Lanner, Laggar Falcon). In R. D. Chancellor and B.-U. Meyburg (eds) *Raptors Worldwide.* WWGBP, Berlin.

Wink, M., Sauer-Gürth, H. and Fuchs, M. 2004e. Phylogenetic relationships in owls based on nucleotide sequences of mitochondrial and nuclear marker genes. In R. D. Chancellor and B.-U. Meyburg (eds) *Raptors Worldwide.* WWGBP, Berlin.

Wink, M., Sauer-Gurth, H., Heidrich, P., Witt, H.-H. and Gwinner, E. 2002b. A molecular phylogeny of Stonechats and related turdids. In Urquhart, E. (ed.) *Stonechats: a Guide to the Genus* Saxicola. Christopher Helm, London.

Wink, M., Sauer-Gürth, H. and Pepler, D. 2004c. Phylogeographic relationships of the Lesser Kestrel *Falco naumanni* in breeding and wintering quarters, inferred from nucleotide sequences of the mitochondrial cytochrome b gene. In R. D. Chancellor and B.-U. Meyburg (eds) *Raptors Worldwide.* WWGBP, Berlin.

Wink, M., Sauer-Gürth, H. and Witt, H.-H. 2004b. Phylogenetic differentiation of the Osprey *Pandion haliaetus* inferred from nucleotide sequences of the mitochondrial cytochrome b gene. In R. D. Chancellor and B.-U. Meyburg (eds) *Raptors Worldwide.* WWGBP, Berlin.

Winker, K. and Pruett, C. L. 2006: Seasonal migration, speciation, and morphological convergence in the avian genus *Catharus* (Turdidae). *Auk* 123: 1052–1068.

Winkler, H., Christie, D. A. and Nurney, D. 1995. *Woodpeckers: a Guide to the Woodpeckers, Piculets and Wrynecks of the World.* Pica Press, Robertsbridge, E Sussex.

Winney, B. J., Litton, C. D., Parkin, D. T. and Feare, C. J. 2001. The sub-specific origin of the inland breeding colonies of the Cormorant *Phalacrocorax carbo* in Britain. *Heredity* 86: 45–53.

Winstanley, D., Spencer, R. and Williamson, K. 1974. Where have all the Whitethroats gone? *Bird Study* 21: 1–14.

Witmer, M. C. 2002. Bohemian Waxwing (*Bombycilla garrulus*). In A. Poole and F. Gill (eds), *The Birds of North America, No. 714.* The Birds of North America, Inc., Philadelphia, PA.

Woodin, M. C. and Michot, T. C.. 2002. Redhead (*Aythya americana*). In A. Poole and F. Gill (eds), *The Birds of North America, No. 695.* The Birds of North America, Inc., Philadelphia, PA.

Worden, J., Crowe, O., Einarsson, O., Gardarsson, A., McElwaine, G. and Rees, E. C. In press. Population size and breeding success of the Icelandic Whooper Swan *Cygnus cygnus*: results of the January 2005 International Census. *Bird Study*

Worden, J., Mitchell, C., Merne, O. and Cranswick, P. 2004. *Greenland Barnacle Geese* Branta leucopsis *in Britain and Ireland: Results of the International Census, Spring 2003.* WWT Report, Slimbridge.

Wotton, S. R., Rylands, K., Grice, P., Smallshire, D. and Gregory, R. 2004. The status of the Cirl Bunting in Britain and the Channel Islands in 2003. *British Birds* 97: 376–384.

Wyllie, I., Dale, L. and Newton, I. 1996. Unequal sex ratio, mortality and pollutant residues in Long-eared Owls in Britain. *British Birds* 89: 429–436.

Wynne-Edwards, V. C. 1962. *Animal Dispersion in Relation to Social Behaviour.* Oliver and Boyd, London.

Yalden, D. W. 1999. *The History of British Mammals.* T and AD Poyser, London.

Yalden, D. W. and Yalden, P. E. 1988. *Golden Plovers and Recreational Disturbance.* Report to Nature Conservancy Council, NCC, Peterborough.

Yang, S. J., Lei, F. M. and Yin, Z. H. 2006. Molecular phylogeny of rosefinches and rose bunting (Passeriformes, Fringillidae, Urocynchramidae). *Acta Zootaxonomica Sinica* 31: 453–458.

Yésou, P. 2002. Trends in systematics. Systematics of *Larus argentatus–cachinnans–fuscus* complex revisited. *Dutch Birding* 64, 271–298.

Yésou, P. and Paterson, A. M. 1999. Puffin Yelkouan et Puffin des Baléares: une ou deux espèces? *Ornithos* 6: 20–31.

Zamora, J., Lowy, E., Ruiz-del-Valle, V., Moscoso, J., Serrano-Vela, J. I., Rivero-de-Aguilar, J. and Arnaiz-Villena, A. 2006. *Rhodopechys obsoleta* (desert finch): a pale ancestor of greenfinches (*Carduelis* spp.) according to molecular phylogeny. *Journal of Ornithology* 147: 448–456, 511–512.

Zink, R. M. 2004. The role of subspecies in obscuring avian biological diversity and misleading conservation policy. *Proceedings of the Royal Society of London B* 271: 561–564.

Zink, R. M. 2005. Natural selection on mitochondrial DNA in *Parus* and its relevance for phylogeographic studies. *Proceedings of the Royal Society of London B* 272: 71–78.

Zink, R. M. 2008. Microsatellite and mitochondrial DNA differentiation in the Fox Sparrow. *Condor* 110: 482–492.

Zink, R. M., Dittman, D. L. and Rootes, W. L. 1991. Mitochondrial variation and the phylogeny of *Zonotrichia*. *Auk* 108: 578–584.

Zink, R. M., Drovetski, S. V., Questiau, S., Fadeev, I. V., Nesterov, E. V., Westberg, M. C. and Rohwer, S. 2003. Recent evolutionary history of the Bluethroat (*Luscinia svecica*) across Eurasia. *Molecular Ecology* 12: 3069–3075.

Zink, R. M., Drovetski, S. V. and Rohwer, S. 2002. Phylogenetic patterns in the Great Spotted Woodpecker *Dendrocops major* across Eurasia. *Journal of Avian Biology* 33: 175–178.

Zink, R. M., Drovetski, S. V. and Rohwer, S. 2006b. Selective neutrality of mitochondrial ND2 sequences, phylogeography and species limits in *Sitta europaea*. *Molecular Phylogenetics and Evolution* 40: 679–686.

Zink, R. M. and Johnson, N. K. 1984.

Evolutionary genetics of flycatchers. 1. Sibling species in the genera *Empidonax* and *Contopus*. *Systematic Zoology* 33: 205–216.

Zink, R. M. and Klicka, J. T. 1990. Genetic variation in the Common Yellowthroat and some allies. *Wilson Bulletin* 102: 514–520.

Zink, R. M., Pavlova, A. Drovetski, S., Wink, M. and Rohwer, S. 2009. Taxonomic status and evolutionary history of the *Saxicola torquata* complex. *Molecular Phylogenetics and Evolution* 52: 769–773.

Zink, R. M., Pavlova, A., Rohwer, S. and Drovetski, S. V. 2006a. Barn Swallows before barns: population histories and intercontinental colonization. *Proceedings of the Royal Society of London B* 273: 1245–1251.

Zink, R. M., Rising, J. D., Mockford, S., Horn, A. G., Wright, J. M., Leonard, M. and Westberg, M. C. 2005. Mitochondrial DNA variation, species limits, and rapid evolution of plumage coloration and size in the Savannah Sparrow. *Condor* 107: 21–28.

Zink, R. M., Rohwer, S., Andreev, A. V. and Dittmann, D. L. 1995. Trans-Beringia comparisons of mitochondrial DNA differentiation in birds. *Condor* 97: 639–649.

Zink, R. M. and Weckstein, J. D. 2003. Recent evolutionary history of the Fox Sparrows (genus: *Passerella*). *Auk* 120: 522–527.

Zino, F., Brown, R. and Biscoito, M. 2008. The separation of *Pterodroma madeira* (Zino's Petrel) from *Pterodroma feae* (Fea's Petrel) (Aves: Procellariidae). *Ibis* 150: 326–334.

Zöckler, C. 2002. Declining Ruff *Philomachus pugnax* populations: a response to global warming? *Wader Study Group Bulletin* 97: 19–29.

Zonfrillo, B. 2002. Puffins return to Ailsa Craig. *Scottish Bird News* 66: 1–2.

Zonfrillo, B. and Monaghan, P. 1995. Rat eradication on Ailsa Craig. In M. L. Tasker (ed.) *Threats to Seabirds*. Proceedings of the Fifth International Seabird Group Conference. Seabird Group, Glasgow.

Zwarts, L., Kamp, J., van der Overdijk, O., Spanje, T., Veldkamp, R., West, R. and Wright, M. 1998. Wader count of the Banc d'Arguin, Mauritania in January/February 1997. *Wader Study Group Bulletin* 86: 53–69.

INDEX